T0343173

DIET AND EXERCISE IN CYSTIC FIBROSIS

DIET AND EXERCISE IN CYSTIC FIBROSIS

Edited by

RONALD ROSS WATSON

University of Arizona, Mel and Enid Zuckerman College of Public Health,
Sarver Heart Center in the School of Medicine, Tucson, AZ

AMSTERDAM • BOSTON • HEIDELBERG • LONDON
NEW YORK • OXFORD • PARIS • SAN DIEGO
SAN FRANCISCO • SINGAPORE • SYDNEY • TOKYO
Academic Press is an imprint of Elsevier

Academic Press is an imprint of Elsevier
32 Jamestown Road, London NW1 7BY, UK
225 Wyman Street, Waltham, MA 02451, USA
525 B Street, Suite 1800, San Diego, CA 92101-4495, USA

British Library Cataloguing-in-Publication Data
A catalogue record for this book is available from the British Library

Library of Congress Cataloging-in-Publication Data
A catalog record for this book is available from the Library of Congress

ISBN: 978-0-12-800051-9

For information on all Academic Press publications
visit our website at elsevierdirect.com

Typeset by TNQ Books and Journals
www.tnq.co.in

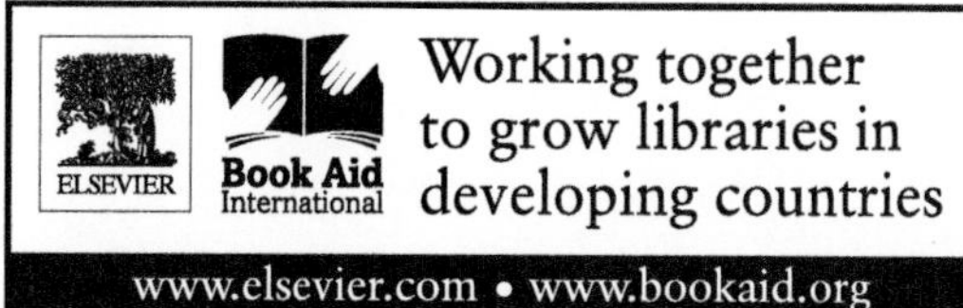

Dedication

The book is dedicated to the strong interest of Mr. Elwood Richards in food and supplements to help those with cystic fibrosis. This was possible due to his lifelong interest in components in plants that provide health promotion and disease prevention for those with genetic diseases. He knows and supports friends with cystic fibrosis and this book is an outgrowth of his concern for individuals.

Contents

D

MANAGEMENT OF DIABETES AND CELIAC DISEASE ASSOCIATED WITH CYSTIC FIBROSIS: ROLE OF NUTRITION AND FOOD

E

NUTRITION AND PULMONARY FUNCTION IN CYSTIC FIBROSIS PATIENTS

F

EXERCISE AND BEHAVIOR IN MANAGEMENT OF CYSTIC FIBROSIS

32. Exercise Testing in CF, the What and How

LARRY C. LANDS, HELGE HEBESTREIT

33. Mechanisms of Exercise Limitation in Cystic Fibrosis: A Literature Update of Involved Mechanisms

H.J. HULZEBOS, M.S. WERKMAN, B.C. BONGERS, H.G.M. ARETS, T. TAKKEN

34. Physical Activity Assessment and Impact

NANCY ALARIE, LISA KENT

35. Motivating Physical Activity: Skills and Strategies for Behavior Change

HEATHER CHAMBLISS

36. Diet, Food, Nutrition, and Exercise in Cystic Fibrosis

ANDREA KENCH, HIRAN SELVADURAI

37. Personalizing Exercise and Physical Activity Prescriptions

MATTHEW NIPPINS

G

FAT AND LIPID METABOLISM IN CYSTIC FIBROSIS

38. The Pancreatic Duct Ligated Pig as a Model for Patients Suffering from Exocrine Pancreatic Insufficiency—Studies of Vitamin A and E Status

A. MÖßELER, T. SCHWARZMAIER, M. HÖLTERSHINKEN, J. KAMPHUES

39. Unsaturated Fatty Acids in Cystic Fibrosis: Metabolism and Therapy

ADAM SEEGMILLER, MICHAEL O'CONNOR

Preface

There is increasing research including clinical trials showing a link between nutrition and food, and many aspects of health in cystic fibrosis. This book brings together many expert basic and clinical researchers to focus on the role of nutrition and diets in disease and therapy of cystic fibrosis. The overall evaluation of the various reviews is that nutrition and food protects and provides health and function to cystic fibrosis with and is important in longevity, cognitive impairment, and lung structure in cystic fibrosis.

In the first section overviews of the role of nutrition and diet on cystic fibrosis (CF) patients are provided. In particular nutrition plays a key role in early life in pregnant women according to Michel and Mueller. Dr. Quick shows that disordered eating changes body image. Screening in neonatal is important to provide optimum nutrition for growth and health. Amanda Leonard describes the various treatment guidelines and risk classifications of children with CF. Torres evaluates the role of nutrition in clinical practice relating to spirometry. Clearly family plays a critical training for CF children at mealtimes according to Hammons, Everhart, and Fiese. As found in many other conditions, the hormone melatonin, available over-the-counter, regulates and treats sleep dysfunction and CF as summarized by Watson. De Monestrol's review shows the effects of age at diagnosis on CF disease progression. Finally a group of Italian researchers describe the role of lactoferrin in lung and airway infection. The very important actions of CF on digestive enzymes and the role of probiotics and foods in treatment of intestinal dysfunction are reviewed. In particular Pohl reviews the actions of celiac disease on CF and its manifestations. Bryon discusses the evidence that eating disorders and disturbed eating attitudes and behaviors have unique manifestations in CF patients. The Starks describe methods of better airway use in breathing for better health. Alcohol use can play an important role in CF function and is reviewed. Finally specialty foods designed for CF children are helpful according to McGuckin.

In the second section the role of vitamin D deficiency as well as replacement by supplement is reviewed from several aspects. Mailhot shows the effects of the disease on vitamin D bioavailability. Singh describes fat soluble vitamin deficiency in general in CF. Robberecht asks and answers the question about the role of light in providing vitamin D supplementation. Beckett, Shaw, and Sathe define pediatric CFs' effects on fat-soluble vitamins in children.

The third section continues this theme by evaluating vitamin deficiency, especially antioxidants and then their supplementation in promoting health in cystic fibrosis. Inflammation is a significant problem in CF patients. Sadowska-Bartosz describes nutritional strategies to reduce inflammation and oxidative stress. Similarly Offenberger describes the importance of vitamin A supplementation as therapy for CF. Many dietary supplementations with natural products contain polyphenols which are very functional. Kubow describes the developing sets of studies showing polyphenols in treatment modalities in CF. To study bacterial infections animal models of CF are used. Ciofu demonstrates the role of oxidative stress and their lessons for CF patients. Two different authors, Jagannath and Kleinman review vitamin K in therapy of CF.

The fourth section describes major chronic diseases, particularly diabetes are associated with CF. It covers the role of nutrition and food in CF. Hameed reviews the role of insulin on body mass and growth in young CF patients. Balzer investigates the published information on low glycemic index through dietary interventions in CF patients. Finally Jimenez reviews the management of insulin resistance common in CF.

The fifth section deals with the critical dysfunctions of the lungs and where food and nutrition may play roles. Probiotics are increasingly used to modify and regulate pulmonary functions. Shalem describes the literature on probiotic supplementation as it relates to CF dysfunctional pulmonary exacerbations. Diabetes, according to the review by Frias and Barrio affects lung function and nutritional status, and thus is open to regulation.

The sixth section deals with exercise in the health of CF patients and is a major component of the book for CF patient health. Exercise is something that is affected by lung dysfunction, but within the reach of most patients to include in their lives and lifestyles. Lands and Hebestreit describe methods to test for exercise efficacy including how and what to test. A team of authors, Hulzebos, Werkman, Bongers, Arets and Takken review and update a discussion of the exercise limitations imposed by CF on the lungs. Alarie reviews exercise by assessing physical activity and then discussing its impact on the physiology and functions of CF patients. Clearly for anyone to exercise effectively requires motivation. Chambliss investigates the literature relative to the skills and strategies for behavior change to affect the level of physical activity. Kench takes a broad overview of foods and diet and their roles in exercise in the special population that CF patients are. Finally Nippens reviews literature personalizing exercise and physical activities and how to define that for the CF patient.

The final section focuses on a critical component of the diet for the health of CF patients, lipids and fats. Mosseler, Schwarzmaier, Höltershinken, and Kamphues work together to describe results from the pig, an excellent animal model for human lipid use. They look at pancreatic duct ligation as a model for patients with pancreatic insufficiency with the focus on fat soluble vitamins E and A. Van Biervliet and Strandvick ask and answer the question as to whether essential fatty acid deficiency in CF patients is due to malabsorption and/or metabolic abnormalities. Understanding of this concern is vital for designing methods to treat fat deficiencies. Bodewes and Wouthuyzen-Bakker review the causes and consequences of persistent fat malabsorption. Finally Paul and Watson look at the role of omega-3 fatty acids which are becoming recognized as health promoting and needed in increased amounts in the diets of most children and adults, with an emphasis on CF patients.

Acknowledgments

The work of Dr Watson's editorial assistant Bethany L. Stevens and the editorial project manager Ms. Shannon Stanton, in communicating with authors and working on the manuscripts, was critical for the successful completion of the book. It is very much appreciated. Support for Ms Stevens' and Dr Watson's work was graciously provided by Natural Health Research Institute www.naturalhealthresearch.org. It is an independent, nonprofit organization that supports science-based research on natural health and wellness. It is committed to informing about scientific evidence on the usefulness and cost-effectiveness of diet, supplements, and a healthy lifestyle to improve health and wellness, and reduce disease. Finally the work of the librarian of the Arizona Health Science Library, Mari Stoddard, was vital and very helpful in identifying the key researchers who participated in the book.

Biography

Dr Ronald Ross Watson has studied the role of bioactive nutrients, dietary supplements, and alternative medicines for 40 years. He has been funded to conduct research by grants provided by the U.S. National Institute of Heart, Lung and Blood, the American Heart Foundation, as well as companies and private foundations; in addition, he conducts research for a small company to study effects of novel dietary supplements to modify age and autoimmune diseases in mice and humans, including obtaining patents for the discoveries. Dr Watson has edited 102 biomedical books on topics including aging, dietary supplements, and the role of nutrients in health and prevention of disease. He graduated from Brigham Young University with a degree in Chemistry in 1966 and later completed his PhD from Michigan State University in 1971 with a focus on Biochemistry. Dr Watson's current appointments are in the School of Medicine and the Department of Nutritional Sciences at the University of Arizona, reflecting a long and distinguished interest in dietary supplements and novel foods in health.

List of Contributors

Nancy Alarie Physiotherapy Department, Montreal Children's Hospital, McGill University Health Centre, Montreal, QC, Canada

H.G.M. Arets Department of Pediatric Respiratory Medicine, Cystic Fibrosis Center, University Medical Center Utrecht, Utrecht, The Netherlands

Fiona S. Atkinson School of Molecular Bioscience, The University of Sydney, NSW, Australia

Ben W.R. Balzer Academic Department of Adolescent Medicine, The Children's Hospital at Westmead, NSW, Australia; Discipline of Paediatrics and Child Health, Sydney Medical School, The University of Sydney, NSW, Australia

Raquel Barrio Diabetes Pediatric Unit, Ramón y Cajal Hospital, Alcalá University, Madrid, Spain

Grzegorz Bartosz Department of Biochemistry and Cell Biology, University of Rzeszów, Rzeszów, Poland; Department of Molecular Biophysics, University of Łódź, Łódź, Poland

Kacie Beckett Cystic Fibrosis Clinic, University of Texas Southwestern, Dallas, TX, USA

Kirstine J. Bell School of Molecular Bioscience, The University of Sydney, NSW, Australia

Francesca Berlutti Department of Public Health and Infectious Diseases, Sapienza University of Rome, Rome, Italy

S. Van Biervliet Department of Pediatric Gastroenterology and Nutrition, Ghent University Hospital, Ghent, Belgium

Frank A.J.A. Bodewes University Medical Center Groningen, University of Groningen, Groningen, The Netherlands

B.C. Bongers Child Development & Exercise Center, Wilhelmina Children's Hospital, University Medical Center Utrecht, Utrecht, The Netherlands

Aaron Robert Brussels Health Sciences Center, School of Medicine, Mel and Enid Zuckerman College of Public Health, University of Arizona, Tucson, AZ, USA

Mandy Bryon Cystic Fibrosis Service, Great Ormond Street Hospital for Children, London, UK

Carol Byrd-Bredbenner Department of Nutritional Sciences, Rutgers University, New Brunswick, NJ, USA

Amy Cantrell Division of Pediatric Endocrinology, Scott and White Hospital, Temple, TX, USA

Angela Catizone Department of Anatomy, Histology, Forensic Medicine, and Orthopedics, Sapienza University of Rome, Rome, Italy

Heather Chambliss

Oana Ciofu Department of International Health, Immunology and Microbiology, Costerton Biofilm Center, Faculty of Health Sciences, University of Copenhagen, Denmark

M. Francisco Rivas Crespo Pediatric Endocrinology, Universidad de Oviedo, Hospital Universitario Central de Asturias, Oviedo, Spain

Isabelle de Monestrol Stockholm CF Center, Karolinska University Hospital, Stockholm, Sweden; Department of Clinical Science, Intervention and Technology, Division of Pediatrics, Karolinska Institutet, Stockholm, Sweden

Dimitri Declercq Cystic Fibrosis Centre, Department of Pediatrics, Ghent University Hospital, Belgium

Ieda Regina L. Del Ciampo University of São Paulo, São Paulo, Brazil

Robin S. Everhart Psychology, Virginia Commonwealth University, Richmond, VA, USA

Barbara H. Fiese Human and Community Development, University of Illinois, Urbana–Champaign, IL, USA

Mary Shannon Fracchia Department of Pediatrics, Massachusetts General Hospital, Boston, MA, USA; Harvard Medical School, Boston, MA, USA

Alessandra Frioni Department of Public Health and Infectious Diseases, Sapienza University of Rome, Rome, Italy

Manyan Fung School of Dietetics and Human Nutrition, McGill University, QC, Canada

Sabina Galiniak Department of Biochemistry and Cell Biology, University of Rzeszów, Rzeszów, Poland

Carlos Bousoño García Pediatric Gastroenterology and Nutrition Unit, Universidad de Oviedo, Hospital Universitario Central de Asturias, Oviedo, Spain

Shihab Hameed Endocrinology Department Sydney Children's Hospital, Randwick, NSW, Australia; School of Women's and Children's Health, University of New South Wales, Kensington, NSW, Australia

Amber J. Hammons Fresno in the Child, Family, and Consumer Sciences Department, California State University, Fresno, CA, USA

Helge Hebestreit Universitaets-Kinderklinik, Josef-Schneider-Strasse 2, Wuerzburg, Germany

Wendy Anne Hermes Department of Nutrition, Byrdine F. Lewis School of Nursing and Health Professionals, Georgia State University, Atlanta, GA, USA

M. Höltershinken Institute of Animal Nutrition, University of Veterinary Medicine Hannover, Foundation, Hannover, Germany

H.J. Hulzebos Child Development & Exercise Center, Wilhelmina Children's Hospital, University Medical Center Utrecht, Utrecht, The Netherlands

Vanitha Jagannath Specialist Pediatrician in American Mission Hospital, Manama, Kingdom of Bahrain

David Gonzalez Jiménez Pediatric Gastroenterology and Nutrition Unit, Universidad de Oviedo, Hospital Universitario Central de Asturias, Oviedo, Spain

J. Kamphues Institute of Animal Nutrition, University of Veterinary Medicine Hannover, Foundation, Hannover, Germany

Andrea Kench Department of Nutrition & Dietetics and Respiratory Medicine, The Children's Hospital Westmead, Westmead, NSW, Australia

Lisa Kent Centre for Health and Rehabilitation Technology (CHaRT), Institute of Nursing and Health Research, University of Ulster, Northern Ireland,UK; Department of Respiratory Medicine, Belfast Health and Social Care Trust, Belfast, Northern Ireland, UK

Ronald E. Kleinman Department of Pediatrics, Massachusetts General Hospital, Boston, MA, USA; Harvard Medical School, Boston, MA, USA

Stan Kubow School of Dietetics and Human Nutrition, McGill University, QC, Canada

Larry C. Lands Montreal Children's Hospital, Pediatric Respiratory Medicine, McGill University, Montreal, QC, Canada

Amanda Leonard Division of Gastroenterology and Nutrition, The Johns Hopkins Children's Center, Baltimore, MD, USA

Amy Lowichik Division of Pediatric Pathology, Primary Children's Medical Center, Salt Lake City, UT, USA

G. Mailhot Department of Nutrition, Université de Montréal, Montréal, QC, Canada; Gastroenterology, Hepatology and Nutrition Unit, CHU Sainte-Justine Research Center, Montréal, QC, Canada

María Martín-Frías Diabetes Pediatric Unit, Ramón y Cajal Hospital, Alcalá University, Madrid, Spain

Megan Elizabeth McGuckin University of Arizona, Department of Immunobiology, USA

Suzanne H. Michel Clinical Assistant Professor, Medical University of South Carolina, Charleston, SC, USA

Alison Morton Regional Adult Cystic Fibrosis Unit and Department of Nutrition and Dietetics, Leeds Teaching Hospitals NHS Trust, St James's Hospital, Leeds, UK

A. Mößeler Institute of Animal Nutrition, University of Veterinary Medicine Hannover, Foundation, Hannover, Germany

Donna H. Mueller Nutrition Sciences Department, Drexel University, Philadelphia, PA, USA

Noor Naqvi School of Dietetics and Human Nutrition, McGill University, QC, Canada

Matthew Nippins Northeastern University, Boston, MA, USA; Massachusetts General Hospital, Wang Ambulatory Care Center, Boston, MA, USA

Michael O'Connor Department of Pediatrics, Division of Allergy, Immunology, and Pulmonology Medicine, Vanderbilt University School of Medicine, Nashville, TN, USA

Holly M. Offenberger College of Science, Mel and Enid Zuckerman College of Public Health, University of Arizona, Tucson, AZ, USA

Gaurav Paul College of Public Health, University of Arizona, Tucson, AZ, USA

John F. Pohl Department of Pediatric Gastroenterology, Primary Children's Medical Center, University of Utah, Salt Lake City, UT, USA

Virginia Quick Eunice Kennedy Shriver National Institute of Child Health and Human Development/National Institutes of Health, Division of Intramural Population Health Research, Bethesda, MD, USA

Gilbert L. Rivera Jr. University of Arizona, College of Public Health, Tucson, AZ, USA

Eddy Robberecht Cystic Fibrosis Centre, Department of Pediatrics, Ghent University Hospital, Belgium

Izabela Sadowska-Bartosz Department of Biochemistry and Cell Biology, University of Rzeszów, Rzeszów, Poland

Donatello Salvatore Cystic Fibrosis Center, Pediatric Center Bambino Gesù Basilicata, Hospital San Carlo, Potenza, Italy

Meghana Sathe Department of Pediatric Gastroenterology and Nutrition, University of Texas Southwestern, Dallas, TX, USA

T. Schwarzmaier Institute of Animal Nutrition, University of Veterinary Medicine Hannover, Foundation, Hannover, Germany

Adam Seegmiller Department of Pathology, Microbiology, and Immunology, Vanderbilt University School of Medicine, Nashville, TN, USA

Hiran Selvadurai Respiratory Medicine, The Children's Hospital Westmead, Westmead, NSW, Australia

Tzippora Shalem Division of Pediatric Gastroenterology and Nutrition, Edmond and Lily Safra Children's Hospital, Tel-Hashomer, Israel; Sackler Faculty of Medicine, Tel-Aviv University, Tel-Aviv, Israel

Karyn Shaw Department of Clinical Nutrition, Children's Medical Center, Dallas, TX, USA

Vijay Karam Singh Health Sciences Center, School of Medicine, Mel and Enid Zuckerman College of Public Health, University of Arizona, Tucson, AZ, USA

Jennifer Stark Buteyko Institute of Breathing and Health (Inc), Fitzroy, VIC, Australia; Buteyko Breathing Educators Association, IN, USA

Russell Stark Buteyko Institute of Breathing and Health (Inc), Fitzroy, VIC, Australia; Buteyko Breathing Educators Association, IN, USA

Katharine S. Steinbeck Academic Department of Adolescent Medicine, The Children's Hospital at Westmead, NSW, Australia; Discipline of Paediatrics and Child Health, Sydney Medical School, The University of Sydney, NSW, Australia

B. Strandvik Department of Biosciences and Nutrition, Karolinska Institutet, Stockholm, Sweden

T. Takken Child Development & Exercise Center, Wilhelmina Children's Hospital, University Medical Center Utrecht, Utrecht, The Netherlands

Vin Tangpricha Division of Endocrinology, Metabolism and Lipids, Department of Medicine, Emory University School of Medicine, Atlanta, GA, USA

Lidia Alice G.M.M. Torres University of São Paulo, São Paulo, Brazil

Piera Valenti Department of Public Health and Infectious Diseases, Sapienza University of Rome, Rome, Italy

Charles F. Verge Endocrinology Department Sydney Children's Hospital, Randwick, NSW, Australia; School of Women's and Children's Health, University of New South Wales, Kensington, NSW, Australia

Henkjan J. Verkade University Medical Center Groningen, University of Groningen, Groningen, The Netherlands

Ronald Ross Watson University of Arizona, Mel and Enid Zuckerman College of Public Health, Sarver Heart Center in the School of Medicine, Tucson, AZ

Batia Weiss Division of Pediatric Gastroenterology and Nutrition, Edmond and Lily Safra Children's Hospital, Tel-Hashomer, Israel; Sackler Faculty of Medicine, Tel-Aviv University, Tel-Aviv, Israel

M.S. Werkman Child Development & Exercise Center, Wilhelmina Children's Hospital, University Medical Center Utrecht, Utrecht, The Netherlands

Sue Wolfe Regional Paediatric Cystic Fibrosis Unit and Department of Nutrition and Dietetics, Leeds Teaching Hospitals NHS Trust, Leeds Children's Hospital, Leeds General Infirmary, Leeds, UK

Marjan Wouthuyzen-Bakker University Medical Center Groningen, University of Groningen, Groningen, The Netherlands

SECTION A

OVERVIEW OF NUTRITION AND DIETS IN CYSTIC FIBROSIS

CHAPTER

1

Nutrition for Pregnant Women Who Have Cystic Fibrosis

Suzanne H. Michel[1], *Donna H. Mueller*[2]

[1]Clinical Assistant Professor, Medical University of South Carolina, Charleston, SC, USA; [2]Nutrition Sciences Department, Drexel University, Philadelphia, PA, USA

1.1 INTRODUCTION

Nutrition before and during pregnancy is fundamental for all women. Determinants of pregnancy outcome for women and their babies include weight and nutrient status before conception and weight gain and overall nutrient stores and intake during gestation [1]. For women who have CF, pregnancy heightens their need for optimal nutrition and is regarded as an essential factor influencing pregnancy and fetal outcome [2].

Median age of survival for all people who have CF is 41.1 years with women having a median age of survival of 38 years [3]. More women with CF are reaching reproductive age, becoming pregnant, and delivering babies. On the basis of the 2012 U.S. Cystic Fibrosis Foundation (CFF) Patient Registry, 249 pregnancies were reported in women who have CF, with 2 live births per 100 women. Of the 249 pregnancies, 145 resulted in live births, 1 was a stillborn event, 23 were spontaneous abortions, and 7 were therapeutic abortions. At the time of data analysis, 71 women had yet to deliver their babies [3]. Although pregnancy increasingly is common among women who have CF, few facts are available about nutrition management. This chapter summarizes evidence related to outcomes for pregnant women who have CF, available nutrition recommendations, and applications by clinicians.

1.2 HISTORICAL PERSPECTIVE

In 1960, at which time the median age of survival for people who had CF was 10 years of age, the first report of pregnancy in a woman who had CF was published. The woman died 6 weeks following delivery [4]. In 1966, Grand et al. evaluated information available for 10 women who became pregnant and concluded that substantive factors to consider during pregnancy should include the following: (1) comprehensive prenatal evaluation and therapy, including cor pulmonale; (2) careful management of electrolytes; (3) optimal nutrition and enzyme therapy; (4) diagnosis of maternal diabetes, proper labor management, and care of the newborn; and (5) assessment of the infant for CF [5]. Cohen et al. described 84 pregnancies. Between conception and 6 months postpartum, there were 10 maternal deaths and significantly increased morbidity, with 60% of the women experiencing pulmonary symptomatology and an increase of 14% for cor pulmonale [6]. Palmer et al. performed a retrospective chart review of eight women who had 11 pregnancies between 1974 and 1981. Fifty-two percent of the women were pancreatic sufficient (PS) and at 95% of ideal body weight. Four had diabetes either before their pregnancy or developed it during pregnancy [7]. Corkey et al. described 11 pregnancies in seven women with 10 live births. Two of the women were pancreatic insufficient (PI) and four of the pregnancies occurred before the women knowing they had CF [8]. In a review of pregnancy occurring between 1963 and 1990, Canny et al. reported outcomes for 25 women with CF who had 38 pregnancies. Two women were diagnosed with CF after their pregnancies. Twelve of the women were PS, one woman developed pancreatitis following delivery, one was using insulin at the time of the pregnancy, and three developed gestational diabetes. Of the 38 pregnancies, 4 were interrupted by abortion, 3 were therapeutic, and 1 was spontaneous. Mean weight gain for 24 of the women for whom the information was available was 10.5 ± 5.7 kg [9].

As more women with CF had babies, reports provided survival data, but few described management techniques, including nutrition management. Case reports and epidemiological studies showed that the

Diet and Exercise in Cystic Fibrosis
http://dx.doi.org/10.1016/B978-0-12-800051-9.00001-8

long-term outcome of women with CF who become pregnant was not worse than that of CF women who did not become pregnant [2,10–15]. Some data indicated that women with better pulmonary function might do better [12,13,16]. Using CFF Patient Registry data, Fitzsimmons et al. found no differences in CF status up to 18 months following delivery [17]. McMullen et al. found no difference in rate of CF decline in women who had babies versus those who did not. No information was provided regarding fetal outcome [2]. In another review of pregnancy, Gilljam et al. found that women who were PS; did not have *Burkholderia cepacia*; and, measured by standardized spirometry, had values of forced expiratory volume at one second (FEV1%) greater than 50% demonstrated better outcome than those who did not. Postdelivery pulmonary decline was similar to those women who did not have babies [12]. In a retrospective chart review of pregnancies occurring between 1990 and 2009 in the United Kingdom, Etherington et al. reported the presence of preexisting cystic fibrosis-related diabetes (CFRD), coupled with a prepregnancy FEV1% less than 60% predicted, were associated with significantly worse outcomes [18]. Thorpe-Beeston et al. reviewed 48 pregnancies in 41 women. There were 2 miscarriages, 44 singleton births, and 2 twin births. Forty-six percent of the deliveries were preterm deliveries and all but two were delivered preterm because of medical concern for maternal well-being. Caesarean sections were performed for 48% of the deliveries. The authors noted that women should receive optimal antenatal counseling to ensure that they understand the implications of their disease on overall survival [13]. Using data from the 1994–2005 Epidemiologic Study of CF, Schechter et al. studied long-term physiologic and functional effects of pregnancy. They concluded that although pregnancy and motherhood do not appear to accelerate disease progression, these two situations may lead to more illness-related visits, pulmonary exacerbations, and a decrease in some domains of quality of life [15].

Early reports of pregnancy described poor outcomes for both the women and their babies [4–7]. The majority of women in the studies with good outcome were PS, with optimal weight and pulmonary function before conception [8,9]. On the basis of the evidence available at the time, a major recommendation was that pregnancy be attempted only by women in excellent health. Reviews of more recent population data and clinical data do not indicate a survival difference in CF women between those who became pregnant and had a baby and those who had not become pregnant [15,17]. On an individual basis, however, women with inadequate weight, inadequate pulmonary function tests, or diabetes before pregnancy may have a more compromised pregnancy outcome for both themselves and their infants [12,13,19,20]. Table 1.1 summarizes information provided in the reviewed papers.

1.3 NUTRITION: REVIEW OF THE LITERATURE

1.3.1 Weight

In all women, with or without CF, prepregnancy weight, rate of weight gain, and total weight gain during pregnancy are crucial measures to follow for both the woman's postpartum health and for her baby. Although historically pregnancy in women who had CF occurred in women who were PS and more nutritionally stable [8,9], more recent studies report pregnancy in both PS and PI women with weight, rather than pancreatic status, considered to be the important factor [2,10].

Two studies provide useful clarification. Cheng et al. reported that 7 of the 16 women he studied had a prepregnancy body mass index (BMI) of $<20\,kg/m^2$. Lower preconception BMI was associated with more hospitalizations and greater need for intravenous antibiotics. Mean weight gain was $10.4 \pm 4.5\,kg$ for the entire group and $11.4 \pm 5.0\,kg$ for women who delivered at term. Compromised maternal weight or pulmonary status resulted in nine inductions and was associated with longer postpartum length of stay as well as with a tendency toward lower fetal weight. All of the women returned to prepregnancy weight by 2 months after delivery [10]. Gilljam et al. studied 92 pregnancies in 54 women. The mean BMI both pre- and postpregnancy was $21.6\,kg/m^2$, with a mean pregnancy weight gain of $8 \pm 5\,kg$. No difference in the 10-year postpartum survival was identified between women with a prepregnancy BMI $<20\,kg/m^2$ or $>20\,kg/m^2$. Yet, PS women had better survival when compared with PI women, with no explanation from the authors as to the cause. Six infants were born before 37 weeks' gestation, four infants were considered small for gestational age, and three were classified low birth weight [12].

The increased nutritional demands of pregnancy may place women with CF at risk [21]. Lau et al. wrote, "Poor nutrition is usually a marker of more severe disease and has been associated with worse outcomes and low birth weights. Severe malnutrition ($BMI < 18\,kg/m^2$) is considered a relative contraindication to pregnancy" [20]. Goss et al. followed women for 12 years after delivery. Although the women who became pregnant had a higher BMI when compared with women who did not become pregnant, after adjusting for demographic differences, there was no difference in survival [11]. Using British registry data of 1148 females, Boyd et al. noted that 24.3% had weight below the 10th percentile, which placed them at risk for infertility, preterm birth, and delivery complications [22]. The women assessed by McMullen et al. had a higher total weight but not a higher percent of ideal body weight when compared with women who did not become pregnant. Following delivery, the ideal body weight was greater when

TABLE 1.1 Summary of Selected Papers Describing Pregnancy in Women Who Have CF

Author	Data Source/ Collection Dates	Subjects[a]	Pancreatic Insufficient (% of Subjects)	Weight Pregravid/ Gravid Gain (Range)	Diabetes Pregravid/ Gravid	Nutrition Support Oral/TF/PN	Infant Gestational Age, Weeks (Range)	Infant Birth Weight, Gms (Range)
Corkey et al. [8]	Clinic charts 1966–1980	7/7/10	29%	NI	NI	NI	40 (38–42)	3469 (2690–4235)
Canny et al. [9]	Clinic charts 1963–1990	25/38/34	52%	95% IBW ± 12/10.4 kg ± 5.7 (71–122%/0.9–24 kg)	1/3	NI	39 ± 2 (31–42)	3240 ± 650 (1100–4500)
Palmer et al. [7]	Clinic charts 1974–1981	8/14/11	50%	90% of recommended/NI	NI	NI	37 (31–40)	2450 (1100–3400)
Gilljam et al. [12]	Clinic charts 1963–1998	54/92/74	71%	NI/8 kg ± 5 (NI)	3/7	NI	40 ± 2 (NI)	3200 ± 0.6
Odegaard et al. [21]	Clinic charts 1977–1998	23/NA/33	91%	BMI 20.9/10.3 kg (16.5–23.7/1.1–26 kg)	2/4	NI/5/9	37.7 (NI)	3200 (NI)
Goss et al. [11]	National registry 1985–1997	680/993/455	58.3%	BMI 20.4 ± 3.1/NI (NI)	30/NI	NI	NI	NI
Boyd et al. [22]	National registry prior to 2001	65/84/76	65%	24% IBW < 10% tile/NI (NI)	19/NI	NI	14 preterm (NI)	NI
McMullen et al. [2]	National database 1995–2003	216/NI/NI	NI	53.5 kg ± 0.6/57.2 kg ± 0.6 (NI)	9.3%/20.8%	NI/42.1%/2.3%	NI	NI
Cheng et al. [10]	Clinic charts 1989–2004	25/43/36	63%	BMI 22.1/10.4 kg (17.7–34.9/1.8–21.4 kg)	5/5	NI/NI/6	37 ± 2 (33–39)	NI (2211–4337)
Lau et al. [20]	Clinic charts 1995–2009	18/20/19	94%	BMI 21/7.7 ± 3.2 kg (18.4–24.6/2.4–15.6 kg)	22%/43%	NI	37 ± 3 (30–41)	2781 ± 604 (1356–3875)
Burden et al. [14]	Clinic charts 2003–2011	12/15/14	NI	BMI 21.4/5.74 kg (18.3–24/1–10 kg)	28.5%/28.5%	NI	38 2 preterm	2970 (2.2–3.8)
Thorpe-Beeston et al. [13]	Clinic charts 1998–2011	41/48/48 2 sets of twins 2 miscarriages	54.2%	BMI 21.9/NI (±3.6)	NI/35.4%	NI	35.9 ± 3.3 46% preterm	2563 (700–3900)
Schechter et al. [15]	National database 1994–2005	119/119/NI	NI	BMI z score −0.23/0.06	NI/14	NI	NI	NI

NI, Not included in paper; IBW, Ideal body weight; BMI, Body mass index; TF, Enteral tube feeding; PN, Parenteral nutrition.
[a]*Number of women/number of pregnancies/number of live births.*
Modified from Michel SH, Mueller DH. Nutrition for pregnant women who have CF. J Acad Nutr Diet 2012;112:1943–1948.

compared with preconception weight [2]. In the group studied by Thorpe-Beeston et al., 25 women (52.1%) were PI, with a mean BMI at conception of 21.9±3.6 kg/m^2. No information was provided describing weight gain throughout the pregnancy [13]. In the retrospective review of 15 pregnancies in 12 women by Burden et al. the mean prepregnancy BMI was 21.4 kg/m^2 (18.3–24) with a mean weight gain of 5.74 kg (1–10) [14].

In a study of preconception counseling, women who had CF and received preconception counseling had significantly greater mean maternal weight gain and significantly heavier babies when compared with women who did not receive counseling [23]. Edenborough et al. suggested that BMI before pregnancy should be within, and preferably at the upper end, of recommended values for women [24]. Chetty et al. noted that increased preterm delivery and poor fetal growth is associated with low BMI; therefore, careful management of maternal weight is important. The authors noted that BMI of at least 90% of recommended is ideal before conception. [25]

1.3.2 Diabetes

In women without CF, elevated glucose levels during pregnancy is associated with adverse perinatal and maternal outcomes [26]. In these women, aggressive treatment of mild gestational diabetes improved outcomes [27]. Work by Hardin et al. elucidated the metabolic basis that predisposes pregnant CF women to early development of diabetes and poor weight gain, thereby requiring greater intake of calories and protein when compared with the non-CF pregnant women. The researchers noted that pregnancy in CF is associated with decreased insulin sensitivity and high hepatic glucose production, as well as decreased insulin secretion [28]. Along with increased protein turnover, response was lower to insulin's anticatabolic effect. McMullen et al. noted the most frequently documented complication was the need for the management of diabetes [2]. Nine percent of the women studied by Tonelli et al. were on diabetes treatment before pregnancy, which increased to 21% during pregnancy [29]. Both Cheng et al. and Gilljam et al. found a high rate of diabetes in the populations they studied [10,12]. In the 10 diabetic mothers followed by Gilljam et al., one infant was born at 31 weeks' gestation and died of sepsis at 18 days of life, two infants were preterm with low birth weight, and one was small for gestational age [12]. Thorpe-Beeston et al. reported that 17 women (35.4%) required insulin, but it is not clear whether the insulin was used before pregnancy or as the result of the pregnancy. Although the women required insulin, the mean birth weight percentile in the series was 31.9. The authors hypothesize that the birth weight may reflect good diabetic control, frequent clinic visits, or other metabolic factors, such as the increased energy needs of CF [13]. In the series described by Burden et al., diabetes was present in 57% (8/14), with preexisting CFRD in 28.5% (4/14) and with diagnosis during pregnancy in 28.5% (4/14) [14]. Schechter et al. reported diabetes at baseline in 11.8% of the women who became pregnant and in 12.5% in the women who did not become pregnant [15].

In women who have CF, information describing the progression of CFRD during pregnancy or the progression of those who develop gestational diabetes is limited. Therefore, it is difficult to determine either maternal or infant risk [10,30]. Data that are available imply a greater risk for both mother and infant when diabetes and poor pulmonary function are present in the mother [31]. As is true for women who do not have CF, either diabetes before pregnancy or gestational diabetes will place the woman and her baby at risk; both situations require careful overall management [13,26]. Balancing the need for maternal weight gain while managing previously existing CFRD or gestational diabetes is challenging and requires the ongoing involvement of a registered dietitian or nutritionist familiar with CF [32].

1.3.3 Vitamins and Minerals

No studies have been conducted specific to the management of vitamin and mineral nutrition for women who have CF, either for those considering pregnancy or during pregnancy. Edenborough et al. suggested folic acid at the dosage recommended for women who do not have CF [31]. Consideration is given to appropriate dosage for women with a previous infant with neural tube defects or women with diabetes [33]. Vitamin A deficiency or excess are teratogenic and associated with adverse reproductive outcomes [34]. Evaluation of serum retinol levels in a small group of pregnant women who had CF found serum levels to be within normal range on usual CF-specific vitamin supplementation [35]. Table 1.2 provides fat-soluble vitamin and zinc content of CF-specific products available for adults at the time this chapter was prepared. Information on the over-the-counter product is provided as a basis of comparison.

1.3.4 Nutrition Support

Pregnancy places an additional nutrition burden on women who have CF. Besides maintaining optimal energy and nutrient intake to meet the demands of CF, women also must meet the increased requirements imposed by pregnancy. Less than optimal prepregnancy weight, coupled with an inability to consume adequate energy for weight gain before pregnancy may be reasons to initiate oral energy supplementation or enteral tube feeding. Enteral tube feeding that is required before pregnancy may be an indicator of the challenges that

TABLE 1.2 Comparison of CF-Specific Vitamin and Mineral Supplements for Adults in the United States, July 2013[a]

MVW Complete Formulation™[b] Softgels (2 Softgels)	AquADEKs®[c] Softgels (2 Softgels)	Centrum®[d] Tablets (2 Tablets)
Vitamin A (IU) beta-carotene (%)		
32,000 88% BC	36,334 92% BC	7000 29% BC
Vitamin E (IU)		
400	300[e,f]	60
Vitamin D (IU)		
3000	1600	800
Vitamin K (mcg)		
1600	1400	50
Zinc (mg)		
20	20	22

[a]*The content of this table was confirmed in July 2013. Products also contain water-soluble vitamins.*
[b]*MVW Complete Formulation Softgels™ is a trademark of MVW Nutritionals, Inc.*
[c]*AquADEKs Softgels® is a registered trademark of Yasoo Health, Inc.*
[d]*Centrum Tablets® is a registered trademark of Wyeth Consumer Care, Inc.*
[e]*Alpha-tocopherol.*
[f]*Product contains mixed tocopherols.*

will face the woman while pregnant [31]. Moreover, during pregnancy, enteral feeding may be necessary for women with inadequate weight gain. Tube feeding combined with the elevated intra-abdominal pressure of the expanding uterus may reveal or exacerbate gastroesophageal reflux. Gastrostomy or jejunostomy tube feeding may be prescribed [19,35,36]. Total parenteral nutrition (TPN) may become necessary [10,19]. Inadequate weight gain, weight loss, intractable nausea, and vomiting were the indications for TPN in the population studied by Cheng et al. Prepregnancy BMI or lung function did not predict who would require TPN. Weight gain was less for those receiving TPN compared with those who did not require it [10]. TPN may be indicated for pregnant women who develop pancreatitis during pregnancy [37,38]. McMullen et al. reported that the percentage of women receiving both oral and parenteral nutritional supplementation increased during pregnancy and continued after pregnancy as well as during lactation [2].

1.4 CLINICAL GUIDANCE

Managing reproductive nutrition in all women includes preconception, pregnancy, and postnatal care. Nourishment, nutrients, food, and eating are intertwined with other aspects of daily life as well as with CF-related daily treatments. Each woman with CF is different. Although everyone awaits pertinent research discoveries, patients and their professional team members use current knowledge and then modify the approaches for personalized care. Medical nutrition therapy for each patient must be individualized, based on the known and unknown scientific facts, and in consideration of the patient's current health status and wishes.

Fortunately, in clinical practice for patients with CF, the standard of practice expects the patient to be at the center of an interdisciplinary team model of care and, in the United States, receiving care at a CFF-accredited center. Such centers have core teams, including a physician, registered nurse, registered dietitian, social worker, and physical or respiratory therapist. When possible before conception and during pregnancy, patients and their families are involved with additional team members in endocrinology, gastroenterology, genetics, obstetrics, psychology, and pharmacology.

1.4.1 Specific Nutrition Actions

Women who have CF can become pregnant and deliver healthy infants, although prematurity and low birth weight are possible [10,29,39]. On the basis of currently available research, pregnancy outcome is unpredictable. Designing nutrition care plans is challenging, because currently no standard of care has been established for the nutrition management of pregnant women who have CF. Consequently, nutritionally, each woman must be assessed and managed individually, before, during, and after pregnancy [10]. The specialty-trained and experienced registered dietitian or nutritionist is essential in preparing all women with CF of childbearing age for possible pregnancy by providing nutrition management directed toward optimal weight, serum vitamin and mineral levels, and overall dietary quality [14].

1.4.2 Assessment and Counseling

For women who have CF, little research evidence currently exists. Thus, it seems reasonable that overall medical nutrition therapy should rely on recommendations established for women without CF and be modified for the specific situations experienced by each individual woman who has CF [1,40]. Evidence suggests that prenatal counseling can result in better nutrition status and pregnancy outcomes [23]. For the woman planning pregnancy or who is already pregnant, the registered dietitian or nutritionist should perform a comprehensive nutrition assessment that focuses on (1) historical and current weight status and (2) historical and current dietary nutrient adequacy that includes both the intake levels of known nutrients, as well as the serum levels of glucose, fat-soluble vitamins, folic acid, and minerals such as iron and calcium.

Other recommended topics for assessment and counseling include the use of alcohol and caffeine; methods to avoid foodborne illness; appropriate intake of fish regarding potential heavy metals and toxins; and use of herbals, either as dietary supplements or as medicines [21]. All women may be at risk for developing pregnancy-exacerbated reflux and constipation or pancreatitis in PS women [24]. A woman who is diagnosed with pancreatitis, gestational diabetes, and other concomitant medical conditions, or who is an adolescent, presents additional nutritional challenges.

1.4.3 Weight, Enzymes, and Energy

The importance of optimal weight for all women during pregnancy has been well documented [40]. Weight before pregnancy determines pregnancy weight gain goals. Women at recommended body weight require less weight gain throughout pregnancy when compared with weight gain goals for women not at the recommended weight [41]. The Institute of Medicine considers "normal weight" for women who do not have CF to be a BMI of 18.5–24.9 kg/m^2 and recommends a total weight gain of 25–35 pounds. If body weight before conception is less than the recommended level, weight gain to the desired level is suggested before pregnancy; otherwise, the advised weight gain goal during pregnancy is greater [42]. Some women find it helpful to follow weight gain on a prenatal weight-gain curve, because this is a visual guide to actual weight gain compared with weight goals [43].

Central to optimal weight gain for women who are PI is the use of pancreatic enzyme replacement therapy (PERT). Pregnant women who are PI must continue to use PERT with all meals, snacks, and beverages containing fat and protein. One concern for some pregnant women with CF is phthalates. As of 2009, the Food and Drug Administration (FDA) required all pancreatic enzyme preparations to undergo all steps of a new drug application [44]. Before 2009, supplemental pancreatic enzymes contained a variety of phthalates, including dibutyl phthalate and diethyl phthalate. Not only were these forms of phthalates found in pancreatic enzyme preparations, but they also were in other medical and household items. These forms were researched and were found to be concerning. Since the legislation was enacted, these forms of phthalates have been removed from a majority of products, including medications [45] as well as the currently available FDA-approved pancreatic enzyme replacement products. With the exception of one enzyme product, currently available products contain hypromellose phthalate (HP), which is used as a coating around the enzyme beads to delay degradation by gastric hydrochloric acid. HP is a phthalic acid ester of hydroxypropyl methylcellulose and was introduced in 1971 as a cellulose derivative for enteric coating and is part of the U.S. National Formulary [46]. To date, no research has indicated a harmful effect of this form of phthalate. As suggested by the CFF, patients concerned about HP in enzymes should talk to their CF health care team, but they should not stop taking PERT [46].

Defining the energy needs of people who have CF is challenging and depends on such variables as level of maldigestion and resultant malabsorption, pulmonary function, pulmonary inflammation, fat-free mass, gender, genetic mutation, age, and other medical complications [47]. A formula that incorporates some of these variables has been published [48]. Energy recommendations specific for pregnant women who have CF remain unavailable, however. Consensus suggests the standardized formula be used as a baseline estimate, to which is added the individual caloric needs from foods and supplements of each woman with CF during pregnancy. The objective measure to be followed clinically is suitable weight gain throughout pregnancy to achieve the weight goal. If the weight gain remains inadequate, then oral, enteral, or parenteral nutrition supplementation may become necessary [31].

1.4.4 Protein

Specific information describing the protein needs of people who have CF is limited, with none available for women with CF who are pregnant. Published data from the CF population suggests that protein intake is correlated strongly with energy intake and that those people who consume adequate calories also consume adequate protein [48,49,50]. Research by Hardin et al. suggested that women who have CF and are pregnant may require higher intakes of both calories and protein when compared with pregnant women without CF [28].

1.4.5 Vitamins and Minerals

Before pregnancy, CF standard practice for women is that serum levels of fat-soluble vitamins are to be assessed routinely with adjustments made with the use of foods and dietary supplements. Levels usually are rechecked following changes to supplemental vitamin prescription [31]. Some special considerations exist. Assessment of serum retinol level and retinol intake should be evaluated in the preconceptional period and during pregnancy. Total retinol intake, including the contribution of beta-carotene, is adjusted based on serum retinol levels, with the goal of maintaining serum levels within the normal reference range [31,35]. Serum vitamin D levels are evaluated and vitamin D is supplemented as indicated by serum results [31]. There are no CF-specific folic acid supplementation recommendations. For non-CF pregnant women, the usual recommendation is 400 mcg daily and 4000–5000 mcg daily if considered to be at high risk for an affected (neural tube defects) pregnancy [31]. Dosing recommendations for the general public have been published and include levels based on risk, such as insulin-dependent diabetes [33]. Mineral nutrition is a challenge for pregnant women who have CF. Without calcium and iron recommendations specific for women who have CF, initial supplementation of these nutrients may be based on recommendations for pregnant women who do not have CF.

1.4.6. Blood Glucose

Blood glucose screening is recommended before pregnancy for women who have CF, but do not have diabetes, and who have not been screened using a 2-h oral glucose tolerance test (OGTT) in the previous 6 months [51]. The test is repeated twice: during gestation weeks 12–16 and again during weeks 24–28. For women diagnosed with gestational diabetes, another 2-h OGTT is recommended 6–12 weeks after the end of pregnancy. Blood glucose levels are monitored based on CFF recommendations [51]. Because the diet during pregnancy should contain sufficient calories to promote optimal weight gain, exogenous insulin may be required [31].

1.5 CONCLUSION

Review of peer-reviewed published research, organizational consensus statements, and clinical judgment opinions strongly indicate the enormous imperative for basic and clinical research studies. Historically and currently, the main nutritional variable identified and studied is weight, or a combination of weight and height as designated as BMI. Because anthropometric indicators only provide superficial information about the nutritional status and nutritional requirements of pregnant women who have CF, clinicians who are involved with people who have CF must become engaged actively in the research process. For example, codifying elements for a case study publication provides the underpinnings for further research endeavors. Some research is the purview of dietitians and nutritionists, whereas some research is better conducted by an interdisciplinary team. Areas to be considered for research include dietary intake and nutrition status in women with CF throughout adolescence and the childbearing years, nutrient needs during pregnancy for women who have CF and for women who experience exacerbation or other medical complications during pregnancy, and optimal methods of nutrition counseling to elucidate the best outcomes for women and their babies. Investigations on reproductive nutrition in all women with CF during all phases of life are an opportunity to advance health care practices for women who have CF and their children.

References

[1] Harvey LB, Ricciotti HA. Nutrition for a healthy pregnancy. AJLM 2013. http://dx.doi.org/10.1177/1559827613498695. Published online July 30, 2013.

[2] McMullen AH, Pasta DJ, Frederick PD, Konstan MW, Morgan WJ, Schechter MS, et al. Impact of pregnancy on women with cystic fibrosis. Chest 2006;129:706–11.

[3] Patient registry: annual data report. Bethesda, Maryland: Cystic Fibrosis Foundation; 2012.

[4] Siegel B, Siegel S. Pregnancy and delivery in a patient with cystic fibrosis of the pancreas: report of a case. Obstet Gynecol 1960;16:438–40.

[5] Grand RJ, Talamo RC, di Sant'Agnese PA, Schwartz RH. Pregnancy in cystic fibrosis of the pancreas. JAMA 1966;195:117–24.

[6] Cohen LF, di Sant'Agnese PA, Friedlander J. Cystic fibrosis and pregnancy: a national survey. Lancet 1980;2:842–4.

[7] Palmer J, Dillon-Baker C, Tecklin JS, Wolfson B, Rosenberg B, Burroughs B, et al. Pregnancy in patients with cystic fibrosis. Ann Intern Med 1983;99:596–600.

[8] Corkey CW, Newth CJL, Corey M, Levison H. Pregnancy in cystic fibrosis: a better prognosis in patients with pancreatic function. Am J Obstet Gynecol 1981;140:737–42.

[9] Canny GJ, Corey M, Livingstone RA, Carpenter S, Green L, Levison H. Pregnancy and cystic fibrosis. Obstet Gynecol 1991;77:850–3.

[10] Cheng EY, Goss CH, McKone EF, Galic V, Debley CK, Tonelli MR, et al. Aggressive prenatal care results in successful fetal outcomes in CF women. J Cyst Fibros 2006;5:85–91.

[11] Goss CH, Rubenfeld GD, Otto K, Aitken ML. The effect of pregnancy on survival in women with cystic fibrosis. Chest 2003;124:1460–8.

[12] Gilljam M, Antoniou M, Shin J, Dupuis A, Corey M, Tullis DE. Pregnancy in cystic fibrosis: fetal and maternal outcome. Chest 2000;118:85–91.

[13] Thorpe-Beeston JG, Madge S, Gyi K, Hodson M, Bilton D. The outcome of pregnancies in women with cystic fibrosis-a single centre experience 1998-2011. BJOG 2013;120:354–61.

[14] Burden C, Ion R, Chung Y, Henry A, Downey DG, Trinder J. Current pregnancy outcomes in women with cystic fibrosis. Eur J Obstet Gynecol Reprod Biol 2012;164:142–5.

[15] Schechter MS, Quittner AL, Konstan MW, Millar SJ, Pasta DJ, McMullen A. Long-term effects of pregnancy and motherhood on disease outcomes of women with cystic fibrosis. Ann Am Thorac Soc 2013;10:213–9.

[16] Edenborough FP, Stableforth DE, Webb AK, Mackenzie WE. Outcome of pregnancy in women with cystic fibrosis. Thorax 1995;50:170–4.

[17] FitzSimmons SC, Fitzpatrick S, Thompson D, Aitkin M, Fiel S, Winnie G, et al. A longitudinal study of the effects of pregnancy on 325 women with cystic fibrosis. Pediatr Pulmonol 1996;(Suppl. 13):99–101S.

[18] Etherington C, Peckham D, Clifton I, Conway S. Pregnancy and motherhood in women with cystic fibrosis: experience and outcome in a regional adult UK centre. Pediatr Pulmonol 2010;45(Suppl. 33):431S.

[19] Whitty JE. Cystic fibrosis in pregnancy. Clin Obstet Gynecol 2010;53:369–76.

[20] Lau EM, Moriarty C, Ogle R, Bye PT. Pregnancy and cystic fibrosis. Paediatr Respir Rev 2010;11:90–4.

[21] Odegaard I, Stray-Pedersen B, Hallberg K, Hannaes OC, Storrosten OT, Johannesson M. Maternal and fetal morbidity in pregnancies of Norwegian and Swedish women with cystic fibrosis. Acta Obstet Gynecol Scand 2002;81:698–705.

[22] Boyd JM, Mehta A, Murphy DJ. Fertility and pregnancy outcomes in men and women with cystic fibrosis in the United Kingdom. Hum Reprod 2004;19:2238–43.

[23] Morton A, Wolfe S, Conway SP. Dietetic intervention in pregnancy in women with CF-the importance of pre-conceptional counseling. Pediatr Pulmonol 1996;(Suppl. 13):315S.

[24] Edenborough FP, Morton A. Cystic fibrosis-A guide for clinicians in reproductive and obstetric medicine. Fetal Matern Med Rev 2010;21:36–54.

[25] Chetty SP, Shaffer BL, Norton ME. Management of pregnancy in women with genetic disorders: part 2: inborn errors of metabolism, cystic fibrosis, neurofibromatosis type 1, and turner syndrome in pregnancy. Obstet Gynecol Surv 2011;66:765–76.

[26] The HAPO Study Cooperative Research Group. Hyperglycemia and adverse pregnancy outcomes. N Engl J Med 2008;358:1991–2002.

[27] Landon MB, Spong CY, Thom E, Carpenter MW, Ramin SM, Casey B, et al. A multicenter, randomized trial of treatment for mild gestational diabetes. N Engl J Med 2009;361:1339–48.

[28] Hardin DS, Rice J, Cohen RC, Ellis KJ, Nick JA. The metabolic effects of pregnancy in cystic fibrosis. Obstet Gynecol 2005;106:367–75.

[29] Tonelli MR, Aitken ML. Pregnancy in cystic fibrosis. Curr Opin Pulm Med 2007;13:537–40.

[30] Milla CE, Billings J, Moran A. Diabetes is associated with dramatically decreased survival in female but not male subjects with cystic fibrosis. Diabetes Care 2005;28:2141–4.

[31] Edenborough FP, Borgo G, Knoop C, Lannefors L, Mackenzie WE, Madge S, et al. Guidelines for the management of pregnancy in women with cystic fibrosis. J Cyst Fibros 2008;7:2S–32S.

[32] McArdle JR. Pregnancy in cystic fibrosis. Clin Chest Med 2011;32:111–20.

[33] Wilson DR, The Genetics Committee of the Society of Obstetricians and Gynaecologists of Canada and The Motherisk Program, The Hospital for Sick Children Toronto. Pre-conceptional vitamin/folic acid supplementation 2007: the use of folic acid in combination with a multivitamin supplement for the prevention of neural tube defects and other congenital anomalies. J Obstet Gynaecol Can 2007;29:1003–13.

[34] World Health Organization. Vitamin A dosage during pregnancy and lactation: recommendations and report of a consultation; 1998. Document NUT/98.4.

[35] Stephenson AL, Robert R, Brotherwood M, Duan B, Tullis E. Vitamin A supplementation and serum vitamin A levels in pregnant women with cystic fibrosis. Pediatr Pulmonol 2008;43(Suppl. 31):420S.

[36] Pereira JL, Velloso A, Parejo J, Serrano P, Fraile J, Carrido M, et al. Percutaneous endoscopic gastrostomy and gastrojejunostomy. Experience and its role in domiciliary enteral nutrition. Nutr Hosp 1998;13:50–6.

[37] Wejda BU, Soennichsen B, Huchzermeyer H, Mayr B, Cirkel U, Dormann AJ. Successful jejunal nutrition therapy in a pregnant patient with apallic syndrome. Clin Nutr 2003;22:209–11.

[38] Virgilis D, Rivkin L, Samueloff A, Picard E, Goldberg S, Faber J, et al. Cystic fibrosis, pregnancy, and recurrent acute pancreatitis. J Pediatr Gastroenterol Nutr 2003;36:486–8.

[39] Sciaky-Tamir Y, Armony S, Elyashar-Earon H, Wilschanski M. Prolonged TPN during pregnancy in a cystic fibrosis patient with chronic pancreatitis. Eur J Obstet Gynecol Reprod Biol 2009;143:61–3.

[40] The American Congress of Obstetricians and Gynecologists. Nutrition during pregnancy, http://www.acog.org/Resources_And_Publications/Committee_Opinions/Committee_on_Obstetric_Practice/Weight_Gain_During_Pregnancy; [accessed 27.08.13].

[41] Winick M. Nutrition in pregnancy and early infancy. Baltimore, MD: Williams and Wilkins; 1989. p. 25–43.

[42] Institute of Medicine. The National Academies of Science. Weight gain during pregnancy: reexamining the guidelines. Report Brief May 2009. www.iom.edu/pregnancyweightgain; [accessed 27.08.13].

[43] Prenatal weight gain chart. www.nal.usda.gov/wicworks/Sharing_Center/NY/prenatalwt_charts.pdf; [accessed 27.08.13].

[44] Food and Drug Administration. Guidance for industry: exocrine pancreatic insufficiency drug products-submitting NDAs. April 2006. www.fda.gov/cder/guidance/index.htm; [accessed 27.08.13].

[45] Food and Drug Administration. Guidance for industry: limiting the use of certain phthalates as excipients in CDER-regulated products. December 2012. [27.08.13] www.fda.gov/Drugs/GuidanceComplianceRegulatoryInformation/Guidances/default.htm.

[46] Cystic Fibrosis Foundation. Phthalates and pancreatic enzymes. March 2012. Cff.org/livingwithcf/stayinghealthy/diet/phthalates; [accessed 27.08.13].

[47] Michel SH, Maqbool A, Hanna MD, Mascarenhas M. Nutrition management of pediatric patients who have cystic fibrosis. Pediatr Clin N Am 2009;56:1123–41.

[48] Ramsey BW, Farrell PM, Pencharz P, The Consensus Committee. Nutritional assessment and management in cystic fibrosis: a consensus report. Am J Clin Nutr 1992;55:108–16.

[49] Kawchak DA, Zhoa H, Scanlin TF, Tomezsko J, Cnaan A, Stallings VA. Longitudinal, prospective analysis of dietary intake in children with cystic fibrosis. J Pediatr 1996;129:119–29.

[50] White H, Morton AM, Peckham DG, Conway SP. Dietary intakes in adult patients with cystic fibrosis-do they achieve guidelines. J Cyst Fibros 2003;3:1–7.

[51] Moran A, Brunzell C, Cohen RC, Katz M, Marshall BC, Onady G, The CFRD Guidelines Committee, et al. Clinical care guidelines for cystic fibrosis-related diabetes. Diabetes Care 2010;33:2697–708.

CHAPTER

2

Disordered Eating and Body Image in Cystic Fibrosis

Virginia Quick[1], *Carol Byrd-Bredbenner*[2]

[1]Eunice Kennedy Shriver National Institute of Child Health and Human Development/National Institutes of Health, Division of Intramural Population Health Research, Bethesda, MD, USA,

[2]Department of Nutritional Sciences, Rutgers University, New Brunswick, NJ, USA

2.1 INTRODUCTION

Cystic fibrosis (CF) is the most common recessively inherited disease caused by defects in a single gene on a chromosome (i.e., CFTR) [1]. This condition occurs in 1 of 2500 live births [2]. The CFTR genetic defect causes increased production of abnormally thick, tenacious mucous secretions that obstruct glands and ducts throughout the body. This severe, progressive, and not-yet-curable disease affects most organs, with effects on the lungs, pancreas, intestines, and liver being particularly profound.

Lung conditions, including recurrent infections, inflammation, damage, and respiratory failure are common among CF patients due to their diminished capacity to clear respiratory secretions. Respiratory conditions in CF patients are treated with antibiotics, bronchodilators, and corticosteroids. Chest physiotherapy (i.e., a procedure in which the patient is clapped repeatedly on the back to free mucus from chest) and regular exercise are critical to clearing respiratory secretions. Treatment of respiratory conditions may cause weight fluctuations that influence body satisfaction of patients. Infections, including pulmonary infections, cause anorexia and increase resting energy expenditure, which elevates energy needs. Pulmonary infections are particularly problematic for CF patients because they exacerbate their already marginal-to-poor nutritional status, thereby further elevating their risk for infections [3]. Thus, an integral part of CF treatment involves optimizing nutritional status through dietary intake and weight gain.

The pancreas is the source of key digestive enzymes critical to carbohydrate, protein, and fat digestion. Blockage of pancreatic ducts by mucous impairs enzyme secretion and causes maldigestion and malabsorption of nutrients. The vast majority of CF patients (i.e., 85–90%) experience digestive absorptive insufficiencies, most notably of dietary fats and fat-soluble vitamins [4]. Pancreatic enzyme replacement therapy reduces the degree of malabsorption; however, many patients continue to poorly absorb nutrients and suffer from associated abdominal pain, bloating, and irregular bowel habits. These common gastrointestinal problems may suppress appetite or cause CF patients to alter dietary intake by purposely avoiding foods that they associate with gastrointestinal distress or reducing overall food intake [5]. Over time, pancreatic insufficiencies also may cause some patients to develop CF-related diabetes, which requires exogenous insulin and daily monitoring of blood glucose levels and often requires further adjustments to dietary intake [6]. Overall, pancreatic insufficiencies, gastrointestinal problems, and CF-related diabetes influence the diet quality [7] and quality of life of CF patients [8].

Abnormal liver function or fat in the liver is common in CF patients. In some instances, advanced liver disease, called CF liver disease (CFLD), occurs. Previous research has found that CFLD onset mainly occurs in the first decade of life with a prevalence rate of 41% among 12-year-olds [9]. CFLD risk factors include a history of meconium ileus (i.e., born with thick secretions blocking the lower part of the small intestine) and pancreatic insufficiency [9]. In CFLD, it is thought that thick bile obstructs the intrahepatic ducts, resulting in inflammation and scarring of the liver. As liver scarring becomes more severe (i.e., cirrhosis), pressure in the vein leading to the liver from the intestinal tract and spleen (the portal vein) rises and complications of portal hypertension can occur. Complications include an enlarged spleen, fluid

Diet and Exercise in Cystic Fibrosis
http://dx.doi.org/10.1016/B978-0-12-800051-9.00002-X

accumulation in the abdomen (i.e., ascites), and bleeding from the dilated veins (i.e., varices) in the esophagus and stomach. There is no known cure for CFLD, but drug therapy may help prevent or reduce related health problems, such as ursodeoxycholic acid therapy (i.e., drug treatment of chronic cholestatic liver disease) [10]. Poor nutritional status also is associated with poor prognosis and survival for adults with CFLD [11].

Maintaining a healthy body weight and optimal nutritional status in people with CF is associated with better health outcomes and survival [12]. Optimal dietary management, however, is an extremely demanding treatment regimen for CF patients. Maintaining a healthy body weight and adequate nutritional status can be difficult because of pancreatic insufficiency, recurrent infections, and associated increased energy needs and anorexia, as well as the regular exercise routines needed to clear lung secretions [2,13]. Long-term treatment includes energy intakes of 120–150% of daily recommended calories for individuals without CF, with 40% of the calories coming from fat [14], coupled with vitamin and mineral supplementation and pancreatic enzyme replacement therapy [2]. Calorie requirements are increased further when there are chest exacerbations (i.e., infections in the lungs). In some instances, oral and enteral tube feedings (i.e., nasogastric or percutaneous endoscopic gastrostomy) may be necessary when those with CF are unable to meet nutrient needs from food sources. Dietary management becomes increasingly complex for individuals who also develop CF-related diabetes.

CF itself along with the difficulty of adhering to the CF dietary regimen may lead to low body weight, short stature, and pubertal delay [15]. In addition, recurrent lung infections contribute to body shape aberrations, such as protruding sternum, rounded shoulders, and clubbed fingers and toes [16]. These outwardly visible changes may precipitate quality of life issues, negative body image cognitions, and unhealthy dietary behaviors in CF patients that interfere with treatment adherence.

2.2 QUALITY OF LIFE

Fortunately, advances in CF management have enabled more CF patients to survive into adulthood, with half of CF children born in the 1990s expected to live into their 40s and older [17,18]. Despite these advancements, the progressive and complex nature of CF, coupled with the highly demanding and intensive dietary and therapeutic treatments, has the potential to affect their quality of life.

Social functioning (e.g., reduced time spent with peers), coping with illness, emotional well-being (e.g., self-esteem, depression), concerns for the future (e.g., shorter life expectancy), interpersonal relationships, body image, career issues, and general health perceptions are all quality-of-life domains that may be affected and are risk factors for disordered eating [19,20]. In particular, girls with CF report illness-related stresses and worries (e.g., emotional strains), treatment discouragement, low self-esteem, and low adherence to some aspects of the CF treatment (e.g., eating high-fat foods, taking medications) [21].

Physical growth (e.g., short stature and low weight) and pubertal development delays may lead to body image concerns [22]. For instance, the visible differences in heights of boys with CF and their healthy peers may prompt those with CF to overcompensate for their shorter stature in other ways, such as by increasing risk-taking behaviors (e.g., use of anabolic steroids) to achieve an appearance similar to that of their peer group [23]. Developing a positive self-image in the presence of CF-associated growth and body shape issues is one of the many challenges CF patients face as they grow into their teen and adult years [13]. Given the rigorous dietary management demands, appearance differences, and typical desire to "fit in", CF patients may be at risk for body image disturbances [24] and disordered eating [25].

2.3 BODY IMAGE

Body image is a complex and multifaceted construct. In general, body image is how an individual perceives (i.e., thoughts and feelings) his or her own body and how the individual perceives that others view him or her [26,27]. Thus, body image is a delimiter of social interactions and has a profound effect on physical health, social interactions, psychological development, and interpersonal relationships [28]. A person may have body dissatisfaction because of a discrepancy between one's real and ideal body image [27]. Young people are at increased risk for body image disturbances that commonly involve body weight and shape dissatisfaction because adolescence is an important developmental period during which body and social changes occur [29]. Other body image concerns among girls and boys, and particularly among people with CF, include physical and appearance attributes, such as facial characteristics, skin appearance, muscularity, fitness, and strength [24,27].

Societal and media messages portraying how male and female bodies "should" appear reveal important gender differences in body image [30]. In most of the industrial world, an emphasis on the functional capacity of the body is apparent for males, whereas in females, the body is judged on aesthetics [31]. Thus, males often want to be lean and muscular, whereas females commonly wish to be thin. Those with CF tend to have difficulty maintaining normal body weights, which likely is why males with CF tend to have greater body dissatisfaction than both females with CF and healthy men [32]. This body dissatisfaction experienced by males with CF is related to their desire to reflect the male body ideal (i.e., lean and muscular) and could elevate their risk for anabolic steroid abuse as a means for achieving the desired muscular shape; however, studies exploring this issue are limited [23].

In a qualitative study, females with CF reported being happy about their thinness and associated thinness with attractiveness [33]. Furthermore, females with CF stated they were not concerned about their low body weight, even though they knew it was dangerous for their health [33]. Although females with CF have less body dissatisfaction than their healthy peers and males with CF [32], thinness is detrimental to their physical health, prognosis, and survival [12]. Indeed, females with CF that value a thin body frame may be less motivated at working toward maintaining an optimal body weight, which may be contributing to the poorer survival rates of females than males with CF [34].

Only two studies on body image that include non-CF controls could be located [5,32]. In general, non-CF controls reported being less satisfied with their facial appearance than CF patients and received less external pressure to eat than CF patients [32]. Females with CF desired a low body weight (i.e., body mass index [BMI] 18–19), which is similar to their actual weight and, thus, were more content with their shape than healthy peers [32]. However, females with CF who perceived themselves as having a higher BMI, restricted calories to a greater extent than healthy peers despite their greater calorie need [5]. Male controls had an accurate view of their body weight and were content with their size [32]. Males with CF wanted to be bigger, however, and thus were more motivated to adhere to nutritional advice than females with CF [5,32]. A recent review on body image among adults and adolescents with CF confirms these findings and suggests that practitioners approach the topic of body image at clinic appointments to ensure that body dissatisfaction does not have a detrimental impact on disease self-management [24].

2.4 EATING BEHAVIORS

Disordered eating and eating disorders commonly are used interchangeably; however, they are distinctively different. Eating disorders, such as anorexia nervosa, bulimia nervosa, and binge eating disorder, are complex mental illnesses that require psychiatric diagnoses based on specific criteria in the *Diagnostic Statistical Manual of Mental Disorders*, 5th edition (*DSM-5*) [35]. Eating disorders often lead to serious physical problems, such as cardiac conditions and osteoporosis, and, in some instances, can become life threatening [36]. Around 4% of women and 1% of men in the United States are affected by eating disorders [37].

On the other hand, disordered eating is far more common than eating disorders [38]. Disordered eating involves an unhealthy relationship with food, causing disturbed eating behaviors such as restrictive eating, dieting, laxative abuse, binge eating, and self-induced vomiting that occurs less frequently (i.e., typically caused by a particular event) or are less severe than those required to meet the full criteria for a psychiatric diagnosis of an eating disorder. When disordered eating becomes persistent and more frequent, this can lead to a psychiatric diagnosis of an eating disorder.

Common disordered eating behaviors and attitudes among CF patients include atypical eating disorder behaviors, such as spitting out chewed food [39], restricting fat intake [40], food avoidance, preoccupation with food, bulimic tendencies [32], and misuse of pancreatic enzyme (i.e., digestive enzyme) replacement therapy [41]. The disordered eating behaviors and cognitions presented in those with CF pose harm to their health [41]. Disordered eating practices and risk factors for young people with CF can be found in Table 2.1 (adapted from a recent review article) [20].

2.4.1 Parental Perspective of CF Children's Eating Behaviors

Parents find it difficult to manage the dietary needs of children with CF [39,45]. Parents may feel strong pressure to ensure that their CF child's eating habits are optimal

TABLE 2.1 Disordered Eating Practices and Risk Factors of Young People with Cystic Fibrosis

Cystic Fibrosis Patients
Types of disordered eating behaviors documented: Bulimic tendencies, misuse of pancreatic enzyme replacement medications for purposes of controlling weight, spitting out chewed food, food avoidance, and food preoccupation [32,39,41].
Potential factors increasing risk of disordered eating
• Delayed growth and onset of puberty [42].
• Low body weight [24,42].
• Increased disease severity [42].
• Preoccupation with disease and dietary management (higher caloric needs in cases in which 40% total calories required are from fat, pancreatic enzyme replacement therapy) [43].
• Pressure to manage weight (usually to gain weight) and exercise regularly to help pulmonary function [42].
• Presence of comorbid conditions (e.g., cystic fibrosis-related diabetes) [6].
• Reduced social functioning (e.g., extended time away from peers) [19].
• Impaired interpersonal relationships (family, friends, significant others) [19].
• Stressful family mealtimes [39].
• Lack of coping skills [19].
• Poor emotional well-being (e.g., self-esteem, depression) [8,19,44].
• Body image disturbances [24,44].
• Concerns for the future (shorter life expectancy) [19].

Adapted from Quick and et al. [20].

as a result of the emphasis health professionals put on the significance of the child adhering closely to the prescribed dietary regimen and need for weight gain [43].

A recent meta-analysis suggests that families with a CF child encounter more difficulties during mealtimes than families without CF children [46]. Parents of CF children have reported that their children's mealtime behaviors are more stressful than parents who do not have a chronically ill child [39]. Problematic mealtime behaviors reported by CF parents are as follows: CF child has a poor appetite, has problems chewing food, is reluctant to eat at mealtimes, takes more than 20 min to finish meals, spits out food [39], and is a picky eater [45]. The etiology of these mealtime feeding problems and disordered eating behaviors are not well understood and can increase over time, thereby placing CF patients at risk for poor health outcomes [39].

2.4.2 Development of Disordered Eating Behaviors

Figure 2.1 is a proposed model for development of disordered eating in the CF population. CF management demands for optimal health and longevity, such as dietary and pulmonary management, medications, and infection or liver treatment, affect weight status [47,48]. For instance, poor dietary management may lead to insufficient weight gain [5], which is perceived favorably by females who want to maintain a slim profile, whereas insufficient weight gain is perceived unfavorably by males who want larger body sizes [33]. Furthermore, insufficient weight gain (i.e., low BMI) is associated with body dissatisfaction in males [44]; in contrast, females prefer a low body weight and fear needing to gain weight to achieve optimal health and ensure survival [33].

Social and environmental influences may potentiate the negative feelings about weight and shape experienced by CF patients. In the general population, negative feelings about weight and shape (i.e., body dissatisfaction) can be influenced by social and environmental influences, such as the media (i.e., sociocultural messages about ideal body forms), peer and family calorie restriction (dieting) behaviors, and weight concerns, and exposure to weight teasing [49–51]. Although little is known about the association between body dissatisfaction and social and environmental influences among CF

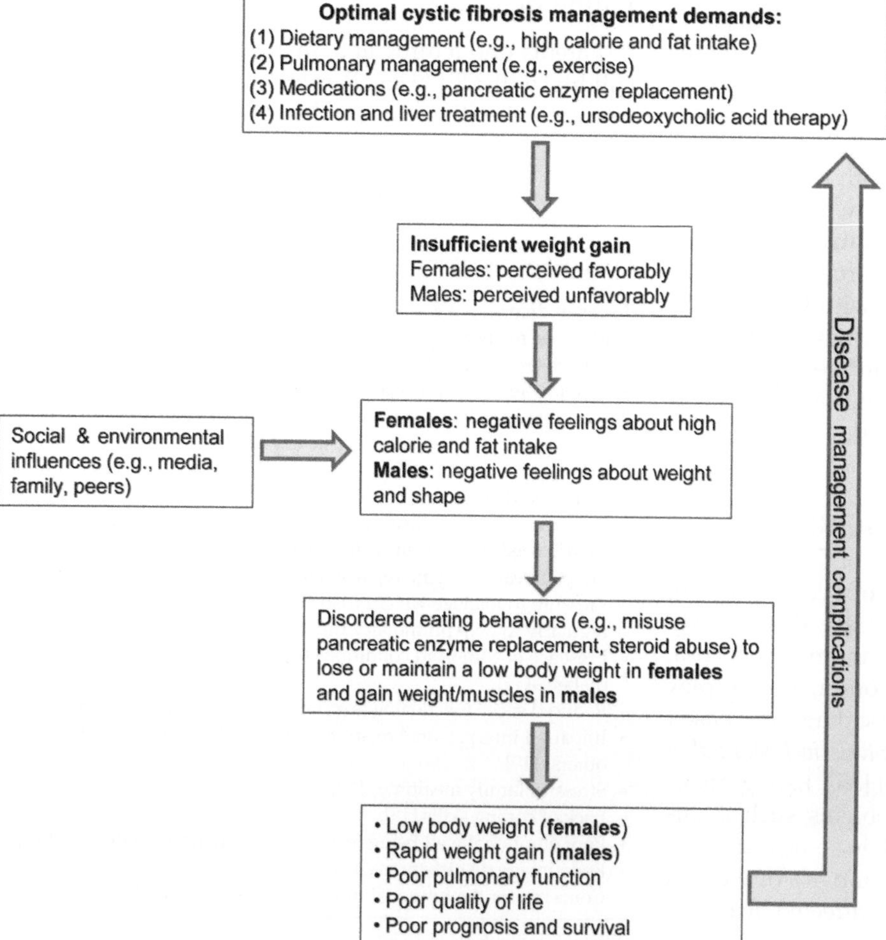

FIGURE 2.1 **Proposed model for development of disordered eating in cystic fibrosis population.**

patients, associations observed in the general population likely also hold true for those with CF, particularly given that these patients are exposed to the same social and environmental influences as their healthy peers.

Negative feelings about weight and shape can precipitate disordered eating behaviors [40,52] that cascade into weakened physical health, diminished quality of life, and poorer prognosis for survival—all of which cause further disease management complications [41]. Research that examines the risk factors supporting and thwarting the progression of disordered eating in those with CF could help practitioners develop treatment interventions to effectively halt the development of disordered eating behaviors and support optimal CF management.

2.4.3 Prevalence of Disordered Eating and Eating Disorders in CF

Disordered eating behaviors in CF patients do occur [41], but little research has examined prevalence and severity, and the limited studies available have conflicting results. Some studies have suggested that CF patients are at increased risk for eating disorders (i.e., Eating Disorders Not Otherwise Specified) [5,52], whereas other studies report that the prevalence rates of eating disorders in those with CF are no different from rates in the general population [32,53,54]. Conflicting results likely are due to differences in the eating disorder risk-screening tools used, along with the age of diagnosis, duration and severity of CF at time of assessment, type of medical care received (e.g., general vs. specialized physician, nutrition counseling, psychological counseling), and salient psychographic characteristics not being taken into account.

The most recent study examining the prevalence of eating disorders and disordered eating in adolescents with CF (N=55) revealed that none of the CF participants met criteria for a diagnosis of anorexia nervosa or bulimia nervosa; however, 53% demonstrated disordered eating attitudes (e.g., high concerns about body weight and shape) and 16% had disordered eating behaviors [53]. Although no formal diagnoses of eating disorders were evident in this study, it is apparent that disordered eating behaviors and attitudes are common in those with CF. Thus, there is a need for health professionals working with CF patients to regularly monitor disordered eating behaviors and attitudes among young people with CF during clinic visits to identify those who may need help early when treatment is most effective [55].

It is possible that having CF lends protection against the development of eating disorders because of the high frequency of contact with health professionals. Indeed, CF patients are monitored constantly for weight and nutritional status by their CF medical team and parents. The high rate of disordered eating behaviors and attitudes, however, may be a function of the level of pressure placed on weight, dietary intake, and exercise faced by an individual with CF [44,53].

2.5 BODY DISSATISFACTION AND DISORDERED EATING: IDENTIFICATION, TREATMENT, AND INTERVENTION STRATEGIES

Body dissatisfaction is associated with disordered eating and eating disorders [56–58], and this dissatisfaction has the potential to influence CF patients' self-management and motivation to adhere to CF treatment. Given that body dissatisfaction is a risk factor for disordered eating and eating disorders, it is imperative that health care providers and CF parents are aware of CF body image concerns. The pressures placed on CF patients to achieve and maintain a high body weight through eating a high-calorie diet runs counter to the social culture of the environment in which they live, where images portrayed in the media of extremely thin women and lean, muscular men are idolized.

2.5.1 Health Care Providers

Regular screening for body dissatisfaction and disordered eating is needed for early detection and treatment. No brief screening tools for disordered eating have been developed and validated for the CF population. Nevertheless, monitoring weight, disease management behaviors, and related cognitions of young people with CF by asking a few questions at regular clinical visits are important. Screening questions could include the following: (1) "How do you feel about your weight and body size? Do you feel your weight is too low, just about right, or too heavy?" (probe to determine why a patient feels this way), and (2) "In the last month, have you done one or more of the following things to manage your weight or shape: taken diet pills or anabolic steroids, skipped meals, self-induced vomiting, restricted food intake, or omitted pancreatic enzyme medications?"

Health care providers can use a brief, eight-item body image attitudes scale for young people with CF to assess body image (e.g., "I am content with my physical appearance") and importance (e.g., "It is important to me that I look good"), and trust (e.g. "I can count on my body") [59] (Table 2.2). All items are answered on a Likert scale ranging from 1 (exactly true) to 5 (not at all true). Some items are reverse scored before summing all of the items for a total score and then standardized for a score range of 0–100. Higher scores represent a more positive body image and a higher importance being placed on aspects of the body. Preliminary evidence indicates that

TABLE 2.2 Brief, Diagnostic Scale for Assessing Body Image Satisfaction in Cystic Fibrosis Patients

Body Satisfaction Scale Items[a]
IMPORTANCE DOMAIN
1. It is important to me that I look good.
2. To train my physical fitness is important to me.
EVALUATION DOMAIN
3. I am content with my physical appearance.
4. Sometimes I feel shamed because of my body.
5. I am proud of my body.
TRUST DOMAIN
6. I can count on my body.
7. I am able to relax easily.
8. I am worried about my physical condition.

Body satisfaction scoring methodology
1. Score each item. Exactly true, mostly true, moderately true, hardly true, and not at all true are scored as 5, 4, 3, 2, and 1, respectively, except for items 4 and 8 that are reverse scored.
2. Sum individual item scores in a domain to create a domain score. Sum all domain scores to create a total score. Higher overall scores indicate greater body satisfaction.
3. To transform total scores to a 0 to 100 point system, use this equation: [((total score − 8)/32) * 100].
[a]*Answer choices for all items are exactly true, mostly true, moderately true, hardly true, not at all true.*
Instrument developed and validated by Wenninger and colleagues [59]. Reprinted by permission of the authors.

the Body Image Scale is reliable and has clinical validity [59]. The Body Image Scale, however, has yet to be cross-validated in a larger sample of participants and norms need to be calculated to facilitate interpretation of scores. By screening for body dissatisfaction and disordered eating, health care providers can gain a better understanding of their patients' psychological well-being and assist patients and families in addressing these problems. Psychological support should be provided as needed, especially at times when young people with CF are emotionally distressed [13].

The CF Foundation recommends using a behavioral approach to address problematic mealtime behaviors that can be integrated into the standard nutrition care plan [14,47,60]. Previous behavioral interventions have shown promise at decreasing problematic eating along with improving weight and caloric intake among children with CF [5,61,62]. Few of these behavioral interventions comprehensively examined eating disorder symptomology and body dissatisfaction. Abbott et al. explored the effects of a nutrition intervention on adult CF patient's perceptions and behaviors concerning body image and eating [5]. Results revealed that those receiving the nutrition intervention were more aware of their actual BMI, engaged in fewer disordered eating behaviors, and were less occupied with food compared with their counterparts who did not receive the nutrition intervention [5]. In addition, those receiving the nutrition intervention wanted to be heavier and have a normal BMI, suggesting improvement in understanding the value of being heavier for survival. Replication of this nutrition intervention in younger persons with CF is warranted. Overall, a nutritional management approach that takes into account the CF patient's attitudes toward eating, shape, and personal appearance should be considered [53]. For instance, emphasizing healthy eating rather than pressuring patients to eat a high calorie and fat diet might be a more acceptable dietary approach for some CF patients.

Interventions aimed at teaching young people with CF that being attractive does not depend on the appearance of their bodies could improve body image as well [24]. Teachable moments from health care providers or parents on media literacy (e.g., rejecting or challenging images and messages that could endanger body image on a regular basis) could help to improve self-image [63].

2.5.2 Parents and Families

Some parents of CF children may too rigidly control the management of their children's condition. This overprotectiveness can interfere with the child's individuation process. Other parents may feel overwhelmed and frustrated with managing their children's condition and become disengaged, which can interfere with optimal disease management.

Optimal treatment strategies and interventions should include the CF patient and family and seek to maintain and improve family functioning. Positive family functioning improves emotional well-being, and psychosocial and medical outcomes for young people who suffer from chronic illnesses such as CF [64]. A meta-analysis revealed that positive family functioning also is related to a higher weight status of children with CF [46].

Frequent family meals also may help to defend against body dissatisfaction and disordered eating because it allows time for parents to "check-in" with their child and provide emotional support when needed [45]. Indeed, CF families report that effectively communicating with the CF team, including asking for help when needed and eating family dinners together had a positive impact on CF management [45]. Parents also mentioned that communicating more openly and honestly with their child about CF is beneficial for the child in learning about their condition and better understanding the importance of adhering to treatment recommendations [45]. Although research is limited on interventions for body image and disordered eating with CF [65], findings suggest that a behavioral family systems approach at improving family functioning may be protective against body dissatisfaction and disordered eating among young people [64,66].

2.5.3 Peer Socialization

Evidence also suggests that encouraging socialization with peers having chronic illnesses may promote lateral or downward social comparisons (i.e., comparing oneself to others who are less fortunate) and enable individuals to develop empathy for others while also gaining a more positive attitude toward their own body and health [67,68]. In addition, young people who surround themselves with others who have positive body images and are understanding and supportive of their illness are more likely to be protected from body dissatisfaction [69].

2.6 CONCLUSION

Health care providers and families need to be aware of the body image concerns and disordered eating behaviors that may be precipitated when coping with CF and regularly monitor patient psychosocial well-being to intervene early. Brief, valid, and reliable disordered eating and body image screening measures specially developed for those with CF are needed to facilitate monitoring and diagnosis of CF patients. Treatment strategies and interventions for preventing negative body image and disordered eating that use a behavioral approach involving the CF patient's family are promising. Future longitudinal studies should explore risk and protective factors of disordered eating and body dissatisfaction in the CF population to increase our understanding of the etiology and course of CF management and treatment.

References

[1] Kerem B, Rommens J, Buchanan J, Markiewicz D, Cox T, Chakravarti A, et al. Identification of the cystic fibrosis gene: genetic analysis. Science 1989;245:1073–80.

[2] Mahan LK, Escott-Stump S. Krause's food, nutrition, & diet therapy. 11th ed. Philadelphia: Saunders; 2004.

[3] Steinkamp G, Wiedemann B. On behalf of the German CFQA group. Relationship between nutritional status and lung function in cystic fibrosis: cross sectional and longitudinal analyses from the German Quality Assurance Project. Thorax 2002;57:596–601.

[4] Borowitz D. Pathophysiology of gastrointestinal complications of cystic fibrosis. Semaine Respir Crit Care Med 1994;15:391–4.

[5] Abbott J, Morton A, Musson H, Conway S, Etherington C, Gee L, et al. Nutritional status, perceived body image and eating behaviors in adults with cystic fibrosis. Clin Nutr 2007;26:91–9.

[6] Brennan A, Gedde D, Gyi K, Baker E. Clinical importance of cystic fibrosis-related diabetes. J Cyst Fibros 2004;3:209–22.

[7] Murphy J, Wooton S. Nutritional management in cystic fibrosis—an alternative perspective in gastrointestinal function. Disabil Rehabil 1998;20:226–34.

[8] Abbott J, Hart A, Morton A, Gee L, Conway S. Health-related quality of life in adults with cystic fibrosis: the role of coping. J Psychosom Res 2008;64:149–57.

[9] Lamireau T, Monnereau S, Martin S, Marcotte J, Winnock M, Alvarez F. Epidemiology of liver disease in cystic fibrosis: a longitudinal study. J Hepatol 2004;41:920–5.

[10] Kappler M, Espach C, Schweiger-Kabesch A, Lang T, Hartl D, Hector A, et al. Ursodeoxycholic acid therapy in cystic fibrosis liver disease. Aliment Pharmacol Ther 2012;36(3):266–73.

[11] Chryssostalis A, Hubert D, Coste J, Kanaan R, Burgel P, Desmazes-Dufeu N, et al. Liver disease in adult patients with cystic fibrosis: a frequent and independent prognostic factor associated with death or lung transplantation. J Hepatol 2011;55(6):1377–82.

[12] Beker L, Russek-Cohen E, Fink R. Stature as a prognostic factor in cystic fibrosis survival. J Am Dietetic Assoc 2001;101:438–42.

[13] Segal T. Adolescence: what the cystic fibrosis team needs to know. J R Soc Med 2008;101:S15–27.

[14] Borowitz D, Baker R, Stallinings V. Consensus report on nutrition for pediatric patients with cystic fibrosis. J Pediatr Gastroenterol Nutr 2002;35:246–59.

[15] Kepron W. Cystic fibrosis: everything you need to know. New York: Firefly Books; 2003.

[16] Harris A, Super M. Cystic fibrosis, the facts. 3rd ed. Oxford: Oxford University Press; 1995.

[17] Dodge J, Morison S, Lewis P, Coles E, Geddes D, Russell G, et al. Incidence, population and survival of cystic fibrosis in the UK (1968–95). Arch Dis Child 1997;77:493–6.

[18] Elborn J, Britton J, Shale D. Cystic fibrosis: current survival and population predictions until the year 2000. Thorax 1991;46:881–5.

[19] Gee L, Abbott J, Conway S, Etherington C, Webb A. Quality of life in cystic fibrosis: the impact of gender, general health perceptions and disease severity. Eur Cyst Fibros Soc 2003;2:206–13.

[20] Quick V, Byrd-Bredbenner C, Neumark-Sztainer D. Chronic illness and disordered eating: a discussion of the literature. Adv Nutr 2013;4:277–86.

[21] Patterson J, Wall M, Berge J, Milla C. Gender differences in treatment adherence among youth with cystic fibrosis: development of a new questionnaire. J Cyst Fibros 2008;7:154–64.

[22] Sawyer S, Rosier M, Phelan P, Bowes G. The self-image of adolescents with cystic fibrosis. J Adolesc Health 1995;16:204–8.

[23] Morris A, Ledson M, Walshaw M. Anabolic steroid use in CF: a two year follow-up report. Paper presented at the meeting of the European Cystic Fibrosis Conference. Valencia; 2010.

[24] Tierney S. Body image and cystic fibrosis: a critical review. Body Image 2012;9:12–9.

[25] Quick V, McWilliams R, Byrd-Bredbenner C. A case-control study of disturbed eating behaviors and related psychographic characteristics in young adults with and without diet-related chronic health conditions. Eat Behav 2012;13(3):207–13.

[26] Wertheim E, Paxton S. Body image development in adolescent girls. In: Cash T, Smolak L, editors. Body image development in adolescent girls. New York: The Guilford Press; 2011. pp. 76–84.

[27] Thompson J, Heinberg L, Altabe M, Tantleff-Dunn S. Exacting beauty: theory, assessment and treatment of body image disturbance. Washington, DC: American Psychological Association; 1999.

[28] Biordi DL, Galon PM. Body image. In: Lubkin I, Larsen P, editors. Body image. 8th ed. Burlington, MA: Jones & Barlett Learning; 2013. pp. 133–59.

[29] Suris J, Michaud P, Viner R. The adolescent with a chronic condition. Part I: developmental issues. Arch Dis Child 2004;89(10):938–42.

[30] Tiggemann M. Sociocultural perspectives on human appearance and body image. In: Cash T, Smolak L, editors. Sociocultural perspectives on human appearance and body image. 2nd ed. New York: The Guilford Press; 2011. pp. 12–9.

[31] Cash T, Pruzinsky T. Body image; a handbook of theory, research & clinical practice. New York: The Guilford Press; 2002.

[32] Abbott J, Conway S, Etherington C, Fitzjohn J, Gee L, Morton A, et al. Perceived body image and eating behavior in young adults with cystic fibrosis and their healthy peers. J Behav Med 2000;23(6):501–17.
[33] Willis E, Miller R, Wyn J. Gendered embodiment and survival for young people with cystic fibrosis. Soc Sci Med 2001;53:1163–74.
[34] Rosenfield M, Davis R, Fitzsimmons S, Pepe M, Ramsey B. Gender gap in cystic fibrosis mortality. Am J Epidemiol 1997;145:794–803.
[35] Diagnostic and statistical manual of mental disorders. 5th ed. Washington DC: American Psychiatric Association; 2013.
[36] Hoek H. Incidence, prevalence and mortality of anorexia and other eating disorders. Curr Opin Psychiatry 2007;19(4):389–94.
[37] Hudson J, Hiripi E, Pope H, Kessler R. The prevalence and correlates of eating disorders in the national comorbidity survey replication. Biol Psychiatry 2007;61(3):348–58.
[38] Grilo C. Eating and weight disorders. New York, NY: Psychology Press; 2006.
[39] Crist W, McDonnell P, Beck M, Gillespie C, Barrett P, Mathews J. Behavior at mealtimes and the young child with cystic fibrosis. Dev Behav Pediatr 1994;15:157–61.
[40] Walters S. Sex differences in weight perception and nutritional behavior in adults with cystic fibrosis. J Hum Nutr Diet 2001;14:83–91.
[41] Pumariega A, Pursell J, Spock A, Jones J. Eating disorders in adolescents with cystic fibrosis. J Am Acad Child Psychiatry 1986;25(2):269–75.
[42] Sawyer S, Drew S, Yeo M, Britto M. Adolescents with a chronic condition: challenges living, challenges treating. Lancet 2007;369:1481–9.
[43] Chase H, Long M, Lavin M. Cystic fibrosis and malnutrition. J Pediatr 1979;95:337–47.
[44] Truby H, Paxton S. Body image and dieting behaviors in cystic fibrosis. Pediatrics 2001;107(6):e902.
[45] Filigno S, Brannon E, Chamberlin L, Sullivan S, Barnett K, Powers S. Qualitative analysis of parent experiences with achieving cystic fibrosis nutrition recommendations. J Cyst Fibros 2012;11: 125–30.
[46] Hammons A, Fiese B. Mealtime interactions in families of a child with cystic fibrosis: a meta-analysis. J Cyst Fibros 2010;9:377–84.
[47] Stallings V, Stark L, Robinson K, Fernanchak A, Quinton H. Evidence-based practice recommendations for nutrition-related management of children and adults with cystic fibrosis and pancreatic insufficiency: results of a systematic review. J Am Diet Assoc 2008;108(5):832–9.
[48] Anthony H, Bines J, Phelan P, Paxton S. Relation between dietary intake and nutritional status in cystic fibrosis. Arch Dis Child 1998;78:443–7.
[49] Quick V, Eisenberg M, Bucchianeri M, Neumark-Sztainer D. Prospective predictors of body dissatisfaction in young adults: 10-year longitudinal findings. Emerg Adulthood 2013;1(4):271–282.
[50] van den Berg P, Paxton S, Keery H, Wall M, Guo J, Neumark-Sztainer D. Body dissatisfaction and body comparison with media images in males and females. Body Image 2007;4:257–68.
[51] Worobey J. Early family mealtime experiences and eating attitudes in normal weight, underweight and overweight females. Eating Weight Disord 2002;7:39–44.
[52] Shearer J, Bryon M. The nature and prevalence of eating disorders and eating disturbance in adolescents with cystic fibrosis. J R Soc Med 2004;97:36–42.
[53] Bryon M, Shearer J, Davies H. Eating disorders and disturbance in children and adolescents with cystic fibrosis. Child Health Care 2008;36:67–77.
[54] Raymond N, Chang P, Crow S, Mitchell J, Bieperink B, Beck M, et al. Eating disorders in patients with cystic fibrosis. J Adolesc 2000;23:359–63.
[55] DeSocio J, O'Toole J, Nemirow S, Lukach M, Magee M. Screening for childhood eating disorders in primary care. Primary Care Companion J Clin Psychiatry 2007;9:16–20.
[56] Mcvey G, Pepler D, Davis R, Flett G, Abdolell M. Risk and protective factors associated with disordered eating during early adolescence. J Early Adolesc 2002;22:75–95.
[57] Neumark-Sztainer D, Wall M, Haines J, Story M, Sherwood N, van den Berg P. Shared risk and protective factors for overweight and disordered eating in adolescents. Am J Prev Med 2007;33(5):350–69.
[58] Quick V, Byrd-Bredbenner C. Disturbed eating behaviors and associated psychographic characteristics of young adults. J Hum Nutr Diet 2013. http://dx.doi.org/10.111/jhn.12060. Advanced publication.
[59] Wenninger K, Weiss C, Wahn U, Staab D. Body image in cystic fibrosis: development of a brief diagnostic scale. J Behav Med 2003;26:81–94.
[60] Borowitz D, Robinson K, Rosenfeld M, Davis S, Sabadosa K, Spear S, et al. Cystic fibrosis foundation evidence-based guidelines for management of infants with cystic fibrosis. J Pediatr 2009;155:S73–93.
[61] Stark L, Opipari L, Spieth L, Jelalian E, Quittner A, Higgins L, et al. Contribution of behavior therapy to dietary treatment in cystic fibrosis: a randomized controlled study within 2-year follow-up. Behav Ther 2003;34:237–58.
[62] Stark L, Mulvihill M, Powers S, Jelalian E, Keating K, Creveling S, et al. Behavioral interventions to improve calorie intake of children with cystic fibrosis: treatment versus wait list control. J Pediatr Gastroenterol Nutr 1996;22(3):240–53.
[63] Manganello J. Health literacy and adolescents: a framework and agenda for future research. Health Educ Res 2008;23(5):840–7.
[64] Cohen M. Families coping with childhood chronic illness: a research review. Fam Syst Health 1999;17:149–64.
[65] Neumark-Sztainer D, Story M, Falkner N, Beuhring T, Resnick M. Disordered eating among adolescents with chronic illness and disability. Arch Pediatr Adolesc Med 1998;152:871–8.
[66] Drotar D. Relating parent and family functioning to the psychological adjustment of children with chronic health conditions: what have we learned? what do we need to know? J Pediatr Psychol 1997;22:149–65.
[67] Meltzer L, Rourke M. Oncology summer camp: benefits of social comparison. Child Health Care 2005;34(4):305–14.
[68] Pinquart M. Body image of children and adolescents with chronic illness: a meta-analytic comparison with healthy peers. Body Image 2013;10:141–8.
[69] Michaud P, Suris J, Viner R. The adolescent with a chronic condition: Epidemiology, developmental issues and health care provision. Geneva, Switzerland: WHO Press, World Health Organization; 2007.

CHAPTER

3

Neonatal Screening and Nutrition/ Growth in Cystic Fibrosis: A Review

Donatello Salvatore

Cystic Fibrosis Center, Pediatric Center Bambino Gesù Basilicata, Hospital San Carlo, Potenza, Italy

3.1 INTRODUCTION

Cystic fibrosis (CF) is a life-shortening, common, recessive genetic disease that is characterized by intestinal malabsorption, impaired growth, and lung disease [1]. Malnutrition and faltering growth are common [2–4]. The nutritional complications of patients with CF generally relate to nutrient imbalance, as nutrient intake for normal function and growth is inadequate to balance losses due to maldigestion/malabsorption or excessive utilization [5]. Optimizing nutritional status is critical in children with CF because malnutrition is associated with poor clinical outcomes [6–10].

CF can now be diagnosed routinely through newborn screening (NBS) by evaluating immunoreactive trypsin levels by using a dried blood specimen collected between 1 and 5 days of age, with the second tier being either a second immunoreactive trypsin or DNA analysis for CF transmembrane conductance regulator mutations [11].

Research over the past two decades has proven that NBS for CF is workable [12–15]; moreover, research focused on possible benefits in the fields of nutrition and growth, in pulmonary disease, and, finally, on survival [11]. This review provides an overview of the effects of NBS on the nutrition and growth of patients with CF.

3.2 REVIEW OF THE LITERATURE

In general, except for patients with meconium ileus (MI), infants with CF are quite healthy and well-nourished at the time of birth, but a large percentage of them (about 90%) will develop pancreatic insufficiency (PI) during the first months of life. Early identification offers the opportunity to provide nutrition intervention, which can lead to an overall better outcome for the infant, thus avoiding such complications as protein calorie malnutrition and vitamin and mineral deficiencies. This theoretical advantages of early diagnosis through NBS provide compelling rationale for this practice [16].

The assessment of the effects of NBS for CF in terms of health benefits is based on three types of surveys: (1) randomized controlled trials (RCTs); (2) evaluation of the outcomes of patients with CF reported by national registries; and (3) observational, nonrandomized comparison between geographical and temporal cohorts of screened and nonscreened patients.

3.3 RCTS STUDIES

The Wisconsin CF Neonatal Screening Project, continued for over two decades with the accumulation of evidence on the benefits of an early diagnosis and potential risks (challenges) of NBS, was the main font of literature on this issue. The first contribution on the relationship between NBS and nutritional status was in 1991, when the Neonatal Screening Project clearly demonstrated that normal growth and biochemical indices of nutritional status can be achieved in the majority of infants with CF when they are diagnosed via NBS and managed with appropriate dietary intervention and supplements [17]. In fact, evaluation of growth revealed that normal patterns could be achieved with mean energy intake values of 115 and 102 kcal/kg body weight at ages 6 and 12 months, respectively. Biochemical assessment demonstrated low α-tocopherol and linoleic acid values at diagnosis in the majority of infants, whereas one-third had indices of protein calorie malnutrition. Essential fatty acid deficiency also was demonstrated in 27% of screened infants. With predigested formula and dietary supplementation, there was improvement in all indices

Diet and Exercise in Cystic Fibrosis
http://dx.doi.org/10.1016/B978-0-12-800051-9.00003-1

of nutritional status, and only a small percentage of patients showed mild biochemical abnormalities at age 12 months.

Later, focusing on patients without MI and with PI, the project showed that anthropometric indices of nutritional status were significantly higher at diagnosis in the screened group, including length/height, weight, and head circumference, presumably thanks to the much earlier age at diagnosis of screened patients (mean ± standard deviation, 12 ± 37 weeks) compared with the standard diagnosis group (72 ± 106 weeks) [18].

During the 13-year follow-up evaluation, analysis of nutritional outcomes revealed significantly greater growth associated with early diagnosis in spite of similar nutritional therapy. Most impressive was that the screened group had a much smaller proportion of patients with weight and height data below the 10th percentile throughout childhood. The odds ratio (OR) for the risk of a weight below the 10th percentile in the control group, compared with the screened group, was 4.12 (95% confidence interval (CI), 1.64–10.38), and the corresponding OR for height was 4.62 (95% CI, 1.70–12.61) [19,20]. These data were further reinforced by accurate description of nutritional advantages of screened patients, whose follow-up evaluations over 16 years showed that all anthropometric indexes of nutritional status (height and weight *z*-scores and percentage of patients below the 10th percentile) were significantly better in the screened subgroup compared with the control subjects. Indeed this observation was particularly significant since, by chance, there was a disproportionate number of pancreatic-sufficient children (with their expected better nutritional status) in the nonscreened group (21% vs. 8%).

Besides the anthropometric parameters, the authors evaluated the biochemical parameters of nutritional status (serum levels of α-tocopherol, retinol, and linoleic acid): the values were similar between the two groups, although in the screening group serum levels increased and subsequently normalized about 1–3 months after diagnosis, in contrast to patients diagnosed by symptoms, whose biochemical evidence of malnutrition persisted for about 1 year after diagnosis [20]. The relevance of these results was strengthened by evidence that prolonged α-tocopherol deficiency during infancy was associated with lower subsequent cognitive performance. Thus, diagnosis via NBS may also benefit the cognitive development of children with CF, particularly in those prone to vitamin E deficiency during infancy [21,22].

Another trial was from United Kingdom (UK) and involved newborns from Wales and West Midland. This study was unable to show any significant clinical difference between the two groups (screened vs. nonscreened) of patients at the ages of 1–4 years, with the exception of shorter time in the hospital during the first year of life for the screened group. However, the lousy attention given to diagnostic procedures in the control group, the lack of treatment protocols, and short and irregular follow-up lower the quality of this study [23].

3.4 STUDIES USING CF REGISTRY DATA

In 2005 the article by Accurso et al. [24] analyzed new diagnoses of CF, identified by the US Cystic Fibrosis Foundation registry between 2000 and 2002, to determine the correlation between mode of diagnosis and complication rates. Nutritional benefits of early diagnosis through NBS vs. symptomatic diagnosis included the observation that infants of the symptomatic group were stunted or wasted at three times the rate of infants with NBS; moreover, no infants in the NBS group suffered hypoproteinemia, compared with 5% of infants in the symptomatic group.

In the same year, a study from the UK CF database [25] showed the benefits of NBS, expressed as the achievement of a significantly greater median height and a reduction in morbidity in screened patients (NBS group; $n = 184$ patients) compared to controls (CG; $n = 950$ patients) matched for age and genotype. Although the differences in weight *z*-scores were not significantly different overall or for any of the age subgroups, the height *z*-scores of those diagnosed via NBS were consistently higher. Overall, the NBS group had a median height *z*-score 0.32 greater than the CG. This difference seems to be primarily accounted for by superior median heights for those aged 6 years and younger, as there was no apparent difference in the 7- to 9-year-old groups. A comparison of the homozygous F508del NBS subpopulation to equivalent subjects who were in the CG gave comparable results to those of the all-genotype study population, ruling out the probability that a milder genotype/phenotype "mix" within the NBS population could account for the apparent improvement in outcome.

The UK CF database further suggests the positive effect of NBS in promoting better nutritional status (expressed as height *z*-score and rate of patients below the 10th percentile for height) and lowering morbidity, in terms of a higher clinical score and reduced long-term therapy, compared to diagnosis by symptoms, especially when it occurs after the first 2 months of life [26]. This study highlights that increasing the window of diagnosis from 2 to 3 months or beyond increases the proportion of patients with poor long-term growth. On the basis of these results, the authors propose, and find full agreement by Farrell [27], that the definition of early diagnosis of CF is redefined to a diagnosis made within 2 months of birth, with respect to the current definition of 3 months of birth, created by Shwachman et al. [28] in the 1970s.

3.5 COHORT OBSERVATIONAL STUDIES

An observational Australian study from South Wales evaluated the effects of human milk or formula feeding on growth during the first 2 years of life of infants identified by NBS [29]. The authors concluded that human milk feeding with appropriate enzyme replacement was suitable for the nutritional management of infants with CF. This study showed that early diagnosis before the onset of malnutrition makes breastfeeding adequate for the nutritional needs of infants with CF.

The same research group subsequently studied 57 children with CF born before NBS was introduced (1978 to mid-1981) and a further 60 children born during the first 3 years of the program (mid-1981–1984). The children were followed up to the age of 10, and the two cohorts were compared on measures of clinical outcome. Age- and sex-adjusted *z*-scores for height and weight were consistently higher in children screened for CF than in those born before screening. At 10 years of age, average differences in *z*-score between groups were 0.4 (95% CI, 0.1–0.8) for weight and 0.3 (95% CI, 0.1–0.7) for height. This translates to an average difference of about 2.7 cm in height and 1.7 kg in weight [30].

Children from these cohorts were further evaluated during adolescence [31]. The previously observed advantage of screened patients was still apparent in adolescence; they displayed statistically better total Shwachman-Kulczycki scores, chest radiograph scores, and lung function, whereas nutritional indexes, although consistent with previous measurements (mean differences in *z*-score were 0.4 for height and 0.3 for weight), did not reach the statistical significance.

In another more recent report, these observations were extended and included a survival analysis of these cohorts into early adulthood [32]. Height, weight, and body mass index (BMI) *z*-scores (all $P<0.01$) were better in the screened group ($n=41$) compared to nonscreened ($n=38$) subjects upon transfer to adult care. Similarly, lung function measurements were consistently better in the screened group, with the difference also becoming more evident over time. These cumulative benefits through childhood were predictors of survival, contributing to an increased survival at age 25 in the screened cohort. The survival analysis in particular showed that each 1.0-kg/m^2 increase in BMI contributed to a 44% (95% CI, 31–55%; $P<0.001$) decrease in risk of death.

In another article from Australia, Neville and Ranganathan [33] studied vitamin levels in infants diagnosed with CF by NBS over a 5-year period. The percentages of infants deficient in vitamins D, E, and A were 37, 16 and 60%, respectively. Vitamin D levels were unrelated to sex, vitamin A or E levels, month of birth, or pancreatic status, whereas vitamin A and E levels were significantly lower in those who were pancreatic insufficient.

Two cohorts of children born in the Netherlands between 1973 and 1979 were followed into early adult life [34]. One group was diagnosed at a mean of 1 month using NBS (screened group (SG)) and the other diagnosed at a mean of 18 months on clinical grounds (clinical group (CG)). In a first article describing the clinical outcomes up to the age of 11 years in the patients in the screened and nonscreened birth cohorts, the authors found, as a nutritional outcome, higher vitamin A levels ($P<0.01$) in the screened patients [34].

The SG was subsequently compared to the CG and to a third birth cohort consisting of patients with CF born in the same region immediately after the closure of the experimental screening program (after-SG) [35]. At the start of the study the height of the patients in the SG was just below the population mean values, and at the end of the study it was around the expected level. The mean weight remained about half a standard deviation below the mean levels throughout the observation period. The patients in the CG and the after-SG groups showed greater growth retardation for both height and weight than those in the SG. The catch-up growth shown by these patients after diagnosis results in similar nutritional status at the end of the observation period, despite whether the diagnosis was made by NBS or clinically. Considerably less catch-up growth was noted in the patients detected by NBS.

Finally, these three cohorts were again reassessed in a third article [36]. Analysis of the longer-term clinical outcomes revealed that over the first 18 years of life, there was no statistically significant difference in *z*-scores for height or weight in the SG and the CG.

The overall results of these studies should be interpreted noting that the patients in the SG and the CG were born in the 1970s, before the introduction of enteric-coated pancreatic enzymes and a fat-rich diet, and a normal growth pattern, such as was observed in later studies, could not be achieved. The patients in the after-SG group were mostly born in the 1980s, when the fat-enriched diet and the availability of enteric-coated pancreatic extracts was established. Moreover, only about half of the patients were followed in a specialized center.

Several studies were provided by US researchers in Denver, Colorado. In 1984 there was the first description of nutritional deficits in the first 20 infants in whom CF was diagnosed by NBS [37]. At a mean age of 5.5 weeks, 50% of patients had decreased their weight percentile and 15 of 17 infants had a triceps skinfold <50th percentile for age.

To evaluate the effect of early PI on growth and nutritional status in CF, 49 infants identified by NBS were studied [38]. PI, determined by increased 72-h fecal fat excretion, was present in 59% of infants at diagnosis. Before initiation of pancreatic enzyme replacement, growth and nutritional status of pancreatic-insufficient ($n=16$) and pancreatic-sufficient ($n=13$) infants were compared. Pancreatic-insufficient infants gained less weight from birth to diagnosis, had decreased triceps skinfold thicknesses, and had lower blood urea nitrogen and albumin levels despite higher gross calorie and protein intake. Fat malabsorption was present in 79% (30/38) and 92% (33/36) of infants tested at 6 and 12 months of age, respectively, indicating that PI persists and increases in frequency throughout infancy. The authors concluded that PI is prevalent in young infants with CF and has a significant effect on growth and nutrition.

The same group of researchers evaluated the biochemical status of vitamins A, D, and E in children with CF identified by the Colorado CF NBS program in the first year of life and then followed prospectively to the age of 10 years [39,40]. Despite supplementation with standard multivitamins and pancreatic enzymes, these studies highlighted that the sporadic occurrence of fat-soluble vitamin deficiency and persistent vitamin deficiency were relatively common, suggesting that frequent and serial monitoring of the serum concentrations of these vitamins is therefore essential in children with CF.

A French observational study [41] compared clinical characteristics at diagnosis and their evolution over a 10-year period in a group of 77 children with CF born during the period 1989–1998 in Britain, where NBS was performed, with a group of 36 nonscreened children with CF born in a neighboring region without NBS, Loire-Atlantique, France. Patients with MI, the prevalence of which was quite different in the two groups (14 in the first group, 10 in the second), were excluded. There were no significant differences in sex ratio, gestational age, anthropometric data at birth, frequency of F508del homozygotes, proportion of pancreatic-insufficient patients, and mean age between the two populations. Z-scores for weight and height were significantly better in the screened population, not only in the first years of life but also at 3 and 5 years old for height and 8 years old for weight. The clinical and radiologic scores were higher among the screened children during the whole period of follow-up, whereas no significant differences in colonization by *Pseudomonas aeruginosa* (PA) or in lung function were found.

Data from the northeast of Italy have been used to compare outcomes among patients with CF diagnosed by NBS compared to those diagnosed by symptoms [42]. A prospective analysis, with a follow-up until adult age (up to 26 years), compared the outcomes of patients with CF who were born during the period 1973–1981 in the Veneto and Trentino Alto Adige regions and diagnosed through NBS with those diagnosed by symptoms in the same districts and in two other neighboring areas that did not perform NBS. The NBS group showed significant advantages compared to the symptoms group in terms of mortality, nutritional parameters, radiological score, and lung function. A second retrospective analysis compared the outcomes of children with CF born during the period 1983–1992 in the Veneto region, where NBS had become universal, with those of patients born in Sicily (South Italy), where NBS for CF was not practiced. The two groups had similar frequencies of severe CF transmembrane conductance regulator mutations and therapeutic management but differed by other factors, including socioeconomic conditions. Screened patients were highly advantaged compared to the nonscreened patients in relation to mortality, nutritional status, radiological score, and prevalence of PA and *Burkholderia cepacia* infections.

In addition to the RCTs, the Wisconsin CF Neonatal Screening Group made some relevant observational studies. The first study tested the hypothesis that sustained high-energy intake (HEN) and normal plasma essential fatty acid status are critical determinants of treatment responsiveness within 2 years after diagnosis of CF. In fact, it is unclear why some patients with CF succeed in recovering from malnutrition and growth faltering after treatment initiation (responders) and others fail to do so (nonresponders) [43]. The first result of this study was that a caloric intake >120% of the estimated requirement for a prolonged period (i.e., HEN) is critical in promoting adequate weight gain. The association between HEN and recovery of weight *z*-score at birth (WtzBR) was observed in both screened patients and patients whose CF was diagnosed conventionally but was stronger in screened patients (91% vs. 56%). The second result was that high plasma linoleic acid (HPLA) are necessary but not sufficient to promote adequate weight gain. In fact, the subgroup of patients who achieved combined HEN and HPLA had the highest response rate of recovering the WtzBR, whereas the response rate in the subgroup of patients who received HEN alone was approximately half of those who received combined HEN and HPLA (40% vs. 83%) in the conventional diagnosis group. Another relevant finding of this study was that the severity of malnutrition at the time of diagnosis has a significant, negative effect on attaining adequate weight gain. So it could be speculated that some threshold effects may exist regarding the effect of malnutrition on the WtzBR, also on the basis of further clinical factors such as lung involvement. This study eventually provided additional evidence to support routine implementation of NBS programs because children with CF who are identified through NBS not only require less extensive nutrition intervention but also respond to it at younger ages than children with CF who are diagnosed conventionally.

The topic of optimal feeding (breast milk, formula, or a combination) for infants with CF and its association to longitudinal growth and pulmonary outcomes while taking into account MI and PI was examined in a cohort of 103 infants who were born in 1994–2006 and identified through the Wisconsin Routine CF NBS Program from diagnosis to 2 years of age [44]. In infants with PI, weight *z*-score declined from birth to 6 months in infants who were exclusively breastfed (ExBF) for ≥2 months, and the number of PA infections through the age of 2 years was fewer in breastfed than in exclusively formula-fed (ExFM) infants but did not differ by the duration of ExBF. This study provided evidence that, compared with ExFM, ExBF <2 months was associated with adequate growth and protected against PA infections during the first 2 years of life in infants with CF who had PI, whereas the benefits of ExBF >2 months in this at-risk population are unclear. However, the authors suggest that prospective studies with a larger sample size, a longer duration of follow-up, and data collection including potential confounding factors are needed to confirm these observations, further evaluate potential risks of ExBF for >2 months on growth faltering and their long-term effects (i.e., whether attenuated growth persists or catch-up growth occurs after 2 years of age), and clarify whether the low incidence of PA infection associated with breastfeeding leads to improved pulmonary outcomes later in life.

The Wisconsin Newborn Screening Project recently studied longitudinal associations between nutritional status and health-related quality of life in 95 children aged 9–19 years [45]. Height and BMI *z*-scores were positively associated with physical functioning and body image. The authors concluded that this study further suggests that early diagnosis through NBS and improved nutrition provide an opportunity to enhance the quality of life and body image perceptions of patients with CF.

A retrospective study of individuals with CF born in Connecticut between 1983 and 1997 evaluated growth, pulmonary function, and bacterial acquisition/colonization data from diagnosis through July 1, 2005, comparing those diagnosed by NBS ($n=34$) to those diagnosed by sweat test after symptoms appeared ($n=21$) [46]. Screened individuals demonstrated greater weight and height for age at diagnosis and through 15 years of age. BMI was higher in screened individuals (21 vs. 18 kg/m^2) at 15 years of age ($P=0.01$). Moreover, at 15 years of age, screened individuals had better pulmonary function than nonscreened individuals; over a 9-year period, from ages 6 to 15, percentage predicted forced expiratory volume in 1 second and forced vital capacity increased by 4% and 13% in screened individuals and declined by 14% and 5%, respectively, in nonscreened individuals. Acquisition/colonization of PA was similar between groups.

An interesting point of view is that derived from the evaluation of the baseline characteristics and factors associated with nutritional and pulmonary status at enrollment in the CF Early *Pseudomonas* Infection Control (EPIC) Observational Cohort [47]. EPIC is a prospective cohort study investigating risk factors for and clinical outcomes associated with early PA acquisition in young children with CF. It is noteworthy that enrolled patients diagnosed by newborn or prenatal screening (17.3 and 2.6% of participants, respectively) had significantly higher average height (8.5 (95% CI, 4.8–12.2); $P<0.0001$) and weight (6.5 (95% CI, 2.6–10.3); $P=0.0009$) percentiles than participants whose diagnosis of CF was determined by symptoms. Moreover, the advantage of NBS is unlikely to simply reflect detection of milder genotypes, as the same proportion of class A genotypes (two mutations in functional class I, II, or III) was present either in infants with CF diagnosed by NBS or in those with CF detected by other means (75% vs. 74%).

Improved nutrition is considered to be associated with better long-term outcomes in CF and is the major proven benefit arising from NBS programs, and the Australian Respiratory Early Surveillance Team for Cystic Fibrosis aimed to describe the relationship of pulmonary inflammation from NBS diagnosis over the first 3 years of life to the clinically relevant outcome of nutritional status [48]. The authors concluded that pulmonary inflammation in infants with CF diagnosed by NBS was associated with worse nutritional status. In particular, increased neutrophil elastase from bronchoalveolar lavage fluid and infection with *Staphylococcus aureus* were associated with lower BMI, whereas age and antistaphylococcal antibiotics were associated with increased BMI. The relationships between pulmonary inflammation and nutrition need more investigation; the observation that infection with *S. aureus* is associated with both pulmonary inflammation and worse nutritional status makes necessary specifically designed studies to determine whether a more aggressive approach of prevention, early detection, and treatment of infection with this organism results in improvement in BMI and other clinical outcomes in the first years of life.

A study from Louisiana evaluated the effect of NBS, introduced in that state in 2007, comparing growth and nutritional parameters in the first 3 years of life of infants with CF diagnosed by NBS ($n=20$) with patients with CF diagnosed by symptoms in a historic cohort from 1971 to the present ($n=45$) [49]. The study pointed out that there was an improvement in the nutritional status shown by a decrease in the percentage of patients below the 50th percentile of weight and height or, in particular, patients below the 10th percentile of weight and height, at 1–3 years of age after the implementation of NBS. There was also a reduction in vitamin deficiency, hospitalizations, and colonization by PA among the screened patients compared to the patients with CF who were diagnosed clinically. This study has several relevant limitations because of the small sample size and unequal number of patients in the two groups and the comparison between patients born and cared for in different decades.

3.6 DISCUSSION

The balance of evidence from RCTs, observational studies, and CF Registry data indicates that diagnosis of CF via NBS programs likely prevents malnutrition during infancy and can enhance nutritional status beyond this period during childhood, adolescence, and until adulthood. In fact, the outcomes of an early diagnosis of CF via NBS related to short- and long-term nutritional status can be assessed by examining anthropometric and biochemical measures, and it is naturally assumed that the earlier institution of pancreatic enzyme replacement therapy and a high-fat diet will result in better nutrition in those children diagnosed via NBS.

The difficulty associated with performing RCTs of NBS is indicated by the lack of such studies, but, largely thanks to the Wisconsin Screening Project, the evidence suggests that both long-term cognitive and growth outcomes may be optimized through NBS and early intervention, whereas it is less compelling and weaker for lung health. However, the relationship between growth and nutritional indexes in early life with pulmonary function also has some evidence, either from cohort studies directly involving infants with CF diagnosed by NBS or from epidemiologic studies in which screened infants are less numerous, but the large number of patients included makes these results quite remarkable [10].

The major source of information for assessing the nutritional benefits (or not) of NBS for CF is available from cohort studies, although the small sample sizes and the short follow-up periods could be relevant limitations of this kind of study. Despite this, data from cohort studies do show consistency for positive effects on nutritional outcomes, with only sporadic exceptions.

Registries reports are limited to studies from the United States and the United Kingdom. These registries provide data pertaining to larger numbers of patient and are of vital importance in providing guidance regarding organizational and therapeutic strategies in the future. Efficacy studies carried out using registry data are obviously retrospective, nonrandomized, observational studies that, as such, are exposed to the risk of many kinds of biases (i.e., year of birth, geographical differences, center bias), although the large number of patients in the registries could sometimes partially compensate for them [50]. In spite of these limitations, data from this source consistently support the concept that children diagnosed as early as possible by NBS, and therefore receiving early and intensive treatment, have a clear advantage in terms of growth and nutrition. In this view, and with the probable availability of future therapies, such as correctors and potentiators [51], leading to significant advantages for lung health, the importance of early diagnosis will be further reinforced, and the role of NBS will most likely be overriding in the organization of CF care around the world.

References

[1] Egan M. Cystic fibrosis. In: Kliegman RM, Stanton BF, St. Geme III JW, Schor NF, Berhrman RE, editors. Nelson textbook of pediatrics. 19th ed. Philadelphia: Saunders Elsevier; 2011. pp. 1481–96.

[2] Lai HC, Kosorok MR, Sondel SA, Chen ST, FitzSimmons SC, Green CG, et al. Growth status in children with cystic fibrosis based on the national cystic fibrosis patient registry data: evaluation of various criteria used to identify malnutrition. J Pediatr 1998;132:478–85.

[3] McNaughton SA, Stormont DA, Shepherd RW, Francis PW, Dean B. Growth failure in cystic fibrosis. J Paediatr Child Health 1999;35:86–92.

[4] Lai HJ, Farrell PM. Nutrition and cystic fibrosis. In: Coulston AM, Rock CL, Monsen ER, editors. Nutrition in the prevention and treatment of disease. 2nd ed. San Diego: Academic Press; 2008. pp. 787–804.

[5] Gaskin KJ. Nutritional care in children with cystic fibrosis: are our patients becoming better? Eur J Clinl Nutr 2013;67:558–64.

[6] Kerem E, Reisman J, Corey M, Canny GJ, Levison H. Prediction of mortality in patients with cystic fibrosis. N Engl J Med 1992;326:1187–91.

[7] Nir M, Lanng S, Johansen HK, Koch C. Long-term survival and nutritional data in patients with cystic fibrosis treated in a Danish centre. Thorax 1996;51:1023–7.

[8] Zemel BS, Jawad AF, FitzSimmons S, Stallings VA. Longitudinal relationship among growth, nutritional status, and pulmonary function in children with cystic fibrosis: analysis of the cystic fibrosis foundation national CF patient registry. J Pediatr 2000;137:374–80.

[9] Beker LT, Russek-Cohen E, Fink RJ. Stature as a prognostic factor in cystic fibrosis survival. J Am Diet Assoc 2001;101:438–42.

[10] Konstan MW, Butler SM, Wohl ME, Stoddard M, Matousek R, Wagener JS, et al. Growth and nutritional indexes in early life predict pulmonary function in cystic fibrosis. J Pediatr 2003;142:624–30.

[11] Wagener JS, Zemanick ET, Sontag MK. Newborn screening for cystic fibrosis. Curr Opin Pediatr 2012;24:329–35.

[12] Hammond KB, Abman SH, Sokol RJ, Accurso FJ. Efficacy of statewide neonatal screening for cystic fibrosis by assay of trypsinogen concentrations. N Engl J Med 1991;325:769–74.

[13] Gregg RG, Simantel A, Farrell PM, Koscik R, Kosorok MR, Laxova A, et al. Newborn screening for cystic fibrosis in Wisconsin: comparison of biochemical and molecular methods. Pediatrics 1997;99:819–24.

[14] Fost N, Farrell PM. A prospective randomized trial of early diagnosis and treatment of cystic fibrosis: a unique ethical dilemma. Clin Res 1989;37:495–500.

[15] Rock MJ, Hoffman G, Laessig RH, Kopish GJ, Litsheim TJ, Farrell PM. Newborn screening for cystic fibrosis in Wisconsin: nine-year experience with routine trypsinogen/DNA testing. J Pediatr 2005;147:S73–7.

[16] Michel SH, Mallowe A. Nutrition management of the infant identified with cystic fibrosis. Top Clin Nutr 2012;27:260–9.

[17] Marcus MS, Sondel SA, Farrell PM, Laxova A, Carey PM, Langhough R, et al. Nutritional status of infants with cystic fibrosis associated with early diagnosis and intervention. Am J Clin Nutr 1991;54:578–85.

[18] Farrell PM, Kosorok MR, Laxova A, Shen G, Koscik RE, Bruns WT, et al. Nutritional benefits of neonatal screening for cystic fibrosis. N Engl J Med 1997;337:963–9.
[19] Farrell PM, Kosorok MR, Rock MJ, Laxova A, Zeng L, Lai H, et al. Early diagnosis of cystic fibrosis through neonatal screening prevents severe malnutrition and improves long-term growth. Pediatrics 2001;107:1–13.
[20] Farrell PM, Lai HJ, Li Z, Kosorok MR, Laxova A, Green CG, et al. Evidence on improved outcomes with early diagnosis of cystic fibrosis through neonatal screening: enough is enough!. J Pediatr 2005;147:S30–6.
[21] Koscik RL, Farrell PM, Kosorok MR, Zaremba KM, Laxova A, Lai HC, et al. Cognitive function of children with cystic fibrosis: deleterious effect of early malnutrition. Pediatrics 2004;113:1549–58.
[22] Koscik RL, Lai HJ, Laxova A, Zaremba KM, Kosorok MR, Douglas JA, et al. Preventing early, prolonged vitamin E deficiency: an opportunity for better cognitive outcomes via early diagnosis through neonatal screening. J Pediatr 2005;147:S51–6.
[23] Chatfield S, Owen G, Ryley HC, Williams J, Alfaham M, Goodchild MC, et al. Neonatal screening for cystic fibrosis in Wales and the West Midlands: clinical assessment after 5 years of screening. Arch Dis Child 1991;66:29–33.
[24] Accurso FJ, Sontag MK, Wagener JS. Complications associated with symptomatic diagnosis in infants with cystic fibrosis. J Pediatr 2005;147:S37–41.
[25] Sims EJ, McCormick J, Mehta G, Mehta A. Neonatal screening for cystic fibrosis is beneficial even in the context of modern treatment. J Pediatr 2005;147:S42–6.
[26] Sims EJ, Clark A, McCormick J, Mehta G, Connett G, Mehta A. Cystic fibrosis diagnosed after 2 months of age leads to worse outcomes and requires more therapy. Pediatrics 2007;119:19–28.
[27] Farrell PM. The meaning of "early" diagnosis in a new era of cystic fibrosis care. Pediatrics 2007;119:156–7.
[28] Shwachman H, Redmond A, Khaw KT. Studies in cystic fibrosis: report of 130 patients diagnosed under 3 months of age over a 20-year period. Pediatrics 1970;46:335–43.
[29] Holliday KE, Allen JR, Waters DL, Gruca MA, Thompson SM, Gaskin KJ. Growth of human milk-fed and formula-fed infants with cystic fibrosis. J Pediatr 1991;118:77–9.
[30] Waters DL, Wilcken B, Irwig L, Van Asperen P, Mellis C, Simpson JM, et al. Clinical outcomes of newborn screening for cystic fibrosis. Arch Dis Child Fetal Neonatal Ed 1999;80:F1–7.
[31] McKay KO, Waters DL, Gaskin KJ. The influence of newborn screening for cystic fibrosis on pulmonary outcomes in New South Wales. J Pediatr 2005;147:S47–50.
[32] Dijk FN, McKay K, Barzi F, Gaskin KJ, Fitzgerald DA. Improved survival in cystic fibrosis patients diagnosed by newborn screening compared to a historical cohort from the same centre. Arch Dis Child 2011;96:1118–23.
[33] Neville LA, Ranganathan SC. Vitamin D in infants with cystic fibrosis diagnosed by newborn screening. J Paediatr Child Health 2009;45:36–41.
[34] Dankert-Roelse JE, te Meerman GJ, Martjin A, ten Kate LP, Knol K. Survival and clinical outcome in patients with cystic fibrosis, with or without neonatal screening. J Pediatr 1989;114:362–7.
[35] Dankert-Roelse JE, te Meerman GJ. Long term prognosis of patients with cystic fibrosis in relation to early detection by neonatal screening and treatment in a cystic fibrosis centre. Thorax 1995;50:712–8.
[36] Merelle ME, Schouten JP, Gerritsen J, Dankert-Roelse JE. Influence of neonatal screening and centralized treatment on long-term clinical outcome and survival of CF patients. Eur Respir J 2001;18:306–15.
[37] Reardon MC, Hammond KB, Accurso FJ, Fisher CD, McCabe ER, Cotton EK, et al. Nutritional deficits exist before 2 months of age in some infants with cystic fibrosis identified by screening test. J Pediatr 1984;105:271–4.
[38] Bronstein MN, Sokol RJ, Abman SH, Chatfield BA, Hammond KB, Hambidge KM, et al. Pancreatic insufficiency, growth, and nutrition in infants identified by NBS as having CF. J Pediatr 1992;120(4 Pt 1):533–40.
[39] Sokol RJ, Reardon MC, Accurso FJ, Stall C, Narkewicz M, Abman SH, et al. Fat-soluble-vitamin status during the first year of life in infants with cystic fibrosis identified by screening of newborns. Am J Clin Nutr 1989;50:1064–71.
[40] Feranchak AP, Sontag MK, Wagener JS, Hammond KB, Accurso FJ, Sokol RJ. Prospective, long-term study of fat-soluble vitamin status in children with cystic fibrosis identified by newborn screen. J Pediatr 1999;135:601–10.
[41] Siret D, Bretaudeau G, Branger B, Dabadie A, Dagorne M, David V, et al. Comparing the clinical evolution of cystic fibrosis screened neonatally to that of cystic fibrosis diagnosed from clinical symptoms: a 10-year retrospective study in a French region (Brittany). Pediatr Pulmonol 2003;35:342–9.
[42] Mastella G, Zanolla L, Castellani C, Altieri S, Furnari M, Giglio L, et al. Neonatal screening for cystic fibrosis: long-term clinical balance. Pancreatology 2001;1:531–7.
[43] Shoff SM, Ahn HY, Davis L, Lai H. Temporal associations among energy intake, plasma linoleic acid, and growth improvement in response to treatment initiation after diagnosis of cystic fibrosis. Pediatrics 2006;117:391–400.
[44] Jadin SA, Wu GS, Zhang Z, Shoff SM, Tippets BM, Farrell PM, et al. Growth and pulmonary outcomes during the first 2 y of life of breastfed and formula-fed infants diagnosed with cystic fibrosis through the Wisconsin Routine Newborn Screening Program. Am J Clin Nutr 2011;93:1038–47.
[45] Shoff SM, Tluczek A, Laxova A, Farrell PM, Lai HJ. Nutritional status is associated with health-related quality of life in children with cystic fibrosis aged 9–19 years. J Cyst Fibros 2013;12:746–53.
[46] Collins MS, Abbott MA, Wakefield DB, Lapin CD, Drapeau G, Hopfer SM, et al. Improved pulmonary and growth outcomes in cystic fibrosis by newborn screening. Pediatr Pulmonol 2008;43:648–55.
[47] Rosenfeld M, Emerson J, McNamara S, Joubran K, Retsch-Bogart G, Graff GR, et al. Baseline characteristics and factors associated with nutritional and pulmonary status at enrollment in the cystic fibrosis EPIC observational cohort. Pediatr Pulmonol 2010;45:934–44.
[48] Ranganathan SC, Parsons F, Gangell C, Brennan S, Stick SM, Sly PD. Evolution of pulmonary inflammation and nutritional status in infants and young children with cystic fibrosis. Thorax 2011;66:408–13.
[49] Venkata JA, Jones KL. Benefits of newborn screening for cystic fibrosis in Shreveport, Louisiana, Cystic Fibrosis Center. J La State Med Soc 2011;163:316–9.
[50] Salvatore D, Buzzetti R, Baldo E, Forneris MP, Lucidi V, Manunza D, et al. An overview of international literature from cystic fibrosis registries 2. Neonatal screening and nutrition/growth. J Cyst Fibros 2010;9:75–83.
[51] Wilschanski M. Novel therapeutic approaches for cystic fibrosis. Discov Med 2013;15:127–33.

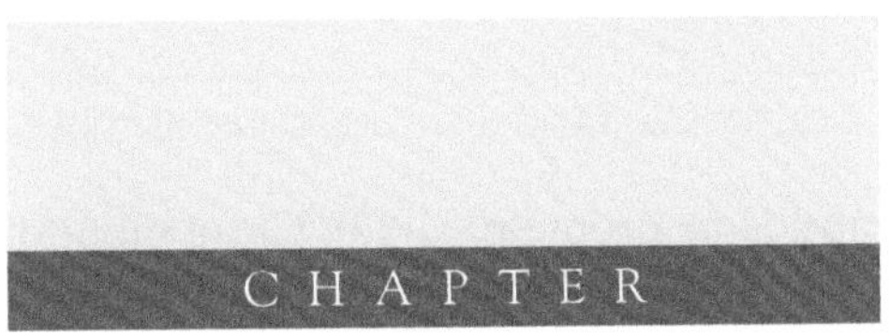

CHAPTER

4

Cystic Fibrosis Nutrition: Outcomes, Treatment Guidelines, and Risk Classification

Amanda Leonard

Division of Gastroenterology and Nutrition, The Johns Hopkins Children's Center, Baltimore, MD, USA

Cystic fibrosis (CF) is a chronic, life-shortening, genetic disease that affects approximately 1 in 3500 live births in the United States [1]. The genetic defect affects a variety of organs including the lungs, pancreas, gastrointestinal tract, and liver. Primary problems include lung disease and pancreatic insufficiency. Poor growth is also a common problem seen in this population [2,3]. Historically, people with CF died during childhood [4]. As treatments have improved, the median life expectancy for individuals with CF continues to increase. In 2013, the median life expectancy in the United States was 41 years old [1].

4.1 NUTRITION AND OUTCOMES

Nutrition goals and recommendations for individuals with CF have evolved over the past 25 years [3,5,6]. Poor nutritional status was accepted and even expected with CF; this is no longer the case [6]. Nutrition is now believed to be a vital component of CF care. The work of Corey and colleagues [5] was the first publication connecting better nutrition and growth to improved survival. In their work they looked at pulmonary function, growth, and survival outcomes at two similarly sized CF centers in Toronto and Boston. Pulmonary function was similar at the two centers, but both growth and survival were better at the Toronto center. At that time, a fat-restricted diet was the standard approach at the Boston center, with the goal of minimizing symptoms of malabsorption. In contrast, Toronto prescribed a high-fat diet, with increased pancreatic replacement enzyme therapy (PERT) dosing. There was a dramatic difference in survival data with the high-fat diet, with 30 years of survival for Toronto versus 21 years for Boston. This result changed the way nutrition was approached in the CF community [2].

Since the work of Corey and colleagues [5], numerous studies have further investigated the relationship between nutritional status, mortality, and pulmonary outcomes [7–17]. In 2006, the North American Cystic Fibrosis Foundation (CFF) first published data that compared nutritional status (as body mass index [BMI]) to lung function (forced expiratory volume in 1 s [FEV_1]) [6]. Figure 4.1 shows the FEV_1 versus BMI for children aged 6–18 years. Figure 4.2 shows FEV_1 versus BMI for adults. Although causality of the relationship was not clear, the data showed a strong association between lung function and nutritional status. Improved nutritional status at ages as young as 3–4 years has been linked with better pulmonary function tests and improved mortality [7,8]. Both wasting and stunting have been shown to be independent predictors of mortality [9,14,17]. Studies also have looked at preventing a decline in pulmonary function as a means to prolong and improve quality of life. Higher FEV_1 as well as better nutritional status have been correlated to a better quality of life [18–20]. Nutritional factors have been associated with a decrease in pulmonary function, as measured by FEV_1 [11]. An evaluation of data from the European Cystic Fibrosis Society (ECFS) Patient Registry showed that BMI had a statistically significant and clinically relevant effect on FEV_1. Patients with a BMI ≥2 standard deviations below the mean had a sixfold increased odds of having severe lung disease [11]. Focusing efforts on good nutrition early in life may improve outcomes.

4.2 ASSESSING WEIGHT CHANGE AND GROWTH IN CF

Consensus documents or guidelines for CF care have been published by a variety of groups including the CFF, the European Cystic Fibrosis Society, and the Australian Cystic Fibrosis Federation [2,6,21,22]. The guidelines

Diet and Exercise in Cystic Fibrosis
http://dx.doi.org/10.1016/B978-0-12-800051-9.00004-3

concur that the goal of nutrition therapy is to promote normal growth. More recent guidelines recommend BMI as a means to assess growth in individuals older than the age of 2 years. In children younger than 2 years old, weight for length should be utilized to assess growth [23]. Close monitoring is essential to promote adequate growth and, during adulthood, weight maintenance. Early detection of problems is vital to prevent further decline [6].

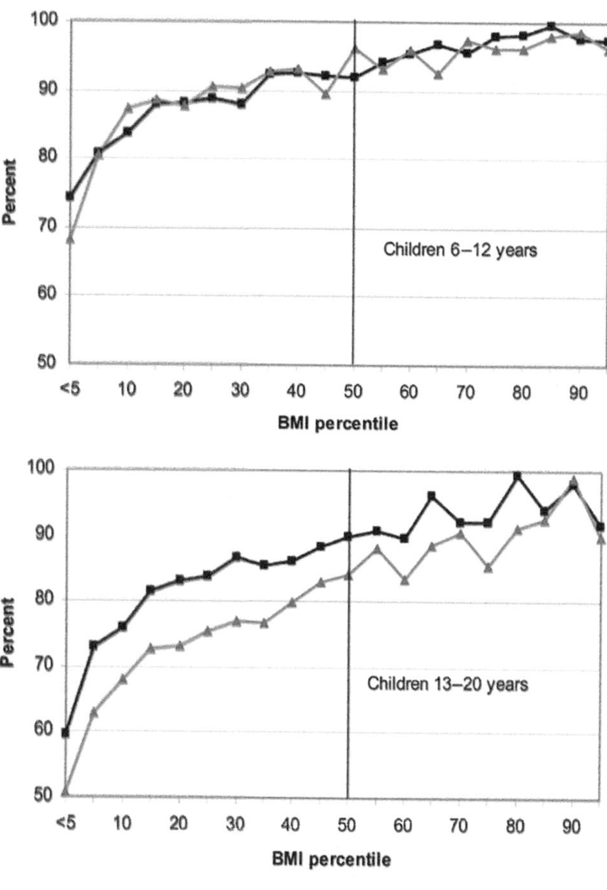

FIGURE 4.1 **Percent predicted forced expiratory volume per 1 s versus body mass index (BMI) percentile in children 6–12 (top) and 13–20 years old (bottom).**

4.2.1 Infants

The recommended goal for children younger than 2 years of age is to achieve the 50th percentile of weight for length by age 2, although achieving those goals earlier is likely beneficial [6,23]. An infant's weight, length, and head circumference should be measured every month until age 6 months and then every 3 months thereafter [23]. Anthropometric data should be plotted on a growth curve. In the United States there are two growth curves available for infants: the Centers for Disease Control and Prevention (CDC) growth curve and the World Health Organization (WHO) growth curve. The American Academy of Pediatrics, the CDC, and the National Institutes of Health recommend using the WHO growth chart in children younger than 2 years old [24]. In addition to following progression on the growth curve, average daily weight gain should be calculated and compared to norms for age [23,25].

4.2.2 Children

Recommendations suggest a goal BMI percentile at or above the 50th percentile for children ages 2–20 years. BMI percentile should be monitored every 3 months and plotted on the CDC growth curve (2–20 years). In addition to BMI percentile, weight and stature should be assessed every 3 months [6,23,26]. Stunting, defined as height for age percentiles below the fifth, has been associated with decreased survival [9,17]. The CFF recommends estimating the genetic potential for height for each patient [2]. This additional information can help determine whether the patient is stunted or has a lower genetic potential for height. Zhang and colleagues [27] investigated this issue and found that without adjusting

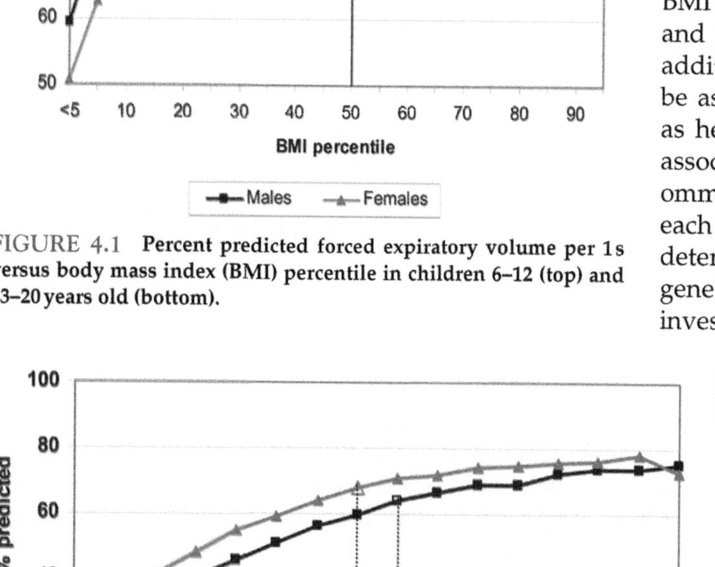

FIGURE 4.2 **Percent predicted forced expiratory volume per 1 s (FEV1) versus body mass index (BMI) percentile in adults.**

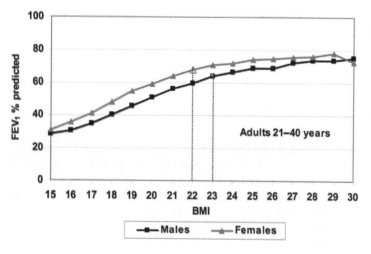

for genetic potential there is a risk of underestimating the prevalence of short stature and potentially biasing the association between lung function and height.

4.2.3 Adults

Recommended BMI goals for adults are 22 kg/m^2 for females and 23 kg/m^2 for males. Weight change should be tracked at each visit, ideally every 3 months [6]. Although there are not guidelines specific to an upper limit for BMI in CF, obesity is becoming more common in the CF population [3,15]. Stephenson and colleagues [15] caution against the potential negative impact of obesity on CF outcomes. In their work, there was little pulmonary benefit found with a BMI >25 kg/m^2. Monitoring growth and weight change is essential to the care of people with CF. Prevention combined with early detection and treatment of growth issues is the best approach [6].

4.3 ENERGY GUIDELINES

4.3.1 Energy

Individuals with CF and pancreatic insufficiency (PI) have increased energy requirements secondary to malabsorption and hypermetabolism [2]. Energy requirements are estimated to be 110–200% of the energy needs of the healthy population [6,7]. The degree of malabsorption, as well as the severity of lung disease, is thought to affect the total calorie needs. A high-calorie, high-fat diet, with approximately 35–40% of total calories from fat, is recommended to achieve the intake necessary to meet these increased caloric needs [2].

Patient-to-patient variations in energy needs are thought to be secondary to the differences in the severity of lung disease, degree of malabsorption, and presence of other comorbid diseases. There is not one specific recommended method to calculate energy needs. The goal is sustained weight gain and growth in children and adolescents and weight gain or maintenance in adults, depending on their nutritional status [6]. Monitoring growth and weight change may be the most useful method of assessing whether caloric needs are met [2,6].

4.4 VITAMINS AND MINERALS

4.4.1 Vitamins

Individuals with CF are at risk for fat-soluble vitamin deficiency (vitamins A, D, E, and K) because of fat malabsorption. There are commercially available vitamins that are specifically designed for individuals who have fat malabsorption, such as individuals with CF. These products have a water-miscible form of the fat-soluble nutrients to make them more easily absorbed [28]. These products are available in liquid, soft-gel, or chewable form. In addition to fat-soluble vitamins, these preparations also contain water-soluble vitamins and zinc [28].

For infants, children, and adults, supplementation with a CF-specific vitamin should be started at diagnosis [2,23]. Infants should have serum vitamin concentrations checked 2–4 months after diagnosis and then annually [23]. For individuals diagnosed after infancy, serum vitamin concentrations should be checked at diagnosis and then annually. When a serum vitamin concentration is low, it is important to assess adherence. If doses are frequently missed, levels should be rechecked once adherence is improved. Despite supplementation, serum concentrations may still be low and individuals may require additional amounts of specific vitamins. CF-specific doses of vitamin preparations should not be increased if the concentration of only one vitamin is low. This could lead to toxicity of other vitamins. Additional supplementation of only the vitamin that is low should be started and levels checked again in 2–4 months [6].

4.4.2 Minerals

4.4.2.1 Sodium

The CF population is at risk for hyponatremia as a result of excess loss of salt through the skin [2,29,30]. This risk increases in hot weather, with fever, or with extensive exercise/outdoor sports. To replace these losses, a high-salt diet is recommended for all patients with CF. Children and adults should add salt to foods and consume high-salt foods [2]. Standard infant feedings (breast milk, formula, and/or baby foods) do not provide adequate salt to meet these needs. Recommendations suggest that infants with CF receive an additional one-eighth teaspoon (13 mEq) of salt daily from birth to 6 months of age and one-quarter teaspoon (26 mEq) of salt for infants >6 months of age, not to exceed 4 mEq/kg body weight [23].

4.4.2.2 Zinc

Zinc deficiency, which is difficult to characterize in CF, also has been linked to decreased pulmonary function and bone disease [31]. Plasma zinc concentrations are not an adequate measure of zinc deficiency, and a deficiency can be present with normal serum zinc concentrations [2,23]. For this reason, it is recommended that empiric zinc supplementation (1 mg elemental zinc/kg/day in divided doses for 6 months) be used for individuals who are not growing well despite adequate calories and PERT [2].

Vitamin A status can be negatively affected by a zinc deficiency. Zinc supplementation, in addition to vitamin A, may improve serum concentrations of vitamin A when a low vitamin A level is noted [2].

4.4.2.3 *Iron*

Iron deficiency is common in children and adults with CF [32–34]. Annual laboratory tests should include hemoglobin and hematocrit to screen for anemia. The etiology of anemia may be multifactorial (i.e., anemia of chronic disease, blood loss, low serum iron). Serum ferritin may be falsely elevated in CF because it is an acute phase reactant. A low serum ferritin concentration indicates deficiency and should be treated [2,35].

4.5 MALABSORPTION

Approximately 85–90% of all individuals with CF have PI [1,36]. PI causes malabsorption of fat, vitamins, and other nutrients. Primary symptoms of PI include greasy, frequent, and malodorous stools. While the majority of individuals with CF have PI, approximately 10–15% are pancreatic sufficient (PS) [1].

Pancreatic status can be determined using a fecal elastase test [6,37]. This test can be conducted at diagnosis to determine pancreatic status [23]. O'Sullivan and colleagues [38] found that the pancreatic status can change over the first year of life. For this reason, they recommend that infants whose initial results are consistent with PS but who do not have a PS mutation have a fecal elastase test repeated at 1 year of age, or sooner if clinically indicated [38]. Once PI has been identified, treatment with PERT should be initiated [2]. The goal of enzyme therapy is to minimize symptoms of malabsorption, promote adequate weight gain and growth, and prevent deficiencies [39].

4.5.1 Pancreatic Replacement Enzyme Dosing

Enzymes should be taken with all meals and snacks. Enzymes can be dosed per kilogram or per gram of fat ingested. Dosing per gram of fat can be cumbersome and is not widely used [39]. Refer to Table 4.1 for PERT dosing guidelines.

Recommended upper limits for enzyme dosing were developed in 1995 after a noted increase in the incidence of fibrosing colonopathy [39]. Fibrosing colonopathy, a submucosal fibrosis of the colon and associated colonic strictures, is associated with exposure to prolonged high doses (>6000 IU lipase/kg meal) of pancreatic enzymes [39,40]. For this reason the recommended upper limit of 2500 IU lipase/kg/meal or 10,000 IU lipase/kg/day was established. This maximum is meant to prevent further disease not maximize absorption [39].

4.5.2 Assessing Absorption

Assessing the adequacy of enzyme therapy can be done subjectively, using stool history and growth, or objectively, using laboratory studies. Although fecal elastase is useful in determining pancreatic status, it does not measure the degree of absorption [37]. The 72-h fecal fat, with the coefficient of absorption calculation, is considered to be the best objective measure of enzyme adequacy [2].

TABLE 4.1 Recommended Dosing Guidelines for Pancreatic Enzyme Therapy Replacement [40]

DOSING BASED ON WEIGHT	
Infants	2000–4000 lipase units/120 mL formula or breast milk
Children <4 years old	Start with 1000 lipase units/kg/meal; titrate as needed Use half the dose for snacks Maximum dose is 2500 lipase units/kg/meal or 10,000 lipase units/kg/day
Children >4 years old and adults	Start with 500 IU lipase units/kg/meal; titrate as needed Use half of the dose for snacks Maximum dose is 2500 lipase units/kg/meal or 10,000 IU lipase units/kg/day
DOSING BASED ON GRAMS OF FAT INGESTED	
	500–4000 IU lipase units/g fat ingested Mean of 1800 lipase units/g fat ingested Maximum dose 4000 lipase units/g fat ingested

4.6 INCREASING INTAKE

It can be difficult for individuals with CF to meet their calorie needs. If caloric intake is inadequate, high-calorie nutritional supplements can be used. Appetite stimulants and enteral feedings are also options to improve caloric intake.

4.6.1 Nutritional Supplements

High-calorie supplements (foods or beverages) can be used to supplement calories. These products should be used as additional calories, not in place of meals [2]. Studies looking at the effect of oral supplements on overall caloric intake and weight gain have not consistently shown that supplements are beneficial [41].

4.6.2 Appetite Stimulants

Appetite stimulants may be useful in CF if the primary cause of malnutrition is inadequate intake secondary to diminished appetite [42]. Other causes of inadequate intake, such as gastrointestinal dysmotility, depression, or eating disorders, should be assessed and ruled out before an appetite stimulant is prescribed [6,42]. There have been case reports and small studies using megestrol acetate, cyproheptadine hydrochloride, and dronabinol in the CF population. Use of these appetite stimulants has potential positive effects, but more research is needed, especially with respect to side effects and long-term outcomes [43,44].

4.6.3 Enteral Feedings

Enteral feedings have been shown to improve nutritional status, as measured by BMI, in both children and adults with CF [45,46]. Enteral feedings should be considered when a patient is not achieving adequate weight gain or meeting a BMI goal with oral intake after other possible causes for malnutrition have been ruled out [2].

Available data does not support one type of enteral access, kind of tube, or method for placement (nasogastric, orogastric, gastrostomy, or jejunostomy). Intact formulas, when used with PERT, are generally well tolerated in the CF population [2]. Formulas with a higher caloric density may be required to meet calorie needs. The CFF recommends nocturnal feedings to promote oral intake during the day, with a goal of 30–50% of energy requirements provided via feeds [2,46]. There is not a great deal of data regarding enzyme administration with tube feedings. Current recommendations suggest giving a meal dose of PERT at the beginning and end of nocturnal feedings, with the addition of a dose midcycle if one is clinically indicated [2].

Administration can be problematic when patients are unable to swallow pills by mouth. There have been case reports of tubes becoming clogged when enzymes are administered through enteral tubes [47]. Some report success in adults when the enzymes beads are mixed with an 8.4% sodium bicarbonate solution and administered through a large-bore (16-French or larger) tube [47]. Crushing the enzyme beads (or nonenteric-coated tablets) and mixing with sodium bicarbonate solution may allow for use with a smaller-bore gastric tube or a jejunal tube [48]. The use of thickened fruit juice to suspend the enzyme beads and allow for successful delivery via enteral tube feedings also has been described [48].

4.7 COMORBID DISEASES/COMPLICATIONS THAT AFFECT NUTRITION

There are multiple comorbidities in the CF population. Many of these conditions can have a negative effect on nutritional status if they are not identified and treated.

4.7.1 Cystic Fibrosis–Related Diabetes

CF-related diabetes (CFRD) occurs in about 20% of children and 40–50% of adults with CF [1,49]. CFRD, which is often asymptomatic, is most commonly diagnosed after the age of 18 but can occur at any age. A decrease in pulmonary and nutritional status begins about 6–24 months before the diagnosis of CFRD. Annual screening with an oral glucose tolerance test should begin at age 10 [49–51] unless symptoms present at an earlier age. Insulin is the treatment of choice once CFRD is diagnosed [49]. As treatment for CFRD has become more aggressive, the difference in nutritional and pulmonary outcomes between those with and without CFRD is becoming less pronounced [52].

4.7.2 Cystic Fibrosis–Related Bone Disease

CF-related bone disease is characterized by decreased bone density and increased low trauma fraction rates [52]. Risk factors include poor nutritional status, lung infection, chronic inflammation, severity of lung disease, PI, vitamin deficiencies, and delayed puberty [53–55]. Screening with dual-energy X-ray absorptiometry of the lumbar spine and hip should begin at age 18 for those without additional risk factors. Screening should begin at 8 years if any of the following risk factors are present: <90% ideal body weight, FEV_1 <50% of predicted, prolonged glucocorticoid use (>90 days/year), delayed puberty, or history of fractures [54]. Adequate vitamin D and calcium intake are critical during development. Interventions include supplementation with calcium, vitamin D, vitamin K (if levels are low), or all three, as well as weight-bearing exercise [54]. Treatment with bisphosphonates is commonly used in adults; recommendations from the ECFS also include guidelines for use in children and adolescents [53].

4.7.3 Cystic Fibrosis–Associated Liver Disease

There are a variety of hepatic abnormalities seen in CF, including neonatal cholestasis, hepatic steatosis, focal biliary cirrhosis, multilobular cirrhosis, and synthetic liver failure [56]. CF-associated liver disease (CFLD) with cirrhosis or portal hypertension is thought to be the most clinically relevant of the abnormalities [57]. About 5–10% of people with CF develop CFLD during the first decade of life. Of these, a small percentage progress to liver failure [58]. Annual laboratory testing of liver enzymes, clinical examination, and ultrasound are recommended to screen for CFLD [58].

4.7.4 Gastrointestinal Disorders

Numerous gastrointestinal disorders, in addition to PI, can be seen in the CF population. Nonpancreatic causes may contribute to poor growth in CF. Diagnosis of these can often be missed if symptoms are thought to be secondary to malabsorption. Disorders in CF include distal intestinal obstructive syndrome, gastroesophageal reflux disease, infectious enteritis, bacterial overgrowth of the small intestine, Crohn's disease, and celiac disease [2,59]. These abnormalities should be considered if a patient is having poor weight gain and/or malabsorptive symptoms despite adequate calories and a sufficient dose of PERT [2].

4.8 NUTRITION SCREENING AND RISK CLASSIFICATION IN A CLINICAL SETTING

Current recommendations suggest that all individuals with CF strive to achieve normal growth [2]. With this recommendation comes the need to assess nutritional status and intervene as necessary to promote optimal growth and weight gain. This task can be daunting in a busy clinical setting. A standardized approach to screening, classification, and intervention can potentially make this task easier and can improve nutrition outcomes [60]. Although recommendations give parameters for clinical evaluation, the specifics of how to approach this task in the clinical setting are up to the clinician.

Nutrition screening is defined by the American Society for Enteral and Parenteral Nutrition as, "a process to identify an individual who is malnourished or who is at risk for malnutrition to determine if a detailed nutrition assessment is indicated" [61]. Nutrition risk screening and classification, as well as a host of chronic diseases, have been studied in the general population [62–68]. However, there is limited data looking at classification and/or screening in CF clinics.

Two recent publications describe screening and nutrition risk classification approaches in a CF clinic [60,69]. McDonald [69] validated a screening tool that assessed three aspects of growth: BMI, weight gain, and linear growth (until adult height was achieved). Points were assigned based on meeting the expected standards for each category (e.g., weight gain greater than or equal to the expected rate for age). Based on the total number of points, patients are assigned a nutrition risk category: no or low risk, moderate risk, or high risk. Focused nutrition screening, along with subsequent nutrition risk category assignment, allowed for identification of and emphasis on those patients who may require more interventions.

Leonard et al. [60] adopted a standardized nutrition risk classification system in an effort to improve nutrition care and eventually nutrition-related outcomes, primarily BMI. The system evaluated weight for age percentile, height for age percentile, BMI percentile, and weight change. Based on this information, a nutrition risk category of optimal, acceptable, at risk, or failure was assigned. Since this work, the use of the term *failure* has been discontinued because of the negative connotation for patients and families (unpublished data). The term *high nutrition risk* is now used. Leonard and colleagues [60] also included a "concerning" category. This category included patients who met the criteria for acceptable or optimal but who had not had adequate gains in the past 3 months. The purpose of this category was to promote early intervention. With the adoption of this classification system, the median clinic BMI went from 35.2% to 42% in a 15-month period (2005–2006). The system is still in use, and the 2012 median clinic BMI was 50% [1,60].

Use of a screening tool and assigning a risk category may help to ensure that patients with the greatest need are evaluated. Despite the fact that there is limited published information on screening and risk classification, many CF centers in the United States have looked at improving nutrition outcomes in conjunction with the CFF quality improvement initiative [70–72]. The use of a standardized approach to nutrition screening and risk classification may improve care and outcomes through focused interventions.

References

[1] Cystic Fibrosis Foundation Patient Registry. 2012 Annual Data Report. Bethesda, MD.

[2] Borowitz D, Baker RD, Stallings V. Consensus report on nutrition for pediatric patients with cystic fibrosis. J Pediatr Gastroenterol Nutr 2002;35(3):246–59.

[3] Panagopoulou P, Maria F, Nikolaou A, Nousia-Arvanitakis S. Prevalence of malnutrition and obesity among cystic fibrosis patients. Pediatr Int 2014;56(1):89–94.

[4] Warwick WJ, Pogue RE, Gerber HU, Nesbitt CJ. Survival patterns in cystic fibrosis. J Chronic Dis 1975;28(11–12):609–22.

[5] Corey M, McLaughlin FJ, Williams M, Levison H. A comparison of survival, growth, and pulmonary function in patients with cystic fibrosis in Boston and Toronto. J Clin Epidemiol 1988;41(6):583–91.

[6] Stallings VA, Stark LJ, Robinson KA, Feranchak AP, Quinton H, Clinical Practice Guidelines on Growth and Nutrition Subcommittee. Evidence-based practice recommendations for nutrition-related management of children and adults with cystic fibrosis and pancreatic insufficiency: results of a systematic review. J Am Diet Assoc 2008;108:832–9.

[7] Yen EH, Quinton H, Borowitz D. Better nutritional status in early childhood is associated with improved clinical outcomes and survival in patients with cystic fibrosis. J Pediatr 2013;162(3):530–5.

[8] Konstan MW, Butler SM, Wohl ME, Stoddard M, Matousek R, Wagener JS, et al. Investigators and Coordinators of the Epidemiologic Study of Cystic Fibrosis. Growth and nutritional indexes in early life predict pulmonary function in cystic fibrosis. J Pediatr 2003;142(6):624–30.

[9] Beker LT, Russek-Cohen E, Fink RJ. Stature as a prognostic factor in cystic fibrosis survival. J Am Diet Assoc 2001;101(4):438–42.

[10] McPhail GL, Acton JD, Fenchel MC, Amin RS, Seid M. Improvements in lung function outcomes in children with cystic fibrosis are associated with better nutrition, fewer chronic pseudomonas aeruginosa infections, and dornase alfa use. J Pediatr 2008;153(6):752–7.

[11] Kerem E, Viviani L, Zolin A, Macneill S, Hatziagorou E, Ellemunter H, et al. Factors associated with FEV1 decline in cystic fibrosis: analysis of the data of the ECFS Patient Registry. Eur Respir J 2014;43(1):125–33.
[12] Lai HJ, Shoff SM, Farrell PM, Wisconsin Cystic Fibrosis Neonatal Screening Group. Recovery of birth weight z score within 2 years of diagnosis is positively associated with pulmonary status at 6 years of age in children with cystic fibrosis. Pediatrics 2009;123(2):714–22.
[13] Peterson ML, Jacobs Jr DR, Milla CE. Longitudinal changes in growth parameters are correlated with changes in pulmonary function in children with cystic fibrosis. Pediatrics 2003;112(3 Pt 1):588–92.
[14] Sharma R, Florea VG, Bolger AP, Doehner W, Florea ND, Coats AJ, et al. Wasting as an independent predictor of mortality in patients with cystic fibrosis. Thorax 2001;56(10):746–50.
[15] Stephenson AL, Mannik LA, Walsh S, Brotherwood M, Robert R, Darling PB, et al. Longitudinal trends in nutritional status and the relation between lung function and BMI in cystic fibrosis: a population-based cohort study. Am J Clin Nutr 2013;97(4):872–7.
[16] Vandenbranden SL, McMullen A, Schechter MS, Pasta DJ, Michaelis RL, Konstan MW, et al. Lung function decline from adolescence to young adulthood in cystic fibrosis. Pediatr Pulmonol 2012;47(2):135–43.
[17] Vieni G, Faraci S, Collura M, Lombardo M, Traverso G, Cristadoro S, et al. Stunting is an independent predictor of mortality in patients with cystic fibrosis. Clin Nutr 2012;3:382–5.
[18] Abbott J, Hurley MA, Morton AM, Conway SP. Longitudinal association between lung function and health-related quality of life in cystic fibrosis. Thorax 2013;68(2):149–54.
[19] Sawicki GS, Rasouliyan L, McMullen AH, Wagener JS, McColley SA, Pasta DJ, et al. Longitudinal assessment of health-related quality of life in an observational cohort of patients with cystic fibrosis. Pediatr Pulmonol 2011;46(1):36–44.
[20] Shoff SM, Tluczek A, Laxova A, Farrell PM, Lai HJ. Nutritional status is associated with health-related quality of life in children with cystic fibrosis aged 9–19 years. J Cyst Fibros 2013;12(6):746–53.
[21] Sinaasappel M, Stern M, Littlewood J, Wolfe S, Steinkamp G, Heijerman HG, et al. Nutrition in patients with cystic fibrosis: a European Consensus. J Cyst Fibros 2002;1(2):51–75.
[22] Anthony H, Collins CE, Davidson G, Mews C, Robinson P, Shepherd R, et al. Pancreatic enzyme replacement therapy in cystic fibrosis: Australian guidelines. J Ped Child Health 2002;35:125–9.
[23] Borowitz D, Robinson KA, Rosenfeld M, Davis SD, Sabadosa KA, Spear SL, et al. Cystic Fibrosis Foundation evidence-based guidelines for management of infants with cystic fibrosis. J Pediatr 2009;155(Suppl. 6):S73–93.
[24] Grummer-Strawn LM, Reinold C, Krebs NF, Centers for Disease Control and Prevention (CDC). Use of World Health Organization and CDC growth charts for children aged 0–59 months in the United States. MMWR Recomm Rep 2010;59(RR-9):1–15.
[25] Guo SM, Roche AF, Fomon SJ, Nelson SE, Chumlea WC, Rogers RR, et al. Reference data on gains in weight and length during the first two years of life. J Pediatr 1991;119(3):355–62.
[26] Lai HJ. Classification of nutritional status in cystic fibrosis. Curr Opin Pulm Med 2006;12(6):422–7.
[27] Zhang Z, Shoff SM, Lai HJ. Incorporating genetic potential when evaluating stature in children with cystic fibrosis. J Cyst Fibros 2010;9(2):135–42.
[28] Papas KA, Sontag MK, Pardee C, Sokol RJ, Sagel SD, Accurso FJ, et al. A pilot study on the safety and efficacy of a novel antioxidant rich formulation in patients with cystic fibrosis. J Cyst Fibros 2008;7(1):60–67.
[29] Laughlin JJ, Brady MS, Eigen H. Changing feeding trends as a cause of electrolyte depletion in infants with cystic fibrosis. Pediatrics 1981;68(2):203–7.
[30] Legris GJ, Dearborn D, Stern RC, Geiss CL, Hopfer U, Douglas JG, et al. Sodium space and intravascular volume: dietary sodium effects in cystic fibrosis and healthy adolescent subjects. Pediatrics 1998;(1 Pt 1):48–56.
[31] Damphousse V, Mailhot M, Berthiaume Y, Rabasa-Lhoret R, Mailhot G. Plasma zinc in adults with cystic fibrosis: Correlations with clinical outcomes. J Trace Elem Med Biol 2014;28(1):60–4.
[32] Ater JL, Herbst JJ, Landaw SA, O'Brien RT. Relative anemia and iron deficiency in cystic fibrosis. Pediatrics 1983;71(5):810–4.
[33] Uijterschout L, Nuijsink M, Hendriks D, Vos R, Brus F. Iron deficiency occurs frequently in children with cystic fibrosis. Pediatr Pulmonol 2013. http://dx.doi.org/10.1002/ppul.22857. [Epub ahead of print].
[34] von Drygalski A, Biller J. Anemia in cystic fibrosis: incidence, mechanisms, and association with pulmonary function and vitamin deficiency. Nutr Clin Pract 2008;23(5):557–63.
[35] Keevil B, Rowlands D, Burton I, Webb AK. Assessment of iron status in cystic fibrosis patients. Ann Clin Biochem 2000;37(Pt 5):662–5.
[36] Couper RT, Corey M, Moore DJ, Fisher LJ, Forstner GG, Durie PR. Decline of exocrine pancreatic function in cystic fibrosis patients with pancreatic sufficiency. Pediatr Res 1992;32(2):179–82.
[37] Borowitz D, Lin R, Baker SS. Comparison of monoclonal and polyclonal ELISAs for fecal elastase in patients with cystic fibrosis and pancreatic insufficiency. J Pediatr Gastroenterol Nutr 2007;44(2):219–23.
[38] O'Sullivan BP, Baker D, Leung KG, Reed G, Baker SS, Borowitz D. Evolution of pancreatic function during the first year in infants with cystic fibrosis. J Pediatr 2013;162(4):808–12.
[39] Borowitz D, Grand RJ, Durie PR. Use of pancreatic enzyme supplements for patients with cystic fibrosis in the context of fibrosing colonopathy. Consensus Committee. J Pediatr 1995;127(5):681–4.
[40] Reichard KW, Vinocur CD, Franco M, Crisci KL, Flick JA, Billmire DF, et al. Fibrosing colonopathy in children with cystic fibrosis. J Pediatr Surg 1997;32(2):237–41.
[41] Smyth RL, Walters S. Oral calories supplements for cystic fibrosis. Cochrane Database Syst Rev 2012;17(10):CD000406.
[42] Nasr S.Z., Drury D.. Appetite stimulants use in cystic fibrosis. Pediatr Pulmonol 2008;43:209–19.
[43] Epifanio M, Marostica PC, Mattiello R, Feix L, Nejedlo R, Fischer GB, et al. A randomized, double-blind, placebo-controlled trial of cyproheptadine for appetite stimulation in cystic fibrosis. J Pediatr 2012;88(2):155–60.
[44] Bradley GM, Carson KA, Leonard AR, Mogayzel Jr PJ, Oliva-Hemker M. Nutritional outcomes following gastrostomy in children with cystic fibrosis. Pediatr Pulmonol 2012;47(8):743–8.
[45] White H, Morton AM, Conway SP, Peckham DG. Enteral tube feeding in adults with cystic fibrosis; patient choice and impact on long term outcomes. J Cyst Fibros 2013;12(6):616–22.
[46] Borowitz D, Baker RD, Stallings V. Consensus report on nutrition for pediatric patients with cystic fibrosis. J Pediatr Gastroenterol Nutr 2002;35(3):246–59.
[47] Nicolo M, Stratton KW, Rooney W, Boullata J. Pancreatic enzyme replacement therapy for enterally fed patients with cystic fibrosis. Nutr Clin Pract 2013;28(4):485–9.
[48] Ferrie S, Graham C, Hoyle M. Pancreatic enzyme supplementation for patients receiving enteral feeds. Nutr Clin Pract 2011;26(3):349–51.
[49] Moran A, Brunzell C, Cohen RC, Katz M, Marshall BC, Onady G, et al. CFRD Guidelines Committee. Clinical care guidelines for cystic fibrosis-related diabetes: a position statement of the American Diabetes Association and a clinical practice guideline of the Cystic Fibrosis Foundation, endorsed by the Pediatric Endocrine Society. Diabetes Care 2010;(12):2697–708.
[50] Moran A, Hardin D, Rodman D, et al. Diagnosis, screening, and management of CFRD: a consensus conference report. J Diabetes Res Clin Pract 1999;45:61–73.

[51] Moran A, Becker D, Casella SJ, Gottlieb PA, Kirkman MS, Marshall BC, et al. Epidemiology, pathophysiology, and prognostic implications of cystic fibrosis-related diabetes: a technical review. Diabetes Care 2010;(12):2677–83.

[52] Moran A, Dunitz J, Nathan B, Saeed A, Holme B, Thomas W. Cystic fibrosis-related diabetes: current trends in prevalence, incidence, and mortality. Diabetes Care 2009;(9):1626–31.

[53] Paccou J, Fardellone P, Cortet B. Cystic fibrosis-related bone disease. Curr Opin Pulm Med 2013;19(6):681–6.

[54] Aris RM, Merkel PA, Bachrach LK, Borowitz DS, Boyle MP, Elkin SL, et al. Consensus conference report: guide to bone health and disease in cystic fibrosis. J Clin Endocrinol Metab 2005;90(3):1888–96.

[55] Boyle MP. Update on maintaining bone health in cystic fibrosis. Curr Opin Pulm Med 2006;12(6):453–8.

[56] Leeuwen L, Fitzgerald DA, Gaskin KJ. Liver disease in cystic fibrosis. Paediatr Respir Rev 2014;15(1):69–74.

[57] Flass T, Narkewicz MR. Cirrhosis and other liver disease in cystic fibrosis. J Cyst Fibros 2012;S1569-1993(12):00227–235.

[58] Debray D, Kelly D, Houwen R, Strandvik B, Colombo C. Best practice guidance for the diagnosis and management of cystic fibrosis-associated liver disease. J Cyst Fibros 2011;10(Suppl. 2):S29–36.

[59] Borowitz D, Gelfond D. Intestinal complications of cystic fibrosis. Curr Opin Pulm Med 2013;(6):676–80.

[60] Leonard A, Davis E, Rosenstein BJ, Zeitlin PL, Paranjape SM, Peeler D, et al. Description of a standardized nutrition classification plan and its relation to nutritional outcomes in children with cystic fibrosis. J Pediatr Psychol 2010;35(1):6–13.

[61] Mueller C, Compher C, Ellen DM, American Society for Parenteral and Enteral Nutrition (A.S.P.E.N.) Board of Directors. A.S.P.E.N. clinical guidelines: nutrition screening, assessment, and intervention in adults. J Parenter Enteral Nutr 2011;35(1):16–24.

[62] Lim SL, Ang E, Foo YL, Ng LY, Tong CY, Ferguson M, et al. Validity and reliability of nutrition screening administered by nurses. Nutr Clin Pract 2013;28(6):730–6.

[63] Beberashvili I, Azar A, Sinuani I, Kadoshi H, Shapiro G, Feldman L, et al. Comparison analysis of nutritional scores for serial monitoring of nutritional status in hemodialysis patients. Clin J Am Soc Nephrol 2013;8(3):443–51.

[64] Holst M, Yifter-Lindgren E, Surowiak M, Nielsen K, Mowe M, Carlsson M, et al. Nutritional screening and risk factors in elderly hospitalized patients: association to clinical outcome? Scand J Caring Sci 2013;27(4):953–61.

[65] Hartman C, Shamir R, Hecht C, Koletzko B. Malnutrition screening tools for hospitalized children. Curr Opin Clin Nutr Metab Care 2012;15(3):303–9.

[66] Wiskin AE, Owens DR, Cornelius VR, Wootton SA, Beattie RM. Paediatric nutrition risk scores in clinical practice: children with inflammatory bowel disease. J Hum Nutr Diet 2012;25(4):319–22.

[67] Joosten KF, Hulst JM. Nutritional screening tools for hospitalized children: methodological considerations. Clin Nutr 2014;33(1):1–5.

[68] Mehta NM, Corkins MR, Lyman B, Malone A, Goday PS, Carney LN, et al. Defining pediatric malnutrition: a paradigm shift toward etiology-related definitions. J Parenter Enteral Nutr 2013;37(4):460–81.

[69] McDonald CM. Validation of a nutrition risk screening tool for children and adolescents with cystic fibrosis ages 2–20 years. J Pediatr Gastroenterol Nutr 2008;46(4):438–46.

[70] Britton LJ, Thrasher S, Gutierrez H. Creating a culture of improvement: experience of a pediatric cystic fibrosis center. J Nurs Care Qual 2008;23(2):115–20.

[71] Schechter MS, Gutierrez HH. Improving the quality of care for patients with cystic fibrosis. Curr Opin Pediatr 2010;22(3):296–301.

[72] Kraynack NC, McBride JT. Improving care at cystic fibrosis centers through quality improvement. Semin Respir Crit Care Med 2009;30(5):547–58.

CHAPTER

5

Clinic, Nutrition, and Spirometry in Cystic Fibrosis

Lidia Alice G.M.M. Torres, Ieda Regina L. Del Ciampo
University of São Paulo, São Paulo, Brazil

5.1 INTRODUCTION

Cystic fibrosis (CF) is the most common recessive autosomal hereditary disease in Caucasians. More than 1500 mutations related to the CF gene have been identified, the first in 1989. The CF gene is a single gene located on the long arm of chromosome 7. It consists of 27 exons, is 250 kb in size, and transcribes a 1480 amino acid protein. The vast majority of mutations involve three or fewer nucleotides and result in amino acid substitution, frame shift, splice site, or nonsense mutations. One common mutation—and the first to be described—is F508del, which is due to the loss of three base pairs encoding a phenylalanine in position 508. It is the most common mutation among Caucasians, but there are approximately 20 mutations with higher incidence, some being characteristic of specific groups. Mutations in the CF gene lead to the production of a defective protein, CFTR (cystic fibrosis transmembrane conductance regulator), which is a chloride channel located in the apical portion of epithelial cells. This defect impairs chloride transport. In turn, this impairment results in a change in the viscosity of the respiratory tract secretions, in the decreased production of pancreatic enzymes due to obstruction of the pancreatic ducts, and in the loss of electrolytes through sweat that are not reabsorbed [1–3]. Some individuals, however, may produce small amounts of protein and may therefore only partially lose protein function. The CFTR mutations can be divided into at least five classes according to the level of function of the transporter: Classes I to III indicate nearly absent or very poor function, and classes IV and V indicate preservation of part of the protein function. This partial function has been one of the forms of treatment being studied, and there are drugs that partially correct the defect, although they are restricted to a few types of mutations [4].

5.2 CLINICAL CHARACTERISTICS

The phenotypic presentation in both the respiratory and gastrointestinal systems may be different in each patient due to enormous genetic variability, complicated further by modifying genes that can alter the expression of the original genes. However, progressive lung impairment is the most common cause of death. A change in the periciliary fluid (PCF) is caused by defective ion and fluid homeostasis. This change increases the gel layer and decreases the liquid layer of the PCF. Maintaining the thickness and hydration of the PCF is important for proper ciliary movement and airway mucus clearance. Another situation that can worsen the condition of the PCF is known as the "low volume hypothesis." The epithelia of the airways in CF patients secrete more sodium than normal through the epithelial sodium channel (ENaC) to compensate for changes in the chlorine metabolism. It is believed that these channels do not function fully in CF patients. Although the contribution of ENaC channels to the pathogenesis of CF is still debated, there are convincing data demonstrating that the rehydration of the PCF can improve mucociliary clearance in CF patients [5].

The cells that suffer from the lack of the transporter are predominantly the bronchial epithelial cells and cells from submucosal glands, which do not affect the interstitium and alveoli at the early stages of the disease [6]. The retained secretion facilitates the establishment of bacteria and promotes the accumulation of neutrophils. The bacteria and neutrophils cause endobronchitis and peribronchial inflammation, and the process can then become chronic and lead to the formation of bronchiectasis. The presence of bacteria and the apoptosis of neutrophils triggers an inflammatory response that increases secretion production and hinders the body's defense even more. This hindering eventually leads to a

Diet and Exercise in Cystic Fibrosis
http://dx.doi.org/10.1016/B978-0-12-800051-9.00005-5

gradual loss of lung function, initially by an obstructive process. After the development of bronchiectasis, the structure of the bronchial walls is destroyed, and purulent material accumulates. At this point, the ventilation process has become a mixed process of obstruction and restriction due to the destruction of pulmonary tissue, which culminates in respiratory failure [5–7]. Therefore, the control of pulmonary infections and improved clearance of bronchial secretions has contributed to decreased morbidity and mortality of CF [5–11].

Regarding infections, *Pseudomonas aeruginosa* is the most important pathogen because chronic colonization by this agent is associated with a worse prognosis and decreased pulmonary function. In these patients, the serial analysis of lung function is critical, as lung function is an important predictor of mortality and survival [8,9]. The control of pulmonary infections is an important part of the health care provided, as are improving the clearance of bronchial secretions, correcting malabsorption, and providing nutritional support [5,10]. These steps are important because morbidity and mortality are caused by chronic infection by *P. aeruginosa* and loss of lung function, in addition to genotype, growth, and hospitalizations [11]. Careful attention must also be paid to pancreatic enzyme replacement therapy and to nutritional supplementation, because both allow most children with CF to achieve normal growth and to reach adulthood in good nutritional condition and with a near-normal height, and they are also essential for maintaining immune status and lung function [12].

The importance of a multidisciplinary approach to these patients has also been stressed as a factor to improve some parameters and maintain the stability of others, while always focusing on the importance of the growth and general well-being of the patients. This approach is much better when performed by a group of specialists [13]. Poor somatic growth and deficient nutritional status, both resulting from pancreatic insufficiency, as well as other effects of the transmembrane conductance regulator (CFTR), most likely affect lung growth, host defense, and the ability to repair lung injury [14].

The pancreatic and intestinal changes are caused by the obstruction of the pancreatic ducts by thick secretions, which begin in the intrauterine period. This obstruction leads to dilation of the secretory ducts and acini and to flattening of the epithelium. It also leads to the gradual formation of cysts, which correspond to dilated ducts, but inflammatory changes are not detected. The pancreatic islets, however, are preserved longer. These pathophysiological changes lead to the reduction or loss of enzyme production, and thus to malabsorption. The pancreatic juice of these patients is reduced in volume and has low enzyme and bicarbonate concentrations, which leads to poor digestion and malabsorption of fats. This malabsorption is one of the predominant factors leading to malnutrition in this group of patients and is related to growth and lung development [15,16].

Cystic fibrosis has been widely studied by authors from various countries. This study has led to greater understanding of its pathophysiology and the advent of new therapeutic modalities, which in turn are reflected in significantly reduced morbidity and increased survival [17]. The treatment of CF patients in specialized centers has been considered critical because it allows various professionals to implement a multidisciplinary assessment. This approach leads to benefits in terms of clinical outcomes, especially regarding nutrition and the proper management of lung disease, which is important because many studies have demonstrated the importance of both nutrition and the proper management of lung disease as prognostic determinants. Thus, it is clear that optimizing growth and nutritional status are fundamental for the effective treatment of patients with cystic fibrosis [18]. Monitoring growth, nutritional status, and lung function, and controlling bacterial infections adequately starting from the early years of life, are crucial in maintaining good lung function and the integrity of the respiratory system. Numerous epidemiological studies have demonstrated the link between growth abnormalities and pulmonary disease in CF patients [19–22].

Another exciting observation is that many pulmonary lesions may be reversible in childhood, as long as the child suffers no exacerbations and continues growing properly. Studies have shown unequivocally that there is a possibility of reversing the changes, even if they are structural. One example is bronchiectasis detected by chest computed tomography in children less than 4 years old with cystic fibrosis, which disappeared later if treated correctly with early bacterial eradication [23,24]. An example of this can be seen in Figure 5.1.

5.3 METHODS OF ASSESSING GROWTH AND PULMONARY FUNCTION

The percentile curves of the body mass index (BMI) for children and the absolute values of the BMI for adults are currently considered good parameters for the detection of malnutrition in children and adults with cystic fibrosis. Appropriate values are BMI ≥ 50th percentile for children, BMI ≥ 22 kg/m^2 for women, and BMI ≥ 23 kg/m^2 for adult men [25]. However, it should be noted that when calculating target height, or the parental target, the correct interpretation of that percentile is necessary, because a patient with such a severe disease is not expected to grow more than he or she is genetically programed to grow [26].

The functional measures that are recommended and most frequently used are FEV1 (forced expiratory volume in 1 s) and FVC (forced vital capacity). These measures show little variability, good reproducibility, proper standardization, and easy implementation, even for children. They have been considered the

First CT

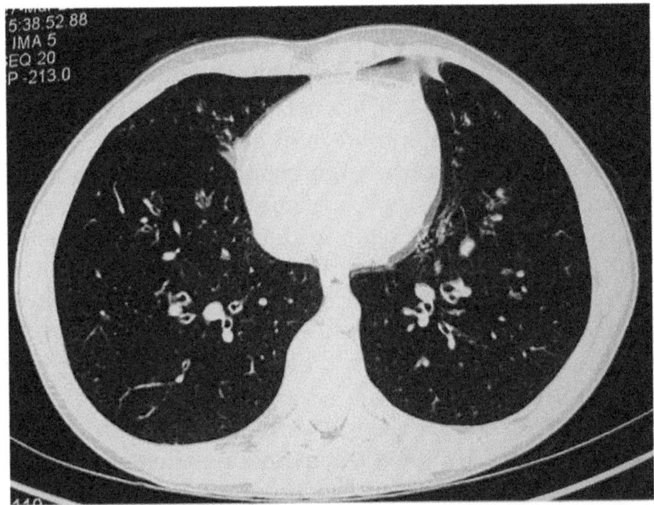

Second CT

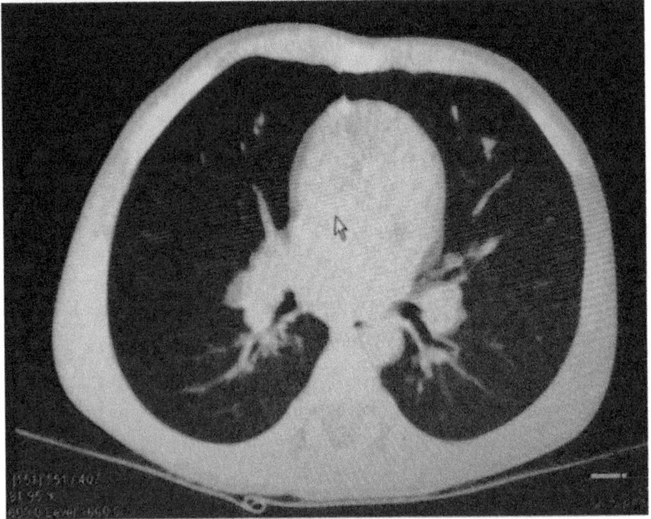

FIGURE 5.1 **Comparison of CT scan of a well-nourished patient with reduction of the number of bronchiectases, after 6 months of treatment.** The second image was taken with reduction of radiation doses.

measures of choice for population-based and follow-up studies. These measures are commonly used in most studies on cystic fibrosis and are used for reference in this chapter [27].

5.4 FACTORS THAT INTERFERE WITH THE GROWTH OF CF PATIENTS

Pancreatic insufficiency is present in 75% of CF patients at birth, in 80–85% by the end of the first year, and in 90% by adulthood. In these patients, pancreatic enzyme replacement therapy is critical. Individuals who do not develop pancreatic insufficiency have a better prognosis because they are able to maintain their nutritional status more easily [28].

The metabolic disorders associated with the excessive loss of sodium and chlorine also favor the onset of dehydration and malnutrition. Especially during the first two years of life, patients should be monitored for the onset of pseudo-Bartter syndrome. This syndrome is characterized by hypokalemia and metabolic alkalosis, which are common in warm climates and can lead to inadequate weight and height growth (failure to thrive) [29,30].

Breastfeeding is the recommendation of choice because it is the best source of nutrition in the first year of life and acts as a protective factor for children with CF. Studies comparing breast milk and formula feeding identified a lower incidence of infection and improved lung function in the first three years of life in the breastfed group [31]. However, it is important to note that breast milk has lower amounts of sodium. Due to excessive sodium ion loss in CF patients, the supplementation of milk or solids with table salt to provide 2–4 mmol/kg/day may be recommended, especially when the loss is excessive, such as during the summer [32,33].

Regarding the replacement of fat-soluble vitamins, studies have shown that supplementation, especially of vitamins A, D, and E, could play a role in decreasing pulmonary exacerbations even in patients without malabsorption. Vitamin D appears to have an effect on lung function and on the ability of the lungs to fight infection, and some studies have shown a positive correlation between vitamin D levels and FEV1.

Higher levels of serum retinol also appear to be positively correlated with FEV1 and negatively correlated with the number of pulmonary exacerbations, even when the values are within the reference range [34–38].

Deficiency of essential fatty acids has also been well described in CF and is characterized by low plasma levels of linoleic acid and docosahexaenoic acid (DHA). Supplementation with these fatty acids appears to provide beneficial anti-inflammatory effects for chronic pulmonary inflammation in CF [39]. In a study conducted in adults, low-dose supplementation of essential fatty acids for a period of one year seems to have improved pulmonary function and decreased both respiratory exacerbations and antibiotic consumption, although more studies are needed in this area before it can be regarded as a recommended treatment [40,41].

5.5 METABOLIC PROCESS IN CYSTIC FIBROSIS: HIGH CALORIE CONSUMPTION

Malnutrition contributes significantly to muscle weakness in patients with any chronic lung disease, and this condition is even more severe in cystic fibrosis. The main cause of this muscle weakness is the significant increase in energy expenditure at rest associated with malnutrition, which can be further worsened by pancreatic insufficiency. This interaction leads to a vicious circle in which there is weight loss caused by increased energy consumption from respiratory effort, and there is lung inflammation, both of which are associated with low oxygen intake and a progressive increase in oxygen demand. This phenomenon will lead to the loss or weakness of the respiratory muscles, similar to what happens in patients with chronic obstructive pulmonary disease (COPD) [42].

As the lung disease worsens for these patients, respiratory effort leads to a hypercatabolic state. In this situation, increased cardiac output is required to maintain the blood flow through capillary beds vasoconstricted by the presence of hypoxia. Respiratory effort gets progressively worse, and energy consumption is increased. All of these factors lead to respiratory muscle fatigue, which seems to be exacerbated by the lungs being hyperinflated. In short, respiratory effort is generally increased in the presence of a parenchymal lung injury leading to hypoxia, and it may be further increased by pulmonary hyperinflation, which originates from secretions obstructing the bronchi. The consequence is that the respiratory muscles have a short recovery time, and later there is an increase in the fatigue of the diaphragm [43].

These changes in respiratory effort, coupled with the presence of infection and tissue hypercatabolism, suggest that nutritional requirements are greatly increased in CF patients, and that the individual needs special metabolic support. The metabolism of 33 children about 5 years old with chronic and severe lung diseases was studied using calorimetry and nitrogen balance. That study showed that the need for energy supplementation was 20% higher for this patient group than for other groups. A negative nitrogen balance with a higher fat oxidation profile was also observed, suggesting that fat is preferentially used for the production of most calories. Thus, fat can be considered a good source of energy production for those patients who have chronic CO_2 retention, which can be worsened by carbohydrate intake. However, it should be noted that fats cause delayed gastric emptying, which may promote frequent gastroesophageal reflux (GERD) in these patients. This higher incidence of GERD can be explained both by chronic lung disease and lung hyperinflation and by the delayed gastric emptying that is characteristic of the disease. Thus, food supplementation in these patients should be increased, but without exaggeration of any of the components [44].

5.6 METABOLIC AND NUTRITIONAL SECONDARY PROCESSES

In addition to inflammation caused by infection, other mechanisms may also be involved in CF, especially in more severe cases. Hypoxia is associated with increased pressure in the pulmonary arteries, leading to epithelial dysfunction, edema, and acute inflammation accompanied by the release of IL1, IL6, interferon gamma, and TNF-alpha. These events have been observed in experimental studies. If hypoxia persists for long periods, vascular proliferation may occur, with increased vessel reactivity. Lastly, pulmonary hypertension may occur, although it is not very common in CF patients. The exposure of the lung to oxidative stress is an important mechanism associated with hypoxia and can be a factor in maintaining and increasing the inflammatory response with the release of more inflammatory mediators, such as interleukins and other factors, thus feeding this vicious circle [45].

An imbalance between food intake and calorie consumption may also be a factor for CF patients and will lead to a reduction in lung function. Metabolic consumption leads to changes in the metabolism of macronutrients, ultimately leading to muscle protein breakdown, which is also a source of calorie intake and total energy expenditure. Therefore, protein supplementation is also recommended [46]. That is, inflammation due to infection or to bronchial obstruction prevents weight gain and proper growth, which, in turn, hinders lung growth in children and the maintenance of FEV1 in adults.

5.7 RELATIONSHIP BETWEEN ENDOCRINE-METABOLIC DISEASES AND LUNG DEVELOPMENT

Cystic fibrosis-related diabetes (CFRD), which occurs in approximately 30% of patients, is associated with reduced lung function, worse nutritional status, and increased mortality rate [47]. A recent study that retrospectively evaluated 34 children with CF showed that there was a more rapid progression of structural lung disease in these individuals compared to individuals with normal glucose tolerance [48]. Moreover, stabilization of the disease state and improvement of both nutritional status and lung function appear to have occurred after adequate therapy. However, it is essential to strictly control the nutritional conditions, because changes in both growth and in nutritional status may precede the diagnosis of CFRD by a few years [49,50].

Another metabolic situation that seems to be involved in the inflammation of cystic fibrosis occurs because CFTR, in addition to its role in maintaining the hydration of secretions and electrolyte balance, also seems to have a regulatory role in inflammatory homeostasis. The absence of CFTR has been implicated in the increased activity of inflammatory mediators such as NF-κB and the decreased function of PPAR-γ, which exerts a protective effect. So these activity changes are additional factors that may be associated with exacerbated inflammatory response and with increased loss of muscle mass and skeletal muscle strength. This weakening, in turn, can increase exercise intolerance and cause muscle atrophy, in addition to interfering with the muscular process of respiration, thus worsening respiratory flows and volumes [51,52].

Among the substances that are involved in the metabolic and endocrine process of CF and are related to pulmonary function, ghrelin has been studied in relation to weight loss and lung function in CF patients. Ghrelin is an important orexigenic that is found in high levels in patients with chronic lung diseases. High ghrelin levels were also observed in patients with cystic fibrosis and were inversely related to FEV1, although this relationship is believed to be an effect of the metabolic process with high energy expenditure rather than its cause [53,54].

5.8 NUTRITION AND LUNG FUNCTION

As was already noted, numerous studies have shown that worsened lung function is associated with malnutrition, even when the infection is not severe. It is not known if low linear height growth is directly related to a decrease in lung function, as height has an important relationship with lung function [55]. Alternatively, progressive lung disease may inhibit appetite and lead to weight loss and poor linear growth, which was the initial thinking on this topic [56].

However, despite the apparent relationship between lung function and weight/height growth, the mechanism is poorly understood. One of the most important studies in this field showed that children who have continuous weight gain during childhood maintain FEV1 at higher levels than children with irregular weight gain. This study also emphasizes that having steady weight gain greater than 200 g/month, plus having a higher weight at the beginning of the study, is related to higher FEV1 values. This result suggests that nutritional loss early in life would be decisive for the rate of lung growth and development and is an important predictor of future lung disease [57].

The importance of nutrition to maintain lung function is a concept that has existed for many years, but it was reinforced by the American Cystic Fibrosis Foundation Guidelines. It was shown that high levels of indices such as weight-age, height-age, and height-to-weight ratios correlate with higher FEV1 and better survival, but only in childhood. One hypothesis that supports this theory is that after adolescence ends, the period of lung growth and development is completed. Therefore, there is no possibility for the recovery of lung function, even when nutritional recovery occurs. According to studies, the best nutritional measure for children older than 2 years would be a BMI above the 50th percentile, because this value had the best direct correlation with the FEV1 levels. For children younger than 2 years, a height/weight curve should be used with the aim of keeping the values at the 50th percentile. For adults, a BMI of 22 for women and 23 for men [18,58] is recommended.

The majority of studies, however, show that increasing food intake in CF patients leads to steady weight gain but not height gain. Moreover, it is well known that there is a relationship between height and pulmonary function in prepubertal patients [55]. Another point to consider is that during adolescence, the child's nutritional needs are increasing, due both to rapid growth and to the progression of lung disease. In addition to increasing the energy demand, the lung disease interferes by decreasing appetite and hence oral intake. The latest recommendations of the Cystic Fibrosis Foundation (CFF) indicate that children under 13 years of age should receive calorie assistance that is 110–200% of the recommendation for normal individuals to maintain normal gain weight. In addition to feeding, this assistance should be accomplished by means of nutritional supplements. However, the Cochrane Foundation examined this supplementation and assessed that it did not differ from the administration of well-balanced common foods with emphatic instructions to ingest them. If there are any nutritional deficiencies in children younger than 13 years, behavioral and nutritional approaches can achieve the necessary gains, although gains were not observed in older individuals [18].

When oral food intake alone is not sufficient for growth or to maintain nutritional status, dietary supplements may be needed. The routes of administration may vary from oral to enteral (including enteral tube feeding) or, more rarely, parenterally. Studies that evaluated nutritional supplementation via gastrostomy in CF patients noted improvement in the height Z-score and in the FEV1 levels from the second year of treatment. However, this approach should only be used when all possibilities for oral administration have been exhausted [59–61].

A final and important aspect of this topic is that not only weight gain but also body composition may be associated with higher lung function in this group of patients. A study by Williams et al. shows that the supplementation of various body building blocks, such as proteins and minerals, is associated with better nutrition and lung function. These authors showed that an increase in lean mass, contrary to all expectations, was not associated with a modification of FEV1 levels. Conversely, having higher fat levels was related to higher levels of lung function in male patients, although they were lower than the normal controls. Additionally, the disease began earlier in girls than in boys in this group, and the girls had a worse prognosis, which most likely reflects their worse nutritional status [62].

5.9 CONCLUSION

Several studies, including recent ones, have highlighted the poor outcomes associated with malnutrition in CF patients [63]. There are several aspects that need to be considered—mainly that it is a chronic disease that requires a multidisciplinary approach—and that confirm the hypothesis that nutrition is critical for maintaining lung function and consequently the survival and quality of life of these patients. The authors note that from breastfeeding on up to the more intricate and intracellular inflammation responses, including responses from cell signaling or orchestrated by cytokines, nutrition is essential for the maintenance, growth, and preservation of pulmonary function in patients with cystic fibrosis.

References

[1] Davis PB, Drumm M, Konstan MW. Cystic fibrosis: state of the art. Am J Respir Crit Care Med 1996;154:1229–56.
[2] Davis PM. Cystic fibrosis since 1938. Am J Respir Crit Care Med 2006;173:475–82.
[3] Riordan JR, Rommens JM, Kerem B, Alon N, Rozmahel R, Grzelczak Z, et al. Identification of the cystic fibrosis gene: cloning and characterization of complementary DNA. Science 1989;245:1066–72.
[4] Orozco L, Chávez M, Saldaña Y, Velázquez R, Carnevale A, González-del Angel A, et al. Cystic fibrosis: molecular update and clinical implications. Rev Invest Clin 2006;58(2):139–52.
[5] Althaus L. ENaC inhibitors and airway re-hydration in cystic fibrosis: state of the art. Curr Mol Pharmacol Mar 2013;6(1):3–12.
[6] Gibson RL, Burns JL, Ramsey BW. Pathophysiology and management of pulmonary infections in cystic fibrosis. State of art. Am J Respir Crit Care Med 2003;168:918–51.
[7] Engelhardt JF, Yankaskas JR, Ernst SA, Yang Y, Marino CR, Boucher RC, et al. Submucosal glands are the predominant site of CFTR expression in the human bronchus. Nat Genet 1992;2:240–8.
[8] Kerem E, Reisman J, Corey M, Canny G, Levison H. Prediction of mortality in patients with cystic fibrosis. N Engl J Med 1992;326:1187–91.
[9] Chimiel JF, Konston MW. Inflammation and anti-inflammatory therapies for cystic fibrosis. Clin Chest Med 2007;28:331–46.
[10] Mc Phail GL, Acton JD, Fenchel MS, Amin RS, Seid M. Improvements in lung function outcomes in children with cystic fibrosis are associated with better nutrition, fewer chronic Pseudomonas aeruginosa infections, and Dornase Alfa Use. J Pediatr 2008;153:752–7.
[11] Sanders DB, Li Z, Laxova A, Rock MJ, Levy H, Collins J, et al. Risk factors for the progression of cystic fibrosis lung disease throughout childhood. Ann Am Thorac Soc Nov 21, 2013. [Epub ahead of print].
[12] Bell SC, Shepherd RW. Optimising nutrition in cystic fibrosis. J Cyst Fibros 2002;1:47–50.
[13] Konstan MW, Butler SM, Wohl MB, Stoddard MS, Matousek R, Wagener J, for the investigators and coordinators of the epidemiologic study of cystic fibrosis, et al. Growth and nutritional indexes in early life predict pulmonary function in cystic fibrosis. J Pediatr 2003;142:624–30.
[14] Torres L, Hernandez JLJ, Almeida GB, Gomide LB, Ambrósio V, Fernandes MIM. Clinical, nutritional and spirometric evaluation of patients with cystic fibrosis after the implementation of multidisciplinary treatment. J Bras Pneumol 2010;36:731–7.
[15] Ratjen F, Döring G. Cystic fibrosis. Lancet 2003;361(9358):681–9.
[16] Kalnins D, Wilschanski M. Maintenance of nutritional status in patients with cystic fibrosis: new and emerging therapies. Drug Des Dev Ther 2012;6:151–61.
[17] World Health Organization and International Cystic Fibrosis (Mucoviscidosis) Association. Implementation of cystic fibrosis services in developing countries: memorandum from a joint WHO/ICF (M)A meeting. Bull World Health Organ 1997;15(1):10.
[18] Stallings V, Stark LJ, Robinson KA, Feranchak AP, Quinton H, Clinical practice guidelines on growth and nutrition subcommittee; AD HOC group. Evidence-based practice recommendations for nutrition-related management of children and adults with cystic fibrosis and pancreatic insufficiency: results of a systematic review. J Am Diet Assoc 2008;108:832–9.
[19] Kraemer R, Rudeberg A, Hadorn HB, Rossi E. Relative underweight in cystic fibrosis and its prognostic value. Acta Paediatr Scand 1978;76:33–7.
[20] Nir M, Lanng S, Johansen HK, Koch C. Long term survival and nutritional data in patients with cystic fibrosis treated in a Danish centre. Thorax 1996;51:1023–7.
[21] Peterson ML, Jacobs Jr DR, Milla CE. Longitudinal changes in growth parameters are correlated with changes in pulmonary function in children with cystic fibrosis. Pediatrics 2003;112(3 Pt 1): 588–92. PMid:12949289.
[22] Stapleton D, Kerr D, Gurrin L, Sherriff J, Sly P. Height and weight fail to detect early signs of malnutrition in children with cystic fibrosis. J Pediatr Gastroenterol Nutr 2001;33(3):319–25.
[23] Davis SD, Fordham LA, Brody AS, Noah TL, Retsch-Bogart GZ, Qaqish BF, et al. Computed tomography reflects lower airway inflammation and tracks changes in early cystic fibrosis. Am J Respir Crit Care Med 2007;175:943–50.
[24] Gaillard EA, Carty H, Heaf D, Smyth RL. Reversible bronchial dilatation in children: comparison of serial high-resolution computer tomography scans of the lungs. Eur J Radiol 2003;47:215–20.
[25] Zhang Z, Lai HC. Comparison of the use of body mass index percentiles and percentage of ideal body weight to screen for malnutrition in children with cystic fibrosis. Am J Clin Nutr 2004;80:982–91.
[26] Tanner JM. The use and abuse of growth standards. In: 2nd ed. Falkner F, Tanner JM, editors. Human growth, vol. 3. New York: Plenum; 1986. pp. 95–112.
[27] Miller MR, Hankinson J, Brusasco V, Burgos F, Casaburi R, Coates A, et al. Standardisation of spirometry. Eur Respir J 2005;26:319–38.
[28] Evans AK, Fitzgerald DA, Mckay KO. The impact of meconium ileus on the clinical course of children with cystic fibrosis. Eur Respir J 2001;18(5):784–9.
[29] Bates CM, Baum M, Quigley R. Cystic fibrosis presenting with hypokalemia and metabolic alkalosis in a previously healthy adolescent. J Am Soc Nephrol 1997;8:352–5.
[30] Ruddy R, Anolik R, Scanlin TF. Hypoelectrolytemia as a presentation and complication of cystic fibrosis. Clin Pediatr 1982;21(6):367–9.
[31] Colombo C, Constantini D, Zazzeron L, Faelli N, Russo MC, Ghisleni D, et al. Benefits of breastfeeding in cystic fibrosis: a single-centre follow-up survey. Acta Paediatr 2007;96(8):1228–32.
[32] Sinaasappel M, Stern M, Litlewood J, Wolfe S, Steinkamp G, Heijerman HG, et al. Nutrition in patients with cystic fibrosis: a European consensus. J Cyst Fibros 2002;1(2):51–75.
[33] Borowitz D, Robinson KA, Rosenfeld M, Davis SD, Sabadosa KA, Spear SL, et al. Cystic fibrosis foundation evidence-based guidelines for management of infants with cystic fibrosis. J Pediatr 2009;155(6 Suppl.):S73–92.
[34] Hakim F, Kerem E, Rivlin J, Bentur L, Stankiewicz H, Bdolach-Abram T, et al. Vitamins A and E and pulmonary exacerbations in patients with cystic fibrosis. J Pediatr Gastroenterol Nutr 2007;45(3):347.

[35] McCauley LA, Thomas W, Laguna TA, Regelman WE, Moran A, Polgreen LE. Vitamin D deficiency is associated with pulmonary exacerbations in children with cystic fibrosis. Ann Am Thorac Soc 2013 oct 1. [EPUB ahead print].

[36] Wolfenden LL, Judd SE, Shah R, Sanyal R, Ziegler TR, Tangpricha V. Vitamin D and bone health in adults with cystic fibrosis. Clin Endocrinol (Oxf) 2008;69:374–81.

[37] Pincikova T, Nilsson K, Moen IE, Karpati F, Fluge G, Hollsing A, et al. Inverse relation between vitamin D and serum total immunoglobulin G in the Scandinavian Cystic Fibrosis Nutritional Study. Eur J Clin Nutr 2011;65:102–9.

[38] Rivas-Crespo MF, González Jiménez D, Acunha Quirós MD, Sojo Aguirre A, Heredia González S, Diaz Martin JJ, et al. High serum retinol and lung function in young patients with cystic fibrosis. J Pediatr Gastroenterol Nutr June 2013;56(6):657–62.

[39] Roulet M, Frascarolo P, Frappaz I, Pilet M. Essential fatty acid deficiency in well nourished young cystic fibrosis patients. Eur J Pediatr 1997;156:952–6.

[40] Olveira G, Olveira C, Acosta E, Espindola F, Garrido-Sanchez L, Garcia-Escpbar E, et al. Fatty acid supplements improve respiratory, inflammatory and nutritional parameters in adults with cystic fibrosis. Asch Bronconeumol Feb 2010;46(2):70–7.

[41] Oliver C, Jahnke N. Omega-3 fatty acids for cystic fibrosis. Cochrane database Syst Rev Aug 10, 2011;8.

[42] Ramires BR, Oliveira EP, Pimental GD, McLellan KCP, Nakato DM, Faganello MM, et al. Resting energy expenditure and carbohydrate oxidations are higher in elderly patients with COPD: a case control study. Nutr J 2012;11:37–43.

[43] Schöni MH, Casaulta-Aebischer. Nutrition and lung function in cystic fibrosis patients. Clin Nutr 2000;19:79–85.

[44] Müller MJ, Bosy-Westphal A. Energy expenditure in children. Curr Opin Clin Nutr Metab Care 2003;6:519–30.

[45] Araneda OM, Tuesta M. Lung oxidative damage by hypoxia. Oxid Med Cell Longev 2012:1–18. Article ID 856918.

[46] Kao CC, Hsu JW-C, Jahoor F. Resting energy expenditure and protein turnover are increased in patients with severe chronic obstructive pulmonary disease. Metabolism 2011;60(10):1449–55.

[47] Sinaasappel M, Stern M, Littlewood J, Wolfe S, Steinkamp G, Heijerman GM, et al. Nutrition in patients with cystic fibrosis: a European consensus. J Cyst Fibros 2002;1:51–75.

[48] Widger J, Ranganathan S, Robinson PJ. Progression of structural lung disease on CT scans in children with cystic fibrosis related diabetes. J Cyst Fibros 2013;12(3):216–21.

[49] Kolouskova S, Zemkova D, Bartosova J, et al. Low-dose insulin therapy in patients with cystic fibrosis and early-stage insulinopenia prevents deterioration of lung function: a 3-year prospective study. J Pediatr Endocrinol Metab 2011;24(7–8):449–54.

[50] Cheung MS, Bridges NA, Prasad SA, Francis J, Carr SB, Suri R, et al. Growth in children with cystic fibrosis-related diabetes. Pediatr Pulmonol 2009;44(12):1223–5.

[51] Dekkers JF, Van der Ent CK, Kalkhoven E, Beekman JM. PPAR-gama as a therapeutic target in cystic fibrosis. Trends Mol Med 2012;18(5):283–91.

[52] Van Der Weert-van Leeuween PB, Prets HGM, Van Der Ent CK, Beekman JM. Infection, inflammation and exercise in cystic fibrosis. Respir Res 2013;14:32–42. J Cyst Fibros. Sep 2008; 7(5):398–402. doi: 10.1016/j.jcf.2008.02.002. Epub 2008 Mar 18.

[53] Cohen RI, Tsang D, Koenig S, Wilson D, McCloskey T, Chandra S. Plasma ghrelin and leptin in adult cystic fibrosis patients. J Cyst Fibros 2008;7(5):398–402.

[54] Monajemzadeh M, Mokhtari S, Motamed F, Sedigheh S, Mohammad THA, Abbasi A, et al. Plasma ghrelin levels in children with cystic fibrosis and healthy. Arch Med Sci 21, 2013;9(1):93–7.

[55] Torres LAGMM, Martinez FE, Manço JC. Correlation between standing height, sitting height, and arm span as an index of pulmonary function in 6–10-year-old children. Pediat Pulmonol 2003;36:202–8.

[56] Zemel B, Kawchak D, Cnann A, Zhao H, Scanlin T, Stallings V. Prospective evaluation of resting energy expenditure, nutritional status, pulmonary function, and genotype in children with cystic fibrosis. Pediatr Res 1996;40:578–86.

[57] Petersen ML, Jacobs DR, Milla C. Longitudinal changes in growth parameters are correlated with changes in pulmonary function in children with cystic fibrosis. Pediatrics 2003; 112:588–92.

[58] Elborn, Bell. Nutrition and survival in cystic fibrosis. Thorax 1996;51:971–2.

[59] Efrati O, Mei-Zahav M, Rivlin J, Kerem E, Blau H, Barak A, et al. Long term nutritional rehabilitation by gastrostomy in Israeli patients with cystic fibrosis: clinical outcome in advanced pulmonary disease. J Pediatr Gastroenterol Nutr 2006;42(2):222–8.

[60] Williams SG, Ashworth F, McAlweenie A, Poole S, Hodson ME, Westaby D. Percutaneous endoscopic gastrostomy feeding in patients with cystic fibrosis. Gut 1999;44(1):87–90.

[61] Walker SA, Gozal D. Pulmonary function correlates in the prediction of long-term weight gain in cystic fibrosis patients with gastrostomy tube feedings. J Pediatr Gastroenterol Nutr 1998;27(1):53–6.

[62] Williams JE, Wells JC, Benden C, Jaffe A, Suri R, Wilson CM, et al. Body composition assessed by the 4-component model and association with lung function in 6-12-y-old children with cystic fibrosis. Am J Clin Nutr Dec 2010;92(6):1332–43.

[63] Kalnius D, Wilschauski M. Maintenance of nutritional status in patients with cystic fibrosis: new and emerging therapies. Drug Des Devel Ther 2012;6:151–61.

CHAPTER

6

Family Mealtimes and Children with Cystic Fibrosis

Amber J. Hammons[1], *Robin S. Everhart*[2], *Barbara H. Fiese*[3]

[1]Fresno in the Child, Family, and Consumer Sciences Department, California State University, Fresno, CA, USA; [2]Psychology, Virginia Commonwealth University, Richmond, VA, USA; [3]Human and Community Development, University of Illinois, Urbana–Champaign, IL, USA

Mealtimes can be particularly challenging for families with a child with cystic fibrosis (CF) because a strict nutrition regimen is required. Nutritional recommendations are that children with CF consume 120–150% of the expected recommended dietary allowance of calories [1]. Unfortunately, few children meet this requirement [2]. Consuming sufficient calories is essential for healthy functioning in children with CF. A review examining nutritional status in relation to pulmonary health found that nutritional status is a strong predictor of pulmonary functioning, reiterating the importance of adequate nutrition in individuals with CF [3]. Similarly, weight gain is related to better lung functioning [4]. Many factors can influence adherence to the demanding nutrition regimen, including family stress. For instance, increased family stress is linked to the consumption of fewer calories in children with CF [5,6]. Because overall family functioning is an important moderator of child health, we briefly review family functioning in families with a child with CF and focus on common mealtime challenges faced by families with a child with CF as an example of how family factors influence child nutritional habits. In addition, research on direct mealtime observations is presented to illustrate how families manage this daily routine that affects an important aspect of daily functioning. We end with a discussion of behavioral and nutritional interventions designed to increase weight gain and caloric intake and decrease feeding difficulties.

Family members are most frequently responsible for the medical and nutritional treatment of young children with CF [7]. Studies have examined caregiving differences in relation to maternal stress between mothers with a child with CF compared to those without children with CF. One such study looked at 40 mothers and their caregiving role strain through the use of phone diaries and an in-home interview [8]. As expected, mothers with a child with CF spent more time on medical care, chores, and child care and less time in play and recreation than the control group [8]. This is not surprising given that, in addition to the intense nutrition regimen required of patients with CF, treatment also consists of time-consuming procedures to clear airways. Indeed, as part of medical care, techniques to clear mucus can take an average of 2h a day [7].

When considering the family management of CF behaviors related to mealtimes, compliance with pancreatic enzyme medication has been identified as a particularly challenging area for families [9]. Pancreatic enzymes often are given up to five times a day with each meal and snack. The repetition and scheduling of such enzyme administration may be burdensome and overwhelming for families. In their study of 37 families with a child with CF between the ages of 6 and 13 years, Modi and Quittner [9] asked parents to report specifically on "what got in the way" of completing disease-specific treatments, including adherence to enzymes. Parents identified forgetting (46% of parents), embarrassment (17%), and oppositional behaviors from the child (11%) as barriers to enzyme adherence.

Given that parents are primarily responsible for the day-to-day treatment decisions related to pediatric illnesses [10], especially among young children, the psychological well-being of parents also has been associated with enzyme adherence in CF. For instance, parent depression was negatively associated with adherence to electronically monitored enzyme adherence in children with CF between 1 and 11 years of age [11]. Caregivers with elevated levels of depression demonstrated rates of

Diet and Exercise in Cystic Fibrosis
http://dx.doi.org/10.1016/B978-0-12-800051-9.00006-7

enzyme adherence that were 11 percentage points lower than rates of adherence for nondepressed caregivers. Furthermore, caregiver depression was associated with poor weight gain among children with CF across a 3-month period [11]. These findings highlight the importance of focusing on parental adjustment in interventions aimed at improving pancreatic enzyme adherence.

In addition to parental functioning, the overall level of functioning within families has important implications for CF treatment behaviors. Warm, supportive, and responsive interactions between family members, as well as lower levels of family conflict, have been associated with higher rates of treatment adherence [12,13]. In Berge and Patterson's review [7] of the CF literature looking at family variables, children with CF have a heightened risk for behavioral problems including anxiety, social withdrawal, and aggression, all of which may affect family functioning. Siblings of the child with CF also received less attention because of the time necessary for medical treatments for the child with CF. These siblings showed more fighting, aggression, and jealousy, likely contributing to parental role strain. In addition, parents with a child with CF were found to have lower levels of marital satisfaction than parents without a child with CF [7]. All of these factors may contribute to the overall level of family functioning.

In their sample of 96 families with a child with CF between the ages of 9 and 16 years, DeLambo et al. [12] found that better family relationship quality was associated with higher rates of adherence to airway clearance and aerosolized medications. Interestingly, disease-specific problem solving was not associated with adherence after controlling for relationship quality. In the absence of positive family relationships, problem-solving skills are unlikely to increase adherence to the CF treatment regimen [12]. Positive family functioning may play a critical role in promoting treatment adherence in pediatric CF and may be an area for future interventions.

In measuring family functioning, researchers often rely on mealtime observations, which can provide an opportunity to observe how families regulate child behavior, delegate roles, express and manage their affect, communicate, and interact with one another [14]. Observational measures have advantages over self-report measures of family functioning in that the whole family is included in the observation and biases related to self-report measures are minimized (e.g., social desirability, recall bias [15]). Cohesive and supportive family environments at mealtime may lead to the consumption of more food and a more enjoyable mealtime [16]. On the other hand, families may experience worse functioning if parental attempts at feeding the child are met with resistance, creating an unpleasant mealtime experience for the family [17].

One team of researchers proposes a cyclical pattern in which parental stress over the child not getting sufficient calories leads to parental coercive behavior for bids to eat, which further leads to a decrease in the child's eating and pleasant interactions with the child, which in turn increases parental stress [18]. From an examination of the studies included in their review, Berge and Patterson [7] concluded that striking a balance between treatment needs and normative family needs is challenging, with many parents focusing on treatment needs more than normative family needs. Further, as parents typically attend to the nutritional needs of the child with CF, less attention may be paid to the social and emotional needs of other family members, including siblings who are not ill [19]. Mealtimes for families with a child with CF often are equated with illness-management tasks and provide an overall picture of how families work together in managing their child's CF.

Previous research incorporating mealtime observations as an assessment of family functioning have found that families with a child with CF often demonstrate lower levels of family functioning across several domains (e.g., affect regulation, task accomplishment) compared to families with non-ill children [17,19–21]. For instance, Janicke et al. [17] reported that in 28 families with a child with CF between the ages of 6 and 13 years, levels of communication, affect management, interpersonal involvement, and behavior control were significantly lower than in a control group of 27 families. A larger percentage of families with a child with CF also had levels of family functioning that fell in the "unhealthy" range. In another study, Mitchell and colleagues [20] found that levels of communication, affect management, interpersonal involvement, behavior control, and role distribution were lower in a group of 33 families with a child with CF. Existing research focused on understanding family adjustment to CF suggests that families with a child with CF may struggle in interactions aimed at meeting the dietary demands of CF, which has implications for other components of the treatment regimen.

In fact, little research has considered how a family's level of functioning, as measured by a mealtime observation, may be associated with treatment adherence in children with CF. A recent dissertation study [22] considered associations between family functioning, as measured by the Mealtime Interaction Coding System (MICS [23]), and treatment adherence, as measured by the Treatment Adherence Rating Scale [24]. This study included 19 children and adolescents between the ages of 8 and 19 years and their families. Children and primary caregivers completed psychosocial measures in a single session; families then were videotaped having dinner together, and the mealtime tapes were coded by trained, reliable coders using the MICS. Study findings suggested that higher levels of family functioning were most commonly associated (at a trend level) with better antibiotic adherence. Results from this pilot study suggest that how well families are able to work together as a team may be most useful in ensuring that children are adhering to their antibiotic treatment. For instance, families may need to plan ahead to have the appropriate antibiotic when needed and to have the child's nebulizer set up correctly in advance. However, given this study's small sample, more research is needed to investigate associations between family functioning and treatment adherence in children and adolescents with CF.

An in-depth meta-analysis examining 10 studies looked at how family mealtimes are affected in families with a child with CF [25]. The meta-analysis revealed that families with a child with CF encounter more challenges during mealtimes and that these challenges have a larger impact on family functioning than for comparison families [25]. Both specific (microanalytic) and global mealtime behaviors were analyzed in the meta-analysis. Specific mealtime behaviors were measured using the Dyadic Interaction Nomenclature for Eating (DINE) [21] and the Mealtime Observation Schedule [26], both of which measure parent and child feeding behaviors. Examples of parent feeding behaviors include frequency of use of commands, prompts, and reinforcements. Child behaviors measured include number of sips and bites, leaving the table, and refusing to eat. In the meta-analysis, the micro behaviors were combined to form an aggregate variable. There were four studies that examined these micro behaviors, and the meta-analysis revealed that a greater number of micro behaviors were present for families with a child with CF than for comparison families. In addition, mealtime length was related to child weight, with longer meals associated with lower weight.

Broad areas of family functioning also were examined globally using the MICS [23] and included affect management, interpersonal involvement, communication, behavior control, and family functioning. Affect management examines the appropriateness of emotions during mealtime. For example, this dimension assesses whether emotions are generally disruptive or attended to in a sensitive and nurturing way [23]. Interpersonal involvement examines family members' interest in one another and one another's activities, whereas communication examines how family members talk with each other and communicate as well as respond to one another's needs, feelings, and interests [23]. The results of the meta-analysis revealed that affect management, interpersonal involvement, and communication were the areas that were most strongly affected in the mealtime context [25]. Family functioning, which is its own dimension and a measure of the quality of the mealtime (not an average of all of the other dimensions), was related to child weight status, with lower family functioning associated with a lower child weight. These dimensions are particularly relevant to mealtimes with a child with CF because the priority is often making sure that the child consumes the necessary calories. Similar to Berge and Patterson's [7] conclusions from their review, speculation as to why meals are more challenging for families with a child with CF included immediate needs of feeding taking precedent over the relational aspects of mealtimes. The results of the meta-analysis led the researchers to encourage those working on interventions to incorporate the broader family context and to take a family systems perspective.

Other studies in this area have focused on interventions designed to increase weight gain. Another meta-analysis, this one looking specifically at interventions, found that behavior interventions were just as effective at increasing weight gain as invasive medical interventions for children with CF [27]. Behavioral interventions have an advantage over medical interventions (e.g., enteral and parenteral nutrition) because they are less invasive and can begin earlier in the disease progression [27]. Behavioral treatments often incorporate social learning theory as a framework to understand how parent–child mealtime interactions can reinforce maladaptive eating behaviors in children with CF. Behavioral interventions in this meta-analytic review included a focus on nutrition as well as parent training in effective behavioral management techniques. Eighteen studies looking at enteral nutrition, parenteral nutrition, oral supplementation, and behavior treatment were included in the meta-analysis [27]. The results revealed that the behavioral studies were effective at increasing weight, yielding a large effect size across studies.

Behavior–nutrition interventions have expanded since the meta-analysis by Jelalian and colleagues [27]. Behavior–nutrition interventions have been found to be effective in improving dietary adherence as well as growth [28]. One study examined the transition from toddlerhood to school-age and nutrition/mealtime challenges encountered by families with children with CF [29]. The study examined eight parents of children with CF (average age of 8 years) using a semistructured interview. This was a follow-up study that took place 5 years after these families participated in a behavior–nutrition treatment study [28]. A thematic approach was used to describe the information extracted from the interviews. One of the themes that emerged regarded nutrition and how parents helped their children to consume the recommended calories. Strategies included using addables (i.e., adding extra things such as toppings, e.g., adding nuts, sugar, or fruit to oatmeal) and spreadables (e.g., butters, sauces) and feeding high-calorie foods [29]. Parents reported high stress over getting their children to eat, with picky eating complicating matters. Parents also discussed their improved behavioral techniques such as focusing on micro behaviors such as the number of bites the child takes instead of focusing on more general behaviors [29].

The transition to school introduces even more unique challenges for parents, such as not being able to monitor what the child eats while away and consequently not knowing how many calories were consumed during lunchtime. An additional concern that parents expressed revolved around the child not having enough time to eat during the allocated lunchtime at school. There were nutrition-related concerns as well as worries surrounding academics, specifically when missing school because of illness [29]. Families including an individual with CF face many difficulties that are unique to the disease itself as well as specific family-level difficulties [28].

The efficacy of behavioral and nutritional interventions has been compared to nutrition education-only programs with regard to changing caloric intake and weight gain [30]. A randomized controlled trial that included 79 children was used to examine changes in the number of calories consumed and amount of weight gain from baseline to after the intervention. Although previous studies have compared behavioral and nutrition interventions to nutritional education-only interventions, this was the first study to use a randomized controlled trial with a large sample and thus was able to detect differences between the groups. The program consisted of separate parent and child groups. Each group received a 90-min session before treatment, five weekly group sessions, and a session after treatment [30]. In addition, each group was followed up five times over the subsequent 2 years. The calorie goal included an increase of total calories by 1000 per day. For the nutrition component, both groups (i.e., behavioral plus nutrition education and nutrition education only) received the same information. The behavioral plus nutrition education followed the Be-In-Charge protocol (for more information on this intervention see Stark et al. [31]). The only difference between the two types of interventions was that the nutrition-only group did not receive any training on behavioral management. The behavioral management component emphasized positive reinforcement and differential attention.

Although researchers hypothesized that the nutritional plus behavioral intervention would be more effective immediately and at a 24-month follow-up, they found that the combined behavioral and nutritional intervention yielded stronger findings in relation to the outcomes (averaged 383 more calories a day and 0.55 kg more in weight) after the 9-week intervention, but significant differences between the two types of interventions were not found 24 months after treatment. In addition, parents in both groups reported being satisfied with the treatment. With regard to estimated energy requirement, both groups were effective at the 24-month follow-up [30].

A related feasibility study [18] adapted the Be-In-Charge intervention [31] by including developmentally appropriate information regarding toddlers. This adaptation was tested using four families. Primary goals included improving diet quality and food intake and increasing child weight (all based on the specific needs of the children in the study). Group sessions consisted of behavioral as well as nutritional components, and a registered dietician on staff created individualized calorie goals for each child. The intervention itself lasted 6 weeks. Parents used food diaries to report child food intake and were taught how to measure food intake through weighing and counting calories. A registered dietician also verified caloric estimates. Each family was asked to record one family dinner each week, for a total of nine family dinners. Mealtime interactions were coded using the DINE [21], through which parent and child eating behaviors were examined. Baseline data were collected across a 2-week period, and then a 12-week follow-up after the intervention also was measured. The program was effective for three of the four children, and these results were sustained at follow-up. The parents believed that the information and materials presented were informative and would be useful to them [18].

To assist professionals working with families with a child with CF, Powers and colleagues [32] designed a tool to give information about children's feeding patterns, food choices, and enzyme usage—information vital for nutritional intervention. This tool provides an individualized report on nutritional intake and was created to ultimately help individuals with CF achieve optimal health [32]. Powers's team designed the CF Individualized NuTritional Assessment of Kids Eating (CF INTAKE), which is essentially a modification of a method developed by Christensen et al. [33] for type I diabetes mellitus.

The CF INTAKE assesses the number of meals consumed per day, enzyme usage, daily caloric intake, and high-fat food choices [32]. Adequacy of this tool was tested using a sample of 91 infants and children with a mean age of 4.5 years. A 3-day weighed diet diary was completed, and parents were instructed on how to measure food. Meals eaten, food intake, and enzyme use were compared to the nutritional recommendations established at the CF Nutritional Consensus Conference and guidelines from nutritional interventions [32]. The CF INTAKE gives scores for missed opportunities (i.e., missed opportunity to optimize dietary intake) and power (i.e., number of times high-fat/high-calorie foods were consumed each day). Missed opportunity scores break down into scores for meals (number of missed meals per day (out of three) and number of missed snacks per day (out of two)), enzymes (taken or not taken when necessary), calories (<120% of recommended daily allowance (RDA) per day), and food choices (low-fat foods consumed in a day divided by total foods per day) [32]. Power scores were separated into a power foods score (assigned one point for every high-fat food consumed) and a power addables score (one point for every high-fat addable/spreadable).

Results revealed that missed meals occurred 10% of the time and enzymes were not taken 15% of the time. In addition, 60% of the children consumed less than 120% of the RDA. Of the foods consumed, 50% of the food eaten were low-fat foods, 33% of the foods eaten each day maximized their calorie intake, and 16% of the foods included the use of addables/spreadables [32]. Parent report revealed a higher frequency of addables/spreadables and power foods in preschool children than in infants and toddlers. The CF INTAKE was used later in two case studies that demonstrated the power of the CF INTAKE in identifying nutritional areas needing attention. Findings from this and subsequent studies highlight the clinical utility of the CF INTAKE measure in improving nutrition among children with CF.

CONCLUSION AND FUTURE DIRECTIONS

Raising a child with a chronic health condition can present many challenges for families. The daily routines of sharing meals, taking medications, and even sending a child to school are complicated for families who have a child with CF. The core elements of family routines associated with healthy outcomes include predictability, positive communication, positive affect regulation, and repetition over time [14,34,35]. This brief review focused on mealtime routines and medical adherence as two essential family practices associated with health and well-being of children with CF. Motivated to see their children thrive and gain weight, parents often are compelled to attend to the microdynamics of the meal, including how many times the child takes a bite of food, forcing the child to eat more, and focusing more on caloric intake than conversation. Given the dietary consequences of not consuming enough calories, these behaviors are understandable. However, the evidence from this review suggests that the behavioral and dietary consequences are intertwined [5,6] to a degree that a focus on positive communication and attention to positive emotions may be more closely aligned to weight gain [16,25]. A challenge for future intervention programming is to convince caregivers that a calm and emotionally responsive approach to mealtimes may actually encourage caloric intake rather than a controlling atmosphere focusing on the food itself. Subtle changes in mealtime conversations that focus on the routine events of the day may reduce stress not only for the child but also for caregivers. Combining mealtime conversation approaches applied to common mealtime challenges such as picky eating (see http://familyresiliency.illinois.edu/MealtimeMinutes.htm) with evidence-based programs such as CF-INTAKE may provide families with easy-to-apply solutions.

Adhering to a complex medical regimen is essential for the health and survival of patients with CF. Adherence in CF is critical in slowing disease progression and preventing lung infections. There is a strong need for intervention programs that can be delivered more readily to families who are unable to come into the laboratory for a long period of time. Families with a child with CF are likely already busy completing daily activities (e.g., family, socializing, school, sports) in addition to the rigorous demands of caring for a child with CF. More than 60% of the eligible families in the study by Hourigan et al. [18] study declined to participate, with some giving time constraints and travel concerns as reasons. Advances in mobile technologies and e-health have the potential to assist patients with CF in self-management. However, these technologies will be effective only to the extent that they also incorporate family factors into their applications [36]. It is clear that a one-size-fits-all approach will not work in the self-management of CF, and current levels of stress, variations in diet, and availability of social support will have to be taken into account.

As new therapies are developed and the life span of patients with CF increases, it is important to develop healthy habits at an early age. Establishing good dietary habits in a supportive family environment may bode well for children as they reach the vulnerable period of adolescence and young adulthood, when adherence and dietary habits often are weakened. For instance, rates of adherence decline when parents are not able to directly monitor treatment adherence [37]. Social influences related to body image and appearance also affect adolescents with CF and can lead to self-imposed weight loss regimens [38]. Future research and intervention efforts to address these developmental transitions are warranted. Although CF is diagnosed in one, and sometimes two or more, member(s) of a family, it is a family affair to manage the disease. The health and well-being of the entire family needs to be considered when planning treatment protocols. Everyone's health will benefit from a whole-family approach to this complex and challenging condition.

References

[1] Borowitz D, Baker RD, Stallings V. Consensus report on nutrition for pediatric patients with cystic fibrosis. J Pediatr Gastroenterol Nutr 2002;35:246–59.

[2] Mackner L, McGrath A, Stark L. Dietary recommendations to prevent and manage chronic pediatric health conditions: adherence, intervention, and future directions. J Dev Behav Pediatr 2001;22(2):130–43.

[3] Milla CE. Association of nutritional status and pulmonary function in children with cystic fibrosis. Curr Opin Pulm Med 2004;10:505–9.

[4] Konstan MW, Butler SM, Wohl ME, Stoddard M, Matousek R, Wagener JS, Investigators & Coordinators of the Epidemiologic Study of Cystic Fibrosis, et al. Growth and nutritional indexes in early life predict pulmonary function in cystic fibrosis. J Pediatr 2003;142:624–30.

[5] Crist W, Mcdonnell P, Beck M, Gillespie CT, Barrett P, Mathews J. Behavior at mealtimes and the young-child with cystic-fibrosis. J Dev Behav Pediatr 1994;15:157–61.

[6] Opipari-Arrigan L, Powers S, Quittner A, Stark L. Mealtime problems predict outcome in clinical trial to improve nutrition in children with CF. Pediatr Pulmonol 2010;45.1:78–82.

[7] Berge J, Patterson J. Cystic fibrosis and the family: a review and critique of the literature. Families Systems Health 2004;22:74–100.

[8] Quittner AL, Opipari LC, Regoli MJ, Jacobsen J, Eigen H. The impact of caregiving and role strain on family-life – comparisons between mothers of children with cystic-fibrosis and matched controls. Rehabil Psychol 1992;37:275–90.

[9] Modi AC, Quittner AL. Barriers to treatment adherence for children with cystic fibrosis and asthma: what gets in the way? J Pediatr Psychol 2006;31:846–58.

[10] Kazak AE, Rourke MT, Crump TA. Families and other systems in pediatric psychology. In: Roberts MC, editor. Handbook of pediatric psychology. third ed. New York: Guilford Press; 2003.

[11] Quittner AL, Barker DH, Geller D, Butt S, Gondor M. 319 Effects of maternal depression on electronically monitored enzyme adherence and changes in weight for children with CF. J Cyst Fibros 2007;6.

[12] DeLambo KE, Ievers-Landis CE, Drotar D, Quittner AL. Association of observed family relationship quality and problem-solving skills with treatment adherence in older children and adolescents with cystic fibrosis. J Pediatr Psychol 2004;29:343–53.

[13] Quittner AL, Tolbert VE, Regoli MJ, Orenstein DM, Hollingsworth JL, Eigen H. Development of the role-play inventory of situations and coping strategies for parents of children with cystic fibrosis. J Pediatr Psychol 1996;21:209–35.

[14] Fiese BH, Foley KP, Spagnola M. Routines and ritual elements in family mealtimes: contexts for child well-being and family identity. New Dir Child Adolesc Dev 2006;111:67–90.

[15] Smyth JM, Stone AA. Ecological momentary assessment research in behavioral medicine. J Happiness Studies 2003;4.

[16] Stanhope R, Wilks Z, Hamill G. Failure to grow: lack of food or lack of love? Prof Care Mother Child 1994;4:234–7.

[17] Janicke DM, Mitchell MJ, Stark LJ. Family functioning in school-age children with cystic fibrosis: an observational assessment of family interactions in the mealtime environment. J Pediatr Psychol 2005;30:179–86.

[18] Hourigan S, Helms S, Christon L, Southam-Gerow M. Improving nutrition in toddlers with cystic fibrosis: feasibility of a behavioral parent-training intervention. Clin Practice Pediatr Psychol 2013;1:235–49.

[19] Spieth LE, Stark LJ, Mitchell MJ, Schiller M, Cohen LL, Mulvihill M, et al. Observational assessment of family functioning at mealtime in preschool children with cystic fibrosis. J Pediatr Psychol 2001;26:215–24.

[20] Mitchell MJ, Powers SW, Byars KC, Dickstein S, Stark LJ. Family functioning in young children with cystic fibrosis: observations of interactions at mealtime. J Dev Behav Pediatr 2004;25:335–46.

[21] Stark LJ, Jelalian E, Powers SW, Mulvihill MM, Opipari LC, Bowen A, et al. Parent and child mealtime behavior in families of children with cystic fibrosis. J Pediatr 2000;136:195–200.

[22] Everhart RS. Family functioning and treatment adherence in children and adolescents with cystic fibrosis. Syracuse University (Order No. 3430679, Syracuse University), ProQuest Dissertations and Theses, 184; 2010.

[23] Dickstein S, Seifer R, Hayden LC, Schiller M, Sameroff AJ, Keitner G, et al. Levels of family assessment: II. Impact of maternal psychopathology on family functioning. J Family Psychol 1998;12:23–40.

[24] Quittner AL, Drotar D, Levers CE. Treatment adherence rating scale: unpublished measure; 1998.

[25] Hammons AJ, Fiese B. Mealtime interactions in families of a child with cystic fibrosis: a meta-analysis. J Cyst Fibros 2010;9:377–84.

[26] Sanders MR, Turner KM, Wall CR, Waugh LM, Tully LA. Mealtime behavior and parent-child interaction: a comparison of children with cystic fibrosis, children with feeding problems, and nonclinic controls. J Pediatr Psychol 1997;22:881–900.

[27] Jelalian E, Stark LJ, Reynolds L, Seifer R. Nutrition intervention for weight gain in cystic fibrosis: a meta analysis. J Pediatr 1998;132:486–92.

[28] Powers SW, Jones JS, Ferguson KS, Piazza-Waggoner C, Daines C, Acton JD. Randomized clinical trial of behavioral and nutrition treatment to improve energy intake and growth in toddlers and preschoolers with cystic fibrosis. Pediatrics 2005;116:1442–50.

[29] Filigno SS, Brannon EE, Chamberlin LA, Sullivan SM, Barnett KA, Powers SW. Qualitative analysis of parent experiences with achieving cystic fibrosis nutrition recommendations. J Cyst Fibros 2012;11:125–30.

[30] Stark LJ, Quittner AL, Powers SW, Opipari-Arrigan L, Bean JA, Duggan C, et al. Randomized clinical trial of behavioral intervention and nutrition education to improve caloric intake and weight in children with cystic fibrosis. Arch Pediatr Adolesc Med 2009;163:915–21.

[31] Stark LJ, Opipari LC, Spieth LE, Jelalian E, Quittner AL, Higgins L, et al. Contribution of behavior therapy to dietary treatment in cystic fibrosis: a randomized controlled study with 2-year follow-up. Behavior Therapy 2003;34:237–58.

[32] Powers SW, Patton SR, Henry R, Heidemann M, Stark LJ. A tool to individualize nutritional care for children with cystic fibrosis: reliability, validity, and utility of the CF individualized NuTritional Assessment of Kids Eating (CF INTAKE). Childrens Health Care 2005;34:113–31.

[33] Christensen NK, Terry RD, Wyatt S, Pichert JW, Lorenz RA. Quantitative assessment of dietary adherence in patients with insulin-dependent diabetes mellitus. Diabetes Care 1983;6:245–50.

[34] Fiese BH, Everhart RS. Medical adherence and childhood chronic illness: family daily management skills and emotional climate as emerging contributors. Curr Opin Pediatr 2006;18:551–7.

[35] Fiese BH. Family routines and rituals. New Haven, CT: Yale University Press; 2006.

[36] Kirk S, Beatty S, Callery P, Gellately J, Milnes L, Pryjmachuk S. The effectiveness of self-care support interventions for children and young people with long-term conditions: a systematic review. Child Care Health Dev 2013;3:305–24.

[37] Modi AC, Marciel KK, Slater SK, Drotar D, Quittner AL. The influence of parental supervision on medical adherence in adolescents with cystic fibrosis: developmental shifts from pre to late adolescence. Childrens Health Care 2008;37:78–92.

[38] Patterson JM, Wall M, Berge J, Milla C. Associations of psychological factors with health outcomes among youth with cystic fibrosis. Pediatr Pulmonol 2009;44:46–53.

CHAPTER

7

Disturbed Sleep Behaviors and Melatonin in Sleep Dysfunction and Treatment of Cystic Fibrosis

Aaron Robert Brussels, Ronald Ross Watson

Health Sciences Center, School of Medicine, Mel and Enid Zuckerman College of Public Health, University of Arizona, Tucson, AZ, USA

7.1 INTRODUCTION

Cystic fibrosis (CF) is the most common lethal genetic disease in Caucasians, with 4–5% of the population carrying the recessive gene and 1 in every 2500 newborns diagnosed. The most common cause of both morbidity and mortality in CF patients is respiratory disease resulting from severe and persistent lung inflammation and infection, as well as continual increase in the severity of lung disease [1].

CF patients experience a shortened life span and diminished quality of life due to their illness. However, developments in antibiotics, mucolytic agents, physiotherapy, and nutritional support have helped to greatly increase their survival [2]. Despite such advances in longevity, CF patients still report inferior quality of life as compared with otherwise-healthy individuals, with poor sleep quality being a chief complaint [1,3].

A study revealed that 40% of the CF patients rated themselves as poor or below-average sleepers, noting a delayed sleep latency, increased sleep disturbances, and generally poor sleep quality [1]. Insufficient sleep and the burden it brings on patient life (i.e., fatigue, daytime sleepiness, cognitive impairment, diminished immune response, etc.) is well understood and likely is detrimental to existing courses of CF treatment [3]. Nighttime oxygen supplementation is an integral part of treatment for patients with severe lung diseases, such as chronic obstructive pulmonary disease. CF patients, however, do not experience the same improvement as chronic obstructive pulmonary disease patients with the addition of oxygen therapy, leaving them with few viable noninvasive options [1]. Because it is such a common complaint in the affected population, understanding the mechanisms behind diminished sleep quality in CF patients is paramount and optimizing sleep quality should be an integral component of treatment [2].

Melatonin is a hormone produced endogenously in the pineal gland that regulates the daily sleep–wake cycle. Substantial evidence suggests that the administration of exogenous melatonin is effective both as a long-term chronobiotic and as a short-term hypnotic. Unregulated in most countries and marketed as a supplement, melatonin is easily available and reasonably affordable [4].

Few CF patients report regular use of sleep aids—prescription or over the counter—in their treatment regiments [1]. Because melatonin is shown to have low toxicity and is widely available, it shows great potential for improving sleep quality in CF patients [4].

7.2 SLEEP DYSFUNCTION IN CYSTIC FIBROSIS

7.2.1 Prevalence in the CF Population

Spier et al. were one of the first to measure the marked difference in sleep quality between CF patients and the healthy population using polysomnography [5]. Individuals with CF regularly experience sleep fragmentation and falls in oxyhemoglobin saturation (SpO_2), with nocturnal cough also proving to be a common and persistent source of disruption [1,5]. CF patients are observed experiencing more frequent minor sleep disturbances (i.e., tossing, restlessness) rather than major sleeping disturbances (i.e., wakefulness) when compared

Diet and Exercise in Cystic Fibrosis
http://dx.doi.org/10.1016/B978-0-12-800051-9.00007-9

with healthy patients. Even in instances in which CF patients and healthy participants experienced the same number of sleep disturbances, CF patients consistently report lower subjective sleep quality [6].

Prevalence of sleep disturbance is not distributed evenly among the CF population. Females report more frequent disturbances than males, although this possibly could be explained by gender differences in physical fitness [1,7]. Patient age also appears to be correlated negatively with sleep quality. As age increases, subjective sleep quality steadily decreases, and sleep disruptions become more frequent [1,8]. This phenomenon can be explained by the increase in disease severity with age. Cough, lung inflammation, and mucus accumulation all are linked to sleep disturbances and worsen with growing age [2,5]. As a result, older CF patients experience more cumulative strain from both sleep loss and lung disease and are at greater risk for serious pulmonary issues.

7.2.2 Physiology of Sleep Dysfunction

Unfortunately, there is not yet a definitive explanation for the marked decrease in sleep quality in the CF population. Theories include mucus-related cough and side effects from β-antagonist medications commonly used to treat CF. Such medications are known to cause tremors and excitability, and when taken before bedtime, would negatively affect sleep [9]. These two explanations, however, fall short of encompassing the widely reported phenomenon of disturbed sleep experienced by CF patients.

A low level of SpO_2—and a resulting state of hypoxemia—is a frequently investigated parameter in sleep studies regarding severe lung disease. Hypoxemia is observed most commonly in patients with moderate to severe lung diseases. CF patients often are observed spending both waking and sleeping hours in some degree of hypoxemic state [10]. Typically, studies show that patients who reported better subjective sleep quality had higher overnight SpO_2 averages, suggesting a correlation between blood oxygenation and sleep quality [9]. It is not uncommon for CF patients to spend upward of 50% of their time sleeping in a hypoxemic state [3]. Patients with normal SpO_2 levels during waking hours are not necessarily exempt from nocturnal hypoxemia: Daytime SpO_2 stability is a poor indicator of the occurrence of nocturnal hypoxemia, suggesting that low SpO_2 levels are a likely source of sleep disruption in CF patients of varying disease severity [2].

Episodes of nocturnal hypoxemia appear to be associated closely with rapid eye movement (REM) sleep and the natural hypoventilation that comes with it. It is suggested that a decrease in tidal volume during REM sleep coupled with the anatomy of the CF lung lead to drops in SpO_2 levels and subsequently lead to sleep disturbances and decreased sleep quality [1,2,10]. The exact mechanism responsible for this correlation, however, is still a subject of debate—that is, whether or not damaged or obstructed lungs are to blame for inefficient gas exchange, if CF patients have weakened respiratory muscles that are further compromised by the brain entering REM, or perhaps some combination of the two [2,10]. Further investigations are necessary before this mechanism is fully understood.

7.2.3 Impact

Both sleep loss and nocturnal hypoxemia are correlated with impaired daytime cognitive function in CF patients [11]. Even when CF patients and normal individuals rate themselves as equally tired, CF patients function at 60% of the control when performing tasks designed to test neurocognitive performance. Additionally, CF patients experience low reports of activation and happiness when compared with normal individuals [8]. This suggests that—in addition to adversely affecting physical health—cumulative sleep loss and persistent hypoxemia have significant negative impacts on both the cognitive function and psychological health of CF patients, and clinically addressing sleep-related issues significantly would improve treatment.

7.2.4 Melatonin

7.2.4.1 Overview

Melatonin (*N*-acetyl-5-methoxytryptomine) is a neurohormone produced in the pineal gland at nighttime, typically as a response to a decrease in ambient light. Melatonin is synthesized through the hydroxylation of tryptophan into 5-hydroxytryptophan, which then is decarboxylated to 5-hydroxytryptamine (serotonin). Serotonin then is converted into the melatonin precursor and metabolite *N*-acetylserotonin via the enzyme *N*-acetyl transferase. Finally, this product is methylated by hydroxyindole-*O*-methyltransferase to produce melatonin [4].

The circadian pacemaker drives the production of melatonin via a multisynaptic pathway involving structures in the brain, spinal cord, and sympathetic nervous system. Patients with injury to this pathway experience significantly decreased sleep quality, strongly suggesting melatonin to be the primary antagonist for sleep in humans [12].

Exogenous melatonin is harvested from the pineal glands of beef cattle or synthesized chemically [4]. Melatonin is sold and regulated as a supplement and is available for purchase without a prescription except in the United Kingdom [4].

7.2.4.2 Effects of Exogenous Melatonin

The acute effects of exogenous melatonin administration include rapid, transient, mild sleep induction;

lowered alertness and body temperature; and alertness lasting for 3–5h after administration [4]. Evidence suggests that exogenous melatonin is capable of crossing the blood–brain barrier and is tied closely to the sleep–wake and temperature cycles [13]. It is proposed that melatonin contributes to sleep by inhibiting brain-related wakefulness systems rather than activating sleep-related structures—as do most prescribed hypnotics—providing a smooth transition into sleep [14].

Administration of exogenous melatonin has been shown widely to promote sleep in healthy individuals. Melatonin is an effective hypnotic at many stages of the sleep–wake cycle: the administration of exogenous melatonin during full daylight hours when serum concentration of endogenous melatonin is at its lowest reliably produces sleep and sedation as determined by an involuntary muscle relaxation test [15–17]. Administration of exogenous melatonin in the evening also provides reliable sleep induction [18,19]. Doses of melatonin given within an hour of the natural onset of sleep as dictated by the circadian rhythm provide the most reliable response. A study showed that participants administered a low dose of melatonin at 8:00 p.m. experienced decreased sleep onset and latency to stage-two sleep relative to a placebo. Furthermore, all patients administered melatonin were able to accurately determine that they received the drug rather than a placebo, suggesting a marked increase in subjective sleep quality [15].

There is no discernable difference in sleep onset and quality between doses designed to mimic natural nighttime serum melatonin concentrations (0.3mg) and purely pharmacologic doses (1.0mg). Administration of any dose of exogenous melatonin results in reduced body temperature and awareness, along with sleepiness. This suggests that even the lowest dose of melatonin can trigger the biochemical mechanism responsible for sleep induction [15]. Although melatonin is shown to be capable of inducing quality sleep during waking and evening hours, supplementation does not further improve sleep in healthy sleepers with endogenous serum concentrations that are above 90% pretreatment [20]. This finding suggests that melatonin is beneficial only to patients with reduced nighttime endogenous melatonin levels that fail to trigger sleep onset [15].

7.2.5 Practicality as a Sleep Aid

The effectiveness of melatonin as a sleep aid is well established. When compared with more conventional hypnotics, such as benzodiazepines, melatonin appears to be superior for the treatment of delayed sleep onset and poor sleep quality, although it fails to treat primary and secondary sleeping disorders as well as most prescribed pharmacologic agents [13]. Benzodiazepines—while effectively inducing sleep—decrease the duration of REM sleep after a single dose, as well as total time spent in slow-wave sleep, diminishing overall sleep quality. Melatonin increases sleep quality by bringing patients to REM sleep sooner and maintaining the phase for a longer duration. Comparatively, melatonin would provide more benefits for individuals suffering from minor sleeping difficulties than most prescription hypnotics [15].

Benzodiazepines also are shown to reduce nighttime melatonin concentrations and increase daytime concentrations, leading to a notable "hangover" that hampers cognitive function. Additionally, benzodiazepines are known to cause rebound insomnia because of their disruption of the sleep–wake cycle. Melatonin supplementation has not been observed to cause a rebound effect, and patients administered melatonin experienced normal wakefulness after sleeping [15]. Single doses of melatonin are more effective in improving sleep in the elderly than single doses of prescription hypnotics [21]. Additionally, prolonged melatonin supplementation is shown to improve sleep quality days or weeks beyond cessation of treatment. Therefore, melatonin as a sleep aid is believed to be safe and practical for long-term use [22].

7.3 MELATONIN AND CF

7.3.1 Sleep Dysfunction

Little research has been done on the efficacy of melatonin for treating sleep dysfunction in CF. De Castro et al. conducted one study on melatonin and its roles in improving sleep and lung health in CF patients through a number of parameters, among them a sleep questionnaire. The design of the study was randomized, double blind, and placebo controlled. The study primarily was focused on nonsleep aspects and concluded that melatonin improves subjective sleep quality, sleep efficiency, and sleep latency, among other things, in CF patients [23].

To further speculate the usefulness of melatonin in treating sleep dysfunction in CF, the etiology of disruptions and other issues in the affected population must be investigated in regard to the sleep-inducing mechanisms of exogenous melatonin. Common issues regarding sleep in CF patients include the following:

- Poor subjective sleep quality [1,5,6,9]
- Sleep variability and frequent disruption [1–3,8]
- Persistent cough and lung inflammation [1,2,5]
- Nocturnal hypoxemia and REM hypoventilation [1–3,6,10]
- Possible side effects of existing treatment measures [9].

Melatonin is known to be useful for the following:

- Increasing subjective sleep quality [4,15]
- Increasing duration of REM sleep [15]

- Decreasing sleep latency [15]
- Promoting and regulating regular sleep–wake patterns [4,12,21,22].

Above all, melatonin is useful for improving subjective sleep quality among CF patients. No data have been collected in regards to sleep efficiency and latency of CF patients treated with melatonin. CF patients, however, have reported increases in subjective sleep quality and wakefulness after sleeping following one night of melatonin supplementation [23]. These increases alone could provide alleviation of sleep dysfunction-related issues among CF patients, including impaired cognitive function, reduced activation, and liable mood, leading to a stark increase in quality of life [8,11]. Improvement in these areas would be beneficial in a greater course of treatment and warrant further investigation.

CF patients report highly variable night-to-night sleeping patterns. Additionally, periods of sleep often are peppered with disruptions including restlessness, cough, and wakefulness [10]. Melatonin has been shown to be effective at restoring, promoting, and maintaining regular sleep–wake patterns among healthy individuals [4,12,21,22]. Its efficacy, however, at doing so in the CF population is not yet known. Melatonin supplementation would likely aid CF patients in maintaining a semiregular sleep–wake pattern by essentially resetting the circadian pacemaker, but it would do little to address the persistent underlying problems that ultimately contribute to dysfunction [4,12]. Even with the resetting of the natural sleep–wake cycle, patients likely still would experience diminished sleep quality due to persistent disruption.

In regards to nocturnal hypoxemia in CF patients, the notion of melatonin supplementation proves troublesome. Low overnight SpO_2 is a chronic, commonly occurring problem among CF patients, resulting in cognitive impairment and a general negative impact on quality of life [2,10]. Furthermore, hypoxemia appears to be correlated tightly to REM sleep onset and duration [3,6]. Melatonin is known to decrease the latency of REM sleep and increase the total duration of REM relative to other phases of the sleep cycle [15]. This poses a potential safety issue concerning melatonin as an option for treating sleep dysfunction in CF: although it may provide an overall benefit regarding sleep quality, melatonin potentially could increase the occurrence and severity of nocturnal hypoxemia among CF patients. Because the exact cause of nocturnal hypoxemia in this population is not yet fully understood, further research must be conducted before the extent of melatonin's potential negative effects can be appreciated fully [2,10].

Melatonin is understood to promote sleep by decreasing activity in brain structures related to wakefulness [14]. CF patients who are treated with β-antagonists as bronchodilators report sleep disruption, tremors, and restlessness [9]. It is speculated that these reports of sleep loss stem from β-antagonists' tendency to inhibit nocturnal melatonin secretion, leading to markedly increased wakefulness [24]. Therefore, supplementation of exogenous melatonin before sleep could help directly reverse sleep dysfunction in CF patients brought by β-blocker treatment by raising serum concentrations of melatonin that otherwise would be depleted from secretion inhibition.

7.3.2 Other Implications

Melatonin also has some indirect benefits in regard to the treatment of sleep dysfunction in CF. Melatonin and its metabolites are potent reactive oxygen and nitrogen scavengers, giving them protective antioxidant properties [1,23]. Patients administered melatonin before sleep experienced a marked decrease in the nitrite concentration of their exhaled breath concentrate (EBC) the following morning. EBC concentrations of isoprotane—another indicator of lung stress—did not decrease to a clinically significant degree after a single dose, although it is speculated that continued use of melatonin could result in significant decreases [23]. In such parameters, nitrite and isoprotane are surrogate measure for lung oxidative stress and inflammation, meaning that exogenous melatonin supplementation has the added benefit of improving lung health with just one dose. The reduction of oxidative stress and the decrease in inflammation results in a generally healthier lung, which potentially could lead to fewer instances of nocturnal cough and improve nocturnal SpO_2 [1,2,5].

Melatonin has been shown to have an in vitro antimicrobial effect, particularly on Gram-negative bacteria [25]. Theoretically, exogenous melatonin supplementation could benefit CF patients by altering the bacterial profile of the lung and reducing the risk or extent of bacterial infection, significantly improving overall health. By extension, this improvement in lung health could lead to an increase in sleep quality by reducing the likelihood of sleep disturbances. However, the lack of in vivo evidence supporting this claim limits its credibility, and further investigation is warranted.

7.4 CONCLUSION

Melatonin shows considerable potential for treating sleep dysfunction in patients with CF. Regardless of individual disease severity, exogenous melatonin supplementation is shown to increase subjective sleep quality and efficiency, as well as decrease sleep latency in the CF population, consequently decreasing the burden of sleep loss and increasing overall patient health.

Melatonin is useful for regulating a normal sleep–wake pattern, which otherwise can be complicated by CF-related disturbances among patients. These benefits are not unique to the CF population, and melatonin does not appear to improve sleep for CF patients any differently than it does for healthy individuals. Reported β-blocker–related sleep loss, however, can be traced back to the inhibition of endogenous melatonin secretion, suggesting that melatonin supplementation may give CF patients additional sleep-promoting benefits. The antimicrobial and antioxidant properties of melatonin show promise in improving the overall health of the CF lung, leading to fewer sleep disturbances and generally better sleep quality. Unfortunately, melatonin potentially could have a negative impact on the health of CF patients by worsening nocturnal hypoxemia through the prolongation of REM sleep, and its long-term safety for treating sleep dysfunction in this population remains unknown.

References

[1] Milross MA, Piper AJ, Norman M, Willson GN, Grunstein RR, Sullivan CE, et al. Subjective sleep quality in cystic fibrosis. Sleep Med 2002;3:205–12.

[2] de Castro-Silva C, de Bruin VM, Cavalcante AG, Bittencourt LR, de Bruin PF. Nocturnal hypoxia and sleep disturbances in cystic fibrosis. Pediatr Pulmonol 2009;44:1143–50.

[3] Yue HJ, Conrad D, Dimsdale JE. Sleep disruption in cystic fibrosis. Med Hypotheses 2008;71:886–8.

[4] Arendt J. Melatonin. Br Med J (Int Ed) 1996;312:1242.

[5] Spier S, Rivlin J, Hughes D, Levison H. The effects of oxygen on sleep, blood gases, and ventilation in cystic fibrosis. Am Rev Resp Dis 1984;129:712–8.

[6] Jankelowitz L, Reid KJ, Wolfe L, Cullina J, Zee PC, Jain M. Cystic fibrosis patients have poor sleep quality despite normal sleep latency and efficiency. Chest 2005;127:1593–9.

[7] Ryujin DT, McCleary JL, Mannebach SC, Samuelson WM, Marshall BC. Gender differences in physical activity and its relationship to physical fitness and pulmonary function in adult cystic fibrosis patients. Pediatr Pulmonol—Suppl 1999;19:444.

[8] Fauroux B, Pepin JL, Boelle PY, Cracowski C, Murris-Espin M, Nove-Josserand R, et al. Sleep quality and nocturnal hypoxaemia and hypercapnia in children and young adults with cystic fibrosis. Arch Dis Child 2012;97:960–6.

[9] Dancey DR, Tullis ED, Heslegrave R, Thornley K, Hanly PJ. Sleep quality and daytime function in adults with cystic fibrosis and severe lung disease. Eur Respir J 2002;19:504–10.

[10] Milross MA, Piper AJ, Norman M, Willson GN, Grunstein RR, Sullivan CE, et al. Night-to-night variability in sleep in cystic fibrosis. Sleep Med 2002;3:213–9.

[11] Amin RS, Burklow K, Hsieh T. Sleep disorders in cystic fibrosis. Pediatr Pulmonol—Suppl 1999;(19):445.

[12] Scheer FA, Czeisler CA. Melatonin, sleep, and circadian rhythms. Sleep Med Rev 2005;9:5–9.

[13] Buscemi N, Vandermeer B, Pandya R, Hooton N, Tjosvold L, Hartling L, et al. Melatonin for treatment of sleep disorders. Summary, evidence report/technology assessment no. 108. (Prepared by the University of Alberta evidence-based practice center, under contract no. 290-02-0023.) AHRQ Publication No. 05-E002-1. Rockville (MD): Agency for Healthcare Research and Quality; November 2004.

[14] Luboshizsky R, Lavie P. Sleep-inducing effects of exogenous melatonin administration. Sleep Med Rev 1998;2:191–202.

[15] Zhdanova IV, Wurtman RJ, et al. Sleep-inducing effects of low doses of melatonin ingested in the evening. Clin Pharmacol Ther 1995;57:552–8.

[16] Zhdanova IV. Melatonin as a hypnotic: pro. Sleep Med Rev 2005;9:51–65.

[17] van den Heuvel CJ, Ferguson SA, Macchi MM, Dawson D. Melatonin as a hypnotic: con. Sleep Med Rev 2005;9:71–80.

[18] Zhdanova IV, Wurtman RJ, Morabito C, Piotrovska VR, Lynch HJ. Effects of low oral doses of melatonin, given 2–4 hours before habitual bedtime, on sleep in normal young humans. Sleep 1996;19:423–31.

[19] Sack RL, Hughes RJ, Edgar DM, Lewy AJ. Sleep-promoting effects of melatonin: at what dose, in whom, under what conditions, and by what mechanisms? Sleep 1997;20:908–15.

[20] James SP, Mendelson WB, Sack DA, Rosenthal NE, Wehr TA. The effect of melatonin on normal sleep. Neuropsychopharmacology 1987;1:41–4.

[21] Haimov I, Lavie P, Laudon M, Herer P, Vigder C, Zisapel N. Melatonin replacement therapy of elderly insomniacs. Sleep 1995;18:598–603.

[22] Scheer FA, Van Montfrans GA, Van Someren EJ, Mairuhu G, Buijs RM. Daily nighttime melatonin reduces blood pressure in male patients with essential hypertension. Hypertension 2004;43:192–7.

[23] de Castro-Silva C, de Bruin VM, Cunha GM, Nunes DM, Medeiros CA, de Bruin PF. Melatonin improves sleep and reduces nitrite in the exhaled breath condensate in cystic fibrosis—a randomized, double-blind placebo-controlled study. J Pineal Res 2010;48:65–71.

[24] Betts TA, Alford C. Beta-blockers and sleep: a controlled trial. Eur J Clin Pharmacol 1985;28:65–8.

[25] Tekbas OF, Ogur R, Korkmaz A, et al. Melatonin as an antibiotic: new insights into the actions of this ubiquitous molecule. J Pineal Res 2008;44:222–6.

CHAPTER

Age at Diagnosis and Disease Progression of Cystic Fibrosis

Isabelle de Monestrol[1,2]

[1]Stockholm CF Center, Karolinska University Hospital, Stockholm, Sweden; [2]Department of Clinical Science, Intervention and Technology, Division of Pediatrics, Karolinska Institutet, Stockholm, Sweden

8.1 AGE AT DIAGNOSIS OF CF

Historically, CF was diagnosed in childhood by clinical symptoms or a family history of CF. During the past decades, some individuals with CF have been diagnosed in adulthood. The majority of patients, however, have classic CF diagnosed in childhood. Neonatal/newborn screening (NBS) programs around the world are influencing the age at diagnosis of CF.

In 1990, the median age at diagnosis in the United States was 7 months, and the mean age at diagnosis was 2.9 years [1]. The majority of cases (70%) were diagnosed during the patients' first year of life; 80% of the cases had been diagnosed by age 4 years and 90% by age 12. The median age at diagnosis was 6 months in 1992 and again in 2002. Since 2010, all 50 states and the District of Columbia have screened all newborns for CF, and in 2012 the median age at diagnosis had decreased to 4 months [2]. More than 61% of people with CF who were diagnosed in 2012 were found by NBS.

In Canada, the median age at diagnosis was 7 months in 1985 and again in 2011, mean age 2.2 and 3.9 years, respectively. In 2011, 50% of patients were diagnosed by age 6 months and 73% by age 2 years [3]. NBS programs exist in only three Canadian provinces. A study from British Columbia, a province without NBS, showed that the median age at diagnosis for patients diagnosed between 1993 and 2005 was 1.2 years for all pediatric patients and 2.2 years for pediatric patients without meconium ileus [4]. The median delay of diagnosis in this study was 0.5 and 0.8 months, respectively.

The median age at diagnosis in the European CF Society Patient Registry Annual Report 2008–2009, comprising almost 19,000 patients from 20 countries, was 6.0 months, and the mean age at diagnosis was 3.6 years [5]. Approximately 45% of patients 5 years or younger had undergone NBS.

In the United Kingdom, the median age at diagnosis was 3 months for patients with complete data in the UK CF Registry Annual Data Report 2012. For children <16 years of age, the median age at diagnosis was 1 month (range 0–15 years), and for adults >16 years of age it was 7 months (range 0–79 years) [6]. Almost 60% (157/274) of all diagnosed patients were identified through NBS.

The Australian CF Registry Report 2012 shows that over 80% of diagnoses in infants were completed by the time the infants were 3 months of age [7]. Approximately 65% of new cases were found by NBS. In a Registry study comparing the CF patients younger than 18 years in 2001 in the UK, USA, France, and Australasia, the Australasian population had the lowest median age at diagnosis (1.8 months) because 70% of them were diagnosed by NBS [8].

8.2 MORBIDITY AT DIAGNOSIS AND INTRODUCTION TO MORBIDITY

The main clinical characteristics in CF are progressive pulmonary obstructive disease and gastrointestinal disease, such as pancreatic insufficiency and liver disease. Pancreatic insufficiency is present in 85% to 90% of patients, resulting in malabsorption of fat and protein [2].

The most common clinical modes of presentation of CF in infancy are failure to thrive and steatorrhea, in areas without NBS approximately between 50% and 90% [9–11]. These symptoms may also be present in children who have been screened for CF. Failure to thrive was seen at diagnosis in 6.2% of children in Australia, where 65.8%

Diet and Exercise in Cystic Fibrosis
http://dx.doi.org/10.1016/B978-0-12-800051-9.00008-0

had been diagnosed after NBS [12]. Respiratory symptoms as a clinical presentation vary—approximately between 40% and 80% in nonscreened areas—and is argued to be higher when diagnosis is delayed [4,9–11]. In Australia, which has had nationwide NBS since 2003, respiratory symptoms are noted as a mode of presentation in 12% to 15% of patients [7,12]. Meconium ileus occurs in 6% to 23% of CF newborns [1,9,12,13]. It is rarely fatal with prompt treatment, and the long-term outcome may not be worse than for other CF patients. Some infants with CF present with prolonged jaundice and elevated liver enzyme test results, which might be secondary to mucus-plugged bile ducts [14].

We studied disease progression in a Stockholm CF Registry study of 119 patients born between 1974 and 2001 in an area without NBS, treated in one center with a constant and intense treatment policy over the years, especially regarding intravenous essential fatty acids and antibiotics [9]. Patients included had a first visit to the center within 6 months from the diagnosis and a minimum of two annual reviews. At diagnosis and at each annual review, we defined the patient as healthy or having symptoms of morbidity. Healthy was defined so that it would be a normal medical status regardless of the medication the patient had taken, for example, normal growth in a patient taking pancreatic enzymes. Nutritional morbidity was defined as weight or height below −0.5 SD from the birth weight or height, both according to Swedish growth charts (if age <18 years) or body mass index <18.5 (if age ≥18 years). Lung morbidity was defined as pulmonary radiologic results showing pathologic changes other than peribronchial thickening, or pathologic scintigraphy of the lungs (if age <7 years) or spirometry results with forced expiratory volume in 1 second (FEV_1) <85% predicted (if age ≥7 years). We chose to look at liver morbidity defined as pathologic ultrasonographic results, for example, heterogeneous echo texture of the liver parenchyma, or gamma glutamyltransferase or alanine transaminase above upper reference levels. Overall morbidity was defined as having symptoms of at least one of the specific morbidities.

In the study, 66% of patients had nutritional morbidity, 21% lung morbidity, and 45% liver morbidity at diagnosis (several morbidities were possible, giving a total of >100%). No symptoms of morbidity was seen in 14% of patients according to the definitions.

To create a scenario to allow patients to move from healthy to morbidity and vice versa, we analyzed disease progression over time by following up the patients from (A) healthy at diagnosis to endpoint morbidity, (B) diagnosis with symptoms of morbidity to endpoint free of morbidity, and (C) free of morbidity to endpoint relapse of morbidity, using time-to-event methodology (Kaplan-Meier curves). A Cox proportional hazards regression model was used to estimate the effect of sex (male vs. female), birth cohort (1974–1984 reference, 1985–1994, 1995–2002), age at diagnosis (0–2, 2–12 reference, 12–24, and >24 months), and length of time span between start of symptoms and diagnosis (0–2 reference, 2–4, 4–12, and >12 months) on outcomes. Possible effect modification between the above-mentioned factors and genotype (severe genotypes consisting of two class I or class II mutations vs. mild genotypes consisting of at least one other mutation being a mild or unknown mutation), colonization with *Pseudomonas aeruginosa* and county of residence (Stockholm vs. non-Stockholm patients) were also evaluated in the regression models.

8.3 NUTRITIONAL MORBIDITY

Pancreatic insufficiency is treated with pancreatic enzyme preparations with every meal. CF patients are recommended to use a high-fat, high-calorie diet and supplementation with fat-soluble vitamins. Nevertheless, CF patients have difficulty maintaining a good nutritional status throughout life. In our study, 40 of 119 patients were in nutritional health at diagnosis; only 10 of 40 patients experienced nutritional morbidity by 10 years of follow-up (Figure 8.1(A)) [9]. The Kaplan-Meier curves give us good insight into the disease progression. For the 79 patients with nutritional morbidity at diagnosis, half were free of nutritional morbidity after 2.0 years, and two thirds became free of nutritional morbidity within the 10 years of follow-up (Figure 8.1(B)). This was after they had received proper CF treatment according to international guidelines. All patients who were healthy/free of morbidity at diagnosis or who became free of morbidity regardless of follow-up time—57 patients in all—were followed up until relapse of morbidity. This relapse in nutritional morbidity came very slowly; only 18 of 57 patients experienced a relapse in nutritional morbidity within 10 years of follow-up (Figure 8.1(C)). Among the patients who recovered and became free of nutritional morbidity, there was a significantly increased risk of relapse to nutritional morbidity for the patients given the diagnosis with >12 months of delay after the start of the first symptoms (HR 5.22, 95% CI: 1.05–25.91, $p=0.12$), for the male patients (HR 2.87, 95% CI: 1.01–8.15, $p=0.04$) and for the birth cohort 1995–2004 (HR 10.08, 95% CI: 1.21–83.80, $p=0.01$). There was no other evidence of any association with the factors investigated regarding nutritional morbidity.

In the Registry study analyzing 12,994 United States and 1220 Australian CF patients aged ≤18 years, different outcomes were assessed by creating models controlling for differences in age, gender, genotype, and diagnosis after NBS [12]. The Australian patients were significantly more likely (65.8% vs. 7.2%; $p<0.001$) to have been diagnosed after NBS and were therefore diagnosed at a younger age than were children in the United States.

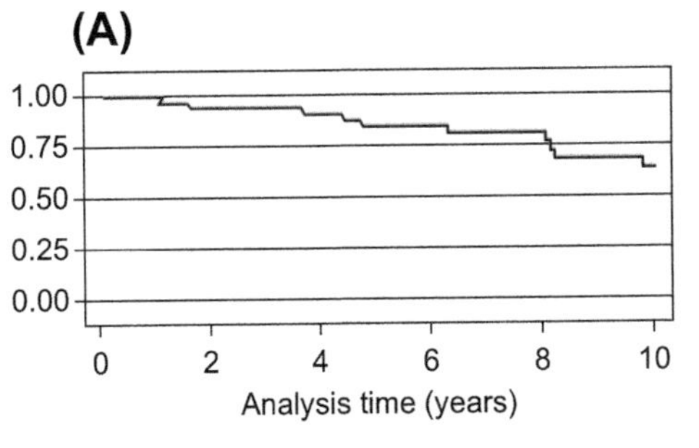

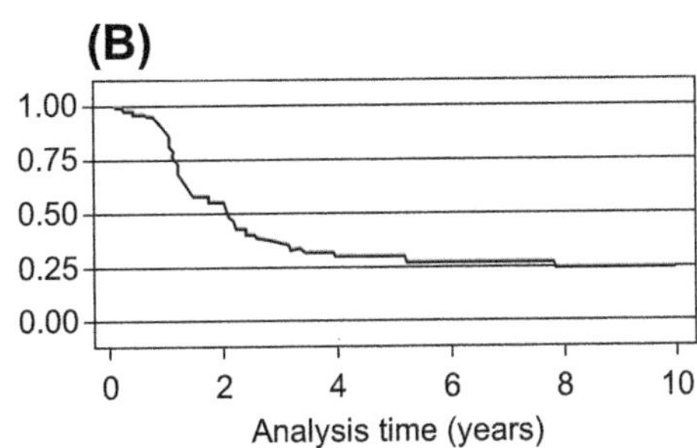

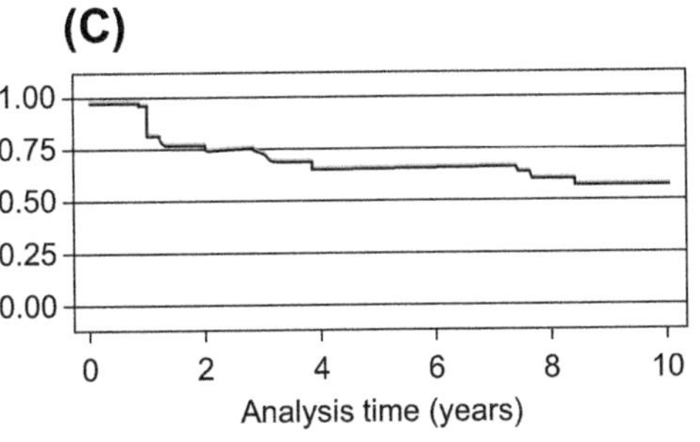

FIGURE 8.1 **Disease progression and nutritional morbidity.** (A) From healthy at diagnosis to morbidity. (B) From diagnosis with morbidity to free of morbidity. (C) From free of morbidity to relapse of morbidity. *Granted copyright permission, see acknowledgment.*

The Australian children had a significantly greater mean height percentile (41.0 vs. 32.6; $p < 0.001$) and weight percentile (43.5 vs. 36.1; $p = 0.028$) than their counterparts in the United States. The children diagnosed with CF after NBS demonstrated a mean body mass index 0.26 greater (95% CI: 0.09–0.43; $p = 0.003$) than did children diagnosed clinically, independently of country.

8.4 LUNG MORBIDITY

The major cause of morbidity and mortality in CF is lung disease, which is a progressive obstructive disease with chronic inflammation and infection, resulting in frequent lung exacerbations and tissue destruction. The end-stage of lung disease is respiratory failure, which is responsible for at least 80% of CF-related deaths [15].

In our study, patients who were healthy in the lungs at diagnosis showed a steady decline, with half of the patients experiencing lung morbidity after 3.4 years (Figure 8.2(A)) [9]. For the patients with lung morbidity at diagnosis, the symptoms were rapidly treated; half of these patients were free of lung morbidity after 1.3 years (Figure 8.2(B)). Relapse was rapid despite CF treatment; half of the patients experienced relapse in lung morbidity after 1.2 years (Figure 8.2(C)). For the patients with healthy lungs at diagnosis, there was a significantly increased risk of morbidity for the birth cohort 1995–2004 (HR 4.17, 95% CI: 1.67–10.39, $p = 0.01$). For the patients with lung morbidity at diagnosis, none of the investigated factors (sex, birth cohort, age at diagnosis, and length of time between start of symptoms and diagnosis) were associated with the outcome free of lung morbidity. There was no evidence of associations with the factors investigated when relapse in lung morbidity were looked at.

The Wisconsin Cystic Fibrosis NBS Project, established in 1984 as a randomized controlled trial, had difficulty proving that low age at diagnosis through NBS had an advantage regarding lung morbidity because of unexpected confounders when a subgroup of screened patients acquired *P. aeruginosa* prematurely during 1985–1990 at an old Wisconsin clinic [16]. In patients 7 years of age, the study compared outcomes using pulmonary function data and quantitative chest radiologic results. In the screened group (56 patients), diagnosis was made at a younger age, 12.4 weeks, than in the control group (47 patients) diagnosed at the age of 95.8 weeks, but the screened group included a significantly greater proportion of patients with delF508 genotypes and pancreatic insufficiency. The first chest radiograph showed significantly fewer abnormalities in the screened group, but over time, the two groups converged, and after 10 years of age the screened patients showed worse chest radiologic results associated with earlier acquisition of *P. aeruginosa*. No differences were detected in any measure of pulmonary dysfunction, which was generally mild in each group. However, in the Registry comparison between children with CF in the United States and in Australia, the children diagnosed after NBS benefited from better lung function [12]. The mean FEV_1 % predicted, adjusted for age, gender, and genotype, was

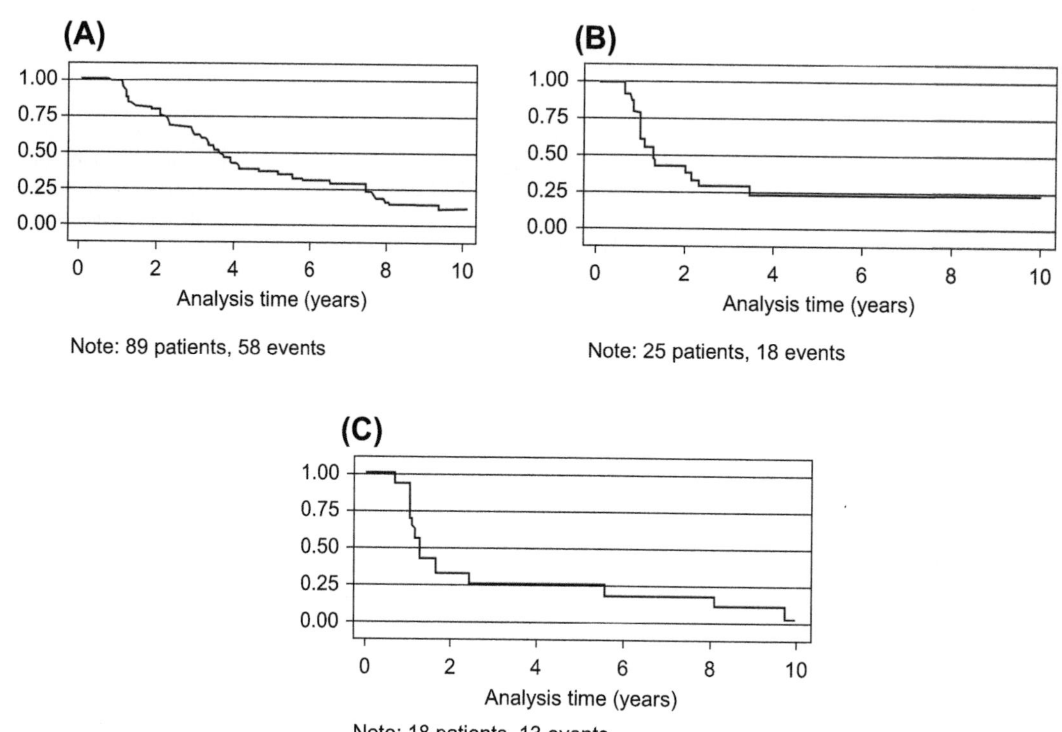

FIGURE 8.2 **Disease progression and lung morbidity.** (A) From healthy at diagnosis to morbidity. (B) From diagnosis with morbidity to free of morbidity. (C) From free of morbidity to relapse of morbidity. *Granted copyright permission, see acknowledgement.*

similar in the two countries ($p=0.80$); all patients diagnosed after NBS had higher mean FEV_1 % predicted (5.3, 95% CI: 3.6–7.0, $p<0.001$).

In a retrospective cohort study in two Dutch CF centers, effects of birth order and age at diagnosis during the first two decades of life in 52 sibling pairs with CF were analyzed both cross-sectionally and longitudinally by the use of Kaplan-Meier curves and modified log-rank tests [11]. The median age at diagnosis was significantly higher in the older siblings than in the younger siblings (3.0 and 0.2 years, respectively, $p<0.0001$). In the younger sibling group, FEV_1 at age 20 years was significantly better in those who had a diagnosis before the age of 6 months (difference 22.9%, 95% CI: 0.1–45.8%, $p<0.05$). This analysis was performed only in patients with a severe CFTR genotype. At the age of 20 years, the FEV_1 in older siblings was 19.4% lower (95% CI: 5.9–32.9%, $p=0.007$) than in younger siblings.

8.5 LIVER MORBIDITY

Liver disease develops in about one third of patients with CF. The characteristics are periportal fibrosis and focal biliary cirrhosis that present before 18 years of age [14]. This may lead to portal hypertension and end-stage liver disease, needing liver transplantation. End-stage liver disease is the single most important nonpulmonary cause of death in CF, accounting for about 2.5% of overall CF mortality. In a Swedish study, advanced fibrosis or cirrhosis was confirmed in about 10% of patients with CF, and 4% of patients had cirrhosis with clinical liver disease [17].

In our study, patients who were healthy in the liver at diagnosis showed a steady decline; half of them experienced liver morbidity after approximately 4.8 years (Figure 8.3(A)) [9]. For the patients with liver morbidity at diagnosis, the symptoms were rapidly treated; the curve showed that half of them were free of liver morbidity after 1.3 years (Figure 8.3(B)). In this age group, liver morbidity was due to increased liver enzymes, inasmuch as ultrasonography was not done at diagnosis. Half of the patients experienced relapse in liver morbidity after 3.4 years (Figure 8.3(C)). For the patients who were healthy in the liver at diagnosis, there was a significantly protective effect for the outcome liver morbidity when diagnosis was given within >4 months, and especially within >12 months, of delay after the first symptoms (4–12 months delay, HR 0.33, 95% CI: 0.11–0.96; >12 months delay, HR 0.22, 95% CI: 0.08–0.61, $p=0.04$). For the patients with liver morbidity at diagnosis, there was a protective effect for the outcome free of liver morbidity for the birth cohort 1995–2004 (HR 3.17, 95% CI: 1.04–9.65, $p=0.02$).

In a case–control study of all children aged 5–18 years with established CF-associated liver disease in Ireland, liver disease was associated with later age at diagnosis

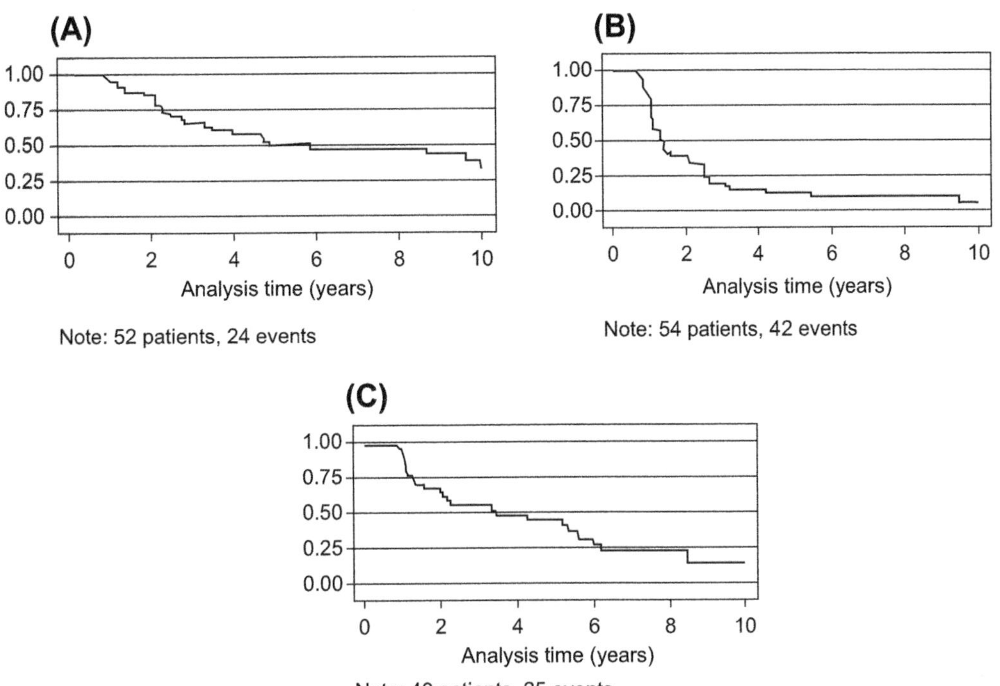

FIGURE 8.3 **Disease progression and liver morbidity.** (A) From healthy at diagnosis to morbidity. (B) From diagnosis with morbidity to free of morbidity. (C) From free of morbidity to relapse of morbidity. *Granted copyright permission, see acknowledgement.*

(>3 months) for all children including those who showed adequate growth and nutrition [18]. Among children given a diagnosis of CF before 3 months, the odds of CF-associated liver disease development were significantly higher only in children who were less than the 10th percentile in height (that is, in poor nutrition).

8.6 OVERALL MORBIDITY

Overall morbidity in our study was a combination of the specific morbidities. The patients who were healthy at diagnosis showed a rapid decline; half of them experienced overall morbidity after 1.4 year (Figure 8.4(A)) [9]. For the patients with overall morbidity at diagnosis, the symptoms declined slowly; the curve showed that half of them became free of overall morbidity after 4.8 years (Figure 8.4(B)). The relapse of morbidity came rapidly despite CF treatment; half of the patients experienced relapse in overall morbidity after 1.2 year (Figure 8.4(C)). For the patients who were healthy in overall morbidity at diagnosis, there was a protective effect when the diagnosis was given with a delay of 12 months or more after the start of the first symptoms compared with the reference group of 0–2 months delay (HR 0.03, 95% CI 0.00–0.50, $p = 0.01$). This might have been because the patients with a milder CF phenotype were in this group.

The most important finding was that for the patients with overall morbidity at diagnosis, there was a reduced chance that they would become free of morbidity if they were over the age of 24 months at diagnosis compared with the reference group of >2–12 months of age (HR 0.14, 95% CI 0.04–0.45, $p < 0.01$), and an increased chance to become free of morbidity for the cohort born from 1995 to 2004 (HR 2.78, 95% CI 1.10–7.01, $p = 0.06$) compared with the reference cohort born from 1974 to 1984. For the patients who had become free of overall morbidity, there was an increased risk of relapse for the cohort born from 1995 to 2004 (HR 4.29, 95% CI 1.29–14.29, $p < 0.01$). There was no evidence of interactions between the risk factors investigated and genotype, colonization with *P. aeruginosa*, or county of residence in any of the types of morbidity.

Sims et al. reported worse outcomes in the patients diagnosed after 2 months of age in a cross-sectional analysis of CF patients who were homozygous delF508, attended specialist CF centers, and were 1 to 10 years of age between 2000 and 2002, identified from the UK Database [19]. The study compared outcomes in, and treatment of, patients clinically diagnosed within the NBS reporting window of 2 months, those presenting after this period, and patients diagnosed by NBS. In two different analyses, NBS was associated with higher z score, higher Shwachman-Kulczyki score, lower likelihood of height <10th percentile, and fewer long-term therapies compared with patients receiving late clinical diagnoses (>2 months). Irrespective of mode of diagnosis, the Shwachman-Kulczyki morbidity score was found to decrease significantly with time (−0.97 per year; 95% CI: 1.41 to −0.53; $p > 0.001$), and increasing age was significantly associated

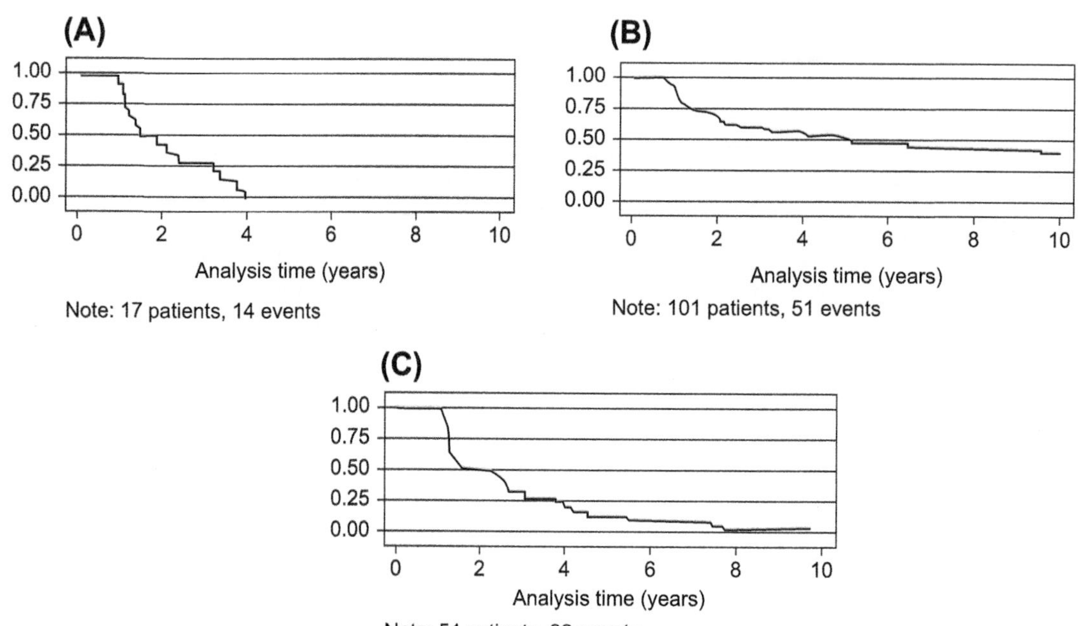

FIGURE 8.4 **Disease progression and overall morbidity.** (A) From healthy at diagnosis to morbidity. (B) From diagnosis with morbidity to free of morbidity. (C) From free of morbidity to relapse of morbidity. *Granted copyright permission, see acknowledgement.*

with a 6% increase per year in long-term therapy requirements (average RR: 1.06; 95% CI: 1.03–1.20). The authors concluded that in the absence of an NBS program, patients with CF diagnosed by symptom presentation after the age of 2 months manifest worse clinical outcomes despite receiving higher levels of long-term therapy for at least the first 10 years. A study from the Colorado CF Registry analyzed the effect of gender and age at diagnosis on disease progression in older CF patients [20]. They found that the population of patients with CF who are older than 40 years is actually composed of two different cohorts that can generally be distinguished by age at diagnosis. Nearly 90% of CF cases are diagnosed in patients before age 10 with the classic clinical features of pancreatic function insufficiency and progressive pulmonary disease. A second cohort is composed of patients diagnosed in adulthood with a variable phenotype and with a later onset of clinically significant symptoms. Patients with an adult diagnosis of CF typically have milder lung disease than do age-matched patients with a childhood diagnosis, and they are predominantly pancreatic sufficient or have recurrent pancreatitis.

8.7 DISCUSSION

Age at CF diagnosis is reported in the CF Registries and Registry studies. When NBS is used, the age at diagnosis is lower than when diagnosis is based on clinical symptoms. In the United States, the median age at diagnosis had decreased to 4 months in 2012 owing to NBS [2]. However, the goal is to diagnose screened infants within 2 months of age to lower the morbidity and to reduce expensive long-term therapy [19]. In the Australasian population, with 70% of CF patients diagnosed by NBS, this was fulfilled [8].

Disease progression over time can be well illustrated with epidemiologic analyzes such as Kaplan-Meier curves. We reported a Registry study in an area without NBS, with the strength of precise longitudinal data from one CF center with a constant treatment policy over the years [9]. The drawbacks were the limited patient numbers and the possibility that the two oldest birth cohorts may have contained less ill patients because the most severely ill patients might never have been included if they were never referred to the center before dying, or they might have been excluded if they were not having a first visit at the center within 6 months of age. In the birth cohort 1995–2004, not all patients may have yet received the diagnosis.

The most common modes of presentation of CF in infancy are failure to thrive, steatorrhea, meconium ileus, and respiratory symptoms. Children diagnosed by NBS have generally fewer symptoms. CF patients diagnosed in adulthood have a variable phenotype, with a later onset of clinical symptoms [20].

Nutritional morbidity is a major CF morbidity. Modern pancreatic enzyme treatments are beneficial. Our study results suggest that nutritional morbidity changes rather slowly, at least more slowly than lung and liver morbidities (Figures 8.1–8.3) [9]. Relapse in nutritional morbidity was much slower than in the other morbidities. However, there was an increased risk of relapse in nutritional morbidity for patients with a diagnostic delay

>12 months. Early diagnosis by NBS has been proved to give nutritional benefits, as seen in the US-Australian Registry study [12] and as further reviewed in the chapter on NBS and on nutrition and growth in CF.

Lung morbidity in CF is complex and multifactorial. There is an advantage for patients without too much lung morbidity at diagnosis; in our study half of the patients who were healthy in the lungs stayed healthy for 3.4 years longer than did the patients who already had symptoms of lung morbidity at diagnosis [9]. The Wisconsin NBS Project had difficulty proving that low age at diagnosis through NBS conferred an advantage regarding lung morbidity [16]. However, better lung function has been shown in children diagnosed early by NBS in the US-Australian Registry study [12] and in several cohort studies [21–23]. In the Dutch sibling study, better lung function was seen in the younger siblings who received a diagnosis within 6 months of age, and in all younger siblings, owing to younger age at diagnosis compared with their older siblings [11]. Several studies have shown a relationship between nutrition and pulmonary function [24] and with pulmonary inflammation [25], showing the complexity of CF disease with multifactorial causes.

Liver morbidity in CF is also complex. An Irish study has found an association with liver disease and age at diagnosis >3 months for children in adequate nutrition [18]. The most important finding in overall morbidity was that patients with symptoms and age above 24 months at diagnosis had a reduced chance of ever becoming free of morbidity, compared with patients diagnosed at a younger age. This speaks in favor of diagnosing CF patients with symptoms as early as possible.

In conclusion, age at CF diagnosis is becoming lower as more countries adopt NBS. A review of the literature shows that early diagnosis gives an advantage regarding CF disease progression. Future prospective studies of children diagnosed by NBS will give more information about early age at diagnosis and CF disease progression.

Acknowledgments

Granted copyright permission, de Monestrol Isabelle et al., Age at diagnosis and disease progression of cystic fibrosis in an area without newborn screening, Pediatric and Perinatal Epidemiology, 25, 298–305, 2011, Blackwell Publishing Ltd, Wiley publication.

References

[1] FitzSimmons SC. The changing epidemiology of cystic fibrosis. J Pediatr January 1993;122(1):1–9.

[2] Cystic Fibrosis Foundation patient registry; 2012. Annual Data Report.

[3] Canadian Cystic Fibrosis Registry; 2011. Annual Report.

[4] Steinraths M, Vallance HD, Davidson AG. Delays in diagnosing cystic fibrosis: can we find ways to diagnose it earlier? Can Fam Physician June 2008;54(6):877–83.

[5] Viviani L, Zolin A, Olesen HV. ECFSPR annual report; 2008–2009.

[6] UK CF registry annual data Report; 2012.

[7] 15th Annual Report from the Australian Cystic Fibrosis Data Registry.

[8] McCormick J, Sims EJ, Green MW, Mehta G, Culross F, Mehta A. Comparative analysis of Cystic Fibrosis Registry data from the UK with USA, France and Australasia. J Cyst Fibros 2005;4(2):115–22.

[9] de Monestrol I, Klint A, Sparén P, Hjelte L. Age at diagnosis and disease progression of cystic fibrosis in an area without newborn screening. Paediatr Perinat Epidemiol May 2011;25(3):298–305.

[10] Farahmand F, Khalili M, Shahbaznejad L, Hirbod-Mobarakeh A, Sani MN, Khodadad A. Clinical presentation of cystic fibrosis at the time of diagnosis: a multicenter study in a region without newborn screening. Turk J Gastroenterol December 2013;24(5):541–5.

[11] Slieker MG, van den Berg JM, Kouwenberg J, van Berkhout FT, Heijerman HG, Van der Ent. Long-term effects of birth order and age at diagnosis: a sibling cohort study. Paediatr Pulmonol 2010;45:601–7.

[12] Martin B, Schechter MS, Jaffe A, Cooper P, Bell SC, Ranganathan S. Comparison of the US and Australian cystic fibrosis registries: the impact of newborn screening. Pediatrics Feb 2012;129(2):e348–55.

[13] Efrati O, Nir J, Fraser D, Cohen-Cymberknoh M, Shoseyov D, Vilozni D, et al. Meconium ileus in patients with cystic fibrosis is not a risk factor for clinical deterioration and survival: the Israeli Multicenter Study. J Pediatr Gastroenterol Nutr February 2010;50(2):173–8.

[14] Colombo C. Liver disease in cystic fibrosis. Curr Opin Pulm Med 2007;13(6):529–36.

[15] O'Sullivan BP, Freedman SD. Cystic fibrosis. Lancet 2009;373(9678):1891–904.

[16] Farrell PM, Li Z, Kosorok MR, Laxova A, Green CG, Collins J, et al. Bronchopulmonary disease in children with cystic fibrosis after early or delayed diagnosis. Am J Respir Crit Care Med November 1, 2003;168(9):1100–8.

[17] Lindblad A, Glaumann H, Strandvik B. Natural history of liver disease in CF. Hepatology 1999;30(5):1151–8.

[18] Corbett K, Kelleher S, Rowland M, Daly L, Drumm B, Canny G, et al. Cystic fibrosis-associated liver disease: a population-based study. J Pediatr 2004;145:327–32.

[19] Sims EJ, Clark A, McCormick J, Mehta G, Connett G, Mehta A, United Kingdom Cystic Fibrosis Database Steering Committee. Cystic fibrosis diagnosed after 2 months of age leads to worse outcomes and requires more therapy. Pediatrics January 2007;119(1):19–28.

[20] Nick JA, Chacon CS, Brayshaw SJ, Jones MC, Barboa CM, St Clair CG, et al. Effects of gender and age at diagnosis on disease progression in long-term survivors of cystic fibrosis. Am J Respir Crit Care Med September 1, 2010;182(5):614–26.

[21] Merelle ME, Schouten JP, Gerritsen J, Dankert-Roelse JE. Influence of NBS and centralized treatment on long-term clinical outcome and survival of CF patients. Eur Respir J 2001;18:306–15.

[22] McKay KO, Waters DL, Gaskin KJ. The influence of newborn screening for cystic fibrosis on pulmonary outcomes in New South Wales. J Pediatr 2005;147:S47–50.

[23] Collins MS, Abbott MA, Wakefield DB, Lapin CD, Drapeau G, Hopfer SM, et al. Improved pulmonary and growth outcomes in cystic fibrosis by newborn screening. Pediatr Pulmonol 2008;43:648–55.

[24] Schoni MH, Casaulta-Aebischer C. Nutrition and lung function in cystic fibrosis patients: review. Clin Nutr 2000;19:79–85.

[25] Pillarisetti N, Williamson E, Linnane B, Skoric B, Robertson CF, Robinson P, Australian Respiratory Early Surveillance Team for Cystic Fibrosis (AREST CF), et al. Infection, inflammation, and lung function decline in infants with cystic fibrosis. Am J Respir Crit Care Med 2011;184:75–81.

CHAPTER

9

The Effects of Caffeine, Alcohol, and Tobacco in Cystic Fibrosis

Gilbert L. Rivera Jr., Ronald Ross Watson

University of Arizona, College of Public Health, Tucson, AZ, USA

KEY POINTS

- Caffeine comes in a variety of sources with varying amounts of the substance.
- Diuretics increase urination, which greatly increases the risk of dehydration in patients with cystic fibrosis.

9.1 INTRODUCTION

Caffeine comes from many sources, but tobacco and alcohol are from a single source (the tobacco plant and ethanol, respectively) and then made available in various forms. Logically, most chronically ill patients should avoid substances that have the potential to worsen their disease or cause significant adverse effects when combined with medications they are currently being treated with. Caffeine is a bitter substance found typically in coffee, tea, soft drinks, chocolate, kola nuts, and some medications. It has several effects on the body's metabolism but is known to stimulate the central nervous system, giving individuals a boost of energy and alertness [1]. Two to four cups of regular coffee a day may not be harmful for some, but for others it may foster the effects of caffeine: jittering of the limbs, difficulty sleeping, headaches, dizziness, tachycardia, and dehydration [1]. With an underlying disease such as cystic fibrosis (CF), it can be very hard on a patient's body, which is already being strained by chronic illness. Tobacco and alcohol are no different in that they cause destruction to the healthy human body let alone one plagued with CF. Smoking tobacco harms nearly every organ in the body, and 87% of lung cancer deaths are attributed to this substance. Other damaging effects of tobacco smoke include lung disease, heart disease, stroke, and even cataracts [2]. Alcohol at high concentrations can take a toll on a healthy individual, as can chronic use. Small amounts of alcohol are generally not beneficial to health, but these amounts typically have no adverse effects on a healthy individual [3]. Individuals who consume alcohol from age 18 and into their early adult years are at an increased risk of liver disease, heart disease, pancreatitis, and cancer. These risks are multiplied for individuals who are chronic consumers of alcohol, those who take prescription medications that interact with alcohol, and those with certain medical conditions [3]. By examining the relationship between the consumption of these staple drugs of American culture by healthy individuals and patients with CF, information can be obtained for future use in the treatment for patients with CF that may help them better understand how to live with CF and what lifestyle choices can determine their survival.

9.2 CYSTIC FIBROSIS OVERVIEW

CF is an inherited chronic disease that affects an estimated 30,000 children and adults in the United States and another 70,000 worldwide [4]. Approximately 1 in every 3,500 babies in the United States are born with CF, and CF primarily affects Caucasians but has the ability to affect all racial and ethnic groups [4]. The disease causes severe damage to the lungs and digestive system that can be life threatening. It is caused by a defective gene that affects the cells that produce mucus, sweat, and digestive juices, causing an abnormal sticky, thick mucus that clogs the lungs, obstructs the normal function of the pancreas, and stops natural enzymes from helping the body break down and absorb nutrients. In a healthy individual, secreted fluids are normally thin, slippery, and act as lubricants; however, in patients with CF these fluids plug up tubes, ducts, and passageways, primarily in the lungs and pancreas. The median survival age for patients with CF is now

Diet and Exercise in Cystic Fibrosis
http://dx.doi.org/10.1016/B978-0-12-800051-9.00009-2

36.8 years due to advancements in medical research and technology [4].

On a cellular level, CF is an autosomal-recessive disorder affecting chloride transport in the pancreas, lung, and other tissues. In the majority of cases, a ΔF508 mutation occurs in the first nucleotide-binding fold (NBF-1) of the cystic fibrosis transmembrane regulator (CFTR) [5]. The CFTR is expressed in airway epithelia on the luminal side of the plasma membrane, where it serves as a phosphorylation-regulated Cl^- channel and a regulator of channels and transporters. Activation of CFTR leads to parallel inhibition of the epithelial Na^+ channel, which is lost when CFTR is absent or dysfunctional [6]. It is postulated that the loss of Cl^- secretion and Na^+ absorption reduces the thickness of the airway surface liquid overlying airway epithelia, resulting in impaired mucociliary clearance. Reduced CFTR-dependent bicarbonate secretion might affect the hydration of the secreted mucus, affecting its physical properties. Since CFTR is also expressed in submucosal glands in the airways, it plays an important role in defending the body from foreign substances that are inhaled [6]. Loss in the function of duct-lining serous cells that contain CFTRs prevents the secretion of mucus and antimicrobial factors by the submucosal glands. These pathological features contribute to the formation of thick, dehydrated mucus that provides a pristine environment for continued growth of bacteria, which ultimately triggers chronic inflammation and organ failure in patients with CF [6].

Patients suffering from CF have various symptoms, depending upon their age. For newborns, symptoms include delayed growth, failure to gain weight normally during childhood, no bowel movements during the first 24–48 hours of life, and salty tasting skin [7]. For individuals who are diagnosed at a later age, any of the following may be symptoms, based on the particular system inundated by the disease: nausea or loss of appetite, weight loss, belly pain from severe constipation, coughing or increased mucus in the sinuses or lungs, fatigue, recurrent episodes of pneumonia, and sinus pressure [7]. Individuals with CF may develop infertility, repeated inflammation of the pancreas (pancreatitis), respiratory symptoms, and clubbed fingers later in life [7]. Since there is currently no cure for CF, treatment for these patients includes (1) antibiotics, which prevent and treat lung and sinus infections, and inhaled medications to open airways; (2) DNAse enzyme therapy to thin mucus; (3) hypertonic saline solutions; (4) oxygen therapy; and (5) alterations in dietary intake, especially with regard to protein and calories, and other therapies relating to increased consumption of antioxidants [7]. Over the years, as more has been learned about CF and treatment options, researchers have been looking into the consumption of other foods and substances to determine how patients with CF are affected compared to individuals who do not have CF. As caffeine, alcohol, and tobacco are further researched in patients with CF, hypotheses can be created to determine whether the typical side effects of these substances affect patients with CF and people without CF in the same or worse ways.

9.3 CAFFEINE AND PATIENTS WITH CF

Caffeine is a white, crystalline xanthine alkaloid and is considered to be a psychoactive drug. Caffeine is believed to work by blocking the adenosine receptors in the brain and other organs, which in turn reduces cellular activity. This action then stimulates nerve cells to release epinephrine, which causes the body to exhibit increased heart rate and blood pressure as well as the other common effects of caffeine. It also increases the levels of dopamine and causes the liver to release glucose [1]. For individuals who become tolerant of the effects of caffeine, headaches may be a withdrawal symptom if they lack it in their body. This is one reason why caffeine is sometimes added to medications that treat headaches. On the other end of the spectrum, individuals who consume too much caffeine and develop intoxication can experience nervousness, increased urination, flushed face, and even hallucinations [1]. According to the US Food and Drug Administration (FDA), the consumption of caffeine in the United States between 2003 and 2008 was approximately 300 mg per person per day [8]. The FDA also reported that solid foods contribute a very small amount of caffeine, whereas the major source of caffeine is coffee, with roughly 64–145 mg in an 8-oz cup of it. Other sources of caffeine include cocoa (11–115 mg); guarana (5%, whereas coffee beans are between 2% and 4.5%); kola nuts (2–3.5%); tea (4%, with a typical cup containing 20–80 mg); taurine, which is an additive in energy drinks; and yerba mate [8].

Patients with CF are at an increased risk for loss of sodium and chloride, two critical components of sweat. These individuals are at a higher risk of losing these vital elements during exercise, fevers, and infections or when exposed to high temperatures [9]. Sodium in the body acts as a sponge and helps maintain bodily fluids wherever it is present, primarily the blood and tissues. As patients with CF sweat out water and salt (sodium and chloride), their bodies are unable to reabsorb the salt. Their bodies do not notice the deficit in sodium and water being lost, which ultimately results in dehydration [9]. This characteristic of patients with CF makes consuming large quantities of caffeine an unwise choice. The diuretic properties of caffeine, which causes frequent urination, can endanger a patient who is already at risk of dehydration based on the underlying pathogenesis of their inherited disease. Some of the characteristics of salt and fluid deficiency mirror those of CF, which help to explain further the complications of the disease and the risk involved with dehydration and caffeine consumption. Hyponatremic dehydration also contributes to sputum that is thicker and more difficult to expel. In addition, thicker secretions from the bowel lead to blockages in the intestines [9].

A study was conducted in 1997 involving children diagnosed with CF and the possibility of enhanced drug metabolism. The study used caffeine as the mechanism to determine the increased or decreased usage of the CYP1A2 metabolic pathway. The results showed that patients with minimal pathophysiological changes in liver and lung function have enhanced metabolism of the pathway involving the CYP1A2 enzyme. The changes in liver function or pulmonary diseases can affect drug metabolism; however, patients with CF often have enhanced clearance of drugs [10a]. Based on this study, it seems that individuals with cystic fibrosis may have an increased amount of medication their body takes in because of the activity of the CYP1A2 enzyme and not specifically related to caffeine. Another study hypothesized that coffee was a food replacement in individuals with chronic pancreatitis [10b]. It was noted that in the later stages of the disease, patients with CF may experience repeated inflammation of the pancreas, which can develop into chronic pancreatitis [7]. Olesen et al. [10b] stated that, due to the pain associated with chronic pancreatitis, individuals often avoid consuming food and replace it with alcohol, tobacco, and coffee. In the study, drug absorption was being researched in patients with chronic pancreatitis, and it showed that these patients may have increased drug absorption due to the pathophysiology of the disease [10b]. The significance of this is again not directly correlated to caffeine and its effect on patients with CF, but on how patients with CF who develop chronic pancreatitis can worsen their disease by consuming coffee to replace meals and suppress their appetite [7]. This can indirectly affect the nutritional efforts of individuals with CF who are lacking nutrient absorption and must consume nutrient-rich foods to nourish their bodies.

9.4 ALCOHOL AND PATIENTS WITH CF

Ethyl alcohol, or ethanol, is the world's most important psychoactive depressant drug and is the intoxicating component of beer, wine, and distilled liquors. It is obtained through the fermentation of various products including yeast, sugars, and starches. As ethanol products are distilled their concentration increases, producing a concoction containing more than 15% alcohol [3]. Depending on the amount of alcohol consumed, the weight of the person, and the strength of the alcohol, the effects will differ. Alcohol is initially consumed for its stimulant effect, the notion that one is "loosened up or relaxed" [11]. But as an individual consumes more alcohol, the depressant effects of this drug present themselves throughout the central nervous system and manifest as slurred language, impaired judgment, and other disturbed perceptions [11].

The absorption of alcohol begins in the stomach but primarily occurs in the small intestine. In the presence of food, the absorption is reduced; however, in the presence of carbonated liquids, the absorption increases [3]. As alcohol is absorbed into the body, it is distributed throughout the bodily fluids and metabolized by the liver at a rate of 0.25 ounces per hour, with 90% metabolized. One sex difference with respect to alcohol is the stomach enzyme that metabolizes alcohol. This enzyme is less active in women than in men, which results in women being more susceptible to the effects of alcohol consumption [3]. As it depresses the central nervous system, alcohol has many effects on the brain. Its mechanism of action is similar to that of barbiturates and benzodiazepines, which enhance the inhibitory effects of gamma-aminobutyric acid (GABA) and the GABA-A receptor. At high doses, it blocks the effects of the excitatory transmitter glutamate and affects dopamine, serotonin, and acetylcholine neurons. Also, alcohol influences peripheral circulation and fluid balance and has hormonal effects. As an individual becomes more and more intoxicated, peripheral blood vessels dilate, causing the user to lose body heat but still feel warm [3]. The diuretic effects of alcohol cause the user to have decreased blood pressure and excessive urination. Excessive urination is caused by the body's inability to release antidiuretic hormone, which prevents the body from retaining fluid. The long-term effects of alcohol are brain tissue loss and intellectual impairment, liver disease, heart disease, cancer, and impaired immunity [3].

According to the Centers for Disease Control and Prevention, two surveys indicated that more than half of the US population drank alcohol in the past 30 days, with approximately 5% of the total population being heavy drinkers and 17% of the population being binge drinkers [12a]. Another study indicated that from 2001 to 2005, there were approximately 80,000 deaths annually that were attributable to excessive alcohol consumption, making it the third leading lifestyle-related cause of death for people in the United States each year [12a]. In the United States overall, chronic alcohol abuse is a major cause of alcoholic hepatitis and chronic pancreatitis, with rates at about 50% and 5%, respectively, which further indicates that only some individuals are susceptible to the development of these diseases [12b]. Interestingly enough, the third leading cause of death in patients with CF is liver disease, which is not necessarily caused by alcohol but is one of the adverse effects of excessive consumption. The prevalence of liver disease among patients with CF is 2–37%, varying largely on liver testing [13]. Since the adverse effects of excessive alcohol consumption include direct toxicity to organs, especially the liver, patients with CF are already at an increased risk because of their condition. Of the percentage of patients with CF with liver disease, up to 3% will see a progression to liver decompensation as a result of abnormal CFTR [14]. Consuming excess amounts of alcohol can be considered a "risky behavior" among the general population; however, individuals who are already suffering from bodily distress are at a greater risk when participating in this behavior. Mc Ewan et al. [14] conducted a study looking into the behaviors of patients with CF and found that of the 599 participants in the study, 77 males (94%) and 98 females (98%) had tried alcohol and 151 participants (83%) continued to drink on a daily basis. In addition, 144 participants (79%) did not drink or drank within the recommended limit, which is significantly higher than that of the general population [14]. The results of this study suggested that individuals with CF who participate in these behaviors are imposing additional health risks with respect to the disease. Risk-taking behavior tended to start at a later age when compared to people without CF and was more prevalent in women [14]. In another study, researchers showed that individual alcohol intake was low in all groups tested but decreased as patients switched from dieting to enteral feeding [15a]. As a result of the complications associated with CF, enteral feeding allows for more nutrients to be absorbed by the body [15a]. Alcohol posses a similar threat to both patients with CF and the general population, but the primary difference here is the current underlying characteristics of the disease that patients with CF already have and that alcohol can greatly increase. The meal replacement hypothesis studied by Olesen et al. [10b] mentions alcohol as another food replacement option for patients with chronic pancreatitis.

Chronic pancreatitis causes irreversible anatomical changes and damage to the pancreas that includes infiltration of inflammatory cells, fibrosis, and calcification with destruction of the glandular structure [15b]. This damage then causes normal digestion and absorption of nutrients to cease altogether. This has become evident in alcoholics who have been consuming alcohol for 8–9 years [15b]. The malabsorption and malnutrition associated with patients with chronic pancreatitis indicates the increased risk for patients with CF who are currently experiencing chronic pancreatitis or who are headed in that direction. Using alcohol as either a meal replacement to prevent the pain associated with chronic pancreatitis or for its other uses poses a significant threat to the long-term survival of patients with CF [10b,15b]. Lifestyle changes are crucial in an attempt to extend the life of patients with CF who are already experiencing the later complications of the disease.

9.5 TOBACCO AND PATIENTS WITH CF

Two main species of tobacco are grown today: *Nicotiana tabacum*, indigenous to South America, and *Nicotiana rustica*, found in the West Indies and eastern North America [2]. Tobacco is found not only in cigarettes but also snuff, chewing tobacco, smokeless tobacco, hookah, and cigars. Similar to caffeine and alcohol, tobacco contains a substance that has significant effects on the body. Nicotine, a naturally occurring liquid alkaloid, is present in tobacco and is the addictive ingredient that quickly builds tolerance and dependence. It is highly toxic, with a lethal dose of 60 mg; cigars contain about twice that amount. Inhalation of nicotine through the use of tobacco products results in the absorption of 90% of what is inhaled. Most of the nicotine absorbed (80–90%) is deactivated by the liver and then excreted by the kidneys. Tolerance is created by an increased use of liver enzyme activity responsible for nicotine deactivation [2]. Nicotine, like caffeine and alcohol, affects the central nervous and circulatory systems. It increases the heart rate and blood pressure; increases the oxygen needs of the heart, decreasing the oxygen-carrying ability of blood and resulting in shortness of breath; reduces hunger; increases blood glucose; and deadens taste buds [2]. Smokers report that nicotine has both a calming and stimulating effect, which may contribute to why users continue to use tobacco and other products containing nicotine despite the possible side effects. The side effects of continued tobacco use are similar but differ slightly based on the forms used. Lung cancer, cardiovascular disease, and chronic obstructive pulmonary disease are the primary adverse effects of continued tobacco use, but others include bad breath, spitting, increased risk of dental disease, and oral cancer. Additional costs of smoking include financial costs, social isolation, physical isolation, increased risk of fires or fire-related injuries, and pollution from the toxins [2].

Tobacco use is another risky behavior many Americans partake in on a daily basis. According to statistics tabulated by the Centers for Disease Control and Prevention, an estimated 45.3 million people, or 19.3% of all individuals 18 years of age and older, smoke cigarettes (21.5% of men and 17.3% of women). Cigarette smoking is the leading cause of preventable death in the United States and accounts for approximately 443,000 deaths each year [16]. In individuals with CF, exposure to smoke irritates the linings of mucosal membranes, which increases coughing and sputum production and increases the risk of bacterial infections and worsening symptoms [14]. Tobacco smoke may also affect the function of CFTR, accelerating the decline in pulmonary function [14]. The study of risky behavior conducted by Mc Ewan et al. [14] reported that 45% and 47% of males and females in her study, respectively, tried cigarette smoking, but only 6% continued smoking throughout the study.

These numbers were significantly lower than the general population, which was roughly at 21%, based on data at the time of the above-mentioned study. Feldman and Anderson [17] analyzed affects caused by cigarette smoke on various systems of the body. One system they looked at was the mucociliary escalator function. In patients with CF, mucociliary function is impaired because of the loss of Cl^- secretion and Na^+ absorption that reduces the thickness of the airway surface liquid overlying airway epithelia [17]. The researchers explained that the primary function of the mucociliary escalator system, which lines the luminal surface of the airways, is to entrap and expel pathogens from the lower airways. The mechanism involves the interaction between proadhesive mucus secreted from goblet cells and submucosal glands operating simultaneously with ciliated respiratory epithelium [17]. As pathogens are inhaled and reach this mucosal layer, they adhere to the viscous luminal gel phase of mucus and are propelled upward toward the larynx, resulting in sputum [17]. When an individual inhales cigarette smoke, the damage it causes to this system is presumed to be caused by the cytotoxic, irritant, and intracellular redox signaling actions that decrease ciliary beating/denudation of ciliated epithelium, mucus hypersecretion/submucosal gland hypertrophy, and squamous cell metaplasia [17]. Furthermore, cigarette smoke has been reported to increase epithelial permeability by compromising the structural integrity of tight junctions of the respiratory epithelium [17].

Roughly 21% of patients with CF have been actively or passively exposed to cigarette smoke. The relationship between smoking and the severity of CF may be apparent in impaired growth, higher rates of lung infections, and a higher frequency of intravenous antibiotic usage [17]. The use of this substance, as well as alcohol and tobacco, can be seen as a risk behavior because of the increases in symptoms among patients with CF [14].

9.6 SUMMARY

The consumption of caffeine may not be as severe as the other two substances, but continued use, or use associated with the general population, can pose threats to patients with CF and should be considered when creating various forms of therapy. The commonality shared by these substances is their exacerbation of preexisting conditions experienced among patients with CF. Caffeine and alcohol's diuretic effect poses a great risk to dehydrated individuals with CF who are typically unaware of their fluid balance. Those who have used tobacco in the past or still do have increased risks of disease from the tobacco itself and the exacerbation of preexisting respiratory symptoms. Similar to what Mc Ewan et al. [14] stated at the conclusion of their study, risk behaviors should be carefully evaluated on an individual basis to determine the possible adverse outcomes of a patient.

The life expectancy of patients with CF has been increasing despite the underlying cellular defects caused by the disease. Patients born today are expected to have a median survival age well into their 60s [18]. The improvements are due to the introduction of pancreas enzymes, better nutrition, and specialized care, and some hypothesize that there are other factors that influence the survival of patients with CF [18]. Other than variances among patients and their genetic makeup related to CFTR genetics, some nongenetic determinants of survival include environmental influences [14,18] such as biological, social, cultural, and health care-related factors and include microorganisms, nutrition, pollutants, and others. The long-term survival of patients with CF can be strongly correlated to maintaining good health during childhood if the patient so happens to be diagnosed early [18]. By instilling good health behaviors in children who are diagnosed at an early age, long-term survival may be achieved. Further research may eventually demonstrate a stronger correlation between the consumption of caffeine, alcohol, and tobacco in patients with CF and reasons why they should be avoided to allow for long-term survival. But based on the current data, it is clear that lifestyle behaviors affect CF symptoms in various ways, often putting the individual at an increased risk above the general population. Individuals using these substances as meal replacements are at the greatest risk of further complications of both CF and chronic pancreatitis. Olesen et al. [10b] also noted that the further along the complications of chronic pancreatitis are, the higher the potential bioavailability of drugs in the body. As patients are being treated with various medications to treat the symptoms of CF, there are a variety of factors that must be considered, especially as they relate to the three substances reviewed herein. Future studies will dictate the course of treatment for patients with CF and what substances they should avoid to prevent further deterioration.

References

[1] Hart C, Ksir C. Caffeine. Drugs Soc Hum Behav 2011;11:257–76.

[2] Hart C, Ksir C. Tobacco. Drugs Soc Hum Behav 2011;10:238–52.

[3] Hart C, Ksir C. Alcohol. Drugs Soc Hum Behav 2011;11:193–226.

[4] Culhane S, George C, Pearo B, Spoded E. Malnutrition in cystic fibrosis: a review. Nutr Clin Pract 2013;28:676. Originally published online October 29, 2013.

[5] Cohen BE, Lee G, Jacobson KA, Kim YC, Huang Z, Sorscher EJ, et al. 8-cyclopentyl-1,3-dipropylxanthine and other xanthines differentially bind to the wild-type and ΔF508 mutant first nucleotide binding fold (NBF-1) domains of the cystic fibrosis transmembrane conductance regulator. American Chemical Society Biochemistry 1997;36.

[6] Conese M, Ascenzioni F, Boyd AC, Coutelle C, De Fino I, de Smedt S, et al. Gene and cell therapy for cystic fibrosis: from bench to bedside. J Cys Fibros 2011;10(Suppl. 2):S114–28.
[7] A.D.A.M. Medical Encyclopedia. Cystic fibrosis. PubMed Health 2012. Web http://www.ncbi.nlm.nih.gov/pubmedhealth/PMH0001167/#adam_000107.disease.symptoms.
[8] Somogyi LP. Caffeine intake by the U.S. population. Food and Drug Administration; 2010.
[9] Queensland Government, Queensland Health. Salt replacement therapy & cystic fibrosis. Queensland: Dietitian/Nutritionist; 2011.
[10a] Parker AC, Pritchard P, Preston T, Smyth RL, Choonara I. Enhanced drug metabolism in young children with cystic fibrosis. Arch Dis Child 1997;77:239–41.
[10b] Olesen AE, Brokjaer A, Fisher IW, Larsen IM. Pharmacological challenges in chronic pancreatitis. World J Gastroenterol 2013.
[11] Foundation for a Drug-Free World. The truth about alcohol: what is alcohol? 2013. http://www.drugfreeworld.org/drugfacts/alcohol.html.
[12a] Centers for Disease Control and Prevention. Alcohol and public health. 2013. http://www.cdc.gov/alcohol/faqs.htm#what Alcohol.
[12b] Blanco P, Salem R, Ollero M, Zaman M, Clutette-Brown J, Freedman S, et al. Ethanol administration to cystic fibrosis knockout mice results in increased fatty acid ethyl ester production. Alcohol Clin Exp Res 2005;29(11).
[13] Nash KL, Allison ME, McKeon D, Lomas DJ, Haworth CS, Bilton D, et al. A single centre experience of liver disease in adults with cystic fibrosis 1995–2006. J Cys Fibros 2008;7:252–7.
[14] Mc Ewan FA, Hodson ME, Simmonds NJ. The prevalence of "risky behaviour" in adults with cystic fibrosis. J Cyst Fibros 2012;11:56–8.
[15a] White H, Morton AM, Peckham DG, Conway SP. Dietary intakes in adult patients with cystic fibrosis—do they achieve guidelines? J Cyst Fibros 2004;3:1–7.
[15b] Rasmussen HH, Irtun O, Olesen SS, Drewes AM, Holst M. Nutrition in chronic pancreatitis. World J Gastroenterol 2013.
[16] Centers for Disease Control and Prevention. Adult cigarette smoking in the United States: current estimate. Smoking & Tobacco Use 2013. http://www.cdc.gov/tobacco/data_statistics/fact_sheets/adult_data/cig_smoking/.
[17] Feldman C, Anderson R. Cigarette smoking and mechanisms of susceptibility to infections of the respiratory tract and other organ systems. J Infect 2013;67:169–84.
[18] Simmonds NJ, MacNeil SJ, Cullinan P, Hodson ME. Cystic fibrosis and survival to 40 years: a case-control study. Eur Respir J 2010;36:1277–83.

CHAPTER

10

Eating Disorders and Disturbed Eating Attitudes and Behaviors Typical in CF

Mandy Bryon

Cystic Fibrosis Service, Great Ormond Street Hospital for Children, London, UK

10.1 POOR NUTRITIONAL STATUS IN CF

Poor nutritional status in cystic fibrosis (CF) has been associated with poor growth and delayed puberty [1]; decreased exercise tolerance [2]; reduced lung function in children and adults [3–5]; and, ultimately, a decreased survival rate [6–8]. There are thought to be many factors contributing to the development of poor nutritional status in CF, including physiological and psychosocial variables. There are several organic disease factors that lead to an increased energy requirement in people with CF, including the malabsorption of nutrients; concurrent disease complications such as diabetes, gastroesophageal reflux, distal intestinal obstruction syndrome, and intestinal inflammation; and recurrent lung infections with the associated increased respiratory and immunological effort. Psychologically, it has been hypothesized that eating disorders (EDs) may be more prevalent in CF due to increased exposure to risk factors [9], including the following: eating difficulties in childhood [10]; low self-esteem [11]; and life-stressors, including the experience of bereavement [12]. Furthermore, increased eating psychopathology has been demonstrated in other conditions where there is a strong emphasis on the monitoring of dietary intake; for example, in diabetes mellitus [13–16]. Young-Hyman & Davis [17] review the risk factors for EDs in diabetes: feelings of loss of control because of required monitoring and reporting of food intake [18] and loss of autonomy because of parental, spousal, or familial concern or vigilance regarding health status [19]. It could be argued that these risk factors may also be applied to the experience and treatment of CF.

10.2 EVIDENCE OF EDs IN CF

Although EDs have been reported in CF, this has generally been at rates equal to or lower than those in the general population, where community prevalence rates have been estimated at 0.66% and 0.16% for anorexia nervosa (AN) in women and men, respectively, and 1.46% and 0.13% for bulimia nervosa (BN) in women and men, respectively [20].

Raymond et al. [21] and White, Miller, Smith, and McMahan [22] administered diagnostic interviews for EDs to samples of adolescents and young adults with CF of both sexes. They found that none of their samples of 58 and 53 participants, respectively, met diagnostic criteria for AN or BN; this prevalence was not significantly different from an age-matched control group in Raymond et al. [21].

Bryon, Shearer, and Davies [23] administered the Child Eating Disorder Examination (CEDE) [24], an adapted version of the gold standard interview for the diagnosis of EDs, the Eating Disorders Examination [25], to a sample of 55 adolescents with CF (males and females). They identified only one adolescent that met criteria for ED not otherwise specified; he met all criteria for a diagnosis of AN, although his body mass index (BMI) was above 17.5.

Pumariega et al. [26], identified 13% of adolescents with CF reached diagnostic criteria for ED in patients attending clinic over a 3 year period. In a general study of adjustment in people with CF, Pearson et al. [27] report 18% of their clinic population to score over significance levels for ED on a questionnaire designed to elicit ED symptoms, although they do not delineate any established diagnosis of ED in the group.

Diet and Exercise in Cystic Fibrosis
http://dx.doi.org/10.1016/B978-0-12-800051-9.00010-9

Single case studies of individuals with CF and EDs have been described: Goldbloom [12] reported on a 24-year-old woman with a bulimic subtype of AN, and Gilchrist and Lenney [28] described a 15-year-old girl with CF and AN. In both cases, there were serious negative health implications. There is as yet a lack of studies examining rates of EDs in individuals with CF surviving into mid- to late adulthood.

10.3 DISTURBED EATING ATTITUDES AND BEHAVIORS IN CF

Disturbed eating attitudes and behaviors (DEABs) refer to disturbances in attitudes related to shape, weight, body image, or food and also to related behaviors, including compensatory strategies (i.e., purging, excessive exercise), food restriction, and bingeing, although not at the level or combination required for a clinical diagnosis of an ED. DEABs are also referred to in the literature using the terms "subclinical eating disorders," "partial syndromes," and "disordered eating."

There has been more evidence of DEABs presenting in people with CF, although often at rates lower than those in the general population. Clearly, any attitude or behavior that may lead to weight loss or the maintenance of low weight in someone with CF could have a serious negative impact on health (greater than in the general population).

Reports of disturbed eating and attitudes in CF have tended to focus on children and adolescents. It could be hypothesized that this is perhaps because difficulties with feeding behaviors in young children with CF have been a clinical challenge. The tendency has been to classify problems with eating in the older age group as EDs.

Bryon et al. [23] reported that 53% of their sample of 11–17 year olds with CF reported some disturbed eating attitudes (positive scores on the CEDE), with 16% displaying disturbed eating behaviors. For example, 44% and 53% of the sample felt that their body shape or weight, respectively, significantly influenced their self-evaluation; 11% "felt fat" (despite none being overweight); and 11% had some fear about weight gain. Regarding behaviors, 16% were either attempting to lose weight or maintain their weight as it was, 5% were using compensatory behaviors such as exercising or misusing pancreatic enzyme medication to facilitate weight loss, and one female reported binge-eating over a 2 month period while trying to restrain her eating. Bryon et al. concluded that the observed rate of attitudinal disturbance was slightly higher than that in the general population (40–47%) [29,30]; however, a healthy control group was not included to enable a direct comparison.

Truby and Paxton [31] administered the Children's Eating Attitude Test (ChEAT) [32] and Children's Body Image Scale (developed for the study) to a sample of 7–12 year olds with CF and found that, although no participants displayed clinically significant scores on the ChEAT, some concerning DEABs were reported. Of the children whose BMI was below the 10th percentile, 75% of girls and 62% of boys did not think they were too thin, and 81% of girls and 68% of boys did not want to get fatter. Children with higher BMIs frequently selected a smaller BMI as their ideal body size, and 3% of females and 8% males in the total sample reported they had ever tried to lose weight. However, all observed body dissatisfaction and ChEAT scores were lower (less disturbed) than a healthy control group of children of the same age.

Raymond et al. [21] reported that the healthy control sample in their study scored significantly higher (more disturbed) on the Drive for Thinness, Perfectionism, and Body dissatisfaction subscales of the Eating Disorder Inventory [33] compared to CF participants, whose mean scores were all below cut-offs for suspected psychopathology.

Behavioral feeding problems have also been reported in children with CF. Many studies have found that parents of children with CF tend to indicate a higher number of and more intense problematic mealtime behaviors compared to parents of healthy control children [10,34–37]. Items endorsed as problematic by the CF parents included the following: "Does not enjoy eating," "Has a poor appetite," "Negotiates food to be eaten," "I get frustrated when feeding," and "I do not feel confident my child eats enough" [34]. Using objective observations, however, some studies have found no significant differences in total problematic mealtime behavior between children with CF and healthy control children [10,35,36], although Sanders et al. [10] found increased noncompliance in children with CF; and Stark et al. [36] found that children with CF engaged in more talk, spent more time away from the table, refused food more, and were more noncompliant toward commands to eat than healthy control children. Children with CF have frequently been found to take longer to eat meals compared to healthy controls [35], and some studies have suggested that this can lead to problematic mealtime behavior, because these behaviors increase with meal length [35,36].

10.4 MEASURES OF DEABs FOR THE CF POPULATION

In their study of adults with CF, Abbott et al. [38] revised the Eating Attitudes Test-26 to reflect its psychometric properties in a CF population; however, this measure does not allow for assessment of CF-specific EABs, as above. Quality-of-life (QOL) measures provide some assessment of EABs. The CF-QOL Questionnaire-2 [39] assesses body image (feeling too small, too thin, and "the way that I look makes life less enjoyable") and the CF Questionnaire-R [40] assesses "enjoyment of eating," "body image," and problems with the following: gaining weight, flatulence, diarrhea, abdominal pain, and "eating problems." However, these measures fail to cover the range of EAB phenomena, both related to EDs and to the CF disease process or treatment regimen that may be observed in this population. Wenninger, Weiss, Wahn, and Staab [41] developed a scale that assesses body image in CF (factor structure of "trust," "evaluation," and "importance") but again does not address other DEABs. A measure of behavioral feeding problems has been adapted for use in CF (The Behavioral Pediatric Feeding Assessment Scale-UK (BPFAS-UK)) [34]. However, this solely assesses child and parent behaviors around eating (i.e., "Takes longer than 20 min to finish a meal," "Eats junky snack food but will not eat at mealtime," "Gets up from the table during meals," "I coax my child to take a bite," "When my child has refused to eat I have to put food in his mouth by force if necessary") and does not assess other cognitions, attitudes, and behaviors around eating that would be relevant for adolescents and adults.

Thus, there appears to be a need for the development of a thorough measure of EABs specifically for, and validated with, the CF population. Such a measure, which would ideally be able to be completed conveniently within a routine clinical encounter, would help guide further assessment and treatment. It may help establish whether further assessment and treatment within a specialist ED team (i.e., if suggesting AN/BN) or in-house management within the CF multidisciplinary team (if suggesting CF-related DEABs) was indicated. It appears that the greatest clinical utility for such a CF-specific measure of EABs would lie in an adolescent-to-adult population, given that (1) individuals in that age group are given increased responsibility for managing their treatment and more independence around eating around adolescence; (2) adolescence represents the period of peak onset and highest incidence of EDs in the general population [42,43]; (3) poor nutritional status is found at all ages in CF and tends to decline from the teenage years onward into adulthood [22]; (4) some measures of pediatric behavioral feeding problems for a younger CF population are already in use (BPFAS-UK) [34]; and (5) a lower age limit of early adolescence could maximize the age range that a single measure could target (i.e., adolescents and adults), whereas it would not be feasible to have a single measure for everyone, including younger children.

Randelsome, Bryon and Evangeli [44] report the development of a CF-specific measure of eating attitudes and behaviors (CFEABs). The measure construction included an expert evaluation of the draft measure's construct validity, a literature review, piloting of the draft measure using a cognitive interviewing procedure, and administration to a large development sample of people with CF attending routine clinics. The final CFEAB measure was psychometrically sound with good internal consistency displayed for total and subscale scores. Items on the measure factor into three subscales suggested as follows: CF-related DEABs, EDs, and healthy appetite. Although the authors caution, as yet, against using the measure to suggest treatment options, it has utility as a clinical indicator of DEABs in CF.

10.5 CONCLUSION

EDs and DEABs are present in people with CF. They contribute to worsened health outcome and present a great challenge to medical management. Methods to help clarify the contributing factors to poor nutritional status are essential, especially those that help indicate when psychological intervention is a key requirement. Research studies are hampered by low participant numbers, and although evidence from published data of prevalence of EDs in CF remains indicative rather than conclusive, reports from clinical teams are that DEABs specific to CF are certainly apparent. With further research, it is hoped that appropriate targeted interventions can be identified to help minimize the negative health consequences associated with poor nutritional status in people with CF.

References

[1] Dodge JA. Malnutrition and age-specific nutritional management in cystic fibrosis. Netherlands J Med 1992;41:127–9.

[2] Marcotte JE, Canny GJ, Grisdale R, Desmond K, Corey M, Zinman R, et al. Effects of nutritional status on exercise performance in advanced cystic fibrosis. Chest 1986;90:375–9.

[3] Peterson ML, Jacobs Jr DR, Milla CE. Longitudinal changes in growth parameters are correlated with changes in pulmonary function in children with cystic fibrosis. Pediatrics 2003;112:588–92.

[4] Steinkamp G, Wiederman B. Relationship between nutritional status and lung function in cystic fibrosis: cross sectional and longitudinal analyses from the German CF quality assurance (CFQA) project. Thorax 2002;57:596–601.

[5] Zemel BS, Jawad AE, Fitzsimmons S, Stallings VA. Longitudinal relationship among growth, nutritional status and pulmonary function in children with cystic fibrosis: analysis of the Cystic Fibrosis Foundation National CF Patient Registry. J Pediatr 2000;137:374–80.

[6] Beker LT, Russek-Cohen E, Fink RJ. Stature as a prognostic factor in cystic fibrosis survival. J Am Diet Assoc 2001;101:438–42.

[7] Corey M, McLaughlan FJ, Williams M, Levison H. A comparison of survival, growth and pulmonary function in patients with cystic fibrosis in Boston and Toronto. J Clin Epidemiol 1998;41:583–91.

[8] Sharma R, Florea VG, Bolger AP, Diehner W, Florea ND, Coats AJ, et al. Wasting as an independent predictor of mortality in patients with cystic fibrosis. Thorax 2001;56:746–50.

[9] Shearer J, Bryon M. The nature and prevalence of eating disorders and eating disturbances in adolescents with cystic fibrosis. J Royal Soc Med 2004;97:36–44.

[10] Sanders M, Turner K, Wall C, Waugh L, Tully L. Mealtime behavior and parent-child interactions: a comparison of children with cystic fibrosis, children with feeding problems and non-clinical controls. J Pediatr Psychol 1997;22:881–900.

[11] Sawyer S, Rosier M, Phelan P, Bowes G. The self-image of adolescents with cystic fibrosis. J Adolesc Heath 1995;16:204–5.

[12] Goldbloom D. Anorexia nervosa and cystic fibrosis: a case report. Int J Eating Disord 1988;7:433–7.

[13] Fairburn CG, Peveler RC, Davis B, Mann JI, Mayou RA. Eating disorders in young adults with insulin dependent diabetes mellitus: a controlled study. Br Med J 1991;303:17–20.

[14] Peveler RC, Fairburn CG, Boller I, Dunger DD. Eating disorders in adolescents with IDDM: a controlled study. Diabet Care 1992;15:1356–60.

[15] Steel JM, Young RJ, Lloyd GG, Clarke BF. Clinically apparent eating disorders in young diabetic women: associations with painful neuropathy and other complications. Br Med J 1987;294:859–62.

[16] Steel JM, Young RJ, Lloyd GG, MacIntyre CCA. Abnormal eating attitudes in young insulin dependent diabetics. Br J Psychiatry 1989;155:515–21.

[17] Young-Hyman DL, Davis CL. Disordered eating behavior in individuals with diabetes: importance of context, evaluation and classification. Diabet Care 2010;33:683–9.

[18] Vamado PJ, Williamson DA, Bentz BG, Ryan DH, Rhodes SK, O'Neil PM, et al. Prevalence of binge eating disorder in obese adults seeking weight loss treatment. Eating Weight Disord 1997;2:117–24.

[19] Surgenor LJ, Horn J, Hudson SM. Links between psychological sense of control and disturbed eating behavior in women with diabetes mellitus: implications for predictors of metabolic control. J Psychosomat Res 2002;52:121–8.

[20] Woodside DB, Garfinkel PE, Lin E, Goering P, Kaplan AS, Goldbloom DS, et al. Comparisons of men with full or partial eating disorders, men without eating disorders, and women with eating disorders in the community. Am J Psychiatry 2001;158:570–4.

[21] Raymond NC, Chang P, Crow SJ, Mitchell JE, Diepernik BS, Beck MM, et al. Eating disorders in patients with cystic fibrosis. J Adolesc 2000;23:359–63.

[22] White T, Miller J, Smith GL, McMahan WM. Adherence and psychopathology in children and adolescence with cystic fibrosis. Eur Child Adolescent Psychiatry 2009;18:96–104.

[23] Bryon M, Shearer J, Davies H. Eating disorders and disturbance in children and adolescents with cystic fibrosis. Child Health Care 2008;37:67–77.

[24] Bryant-Waugh RJ, Cooper PJ, Taylor CL, Lask BD. The use of the eating disorder examination with children: a pilot study. Int J Eating Disord 1996;19:391–7.

[25] Fairburn CG, Cooper Z. The eating disorder examination. In: Fairburn CG, Wilson GT, editors. Binge eating: nature, assessment and treatment. 12th ed. New York: Guildford Press; 1993. p. 317–60.

[26] Pumariega A, Pursell J, Spock A, Jones J. Eating disorders in adolescents with cystic fibrosis. J Am Acad Child Adolescent Psychiatry 1986;25:269–75.

[27] Pearson D, Pumareiga A, Seilheimer D. The development of psychiatric symptomatology in patients with cystic fibrosis. J Am Acad Child Adolescent Psychiatry 1991;30:290–7.

[28] Gilchrist F, Lenney W. Distorted body image and anorexia complicating cystic fibrosis in an adolescent. J Cystic Fibros 2008;7:437–9.

[29] Childress A, Brewerton T, Hodges E, Jarrell M. The kids' eating disorder survey (KEDS): a study of middle school children. J Am Acad Child Adolescent Psychiatry 1993;32:843–50.

[30] Serdula M, Collins M, Williamson D, Auda R, Pamuk E, Byers T. Weight control practices of UD adolescents and adults. Ann Intern Med 1993;119:667–71.

[31] Truby H, Paxton SJ. Body image and dieting behavior in cystic fibrosis. Pediatrics 2001;107:E92.

[32] Maloney MJ, McGuire JB, Daniels SR. Reliability testing of a children's version of the eating attitude test. J Am Acad Child Adolescent Psychiatry 1988;27:541–3.

[33] Garner DM, Garfinkel PE. Eating disorders inventory-2: professional manual. Odessa, FL: Psychological Assessment Resources Inc; 1983.

[34] Duff AJA, Wolfe SP, Dickson C, Conway SP, Brownlee KG. Feeding behavior problems in children with cystic fibrosis in the UK: prevalence and comparison with healthy controls. J Paediatr Gastroenterol 2003;36:443–7.

[35] Powers SW, Patton SR, Byars KC, Mitchell MJ, Jelalian E, Mulvihill MM, et al. Mealtime behaviors in families of infant and toddlers with cystic fibrosis. J Cystic Fibros 2002;4:175–82.

[36] Stark LJ, Jelalian E, Mulvihill MM, Powers SW, Bowen AM, Spieth LE, et al. Eating in pre-school children with cystic fibrosis and healthy peers: a behavioral analysis. Pediatrics 1995;95:210–5.

[37] Ward C, Massie J, Glazner J, Sheehan J, Canterford L, Armstring D, et al. Problem behaviors and parenting in preschool children with cystic fibrosis. Arch Disease Childhood 2009;94:341–7.

[38] Abbott J, Conway S, Etherington C, Fitzjohn J, Gee L, Morton A, et al. Perceived body image and eating behavior in young adults with cystic fibrosis and their healthy peers. J Behavior Med 2000;23:501–17.

[39] Gee L, Abbott J, Conway SP, Etherington C, Webb AK. Development of a disease specific health related quality of life measure for adults and adolescents with cystic fibrosis. Thorax 2000;55:946–54.

[40] Quittner AL. CFQ – cystic fibrosis questionnaire – revised. User Manual, English Version, Available from the author. FL: University of Miami; 2003.

[41] Wenninger K, Weiss C, Wahn U, Staab D. Body image in cystic fibrosis – development of a brief diagnostic scale. J Behavior Med 2003;26:81–94.

[42] Garfinkel PE, Lin E, Goerig P, Spegg C, Goldbloom D, Kennedy S, et al. Bulimia nervosa in a Canadian community sample: prevalence and comparison of subgroups. Am J Psychiatry 1995;152:1052–8.

[43] Hudson JI, Hiripi E, Harrison G, Kessler R. The epidemiology of eating disorders: results from the national co-morbidity survey replication. In: Paper presented at the 11th annual meeting of the Eating Disorders Research Society, Toronto, Canada; 2005.

[44] Randlesome K, Bryon M, Evangeli M. Developing a measure of eating attitudes and behaviors in cystic fibrosis. J Cystic Fibros 2012;12:15–21.

SECTION B

VITAMIN D DEFICIENCY AND SUPPLEMENTATION IN GROWTH AND HEALTH IN CHILDREN WITH CYSTIC FIBROSIS

CHAPTER

11

Vitamin D Bioavailability in Cystic Fibrosis

G. Mailhot[1,2]

[1]Department of Nutrition, Université de Montréal, Montréal, QC, Canada; [2]Gastroenterology, Hepatology and Nutrition Unit, CHU Sainte-Justine Research Center, Montréal, QC, Canada

11.1 INTRODUCTION

Vitamin D is undoubtedly the fat-soluble vitamin that has gained the most attention in recent years. Its high profile is attributed to its broad involvement in various biological functions that go beyond its classical role as a calcium- and phosphate-regulating agent. With a prevalence of up to 95% in certain subgroups of cystic fibrosis (CF) patients, vitamin D insufficiency is by far the most widespread nutritional inadequacy [1]. Vitamin D insufficiency has been reported in both children and adults with cystic fibrosis as well as in presymptomatic children and in infants before diagnosis [1–4]. However, supplementation regimens aimed at correcting suboptimal vitamin D status have failed to achieve complete normalization, thereby suggesting that vitamin D bioavailability may be altered in the setting of this life-limiting pathology. Given the wide range of actions of this vitamin, an adequate supply of this vitamin to overcome vitamin D insufficiency, and more important, maintaining an optimal status over the long term, is of utmost importance.

This chapter will focus on all aspects of vitamin D bioavailability—from its endogenous synthesis and intestinal absorption to its tissue metabolism and utilization in relation to cystic fibrosis. The importance of vitamin D with regard to tissues, organs, and systems affected by cystic fibrosis will be particularly underscored.

11.2 VITAMIN D STATUS IN CYSTIC FIBROSIS

25-Hydroxyvitamin D is considered a measure of total exposure (sun, food, and supplements) and, to some extent, of the vitamin D stored and metabolized throughout the body. It represents the metabolite universally accepted to clinically assess vitamin D status. As opposed to 25-hydroxyvitamin D and despite being a biologically active form, 1,25-dihydroxyvitamin D exhibits a very short half-life (4–15 h) and is thus not considered a good marker of vitamin D status. Furthermore, despite 25-hydroxyvitamin D levels indicative of insufficiency, circulating 1,25-dihydroxyvitamin D is subject to strict homeostatic regulation and may remain normal or even elevated.

Genetic variants may influence not only 25-hydroxyvitamin D baseline levels but also the increment in response to a given dose of vitamin D [5–8]. A genome-wide association study with more than 30,000 Caucasians revealed that four genes contribute to the large interindividual variability in 25-hydroxyvitamin D levels [8]. These genes encode for proteins or enzymes involved in vitamin D metabolism: dehydrocholesterol reductase, responsible for the availability of the vitamin D_3 precursor 7-dehydrocholesterol in the skin; vitamin D-binding protein (DBP), the main serum carrier of vitamin D; and CYP2R1 and CYP24A1, two cytochromes involved respectively in the conversion of vitamin D into 25-hydroxyvitamin D and the degradation of 25-hydroxyvitamin D and 1,25-dihydroxyvitamin D. Whether such variants should be taken into consideration when prescribing vitamin D supplements is currently unknown.

The cut-off value for vitamin D deficiency (25-hydroxyvitamin D levels <25 nmol/L) appears consensual among the scientific community. Conversely, definitions of vitamin D insufficiency and adequacy still represent a matter of debate. A number of clinical guidelines and position statements as well as dietary recommendations have been issued to define the optimal vitamin D status and to explain how to reach and maintain vitamin D sufficiency [9,10]. However, the lack of consensus between two important actors, namely the Institute of Medicine (IOM) and the Endocrine Society, has led to considerable confusion among practitioners. It should be underscored that both reports are intended for different populations: the general population

Diet and Exercise in Cystic Fibrosis
http://dx.doi.org/10.1016/B978-0-12-800051-9.00011-0

within the IOM, and individuals at risk of deficiency within the Endocrine Society. The IOM recognizes that 25-hydroxyvitamin D levels greater than 50nmol/L are indicative of vitamin D sufficiency, and are associated with parathormone (PTH) normalization, reduced risk of osteomalacia, and optimal bone and muscle function [10]. In order to enhance intestinal absorption of calcium, prevent secondary hyperparathyroidism, and maximize the non-skeletal effects of vitamin D, the Endocrine Society claims that vitamin D sufficiency is achieved with 25-hydroxyvitamin D levels greater than 75nmol/L [9]. In their most recent recommendations, the Cystic Fibrosis Foundation adopted the same cut-off point as the Endocrine Society for vitamin D insufficiency [11]. Conversely, European guidelines for cystic fibrosis bone disease have set the cut-off point of vitamin D insufficiency at 50nmol/L, arguing that while PTH suppression may be warranted in certain cases, PTH levels that are too low may block bone formation, whereby they become potentially detrimental to bone health in conditions such as cystic fibrosis [12]. In adults with cystic fibrosis, West et al. attempted to determine the threshold level of 25-hydroxyvitamin D below which the risk of elevated PTH levels increased. They showed that PTH suppression below the level associated with bone loss was observed with 25-hydroxyvitamin D levels above 87.5nmol/L [13].

Solomons et al. were among the first to compare serum 25-hydroxyvitamin D in a group of children with cystic fibrosis to that of healthy controls [14]. They found significantly lower 25-hydroxyvitamin D levels in the cystic fibrosis population compared to controls, notwithstanding disease management practices including multivitamin supplementation and pancreatic enzyme replacement. Observational studies aimed at characterizing vitamin D status of cystic fibrosis populations are difficult to compare due to the varying threshold used to define vitamin D insufficiency. Studies using a threshold of less than 37.5nmol/L reported a prevalence between 20 and 50% [3,15–17] whereas the selection of a higher cut-off point (>75nmol/L) considerably boosted the prevalence of suboptimal vitamin D status, which was found to be as high as 95% [1]. Vitamin D insufficiency prevalence seems to increase with age, but the reasons underlying this observation are unclear. Possible contributing factors include poor compliance with vitamin supplementation and a decrease in vitamin D intake resulting from changes in dietary habits, such as decreased milk and increased soda consumption. In support of this, Grey et al. showed that none of the children aged less than 12years in their sample had serum 25-hydroxyvitamin D below 40nmol/L [18]. Similarly, Green et al. reported that patients aged more than 12years old were nearly four times more likely to exhibit vitamin D deficiency than patients aged less than 5years old [19]. In contrast, Feranchak et al. found that 22.5% of newborns with cystic fibrosis had subnormal 25-hydroxyvitamin D levels, a rate that fell to 2.6% by 10years of age [3].

Inconsistent results were obtained from intervention trials designed to normalize vitamin D status through various supplementation regimens. Earlier work mainly focused on the administration of vitamin D_2 with doses up to 50,000IU administered triweekly to adults and children with cystic fibrosis. Most of the studies, but not all [20], failed to demonstrate normalization of vitamin D status with these regimens [19,21]. More recently, Khazai et al. compared the effectiveness of 50,000IU of vitamin D_2 and D_3 given once a week for 12weeks in adults with cystic fibrosis [22]. They found vitamin D_3 to be more effective than vitamin D_2 in correcting vitamin D insufficiency. Interestingly, they also documented the effect of both supplementation regimens on total serum 25-hydroxyvitamin D concentration as well as on the 25-hydroxyvitamin D_3 and 25-hydroxyvitamin D_2 fractions. They reported lower 25-hydroxyvitamin D_3 after vitamin D_2 therapy, which had attenuated the overall increase in total 25-hydroxyvitamin D concentration. This decrease may be secondary to the increased catabolism of 25-hydroxyvitamin D_3 through upregulation of the catabolic enzyme 24-hydroxylase or through competition between vitamin D_2 and D_3 for the liver 25-hydroxylase. These findings provide a biological explanation whereby earlier studies with high doses of vitamin D_2 failed to demonstrate a significant correction of suboptimal 25-hydroxyvitamin D levels.

There are few follow-up studies that have examined the long-term effects of vitamin D supplementation in cystic fibrosis. Vitamin D_2 supplementation failed to show maintenance of the 25-hydroxyvitamin D level among cystic fibrosis children in the desired range [23]. Conversely, Shepherd et al. documented the efficacy of a vitamin D_3 stoss therapy on serum 25-hydroxyvitamin D over a 12-month period in vitamin D-insufficient children with cystic fibrosis [24]. The stoss therapy consisted of the administration of a single oral high dose of vitamin D_3 per a well-defined protocol, based on the 25-hydroxyvitamin D level and age, followed by oral maintenance of vitamin D supplementation (400IU in children less than one year and 800IU in those older than one year). The stoss-treated group experienced a significant and sustained increase in their serum level of 25-hydroxyvitamin D measured at 1, 3, 6, and 12months posttreatment compared to controls.

Chronic maintenance of high levels of 25-hydroxyvitamin D has thus not been extensively studied in cystic fibrosis despite the inherent risks of adverse effects. High circulating 25-hydroxyvitamin D_3 may cause hypercalciuria, a well-known risk factor for nephrolithiasis. The rate of kidney stones is approximately threefold higher in patients with cystic fibrosis [25]. Although hyperoxaluria has been proposed to account for the increased rate of lithiasis in these patients, close monitoring should be instituted when patients are given high doses of vitamin D [26,27].

11.3 PHOTOSYNTHESIS OF VITAMIN D_3

Thus far, the most beneficial and inexpensive way of getting vitamin D remains short, casual sun exposure. Upon exposure to the ultraviolet B (UVB) portion of sunlight, the precursor 7-dehydrocholestrol, present in the cell membrane of both skin keratinocytes and fibroblasts, undergoes photolysis to generate previtamin D_3. Previtamin D_3 is promptly converted to vitamin D_3 by a nonenzymatic process that occurs in the plasma membrane. Alterations in membrane phospholipid composition, previously documented in CF patients [28], may therefore affect the conversion of previtamin D_3 to vitamin D_3. From the plasma membrane, vitamin D_3 diffuses to the extracellular fluid space where it is attracted to the DBP and thus enters the dermal capillary bed. It has been previously reported that the absence of the cystic fibrosis transmembrane conductance regulator (CFTR) affects intracellular lipid metabolism [29,30]; however, its impact on the cellular metabolism of fat-soluble vitamins has not been investigated as of yet. The CFTR channel is localized in human epidermis [31] and although its role is still unclear, its dysfunction may affect skin biosynthesis of vitamin D_3. The only study that had reported the relationship between 25-hydroxyvitamin D levels and the amount of UVB exposure in a rather heterogeneous cohort of CF patients found that 25-hydroxyvitamin D concentration responded to the amount of sunshine in the preceding months and that this response did not differ from that of healthy control subjects [32]. These findings might thus suggest no discernible difference in the rate of photoproduction of vitamin D_3 among CF patients.

Despite the inherent risk of UVB radiation, controlled ultraviolet light therapy may represent an attractive alternative for patients who suffer from vitamin D malabsorption. A portable UV tanning lamp may constitute an alternate method for providing UVB radiation during the winter months in northern latitudes. Khazai et al. compared UVB therapy to D_2 or D_3 oral supplementation in a group of clinically stable adults with cystic fibrosis as a means to correct vitamin D insufficiency [22]. The UVB regimen was the least effective in raising 25-hydroxyvitamin D levels compared to oral supplementation. This finding is attributed to the poor compliance of the unsupervised UV therapy group. When non-adherent patients were excluded from the analyses (leaving a total of only five patients), the rise in 25-hydroxyvitamin D was comparable to those treated with D_2. Nonetheless, subjects exposed only part of their backs to the sun lamp for 3–15 min, five times a week depending on their skin types. One would have expected a better response with whole body tanning as well as with increased frequency and duration of exposure. Indeed, Gronowitz et al. found a better increase in 25-hydroxyvitamin D in a more substantial sample of CF patients after an 8 week and 24 week whole body exposure to UVB radiation [33]. These data further indicate no impairment in the cutaneous synthesis of vitamin D_3 among CF patients.

Diminished exposure to sunlight is thought to be among the causes of low vitamin D status in CF patients. The intake of antibiotics and antifungals, two classes of drugs found to increase photosensitivity and which use is prevalent among cystic fibrosis patients, represents another incentive for avoiding sun exposure. Nearly half of cystic fibrosis adults reported phototoxicity symptoms within 24 h of ciprofloxacin ingestion [34] whereas the incidence of photosensitivity approaches 15% with the use of voriconazole [35]. Moreover, these patients are instructed to use sunscreens in order to prevent the adverse effects of these drugs [34]. Sunscreens absorb solar UVB radiation before it penetrates skin layers and reduce the amount of UVB reaching the vitamin D_3 precursor, 7-dehydrocholesterol. Application of sunscreen with a sun protection factor (SPF) of 8 or more completely or partially blocks the cutaneous production of vitamin D_3 [36]. However, there are insufficient data regarding the level of sun exposure or outdoor activity in CF patients. The challenge in accurately estimating such variables resides in the lack of a standardized and validated questionnaire [37].

11.4 VITAMIN D DIGESTION AND ABSORPTION

Oral intake represents the other alternative to sunshine to provide sufficient vitamin D. Vitamin D intake is usually not problematic for cystic fibrosis patients as they display intakes far exceeding the recommendations for age-matched healthy individuals [1,38,39]. However, despite these high intakes, they remain chronically depleted of vitamin D, suggesting possible alterations in either the digestion or absorption of vitamin D.

Similar to lipids and other fat-soluble vitamins, vitamin D requires incorporation into bile acid micelles in order to reach the enterocyte membrane. Higher micellar concentration of sodium taurocholate has been shown to decrease the rate of vitamin D_3 absorption by increasing its solubilization within the micelle, thereby creating a shift from the monomeric to the micellar form of the vitamin [40]. Given that vitamin D preferentially enters the enterocyte membrane in its monomeric form, an increased concentration of sodium taurocholate would lead to a reduced absorption of the vitamin. As such, total bile acid secretion is reduced in cystic fibrosis [41]. However, despite this low secretion, impaired water secretion gives rise to higher concentrations of bile salts in the hepatobiliary secretions. Moreover, bile acid malabsorption and the concomitant increase in fecal excretion have been documented in cystic fibrosis [42,43]. Such large losses are accompanied by disturbances in biliary composition, reduction in bile acid pool, and an interruption of the rather efficient enterohepatic recycling of biliary acids. Altogether, these alterations may contribute to vitamin D malabsorption through the disruption in micelle formation or composition.

Reduction in intraluminal pH represents another factor that influences the rate of absorption of vitamin D_3. Hollander et al. demonstrated that increasing the hydrogen ion concentration in the perfusate of both the proximal and distal intestinal segments of rats raised the rate of vitamin D absorption [40]. Changes in hydrogen ion concentration would decrease the charge at the surface of the micelle and the absorptive cell membrane, thus leading to a reduction in the resistance to diffusion of the micelles toward the cell membrane. This observation is of relevance for cystic fibrosis populations as CFTR dysfunction has been related to abnormal output of bicarbonate, therefore leading to luminal acidification and precipitation of biliary salts and digestive enzymes. In addition, studies on intestinal biopsies from cystic fibrosis patients have observed the presence of a thick mucinous layer covering the brush border membrane and areas of the microvilli [44]. This abnormally high mucus production may act as a physical barrier for nutrients, thereby preventing them from reaching the surface of the enterocyte.

It has been generally assumed that vitamin D absorption occurs through a passive nonsaturable diffusion process [40]. However, evidence has emerged in the past years regarding the involvement of membrane transporters such as scavenger receptor class B Type I (SR-BI), cluster determinant 36 (CD36), and Niemann-Pick C1-Like 1 (NPC1L1) in vitamin D_3 absorption [45]. These proteins transport cholesterol, displaying structural homology with vitamin D_3. Therefore, it has been suggested that cholesterol and vitamin D_3 share common absorption pathways whereby they may compete with each other for their absorption [46]. Interestingly, it was reported that the expression of CD36 was modulated in CFTR-depleted intestinal cells [30]. Moreover, microarray analysis of intestinal tissue from *Cftr*-null mice from two distinct genetic backgrounds revealed a decrease in the expression of genes encoding membrane transporters involved in the movement of cholesterol and fatty acids [47,48]. Despite some evidence suggesting a possible correlation between CFTR and vitamin D absorptive pathways, no studies have specifically explored this relationship as of yet. However, the possible role of CFTR in the intestinal handling of lipids has been recently uncovered [29,30]. Fatty acid uptake, transport, metabolism, and secretion have been studied in CFTR-depleted intestinal cells. The phenotype of these cells was the exact opposite of that observed in vivo with demonstration of increased fatty acid uptake as well as enhanced formation and secretion of lipoproteins. These findings dismiss the involvement of CFTR in cystic fibrosis malabsorption but underscore its possible role in the regulation of intestinal nutrient transport and metabolism.

Given that vitamin D is a fat-soluble vitamin, it is logical to assume that the presence of luminal fatty acids would modulate vitamin D absorption. In vitro and in vivo studies in rats and humans have indeed confirmed this assumption. In rats, Hollander et al. revealed that the addition of physiological doses of oleic acid or linoleic acid to the perfusate of intestinal segments decreased the rate of absorption of vitamin D. These findings were explained by an alteration in the structure of the micelles resulting in an expansion of their size and a reduction in the diffusion rate of the vitamin to the enterocyte membrane [40]. In vitro studies, using the intestinal cell line Caco-2, have shown fatty acids to interact in different ways with vitamin D_3 absorption: first on micellar properties by influencing their electric charge and thereafter on vitamin D_3 cellular uptake and secretion [49]. In contrast to the study of Hollander, polyunsaturated fatty acids were less efficient in promoting vitamin D_3 absorption than monounsaturated acids in Caco-2 cells. However, this effect was seen only with single fatty acids and was abrogated in the presence of fatty acid mixtures. These findings were consistent with a clinical trial reporting that intake of monounsaturated fatty acids was positively correlated to the 25-hydroxyvitamin D response to a moderate dose of vitamin D (700 IU/day) whereas polyunsaturated fatty acid intake was inversely associated [50]. Although previous data showed that the presence of fatty acids might impact on vitamin D intestinal absorption, it is currently impossible to extrapolate such findings to cystic fibrosis despite the fact that these patients have fat intakes, regardless of the fatty acid source, far exceeding those of healthy controls [51].

Pancreatic insufficiency is a common feature among cystic fibrosis patients and results primarily from the obstruction of pancreatic small ducts by thick and sticky secretions. Defective fluid secretion leads to high protein concentration in the pancreatic juice and protein precipitation within the pancreatic ducts. These events induce pancreatic atrophy and fibrosis, thus impairing the pancreatic release of digestive enzymes, including the lipase. Moreover, the inappropriately acidic pH levels found in the small intestine of cystic fibrosis patients further inactivate the pancreatic lipase. All such alterations may contribute to lipid and fat-soluble vitamin malabsorption.

The process of vitamin D absorption has not been extensively studied in individuals with cystic fibrosis. To date, only two studies, carried out exclusively in adults, reported the kinetics of circulating vitamin D and its metabolite 25-hydroxyvitamin D following the ingestion of one single dose of vitamin D_2 [7,52]. In a pharmacokinetic study where serum vitamin D_2 and 25-hydroxyvitamin D levels were assessed over a 36h period following the intake of 2500µg of vitamin D_2, Lark et al. found that cystic fibrosis subjects absorbed only 45% of the amount absorbed by controls. Moreover, the increase in the mean 25-hydroxyvitamin D concentration from baseline to 36h was not significant in the cystic fibrosis group compared to the control group, indicating poor conversion of the parent compound or, conversely, increased metabolic clearance of 25-hydroxyvitamin D [7]. Interestingly, a wide interindividual variability in the absorption curves was noted in the cystic fibrosis group where subjects absorbed virtually none of the administered dose whereas others displayed an absorption curve exceeding that of the controls. To circumvent intestinal malabsorption, the same team administered 500,000IU of vitamin D_2 intramuscularly to cystic fibrosis patients in order to distinguish differences in the rate of vitamin D metabolism from variations in the absorption [53]. This dose of D_2 failed to increase fractional calcium absorption and 25-hydroxyvitamin D levels 2weeks postinjection, thus confirming the assumption that vitamin D metabolism may be altered in these patients. Since the studies cited previously investigated vitamin D_2 and because this compound is metabolized in a different manner than vitamin D_3, we cannot firmly conclude, based on these studies, that vitamin D_3 absorption or metabolism is impaired in cystic fibrosis patients. Pharmacokinetic or tracer studies using vitamin D_3 will allow drawing conclusions on this unresolved matter. However, vitamin D_3 administered to cystic fibrosis adults was found to be more effective in raising 25-hydroxyvitamin D levels than vitamin D_2, which argues against the presence of cystic fibrosis-related intrinsic defects in D_3 absorption or metabolism [22].

11.5 VITAMIN D STORAGE

Photoproduced and oral vitamin D_3 can be stored in the adipose tissue, although substantial amounts are also found in the muscle. Measurement of adipose and muscle vitamin D_3 represents a technical challenge that greatly limits their assessment. However, at intakes corresponding to those found in the contemporary world, vitamin D_3 is promptly hydroxylated by the hepatic 25-hydroxylase in the circulating metabolite 25-hydroxyvitamin D and does not accumulate considerably in fat depots. Pharmacokinetic studies reported that serum vitamin D_3 had to exceed 10–15nmol/L to saturate the hepatic hydroxylase and be stored in adipose tissue. This situation is unlikely in cystic fibrosis patients as these patients experience chronically low levels of 25-hydroxyvitamin D, which stimulates the 25-hydroxylase through a feedback mechanism, thereby preventing the accumulation and storage of excess vitamin D_3. Moreover, the presence of depleted fat stores represents another factor limiting the storage of vitamin D_3 in these patients. At present, assessment of serum and adipose concentration of vitamin D has never been realized in cystic fibrosis patients.

At first glance, obesity and cystic fibrosis do not appear to share many similarities. However, advances in the care and knowledge of cystic fibrosis patients have considerably extended their lifespan and have resulted in a shift in the distribution of body mass index of cystic fibrosis adults, thereby bringing about the emergence of a new subset of overweight and obese patients. Recently, Stephenson et al. reported the proportion of overweight and obese subjects to have increased from 7% in the 1980s to 18.4% in their most recent cohort of cystic fibrosis adults [54]. Non-cystic fibrosis obese individuals display reduced concentrations of 25-hydroxyvitamin D and a high prevalence of vitamin D insufficiency. Despite the fact that this has been documented for several years now, the reasons why the obese are prone to vitamin D insufficiency are still unknown. It has been hypothesized that obese people exhibit an enhanced clearance of vitamin D secondary to the increased uptake and storage in the fat compartments [55]. Lately, reduced level of 25-hydroxyvitamin D in obese women has been attributed to a dilution effect arising from differences in blood volume between thin and obese individuals [56]. To date, the vitamin D status of overweight/obese CF patients has not been characterized.

11.6 VITAMIN D TRANSPORT

Lipophilic hormones are largely bound to serum proteins for their extracellular transport. Vitamin D is no exception to this rule as its metabolites circulate bound to the group-specific component of serum (Gc-globulin) DBP, and, to a lesser extent, albumin and lipoproteins. DBP exhibits the highest affinity for 25-hydroxyvitamin D whereas its affinity for 1,25-dihydroxyvitamin D is about 10- to 100-fold lower. Vitamin D has an even much lower affinity, with D_2 having a slightly less affinity than D_3. However, because of the high affinity of DBP for their ligands, most vitamin D metabolites are bound to their binding protein and only a small fraction is either unbound (or free) or bound to nonspecific low-affinity carriers such as albumin. This unbound fraction is considered the biologically active one whereas the protein-bound ligands play a role as a circulatory reservoir to sequester an appropriate level of metabolites and to allow local delivery of free ligands to the target tissues.

Decreased levels of DBP have been described in cystic fibrosis, possibly due to impaired glycosylation of the protein [57,58]. According to the free hormone hypothesis, one would assume that decreased DBP levels would increase the amount of free 25-hydroxyvitamin D and 1,25-dihydroxyvitamin D and possibly enhance the actions of vitamin D. This hypothesis is further corroborated by studies performed in DBP-null mice. These mice displayed normal calcium homeostasis despite undetectable serum levels of 1,25-dihydroxyvitamin D [59]. In contrast, tissue levels of 1,25-dihydroxyvitamin D were normal as well as the expression of the major intestinal vitamin D-dependent genes. Whether reduced levels of DBP affect the distribution and availability of vitamin D metabolites to its target tissues as well as its subsequent biological role in the context of cystic fibrosis remains, however, an open question.

11.7 VITAMIN D METABOLISM

11.7.1 Liver Metabolism

As depicted in Fig. 11.1, circulating vitamin D is rapidly taken up by hepatocytes where it undergoes hydroxylation by members of the cytochrome P450 family to generate the most abundant metabolite of vitamin D, 25-hydroxyvitamin D. The liver is also the site of production of the vitamin D carrier proteins albumin and DBP and plays a role in the biliary elimination of vitamin D. Hepatobiliary diseases such as those afflicting individuals with cystic fibrosis are thus likely to affect vitamin D status. Liver dysfunction can contribute to the low level of 25-hydroxyvitamin D by means of two mechanisms: decreased hepatic uptake of the parent compound, vitamin D_3, and impairment of 25-hydroxylase activity. In an animal model of chemically induced cirrhosis, hepatic extraction of vitamin D_3 was decreased owing to a reduction in the overall hepatocytic mass in association with a probable reduction in the surface area of exchange [60]. The authors speculated that the hepatocytes remaining in the diseased liver parenchyma conserved their intrinsic ability to metabolize vitamin D_3 into 25-hydroxyvitamin D_3. This unaltered capacity of metabolizing vitamin D_3 also suggested that the hepatic 25-hydroxylase is a sturdy enzyme capable of functioning despite the presence of collagen infiltration and cirrhosis. Another study revealed the 25-hydroxylase to function almost normally in the presence of cholestasis, further strengthening the observation that this enzyme is robust [61]. Focal biliary cirrhosis remains the pathognomonic feature of cystic fibrosis hepatic complications. Its pathogenesis is not well characterized, although defective CFTR in the cholangiocytes is thought to play a role. Lack of CFTR

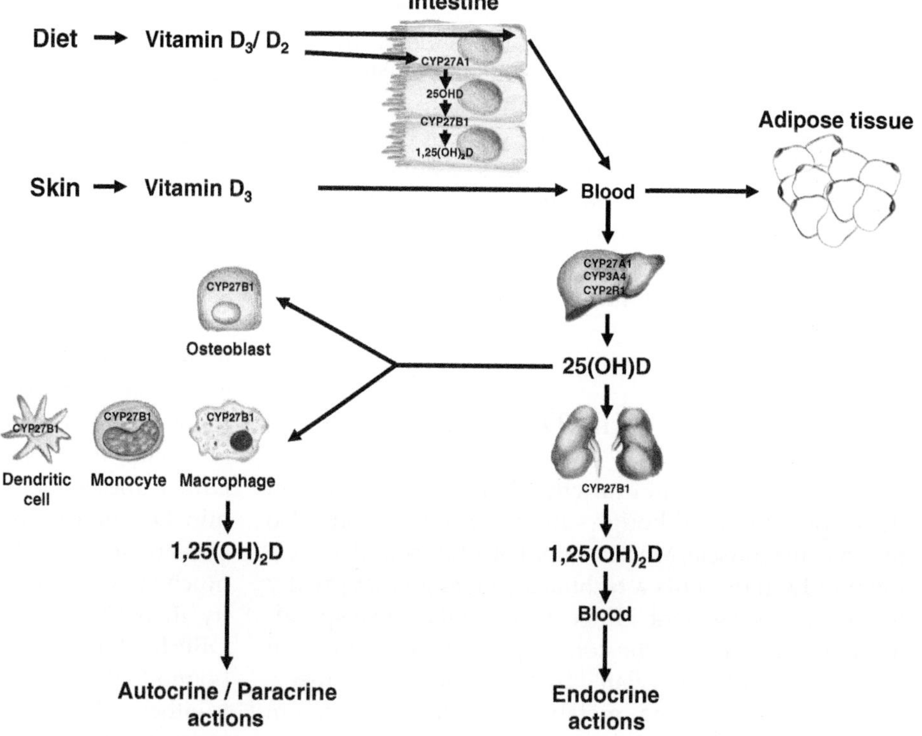

FIGURE 11.1 **Overview of vitamin D metabolism.** Vitamin D_3 is produced through the skin upon exposure to ultraviolet B radiation. Vitamin D_3 along with D_2 can also be provided through diet or supplements. Vitamin D_3/D_2 are biologically inactive. Skin-derived D_3 and dietary vitamin D join the blood circulation and are carried by the high-affinity D-binding protein (DBP). How much dietary vitamin D is taken up and used in situ by the intestine following its absorption is currently unknown. DBP carries vitamin D to the liver where it undergoes cytochrome-mediated hydroxylation on carbon-25 to generate the metabolite 25-hydroxyvitamin D (25OHD). Vitamin D can also be transported to adipose tissue for storage, although the kinetics remain largely unknown. Circulating 25OHD is further hydroxylated on carbon 1 in the kidney and in extrarenal tissues to produce the bioactive form, $1,25(OH)_2D$. $1,25(OH)_2D$ produced in the kidneys is released into circulation to exert endocrine actions whereas $1,25(OH)_2D$ synthesized extrarenally is used locally in an autocrine and/or paracrine fashion.

expression on the apical membrane of the cholangiocytes coupled with the altered biliary transport leads to toxic accumulation of bile acids including taurocholic acid. Taurocholic acid stimulates the expression of a key fibrogenic chemokine, monocyte chemotaxis protein-1 (MCP-1), in hepatocytes and cholangiocytes. MCP-1, in turn, promotes the chemotaxis of the hepatic stellate cells to the peribiliary regions where they are activated into fibrogenic myofibroblast-like cells. These activated cells produce excessive amounts of collagen, thus leading to fibrosis. The pathogenic sequence of events leading to cystic fibrosis focal biliary cirrhosis may indirectly affect hepatic uptake and activation of vitamin D, although this has never been investigated thus far. However, the absence of CFTR immunoreactivity on hepatocytes or other nonparenchymal hepatic cells suggests no direct involvement of this channel in the hepatic handling of vitamin D.

The identity of the hepatic P450 cytochromes that hydroxylate the parent compound vitamin D onto its carbon 25 is not entirely resolved. Many candidates have been proposed based primarily on in vitro studies: CYP2C11, CYP2D25, CYP27A1, CYP3A4, CYP2R1, and CYP2J2/3 [62]. Among them, mitochondrial CYP27A1 and microsomal CYP2R1 have been shown to 25-hydroxylate vitamin D in vivo. CYP27A1 is a low-affinity, high-capacity vitamin D_3 25-hydroxylase also involved in bile acid and cholesterol metabolism. Surprisingly, ablation of this gene in mice did not alter vitamin D metabolism, thereby suggesting that CYP27A1 plays a minor role in 25-hydroxyvitamin D synthesis in vivo. To the contrary, CYP2R1 hydroxylates both vitamin D_2 and D_3 equally well and much more efficiently than CYP27A1. The activity of this cytochrome has not been linked to any other biological functions apart from vitamin D metabolism. Evidence of CYP2R1 being a vitamin D 25-hydroxylase came from the identification of a mutation in the exon 2 of the gene *cyp2r1* in Nigerian families who had vitamin D deficiency and symptoms of vitamin D-dependent rickets [63]. This mutation results in the amino acid substitution of proline for leucine in the CYP2R1 protein, which abolishes its 25-hydroxylase activity. Other mutations were also reported recently on two siblings from a Saudi Arabian family [64]. Recently, *cyp2r1*$^{-/-}$ and *cyp27a1/cyp2r1* double knockout mouse models had been created [65]. The single knockout mice displayed a 50% reduction in serum 25-hydroxyvitamin D_3 whereas the double knockout presented a 25-hydroxyvitamin D_3 level similar to that of the *cyp2r1*$^{-/-}$ mice. The observation that *cyp2r1* and *cyp27a1* ablation did not entirely abolish the synthesis 25-hydroxyvitamin D suggests the presence of other vitamin D 25-hydroxylases. Moreover, the fact that the CYP2R1-associated vitamin D deficiency is corrected by the administration of D_2 may suggest the presence of an alternate D_2 25-hydroxylase pathway that may involve CYP3A4. CYP3A4 is a broad-spectrum cytochrome involved in xenobiotic metabolism and responsible for the detoxification of drugs. It is thought to be involved in the activation as well as the clearance of vitamin D therapeutics, although it is unclear whether it functions in vivo [62].

In an attempt to investigate the altered drug biotransformation that occurred in cystic fibrosis patients, the activity of several P450 cytochromes was assessed in this population [66]. Although this study did not specifically report CYP27A1 and CYP2R1 activity, several cytochrome activities were found to be normal in the context of cystic fibrosis, including CYP3A4. In the absence of additional evidence, one could rule out the involvement of aberrant 25-hydroxylase activities as a contributing factor to low 25-hydroxyvitamin D levels in cystic fibrosis patients.

Mitochondrial dysfunction has been recognized as a feature of cystic fibrosis and is suggested to contribute to the observed lung oxidative stress. In the absence of CFTR, mice exhibited depleted lung mitochondrial glutathione, which resulted in increased levels of mitochondrial reactive oxygen species and oxidative alterations to mitochondrial DNA [67]. Furthermore, impaired mitochondrial calcium homeostasis, changes in mitochondrial morphology, and altered mitochondrial respiration were documented in various models of cystic fibrosis [68–71]. Collectively, these data indicate that mitochondrial dysfunction is an important component of cystic fibrosis, which in turn, may affect the activity of various mitochondrial enzymes, including the 25-hydroxylases, and have an impact on vitamin D metabolism. However, to date, no study has specifically addressed these questions.

11.7.2 Renal Metabolism

The liver metabolite 25-hydroxyvitamin D is transported to the kidney, where a second hydroxylation step takes place to generate the bioactive metabolite, 1,25-dihydroxyvitamin D. In contrast to the 25-hydroxylation process, there is one single enzyme responsible for renal hydroxylation: the 1-alpha hydroxylase, a mitochondrial mixed-function oxidase with cytochrome P450 activity. This enzyme is under the tight control of calcium, phosphate, fibroblast growth factor 23 (FGF23), PTH, and 1,25-dihydroxyvitamin D itself. The 1,25-dihydroxyvitamin D produced by the kidneys is systemically distributed to organs and tissues to exert endocrine actions. Accumulating evidence has shown that the CFTR interacts with a wide variety of proteins, including transporters, ion channels, receptors, enzymes, and scaffold proteins [72]. Most of these interactions occur between the terminal tails (carboxyl or amino) of CFTR proteins and its binding partners, either directly or through PDZ-domain-containing proteins. Interestingly, the PTH receptor, PTHR1, exhibits a PDZ-binding motif, allowing it to interact with PDZ-containing proteins, including the Na/H exchanger regulatory factor 1, which likewise interacts with CFTR. The interaction between PTHR1 and PDZ-containing partners influences PTHR1 relocation and modulates PTH signaling. It could therefore be extrapolated that CFTR dysfunction may disrupt the macromolecular complex involving CFTR and Na/H regulatory protein 1, which, in turn, may impair PTHR1/PTH signaling. The direct interaction between CFTR and PTH has never been documented, but an interesting observation made on *Cftr*-null mice lends weight to this assumption [73]. Osteoblasts isolated from *Cftr*-knockout mice exhibited reduction in PTH-induced PTHR1 internalization as well as PTH-induced Wnt signaling, thereby suggesting some state of PTH resistance at least in the bone cells [73,74]. Whether the same observation also applies to renal PTH signaling and whether this may secondarily impact on vitamin D metabolism remains to be determined.

The kidney is also the main site for vitamin D metabolite catabolism. The first step in this catabolic pathway involves the mitochondrial enzyme 24-hydroxylase, which is responsible for the conversion of 25-hydroxyvitamin D and 1,25-dihydroxyvitamin D in 24,25 dihydroxyvitamin D and 1,24,25(OH)$_3$D, respectively. Owing to the presence of multiple vitamin D receptor (VDR) responsive elements on its promoter, 24-hydroxylase expression is strongly induced by 1,25-dihydroxyvitamin D itself. This VDR-mediated transcriptional response ensures attenuation of the 1,25-dihydroxyvitamin D biological actions inside the target cells. 1,25-Dihydroxyvitamin D levels are rarely decreased in cystic fibrosis, and 24-hydroxylase expression or activity has never been documented, suggesting normal renal catabolism of 25-hydroxyvitamin D and 1,25-dihydroxyvitamin D.

11.7.3 Peripheral Metabolism

The discovery that vitamin D metabolism is not restricted to the liver and kidney resulted in the broad diversification of the physiological actions of vitamin D. The 1-alpha hydroxylase is expressed in various cells and tissues, including, among others, the keratinocytes, placenta, immune cells, intestinal tissue, prostate tissue, brain cells, β-cells, muscle cells, and breast tissue. Some of these tissues, such as the intestine and the colon, express both 25- and 1-alpha hydroxylase [75,76]. There is still some debate on whether the extrarenal 1-alpha hydroxylase activity contributes to the circulating levels of 1,25-dihydroxyvitamin D and whether locally generated 1,25-dihydroxyvitamin D remains local. The regulation of the extrarenal 1-alpha hydroxylase is also poorly understood. In contrast with renal 1-alpha hydroxylase, the peripheral synthesis of 1,25-dihydroxyvitamin D is not influenced by calcium, phosphate, or PTH, whereas fibroblast growth factor 23 (FGF23) appears to inhibit both the renal and extrarenal 1-alpha hydroxylase expression [77,78]. Most of the current knowledge on factors regulating peripheral 1-alpha hydroxylase activity arises from work done in monocytes and macrophages. Cytokines, such as IFN-γ, TNF-α, and IL-2, as well as the bacterial product lipopolysaccharide (LPS), upregulate the macrophage 1-alpha hydroxylase activity. Extrarenal metabolism of vitamin D has not been explored in cystic fibrosis as of yet. However, the observation that macrophages from cystic fibrosis patients displayed enhanced secretion of some proinflammatory cytokines provides a good basis for future studies in that area [79]. Another key determinant to the sufficient peripheral production of 1,25-dihydroxyvitamin D is the bioavailability of 25-hydroxyvitamin D, thereby accentuating the importance of maintaining a sufficient concentration of 25-hydroxyvitamin D to ensure optimal synthesis and autocrine/paracrine actions of 1,25-dihydroxyvitamin D. However, increased expression or enzyme activity was reported under certain pathological conditions such as sarcoidosis, Crohn's disease, and uremia [80–82]. In Crohn's disease, enhanced 1-alpha hydroxylase activity was proposed as a feedback mechanism to endogenous inflammation [80]. Chronic inflammation—both systemic and extrapulmonary—is a characteristic feature of cystic fibrosis; therefore the possibility that 1-alpha hydroxylase activity might be differentially modulated in cystic fibrosis warrants further consideration [83–86].

11.8 VITAMIN D EXCRETION

The discovery of megalin, a member of the low-density lipoprotein receptor family, challenges the free hormone hypothesis, which states that steroids, including the vitamin D metabolite 25-hydroxyvitamin D, crosses the cell membrane by passive diffusion, whereas steroids bound to plasma proteins are believed to be inactive. In fact, it has been shown that steroid uptake may occur through endocytosis of steroid–carrier complexes. Megalin mediates the uptake of DBP-bound 25-hydroxyvitamin D in the renal proximal tubule along with its partner cubilin and the cytosolic adapter protein disabled-2 (Dab2) [87]. The DBP-25-hydroxyvitamin D complex uptake is followed by the intracellular release of the steroid and its mitochondrial conversion to the active form 1,25-dihydroxyvitamin D, which is subsequently released in the blood. Megalin and cubilin mediate the uptake of protein-bound 25-hydroxyvitamin D in other tissues such as the mammary cells, which underscores the importance of such proteins for the extrarenal production of bioactive 1,25-dihydroxyvitamin D [88]. Absence or reduced expression of megalin, cubilin, or Dab2 in mice, dogs, and humans results in exaggerated urinary loss of DBP and 25-hydroxyvitamin D and decreased levels of serum 25-hydroxyvitamin D and 1,25-dihydroxyvitamin D [87,89].

Despite the absence of a clear renal phenotype in cystic fibrosis, the CFTR is expressed on the apical membrane of the renal proximal tubule and in the subcellular fractions of the endosome where it is thought to regulate processing and trafficking of other cellular proteins [90]. *Cftr*-null mice indeed exhibited reduced protein expression of cubilin in the proximal tubule, although cubilin gene expression, distribution pattern, or glycosylation was unaltered by the absence of CFTR. Decreased cubilin expression has been attributed to its increased urinary excretion, thus bringing about a defective cubilin-mediated endocytosis with urinary loss of its low molecular weight ligands, including transferrin. To corroborate their findings, the same authors compared the urinary concentrations of low molecular weight proteins (transferrin, albumin, β_2-microglobulin, and Clara cell protein (CC16)) among stable cystic fibrosis patients, control subjects, and individuals with chronic asthma to rule out the influence of chronic lung inflammation on renal function. The urinary concentrations of albumin, β_2-microglobulin and CC16 were increased nearly two-fold in cystic fibrosis patients, whereas transferrinuria exhibited a 30-fold increase compared to asthma controls. The urinary concentrations of DBP or 25-hydroxyvitamin D were unfortunately not assessed in this study but, nevertheless, these findings raise the possibility that the compromised vitamin D status of cystic fibrosis patients may be related to increased excretion of the complex DBP-25-hydroxyvitamin D.

11.9 VITAMIN D ACTION

DBP-bound 1,25-dihydroxyvitamin D is delivered systemically to target organs where it is internalized. After its cellular internalization, the ligand binds to the ubiquitous VDR. The liganding of VDR triggers the formation of a heterodimer between the VDR and its partner, the retinoid-X receptor (RXR). The liganded VDR-RXR heterodimer recognizes specific DNA sequences called vitamin D response elements located in the promoter region of vitamin D target genes. The ubiquitous distribution of the VDR implies that 1,25-dihydroxyvitamin D exerts pleiotropic actions. However, very few studies have investigated the mechanism of the bioactive form of vitamin D in cystic fibrosis. Aris et al. reported that, following the ingestion of a high-calcium meal, fractional absorption of calcium is impaired in cystic fibrosis patients despite normal 1,25-hydroxyvitamin D levels [91]. Given that active intestinal transport of calcium relies on vitamin D-dependent proteins, it remains tempting to speculate on the presence of an intestinal resistance to the hormonal action of 1,25-dihydroxyvitamin D. Accordingly, the oral administration of 1,25-dihydroxyvitamin D to cystic fibrosis patients resulted in a lesser increase in the fractional absorption of calcium compared to control subjects [92]. Conversely, impaired calcium absorption may also result from other factors such as fat saponification within the gut lumen, chronic intestinal inflammation, or the absence of CFTR, thereby making it difficult to ascertain whether intestinal resistance to 1,25-dihydroxyvitamin D is present or not in cystic fibrosis patients. A study performed in cystic fibrosis children, who were supplemented with a placebo or 1 g of calcium combined or not with 2000 IU of vitamin D_3 for a period of 6 months, failed to show any effect on true calcium absorption [93]. One of the strengths of this study is that they used the dual stable isotope technique, known as the gold standard approach to measure calcium absorption. Surprisingly, despite a large increase in their calcium intake, the children who received the supplemental calcium did not experience changes in urinary excretion of calcium or bone mineral accretion at the end of the treatment period. The authors speculated that the steatorrhea, commonly seen in cystic fibrosis patients, may have increased the endogenous fecal excretion of calcium.

Cathelicidin is an antimicrobial peptide produced by neutrophils and macrophages that is involved in innate immunity. Its expression is positively regulated by 1,25-dihydroxyvitamin D_3 in keratinocytes, neutrophils, and monocytes as well as in various human cell lines [94]. To date, the only study that compared the 1,25-dihydroxyvitamin D_3 induction of cathelicidin expression between bronchial cells harboring the delta F508 mutation and normal cells failed to show a difference, thereby suggesting that, in this particular cell model, 1,25-dihydroxyvitamin D_3 appears to be fully effective [95].

11.10 IMPORTANCE OF VITAMIN D FOR CYSTIC FIBROSIS PATIENTS

Given the broad involvement of vitamin D in a range of biological functions, a chronic suboptimal vitamin D status may perturb the functioning of several systems, which in turn may impact the cystic fibrosis morbi-mortality.

11.10.1 Lungs

Pulmonary function and pulmonary exacerbations are by far the main determinants of morbidity and mortality in cystic fibrosis. The involvement of vitamin D in lung physiology has been pointed out in epidemiological studies. The largest evidence comes from the Third

National Health and Nutrition Examination Survey (NHANES III), which demonstrated, in healthy adults, a positive relationship between 25-hydroxyvitamin D levels and lung function parameters, forced expiratory volume in 1 s, and forced vital capacity (FVC and FEV_1) [96]. The same trend has been observed in adults with cystic fibrosis [97,98]. In healthy children, studies mainly focused on the association between serum 25-hydroxyvitamin D levels and upper respiratory tract infections [99–102]. Collectively, these studies have documented that vitamin D deficiency is associated with a higher risk and severity of such infections. Recently, McCauley et al. retrospectively studied the relationship between vitamin D and lung function, pulmonary exacerbations and *Pseudomonas aeruginosa* infection in children with cystic fibrosis [103]. The children were classified according to their vitamin D status: sufficient (≥75 nmol/L), insufficient (50–74 nmol/L), and deficient (<50 nmol/L). They found that children with vitamin D deficiency had a 3 times higher rate of pulmonary exacerbations than those with vitamin D sufficiency. The incidence of *P. aeruginosa* infection was, however, not significantly different between the three vitamin D groups. Furthermore, a significant positive relationship was documented between 25-hydroxyvitamin D levels and FEV_1 in the adolescents. Severity of lung disease is often associated with reduced outdoor activity and hence sun exposure, which may, in turn, influence serum 25-hydroxyvitamin D. It is therefore impossible to determine whether low 25-hydroxyvitamin D levels are either the cause or the consequence of reduced lung function.

The underlying causal mechanism between vitamin D and lung function remains unknown, but mice studies have provided some insight into the direct role of vitamin D in lung function [104]. Offspring of vitamin D-deficient and -replete dams were studied at two weeks of age for somatic growth, and lung volume, structure and function. Mice born to vitamin D-deficient mothers exhibited reduced lung function, which was explained by a decrease in lung size and volume. Interestingly, the effect on lung size was not accompanied by a reduction in somatic growth. This study has opened interesting research questions such as whether the changes observed in the vitamin D-deficient mice resulted from the maternal vitamin D depletion or from the pups' own vitamin status.

Thus far, vitamin D supplementation of cystic fibrosis patients undergoing lung transplantation has not been studied. Interestingly, daily administration of 800 IU vitamin D_3 within the first month following liver transplantation had resulted in fewer episodes of rejection [105]. These data support the need to investigate the impact of vitamin D supplementation in cystic fibrosis patients subjected not only to lung but also liver transplantation. Reports on vitamin D status in near-transplant patients with end-stage chronic lung diseases showed the prevalence of vitamin D insufficiency to be as high as 80% [106]. Furthermore, vitamin D-insufficient recipients displayed worse clinical outcomes and prognosis than vitamin D-sufficient patients, with more episodes of acute cellular rejection, a doubling in the rejection rate, an increase in the number of infectious episodes, and a mortality rate 5 times higher 1 year post-transplant. These findings were also reported in a subset of cystic fibrosis patients at the time of lung transplantation [107].

11.10.2 Innate Immune System

The range of non-classical actions of vitamin D was extended to include actions on cells from the immune system. Immune cells, including macrophages and dendritic cells, exhibit 1α-hydroxylase (CYP27B1) activity and express the VDR. Therefore, they have the capacity of producing the bioactive metabolite, 1,25-dihydroxyvitamin D, and eliciting auto- or paracrine actions. The innate immune system is the first line of defense against pathogens. Unlike the adaptive immune system, the innate immune system defends the host in a nonspecific manner. Vitamin D response elements have been identified on the promoter of genes encoding antimicrobial peptides: CAMP (cathelicidin, LL-37) and DEFB2 (β-defensin-2). In vitro studies on cystic fibrosis and non-cystic fibrosis bronchial cells showed that 1,25-dihydroxyvitamin D_3 enhanced the expression of the antimicrobial peptide cathelicidin, thus exhibiting activity against *P. aeruginosa*—the predominant pathogen in cystic fibrosis. However, a small clinical trial investigating the impact of one single dose of vitamin D_3 (250,000 IU) given to cystic fibrosis adults on circulating LL-37 failed to find any effect of vitamin D compared to the placebo group [108]. In contrast, this large bolus of vitamin D caused a reduction in the level of two proinflammatory cytokines, TNF-α and IL-6. Considering the short duration of this study (12 weeks), further investigations are needed to confirm these results. Collectively, these data support a role for vitamin D in the control of bacterial colonization and possibly chronic inflammation.

Unlike the renal 1-alpha hydroxylase, CYP27B1 expression in the immune system is not regulated by calcium homeostasis but rather subjected to modulation by immune inputs. As an example, LPS-activated macrophages exhibit increased expression of CYP27B1 and VDR [109]. In addition, VDR-driven responses were strongly attenuated in macrophages cultured in serum from vitamin D-depleted individuals, a defect associated with increased susceptibility to *Mycobacterium tuberculosis* that may be overcome by vitamin D supplementation. These observations provide evidence that the magnitude of immune responses is dependent on circulating 25-hydroxyvitamin D levels, which may be particularly relevant for cystic fibrosis patients known to present a high rate of vitamin D insufficiency.

Autophagy is a key component of mechanisms of pathogen capture and killing. Defective autophagy has been linked to various disorders, including cystic fibrosis [110,111]. Recent evidence has emerged regarding the involvement of 1,25-dihydroxyvitamin D_3 in the enhancement of autophagy through induction of cathelicidin [112]. Although novel, the association between autophagy and vitamin D has never been explored in the context of cystic fibrosis.

11.10.3 Adaptive Immune System

The adaptive immune system is primarily induced by cells specialized in antigen processing and presentation, in particular dendritic cells, and is mediated by cells specialized in antigen recognition—namely B and T lymphocytes. It has been shown that 1,25-dihydroxyvitamin D could modulate a wide variety of immune cells involved in the adaptive immune responses. Equally important is the capacity of 1,25-dihydroxyvitamin D_3 produced by macrophages, dendritic cells, and B and T lymphocytes to bind to the VDR expressed in all these cells to exert auto/paracrine actions [113–117]. 1,25-Dihydroxyvitamin D influences dendritic cell differentiation, maturation, and function, and favors the induction of dendritic cells with tolerogenic properties leading to the development of regulatory T cells (Treg). The release of IL-12 and IL-23, two cytokines involved in T-helper (Th)1 and Th17 induction, is decreased by 1,25-dihydroxyvitamin D_3 whereas the production of IL-10 is enhanced, resulting in the induction of Treg and Th2 immune responses [118,119]. Although 1,25-dihydroxyvitamin D_3 can modulate dendritic cells to direct T cell responses, it can also target T lymphocytes directly. 1,25-Dihydroxyvitamin D_3 inhibits the secretion of the Th1-signature cytokines, IL-2 and interferon-γ, as well as the secretion of IL-17 and development of Th17 cells. 1,25-Dihydroxyvitamin D_3 may favor deviation to the Th2 pathway by increasing the production of the Th2-cytokines IL-4, IL-5, and IL-10 [120].

Fungi (especially *Aspergillus fumigatus*) emerge as recognized colonizers and potential pathogens in the airways of cystic fibrosis patients. *A. fumigatus* is a known cause of allergic bronchopulmonary aspergillosis (ABPA), which occurs in 4–15% of all cystic fibrosis patients [121]. However, not all *A. fumigatus*-colonized patients develop ABPA, suggesting the involvement of other etiological factors. Kreindler et al. assessed serum levels of fat-soluble vitamins in ABPA-positive and -negative cystic fibrosis patients [122]. Only 25-hydroxyvitamin D differed between the two groups, with ABPA-positive patients having significantly lower levels compared to ABPA-negative individuals. In vitro studies have shown that the addition of a pharmacological dose of 1,25-dihydroxyvitamin D_3 to CD4+ T cells isolated from ABPA-positive patients significantly attenuated *A. fumigatus*-induced Th2 cytokine responses, namely increases in IL-5 and IL-13. The effect of 1,25-dihydroxyvitamin D_3 was found to be through the dendritic cells.

Interestingly, VDR is not constitutively expressed in primary or inactivated B-lymphocytes [123]. Cellular activation is required to upregulate VDR expression and allows for the 1,25-dihydroxyvitamin D_3 actions. In vitro treatment of activated B cells with 1,25-dihydroxyvitamin D_3 led to inhibition of B cell proliferation, plasma cell generation, as well as immunoglobulin E (IgE) and IgG secretion [114,124]. 25-Hydroxyvitamin D_3 exhibited similar effects on B cells, albeit at a much higher level than the bioactive metabolite, indicating that B cells are endowed with 1-alpha hydroxylase activity and emphasizing that low 25-hydroxyvitamin D levels may have a significant impact on these cells. In support of this assumption is the negative association found between serum 25-hydroxyvitamin D as well as vitamin D intake, and serum IgG concentration in a cystic fibrosis cohort [125]. These findings are of relevance in regard to the positive correlation between total IgG levels and severity of lung disease documented in children and adults with cystic fibrosis [126,127].

11.10.4 Pancreas

Norman et al. were among the first to report that pancreases isolated from vitamin D-deficient rats exhibited a reduction in insulin secretion and that both the early and late insulin secretory phases were impaired [128,129]. The importance of vitamin D for the pancreatic beta cell is further demonstrated by the presence of the VDR, the expression of the 1-alpha hydroxylase, and the localization of a vitamin D response element in the promoter of the human insulin receptor gene [130–132]. 1,25-Dihydroxyvitamin D_3 promotes the transcription of the insulin receptor gene and enhances insulin-mediated glucose transport in vitro [133]. Its actions vary depending on the type of diabetes. Vitamin D has been suggested to act via immunomodulatory effects in type I diabetes whereas it increases insulin secretion and its peripheral sensitivity in type II diabetes.

Cystic fibrosis-related diabetes (CFRD) is a condition that affects an increasing number of cystic fibrosis patients and is considered a clinical entity distinct from that of Type I and II diabetes. CFRD is associated with a decline in lung function and a nearly 6-fold greater mortality rate [134]. In the United States, CFRD prevalence ranges from 2% in children to 19% in adolescents and 40–50% in adults [135]. It results primarily from an impaired first-phase insulin secretion, the etiology of which is poorly understood. Loss of beta cells secondary to pancreatic fibrosis is thought to be one cause of CFRD.

Insulinopenia can be further aggravated by the development of peripheral and hepatic insulin resistance. Only one observational study reported the association between vitamin D and CFRD [136]. In a large cohort of Scandinavian patients with cystic fibrosis, the degree of vitamin D insufficiency and serum 25-hydroxyvitamin D levels below 30 and 50 nmol/L were shown to be associated with higher HbA1c concentration. Interestingly, these associations were demonstrated in the pediatric subpopulation only, suggesting that vitamin D might be more beneficial during childhood. Furthermore, the degree of vitamin D insufficiency and 25-hydroxyvitamin D concentration below 30 nmol/L were found to be significant independent predictors for CFRD diagnosis after controlling for lung function among other factors. These findings support the hypothesis that vitamin D insufficiency may play a role in the pathogenesis of impaired glucose tolerance and CFRD; however, only interventional trials with vitamin D will provide the strongest evidence to support this hypothesis.

11.10.5 Intestinal System

The gastrointestinal system is responsible for digestion, absorption, transport, and secretion of nutrients. In addition, it plays a role in detoxifying pathogens and toxic compounds and exhibits non-negligible metabolic activity. It forms a barrier between the external and internal milieu and is in contact with cells from the immune system, the intestinal microbiota, and the systemic circulation. Thus, it is an organ whereby external and internal influences converge. Furthermore, the intestine is exposed to significantly higher amounts of vitamin D compounds than any other organ and, under certain circumstances, such as in cystic fibrosis, pharmacological amounts of vitamin D. Given this unique situation, the intestine should be fully capable of metabolizing vitamin D and responds to its needs through an autocrine manner. Indeed, it was shown in human and rat intestinal explants, as well as in the intestinal cell line Caco-2, that the enzymatic machinery required for the metabolism of the parent compound vitamin D_3 into its metabolites 25-hydroxyvitamin D_3 and 1,25-dihydroxyvitamin D_3 is present [75,76,137]. However, the expression of 1-alpha hydroxylase in colonic cell models appears to be highly dependent on the proliferation and differentiation status of the cell [138]. Similar to other extrarenal sites, regulation of the intestinal 1-alpha hydroxylase is not under the tight control of the PTH/calcium axis. It depends largely on the ambient concentration of 25-hydroxyvitamin D and is not influenced by the level of 1,25-dihydroxyvitamin D. Despite such observations, the exact proportion of ingested vitamin D metabolized in situ and the role of the locally synthesized 1,25-dihydroxyvitamin D_3 are not well characterized. Pharmacological amounts of 1,25-dihydroxyvitamin D_3 (in the nanomolar range) exert anticarcinogenic properties at the intestinal level by suppressing cell proliferation and promoting cell differentiation and apoptosis [139]. These observations are interesting in light of the pharmacological consumption of vitamin D by the cystic fibrosis population. It should be stressed that adequate amounts of calcium in the cell medium are required to optimize the actions of 1,25-dihydroxyvitamin D_3. Indeed, cells grown in low calcium medium (0.25 mM) exhibited only a 20% reduction in cell growth upon treatment with 10 nM of 1,25-dihydroxyvitamin D_3 whereas cells cultured in a normal calcium medium (1.8 mM) and treated with the same concentration of 1,25-dihydroxyvitamin D_3 displayed a reduction of cell growth of 60% [140]. Key targets of the anti-mitogenic action of 1,25-dihydroxyvitamin D_3 are cyclin D1 and c-myc proto-oncogene whose expression is strongly inhibited by the hormone. Furthermore, 1,25-dihydroxyvitamin D_3 induces the expression of the cyclin-dependent kinase inhibitors p21/Waf1 and p27Kip1, which are both involved in G_1 phase arrest. Another anti-proliferative mechanism involves a reduction in the number of the mitogenic epidermal growth factor (EGF)-bound EGF receptor by 1,25-dihydroxyvitamin D_3 [139,141]. These observations may be relevant for cystic fibrosis patients given that two separate studies reported higher than expected rates of digestive cancers in large cystic fibrosis cohorts despite the overall risk of cancer being comparable to that in the general population [142–144].

Microarray studies showed that 1,25-dihydroxyvitamin D_3 significantly induced the transcription of the pattern recognition receptor nucleotide-binding oligomerization domain containing 2/caspase recruitment domain-containing protein (NOD2/CARD15) in various primary cells and cell lines [145]. NOD2 is an intracellular protein that recognizes the bacterial product muramyl peptide and is involved in bacterial clearance. A downstream effect of NOD2 induction is the stimulation of NF-κB transcription factor function, which, in turn, promotes the expression of the antimicrobial peptide defensin β2 (DEFB2/HBD2). Pretreatment with 1,25-dihydroxyvitamin D_3 synergistically induced the NF-κB expression and function as well as DEFB2/HBD2 expression. These synergistic effects were observed when cells were challenged with muramyl peptide. These observations are highly relevant for Crohn's disease as attenuated expression of NF-κB and DEFB2/HBD2 in response to 1,25-dihydroxyvitamin D_3 and muramyl peptide was found in macrophages from Crohn's patients homozygous from non-functional NOD2 variants [145]. These data provide a molecular link between the attenuation of NF-κB and DEFB2/HBD2 response observed in this inflammatory condition and the chronic state of vitamin D insufficiency experienced by Crohn's patients. The regulation of NOD2 by 1,25-dihydroxyvitamin D_3 provides evidence that vitamin D sufficiency is required for optimal innate immune responses. Noteworthy, the prevalence rate of inflammatory bowel disease is also high in cystic fibrosis populations. It was reported to be seven times that of the general population and was accounted for by Crohn's disease whose prevalence was 17 times that of controls [146]. In support of this, cystic fibrosis patients exhibit signs of intestinal inflammation of unknown etiology [86]. Whether vitamin D insufficiency may impact on the intestinal inflammation seen in cystic fibrosis remains to be determined.

11.10.6 Bone

Cystic fibrosis-related bone disease (CFBD) is characterized by the presence of low bone density, kyphosis, and an increased rate of fragility fractures. Low bone mineral density (BMD) develops insidiously during childhood and its prevalence increases with age. Osteopenia and osteoporosis strike, respectively, 28 and 20% of CF children and 38 and 24% of CF adults [147,148]. However, with a prevalence approaching 85%, CF patients with end-stage disease remain by far the group most afflicted by CFBD [149]. Furthermore, these patients present a high incidence of thoracic fractures, the most debilitating complication of CFBD [150]. Thoracic fractures seriously undermine the patient's quality of life as they compromise the person's ability to perform cough and airway clearance techniques, thereby resulting in reduced respiratory capacity, pulmonary exacerbations, and ultimately acceleration of the course of CF that may lead to fatal outcomes [151]. Malnutrition including vitamin D deficiency, glucocorticoid therapy, chronic inflammation, hypogonadism, and lack of weight-bearing physical activity were suggested as contributing factors, although their exact contribution is still unclear. Studies achieved on *Cftr*-null mice have provided some clues on the etiology of CFBD. These mice display a bone phenotype similar to that observed in CF patients, but in the absence of overt lung disease, pancreatic insufficiency, or therapeutic steroid exposure, therefore incriminating other causative factors [152,153]. Like CF patients, these mice exhibit a severe intestinal phenotype, thus giving rise to intestinal obstruction, malabsorption, and a failure to thrive, which may significantly affect the absorption of bone-related nutrients [48]. On the contrary, histomorphometric analysis of bone biopsies from cystic fibrosis patients failed to find features of vitamin D deficiency osteomalacia, suggesting that CFBD more likely results from the contribution of other factors [154,155]. Another mechanism accounting for low bone mass in cystic fibrosis is secondary hyperparathyroidism caused by suboptimal vitamin D status, which may promote osteoclastic bone resorption and prevent cystic fibrosis patients from reaching their genetically predetermined peak bone mass. Finally, the presence of CFTR was reported in the osteoblasts and, less convincingly, in the osteoclasts, thus suggesting that it might play a physiological role [74,156,157].

The role of vitamin D in bone health is multifaceted. Its involvement in bone mineralization and bone mass maintenance arises in part from its role in maintaining serum calcium and phosphate levels. The recent findings that bone cells are endowed with the cellular machinery required for the conversion of the circulating vitamin D metabolite 25-hydroxyvitamin D into its bioactive form, 1,25-dihydroxyvitamin D, have led to the emergence of additional, yet uncharacterized, more local roles for this vitamin in controlling cellular differentiation and bone remodeling [158,159]. Suboptimal concentration of 25-hydroxyvitamin D may therefore have an impact on the auto/paracrine actions of vitamin D on bone by reducing the amount of endogenously produced 1,25-dihydroxyvitamin D. The vitamin D action on the skeleton has been recently reexamined following the publication of a meta-analysis showing little benefit of vitamin D supplementation on bone density. In this meta-analysis including 23 studies and more than 4000 participants, vitamin D supplementation of post-menopausal women for a mean duration of two years resulted in no change of BMD at four anatomical sites [160]. These data are further supported by a study done in mice in which the VDR was conditionally deleted in the intestine or the bone [161]. In the intestinal VDR conditional knockout mice, calcium absorption was impaired and the 1,25-dihydroxyvitamin D levels secondary elevated. This increase in circulating 1,25-dihydroxyvitamin D stimulated bone resorption and decreased bone mineralization. In contrast, the selective loss of VDR in bone increased bone mass and enhanced bone mineralization. Collectively, these data suggest that maintenance of normocalcemia takes precedence over accretion of bone mass and skeletal integrity. In other words, bone resorption is stimulated and bone mineralization suppressed until calcium sufficiency is reached even if these adaptations result in osteopenia and fractures. These findings are particularly relevant given the impaired fractional absorption of calcium documented in cystic fibrosis subjects despite pancreatic enzyme supplementation [91]. Preservation of normocalcemia can thus be achieved at the expense of the skeleton.

The association between vitamin D and bone health has been documented in CF. Donovan et al. demonstrated that end-stage CF adults with 25-hydroxyvitamin D levels below 25 nmol/L displayed significantly lower bone mass than patients with 25-hydroxyvitamin D above that level [2]. Consistent with this, Grey et al. revealed that reduced bone mass in CF children was associated with insufficient vitamin D and K status [1].

However, other studies have failed to find a correlation between low 25-hydroxyvitamin D levels and low bone mass in CF patients [17,162]. The data of Stephenson et al., who reported that CF patients with fractures displayed reduced 25-hydroxyvitamin D levels compared to CF patients without fractures, suggest that vitamin D insufficiency may have an impact on bone quality [163]. In support of this, Priemel et al. carried out histomorphometric analysis on 675 post-mortem iliac crest biopsies coming from individuals without any overt disease and found pathologic mineralization defects in individuals with a serum 25-hydroxyvitamin D below 75 nmol/L [164]. Furthermore, Busse et al. recently reported impaired bone quality in vitamin D-deficient otherwise healthy subjects [165]. In this study, they showed, very elegantly, that vitamin D deficiency is associated with aging of bone tissue and compromised structural integrity and resistance to fracture.

11.10.7 Pregnancy

Advancements in treatments and care have considerably improved the survival of women affected by cystic fibrosis. The proportion of cystic fibrosis women conceiving and carrying a pregnancy to term has increased in the recent years and is likely to keep growing in the future. Cystic fibrosis manifestations represent a burden for both the maternal and fetal health. Alternatively, pregnancy can also adversely affect the progression of the disease. Recurrent lung infection, chronic inflammation, use of medication including antibiotics, poor nutritional status, and comorbidities such as diabetes and liver disease may all have a negative impact on pregnancy as well as postnatal outcomes. On the other hand, a recent case–control study comparing matched pregnant and non-pregnant women with cystic fibrosis revealed that women who became pregnant did not experience worse outcomes than non-pregnant women [166]. Pregnancy did not result in the deterioration of the nutritional status or pulmonary function either at the end of pregnancy or 1-year post-partum. These findings might reflect the recent instauration of a multidisciplinary approach characterized by frequent follow-ups with a nutritionist and high-risk obstetrics/gynecology team as well as the aggressive treatment of infections during pregnancy [167,168].

Vitamin D insufficiency has been widely documented in non-cystic fibrosis women but has yet to be studied in women afflicted with this disease. Maternal vitamin D status is important for the mother but also for fetal and child growth, bone maturation and mineralization, as well as for neonatal handling of calcium. In mothers, low vitamin D intake and low 25-hydroxyvitamin D levels were found to be associated with increased risk of preeclampsia [169–171]. Furthermore, preeclamptic women had lower serum and umbilical cord concentration of the bioactive metabolite 1,25-dihydroxyvitamin D than normotensive women [172]. To date, the incidence of preeclampsia in pregnant women with cystic fibrosis has not been documented despite the recent uncovering of a possible role of CFTR in that respect [173].

Women with suboptimal vitamin D status are at increased risk of delivering a vitamin D-deficient child secondary to the reduced diffusion of 25-hydroxy vitamin D across the placenta. Maternal vitamin D deficiency can affect neonatal calcium homeostasis. Vitamin D deficiency during pregnancy leads to secondary hyperparathyroidism, which in turn may cause transitory hyperparathyroidism, hypocalcemia, and even tetany in the neonate [174]. Vitamin D supplementation of the mother during pregnancy has been shown to reduce the incidence of neonatal hypocalcemia [175,176]. In addition, maternal vitamin D deficiency may lead to impaired fetal growth, low birth weight, and reduced bone mineralization, although reports are more conflicting [177–179]. In vitro studies performed on peripheral blood monocytes cultured in medium supplemented with vitamin D deficient cord blood demonstrated reduced toll-like receptor (TLR)-4 and TLR-2 induction of the antimicrobial peptide cathelicidin as opposed to monocytes cultured with vitamin D sufficient cord blood [180]. These data suggest that host susceptibility to pathogens in newborns may be influenced by the 25-hydroxyvitamin D concentration of the cord blood.

In children, maternal vitamin D status has been linked to bone mass at 9 years of age, muscle strength at 4 years, and the onset of non-skeletal diseases, including Type I diabetes and asthma [181–183]. Moreover, studies done in a vitamin D deficient rodent model underscore the possible involvement of vitamin D in brain development of the offspring [184].

Vitamin D status of pregnant women with cystic fibrosis and their offspring has not been characterized as of yet. Given the potential detrimental impact of a suboptimal vitamin D status on the maternal and child health, close attention should be given to the monitoring of vitamin D status in cystic fibrosis women experiencing a pregnancy.

11.10.8 Depression

Cystic fibrosis population represents a unique population at risk of depression and vitamin D deficiency, thus suggesting a possible link between these two entities. Preliminary data from The International Depression/Anxiety Epidemiological Study findings indicated that 22% of adolescents with cystic fibrosis had elevated depression scores whereas 30% exhibited elevated symptoms of anxiety [185]. In cystic fibrosis patients, depression and anxiety are associated with a lower quality of life, reduced respiratory function, and decreased adherence to therapeutic regimen [186–189].

VDR and 1-alpha hydroxylase are expressed in multiple areas of the brain involved in the pathophysiology of depression. These areas include the prefrontal cortex, the hippocampus, the cingulate gyrus, the thalamus, the hypothalamus, and the substantia nigra. Vitamin D metabolites are known to cross the blood–brain barrier, and they have been identified in the cerebrospinal fluid of humans. Noteworthy, VDR knockout mice exhibit symptoms that overlap with symptoms of depression [190,191]. To date, only one study investigated the association between vitamin D insufficiency and depression in a cystic fibrosis population. In a small cohort of 38 children with cystic fibrosis, Smith et al. found that lower levels of 25-hydroxyvitamin D were associated with increasing depressive symptoms [189]. However, these findings must be interpreted with caution, as the authors did not assess potential confounding variables well known to affect 25-hydroxyvitamin D levels such as physical activity, geographical location, vitamin D intake, sunlight exposure, or season. In fact, any apparent association between 25-hydroxyvitamin D levels and depression may simply reflect the fact that depressed individuals are less likely to do outdoor activities, thus greatly limiting their access to an important source of vitamin D.

Thus far, no randomized controlled trial has been conducted in cystic fibrosis populations to assess the impact of vitamin D supplementation on mental health. However, in a large community-based sample of non-cystic fibrosis women receiving a moderate dose of vitamin D (800 IU) with calcium supplementation, no association was found between vitamin D supplementation and any mental health outcomes [192]. Similar conclusions were reached with different populations and various dosage regimens of vitamin D [193–195].

11.11 CONCLUSION AND FUTURE DIRECTIONS

To date, just a handful of studies have directly addressed vitamin D bioavailability in cystic fibrosis. Although much of the current available data remain indirect, they point to alterations in certain aspects of vitamin D bioavailability in this life-limiting condition. The following points should be emphasized (1) vitamin D insufficiency is highly prevalent in cystic fibrosis populations and 25-hydroxyvitamin D levels should be closely monitored at least once a year; (2) cystic fibrosis is a multisystem condition that affects tissues and organs known to metabolize and use vitamin D, therefore accentuating the vulnerability of these patients to vitamin D insufficiency; and (3) cystic fibrosis patients experience several complications, including chronic inflammation and oxidative stress, for which the impact on vitamin D metabolism has never been investigated. Finally, it should be kept in mind that, given its adverse outcomes, vitamin D insufficiency may represent an important issue for cystic fibrosis patients, as it may constitute an additional burden for an already vulnerable population. Therefore, the management of suboptimal vitamin D status should seek two goals: restoration of serum 25-hydroxyvitamin D levels above 75 nmol/L through a tailored supplementation regimen, with ideally vitamin D_3 and, more important, long-term maintenance of this serum level without overt fluctuations. Given the wide interindividual variability in response to vitamin D supplementation coupled with the influence of genetics, environmental factors, and other factors pertaining to the disease (e.g., steatorrhea, malabsorption), individualized dosage, regular monitoring, and, above all, patient compliance are required. Studies aimed at exploring factors such as dosing schedule, vitamin D formulation, and the influence of nutritional components, pancreatic status, medication, and CFTR mutations on vitamin D absorption and metabolism would all contribute to improving our knowledge and strengthening the recommendations. The risk of long-term maintenance of high 25-hydroxyvitamin D levels in cystic fibrosis patients should also be addressed. According to recent Cystic Fibrosis Foundation guidelines, 25-hydroxyvitamin D levels should not exceed 250 nmol/L [11]. For these reasons, alternative therapeutic options such as the use of vitamin D analogs should be tested on cystic fibrosis patients. The Cystic Fibrosis Foundation guidelines do recommend the use of analogs in difficult-to-treat vitamin D deficient patients only in consultation with an expert in vitamin D therapy [11]. These analogs may correct vitamin D insufficiency in the absence of hypercalcemia or hypercalciuria. Interventional trials investigating the efficiency of these analogs in cystic fibrosis point to promising research avenues.

References

[1] Grey V, Atkinson S, Drury D, Casey L, Ferland G, Gundberg C, et al. Prevalence of low bone mass and deficiencies of vitamins D and K in pediatric patients with cystic fibrosis from 3 Canadian centers. Pediatrics November 2008;122(5):1014–20.

[2] Donovan DSJ, Papadopoulos A, Staron RB, Addesso V, Schulman L, McGregor C, et al. Bone mass and vitamin D deficiency in adults with advanced cystic fibrosis lung disease. Am J Respir Crit Care Med June 1998;157(6 Pt 1):1892–9.

[3] Feranchak AP, Sontag MK, Wagener JS, Hammond KB, Accurso FJ, Sokol RJ. Prospective, long-term study of fat-soluble vitamin status in children with cystic fibrosis identified by newborn screen. J Pediatr November 1999;135(5):601–10.

[4] Neville LA, Ranganathan SC. Vitamin D in infants with cystic fibrosis diagnosed by newborn screening. J Paediatr Child Health January 2009;45(1–2):36–41.

[5] Ahn J, Yu K, Stolzenberg-Solomon R, Simon KC, McCullough ML, Gallicchio L, et al. Genome-wide association study of circulating vitamin D levels. Hum Mol Genet July 1, 2010;19(13):2739–45.
[6] Fu L, Yun F, Oczak M, Wong BY, Vieth R, Cole DE. Common genetic variants of the vitamin D binding protein (DBP) predict differences in response of serum 25-hydroxyvitamin D [25(OH)D] to vitamin D supplementation. Clin Biochem July 2009;42(10–11):1174–7.
[7] Lark RK, Lester GE, Ontjes DA, Blackwood AD, Hollis BW, Hensler MM, et al. Diminished and erratic absorption of ergocalciferol in adult cystic fibrosis patients. Am J Clin Nutr March 2001;73(3):602–6.
[8] Wang TJ, Zhang F, Richards JB, Kestenbaum B, van Meurs JB, Berry D, et al. Common genetic determinants of vitamin D insufficiency: a genome-wide association study. Lancet July 17, 2010;376(9736):180–8.
[9] Holick MF, Binkley NC, Bischoff-Ferrari HA, Gordon CM, Hanley DA, Heaney RP, et al. Evaluation, treatment, and prevention of vitamin D deficiency: an endocrine society clinical practice guideline. J Clin Endocrinol Metab July 2011;96(7):1911–30.
[10] Ross AC, Taylor CL, Yaktine AL, Del Valle HB. Dietary reference intakes for calcium and vitamin D. National Academies Press; 2010.
[11] Tangpricha V, Kelly A, Stephenson A, Maguiness K, Enders J, Robinson KA, et al. An update on the screening, diagnosis, management, and treatment of vitamin D deficiency in individuals with cystic fibrosis: evidence-based recommendations from the cystic fibrosis foundation. J Clin Endocrinol Metab April 2012;97(4):1082–93.
[12] Sermet-Gaudelus I, Bianchi ML, Garabédian M, Aris RM, Morton A, Hardin DS, et al. European cystic fibrosis bone mineralisation guidelines. J Cyst Fibros June 2011;10(Suppl. 2):S16–23.
[13] West NE, Lechtzin N, Merlo CA, Turowski JB, Davis ME, Ramsay MZ, et al. Appropriate goal level for 25-hydroxyvitamin D in cystic fibrosis. Chest March 10, 2011.
[14] Solomons NW, Wagonfeld JB, Rieger C, Jacob RA, Bolt M, Horst JV, et al. Some biochemical indices of nutrition in treated cystic fibrosis patients. Am J Clin Nutr April 1981;34(4):462–74.
[15] Congden PJ, Bruce G, Rothburn MM, Clarke PC, Littlewood JM, Kelleher J, et al. Vitamin status in treated patients with cystic fibrosis. Arch Dis Child September 1981;56(9):708–14.
[16] Conway SP, Morton AM, Oldroyd B, Truscott JG, White H, Smith AH, et al. Osteoporosis and osteopenia in adults and adolescents with cystic fibrosis: prevalence and associated factors. Thorax September 2000;55(9):798–804.
[17] Haworth CS, Selby PL, Webb AK, Dodd ME, Musson H, Mc LNR, et al. Low bone mineral density in adults with cystic fibrosis. Thorax November 1999;54(11):961–7.
[18] Grey V, Lands L, Pall H, Drury D. Monitoring of 25-OH vitamin D levels in children with cystic fibrosis. J Pediatr Gastroenterol Nutr March 2000;30(3):314–9.
[19] Green D, Carson K, Leonard A, Davis JE, Rosenstein B, Zeitlin P, et al. Current treatment recommendations for correcting vitamin D deficiency in pediatric patients with cystic fibrosis are inadequate. J Pediatr October 2008;153(4):554–9.
[20] Boas SR, Hageman JR, Ho LT, Liveris M. Very high-dose ergocalciferol is effective for correcting vitamin D deficiency in children and young adults with cystic fibrosis. J Cyst Fibros July 2009;8(4):270–2.
[21] Boyle MP, Noschese ML, Watts SL, Davis ME, Stenner SE, Lechtzin N. Failure of high-dose ergocalciferol to correct vitamin D deficiency in adults with cystic fibrosis. Am J Respir Crit Care Med July 15, 2005;172(2):212–7.
[22] Khazai NB, Judd SE, Jeng L, Wolfenden LL, Stecenko A, Ziegler TR, et al. Treatment and prevention of vitamin D insufficiency in cystic fibrosis patients: comparative efficacy of ergocalciferol, cholecalciferol, and UV light. J Clin Endocrinol Metab June 2009;94(6):2037–43.
[23] Green DM, Leonard AR, Paranjape SM, Rosenstein BJ, Zeitlin PL, Mogayzel PJJ. Transient effectiveness of vitamin D2 therapy in pediatric cystic fibrosis patients. J Cyst Fibros March 2010;9(2):143–9.
[24] Shepherd D, Belessis Y, Katz T, Morton J, Field P, Jaffe A. Single high-dose oral vitamin D(3) (stoss) therapy - a solution to vitamin D deficiency in children with cystic fibrosis? J Cyst Fibros September 18, 2012.
[25] Gibney EM, Goldfarb DS. The association of nephrolithiasis with cystic fibrosis. Am J Kidney Dis July 2003;42(1):1–11.
[26] Hoppe B, von Unruh GE, Blank G, Rietschel E, Sidhu H, Laube N, et al. Absorptive hyperoxaluria leads to an increased risk for urolithiasis or nephrocalcinosis in cystic fibrosis. Am J Kidney Dis September 2005;46(3):440–5.
[27] Sidhu H, Hoppe B, Hesse A, Tenbrock K, Brömme S, Rietschel E, et al. Absence of *Oxalobacter formigenes* in cystic fibrosis patients: a risk factor for hyperoxaluria. Lancet September 26, 1998;352(9133):1026–9.
[28] Lepage G, Levy E, Ronco N, Smith L, Galeano N, Roy CC. Direct transesterification of plasma fatty acids for the diagnosis of essential fatty acid deficiency in cystic fibrosis. J Lipid Res October 1989;30(10):1483–90.
[29] Mailhot G, Ravid Z, Barchi S, Moreau A, Rabasa-Lhoret R, Levy E. CFTR knockdown stimulates lipid synthesis and transport in intestinal caco-2/15 cells. Am J Physiol Gastrointest Liver Physiol December 2009;297(6):G1239–49.
[30] Mailhot G, Rabasa-Lhoret R, Moreau A, Berthiaume Y, Levy E. CFTR depletion results in changes in fatty acid composition and promotes lipogenesis in intestinal Caco 2/15 cells. PLoS One 2010;5(5):e10446.
[31] Sato F, Soos G, Link C, Sato K. Cystic fibrosis transport regulator and its mRNA are expressed in human epidermis. J Invest Dermatol December 2002;119(6):1224–30.
[32] Robberecht E, Vandewalle S, Wehlou C, Kaufman JM, De Schepper J. Sunlight is an important determinant of vitamin D serum concentrations in cystic fibrosis. Eur J Clin Nutr January 19, 2011.
[33] Gronowitz E, Larko O, Gilljam M, Hollsing A, Lindblad A, Mellstrom D, et al. Ultraviolet B radiation improves serum levels of vitamin D in patients with cystic fibrosis. Acta Paediatr May 2005;94(5):547–52.
[34] Tolland JP, Murphy BP, Boyle J, Hall V, McKenna KE, Elborn JS. Ciprofloxacin-induced phototoxicity in an adult cystic fibrosis population. Photodermatol Photoimmunol Photomed October 2012;28(5):258–60.
[35] Hilliard T, Edwards S, Buchdahl R, Francis J, Rosenthal M, Balfour-Lynn I, et al. Voriconazole therapy in children with cystic fibrosis. J Cyst Fibros December 2005;4(4):215–20.
[36] Matsuoka LY, Ide L, Wortsman J, MacLaughlin JA, Holick MF. Sunscreens suppress cutaneous vitamin D3 synthesis. J Clin Endocrinol Metab June 1987;64(6):1165–8.
[37] McCarty CA. Sunlight exposure assessment: can we accurately assess vitamin D exposure from sunlight questionnaires?. Am J Clin Nutr April 2008;87(4). 1097S–101S.
[38] Rovner AJ, Stallings VA, Schall JI, Leonard MB, Zemel BS. Vitamin D insufficiency in children, adolescents, and young adults with cystic fibrosis despite routine oral supplementation. Am J Clin Nutr December 2007;86(6):1694–9.
[39] Gordon CM, Anderson EJ, Herlyn K, Hubbard JL, Pizzo A, Gelbard R, et al. Nutrient status of adults with cystic fibrosis. J Am Diet Assoc December 2007;107(12):2114–9.
[40] Hollander D, Muralidhara KS, Zimmerman A. Vitamin D-3 intestinal absorption in vivo: influence of fatty acids, bile salts, and perfusate ph on absorption. Gut April 1978;19(4): 267–72.
[41] Weizman Z, Durie PR, Kopelman HR, Vesely SM, Forstner GG. Bile acid secretion in cystic fibrosis: evidence for a defect unrelated to fat malabsorption. Gut September 1986;27(9):1043–8.

[42] Weber AM, Roy CC. Bile acid metabolism in children with cystic fibrosis. Acta Paediatr Scand Suppl 1985;317:9–15.

[43] O'Brien S, Mulcahy H, Fenlon H, O'Broin A, Casey M, Burke A, et al. Intestinal bile acid malabsorption in cystic fibrosis. Gut August 1993;34(8):1137–41.

[44] Freye HB, Kurtz SM, Spock A, Capp MP. Light and electron microscopic examination of the small bowel of children with cystic fibrosis. J Pediatr April 1964;64:575–9.

[45] Reboul E, Goncalves A, Comera C, Bott R, Nowicki M, Landrier JF, et al. Vitamin D intestinal absorption is not a simple passive diffusion: evidences for involvement of cholesterol transporters. Mol Nutr Food Res January 31, 2011.

[46] Goncalves A, Gleize B, Bott R, Nowicki M, Amiot MJ, Lairon D, et al. Phytosterols can impair vitamin D intestinal absorption in vitro and in mice. Mol Nutr Food Res September 2011;55(Suppl. 2):S303–11.

[47] Norkina O, Kaur S, Ziemer D, De Lisle RC. Inflammation of the cystic fibrosis mouse small intestine. Am J Physiol Gastrointest Liver Physiol June 2004;286(6):G1032–41.

[48] Canale-Zambrano JC, Poffenberger MC, Cory SM, Humes DG, Haston CK. Intestinal phenotype of variable-weight cystic fibrosis knockout mice. Am J Physiol Gastrointest Liver Physiol July 2007;293(1):G222–9.

[49] Goncalves A, Gleize B, Roi S, Nowicki M, Dhaussy A, Huertas A, et al. Fatty acids affect micellar properties and modulate vitamin D uptake and basolateral efflux in caco-2 cells. J Nutr Biochem October 2013;24(10):1751–7.

[50] Niramitmahapanya S, Harris SS, Dawson-Hughes B. Type of dietary fat is associated with the 25-hydroxyvitamin D3 increment in response to vitamin D supplementation. J Clin Endocrinol Metab October 2011;96(10):3170–4.

[51] Maqbool A, Schall JI, Gallagher PR, Zemel BS, Strandvik B, Stallings VA. Relation between dietary fat intake type and serum fatty acid status in children with cystic fibrosis. J Pediatr Gastroenterol Nutr November 2012;55(5):605–11.

[52] Lo CW, Paris PW, Clemens TL, Nolan J, Holick MF. Vitamin D absorption in healthy subjects and in patients with intestinal malabsorption syndromes. Am J Clin Nutr October 1985;42(4):644–9.

[53] Ontjes DA, Lark RK, Lester GE, Hollis BW, Hensler MM, Aris RM. Vitamin D depletion and replacement in patients with cystic fibrosis. In: Norman AW, Bouillon R, Thomasset M, editors. Vitamin D endocrine system: structural, biological, genetic and clinical aspects. Riverside (CA): University of California, Printing and Reprographics; 2000. pp. 893–6.

[54] Stephenson AL, Mannik LA, Walsh S, Brotherwood M, Robert R, Darling PB, et al. Longitudinal trends in nutritional status and the relation between lung function and BMI in cystic fibrosis: a population-based cohort study. Am J Clin Nutr April 2013;97(4):872–7.

[55] Wortsman J, Matsuoka LY, Chen TC, Lu Z, Holick MF. Decreased bioavailability of vitamin D in obesity. Am J Clin Nutr September 2000;72(3):690–3.

[56] Gallagher JC, Yalamanchili V, Smith LM. The effect of vitamin D supplementation on serum 25(OH)D in thin and obese women. J Steroid Biochem Mol Biol July 2013;136:195–200.

[57] Speeckaert MM, Wehlou C, Vandewalle S, Taes YE, Robberecht E, Delanghe JR. Vitamin D binding protein, a new nutritional marker in cystic fibrosis patients. Clin Chem Lab Med 2008;46(3):365–70.

[58] Coppenhaver D, Kueppers F, Schidlow D, Bee D, Isenburg JN, Barnett DR, et al. Serum concentrations of vitamin d-binding protein (group-specific component) in cystic fibrosis. Hum Genet 1981;57(4):399–403.

[59] Zella LA, Shevde NK, Hollis BW, Cooke NE, Pike JW. Vitamin d-binding protein influences total circulating levels of 1,25-dihydroxyvitamin D3 but does not directly modulate the bioactive levels of the hormone in vivo. Endocrinology July 2008;149(7):3656–67.

[60] Plourde V, Gascon-Barre M, Coulombe PA, Vallieres S, Huet PM. Hepatic handling of vitamin D3 in micronodular cirrhosis: a structure-function study in the rat. J Bone Miner Res August 1988;3(4):461–71.

[61] Plourde V, Gascon-Barre M, Willems B, Huet PM. Severe cholestasis leads to vitamin D depletion without perturbing its C-25 hydroxylation in the dog. Hepatology November 1988;8(6):1577–85.

[62] Zhu J, DeLuca HF, Vitamin D. 25-hydroxylase - four decades of searching, are we there yet? Arch Biochem Biophys July 1, 2012;523(1):30–6.

[63] Cheng JB, Levine MA, Bell NH, Mangelsdorf DJ, Russell DW. Genetic evidence that the human CYP2R1 enzyme is a key vitamin D 25-hydroxylase. Proc Natl Acad Sci USA May 18, 2004;101(20):7711–5.

[64] Al Mutair AN, Nasrat GH, Russell DW. Mutation of the CYP2R1 vitamin D 25-hydroxylase in a Saudi Arabian family with severe vitamin D deficiency. J Clin Endocrinol Metab October 2012;97(10):E2022–5.

[65] Zhu JG, Ochalek JT, Kaufmann M, Jones G, Deluca HF. CYP2R1 is a major, but not exclusive, contributor to 25-hydroxyvitamin D production in vivo. Proc Natl Acad Sci USA September 24, 2013;110(39):15650–5.

[66] Wang JP, Unadkat JD, McNamara S, O'Sullivan TA, Smith AL, Trager WF, et al. Disposition of drugs in cystic fibrosis. VI. In vivo activity of cytochrome P450 isoforms involved in the metabolism of (R)-warfarin (including P450 3A4) is not enhanced in cystic fibrosis. Clin Pharmacol Ther May, 1994;55(5):528–34.

[67] Velsor LW, Kariya C, Kachadourian R, Day BJ. Mitochondrial oxidative stress in the lungs of cystic fibrosis transmembrane conductance regulator protein mutant mice. Am J Respir Cell Mol Biol November 2006;35(5):579–86.

[68] Antigny F, Girardin N, Raveau D, Frieden M, Becq F, Vandebrouck C. Dysfunction of mitochondria ca2+ uptake in cystic fibrosis airway epithelial cells. Mitochondrion July 2009;9(4):232–41.

[69] Kelly-Aubert M, Trudel S, Fritsch J, Nguyen-Khoa T, Baudouin-Legros M, Moriceau S, et al. GSH monoethyl ester rescues mitochondrial defects in cystic fibrosis models. Hum Mol Genet July 15, 2011;20(14):2745–59.

[70] Kelly M, Trudel S, Brouillard F, Bouillaud F, Colas J, Nguyen-Khoa T, et al. Cystic fibrosis transmembrane regulator inhibitors CFTR(inh)-172 and GlyH-101 target mitochondrial functions, independently of chloride channel inhibition. J Pharmacol Exp Ther April 2010;333(1):60–9.

[71] Shapiro BL. Evidence for a mitochondrial lesion in cystic fibrosis. Life Sci 1989;44(19):1327–34.

[72] Li C, Naren AP. Macromolecular complexes of cystic fibrosis transmembrane conductance regulator and its interacting partners. Pharmacol Ther November 2005;108(2):208–23.

[73] Mechanisms of reduced osteoblast activity and enhanced osteoclastogenesis in CF bone disease. Pediatr Pulmonol 2009.

[74] Stalvey MS, Clines KL, Havasi V, McKibbin CR, Dunn LK, Chung WJ, et al. Osteoblast CFTR inactivation reduces differentiation and osteoprotegerin expression in a mouse model of cystic fibrosis-related bone disease. PLoS One 2013;8(11):e80098.

[75] Theodoropoulos C, Demers C, Mirshahi A, Gascon-Barre M. 1,25-dihydroxyvitamin D(3) downregulates the rat intestinal vitamin D(3)-25-hydroxylase CYP27A. Am J Physiol Endocrinol Metab August 2001;281(2):E315–25.

[76] Theodoropoulos C, Demers C, Delvin E, Menard D, Gascon-Barre M. Calcitriol regulates the expression of the genes encoding the three key vitamin D3 hydroxylases and the drug-metabolizing enzyme CYP3A4 in the human fetal intestine. Clin Endocrinol (Oxf) April 2003;58(4):489–99.

[77] Bacchetta J, Sea JL, Chun RF, Lisse TS, Wesseling-Perry K, Gales B, et al. Fibroblast growth factor 23 inhibits extrarenal synthesis of 1,25-dihydroxyvitamin D in human monocytes. J Bone Miner Res January 2013;28(1):46–55.

[78] Perwad F, Zhang MY, Tenenhouse HS, Portale AA. Fibroblast growth factor 23 impairs phosphorus and vitamin D metabolism in vivo and suppresses 25-hydroxyvitamin d-1alpha-hydroxylase expression in vitro. Am J Physiol Renal Physiol November 2007;293(5):F1577–83.

[79] Simonin-Le Jeune K, Le Jeune A, Jouneau S, Belleguic C, Roux PF, Jaguin M, et al. Impaired functions of macrophage from cystic fibrosis patients: CD11b, TLR-5 decrease and scd14, inflammatory cytokines increase. PLoS One 2013;8(9):e75667.

[80] Abreu MT, Kantorovich V, Vasiliauskas EA, Gruntmanis U, Matuk R, Daigle K, et al. Measurement of vitamin D levels in inflammatory bowel disease patients reveals a subset of Crohn's disease patients with elevated 1,25-dihydroxyvitamin D and low bone mineral density. Gut August 2004;53(8):1129–36.

[81] Adams JS, Sharma OP, Gacad MA, Singer FR. Metabolism of 25-hydroxyvitamin D3 by cultured pulmonary alveolar macrophages in sarcoidosis. J Clin Invest November 1983;72(5):1856–60.

[82] Dusso AS, Finch J, Brown A, Ritter C, Delmez J, Schreiner G, et al. Extrarenal production of calcitriol in normal and uremic humans. J Clin Endocrinol Metab January 1991;72(1):157–64.

[83] Abu-El-Haija M, Ramachandran S, Meyerholz DK, Abu-El-Haija M, Griffin M, Giriyappa RL, et al. Pancreatic damage in fetal and newborn cystic fibrosis pigs involves the activation of inflammatory and remodeling pathways. Am J Pathol August 2012;181(2):499–507.

[84] Levy E, Gurbindo C, Lacaille F, Paradis K, Thibault L, Seidman E. Circulating tumor necrosis factor-alpha levels and lipid abnormalities in patients with cystic fibrosis. Pediatr Res August 1993;34(2):162–6.

[85] Nixon LS, Yung B, Bell SC, Elborn JS, Shale DJ. Circulating immunoreactive interleukin-6 in cystic fibrosis. Am J Respir Crit Care Med June 1998;157(6 Pt 1):1764–9.

[86] Werlin SL, Benuri-Silbiger I, Kerem E, Adler SN, Goldin E, Zimmerman J, et al. Evidence of intestinal inflammation in patients with cystic fibrosis. J Pediatr Gastroenterol Nutr September 2010;51(3):304–8.

[87] Nykjaer A, Dragun D, Walther D, Vorum H, Jacobsen C, Herz J, et al. An endocytic pathway essential for renal uptake and activation of the steroid 25-(OH) vitamin D3. Cell February 19, 1999;96(4):507–15.

[88] Rowling MJ, Kemmis CM, Taffany DA, Welsh J. Megalin-mediated endocytosis of vitamin D binding protein correlates with 25-hydroxycholecalciferol actions in human mammary cells. J Nutr November 2006;136(11):2754–9.

[89] Nykjaer A, Fyfe JC, Kozyraki R, Leheste JR, Jacobsen C, Nielsen MS, et al. Cubilin dysfunction causes abnormal metabolism of the steroid hormone 25(OH) vitamin D(3). Proc Natl Acad Sci USA November 20, 2001;98(24):13895–900.

[90] Jouret F, Bernard A, Hermans C, Dom G, Terryn S, Leal T, et al. Cystic fibrosis is associated with a defect in apical receptor-mediated endocytosis in mouse and human kidney. J Am Soc Nephrol March 2007;18(3):707–18.

[91] Aris RM, Lester GE, Dingman S, Ontjes DA. Altered calcium homeostasis in adults with cystic fibrosis. Osteoporos Int 1999;10(2):102–8.

[92] Brown SA, Ontjes DA, Lester GE, Lark RK, Hensler MB, Blackwood AD, et al. Short-term calcitriol administration improves calcium homeostasis in adults with cystic fibrosis. Osteoporos Int June 2003;14(5):442–9.

[93] Hillman LS, Cassidy JT, Popescu MF, Hewett JE, Kyger J, Robertson JD. Percent true calcium absorption, mineral metabolism, and bone mineralization in children with cystic fibrosis: effect of supplementation with vitamin D and calcium. Pediatr Pulmonol August 2008;43(8):772–80.

[94] Wang TT, Nestel FP, Bourdeau V, Nagai Y, Wang Q, Liao J, et al. Cutting edge: 1,25-dihydroxyvitamin D3 is a direct inducer of antimicrobial peptide gene expression. J Immunol September 1, 2004;173(5):2909–12.

[95] Yim S, Dhawan P, Ragunath C, Christakos S, Diamond G. Induction of cathelicidin in normal and CF bronchial epithelial cells by 1,25-dihydroxyvitamin D(3). J Cyst Fibros November 30, 2007;6(6):403–10.

[96] Black PN, Scragg R. Relationship between serum 25-hydroxyvitamin d and pulmonary function in the third national health and nutrition examination survey. Chest December 2005;128(6):3792–8.

[97] Stephenson A, Brotherwood M, Robert R, Atenafu E, Corey M, Tullis E. Cholecalciferol significantly increases 25-hydroxyvitamin D concentrations in adults with cystic fibrosis. Am J Clin Nutr May 2007;85(5):1307–11.

[98] Wolfenden LL, Judd SE, Shah R, Sanyal R, Ziegler TR, Tangpricha V. Vitamin D and bone health in adults with cystic fibrosis. Clin Endocrinol (Oxf) September 2008;69(3):374–81.

[99] Belderbos ME, Houben ML, Wilbrink B, Lentjes E, Bloemen EM, Kimpen JL, et al. Cord blood vitamin D deficiency is associated with respiratory syncytial virus bronchiolitis. Pediatrics June 2011;127(6):e1513–20.

[100] Ginde AA, Mansbach JM, Camargo CA. Association between serum 25-hydroxyvitamin D level and upper respiratory tract infection in the third national health and nutrition examination survey. Arch Intern Med February 23, 2009;169(4):384–90.

[101] Karatekin G, Kaya A, Salihoğlu O, Balci H, Nuhoğlu A. Association of subclinical vitamin D deficiency in newborns with acute lower respiratory infection and their mothers. Eur J Clin Nutr April 2009;63(4):473–7.

[102] McNally JD, Leis K, Matheson LA, Karuananyake C, Sankaran K, Rosenberg AM. Vitamin D deficiency in young children with severe acute lower respiratory infection. Pediatr Pulmonol October 2009;44(10):981–8.

[103] McCauley LA, Thomas W, Laguna TA, Regelmann WE, Moran A, Polgreen LE. Vitamin D deficiency is associated with pulmonary exacerbations in children with cystic fibrosis. Ann Am Thorac Soc October 1, 2013.

[104] Zosky GR, Berry LJ, Elliot JG, James AL, Gorman S, Hart PH. Vitamin D deficiency causes deficits in lung function and alters lung structure. Am J Respir Crit Care Med February 4, 2011.

[105] Bitetto D, Fabris C, Falleti E, Fornasiere E, Fumolo E, Fontanini E, et al. Vitamin D and the risk of acute allograft rejection following human liver transplantation. Liver Int 2010;30(3):417–44.

[106] Lowery EM, Bemiss B, Cascino T, Durazo-Arvizu RA, Forsythe SM, Alex C, et al. Low vitamin D levels are associated with increased rejection and infections after lung transplantation. J Heart Lung Transplant July 2012;31(7):700–7.

[107] Lowery EM, Landmeier MW, Alex CG, Forsythe SM. Effect of serum vitamin D 25(OH) levels on outcome in patients with cystic fibrosis undergoing lung transplantation. Am J Respir Crit Care Med 2010;181:A4324.

[108] Grossmann RE, Zughaier SM, Liu S, Lyles RH, Tangpricha V. Impact of vitamin D supplementation on markers of inflammation in adults with cystic fibrosis hospitalized for a pulmonary exacerbation. Eur J Clin Nutr September 2012;66(9):1072–4.

[109] Liu PT, Stenger S, Li H, Wenzel L, Tan BH, Krutzik SR, et al. Toll-like receptor triggering of a vitamin d-mediated human antimicrobial response. Science March 24, 2006;311(5768):1770–3.

[110] Luciani A, Villella VR, Esposito S, Brunetti-Pierri N, Medina D, Settembre C, et al. Defective CFTR induces aggresome formation and lung inflammation in cystic fibrosis through ROS-mediated autophagy inhibition. Nat Cell Biol September 2010;12(9):863–75.

[111] Luciani A, Villella VR, Esposito S, Brunetti-Pierri N, Medina DL, Settembre C, et al. Cystic fibrosis: a disorder with defective autophagy. Autophagy January 2011;7(1):104–6.
[112] Yuk JM, Shin DM, Lee HM, Yang CS, Jin HS, Kim KK, et al. Vitamin D3 induces autophagy in human monocytes/macrophages via cathelicidin. Cell Host Microbe September 17, 2009;6(3):231–43.
[113] Cadranel J, Garabedian M, Milleron B, Guillozo H, Akoun G, Hance AJ. 1,25(OH)2D2 production by T lymphocytes and alveolar macrophages recovered by lavage from normocalcemic patients with tuberculosis. J Clin Invest May 1990;85(5):1588–93.
[114] Chen S, Sims GP, Chen XX, Gu YY, Lipsky PE. Modulatory effects of 1,25-dihydroxyvitamin D3 on human B cell differentiation. J Immunol August 1, 2007;179(3):1634–47.
[115] Hewison M, Freeman L, Hughes SV, Evans KN, Bland R, Eliopoulos AG, et al. Differential regulation of vitamin D receptor and its ligand in human monocyte-derived dendritic cells. J Immunol June 1, 2003;170(11):5382–90.
[116] Overbergh L, Decallonne B, Valckx D, Verstuyf A, Depovere J, Laureys J, et al. Identification and immune regulation of 25-hydroxyvitamin d-1-alpha-hydroxylase in murine macrophages. Clin Exp Immunol April 2000;120(1):139–46.
[117] Sigmundsdottir H, Pan J, Debes GF, Alt C, Habtezion A, Soler D, et al. DCs metabolize sunlight-induced vitamin D3 to 'program' T cell attraction to the epidermal chemokine CCL27. Nat Immunol March 2007;8(3):285–93.
[118] Pedersen AW, Holmstrøm K, Jensen SS, Fuchs D, Rasmussen S, Kvistborg P, et al. Phenotypic and functional markers for 1alpha,25-dihydroxyvitamin D(3)-modified regulatory dendritic cells. Clin Exp Immunol July 2009;157(1):48–59.
[119] Penna G, Amuchastegui S, Giarratana N, Daniel KC, Vulcano M, Sozzani S, et al. 1,25-dihydroxyvitamin D3 selectively modulates tolerogenic properties in myeloid but not plasmacytoid dendritic cells. J Immunol January 1, 2007;178(1):145–53.
[120] Hewison M. Vitamin D and the immune system: new perspectives on an old theme. Endocrinol Metab Clin North Am June 2010;39(2):365–79. table of contents.
[121] Stevens DA, Moss RB, Kurup VP, Knutsen AP, Greenberger P, Judson MA, et al. Allergic bronchopulmonary aspergillosis in cystic fibrosis–state of the art: Cystic Fibrosis Foundation consensus conference. Clin Infect Dis October 1, 2003;37(Suppl. 3):S225–64.
[122] Kreindler JL, Steele C, Nguyen N, Chan YR, Pilewski JM, Alcorn JF, et al. Vitamin D3 attenuates th2 responses to *Aspergillus fumigatus* mounted by CD4+ T cells from cystic fibrosis patients with allergic bronchopulmonary aspergillosis. J Clin Invest September 1, 2010;120(9):3242–54.
[123] Morgan JW, Morgan DM, Lasky SR, Ford D, Kouttab N, Maizel AL. Requirements for induction of vitamin d-mediated gene regulation in normal human B lymphocytes. J Immunol October 1, 1996;157(7):2900–8.
[124] Chen WC, Vayuvegula B, Gupta S. 1,25-dihydroxyvitamin D3-mediated inhibition of human B cell differentiation. Clin Exp Immunol September 1987;69(3):639–46.
[125] Pincikova T, Nilsson K, Moen IE, Karpati F, Fluge G, Hollsing A, et al. Inverse relation between vitamin D and serum total immunoglobulin G in the Scandinavian cystic fibrosis nutritional study. Eur J Clin Nutr January 2011;65(1):102–9.
[126] Levy H, Kalish LA, Huntington I, Weller N, Gerard C, Silverman EK, et al. Inflammatory markers of lung disease in adult patients with cystic fibrosis. Pediatr Pulmonol March 2007;42(3):256–62.
[127] Wheeler WB, Williams M, Matthews WJJ, Colten HR. Progression of cystic fibrosis lung disease as a function of serum immunoglobulin G levels: a 5-year longitudinal study. J Pediatr May 1984;104(5):695–9.
[128] Kadowaki S, Norman AW. Dietary vitamin D is essential for normal insulin secretion from the perfused rat pancreas. J Clin Invest March 1984;73(3):759–66.
[129] Norman AW, Frankel JB, Heldt AM, Grodsky GM. Vitamin D deficiency inhibits pancreatic secretion of insulin. Science August 15, 1980;209(4458):823–5.
[130] Bland R, Markovic D, Hills CE, Hughes SV, Chan SL, Squires PE, et al. Expression of 25-hydroxyvitamin D3-1alpha-hydroxylase in pancreatic islets. J Steroid Biochem Mol Biol May 2004;89–90(1–5):121–5.
[131] Lee S, Clark SA, Gill RK, Christakos S. 1,25-dihydroxyvitamin D3 and pancreatic beta-cell function: vitamin D receptors, gene expression, and insulin secretion. Endocrinology April 1994;134(4):1602–10.
[132] Maestro B, Dávila N, Carranza MC, Calle C. Identification of a vitamin D response element in the human insulin receptor gene promoter. J Steroid Biochem Mol Biol February 2003;84(2–3):223–30.
[133] Calle C, Maestro B, García-Arencibia M. Genomic actions of 1,25-dihydroxyvitamin D3 on insulin receptor gene expression, insulin receptor number and insulin activity in the kidney, liver and adipose tissue of streptozotocin-induced diabetic rats. BMC Mol Biol 2008;9:65.
[134] Moran A, Hardin D, Rodman D, Allen HF, Beall RJ, Borowitz D, et al. Diagnosis, screening and management of cystic fibrosis related diabetes mellitus: a consensus conference report. Diabetes Res Clin Pract August 1999;45(1):61–73.
[135] Moran A, Dunitz J, Nathan B, Saeed A, Holme B, Thomas W. Cystic fibrosis-related diabetes: current trends in prevalence, incidence, and mortality. Diabetes Care September 2009;32(9): 1626–31.
[136] Pincikova T, Nilsson K, Moen IE, Fluge G, Hollsing A, Knudsen PK, et al. Vitamin D deficiency as a risk factor for cystic fibrosis-related diabetes in the Scandinavian cystic fibrosis nutritional study. Diabetologia December 2011;54(12):3007–15.
[137] Cross HS, Peterlik M, Reddy GS, Schuster I. Vitamin D metabolism in human colon adenocarcinoma-derived Caco-2 cells: expression of 25-hydroxyvitamin D3-1alpha-hydroxylase activity and regulation of side-chain metabolism. J Steroid Biochem Mol Biol May 1997;62(1):21–8.
[138] Bareis P, Kállay E, Bischof MG, Bises G, Hofer H, Pötzi C, et al. Clonal differences in expression of 25-hydroxyvitamin D(3)-1alpha-hydroxylase, of 25-hydroxyvitamin D(3)-24-hydroxylase, and of the vitamin D receptor in human colon carcinoma cells: effects of epidermal growth factor and 1alpha,25-dihydroxyvitamin D(3). Exp Cell Res June 10, 2002;276(2):320–7.
[139] Tong WM, Hofer H, Ellinger A, Peterlik M, Cross HS. Mechanism of antimitogenic action of vitamin D in human colon carcinoma cells: relevance for suppression of epidermal growth factor-stimulated cell growth. Oncol Res 1999;11(2):77–84.
[140] Cross HS, Pavelka M, Slavik J, Peterlik M. Growth control of human colon cancer cells by vitamin D and calcium in vitro. J Natl Cancer Inst September 2, 1992;84(17):1355–7.
[141] Tong WM, Kállay E, Hofer H, Hulla W, Manhardt T, Peterlik M, et al. Growth regulation of human colon cancer cells by epidermal growth factor and 1,25-dihydroxyvitamin D3 is mediated by mutual modulation of receptor expression. Eur J Cancer December 1998;34(13):2119–25.
[142] Maisonneuve P, FitzSimmons SC, Neglia JP, Campbell 3rd PW, Lowenfels AB. Cancer risk in nontransplanted and transplanted cystic fibrosis patients: a 10-year study. J Natl Cancer Inst March 5, 2003;95(5):381–7.
[143] Maisonneuve P, Marshall BC, Knapp EA, Lowenfels AB. Cancer risk in cystic fibrosis: a 20-year nationwide study from the united states. J Natl Cancer Inst January 16, 2013;105(2):122–9.

[144] Neglia JP, FitzSimmons SC, Maisonneuve P, Schoni MH, Schoni-Affolter F, Corey M, et al. The risk of cancer among patients with cystic fibrosis. Cystic fibrosis and cancer study group. N Engl J Med February 23, 1995;332(8):494–9.

[145] Wang TT, Dabbas B, Laperriere D, Bitton AJ, Soualhine H, Tavera-Mendoza LE, et al. Direct and indirect induction by 1,25-dihydroxyvitamin D3 of the NOD2/card15-defensin beta2 innate immune pathway defective in Crohn's disease. J Biol Chem January 22, 2010;285(4):2227–31.

[146] Lloyd-Still JD. Crohn's disease and cystic fibrosis. Dig Dis Sci April 1994;39(4):880–5.

[147] Paccou J, Zeboulon N, Combescure C, Gossec L, Cortet B. The prevalence of osteoporosis, osteopenia, and fractures among adults with cystic fibrosis: a systematic literature review with meta-analysis. Calcif Tissue Int January 2010;86(1):1–7.

[148] Sermet-Gaudelus I, Souberbielle JC, Ruiz JC, Vrielynck S, Heuillon B, Azhar I, et al. Low bone mineral density in young children with cystic fibrosis. Am J Respir Crit Care Med May 1, 2007;175(9):951–7.

[149] Aris RM, Neuringer IP, Weiner MA, Egan TM, Ontjes D. Severe osteoporosis before and after lung transplantation. Chest May 1996;109(5):1176–83.

[150] Aris RM, Renner JB, Winders AD, Buell HE, Riggs DB, Lester GE, et al. Increased rate of fractures and severe kyphosis: sequelae of living into adulthood with cystic fibrosis. Ann Intern Med February 1, 1998;128(3):186–93.

[151] Latzin P, Griese M, Hermanns V, Kammer B. Sternal fracture with fatal outcome in cystic fibrosis. Thorax July 2005;60(7):616.

[152] Dif F, Marty C, Baudoin C, de Vernejoul MC, Levi G. Severe osteopenia in CFTR-null mice. Bone September 2004;35(3):595–603.

[153] Haston CK, Li W, Li A, Lafleur M, Henderson JE. Persistent osteopenia in adult cystic fibrosis transmembrane conductance regulator-deficient mice. Am J Respir Crit Care Med February 1, 2008;177(3):309–15.

[154] Elkin SL, Vedi S, Bord S, Garrahan NJ, Hodson ME, Compston JE. Histomorphometric analysis of bone biopsies from the iliac crest of adults with cystic fibrosis. Am J Respir Crit Care Med December 1, 2002;166(11):1470–4.

[155] Haworth CS, Webb AK, Egan JJ, Selby PL, Hasleton PS, Bishop PW, et al. Bone histomorphometry in adult patients with cystic fibrosis. Chest August 2000;118(2):434–9.

[156] Le Heron L, Guillaume C, Velard F, Braux J, Touqui L, Moriceau S, et al. Cystic fibrosis transmembrane conductance regulator (CFTR) regulates the production of osteoprotegerin (OPG) and prostaglandin (PG) E2 in human bone. J Cyst Fibros January 2010;9(1): 69–72.

[157] Shead EF, Haworth CS, Condliffe AM, McKeon DJ, Scott MA, Compston JE. Cystic fibrosis transmembrane conductance regulator (CFTR) is expressed in human bone. Thorax July 2007;62(7):650–1.

[158] Kogawa M, Findlay DM, Anderson PH, Ormsby R, Vincent C, Morris HA, et al. Osteoclastic metabolism of 25(OH)-vitamin D3: a potential mechanism for optimization of bone resorption. Endocrinology October 2010;151(10):4613–25.

[159] van Driel M, Koedam M, Buurman CJ, Hewison M, Chiba H, Uitterlinden AG, et al. Evidence for auto/paracrine actions of vitamin D in bone: 1alpha-hydroxylase expression and activity in human bone cells. FASEB J November 2006;20(13):2417–9.

[160] Reid IR, Bolland MJ, Grey A. Effects of vitamin D supplements on bone mineral density: a systematic review and meta-analysis. Lancet October 10, 2013.

[161] Lieben L, Masuyama R, Torrekens S, Van Looveren R, Schrooten J, Baatsen P, et al. Normocalcemia is maintained in mice under conditions of calcium malabsorption by vitamin D-induced inhibition of bone mineralization. J Clin Invest May 1, 2012;122(5): 1803–15.

[162] Elkin SL, Fairney A, Burnett S, Kemp M, Kyd P, Burgess J, et al. Vertebral deformities and low bone mineral density in adults with cystic fibrosis: a cross-sectional study. Osteoporos Int 2001;12(5):366–72.

[163] Stephenson A, Jamal S, Dowdell T, Pearce D, Corey M, Tullis E. Prevalence of vertebral fractures in adults with cystic fibrosis and their relationship to bone mineral density. Chest August 2006;130(2):539–44.

[164] Priemel M, von Domarus C, Klatte TO, Kessler S, Schlie J, Meier S, et al. Bone mineralization defects and vitamin D deficiency: histomorphometric analysis of iliac crest bone biopsies and circulating 25-hydroxyvitamin D in 675 patients. J Bone Miner Res February 2010;25(2):305–12.

[165] Busse B, Bale HA, Zimmermann EA, Panganiban B, Barth HD, Carriero A, et al. Vitamin D deficiency induces early signs of aging in human bone, increasing the risk of fracture. Sci Transl Med July 10, 2013;5(193):193ra88.

[166] Ahluwalia M, Hoag JB, Hadeh A, Ferrin M, Hadjiliadis D. Cystic fibrosis and pregnancy in the modern era: a case control study. J Cyst Fibros September 6, 2013.

[167] Edenborough FP, Borgo G, Knoop C, Lannefors L, Mackenzie WE, Madge S, et al. Guidelines for the management of pregnancy in women with cystic fibrosis. J Cyst Fibros January 2008;7(Suppl. 1):S2–32.

[168] Cheng EY, Goss CH, McKone EF, Galic V, Debley CK, Tonelli MR, et al. Aggressive prenatal care results in successful fetal outcomes in CF women. J Cyst Fibros May 2006;5(2):85–91.

[169] Bodnar LM, Catov JM, Simhan HN, Holick MF, Powers RW, Roberts JM. Maternal vitamin D deficiency increases the risk of preeclampsia. J Clin Endocrinol Metab September 2007;92(9):3517–22.

[170] Haugen M, Brantsaeter AL, Trogstad L, Alexander J, Roth C, Magnus P, et al. Vitamin D supplementation and reduced risk of preeclampsia in nulliparous women. Epidemiology September 2009;20(5):720–6.

[171] Tabesh M, Salehi-Abargouei A, Tabesh M, Esmaillzadeh A. Maternal vitamin D status and risk of pre-eclampsia: a systematic review and meta-analysis. J Clin Endocrinol Metab August 2013;98(8):3165–73.

[172] Halhali A, Tovar AR, Torres N, Bourges H, Garabedian M, Larrea F. Preeclampsia is associated with low circulating levels of insulin-like growth factor I and 1,25-dihydroxyvitamin D in maternal and umbilical cord compartments. J Clin Endocrinol Metab May 2000;85(5):1828–33.

[173] Castro-Parodi M, Levi L, Dietrich V, Zotta E, Damiano AE. CFTR may modulate AQP9 functionality in preeclamptic placentas. Placenta July 2009;30(7):642–8.

[174] Purvis RJ, Barrie WJ, MacKay GS, Wilkinson EM, Cockburn F, Belton NR. Enamel hypoplasia of the teeth associated with neonatal tetany: a manifestation of maternal vitamin-d deficiency. Lancet October 13, 1973;2(7833):811–4.

[175] Cockburn F, Belton NR, Purvis RJ, Giles MM, Brown JK, Turner TL, et al. Maternal vitamin D intake and mineral metabolism in mothers and their newborn infants. Br Med J July 5, 1980;281(6232):11–4.

[176] Delvin EE, Salle BL, Glorieux FH, Adeleine P, David LS. Vitamin D supplementation during pregnancy: effect on neonatal calcium homeostasis. J Pediatr August 1986;109(2):328–34.

[177] Thiex NW, Kalkwarf HJ, Specker BL. Vitamin D metabolism in pregnancy and lactation. In: Feldman D, Pike JW, Adams JSM, editors. Vitamin D. Amsterdam; Boston: Academic Press; 2011. pp. 679–94.

[178] Ioannou C, Javaid MK, Mahon P, Yaqub MK, Harvey NC, Godfrey KM, et al. The effect of maternal vitamin D concentration on fetal bone. J Clin Endocrinol Metab November 2012;97(11):E2070–7.

[179] Viljakainen HT, Saarnio E, Hytinantti T, Miettinen M, Surcel H, Mäkitie O, et al. Maternal vitamin D status determines bone variables in the newborn. J Clin Endocrinol Metab April 2010;95(4):1749–57.

[180] Walker VP, Zhang X, Rastegar I, Liu PT, Hollis BW, Adams JS, et al. Cord blood vitamin D status impacts innate immune responses. J Clin Endocrinol Metab April 6, 2011.

[181] Harvey NC, Moon RJ, Sayer AA, Ntani G, Davies JH, Javaid MK, et al. Maternal antenatal vitamin D status and offspring muscle development: findings from the Southampton women's survey. J Clin Endocrinol Metab October 31, 2013.

[182] Javaid MK, Crozier SR, Harvey NC, Gale CR, Dennison EM, Boucher BJ, et al. Maternal vitamin D status during pregnancy and childhood bone mass at age 9 years: a longitudinal study. Lancet January 7, 2006;367(9504):36–43.

[183] Lapillonne A. Vitamin D deficiency during pregnancy may impair maternal and fetal outcomes. Med Hypotheses January 2010;74(1):71–5.

[184] Feron F, Burne TH, Brown J, Smith E, McGrath JJ, Mackay-Sim A, et al. Developmental vitamin D3 deficiency alters the adult rat brain. Brain Res Bull March 15, 2005;65(2):141–8.

[185] Quittner AL, Cruz I, Blackwell LS, Schechter MS. The international depression and anxiety epidemiological study (TIDES): preliminary results from the united states. J Cyst Fibros 2010;9:S95.

[186] Havermans T, Colpaert K, Dupont LJ. Quality of life in patients with cystic fibrosis: association with anxiety and depression. J Cyst Fibros November 2008;7(6):581–4.

[187] Quittner AL, Barker DH, Snell C, Grimley ME, Marciel K, Cruz I. Prevalence and impact of depression in cystic fibrosis. Curr Opin Pulm Med November 2008;14(6):582–8.

[188] Riekert KA, Bartlett SJ, Boyle MP, Krishnan JA, Rand CS. The association between depression, lung function, and health-related quality of life among adults with cystic fibrosis. Chest July 2007;132(1):231–7.

[189] Smith BA, Modi AC, Quittner AL, Wood BL. Depressive symptoms in children with cystic fibrosis and parents and its effects on adherence to airway clearance. Pediatr Pulmonol August 2010;45(8):756–63.

[190] Kalueff AV, Lou YR, Laaksi I, Tuohimaa P. Increased anxiety in mice lacking vitamin D receptor gene. Neuroreport June 7, 2004;15(8):1271–4.

[191] Kalueff AV, Keisala T, Minasyan A, Kuuslahti M, Miettinen S, Tuohimaa P. Behavioural anomalies in mice evoked by "Tokyo" disruption of the vitamin D receptor gene. Neurosci Res April 2006;54(4):254–60.

[192] Dumville JC, Miles JN, Porthouse J, Cockayne S, Saxon L, King C. Can vitamin D supplementation prevent winter-time blues? A randomised trial among older women. J Nutr Health Aging 2006;10(2):151–3.

[193] Dean AJ, Bellgrove MA, Hall T, Phan WM, Eyles DW, Kvaskoff D, et al. Effects of vitamin D supplementation on cognitive and emotional functioning in young adults–a randomised controlled trial. PLoS One 2011;6(11):e25966.

[194] Kjærgaard M, Waterloo K, Wang CE, Almås B, Figenschau Y, Hutchinson MS, et al. Effect of vitamin D supplement on depression scores in people with low levels of serum 25-hydroxyvitamin D: nested case-control study and randomised clinical trial. Br J Psychiatry November 2012;201(5):360–8.

[195] Sanders KM, Stuart AL, Williamson EJ, Jacka FN, Dodd S, Nicholson G, et al. Annual high-dose vitamin D3 and mental well-being: randomised controlled trial. Br J Psychiatry May 2011;198(5):357–64.

CHAPTER

12

Vitamin D in Cystic Fibrosis

Wendy Anne Hermes[1], *Vin Tangpricha*[2]

[1]Department of Nutrition, Byrdine F. Lewis School of Nursing and Health Professionals, Georgia State University, Atlanta, GA, USA; [2]Division of Endocrinology, Metabolism and Lipids, Department of Medicine, Emory University School of Medicine, Atlanta, GA, USA

12.1 INTRODUCTION

Vitamin D is essential for the maintenance of skeletal and overall health. Deficiencies in vitamin D are extremely common in the cystic fibrosis (CF) population. The purpose of this chapter is to provide a brief overview of the function of vitamin D and to summarize the current Cystic Fibrosis Foundation (CFF) guidelines on vitamin D pertaining to general health in CF. There are two naturally occurring forms of vitamin D available for supplementation. Vitamin D_2, ergocalciferol, is found in fungi such as mushrooms or yeast. Vitamin D_3, cholecalciferol, is produced in the skin upon exposure to UV light and found in high amounts in fish oils. Both forms of vitamin D (D_2 and D_3) can be used in supplements or to fortify foods. Vitamin D (D_2 or D_3) undergoes two sequential hydroxylations by the liver and kidney to form 1,25-dihydroxyvitamin D (1,25$(OH)_2$D), the hormonal form of vitamin D. Serum 25(OH)D, produced by the first hydroxylation of vitamin D by the liver, is the major circulating form of vitamin D in blood and does not fluctuate much on a daily basis given its long half-life, ~10–14 days. The CFF defines vitamin D deficiency as a serum 25(OH)D concentration <20 ng/mL and vitamin D insufficiency as serum 25-hydroxyvitamin D (25(OH)D) between 21 and 29 ng/mL [1]. The Endocrine Society, the CFF, and the Institute of Medicine recommend the measurement of serum 25(OH)D as the marker for vitamin D status [1,2].

12.2 PREVALENCE OF DEFICIENCY IN CF

Vitamin D deficiency may be an overlooked epidemic among both children and adults in the United States [3]. It is commonly assumed that vitamin D is present in most foods, which perhaps has led to the false sense of security that most healthy adults and children are vitamin D sufficient. The prevalence of vitamin D insufficiency in CF has been reported as high as 90% as compared to 20% in healthy populations in similar geographic areas or as compared to the National Health Examination Survey (NHANES III) [4–6]. In the United States, there has been an increase in the rates of vitamin D deficiency, likely due to decreasing sunlight exposure and intake of vitamin D containing foods [6,7]. Vitamin D intake below the estimated average requirement of 400 IU/day is common in both supplement users and non-supplement users, indicating an overall inadequacy in the perceived need for vitamin D consumption [8].

The prevalence of vitamin D deficiency in CF is well documented across the lifespan [2]. Vitamin D status begins to decline during adolescence, and some study centers report insufficiency between 81% and 90% by adulthood [9–11]. Children with CF are diagnosed with vitamin D deficiency more frequently than healthy children, possibly accounting for decreased bone mineral density (BMD) later in life [12].

Production of vitamin D in the skin upon exposure to sunlight is one of the primary sources of vitamin D in the human body. While the majority of healthy populations get 95% of vitamin D from sunlight, exposure to sunlight is reduced in CF. Exercise has been used as a marker of sunlight exposure in healthy people since activity is usually conducted outdoors [6]. However, individuals with CF often have frequent infections requiring hospitalization and are often prescribed antibiotics causing photosensitivity, which further limits exposure to sunlight [13]. Antibiotic regimens containing chlorines or fluoroquinolones commonly cause UV-induced skin rashes [14]. Furthermore, such photosensitivity often requires the use of sunscreens, which reduces UV sun exposure, thus limiting production of vitamin D in the skin [14–16]. Therefore, formation of vitamin D in the skin in individuals with CF is reduced compared to individuals

Diet and Exercise in Cystic Fibrosis
http://dx.doi.org/10.1016/B978-0-12-800051-9.00011-0

without CF and tends to be the primary reason for decreased vitamin D status [2,17]. CF centers located in the higher latitudes are at additional risk for deficiency due to decreased UVB reaching the Earth's surface for cutaneous vitamin D production [18].

Malabsorption of fat-soluble vitamins due to pancreatic insufficiency is perhaps the most significant barrier for nutrient absorption in CF. Pancreatic scarring and fibrosis causes blockage of digestive enzyme secretion, limiting the absorption of fat-soluble vitamins [19]. Even in the presence of enzymes, research has revealed a reduced level of vitamin D absorption in CF. Lark et al. demonstrated that patients with CF absorbed only approximately half of the amount of vitamin D administered compared to healthy controls [20].

Although malabsorption and low sun exposure are primary contributors to vitamin D deficiency in CF, more than 20 studies highlight other possible reasons for low vitamin D metabolism [16]. Inadequate intake of vitamin D fortified foods and supplements, in addition to pancreatic exocrine function, limit vitamin D bioavailability [12,21]. Impaired hepatic hydroxylation, accelerated excretion of vitamin D, and decreased expression of vitamin D binding protein may also contribute to deficiencies [16,20,22].

12.3 IMPACT OF VITAMIN D DEFICIENCY IN CF

The role of vitamin D in skeletal health in the general population and CF is well established. Several CF centers have established vitamin D deficiency as a risk factor for low BMD in patients with CF and cite preventative supplementation [11,23–26]. However, despite the strong association linking vitamin D and skeletal health, there have been few random controlled trials (RCTs) proving that low BMD is a direct outcome of a vitamin D deficiency [27]. A 2010 Cochrane Review outlined only two RCTs (Popescu *data only*, 1998 and Haworth, 2004) illustrating a BMD outcome after a vitamin D supplement with placebo [27–29]. Because of its inextricable role in calcium and parathyroid hormone homeostasis and bone remodeling, it is difficult to either exclude or definitely prove vitamin D's true magnitude in bone density. The current association between CF, vitamin D, and skeletal health has primarily been studied through retrospective and cross-sectional studies assessing BMD directly while also looking at vitamin D deficiencies in the samples [11,30]. For example, half of the subjects in Elkin et al. had a mean serum 25(OH)D concentration of 11 ng/mL and 36% had a serum concentration of 25(OH)D <10 ng/mL. From this population, 27% also had at least one vertebral fracture and 11% had two vertebral fractures over a 2-year period [30]. This study demonstrated an association between low serum 25(OH)D and the increased risk for fracture, but does not clearly prove the cause of these bone density related fractures.

Lung function and the ability to fight infection are both important in the CF population as a predictor of morbidity and mortality [31,32]. Decreased serum 25(OH)D is associated with decreased lung function as measured by Forced Expiratory Volume in one second (FEV1), as well as a higher frequency of bacterial colonization from bacteria such as *Pseudomonas aeruginosa* [33]. Low vitamin D status is associated with an increased number of hospital stays and ultimately increased mortality in CF [2,16,31,34]. One cross-sectional study examined incidence of the first *P. aeruginosa* infection and percent predicted lung function in children aged 8, 12, and 16 years. The study concluded that higher serum 25(OH)D levels in children with CF were associated with fewer pulmonary exacerbations, and in adolescents, higher FEV1 [32]. This conclusion supported previous NHANES III data demonstrating a positive association between vitamin D status and lung function in non-CF populations [7].

Vitamin D status may also play a role in cystic fibrosis-related diabetes (CFRD). Diabetes develops in up to 25% of young adults, and in up to 50% in later adulthood [35]. CFRD has also been associated with decreases in BMD and risk for fractures [36]. Research in the non-CF population suggests that vitamin D may have a positive effect on insulin sensitivity in that it may help lower serum glucose and mediate insulin sensitivity [37–41]. However, more RCTs are needed in the CF population to more clearly define this role. One RCT in CF, Pincikova et al., studied CF patients from Sweden, Norway, and Denmark and demonstrated a positive association between glycosylated hemoglobin (HbA_{1C}) and vitamin D insufficiency [42] suggesting a link between vitamin D deficiency and pancreatic endocrine insufficiency in CF. Inflammatory markers such as C-reactive protein, IL-1, IL-6, and tumor necrosis factor-α (TNF-α) are elevated in CFRD as well as insulin resistance, Type 2 diabetes, chronic kidney disease, and metabolic syndrome [37]. Pincikova et al. suggest that vitamin D supplementation improves insulin secretion and sensitivity by reducing inflammation [42].

Vitamin D and inflammation are strongly associated with CF infections. A 2011 study of CF respiratory epithelial cell lines exposed to *P. aeruginosa* found a decrease in inflammatory cytokines IL-6 and IL-8 when exposed to the hormonal form of vitamin D, $1,25(OH)_2D$ [43]. Vitamin D may upregulate antimicrobial peptides, namely cathelicidin, to enhance clearance of bacteria at various barrier sites and in immune cells [44,45]. A 2009 review examined literature regarding clinical evidence for vitamin D as a modulator of the innate and adaptive immune system [45]. Cathelicidin exhibits a bacterial killing response in humans and is expressed by epithelial cells of respiratory tract. In people with reduced cathelicidin, bacteria accumulation increases inflammatory responses [45]. A randomized placebo controlled trial in adults with CF comparing vitamin D versus placebo found a statistically significant 50.4% reduction in TNF-α and a trend for a reduction in IL-6 in CF patients given a bolus dose of 250,000 IU of vitamin D_3 [31]. Therefore, vitamin D may have multiple effects on the immune system in CF by enhancing innate immunity by increasing antimicrobial peptides and by dampening inflammation.

12.4 VITAMIN D TREATMENTS IN CF

Current CFF guidelines recommend all people with CF to be assessed for vitamin D deficiency following the winter months and to continue a supplement that enables them to maintain a serum 25(OH)D concentrations of at least 30 ng/mL (75 nmol/L) [2]. The CFF recommends vitamin D_3 as opposed to vitamin D_2 as the ideal formulation of vitamin D for supplementation [1,2]. A 2011 review determined that while there is variation in absorption of vitamin D in the general population, the vitamin D_3 formulation may be more efficiently absorbed and better maintains blood levels above 20 ng/mL in CF as opposed to vitamin D_2 [2]. Only one study has compared the effectiveness of D_2 and D_3 together in CF, with a weekly dose of 50,000 IU, finding a more significant increase in serum 25(OH)D with vitamin D_3 as opposed to vitamin D_2 [4].

High-dose therapy (stoss) vitamin D therapy has been examined in patients with CF. Stoss therapy, which may involve doses between 100,000 and 600,000 IU vitamin D, has been effective in treating children with rickets, kidney disease, and CF as well as osteoporosis prevention in adults [4,46–48]. A retrospective chart review of children ages 3–12 years tested the safety of stoss therapy using doses above 100,000 and found that the treatment group maintained vitamin D levels 6 ng/mL higher than at baseline even 12 months after treatment [9]. Adult high-dose studies are currently underway because daily doses up to 1700 IU have been ineffective at maintaining serum 25(OH)D above 30 ng/mL [28,49,50].

There is no current recommendation as to the vehicle substance to be formulated with the vitamin D supplement since there have been few studies that directly compare an oil-based with a powder-based vehicle. A systematic review suggested that oil may be superior to powder formulations; however, there have been no studies conducted in patients with CF [34]. Given the tendency toward fat malabsorption in CF, a comparison of the optimal vehicle substance is warranted in this population [2]. Alternatively, vitamin D made from the skin does not travel through the gastrointestinal tract and therefore is not inhibited by issues of malabsorption. Use of sunlamps, while not well-studied, may be an alternative for those with particular difficulty maintaining a sufficient vitamin D status [4,17].

12.5 CONCLUSION

Vitamin D deficiency continues to be a health concern for both CF and non-CF populations. Decreased exposure to sunlight and sunscreen use, inadequate intake of vitamin D in the diet, and malabsorption tend to be the primary reasons for low serum 25(OH)D levels in the CF population. Supplementation with vitamin D raising serum 25(OH)D above 30 ng/mL appears to be associated with decreased hospitalizations while improving overall health and mortality. However, large randomized controlled trials confirming these findings are necessary to better inform treatment goals for vitamin D in CF.

References

[1] Holick MF, Binkley NC, Bischoff-Ferrari HA, Gordon CM, Hanley DA, Heaney RP, et al. Evaluation, treatment, and prevention of vitamin D deficiency: an Endocrine Society clinical practice guideline. J Clin Endocrinol Metab 2011;96(7):1911–30.

[2] Tangpricha V, Kelly A, Stephenson A, Maguiness K, Enders J, Robinson KA, et al. An update on the screening, diagnosis, management, and treatment of vitamin D deficiency in individuals with cystic fibrosis: evidence-based recommendations from the Cystic Fibrosis Foundation. J Clin Endocrinol Metab 2012;97(4):1082–93.

[3] Holick MF. Vitamin D: a millenium perspective. J Cell Biochem 2003;88(2):296–307.

[4] Khazai NB, Judd SE, Jeng L, Wolfenden LL, Stecenko A, Ziegler TR, et al. Treatment and prevention of vitamin D insufficiency in cystic fibrosis patients: comparative efficacy of ergocalciferol, cholecalciferol, and UV light. J Clin Endocrinol Metab 2009;94(6):2037–43.

[5] Ginde AA, Liu MC, Camargo Jr CA. Demographic differences and trends of vitamin D insufficiency in the US population, 1988–2004. Arch Intern Med 2009;169(6):626–32.

[6] Ganji V, Zhang X, Tangpricha V. Serum 25-hydroxyvitamin D concentrations and prevalence estimates of hypovitaminosis D in the U.S. population based on assay-adjusted data. J Nutr 2012;142(3):498–507.

[7] Black PN, Scragg R. Relationship between serum 25-hydroxyvitamin d and pulmonary function in the third national health and nutrition examination survey. Chest 2005;128(6):3792–8.

[8] Black LJ, Walton J, Flynn A, Kiely M. Adequacy of vitamin D intakes in children and teenagers from the base diet, fortified foods and supplements. Public Health Nutr 2013:1–11.

[9] Shepherd D, Belessis Y, Katz T, Morton J, Field P, Jaffe A. Single high-dose oral vitamin D(3) (stoss) therapy – a solution to vitamin D deficiency in children with cystic fibrosis? J Cyst Fibros 2012.

[10] Stephenson A, Brotherwood M, Robert R, Atenafu E, Corey M, Tullis E. Cholecalciferol significantly increases 25-hydroxyvitamin D concentrations in adults with cystic fibrosis. Am J Clin Nutr 2007;85(5):1307–11.

[11] Wolfenden LL, Judd SE, Shah R, Sanyal R, Ziegler TR, Tangpricha V. Vitamin D and bone health in adults with cystic fibrosis. Clin Endocrinol 2008;69(3):374–81.

[12] Rovner AJ, Stallings VA, Schall JI, Leonard MB, Zemel BS. Vitamin D insufficiency in children, adolescents, and young adults with cystic fibrosis despite routine oral supplementation. Am J Clin Nutr 2007;86(6):1694–9.

[13] Burdge DR, Nakielna EM, Rabin HR. Photosensitivity associated with ciprofloxacin use in adult patients with cystic fibrosis. Antimicrob Agents Chemother 1995;39(3):793.

[14] Moore DE. Drug-induced cutaneous photosensitivity: incidence, mechanism, prevention and management. Drug Saf Int J Med Toxicol Drug Exp 2002;25(5):345–72.

[15] Cox NS, Alison JA, Holland AE. Interventions for promoting physical activity in people with cystic fibrosis. In: Cochrane database of systematic reviews. John Wiley & Sons, Ltd; 1996. Available at: http://onlinelibrary.wiley.com.ezproxy.gsu.edu/doi/10.1002/14651858.CD009448.pub2/abstract. [accessed 03.01.14].

[16] Hall WB, Sparks AA, Aris RM. Vitamin d deficiency in cystic fibrosis. Int J Endocrinol 2010;2010:218691.

[17] Chandra P, Wolfenden LL, Ziegler TR, Tian J, Luo M, Stecenko AA, et al. Treatment of vitamin D deficiency with UV light in patients with malabsorption syndromes: a case series. Photodermatol Photoimmunol Photomed 2007;23(5):179–85.

[18] Anon. UV-B monitoring and research program at Colorado State University. Available at: http://uvb.nrel.colostate.edu/UVB/da_Erythemal.jsf [accessed 07.01.14].

[19] Gilbert-Barness E. Metabolic diseases: foundations of clinical management, genetics, and pathology. Eaton Publisher; 2000.

[20] Lark RK, Lester GE, Ontjes DA, Blackwood AD, Hollis BW, Hensler MM, et al. Diminished and erratic absorption of ergocalciferol in adult cystic fibrosis patients. Am J Clin Nutr 2001;73(3):602–6.

[21] Couper RT, Corey M, Moore DJ, Fisher LJ, Forstner GG, Durie PR. Decline of exocrine pancreatic function in cystic fibrosis patients with pancreatic sufficiency. Pediatr Res 1992;32(2):179–82.

[22] Aris RM, Lester GE, Dingman S, Ontjes DA. Altered calcium homeostasis in adults with cystic fibrosis. Osteoporos Int 1999;10(2):102–8.

[23] Douros K, Loukou I, Nicolaidou P, Tzonou A, Doudounakis S. Bone mass density and associated factors in cystic fibrosis patients of young age. J Paediatr Child Health 2008;44(12):681–5.

[24] Greer RM, Buntain HM, Potter JM, Wainwright CE, Wong JC, O'Rourke PK, et al. Abnormalities of the PTH-vitamin D axis and bone turnover markers in children, adolescents and adults with cystic fibrosis: comparison with healthy controls. Osteoporos Int 2003;14(5):404–11.

[25] Grey V, Atkinson S, Drury D, Casey L, Ferland G, Gundberg C, et al. Prevalence of low bone mass and deficiencies of vitamins D and K in pediatric patients with cystic fibrosis from 3 Canadian centers. Pediatrics 2008;122(5):1014–20.

[26] Hecker TM, Aris RM. Management of osteoporosis in adults with cystic fibrosis. Drugs 2004;64(2):133–47.

[27] Ferguson JH, Chang AB. Vitamin D supplementation for cystic fibrosis. Cochrane Database Syst Rev 2012;4:CD007298.

[28] Haworth CS, Jones AM, Adams JE, Selby PL, Webb AK. Randomised double blind placebo controlled trial investigating the effect of calcium and vitamin D supplementation on bone mineral density and bone metabolism in adult patients with cystic fibrosis. J Cyst Fibros 2004;3(4):233–6.

[29] Hillman LS, Cassidy JT, Popescu MF, Hewett JE, Kyger J, Robertson JD. Percent true calcium absorption, mineral metabolism, and bone mineralization in children with cystic fibrosis: effect of supplementation with vitamin D and calcium. Pediatr Pulmonol 2008;43(8):772–80.

[30] Elkin SL, Fairney A, Burnett S, Kemp M, Kyd P, Burgess J, et al. Vertebral deformities and low bone mineral density in adults with cystic fibrosis: a cross-sectional study. Osteoporos Int 2001;12(5):366–72.

[31] Grossmann RE, Zughaier SM, Liu S, Lyles RH, Tangpricha V. Impact of vitamin D supplementation on markers of inflammation in adults with cystic fibrosis hospitalized for a pulmonary exacerbation. Eur J Clin Nutr 2012;66(9):1072–4.

[32] McCauley LA, Thomas W, Laguna TA, Regelmann WE, Moran A, Polgreen LE. Vitamin D deficiency is associated with pulmonary exacerbations in children with cystic fibrosis. Ann Am Thorac Soc 2013.

[33] Chalmers JD, McHugh BJ, Docherty C, Govan JRW, Hill AT. Vitamin-D deficiency is associated with chronic bacterial colonisation and disease severity in bronchiectasis. Thorax 2013;68(1):39–47.

[34] Grossmann RE, Tangpricha V. Evaluation of vehicle substances on vitamin D bioavailability: a systematic review. Mol Nutr Food Res 2010;54(8):1055–61.

[35] Middleton PG, Wagenaar M, Matson AG, Craig ME, Holmes-Walker DJ, Katz T, et al. Australian standards of care for cystic fibrosis-related diabetes. Respirology 2014:185–92.

[36] Aris RM, Merkel PA, Bachrach LK, Borowitz DS, Boyle MP, Elkin SL, et al. Guide to bone health and disease in cystic fibrosis. J Clin Endocrinol Metab 2005;90(3):1888–96.

[37] Alvarez JA, Zughaier SM, Law J, Hao L, Wasse H, Ziegler TR, et al. Effects of high-dose cholecalciferol on serum markers of inflammation and immunity in patients with early chronic kidney disease. Eur J Clin Nutr 2013;67(3):264–9.

[38] Bachali S, Dasu K, Ramalingam K, Naidu JN. Vitamin d deficiency and insulin resistance in normal and type 2 diabetes subjects. Indian J Clin Biochem 2013;28(1):74–8.

[39] Ewald N, Hardt PD. Diagnosis and treatment of diabetes mellitus in chronic pancreatitis. World J Gastroenterol 2013;19(42):7276–81.

[40] Robertson J, Macdonald K. Prevalence of bone loss in a population with cystic fibrosis. Br J Nurs 2010;19(10):636–9.

[41] Zheng J-S, Parnell LD, Smith CE, Lee Y-C, Jamal-Allial A, Ma Y, et al. Circulating 25-hydroxyvitamin D, IRS1 variant rs2943641, and insulin resistance: replication of a gene-nutrient interaction in 4 populations of different ancestries. Clin Chem 2014;60(1):186–96.

[42] Pincikova T, Nilsson K, Moen IE, Fluge G, Hollsing A, Knudsen PK, et al. Scandinavian Cystic Fibrosis Study Consortium (SCFSC). Vitamin D deficiency as a risk factor for cystic fibrosis-related diabetes in the Scandinavian Cystic Fibrosis Nutritional Study. Diabetologia 2011;54(12):3007–15.

[43] McNally P, Coughlan C, Bergsson G, Doyle M, Taggart C, Adorini L, et al. Vitamin D receptor agonists inhibit pro-inflammatory cytokine production from the respiratory epithelium in cystic fibrosis. J Cyst Fibros 2011;10(6):428–34.

[44] Herscovitch K, Dauletbaev N, Lands LC. Vitamin D as an anti-microbial and anti-inflammatory therapy for Cystic Fibrosis. Paediatr Respir Rev 2013.

[45] Kamen DL, Tangpricha V. Vitamin D and molecular actions on the immune system: modulation of innate and autoimmunity. J Mol Med 2010;88(5):441–50.

[46] Belostotsky V, Mughal Z, Webb NJA. A single high dose of ergocalciferol can be used to boost 25-hydroxyvitamin D levels in children with kidney disease. Pediatr Nephrol 2009;24(3):625–6.

[47] Cesur Y, Caksen H, Gündem A, Kirimi E, Odabaş D. Comparison of low and high dose of vitamin D treatment in nutritional vitamin D deficiency rickets. J Pediatr Endocrinol Metab 2003;16(8):1105–9.

[48] Diamond TH, Ho KW, Rohl PG, Meerkin M. Annual intramuscular injection of a megadose of cholecalciferol for treatment of vitamin D deficiency: efficacy and safety data. Med J Aust 2005;183(1):10–2.

[49] Boyle MP, Noschese ML, Watts SL, Davis ME, Stenner SE, Lechtzin N. Failure of high-dose ergocalciferol to correct vitamin D deficiency in adults with cystic fibrosis. Am J Respir Crit Care Med 2005;172(2):212–7.

[50] Kelly EMR, Pencharz P, Tullis E. Effect of vitamin D supplementation on low serum 25-hydroxyvitamin D in adults with cystic fibrosis. Pediatr Pulmonol 2002;34(Suppl. 24):344.

CHAPTER

13

Specialty Foods for Children with Cystic Fibrosis

Megan Elizabeth McGuckin[1], *Ronald Ross Watson*[2]

[1]University of Arizona, Department of Immunobiology, USA; [2]University of Arizona, Mel and Enid Zuckerman College of Public Health, Tucson, AZ, USA

13.1 CYSTIC FIBROSIS

Cystic fibrosis (CF) is a chronic, heritable disease that mainly affects the lungs and digestive system. A mutation of the Cystic Fibrosis Transmembrane Regulator (CFTR) gene causes the body to produce unusually thick, sticky mucus. This mucus can clog the lungs, leading to life-threatening lung infections, and obstruct the pancreas, stopping natural enzymes from helping the body break down and absorb food. The CFTR gene provides instructions for making a channel that transports negatively charged particles called chloride ions in and out of cells [1]. Chloride has important functions in cells; for example, the flow of chloride ions helps control the movement of water in tissues, which is necessary for the production of thin, freely flowing mucus. Mutations in the CFTR gene disrupt the function of the chloride channels, preventing them from regulating the flow of chloride ions and water across cell membranes. As a result, cells that line the passageways of the lungs, pancreas, and other organs produce mucus that is unusually thick and sticky. This mucus clogs the airways and various ducts, causing the characteristic signs and symptoms of cystic fibrosis [2]. Symptoms of CF include delayed growth, cough, sputum production, wheeze, chest tightness, difficulty breathing/shortness of breath, fever, and repeated inflammation of the pancreas (pancreatitis) [3]. CF affects about 70,000 children worldwide without a cure, so there are major focuses on reducing symptoms in children with foods, supplements, and nutrients.

13.2 NUTRITIONAL AND GROWTH PROBLEMS ASSOCIATED WITH CYSTIC FIBROSIS

Children with CF are diagnosed in early childhood by poor growth and lack of development. Weight loss and difficulty gaining or maintaining weight are common problems for many people of all ages who have CF [4]. This is due to the mutation on the CFTR gene, which regulates negatively charged chloride ions in cells that allow the movement of water, which creates thin, free-flowing mucus. However, the mutation on the CFTR gene disrupts the function of the chloride channels, which essentially disrupts the body's ability to produce thin mucus [2]. Cells that line organs, such as the pancreas, produce mucus that is unusually thick and sticky [2]. As a result, the pancreas is unable to release enzymes that aid in digestion, and hormones such as insulin. Because of the inability of the pancreas to release sufficient digestive enzymes, children with CF have very reduced absorption of proteins, vitamin A, vitamin C, vitamin E, zinc, omega 3 fatty acids, docosahexaenoic acid, garlic, ginseng, and curcumin, all of which may help absorption by supplementation [5]. The reduced pancreatic enzymes usually cause digestive problems in people with CF, and can lead to diarrhea, malnutrition, poor growth, and weight loss [2]. CF patients also have an inability to retain salt. A common indication of CF in infants is that they may have a salty frosting on their skin or their skin tastes salty. They also may lose abnormally large amounts of body salt when they sweat on hot days [6]. This is due to altered epithelial cell function in the sweat glands. This genetic

Diet and Exercise in Cystic Fibrosis
http://dx.doi.org/10.1016/B978-0-12-800051-9.00013-4

mutation not only affects the *dietary* needs of children with CF, but it also affects their *basic daily* requirements.

13.3 CYSTIC FIBROSIS-RELATED DIABETES MELLITUS

As a result of the thick, sticky mucus, the pores and ducts of the pancreas become clogged. This prevents not only the release of digestive enzymes but also the release of hormones like insulin. When the mucus blocks the ducts of the pancreas, production of insulin is halted. In adolescence or adulthood, a shortage of insulin can cause a form of diabetes known as cystic fibrosis-related diabetes mellitus (CFRDM) [2]. CF patients usually develop this type of diabetes in their 20s, and more develop the disease after 30 years of age [7]. CFRDM has features of both type I diabetes (insulin-dependent diabetes) and type II diabetes (noninsulin-dependent diabetes). People with CF are unable to make enough insulin, as a result of scarring and excess, thick mucus on the pancreas. Insulin resistance develops in CFRDM [8]. Insulin resistance can develop as a result of chronic infections, high levels of cortisol (a hormone that the body secretes in response to stress), and frequent exposure to corticosteroids, which are anti-inflammatory drugs sometimes used in the treatment of lung conditions that mimic the action of cortisol [8]. Treatment of CFRDM includes a combination of insulin, exercise, and diet. The diet for people with CFRDM is different than the calorie-restricted diet typically prescribed for people with other types of diabetes. Despite their diabetes, people with CF must maintain a high-calorie, high-fat diet and compensate by adjusting insulin doses [8]. This treatment allows sugars and proteins to move from the blood into the body's cells. It is used for energy and to build muscle. Keeping blood glucose levels at a normal level (or near normal levels) helps in weight gain and the production of more energy. It also lowers the risk of problems that diabetes can cause [9]. So, due to the effects of the mutation on the CFTR gene, people with CF are forced to alter their diets—regardless of the presence or lack of CFRDM.

The ducts in the pancreas of children with CF become clogged with thick, sticky mucus, reducing the released enzymes necessary for digestion. Not only is the body unable to digest sufficient essential nutrients, but also the body cannot adequately collect and use these essential nutrients for growth and daily functions. Without these essential digestive enzymes, the body cannot absorb key nutrients such as carbohydrates, proteins, and fats, the three parts of food that supply calories [10]. Thus, patients with CF require specific diets, such as high-calorie diets, high-fat diets, and several other diets.

13.4 SPECIALTY FOODS AND DIETS

For children with CF, there are options available to cope with the nutritional problems associated with thick, sticky mucus covering the pancreas. Some options include supplemental vitamins, enzyme supplements, gastrostomy tube (G-tube) feeding, and nasogastric (NG) tube feeding [11]. A change in the diet may have beneficial results without these invasive techniques (see Table 13.1).

13.4.1 Increased Fat and Calorie Diets

One of the major issues for children with CF is the amount of fat present in their diet. The fat present in foods is important for normal weight gain and growth in children. A benefit of a diet high in fat is that it allows a person with CF to absorb more fat-soluble vitamins, which are required for a number of important functions in the human body. Due to the lack of digestive enzymes, there are precautions that children with CF must take into consideration before starting a high-fat diet. First, fatty foods may require an extra dose of pancreatic enzymes to aid in digestion. Children are less likely to eat fats because of this need for extra enzymes [12]. Children with CF tend to become dependent on saturated fats, potentially increasing the risk of cardiovascular disease [13]. Macronutrient intakes do not change significantly in the population of CF children, but there is a consistent imbalance of fat sources, with overdependence on saturated fats, which, in the context of increased survival in CF may potentially increase risk of cardiovascular disease [13]. Thus, by eating diets high in bad fats, children with CF have a higher chance of becoming more ill. Moreover, children with CF are highly encouraged to consume diets that are high in calories. Children with CF need more calories to compensate for the low absorption of nutrients; more specifically, they need to increase their body mass index (BMI) to greater than 25 [14]. The extra calories in this type of diet help them meet the greater energy needed for breathing [15]. High-calorie diets are good for normal growth and development, as well as for gastrointestinal comfort [16]. Unfortunately, a high-calorie diet that contains many carbohydrates may increase the risk of diabetes, where the prevalence of diabetes in the population of people with cystic fibrosis is about 36% already [17,18]. Alternative foods have been researched as options for children with CF.

13.4.2 Breastfeeding and Whey Protein

One major concern of mothers who have infants with CF is the decision on breastfeeding. Is it beneficial for the specific dietary needs of CF patients? Human breast milk has two main proteins, whey and casein, both of which allow easy digestion. Exclusive breastfeeding for infants with CF does not compromise growth, and it provides a respiratory

TABLE 13.1 Fruits and Vegetables for Children with Cystic Fibrosis: Benefits and Risks

Diet	Benefits	Risks
Carrot juice	High amount of bioaccessible provitamin A	High sugar content
Broccoli juice	Substantial amount of health-promoting compounds: vitamins, glucosinolates, phenolic compounds, and dietary essential minerals	Possible genotoxic activities
Tomato juice	Contains lycopene, an antioxidant that protects cells and organs from oxygen damage High in vitamins	Extra dose of pancreatic enzymes for digestion
Cucumber juice	Reversed changes in serum glucose, insulin, total cholesterol, triglyceride Possible role in ameliorating diabetes mellitus	Low cellular antioxidant activity levels
Pineapple juice	Enzyme bromelaine: promotes good digestion Thiamine improves glucose tolerance	High amounts of natural sugar
Acai berry	Reduces the negative effects of a high-fat diet Reduced levels of metabolic disease risk High nutritional value Contains antioxidant compounds	Does not significantly alter the nutrient tract digestibility Protein levels decline
Watermelon	Increases arginine availability Reduces serum concentrations of cardiovascular risk factors Improves glycemic control Ameliorates vascular dysfunction in type II diabetes	Must have specific genetic variants of β-carotene 15,15′-monooxygenase 1
Celery	Highly toxic toward fungi and bacteria Displays neurotoxic, anti-inflammatory and antiplatelet-aggregatory effects	Protects the gastric mucosa Suppresses the basal gastric secretion
Raspberries	Increase the metabolism in the body's fat cells Tiliroside improves insulin balance, and blood sugar balance	Rheosmin decreases the activity of a fat-digesting enzyme
Strawberries	Lowers blood glucose levels Anti-inflammatory and blood glucose-regulating capacity Contains flavonoids: potent antioxidant power	High in sugar
Blueberries	Able to decrease cholesterol Low in sugar	Blueberries and diabetes medicines can lead to extremely low blood sugar levels
Cranberries	Provides novel cardiovascular benefits	Commercially processed products contain lower levels of polyphenols
Avocado	Contains fats, fiber, vitamins, and potassium Low added sugar Bioactive compounds: monounsaturated fatty acids and sterols	Extra dose of pancreatic enzymes Genitourinary ailments, inflammatory ailments in respiratory tract, gastrointestinal tract ailments

benefit [19]. Although breast milk acts as a source for relevant nutritional items, it may be inadequate in caloric density, protein, essential fatty acids, and sodium to meet the increased requirements of CF infants [20]. Therefore, a food supplement like whey protein, rich in sulfhydryl groups, is recognized for its ability to increase glutathione and reduce oxidative stress. Supplementation with whey increased intracellular glutathione levels in patients with CF [21]. Pressurized whey supplementation in children and adults with CF could have significant nutritional and anti-inflammatory benefits [21]. Whey protein improves glucose levels and insulin response, promotes a reduction in blood pressure and arterial stiffness, and improves lipid profile [22]. The main side effects of consumption of whey protein include bloating, and 43% of participants experienced nausea and some vomiting [23]. In a study on whey protein, 27 patients with CF were treated with 20 g of whey protein per day. Anthropometric measures, pulmonary function, serum C-reactive protein, whole blood glutathione, and whole blood IL-8 and IL-6 responses to phytohemagglutinin stimulation were measured at baseline and at one month. Children showed improvement in lung function. For both children and adults, enhancements in nutritional status, as assessed by BMI, were observed [21].

13.4.3 Juices

A major problem that comes with a change in diet for children is the success of the child actually eating the new foods. One way to solve this problem is to use juices.

For children with CF, not only are the juices healthy, but they are also easy to eat and tasty. Carrot juice contains a lot of beta-carotene, vitamins, and minerals, providing a great amount of bioaccessible provitamin A [24]. Carrot juice is also rich in potassium, calcium, sodium, magnesium, iron, and phosphorus. On the other hand, people with diabetes should avoid drinking carrot juice, due to the high sugar content—it has about an 85 on the glycemic index [25,26]. Broccoli juice contains a substantial amount of health-promoting compounds such as vitamins, glucosinolates, phenolic compounds, and dietary essential minerals [27]. These compounds are important, not only in children with CF, but in healthy people as well. Broccoli consumption mediates a variety of physiological functions, including acting as an antioxidant, regulating enzymes, and controlling apoptosis and the cell cycle [28]. In vitro and experimental animal studies indicate that broccoli, its extracts, and the glucosinolate-derived degradation products have genotoxic activities; however, the relevance of the genotoxic activities to human health is not yet known. A quantitative comparison of the benefit and risk of broccoli consumption shows that the benefit from intake in modest quantities and in processed form outweighs the potential risks [29]. Tomato juice has a major organic compound called lycopene. It acts as an antioxidant by protecting cells and organs from oxygen damage. Tomato juice is also very high in vitamins that are essential to a CF diet [30]. However, tomato juice can require another dose of pancreatic enzymes to aid in digestion [12]. Cucumber juice can also be beneficial to children with CF. In an experiment, cucumber peels were fed to male mice in doses of 250 and 500 mg for 15 days. The alterations in serum glucose and hepatic lipid peroxidation were measured. Cucumber juice is high in vitamin A, and cucumber juice's effects on alterations in serum glucose, insulin, triiodothyronine, thyroine, total cholesterol, triglyceride, high-density lipoprotein, low-density lipoprotein, hepatic lipid peroxidation, superoxide dismutase, and catalase were studied [31]. The cucumber extracts nearly reversed most of these changes introduced by alloxan, suggesting its possible role in ameliorating diabetes mellitus [31]. On the other hand, cucumbers had one of the lowest antioxidant activity levels of all the fruits and vegetables tested [32]. A very sweet juice that children may enjoy is pineapple juice. Pineapple juice contains an enzyme, bromelaine, which promotes good digestion by helping the body break down proteins [33]. Also, supplementation with thiamine has been found to improve glucose tolerance in patience with hyperglycemia [34]. Unfortunately, like many fruit juices, pineapple juice contains high amounts of natural sugar, called fructose. When consuming pineapple juice, a large amount of sugar—about 35% more sugar—becomes available for absorption immediately after consumption, with little digestion required [35,36]. This would be a risky food for children who are prone to CFRDM.

13.4.4 Watermelon and Celery

Watermelons work to increase the availability of arginine by reducing serum concentrations of cardiovascular risk factors. They also improve glycemic control and ameliorate vascular dysfunction in obese animals with type II diabetes [37]. Unfortunately, the bioavailability of carotenoids found in watermelon involve specific genetic variants of β-carotene 15,15′-monooxygenase 1. This means that the ability of a child with CF to digest and process some of the nutrients in watermelon is limited—the nutrients must be in the correct form in order for a child with CF to digest them properly. Celery is highly toxic toward fungi, bacteria, and mammalian cells. It also displays neurotoxic, anti-inflammatory and antiplatelet-aggregatory effects [38]. Due to its anti-inflammatory effects, celery is a good choice for CF children. Celery protects the gastric mucosa and suppresses the basal gastric secretion in rats; thus it promotes the production of mucus [39]. Celery would be counteractive in children with CF, as their glands already produce excess mucus.

13.4.5 Berries

Raspberries' phytonutrient rheosmin increases the metabolism in the body's fat cells. Also, by activating adiponectin, the tiliroside in raspberries can help improve insulin balance, blood sugar balance, and blood fat, thus making raspberries great for children at risk for CFRDM [17]. On the other hand, in some cases, rheosmin can decrease the activity of a fat-digesting enzyme released by the pancreas called pancreatic lipase. This decrease in enzyme activity may result in less digestion and absorption of fat [17]. There are several health factors in strawberries, with the most abundant of these being ellagic acid and certain flavonoids: anthocyanin, catechin, quercetin, and kaempferol. These compounds have potent antioxidant power, helping lower risk of cardiovascular events by inhibition of LDL-cholesterol oxidation, promotion of plaque stability, improved vascular endothelial function, and decreased tendency for thrombosis [40]. In an experiment researching the anti-inflammatory and blood glucose-regulating capacity of strawberries with mice, blood glucose values were approximately 6.5% lower in the supplemented mice [41]. These results of low blood glucose levels encourage the consumption of strawberries for children with CF who are also at a risk for CFRDM. However, strawberries are also a fruit that is high in sugar content, about 217 mmol in the vacuole alone [42,43].

Blueberries are extremely low on the glycemic index, meaning they are low in sugar. This makes blueberries a great choice for those with diabetes or those at risk for diabetes [44]. In a double-blinded, randomized, and placebo-controlled clinical study, the bioactives in blueberries increased the participants' insulin sensitivity [45]. Because blueberries are so low in sugar, mixing blueberries and diabetes medications can lead to extremely low insulin levels; in fact, a study found that consumption of blueberries resulted in a lower insulin response [46].

Acai berries contain anthocyanins and flavonoids, which are powerful antioxidants that help defend the body [47]. This berry also has the ability to reduce the negative effects of a high-fat diet in laboratory studies on flies [47]. Consumption of acai fruit pulp can reduce the levels of selected markers of metabolic disease risk [48]. Acai berries have high nutritional values and contain antioxidant compounds [49]. These antioxidant compounds were tested by in vitro antioxidant assays using a macrophage model and in vivo hypolipidemoc activity using zebra fish. In assays in vitro, all extracts demonstrated potent ferric ion reductive capacity, radical-scavenging activity, and inhibition of low-density lipoprotein oxidation at a final concentration of 0.1 mg/mL; the extracts could also abrogate fructose-mediated protein gycation and mildly inhibit cholesteryl ester transfer protein [50]. However, increased consumption of acai berries does not significantly alter the nutrient tract digestibility, and protein levels decline [51]. The decline in protein means the children do not need an extra dose of pancreatic enzymes to aid in digestion, with less needed protein in the overall diet.

Cranberries are another great fruit to consider adding to a child's diet. They have properties that provide novel cardiovascular benefits, improving both vascular function and cholesterol profiles [52]. Processed cranberry products contain significantly lower levels of polyphenols and higher levels of artificial sugar. The commercial processing can alter the levels of biologically active flavonoids (especially anthocyanins and proanthocyanidins) [53].

13.4.6 Avocado

Lastly, avocado is a well-known food, which is a great addition to a CF diet. Avocados contain monounsaturated and polyunsaturated fats, dietary fiber, vitamins E, K, magnesium, and potassium, and lower levels of sugar. There is no significant increase in calories or sodium in this food [54]. Avocados are also a good source of bioactive compounds such as unsaturated fatty acids and sterols [55]. However, because of the high amount of fats in avocado, an extra dose of pancreatic enzymes may be needed to aid in digestion; it is recommended to start out taking 500 units lipase/kg/meal until fat malabsorption is corrected [12]. Also, avocados have been associated with genitourinary aliments, inflammatory ailments in the respiratory tract, and gastrointestinal tract ailments [56]. These factors make it a risky choice for children with CF.

13.5 CONCLUSION

Cystic fibrosis, a heritable disease that mainly affects the lungs and the digestive system, is a result of a mutation of the CFTR. This mutation causes the body to produce unusually thick mucus that can clog the lungs and obstruct the pancreas, thus preventing natural enzymes from breaking down and absorbing essential nutrients. Because of these functional issues, children with CF have difficulty maintaining their weight and are unable to absorb proteins and vitamins. Also, as a result of the thick sticky mucus on the pancreas, insulin is unable to perform its job, so children with CF are also at a high risk for CFRDM. It is due to these vital side effects of the mutation of the CFTR gene that the diet of children with CF is so important.

Of the specialty foods discussed, one of the major contributors to the health of children with CF would be whey protein. Not only does whey protein have important anti-inflammatory benefits but it also improves glucose levels, insulin response, and lipid profile [21]. Because children with CF are at such a high risk for infections, including whey protein in their diet can help their immune system protect them. Also, for children who have CFRDM, the improved insulin response and glucose levels are a huge benefit of this specialty food.

Although calories and proteins are a huge factor in the diets of children with CF, fruits provide many of the essential vitamins and nutrients that are important for normal growth and functioning of the body. A major fruit to consider is the acai berry. Acai berries contain anthocyanins and flavonoids, which are powerful antioxidants that help defend the body [47]. This berry also has the ability to reduce the negative effects of a high-fat diet [47]. Another specialty fruit is the blueberry, which has bioactives that can improve insulin sensitivity [45]. This is a huge factor for children with CFRDM—not only are blueberries an easy snack, but they have monumental effects for controlling a major side effect of CF. Pineapples are another great fruit for children with CF. Pineapples contain an enzyme called bromelaine, which promotes digestion by helping the body break down proteins [33]. This is a huge factor for children with CF—this enzyme could help the body compensate for the lack of release of natural enzymes by the pancreas.

Vegetables are another great source for vitamins for children with CF. The main vegetable to consider adding to a child's diet is broccoli juice. This super vegetable contains a substantial amount of health-promoting compounds such as vitamins, glucosinolates, phenolic compounds, and dietary essential minerals [27]. These components function to mediate a variety of physiological pathways in the body, such as antioxidant-regulating enzymes, and controlling apoptosis in the cell cycle [28]. Minor changes in the diets for children with CF can have incredible results—just by adding a few different fruits and vegetables, and more protein, the daily lives of children with CF can be improved.

References

[1] Billet A, Hanrahan JW. The secret life of CFTR as a calcium-activated chloride channel. National Center for Biotechnology Information. U.S. National Library of Medicine; Aug 19, 2013. Web Sep 07, 2013.

[2] Gaskin KJ. Nutritional care in children with cystic fibrosis: are our patients becoming better?. National Center for Biotechnology Information. U.S. National Library of Medicine; Mar 6, 2013. Web Aug 15, 2013.

[3] Goss CH, Edwards TC, Ramsey BW, Aitken ML, Patrick DL. Patient-reported respiratory symptoms in cystic fibrosis. J Cyst Fibros 2009. Sep 8, 2013.

[4] Pope J. How cystic fibrosis affects the pancreas and the digestive system. University of Michigan Health System; June 15, 2011. Web June 24, 2013.

[5] Braga SF, Almgren MM. Complementary therapies in cystic fibrosis: nutritional supplements and herbal products. Department of Pharmacy Practice, National Center for Biotechnology Information. U.S. National Library of Medicine; Feb 2013. Web Sep 8, 2013.

[6] Traeger N, Shi Q, Dozor AJ. Relationship between sweat chloride, sodium, and age in clinically obtained samples. National Center for Biotechnology Information. U.S. National Library of Medicine; Aug 2, 2013. Web Sep 8, 2013.

[7] Moran A, Pekow P, Grover P, Zorn M, Slovis B, Pilewski J, et al. Insulin therapy to improve BMI cystic fibrosis-related diabetes without fasting hyperglycemia: results of the cystic fibrosis related diabetes therapy trial. Diabetes Care, National Center for Biotechnology Information. U.S. National Library of Medicine; Oct 2009. Web Sep 8, 2013.

[8] Kelly A, Moran A. Update on cystic fibrosis-related diabetes. National Center for Biotechnology Information. U.S. National Library of Medicine; Apr 12, 2013. Web Sep 8, 2013.

[9] Brunzell C, Hardin DS, Moran A, Schindler T. Managing cystic fibrosis-related diabetes. Cystic Fibrosis Foundation; 2011. Sep 8, 2013.

[10] Tomezsko JL, Stallings VA, Scanlin TF. Dietary intake of healthy children with cystic fibrosis compared with normal control children. Department of Pediatrics, National Center for Biotechnology Information. U.S. National Library of Medicine; Oct 1992. Web Sep 8, 2013.

[11] Potter E, McColley S. Supporting nutrition: understanding tube-feeding. Children's Memorial Hospital; 2005. Web July 22, 2013.

[12] Goodin B. Nutrition issues in cystic fibrosis. Nutrition issues in gastroenterology, series number 27; May 2005. Pdf June 24, 2013.

[13] Smith C, Winn A, Seddon P, Ranganathan S. A fat lot of good: balance and trends in fat intake in children with cystic fibrosis. National Center for Biotechnology Information. U.S. National Library of Medicine; Nov 25, 2011. Web Aug 15, 2013.

[14] Stephenson AL, Mannik LA, Walsh S, Brotherwood M, Robert R, Darling PB, et al. Longitudinal trends in nutritional status and the relation between lung function and BMI in cystic fibrosis: a population-based cohort study. National Center for Biotechnology Information. U.S. National Library of Medicine; Feb 6, 2011. Web Nov 16, 2013.

[15] Matel JL, Milla CE. Nutrition in cystic fibrosis. National Center for Biotechnology Information. U.S. National Library of Medicine; Oct 2009. Web Aug 15, 2013.

[16] Proesmans M, De Boeck K. Evaluation of dietary fiber intake in Belgian children with cystic fibrosis: is there a link with gastrointestinal complaints?. National Center for Biotechnology Information. U.S. National Library of Medicine; Nov 2002. Web Aug 15, 2013.

[17] Jeong JB, Jeong HJ. Rheosmin, a naturally occurring phenolic compound inhibits LPS-induced iNOS and COX-2 expression in RAW264.7 cells by blocking NF-kappa B activation pathway. Food Chem Toxicol Aug–Sep 2010;48(8–9):2148–53. Web July 15, 2013.

[18] Somerville R, Lackson A, Zhou S, Fletcher C, Fitzpatrick P. Non-pulmonary chronic disease in adults with cystic fibrosis: analysis of data from the cystic fibrosis registry. National Center for Biotechnology Information. U.S. National Library of Medicine; June 2013. Web Nov 16, 2013.

[19] Jadin SA, Wu Gs, Zhang Z, Shoff SM, Tippets BM, Farrell PM, Miller T, Rock MJ, Levy H, Lai HJ. Growth and pulmonary outcomes during the first 2 y of life of breastfed and formula-fed infants diagnosed with cystic fibrosis through the Wisconsin Routine Newborn Screening Program. Department of Nutritional Sciences, College of Agriculture and Life Sciences, University of Wisconsin, Madison, WI 53706, USA.

[20] Lai HC. UW Department of Nutritional Sciences: First Study; Nov 16, 2012. Web June 24, 2013.

[21] Lands LC, Iskandar M, Beaudoin N, Meehan B, Dauletbaev N, Berthiuame Y. Dietary supplementation with pressurized whey in patients with cystic fibrosis. Montréal (QC, Canada): Division of Pediatric Respiratory Medicine, Montreal Children's Hospital-McGill University Health Centre.

[22] Pal S, Radavelli-Bagatini S. The effects of whey protein on cardiometabolic risk factors. National Center for Biotechnology Information. U.S. National Library of Medicine; Apr 2013. Web Aug 15, 2013.

[23] Jeloka TK, Dharmatti G, Jamdade T, Pandit M. Are oral protein supplements helpful in the management of malnutrition in dialysis patients?. National Center for Biotechnology Information. U.S. National Library of Medicine; Jan 2013. Web Aug 15, 2013.

[24] Courraud J, Berger J, Cristol JP, Avallone S. Stability and bioaccessibility of different forms of carotenoids and vitamin A during in vitro digestion. National Center for Biotechnology Information. U.S. National Library of Medicine; Jan 15, 2013. Web Aug 15, 2013.

[25] Sinchaipanit P, Kerr WL, Chamchan R. Effect of sweeteners and hydrocolloids on quality attributes of reduced-calorie carrot juice. National Center for Biotechnology Information. U.S. National Library of Medicine; Apr 12, 2013. Web Aug 15, 2013.

[26] Donaldson Michael. Let's juice! The glycemic index of carrot juice and controlling blood glucose levels. Hallelujah Acres Foundation. Web Nov 14, 2013.

[27] Ares AM, Nozal MJ, Bernal J. Extraction, chemical characterization and biological activity determination of broccoli health promoting compounds. National Center for Biotechnology Information. U.S. National Library of Medicine; July 16, 2013. Web Aug 15, 2013.

[28] Mukherjee S, Das DK. Health benefits of broccoli. In: International symposium on human health effects of fruits and vegetables; 2007. Web Nov 15, 2013.

[29] Latté KP, Appel KE, Lampen A. Health benefits and possible risks of broccoli – an overview. National Center for Biotechnology Information. U.S. National Library of Medicine; Dec 2011. Web Aug 15, 2013.

[30] Fröhlich K, Kaufmann K, Bitsch R, Böhm V. Effects of ingestion of tomatoes, tomato juice and tomato purée on contents of lycopene isomers, tocopherols and ascorbic acid in human plasma as well as on lycopene isomer pattern. National Center for Biotechnology Information. U.S. National Library of Medicine; Apr 2006. Web Aug 15, 2013.

[31] Dixit Y, Kar A. Protective role of three vegetable peels in alloxan induced diabetes mellitus in male mice. National Center for Biotechnology Information. U.S. National Library of Medicine; Sep 2010. Web Aug 15, 2013.

[32] Song W, Derito CM, Liu MK, He X, Dong M, Liu RH. Cellular antioxidant activity of common vegetables. National Center for Biotechnology Information. U.S. National Library of Medicine; June 9, 2010. Web Aug 15, 2013.

[33] Roxas M. The role of enzyme supplementation in digestive disorders. National Center for Biotechnology Information. U.S. National Library of Medicine; Dec 2008. Web Aug 15, 2013.

[34] Shahmiri AF, Soares MJ, Zhao Y, Sherriff J. High-dose thiamine supplementation improves glucose tolerance in hyperglycemic individuals: a randomized, double-blind cross-over trial. National Center for Biotechnology Information. U.S. National Library of Medicine; May 29, 2013. Web Aug 15, 2013.
[35] Aziz MG, Michlmayr H, Kulbe KD, Del Hierro AM. Biotransformation of pineapple juice sugars into dietetic derivatives by using a cell free oxidoreductase from *Zymomonas mobilis* together with commercial invertase. National Center for Biotechnology Information. U.S. National Library of Medicine; Jan 5, 2011. Web Aug 15, 2013.
[36] Crowe KM, Murray E. Deconstructing a fruit serving: comparing the antioxidant density of select whole fruit and 100% fruit juices. National Center for Biotechnology Information. U.S. National Library of Medicine; Oct 2013. Web Nov 16, 2013.
[37] Wu G, Collins JK, Perkins-Veazie P, Siddig M, Dolan KD, Heaps CL, et al. Dietary supplementation with watermelon pomace juice enhances arginine availability and ameliorates the metabolic syndrome in Zucker diabetic fatty rats. National Center for Biotechnology Information. U.S. National Library of Medicine; Dec 2007. Web Aug 15, 2013.
[38] Christensen LP, Brandt K. Bioactive polyacetylenes in food plants of the Apiaceae family: occurrence, bioactivity and analysis. National Center for Biotechnology Information. U.S. National Library of Medicine; June 7, 2006. Web Aug 15, 2013.
[39] Al-Howiriny T, Alsheikh A, Alqasoumi S, Al-Yahya M, El Tahir K, Rafatullah S. Gastric antiulcer, antisecretory and cytoprotective properties of celery (*Apium graveolens*) in rats. National Center for Biotechnology Information. U.S. National Library of Medicine; July 2010. Web Aug 15, 2013.
[40] Hannum SM. Potential impact of strawberries on human health: a review of the science. Nutritional sciences. University of Illinois; 2004. Web July 15, 2013.
[41] Parelman MA, Storms DH, Dirschke CP, Huang L, Zunino SJ. Dietary strawberry powder reduces blood glucose concentrations in obese and lean C57BL/6 mice, and selectively lowers plasma C-reactive protein in lean mice. National Center for Biotechnology Information. U.S. National Library of Medicine; Nov 28, 2012. Web Aug 15, 2013.
[42] Edirisinghe I, Banaszewski K, Cappozzo J, Sandhya K, Ellis CL, Tadapaneni R, et al. Strawberry anthocyanin and its association with postprandial inflammation and insulin. National Center for Biotechnology Information. U.S. National Library of Medicine; Sep 2011. Web Aug 15, 2013.
[43] John OA, Yamaki S. Sugar content, compartmentation, and efflux in strawberry tissue. J Am Soc Hortic Sci Sep 1994. Web Nov 17, 2013.
[44] Khanal RC, Howard LR, Wilkes SE, Rogers TJ, Prior RL. J Med Food Sep 2012;15(9):802–10.
[45] Stull AJ, Cash KC, Johnson WD, Champagne CM, Cefalu WT. Bioactives in blueberries improve insulin sensitivity in obese, insulin-resistant men and women. American Society for Nutrition; 2010. Web July 15, 2013.
[46] Yvonne GE, Bjorck IME. A bilberry drink with fermented oatmeal decreases postprandial insulin demand in young healthy adults. Nutr J 21 May 2011. Web Aug 15, 2013.
[47] Liedo P, Carey JR, Ingram DK, Zou S. The interplay among dietary fat, sugar, protein and açai (*Euterpe oleracea* Mart.) pulp in modulating lifespan and reproduction in a tephritid fruit fly. National Center for Biotechnology Information. U.S. National Library of Medicine; July 2012. Web Aug 15, 2013.
[48] Udani JK, Singh BB, Singh VJ, Barrett ML. Effects of açai (*Euterpe oleracea* Mart.) berry preparation on metabolic parameters in a healthy overweight population: a pilot study. National Center for Biotechnology Information. U.S. National Library of Medicine; May 12, 2011. Web Aug 15, 2013.
[49] Neia S, Elba S. Characterization of the acai or manaca (*Euterpe oleracea* Mart.): a fruit of the Amazon. National Center for Biotechnology Information. U.S. National Library of Medicine; Mar 2007. Web Aug 15, 2013.
[50] Kim JY, Hong JH, Jung HK, Jeong YS, Cho KH. Grape skin and loquat leaf extracts and acai puree have potent anti-atherosclerotic and anti-diabetic activity in vitro and in vivo in hypercholesterolemic zebrafish. National Center for Biotechnology Information. U.S. National Library of Medicine; June 28, 2012. Web Nov 13, 2013.
[51] Gomez DI, Véras RM, Alves KS, Detmann E, Oliveria LR, Dos Santos RB, et al. Performance and digestibility of growing sheep fed with açai seed meal-based diets. National Center for Biotechnology Information. U.S. National Library of Medicine; May 11, 2012. Web Sep 8, 2013.
[52] Yung LM, Tian XY, Wong WT, Leung FP, Yung LH, Chen ZY, et al. Chronic cranberry juice consumption restores cholesterol profiles and improves endothelial function in ovariectomized rats. National Center for Biotechnology Information. U.S. National Library of Medicine; July 27, 2012. Web Aug 15, 2013.
[53] Grace MH, Massey AR, Mbeunkui F, Yousef GG, Lila MA. Comparison of health-relevant flavonoids in commonly consumed cranberry products. National Center for Biotechnology Information. U.S. National Library of Medicine; July 2, 2012. Web Aug 15, 2013.
[54] Fulgoni 3rd VL, Dreher M, Davenport AJ. Avocado consumption is associated with better diet quality and nutrient intake, and lower metabolic syndrome risk in US adults: results from the National Health and Nutrition Examination Survey (NHANES) 2001–2008. National Center for Biotechnology Information. U.S. National Library of Medicine; Jan 2, 2013. Web Aug 15, 2013.
[55] Duester KC. Avocado fruit is a rich source of beta-sitosterol. National Center for biotechnology information. U.S. National Library of Medicine; Apr 2001. Web Aug 15, 2013.
[56] Rodríguez-Fragoso L, Martiínez-Arismendi JL, Orozco-Bustos D, Reyes-Esparza J, Torres E, Burchiel SW. Potential risks resulting from fruit/vegetable-drug interactions: effects on drug-metabolizing enzymes and drug transporters. National Center for Biotechnology Information. U.S. National Library of Medicine; May 2011. Web Aug 15, 2013.

Further Reading

Wang TT, Edwards AJ, Clevidence BA. Strong and weak plasma response to dietary carotenoids identified by cluster analysis and linked to beta-carotene 15,15′-monooxygenase 1 single nucleotide polymorphisms. National Center for Biotechnology Information. U.S. National Library of Medicine; Aug 24, 2013. Web Sep 8, 2013.
Kawchak DA, Zhoa H, Scanlin TF, Tomezsko JL, Cnaan A, Stallings VA. Longitudinal, prospective analysis of dietary intake in children with cystic fibrosis. J Pediatr 1996;129:119–29.

CHAPTER

14

Fat-Soluble Vitamin Deficiency in Cystic Fibrosis

Vijay Karam Singh, Ronald Ross Watson

Health Sciences Center, School of Medicine, Mel and Enid Zuckerman College of Public Health, University of Arizona, Tucson, AZ, USA

14.1 INTRODUCTION

Cystic fibrosis (CF) is a hereditary disorder that can lead to major genetic dysfunction. Reduction of fats and fat-soluble vitamins (A, D, E, and K) occurs and can cause deficits of some of these vitamins [1]. Many deficiencies can occur with the reduction of these vitamins in patients with CF. The mutation rates also have been based on testing of environmental factors and genotypes [1].

CF is one of the most common genetic disorders within the Caucasian population [2]. One in every three North Americans, one in every 300 African Americans, and one in every 15 Asian Americans will have CF [3]. CF also presents itself in many different ways and varies from patient to patient. Even with the proper treatment and therapies, patients with CF frequently die within the first few years of life [3]. The basic longevity of individuals who carry this disorder is generally very low, ranging from between 37 and 50 years of age [3]. There is a correlation between patients with CF and deficiencies in fat-soluble vitamins.

14.2 CF GENETICS

CF is caused in patients because of how mutations present on chromosome 7 [4]. Mutations in the CF transmembrane conductance regulator (CFTR) protein, a chloride channel in the epithelial cell membrane, cause CF to occur [4]. CFTR is significant for the overall regulation of the body's sweat, digestive fluids, and mucus processes. This protein is also important in regulating the transport of sodium and chloride ions along the epithelial membrane. Many people without CF have two working copies of the *CFTR* gene; however, both of the copies do not have to be present for the CFTR protein to exist in the body [4]. This disorder's recessive nature requires replication to occur because of the absence of both functional copies of the CFTR protein. With a mutation present in the proteins, CF begins to develop throughout the body, which results in autosomal recessive inheritance [5]. Major symptoms that occur with this disorder include poor breathing, pressure on the face, poor growth, liver disease, and pancreatic issues [5]. The CFTR protein is used as an ion channel pump that regulates chloride ions in and out of the cell. It acts as a transporter, which mediates the chloride channels and is characterized under the adenosine triphosphate–binding cassette transporter family of membrane transport proteins [6].

CF is one of the most widespread autosomal recessive disorders among Caucasian populations. CF can be classified under six different categories depending on the function of the *CFTR* mutation [7]. The class I mutation causes the absence or partial absence of the protein. The class II mutation causes a lack of maturation and processing of the ion pumps. Class III mutations disrupt the ion channels of the CFTR pumps and cause them to become dysfunctional. Class IV mutations lead to altered channel conductance. Class V mutations affect the stability of messenger RNA processing and protein coding. Class VI alters the overall stability of the CFTR protein, leading to unpredictable ion pumping [8].

The phenotypes typically vary among patients with CF, depending on what class of the disorder they have [7]. In most cases, classes I, II, III, and IV are generally found to cause pancreatic insufficiency and premature mortality [8]. The phenotypic expression is dependent on the dominance of these classes with the genotype as well as the environmental conditions of the patients who carry it [7]. The *CFTR* mutation has a more profound

Diet and Exercise in Cystic Fibrosis
http://dx.doi.org/10.1016/B978-0-12-800051-9.00014-6

impact than the CF mutation alone since it is the one that causes the disorder to persist in the victim's body [7]. The regulation of the CFTR pump is dependent on the concentration of chloride ions. Attempts at examining gene therapy have been done by cloning of the *CFTR* gene [9]. The early studies and identification of the *CFTR* cloning gene were able to restore the chloride channels in the CFTR pumps [9]. The limitations toward this issue are mainly due to how the host and the vector vary among patients [10]. The lungs of innate hosts are a factor when analyzing the *CFTR* gene because of the secretion of the chloride channels throughout the pumps [10].

The lungs have very convoluted defense mechanisms that are able to protect the host from pathogens and infections. The defense mechanisms in the lungs have to be bypassed for the *CFTR* gene to be received and read [9]. When CF is processing, there is a mucus adherent to the surface of the epithelial tissue that causes inflammation at the site of the infection [11]. Inflammation can also occur with the presence or absence of the infection, depending on the class of CF the patient has [11]. This shows that gene transfer occurs throughout the patient's life span. The lungs are shown to have a strong adaptive immunity toward respiratory viruses and pathogens [12]. The signal response is intact after the signaling and replication of the *CFTR* gene, which causes an antibody response to viral gene transfer vectors [12].

14.3 SIGNS AND SYMPTOMS

The major signs and symptoms that show CF is present include salty tasting skin, poor growth, and poor weight gain, depending on the age at which the disorder develops [13]. Disregarding proper food intake, normal growth and weight still cause problems as people with the disorder age. Males have been noted to become infertile because of the absence of the reproductive organs. Most symptoms occur during younger ages, and the accumulation of mucus in children cause more action to be taken to avoid it [14]. As children begin to age progressively, they must exercise for the mucus in the body to go away, as well as an absence of ciliated epithelial tissue. Loss of the ciliated epithelial tissue can cause lack of maintenance and sinus pains for patients [14].

People with CF and pancreatic insufficiency are at risk of fat-soluble vitamin deficiency when the vitamins are absorbed with fat [15]. Abdominal pain is a very common symptom for patients with CF [16]. Abdominal pain can also decrease overall nutritional status because a patient does not get the necessary amount of food and nutrients. The most common reasons for how abdominal pain occurs in CF are constipation and distal intestinal obstruction syndrome [17]. Patients are at risk of distal intestinal obstruction if they had meconium ileus at birth. Prevention of abdominal pain for patients with CF can be controlled through constant therapy sessions to maintain the abdominals [16]. Distal intestinal obstruction syndrome is especially found in patients who have the severe genotype.

Distal intestinal obstruction syndrome is defined as an acute complete or incomplete obstruction, whereas constipation is defined as a progressive fecal impaction in the colon [17]. Overall, distal intestinal obstruction syndrome was found to be the combination of complete intestinal obstruction [16]. This was found by the signs of vomiting various nauseous materials along with fluid levels in the small intestine, as well as fecal matter in the ileocucum with abdominal pain and distention. Incomplete distal intestinal obstruction syndrome was described as the mix of overall days of abdominal pain and the distension of and fecal mass in the ileocucum [17].

14.4 NUTRITION OVERVIEW FOR CF

Nutrition influences disease status in patients who carry CF genes [18]. Advances in CF have increased the life span of patients with this very common disease [19]. CF is a disease known for its gastrointestinal and nutritional derangements. Damage done to the acinar pancreatic tissue, pancreatic ductular obstruction, and lack of enzymatic activity leads to malabsorption of fats [20]. A small section of patients with CF carrying a milder *CFTR* mutation have preserved pancreatic enzymatic activity, preventing malabsorption early in life. Nevertheless, these patients are still at risk of losing pancreatic function over time. Nutrition plays a major role in CF overall [21]. Patients with CF with a high nutritional intake have a higher survival advantage than those who do not. Many factors contribute to this impaired nutritional status: pancreatic insufficiency, chronic malabsorption, recurrent sinopulmonary infections, chronic inflammations, and increased energy expenditure [20]. Progressive lung disease thus increases calorie requirements by increasing breathing function of patients. Treatment programs that require a higher caloric intake and nutritional management for patients with CF have been reported to have better outcomes. Nutritional status is strongly correlated with pulmonary function along with survival in patients with CF [21]. Attainment of customary growth patterns during childhood and maintenance of nutritional status in adulthood, shows how the CF is represented depending on the health of the patient. Guidelines on energy intake, pancreatic enzyme replacement therapy, and fat-soluble vitamin supplementation are very important in practice for patients with CF [21].

Over the past 30years, major progress has been made in the understanding of CF. Advances in this research have been done through improved antibiotic regimens as well as lung clearance therapy for enhanced survival [22]. A key statistic also has been accredited by how there have been improvements in the median survival rate by 10–20 years [21]. Major achievements were obtained with adherence to a high-fat, high-energy diet rather than a low-fat diet, which was followed up with linear growth failure. A high-fat diet can better control malabsorption because of microspheric pancreatic enzyme replacement therapy [21] and attention to adequate fat-soluble supplementation, and newborn screening has ensured that at least 80–90% of children who have CF will attain a better health in the near future.

Poor clinical outcomes in CF are associated with malnutrition and low overall vitamin intake [23]. Growth and development must be achieved for patients with CF and can be done through nutritional counseling and daily intake. Prevention and early recognition of growth failure is the key to a successful nutritional intervention. Major advances in nutritional management have been one main factor of improved survival of patients with CF for the past several decades [23]. Malnutrition and lung disease are interrelated [24]. The overall care for patients with CF involves attention to both nutritional and lung function status to improve health outcomes [25]. An emphasis on good nutritional status and its association with improved lung function parameters offers patients with CF more adherence to nutritional therapy, enzymes, vitamins, and a balanced, high-energy diet [26]. The North American CF Foundation Consensus Committee on Nutrition recommends that patients with CF include appropriate evaluation of nutritional status at all ages and a diet that is age-appropriate, with ample energy to meet the needs for overall growth stages and weight gain [27]. A diet comprising 35–40% of calories from fat is recommended to meet energy demands for patients with CF [28].

Breast milk is still a recommended part of nutrition for infants in their first year. Breastfeeding is shown to be very protective for infants with CF. Breastfed compared with formula-fed infants carrying CF have better lung function and a reduced incidence of infections within the first 3 years of life [29]. Breast milk provides overall nutritional support for an infant with CF for the first 4–6 months of life. Supplemental energy may sometimes be required for infants to fortify the developing body, which is why sometimes breast milk is combined with formula. After the first 4–6 months, a child slowly begins to add solid foods, where more additional fats can be added for increased energy for babies [30]. For newborns, breast milk is always the main recommendation for nutrition because of its iron and zinc content.

14.5 FAT-SOLUBLE VITAMINS

Patients with CF are routinely given water-soluble and fat-soluble vitamin supplements [31]. The organization of fat-soluble vitamins is readily vindicated by the fact that such patients often have severe fat malabsorption and subsequent fat-soluble vitamin malabsorption. Dietary intake in children with CF is often well maintained because of an increased appetite, and absorption of water-soluble vitamins would be expected to be common so that severe deficiency of water-soluble vitamins seems unlikely [32]. The fat-soluble vitamins that are correlated with CF include A, E, D, and K [31].

Fat-soluble vitamin supplements are required by pancreatic-sufficient patients with CF [33] because of the constant mild to moderate fat malabsorption that occurs, notwithstanding oral enzyme therapy [34]. There is evidence that supplementation of fat-soluble vitamins in the pancreatic-sufficient patients may be associated with a decreased incidence of pulmonary exacerbations that are due to the antioxidant effects of the vitamin compounds [35]. Vitamin D supplements are recommended for patients who have pancreatic issues. The amount of vitamin K that is present in patients has not been analyzed and is still in need of research for those with CF since there is no adequate amount noted yet. Toxicity of fat-soluble vitamins is infrequent for patients who have CF [33].

A decrease in bone density can be present in patients with CF at a very young age [36]. This occurs especially with a reduction in bone mineral levels throughout the body. There are many factors that can affect bone health of both healthy individuals and individuals with CF. Bone mineral levels can be dependent on overall calcium levels, vitamin D and K intake, pulmonary infection, exercise, and the growth of the CFTR protein. Patients with pancreatic deficiency are more exposed to malabsorption of fat-soluble vitamins. Vitamin supplementation is very important for patients with CF for it can cause symptoms to become more prevalent over time [37].

14.6 VITAMIN A DEFICIENCY

Retinoid acid, the bioactive metabolite of vitamin A, is an effective signaling molecule in the brains of growing and developing animals, regulates numerous gene products, and modulates neurogenesis, neuronal survival, and synaptic plasticity [38]. Vitamin A is a fat-soluble micronutrient that is converted into retinoic acid at or near the site of activity in the body for use as a transcriptional regulator. Vitamin A deficiency is globally known as one of the most common forms of malnutrition in the human population, resulting in ocular disorders, immunosuppression, and impaired growth [39].

Ninety percent of patients with CF have pancreatic insufficiency, which is caused by vitamin A deficiency, and is a growing problem due to fat malabsorption [40]. Vitamin A is a fat-soluble vitamin that can help with normal vision, gene expression, epithelial integrity, growth, and immune function in those who take high concentrations of it [41]. Vitamin A comprises a family of pro-vitamin A carotenoids, which help increase immunity and strength in those that consumes it. Studies of vitamin A deficiency have noted that it causes night blindness, xerophthalmia, and immune dysfunction when correlated with pancreatic insufficiency [42]. Taking high concentrations of vitamin A with serum and other foods that carry fat-soluble vitamins will lower symptoms among patients with CF.

14.7 VITAMIN E DEFICIENCY

A variety of vitamin E deficiency syndromes are regularly produced in lower animals. However, humans, except the premature infant have not been shown to develop symptoms during reduced tocopherol intake [43]. The usual dietary intake of tocopherol in the United States is adequate in order to maintain the normal plasma levels within children, infants, and adults with unimpaired gastrointestinal function [44]. Biochemical evidence has shown that vitamin E deficiency has been discovered in CF patients with malabsorption of various origins [45] and in premature infants. Vitamin E is composed mainly of eight different fat-soluble nutrients. The amount of vitamin E that is found in western diets varies depending on region and availability [46].

Studies of human vitamin E deficiency are limited to two populations within developed countries: premature infants and patients with intestinal malabsorption [44]. Patients with CF are particularly suitable for evaluation of the quantity and overall properties of vitamin E deficiency in a human system. Symptoms emerge more often with patients who do not have any sort of dairy intake. With no dairy intake, loss of nutrition and vitamin E levels are more prevalent [44]. Malabsorption also becomes an issue for patients with CF in this state. Studies noted that increased intake of vitamin E was able to prevent the symptoms of CF to occur in patients. These studies further showed that increased uptake of vitamin E was dependent on the patient and how well they are able to digest the supplements [43]. It is also dependent on how the patient digests the dietary nutrients found in the triglycerides of vitamin E [47].

There were many key findings and observations given for patients with CF with vitamin E deficiency. Patients with CF have problems with pancreatic insufficiency when there is a lack of vitamin E present. One of the main constituents that cause this issue is a lack of the proper amounts of plasma α-tocopherol [44]. Plasma α-tocopherol is a mixture of nutrients that are typically found in egg yolks, germ oils, and leafy vegetables and constitute the compound of vitamin E. Plasma α-tocopherol concentrations are an integral part of the diet of a patients with CF since it establishes part of the dietary intake for those with the disorder [43]. The duration of the symptoms of vitamin E deficiency is dependent on the concentration of α-tocopherol and other vitamin E supplements that were given to patients with CF [44]. The correlation of the duration of no vitamin supplement as well as the concentration of the supplement is a factor.

14.8 VITAMIN K DEFICIENCY

On rare occasions, CF is able to discern itself as a clotting disorder. Vitamin K is generally absorbed through breast milk, baby formula, and eventually various solid foods [48]. This consumption is halted in young patients with CF. Newborn babies are very susceptible to vitamin K malabsorption because of how little vitamin K is delivered via the placenta, which gives children very limited dietary resources for absorbing vitamin K while aging [48].

Patients with CF are at risk of having coagulation abnormalities when vitamin K concentrations become reduced [49]. There are four reasons why the deficiencies to occur. Fat malabsorption due to pancreatic insufficiency can cause vitamin K concentrations to be lowered in patients with CF. Liver disease can also cause vitamin K deficiency with bile-salt deficiency [50]. Long-term use of antibiotics can also cause vitamin K deficiency, as well as resection of the small gastrointestinal bowel. All four of these reasons lead to malabsorption in patients with CF. Vitamin K deficiency in patients with CF leads to multiple symptoms and issues that can eventually lead to death [51].

14.9 VITAMIN D DEFICIENCY

Vitamin D is involved with the regulation of calcium and phosphate groups in the body [52]. Poor uptake or deficiency of vitamin D in the body due to malabsorption can lead to osteoporosis. Also, people who have CF often develop clubbing of their fingers and toes due to this chronic illness as well as low oxygen in their tissues [52].

The anti-inflammatory treatments that are currently given have been shown to cause adverse effects in deficiency [53]. There are many current studies that indicate the antimicrobial and anti-inflammatory benefits of vitamin D [52]. Since patients with CF are shown to have deficiencies of vitamin D, it has been noted that having an intake of vitamin D can help increase therapeutic strategies for patients. The serum of vitamin D that was tested also was noted to have had a positive correlation with lung function and a negative correlation with airway inflammation for patients with CF [53]. The purpose of vitamin D in the body is directly correlated with metabolism. The decay of the metabolism in patients with CF is also a developmental symptom that occurs when there is less fat-soluble vitamin intake [53]. High doses of vitamin D supplementation allow for more proper bone function and stabilize the skeletal system of patients with CF.

Osteoporosis as well as bone fractures are found to be more and more evident in adults and children with CF [54]. Vitamin D deficiency affects bone mineral density in patients as well as the severity of severe pulmonary disease. Bone size is drastically reduced in patients with CF with the deficiency. The bone density of patients with CF was reduced in the spinal region and was attributed to factors such as age, weight, and body mass index. The increased life span of patients is attributed to reduced bone mass index because the longevity of CF causes bones to weaken [55]. It was summarized that vitamin D deficiency with pulmonary disease causes more severe osteoporosis and bone fractures. Vitamin D therapy has now become more apparent for patients because more concentrations of nutritional supplements can be given to them to alleviate bone density [54].

14.10 TREATMENT AND THERAPY

DNA sequencing is a fast and growing field that is becoming more and more apparent with the needs of personalized medicine. There are many genomic approaches being used to achieve more personalized analysis and treatments for CF [56]. Lungs that contain CF have an extreme internal environment that contains many bacteria, fungi, and viruses that can be present throughout space and time. Depending on the health status of the patients being tested, the metabolic potential of these species vary [56].

Therapy for CF is mainly focused on limiting the number of microbial species that are present within the host's immune system and lungs [57]. The immune response of the host is dictated through the use of airway clearance techniques, broad-spectrum antibiotics, and treatments that break down the endobronchial biofilm. The variety of antibiotics are used mainly for killing individual microbial strains [57]. Specific antibiotics are needed for these individual strains. Therapies that are currently used are dependent on the overall ecological and evolutionary pressures within the environment of CF [57]. Development of newer strategies are needed for patients that have CF since it can help minimize the number of microbial communities that are present within the immune system.

Liver disease is a symptom that can occur in those who carry CF. At this point of research, the main way that liver disease can be alleviated is by increasing bile flow in the body with ursodeoxycholic acid (UDCA) [58]. UDCA is also able to allow bicarbonate secretion throughout the biliary tract [59]. The early symptoms that are found for those that have CF are severe problems with protein and fat malabsorption with pancreatic insufficiency. Exogenous pancreatic enzymes have been a new innovation used for improving the nutritional intake of patients with CF [60]. This technique is done by taking out the amylase, lipase, and protease that the CF pancreas is not able to produce normally.

14.11 CF AND VITAMIN SUPPLEMENTATION

CF and fat-soluble vitamins directly correlate with each other [23]. Vitamins are needed for proper growth and function as patients with CF age. Patients with CF need extra fat-soluble vitamins to ensure proper function and nutritional intake. Poor clinical outcomes in CF are found in patients having malabsorption [23]. Patients with CF need attention given toward lung function to have the best outcomes available. Patients with CF emphasize a good nutritional status to increase nutrition for certain enzymes, vitamins, and other nutritional supplements [23]. Proper intake of vitamins and additional supplements allow better maintenance of the disorder among patients, and fewer symptoms can occur. The role of these vitamins with CF is in accordance with the symptoms that occur in each patient when the specific vitamin is deficient [23]. The fat-soluble vitamins being studied in these cases each have their own respective correlation with CF.

Vitamin A was used as a neurological signaling molecule to develop brain growth and maintenance [38]. It is also used as a gene regulator, which also helped with the brain's neurological development. With CF, pancreatic insufficiency becomes a problem with reduced vitamin A consumption. With pancreatic insufficiency [39], patients with CF will have problems with digesting the appropriate nutrients to have proper growth and development. Vitamin E also is a similar factor among patients with CF [43]. Nutrient intake will be lowered as a part of decreasing vitamin E intake in patients with CF, and less energy output will occur. The triglyceride levels that are given with vitamin E and tocopherol are reduced and fewer nutrients are given.

Vitamin D is used for calcium regulation and is mainly focused on infants fed with breast milk. With CF, patients can have thinner bone mass, which can cause osteoporosis to occur [52]. Low oxygen levels and tissue mass can also occur with lack of vitamin D in patients with CF. Like vitamin D, infants are more susceptible to vitamin K deficiencies because of breast milk, but they can also cause poor malabsorption in patients with chronic illnesses as well [48].

The main goal of patients with CF is gaining proper function and growth. This can be done is through surveillance of dieting, expert advice, and suitable nutritional intake [23].

References

[1] Jagannath VA, Fedorowicz Z, Thaker V, Chang AB. Vitamin K supplementation for cystic fibrosis. Cochrane Collab 2013:1.

[2] Hammond KB, Abman SH, Sokol RJ, Accurso FJ. Efficacy of statewide neonatal screening for cystic fibrosis by assay of trypsinogen concentrations. N Engl J Med 1991;(329):769–74.

[3] Boat T. Genetic testing for cystic fibrosis. Natl Hum Genome Res 1997;(1):21.

[4] Rowe SM, Miller S, Sorscher EJ. Cystic fibrosis. N Engl J Med 1992;(19):1.

[5] Parisi GF, Dio GD, Leonardi S. Liver disease in cystic fibrosis: an update. Hepat Mon 2013;(8):1–6.

[6] Linsdell P. Functional architecture of the CFTR chloride channel. Informa Healthcare; 2013. p. 1–19.

[7] Parisi GF, Dio GD, Franzonello C, Gennaro A, Rotolo N, Lionetti E, et al. Liver disease in cystic fibrosis: an update. Hepat Mon 2013;(8):1–6.

[8] Haardt M, Benharouga M, Lechardeur D, Kartner N, Lukacs GL. C-terminal truncations destabilize the cystic fibrosis transmembrane conductance regulator without impairing its biogenesis. A novel class of mutation. J Biol Chem 1999;(31):1.

[9] Kreindler J. Cystic fibrosis: exploiting its genetic basis in the hunt for new therapies. Pharmacol Ther 2010;(2):1–5.

[10] Griesenbach U, Alton EW. Gene transfer to the lung: lessons learned from more than 2 decades of CF gene therapy. Adv Drug Delivery Rev 2009;61(2):128–39.

[11] Bartlett JA, Fischer AJ, McCray Jr PB. Innate immune functions of the airway epithelium, vol. 15. Karger: Medical and Scientific Publishers; 2008. p. 147–163.

[12] Kohlmeier JE, Woodland DL. Immunity to respiratory viruses. Annu Rev Immunol 2009;27(1):61.

[13] Mishra A, Greaves R, Massie J. The relevance of sweat testing for the diagnosis of cystic fibrosis in the genomic era. Clin Biochem 2005;26(4):135–53.

[14] Mitchell RS, Kumar V, Robbins SL, Abbas Abul K, Fausto N. Robbins basic pathology. Elsevier; 2007. p. 1–3.

[15] Bonifant CM, Shevill E, Chang AB. Vitamin a supplementation for cystic fibrosis. Cochrane Cyst Fibros Genet Disord Group 2012:1.

[16] Kalnins D, Wilschanski M. Maintenance of nutritional status in patients with cystic fibrosis: new and emerging therapies. Drug Des Devel Ther 2012;6:151–61.

[17] Colombo C, Ellemunter H, Houwen R, Munck A, Taylor C, Wilschanski M. Guidelines for the diagnosis and management of distal intestinal obstruction syndrome in cystic fibrosis patients, vol. 10. Elsevier; 2011(2): 24–28.

[18] Gordon CM, Anderson EJ, Herlyn K, Hubbard JL, Pizzo A, Gelbard R, et al. Nutrient status of adults with cystic fibrosis. J Am Diet Assoc 2007;107(12):2114–9.

[19] Ramsey BW, Farrell PM, Pencharz P. Nutritional assessment and management in cystic fibrosis: a consensus report. Am J Clin Nutr 1992;55(1):108–16.

[20] Matel JL, Milla CE. Nutrition in cystic fibrosis. Thieme E-J 2009;30(5):579–86.

[21] Munck A. Nutritional considerations in patients with cystic fibrosis. NCBI 2010;4(1):47.

[22] Gaskin KJ. Nutritional care in children with cystic fibrosis: are our patients becoming better, vol. 67. NPG: Nature Publishing Group; 2013(5); p. 558.

[23] Kalnins D, Wilschanski M. Maintenance of nutritional status in patients with cystic fibrosis: new and emerging therapies. Dove Press; 2012(1); p. 151–61.

[24] Corey M, McLaughlin FJ, Williams M, Levison H. A comparison of survival, growth, and pulmonary function in patients with cystic fibrosis in Boston and Toronto, vol. 41. Elsevier; 2012(6); p. 583–91.

[25] Kerem E, Reisman J, Corey M, Canny GJ, Levison H. Prediction of mortality in patients with cystic fibrosis. N Engl J Med 1992;326(18):1187–91.

[26] Lai HJ. Classification of nutritional status in cystic fibrosis, vol. 12. Wolters Kluwer; 2006(6); p. 422–7.

[27] Borowitz D, Baker RD, Stallings V. Classification of nutritional status in cystic fibrosis, vol. 12. Wolters Kluwer; 2006(6); p. 422–7.

[28] Sinaasappel M, Stern M, Littlewood J. Nutrition in patients with cystic fibrosis: a European consensus, vol. 1. Elsevier; 2012(2); p. 51–75.

[29] Colombo C, Costantini D, Zazzeron L. Benefits of breastfeeding in cystic fibrosis: a single-centre follow-up survey, vol. 96. Wiley; 2007(8); p. 1228–32.

[30] Borowitz D, Robinson KA, Rosenfeld M. Cystic fibrosis foundation evidence-based guidelines for management of infants with cystic fibrosis, vol. 155. Elsevier; 2009(6); p. 73–93.

[31] Congden PJ, Bruce G, Rothburn MM, Clarke PCN, Littlewood JM, Kelleher J, et al. Vitamin status in treated patients with cystic fibrosis. Arch Dis Child 1981;56(1):708–14.

[32] Kopel FB. Gastrointestinal manifestations of cystic fibrosis. Gastroenterology 1972;62(1):483–91.

[33] Kalnins D, Wilschanski M. Maintenance of nutritional status in patients with cystic fibrosis: new and emerging therapies. Dove Press: Drug Des Dev Ther 2012;6(1):151–61.

[34] Feranchak AP, Sontag MK, Wagener JS, Hammond KB, Accurso FJ, Sokol RJ. Prospective, long-term study of fat-soluble vitamin status in children with cystic fibrosis identified by newborn screen. J Pediatr 1993;135(5):601–10.

[35] Hakim F, Kerem E, Rivlin J, Bentur L, Stankiewicz H, Bdolach-Abram T, et al. Vitamins A and E and pulmonary exacerbations in patients with cystic fibrosis. J Pediatr Gastroenterol Nutr 2007;45(3):347–53.

[36] Bianchi ML, Romano G, Saraifoger S, Costantini D, Limonta C, Colombo C. BMD and body composition in children and young patients affected by cystic fibrosis. J Bone Mine Res 2006;21(3):388–96.

[37] Ramsey BW, Farrell PM, Pencharz P. Nutritional assessment and management in cystic fibrosis: a consensus report: the consensus committee. Am J Clin Nutr 1992;55(1):108–16.

[38] Olson CR, Mello CV. Significance of vitamin A to brain function, behavior and learning. Mol Nutr Food Res 2010;54(4):489–95.

[39] Underwood BA, Arthur P. The contribution of vitamin A to public health. FASEB J 1996;10(1):1040–8.

[40] Graham-Maar RC, Schall JI, Stettler N, Zemel BS, Stallings VA. Elevated vitamin A intake and serum retinol in preadolescent children with cystic fibrosis. Am J Clin Nutr 2006;84(1):174–82.

[41] Food and Nutrition Board, Institute of Medicine. Dietary reference intakes for vitamin A, vitamin K, arsenic, boron, chromium, copper, iodine, iron, manganese, molybdenum, nickel, silicon, vanadium, and zinc, vol. 1. Washington (DC): National Academy Press; 2001 (1); p. 82–161.

[42] Congden PJ, Bruce G, Rothburn MM, Clarke PC, Littlewood JM, Kelleher J, et al. Vitamin status in treated patients with cystic fibrosis. Arch Dis Child 1981;56(9):708–14.

[43] Farrell PM, Biere JG, Fratantoni JF. The occurrence and effects of human vitamin E deficiency. J Clin Investig 1977;60(1):233–41.

[44] Dimitrov NA, Meyer-Leece C, McMillan J, Gilliland D, Perloff M, Malone W. Plasma α-tocopherol concentrations after supplementation with water- and fat-soluble vitamin E. N Engl J Med 1983;308(1):1.
[45] Binder HJ, Hertin DC, Vurst V, Finch SC, Spiro HM. Tocopherol deficiency in man. N Engl J Med 1965;273(1):1289–97.
[46] Bieri JO, Corash L, Hubbard VS. Medical uses of vitamin E. N Engl J Med 1983;308(1):1063–71.
[47] Argao EA, Heubi Je, Hollis BW, Tsang RC. D-Alpha-tocopheryl polyethylene glycol-1000 succinate enhances the absorption of vitamin D in chronic cholestatic liver disease in infancy and childhood. Pediatr Res 1992;31(1):146–50.
[48] Reaves J, Wallace G. Unexplained bruising: weighing the pros and cons of possible causes. Consult Pediatr 2010;9(1):201–2.
[49] Durie PR. Vitamin K and the management of patients with cystic fibrosis. Can Med Assoc J 1994;151(7):933–6.
[50] Lane PA, Hathaway WE. Vitamin K in infancy. J Pediatr 1985;106(1):351–9.
[51] Corrigan J, Taussig LM, Beckerman R. Factor II (prothrombin) coagulant activity and immunoreactive protein: detection of vitamin K deficiency and liver disease in patients with cystic fibrosis. J Pediatr 1981;99(1):254–6.
[52] Haworth C, Selby P, Webb A, Dodd M, Musson H, Niven R, et al. Low bone mineral density in adults with cystic fibrosis. Thorax 1999;54(11):961–7.
[53] Herscovitch K, Dauletbaev N, Lands LC. Vitamin D as an antimicrobial and anti-inflammatory therapy for cystic fibrosis. Paediatr Respir Rev 2013;1(13):1.
[54] Donovan DS, Papadopoulos A, Staron RB, Addesso V, Schulman V, McGregor C, et al. Bone mass and vitamin D deficiency in adults with advanced cystic fibrosis lung disease. Am J Respir Crit Care Med 1982;1(1):1892–9.
[55] Corey M, McLaughlin M, Williams M, Levison H. A comparison of survival, growth, and pulmonary function in patients with cystic fibrosis in Boston and Toronto. J Clin Epidemiol 1988;41(1):583–91.
[56] Lim YW, Evangelista 3rd JS, Schmieder R, Bailey B, Haynes M, Furlan M, et al. Clinical insights from metagenomic analysis of cystic fibrosis sputum. J Clin Microbiol 2013:1.
[57] Conrad D, Haynes M, Salamon P, Rainey PB, Youle M, Rohwer F. Cystic fibrosis therapy: a community ecology perspective. Am J Respir Cell Mol Biol 2013;48(2):150–6.
[58] Parisi GF, Dio GD, Franzonello C, Gennaro A, Rotolo N, Lionetti E, et al. Liver disease in cystic fibrosis: an update. Hepat Mon 2013;13(8):1.
[59] Sokol RJ, Durie PR. Recommendations for management of liver and biliary tract disease in cystic fibrosis. Cystic Fibrosis Foundation Hepatobiliary Disease Consensus Group. J Pediatr Gastroenterol Nutr 1998;28(1):1–13.
[60] Kreindler JL. Cystic fibrosis: exploiting its genetic basis in the hunt for new therapies. Pharmacol Ther 2010;125(2):219–29.

CHAPTER

15

Can Light Provide a Vitamin D Supplement in Cystic Fibrosis?

Dimitri Declercq, Eddy Robberecht

Cystic Fibrosis Centre, Department of Pediatrics, Ghent University Hospital, Belgium

15.1 INTRODUCTION

The markedly growing number of publications on bone health in cystic fibrosis (CF) illustrates the increasing attention paid to the subject. This is the result of several developments. As the proportion of adults with CF has increased over the past 30 years, so has the occurrence of low-impact bone fractures, mainly of the ribs and vertebrae. In parallel, refined techniques have been developed to demonstrate the high frequency of low bone mineral density (BMD). Based on a meta-analysis, the pooled prevalence of osteoporosis in adults with CF is 23.5% (95% confidence interval, 16.6–31.0) [1].

It became generally accepted that disorders of bone health are the sum of many contributing factors, such as CF transmembrane regulator dysfunction, reduced weight-bearing activity or lack of physical activity, elevated cytokine production caused by chronic inflammation, severity of lung disease, delayed pubertal maturation, corticosteroid therapy, essential fatty acid deficiency, CF-related diabetes, immunosuppressive therapy after lung transplantation, negative calcium balance, malabsorption due to exocrine pancreatic insufficiency with poor nutritional status, and decreased serum concentrations of vitamins K and D [2,3]. Unfortunately, vitamin D deficiency is often primarily blamed because most patients with CF suffer from fat malabsorption, and vitamin D is fat soluble [2,4]. As a natural consequence, all nutritional guidelines insist on preventing this deficiency by means of a daily oral supplement [2,5,6].

15.2 VITAMIN D DEFICIENCY IN CF

Studies from non-CF populations show that vitamin D deficiency in early childhood may present as symptomatic hypocalcemia with tetany, seizures, or myopathy or as delayed closure of the anterior fontanelle. In children, vitamin D deficiency can also result in a range of bone deformations such as rickets [7]. In adults, vitamin D deficiency can cause secondary hyperparathyroidism, which may result in osteomalacia, whereby fractures can easily occur and heal poorly [8]. Besides the skeletal effects of vitamin D, studies have suggested a relation between serum 25-hydroxy (25(OH)) vitamin D and the cardiovascular system, the immune system, and glucose metabolism [9–14].

Vitamin D deficiency is a common and well-described phenomenon in the pediatric and adult populations with CF [1,15–18]. In contrast with the non-CF population, the presentation of vitamin D deficiency in the CF population seems more subtle. Several studies of CF populations have shown high levels of parathyroid hormone (PTH) and defects in bone mineralization and bone formation [19,20]. Vitamin D deficiency has mainly been studied in relation to CF-related bone disease, but the complete impact of vitamin D deficiency on extraskeletal effects is yet to be determined [16].

Low serum 25(OH) vitamin D concentrations in CF seem multifactorial and inherent to the pathophysiology. Serum 25(OH) vitamin D is the best biochemical marker of vitamin D status, representing not only the photoproduced and oral intake of vitamin D but also a notion of the vitamin D stored and metabolized throughout the body [21].

Because of exocrine pancreatic insufficiency, which is present in 85–90% of the CF population, the absorption of vitamin D and other fat-soluble vitamins can be impaired. In addition, suboptimal intake of foods containing vitamin D and oral vitamin D supplements can facilitate vitamin D deficiency. Despite the attention that is paid to routine supplementation and pancreatic

Diet and Exercise in Cystic Fibrosis
http://dx.doi.org/10.1016/B978-0-12-800051-9.00015-8

enzyme replacement therapy, there is only some limited improvement in serum 25(OH) vitamin D concentrations, but the thresholds are not met by all patients [22]. In addition to malabsorption, indirect data suggest that vitamin D metabolism in CF is impaired, with accelerated excretion of vitamin D that possibly occurs through enterohepatic dumping [4]. Most 25(OH) vitamin D is carried by vitamin D binding protein (DBP). Patients with CF tend to have decreased concentrations of DBP, which can influence vitamin D concentrations and metabolism [23,24]. Besides several food sources and supplementation, 25(OH) vitamin D_3 can be synthesized by the skin from exposure to sunlight. In the healthy population, about 90% of the vitamin D requirement is delivered by sunlight. The problem of decreased levels of vitamin D can be exacerbated by the fact that patients with CF can have a low percentage of body fat and thus have fewer possibilities for storing vitamin D [16]. Regardless of the etiology of vitamin D deficiency in CF, the most important (and most studied) intervention is routine oral supplementation of vitamin D. Because of the increased attention given to this deficiency, data from studies in large CF centers suggest a trend toward higher serum 25(OH) vitamin D concentrations. However, oral supplementation seems insufficient, which is shown by the persistent existence of low serum 25(OH) vitamin D concentrations [16].

In the past 10 years, preventing vitamin D deficiency by means of a daily oral supplement became the subject of dissension between the American and European CF communities, causing confusing controversies over optimal vitamin D concentrations and supplementation regiments. It is mainly the lower threshold that forms the matter of debate. Based on available pediatric data, the European CF Bone Health Working Group [20] recommends 20 ng/mL (50 nmol/L). In contrast, the North American CF Bone Health Consensus pays more attention to extraskeletal health benefits such as enhanced immune function, and protective properties against the development of diabetes, cancer, and cardiovascular disease [25] that are attributed to vitamin D serum concentrations above 30 ng/mL (75 nmol/L), which they therefore adopt as the minimum concentration to be achieved [26]. At this moment, however, no evidence is available to prove that this level is beneficial for BMD, biochemical markers of bone metabolism, or prevention of fractures in people with CF.

Against this background it is easy to understand that serum concentrations of vitamin D in people with CF is not unequivocal; they are called deficient by some authors [17,22,27], whereas others disagree [18,28]. The level of 30 ng/mL (75 nmol/L) is not even reached by the majority of the healthy population. No consensus has yet been reached regarding the serum concentration of 25(OH) vitamin D that is required to optimize bone mineralization in children, adolescents, and young adults with or without CF. As mentioned earlier, it is clear that most people with CF require vitamin D supplements and have lower 25(OH) vitamin D concentrations and higher PTH levels than healthy controls. Cross-sectional studies, however, show no clear causal association between BMD and the vitamin D status of patients with CF [29]. In addition, interventional studies performed to date show no effect on bone mineralization and bone turnover markers [30,31]. Therefore, current data show no evidence to demonstrate a beneficial skeletal effect of levels above 30 ng/mL (75 nmol/L) in people with CF. The 30 ng/mL (75 nmol/L) cut-off proposed by Holick and colleagues is designed to keep PTH levels as low as possible because elevated PTH levels may increase bone resorption, with demineralization resulting in adults and elderly [32]. However, this may not be the case in all situations, as has been demonstrated in patients with chronic renal insufficiency in whom PTH levels that are too low block bone formation and induce so-called adynamic bone disease. PTH levels that are too low may be detrimental in children and young adults; it has been suggested that PTH levels in the high normal range may favor bone formation [19]. In addition, the long-term consequences of maintaining 25(OH) vitamin D concentrations above 30 ng/mL (75 nmol/L) have not been explored in depth in large groups of children or young adults [20].

15.3 PREVENTION AND TREATMENT OF VITAMIN D DEFICIENCY IN CF

As stated earlier, the threshold for vitamin D deficiency is still under debate; this threshold has a direct influence on the recommended oral supplementation to correct serum 25(OH) vitamin D concentrations. The absence of good-quality clinically applicable intervention studies makes it difficult to formulate effective recommendations to correct vitamin D deficiencies. Therefore, it is advised that the recommendations proposed by relevant CF guidelines be followed when treating vitamin D deficiency [33].

There are two forms of vitamin D available for supplementation, namely ergocalciferol (vitamin D_2) and cholecalciferol (vitamin D_3). In the general population there is an existing debate about the form that is most effective to correct serum 25(OH) vitamin D concentrations. Based on data available in children and adults with CF, vitamin D_2 seems to be less effective in correcting the serum values, possibly because of less bioavailability after oral ingestion. This was directly shown in a randomized controlled trial of 28 adults who were given weekly oral supplements of 50,000 IU vitamin D_3 and vitamin D_2 over a period of 12 weeks. Only 60% of the vitamin D_2 group achieved a level of 30ng/mL (75nmol/L), in contrast to the vitamin D_3 group, in which all patients reached that threshold [34]. Based on the current evidence, the CF Foundation and the European CF Bone Health Working Group prefer supplementation with vitamin D_3 over vitamin D_2 [20,26]. According to the CF Foundation recommendations (consensus recommendation), whether a patient takes the supplement once daily or takes the equivalent once a week can be based on the medical burden of the CF therapy, financial issues, and the preference of the patient. However, high weekly doses that are inadvertently continued or a lack of adherence can make hypervitaminosis D a possible issue, are negative evidence for using a weekly regimen [26].

European guidelines [20] advise a starting dose of 1000–2000IU vitamin D_2 or D_3/day in infants. In children >1year old and adults, the starting dose of vitamin D_2 or D_3 is 1000–5000IU/day. It is advised that the doses should be adjusted to maintain a 25(OH) vitamin D concentration above the deficiency threshold of 20ng/mL (50nmol/L). There is a preference for the use of vitamin D_3 over vitamin D_2.

The American guidelines [26] use other supplementation recommendations to maintain the threshold of 30ng/mL (75nmol/L). All recommendations are consensus recommendations. In the management of vitamin D status in all infants with CF (birth to 1year old), an initial dose of 400–500IU vitamin D_3 per day is recommended. Depending on serum 25(OH) vitamin D, the dose should be adjusted. If the serum 25(OH) vitamin D concentration is at least 20ng/mL (≥50nmol/L) but <30ng/mL (<75nmol/L) with adherence to the prescribed regimen, it is recommended that the dose be increased to 800–1000IU/day. Infants with a serum 25(OH) vitamin D concentration <20ng/mL (<50nmol/L) or a persistent serum level of at least 20ng/mL (≥50nmol/L) but <30ng/mL (<75nmol/L) may have the dose increased to a maximum of 2000IU/day. If patients are still unable to reach the target after treatment with 2000IU vitamin D_3/day, it is recommended that the patient be referred to a specialist with expertise in vitamin D therapy. Patients with a serum 25(OH) vitamin concentration <10ng/mL (<25nmol/L) should also be referred to a specialist with expertise in vitamin D therapy.

Starting with a dose of 800–1000IU vitamin D_3/day is recommended in all children with CF who are >1–10years old. If the patient reaches a serum 25(OH) vitamin D concentration of at least 20ng/mL (≥50nmol/L) but <30ng/ml (<75nmol/L) with adherence to the prescribed regimen, it is recommended that the dose be increased to 600–3000IU/day. If the serum 25(OH) vitamin D concentration is <20ng/mL (<50nmol/L) or is persistent at a concentration of at least 20ng/mL (≥50nmol/L) but <30ng/mL (<75nmol/L) with adherence to the prescribed regimen, then it is recommended that the dose of vitamin D_3 be increased to a maximum of 4000IU/day. If the desired threshold of 30ng/mL (75nmol/L) with a daily dose of 4000IU/day cannot be reached, referring the patient to a specialist with expertise in vitamin D therapy is advised.

For children older than the age of 10years and adults, the CF Foundation recommends that all individuals with CF be treated with an initial dose of 800–2000IU vitamin D_3/day. In patients with a serum 25(OH) vitamin D concentration of at least 20ng/mL (≥50nmol/L) but <30ng/mL (<75nmol/L) with adherence to the prescribed regimen, it is recommended that the dose of vitamin D_3 be increased to 1600–6000IU/day. If the patient has a serum 25(OH) vitamin D concentration <20ng/mL or a persistent concentration of at least 20ng/mL (≥50nmol/L) but <30ng/mL (<75nmol/L) with adherence to the prescribed regimen, it is recommended that the dose of vitamin D_3 be increased to a maximum of 10,000IU/day. If the latter treatment (with adherence to the prescribed regimen) fails to reach a serum 25(OH) vitamin D concentrations of at least 30ng/mL (≥75nmol/L), the patient should be referred to a specialist with expertise in vitamin D therapy. Table 15.1 summarizes the recommendations. Based on these guidelines, it can be suggested that a serum 25(OH) vitamin D concentration of >30ng/mL (75nmol/L) is optimal and >20ng/mL (50nmol/L) be considered as the lower threshold. Based on the amounts of international units that are necessary to achieve a threshold of at least 20ng/mL (≥50nmol/L) or 30ng/mL (≥75nmol/L), it is obvious that supplements are needed in addition to a regular intake of food sources containing vitamin D. Foods (per 100g) that are rich in vitamin D_3 are fatty fish such as eel (3348IU), herring (540IU), salmon (628IU), and trout (488IU), egg yolks (1180IU); and fortified whole milk (100IU). Meats contain between 4 and 60IU/100g, with the exception of, for instance, liver and kidney (60–100IU/100g). Cheese contains between 8 and 40IU/100g. In several countries, margarines are fortified with vitamin D_3 [35]. These data show that it is not possible for people with CF to achieve sufficient intake of vitamin D via foods.

The CF Foundation recommends that 25(OH) vitamin D concentrations be checked annually, preferably at the end of the winter [26]. This recommendation is based on the fact that serum 25(OH) vitamin D concentrations are season-dependent [36–38]. The serum 25(OH) vitamin D is expected to be at its minimum at the end of the winter. If the threshold is reached at that point, then it can be expected that, under the current regimen, a desired serum level is achieved at all times during the year. When vitamin D therapy is adjusted to the serum 25(OH) vitamin D concentrations, it is advised that vitamin D status be reassessed after 3months because of the 2- to 3-week half-life of serum 25(OH) vitamin D [26].

TABLE 15.1 Summary of the European and Cystic Fibrosis Foundation (CFF) Recommendations for Supplementing Vitamin D Based on the Threshold and Serum Values

	European Guidelines	
Age Group	**Serum 25(OH) Vitamin D**	**Supplementation of Vitamin D_3**
	Target >20 ng/mL	
Infants (birth to 1 year)		Start dose = 1000–2000 IU/day
>1 year old		Start dose = 1000–5000 IU/day
CFF GUIDELINES		
	Target ≥30 ng/mL	
Infants (birth to 1 year)		Start dose = 400–500 IU/day
	20–30 ng/mL	Increase the dose to 800–1000 IU/day
	<20 ng/mL or persistent ≥20 ng/mL <30 ng/mL	Increase the dose to a maximum of 2000 IU/day
	<10 ng/mL	Refer to a specialist/endocrinologist
Children (1–10 years)		Start dose = 800–1000 IU/day
	20–30 ng/mL	Increase the dose to 1600–3000 IU/day
	<20 ng/mL or persistent ≥20 ng/mL <30 ng/mL	Increase the dose to a maximum of 4000 IU/day
	<30 ng/mL after treatment with 4000 IU vitamin D_3/day	Refer to a specialist/endocrinologist
Older children and adults		Start dose = 800–2000 IU/day
	20–30 ng/mL	Increase the dose to 1600–6000 IU/day
	<20 ng/mL or persistent ≥20 ng/mL <30 ng/mL	Increase the dose to a maximum of 10000 IU/day
	<30 ng/mL after treatment with 10000 IU vitamin D_3/day	Refer to a specialist/endocrinologist

15.4 ULTRAVIOLET B RADIATION IN THE MANAGEMENT OF VITAMIN D DEFICIENCY IN CF

It is a dubious line of reasoning to hold fat malabsorption solely responsible for vitamin D insufficiency since in healthy people only a minor part (±15%) of serum 25(OH) vitamin D is of nutritional origin [39]. In moderately sunny geographical regions, the majority (up to 90%) of the serum 25(OH) vitamin D concentration is determined by the dermal supply [16]. In less sunny regions, dermal vitamin D production is expected to be high during the sunnier months and low for the rest of the year [40]. In regions with a temperate climate, the customary annual control of serum vitamin D concentration, as a result, highly depends on the moment of the control, which is for practical reasons mostly spread out over the year. Results in subsequent years are consequently rather similar in the same individuals: low after the darker months of the year and normal in the others.

In an attempt to understand the relative importance of different vitamin D sources, more specifically the influence of sunlight, a retrospective study looked at 474 available serum 25(OH) vitamin D concentrations from the annual follow-up visits of 141 patients, collected over 4 consecutive years [37]. They were compared to values of healthy controls in the same age group and who lived at the similar geographical latitude (50° N). Vitamin D serum values also were related to perennial ultraviolet B (UVB) exposure. Patients consumed 800 IU vitamin D_3 daily. Ranked per month, 25(OH) vitamin D concentrations depicted a curve strikingly parallel to the amount of UVB exposure in the preceding months. A significant difference exists between 25(OH) vitamin D concentrations in the months with high UVB exposure (May–October) and the months with low UVB exposure (November–April) (Table 15.2) but not with healthy controls in the same period (Figure 15.1). A fraction of the CF population has 25(OH) vitamin D concentrations <20 ng/mL (Table 15.3). However, there is an important influence of the varying exposure to UVB light from the sun over the different years. The serum 25(OH) vitamin D concentrations were compared to reference values from healthy peers. No significant difference was found in comparison to local controls during the months with low UVB exposure (CF patients: 18.0 ng/mL (interquartile range (IQR), 10.2–24.5 ng/mL); control group: 17.2 ng/mL (IQR, 12.7–19.3 ng/mL); $p=0.60$). During the months with high UVB exposure,

TABLE 15.2 Median 25(OH) Vitamin D Concentrations versus Ultraviolet B (UVB) Exposure

	25(OH) Vitamin D Concentrations (ng/mL)				
Year	**Patients (n)**	**Months with High UVB Exposure[a]**	**Patients (n)**	**Months with Low UVB Exposure[b]**	**p Value**
2005	70	27.9 (19.3–35.0)	59	17.0 (11.8–26.4)	<0.001
2004	72	28.5 (21.7–35.0)	56	22.5 (14.7–28.9)	<0.001
2003	71	28.5 (18.6–35.1)	56	18.6 (11.2–23.4)	<0.001
2002	59	21.7 (14.0–27.5)	40	11.0 (8.0–19.6)	<0.001

Data are median (interquartile range).
[a]*May to October.*
[b]*November to April.*

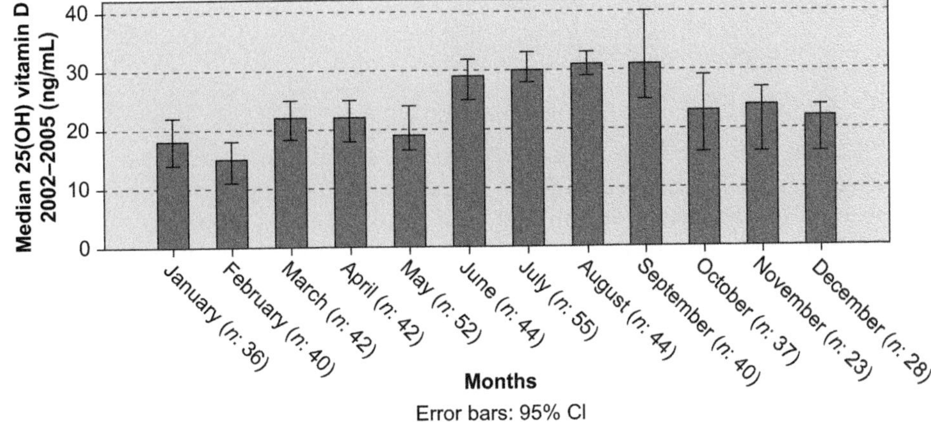

FIGURE 15.1 Median 25(OH) vitamin D serum concentrations per month in the period of 2002–2005 in patients with cystic fibrosis. Median 25(OH) vitamin D serum concentrations per month depicted an *S*-shaped convex curve from May to October (months with high ultraviolet B (UVB) exposure) and a concave curve in the subsequent period from November to April (months with low UVB exposure). The minimum value is found in February. CI, confidence interval.

TABLE 15.3 Percentage of 25(OH) Vitamin D Concentrations <20 ng/mL in 2002, 2003, 2004, and 2005

	25(OH) Vitamin D <20 ng/mL (%)	
Year	**Months with High UVB Exposure**	**Months with Low UVB Exposure**
2002	38.9	77.5
2003	24.6	51.4
2004	19	46.2
2005	19	64.3

serum 25(OH) vitamin D concentrations were not statistically inferior to values from a reported group of patients with comparable age and geographical location [41].

The comparison between the variation in 25(OH) vitamin D serum concentrations in the CF group over the 4 years and the amount of UVB exposure in the preceding months showed a remarkably manifest correlation. It was clear that median 25(OH) vitamin D serum concentrations ran parallel to the amount of UVB exposure in the 2 or 3 preceding months ($p < 0.001$). It also was evident that in a year with more UVB exposure during the summer—such as 2003—25(OH) vitamin D serum concentrations were higher than those recorded in a year with a summer with less UVB exposure (Figure 15.2).

The curve of serum 25(OH) vitamin D concentrations and that of varying intensity of UVB light from the sun run a parallel course, with a time-lag of approximately 2 months, as was also described in healthy people [42]. A maximum is reached in late summer and a minimum at the end of the winter. Higher serum 25(OH) vitamin D concentrations in September correlated with higher serum 25(OH) vitamin D concentrations in March of the subsequent year. This is explained by the storage of 25(OH) vitamin D_3 from dermal sun exposure during the months with high UVB exposure and consumption during the months with less UVB exposure, resulting in a progressive depletion of vitamin D_3 [40,41]. It is generally accepted that the skin is the major source of vitamin D;

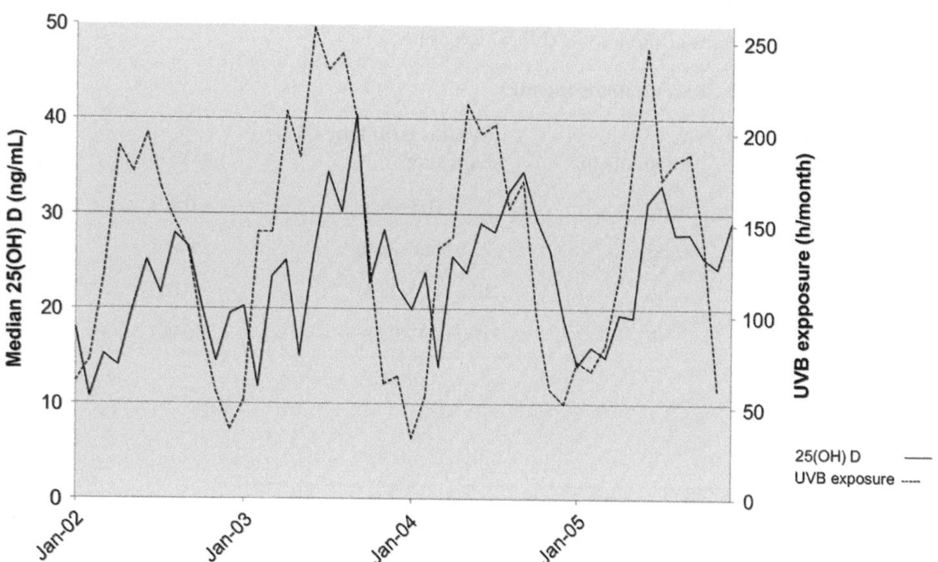

FIGURE 15.2 Median 25(OH) vitamin D serum concentration (solid line) and ultraviolet B exposure (dotted line) per month through the consecutive years 2002–2005 in patients with cystic fibrosis. It is clear that median 25(OH) vitamin D concentrations run parallel to the amount of sunshine in the 2 or 3 preceding months ($p<0.001$).

probably more than 85% of vitamin D is obtained through exposure to sunlight [39]. Exposure of 6% of the body's surface to one minimal erythema dose of sunlight is equivalent to the oral administration of 600–1000 IU of vitamin D. This means that in conditions of sunshine, mild exposure of hands, arms, face, or back to the sun two to three times per week would comply with vitamin D recommendations [8,36].

Because of the problem of malabsorption in CF, it is suggested that patients with CF can benefit from UVB exposure, especially during the winter months in which vitamin D cannot be produced. Indoor tanning machines emit the same spectrum of UVB radiation as sunlight, and as a consequence result in the dermal production of vitamin D_3. In a study by Chandra et al. [43], the Sperti Del Sol sunlamp was used to correct vitamin D status. Vitamin D deficiency was defined as a serum 25(OH) vitamin D concentration <30 ng/mL (<75 nmol/L). Five patients with CF were included. Exposure to UVB radiation was determined by the skin type of the patient. All patients were asked to start with a 3-min session and to increase the time by 1 min/week until they reached their target time. This occurred during five sessions per week for 8 weeks. At the end of 8 weeks the serum 25(OH) vitamin D concentrations were slightly improved (6 ng/mL) and tended toward statistically significant. At the end of 8 weeks only one patient had still a severe vitamin D deficiency, defined as a serum 25(OH) vitamin D concentration <20 ng/mL (<50 nmol/L). The weaknesses of this study were the small size, which limits the power of the data analysis, and the lack of control of compliance. In the study by Khazai et al. [34] the relative efficacy of vitamin D_2, vitamin D_3, and UVB radiation were compared. Thirty patients with CF were randomized to receive 50,000 IU vitamin D_2, 50,000 IU vitamin D_3, or, depending on their skin type, 3–10 min of exposure to UVB radiation five times a week using a portable indoor UV tanning lamp (Sperti Del Sol). The duration of the study was 12 weeks. All patients who received a vitamin D_3 supplement reached vitamin D sufficiency, defined as serum 25(OH) vitamin D >30 ng/mL (>75 nmol/L), compared to those who were treated with vitamin D_2 (60%) and 22% of those treated with UVB radiation. An improvement in serum 25(OH) vitamin D concentration did not occur, probably because of poor adherence to UV therapy. The authors concluded that an oral supplement of 50,000 IU vitamin D_3 once a week is effective in increasing serum 25(OH) vitamin D and lower PTH values in vitamin D deficient patients with CF. Gronowitz et al. [44] studied in a case–control study the effect of total body exposure to UVB radiation during the dark season on serum 25(OH) vitamin D concentrations. The study was conducted for 6 months with 30 patients with CF. Adherence to the treatment was encouraged by frequent telephone contact. Nevertheless, problems with erythema and skin tenderness occasionally caused interruption of the therapy. As a consequence, the adherence to the treatment varied over time. At the end of the study period, however, the patients in the UVB radiation group showed a significant increase in serum 25(OH) vitamin D concentration, from 23.8 to 50.4 ng/mL. The control group did not show a significant increase.

The CF Foundation states that no recommendation can be made for the use of UV lamps in the management of vitamin D deficiency in all individuals with CF because there is a concern that the type of UV light (UVA versus UVB) may not be standardized among all UV lamp devices. Based on the limited availability of data on the use of UVB radiation to correct vitamin D deficiency in patients with CF, it is difficult to recommend an optimal dose and duration of UVB exposure. Further research is needed to tackle this problem as well as to determine the long-term benefits of correcting vitamin D status by UVB exposure via UV lamps on skeletal health and the beneficial extraskeletal effects [26]. Because exposure to a reasonable amount of UVB radiation is feasible in sunny climates, whereas in regions with less sunny climates it is only possible during some months of the year, supplements remain imperative.

15.5 CONCLUSION

People with CF are at risk of vitamin D deficiency, which seems to be a multifactorial issue because of the underlying pathophysiology of the disease. As a consequence, this deficiency may increase the risk of abnormalities in bone metabolism. Studies of beneficial extraskeletal effects in patients with CF, such as the immunomodulatory and glucose-lowering effects of vitamin D, may be of potential for the management of CF. However, a careful approach is recommended because some data represent only in vitro findings or they may present only associations lacking causality. Well-designed, long-term, and well-powered trials with clearly defined outcomes are crucial before any further recommendations can be made.

References

[1] Paccou J, Zeboulon N, Combescure C, Gossec L, Cortet B. The prevalence of osteoporosis, osteopenia, and fractures among adults with cystic fibrosis: a systematic literature review with meta-analysis. Calcif Tissue Int 2010;86(1):1–7.

[2] Aris RM, Merkel PA, Bachrach LK, Borowitz DS, Boyle MP, Elkin SL, et al. Guide to bone health and disease in cystic fibrosis. J Clin Endocrinol Metab 2005;90(3):1888–96.

[3] Paccou J, Fardellone P, Cortet B. Cystic fibrosis-related bone disease. Curr Opin Pulm Med 2013;19(6):681–6.

[4] Lark RK, Lester GE, Ontjes DA, Blackwood AD, Hollis BW, Hensler MM, et al. Diminished and erratic absorption of ergocalciferol in adult cystic fibrosis patients. Am J Clin Nutr 2001;73(3):602–6.

[5] Borowitz D, Baker RD, Stallings V. Consensus report on nutrition for pediatric patients with cystic fibrosis. J Pediatr Gastroenterol Nutr 2002;35(3):246–59.

[6] Sinaasappel M, Stern M, Littlewood J, Wolfe S, Steinkamp G, Heijerman HG, et al. Nutrition in patients with cystic fibrosis: a European consensus. J Cyst Fibros 2002;1(2):51–75.

[7] Wharton B, Bishop N. Rickets. Lancet 2003;362(9393):1389–400.

[8] Holick M. Vitamin D in health and prevention of metabolic bone disease. In: Rosen C, editor. Osteoporosis. Current clinical practice. Humana Press; 1996. pp. 29–43.

[9] Baeke F, Takiishi T, Korf H, Gysemans C, Mathieu C. Vitamin D: modulator of the immune system. Curr Opin Pharmacol 2010;10(4):482–96.

[10] Takiishi T, Gysemans C, Bouillon R, Mathieu C. Vitamin D and diabetes. Endocrinol Metab Clin N Am 2010;39(2):419–46. table of contents.

[11] Schwalfenberg GK. A review of the critical role of vitamin D in the functioning of the immune system and the clinical implications of vitamin D deficiency. Mol Nutr Food Res 2011;55(1):96–108.

[12] Charan J, Goyal JP, Saxena D, Yadav P. Vitamin D for prevention of respiratory tract infections: a systematic review and meta-analysis. J Pharmacol Pharmacother 2012;3(4):300–3.

[13] Elamin MB, Abu Elnour NO, Elamin KB, Fatourechi MM, Alkatib AA, Almandoz JP, et al. Vitamin D and cardiovascular outcomes: a systematic review and meta-analysis. J Clin Endocrinol Metab 2011;96(7):1931–42.

[14] George PS, Pearson ER, Witham MD. Effect of vitamin D supplementation on glycaemic control and insulin resistance: a systematic review and meta-analysis. Diabet Med 2012;29(8):e142–50.

[15] Wolfenden LL, Judd SE, Shah R, Sanyal R, Ziegler TR, Tangpricha V. Vitamin D and bone health in adults with cystic fibrosis. Clin Endocrinol 2008;69(3):374–81.

[16] Hall WB, Sparks AA, Aris RM. Vitamin d deficiency in cystic fibrosis. Int J Endocrinol 2010;2010:218691.

[17] Donovan Jr DS, Papadopoulos A, Staron RB, Addesso V, Schulman L, McGregor C, et al. Bone mass and vitamin D deficiency in adults with advanced cystic fibrosis lung disease. Am J Respir Crit Care Med 1998;157(6 Pt 1):1892–9.

[18] Chavasse RJ, Francis J, Balfour-Lynn I, Rosenthal M, Bush A. Serum vitamin D levels in children with cystic fibrosis. Pediatr Pulmonol 2004;38(2):119–22.

[19] Chapelon E, Garabedian M, Brousse V, Souberbielle JC, Bresson JL, de Montalembert M. Osteopenia and vitamin D deficiency in children with sickle cell disease. Eur J Haematol 2009;83(6):572–8.

[20] Sermet-Gaudelus I, Bianchi ML, Garabedian M, Aris RM, Morton A, Hardin DS, et al. European cystic fibrosis bone mineralisation guidelines. J Cyst Fibros 2011;10(Suppl. 2):S16–23.

[21] Mailhot G. Vitamin D bioavailability in cystic fibrosis: a cause for concern? Nutr Rev 2012;70(5):280–93.

[22] Rovner AJ, Stallings VA, Schall JI, Leonard MB, Zemel BS. Vitamin D insufficiency in children, adolescents, and young adults with cystic fibrosis despite routine oral supplementation. Am J Clin Nutr 2007;86(6):1694–9.

[23] Coppenhaver D, Kueppers F, Schidlow D, Bee D, Isenburg JN, Barnett DR, et al. Serum concentrations of vitamin D-binding protein (group-specific component) in cystic fibrosis. Hum Genet 1981;57(4):399–403.

[24] Speeckaert MM, Wehlou C, Vandewalle S, Taes YE, Robberecht E, Delanghe JR. Vitamin D binding protein, a new nutritional marker in cystic fibrosis patients. Clin Chem Lab Med 2008;46(3):365–70.

[25] Papandreou D, Malindretos P, Karabouta Z, Rousso I. Possible health implications and low vitamin D status during childhood and adolescence: an updated mini review. Int J Endocrinol 2010;2010:472173.

[26] Tangpricha V, Kelly A, Stephenson A, Maguiness K, Enders J, Robinson KA, et al. An update on the screening, diagnosis, management, and treatment of vitamin D deficiency in individuals with cystic fibrosis: evidence-based recommendations from the cystic fibrosis foundation. J Clin Endocrinol Metab 2012;97(4):1082–93.

[27] Mortensen LA, Chan GM, Alder SC, Marshall BC. Bone mineral status in prepubertal children with cystic fibrosis. J Pediatr 2000;136(5):648–52.

[28] Buntain HM, Greer RM, Schluter PJ, Wong JC, Batch JA, Potter JM, et al. Bone mineral density in Australian children, adolescents and adults with cystic fibrosis: a controlled cross sectional study. Thorax 2004;59(2):149–55.

[29] Haworth CS, Selby PL, Webb AK, Dodd ME, Musson H, Mc LNR, et al. Low bone mineral density in adults with cystic fibrosis. Thorax 1999;54(11):961–7.

[30] Haworth CS, Jones AM, Adams JE, Selby PL, Webb AK. Randomised double blind placebo controlled trial investigating the effect of calcium and vitamin D supplementation on bone mineral density and bone metabolism in adult patients with cystic fibrosis. J Cyst Fibros 2004;3(4):233–6.

[31] Hillman LS, Cassidy JT, Popescu MF, Hewett JE, Kyger J, Robertson JD. Percent true calcium absorption, mineral metabolism, and bone mineralization in children with cystic fibrosis: effect of supplementation with vitamin D and calcium. Pediatr Pulmonol 2008;43(8):772–80.

[32] Holick MF. Vitamin D status: measurement, interpretation, and clinical application. Ann Epidemiol 2009;19(2):73–8.

[33] Ferguson JH, Chang AB. Vitamin D supplementation for cystic fibrosis. Cochrane Database Syst Rev 2012;4:CD007298.

[34] Khazai NB, Judd SE, Jeng L, Wolfenden LL, Stecenko A, Ziegler TR, et al. Treatment and prevention of vitamin D insufficiency in cystic fibrosis patients: comparative efficacy of ergocalciferol, cholecalciferol, and UV light. J Clin Endocrinol Metab 2009;94(6):2037–43.

[35] Belgische voedingsmiddelentabel [Internet]. NUBEL vzw; 2009 [cited 18 January 2014].

[36] Holick MF. Biologic effects of light: historical and new perspectives. Biol Eff Light 1998 1999:11–32.

[37] Robberecht E, Vandewalle S, Wehlou C, Kaufman JM, De Schepper J. Sunlight is an important determinant of vitamin D serum concentrations in cystic fibrosis. Eur J Clin Nutr 2011;65(5):574–9.

[38] Tangpricha V, Turner A, Spina C, Decastro S, Chen TC, Holick MF. Tanning is associated with optimal vitamin D status (serum 25-hydroxyvitamin D concentration) and higher bone mineral density. Am J Clin Nutr 2004;80(6):1645–9.

[39] Heaney RP, Davies KM, Chen TC, Holick MF, Barger-Lux MJ. Human serum 25-hydroxycholecalciferol response to extended oral dosing with cholecalciferol. Am J Clin Nutr 2003;77(1):204–10.

[40] Rapuri PB, Kinyamu HK, Gallagher JC, Haynatzka V. Seasonal changes in calciotropic hormones, bone markers, and bone mineral density in elderly women. J Clin Endocrinol Metab 2002;87(5):2024–32.

[41] Guillemant J, Le HT, Maria A, Allemandou A, Peres G, Guillemant S. Wintertime vitamin D deficiency in male adolescents: effect on parathyroid function and response to vitamin D_3 supplements. Osteoporos Int 2001;12(10):875–9.

[42] Need AG, Morris HA, Horowitz M, Nordin C. Effects of skin thickness, age, body fat, and sunlight on serum 25-hydroxyvitamin D. Am J Clin Nutr 1993;58(6):882–5.

[43] Chandra P, Wolfenden LL, Ziegler TR, Tian J, Luo M, Stecenko AA, et al. Treatment of vitamin D deficiency with UV light in patients with malabsorption syndromes: a case series. Photodermatol Photoimmunol Photomed 2007;23(5):179–85.

[44] Gronowitz E, Larko O, Gilljam M, Hollsing A, Lindblad A, Mellstrom D, et al. Ultraviolet B radiation improves serum levels of vitamin D in patients with cystic fibrosis. Acta Paediatr 2005;94(5):547–52.

CHAPTER

16

Pediatric Cystic Fibrosis and Fat-Soluble Vitamins

Kacie Beckett[1], *Karyn Shaw*[2], *Meghana Sathe*[3]

[1]Cystic Fibrosis Clinic, University of Texas Southwestern, Dallas, TX, USA; [2]Department of Clinical Nutrition, Children's Medical Center, Dallas, TX, USA; [3]Department of Pediatric Gastroenterology and Nutrition, University of Texas Southwestern, Dallas, TX, USA

16.1 INTRODUCTION

Cystic Fibrosis (CF) is a genetic disorder that affects approximately 70,000 people worldwide. CF is caused by a mutation to the Cystic Fibrosis conductance Transmembrane Regulator (CFTR) gene, resulting in thick, sticky mucus that resides in the lungs, pancreas, intestines and liver. The mucus can cause chronic lung infections as well as exocrine pancreatic insufficiency [1]. There are over 1000 different mutations in the CFTR gene, some more severe than others; therefore, patients' respiratory and gastrointestinal symptoms can vary greatly depending on which genotype and phenotype they have [1,2]. Patients with CF are either pancreatic insufficient (PI) or pancreatic sufficient (PS). Typically the more severe mutations, type 1-3, result in PI and the milder mutations, type 4 and 5, result in PS. The 72-h fecal fat test is the gold standard to determine pancreatic function; however, the fecal elastase-1 assay is more easily obtained, as it only requires a single stool sample. A fecal elastase of <100 µg/g of stool indicates severe PI, 100–200 µg/g mild to moderate PI, and >200 µg/g indicates normal pancreatic function. Approximately 85–90% of individuals with CF are PI, placing them at increased risk of fat malabsorption, which can lead to fat-soluble vitamin (vitamins A, D, E, and K) deficiency even after the initiation of pancreatic enzyme replacement therapy (PERT). 10–15% of individuals with CF are PS and are at a much lower risk of malabsorption and fat soluble vitamin deficiency. However, PS patients are at an increased risk for the development of recurrent, chronic pancreatitis because of mucus plugging, resulting in the decline of exocrine function and the development of PI. It is important to routinely ask PS patients about their stool frequency and appearance as well as monitor weight trends, assess annual fat-soluble vitamin levels, and repeat the fecal elastase annually to ensure that any changes in pancreatic function can be detected early [1–3].

16.2 VITAMIN A

Vitamin A is a fat-soluble vitamin that can be found in two forms in the diet: preformed (retinol) and provitamin A (carotenoids). Retinol is found in animal-based foods such as dairy products, fish oils, and liver, and carotenoids are found in most orange or yellow fruits and vegetables. Approximately 90% of ingested retinol is absorbed in the body. Carotenoids must be metabolized into retinol, with some remaining unmetabolized in the form of β-carotene; therefore, only about 3% of carotenoids are absorbed. Once retinol is absorbed in the gastrointestinal tract, it requires bile acids for emulsification and micelle formation before being absorbed into enterocytes. Once inside the cell, retinol is directed into chylomicrons that enter the lymphatic system and become stored in the liver. Retinol binding protein (RBP) is created in the liver and binds to retinol along with prealbumin to then carry it to other tissues such as eye, bone marrow, lung, kidney, and adipose tissues [4,5].

Vitamin A plays an important role in several biological processes, including vision, bone health, cell differentiation, immunity, pulmonary function, and reproduction [1,4]. It is recommended to check serum retinol levels at least annually. Normal retinol levels are >20 µg/dL for patients 0–6 months of age and 30–80 µg/dL for patients >6 months of age [4,6,7]. Vitamin A deficiency is defined

Diet and Exercise in Cystic Fibrosis
http://dx.doi.org/10.1016/B978-0-12-800051-9.00016-X

as a serum retinol level <20 μg/dL (0.7 μmol/L), or a retinol-to-retinol binding protein (RBP) ratio <0.8 mol/mol, and has been identified in 10–40% of CF patients. It is important to note that RBP is a negative acute phase reactant protein and, therefore, RBP levels and retinol levels will decrease during times of inflammation and infection. Therefore, it is important to avoid checking retinol and RBP levels during acute illness [1].

Vitamin A deficiency is thought to occur in CF because of a deficiency of bile acids and pancreatic enzymes, as well as compromised liver function, which can deplete RBP stores. Vitamin A deficiency can present with night blindness, Bitot's spots, xerophthalmia, corneal xerosis, keratomalacia, anorexia, anemia, leukopenia, hyperkeratosis, phrynoderma, depressed helper T-cell activity, and impaired mucus secretion [4]. The 2002 CF Guidelines recommend 1,500–10,000 IU vitamin A for all patients with CF, depending on their age [1,4,8,9]. Typically, an oral miscible form is used, as it is more easily absorbed in CF patients with PI. However, case studies and various literature suggest that some CF patients may require additional vitamin A supplementation, with documented doses of up to 25,000–50,000 IU Vitamin A daily [6,10–12]. The risk of vitamin A toxicity does increase with long-term exposure to higher vitamin A supplementation. Traditionally, it was thought that vitamin A toxicity occurred at doses >100,000 IU/day; however, it is now known that the upper limit to cause toxicity varies from patient to patient. Although some literature seems to suggest that toxicities can occur with moderately high doses of 20,000–45,000 IU/day, other case studies and literature have shown vitamin A toxicity in patients to be anywhere from 5000 to 1,400,000 IU/day [4–6,9,11]. Retinol is more potent and has a higher rate of toxicity than beta carotene.

Vitamin A toxicity can cause pseudotumor cerebri, photophobia, alopecia, anorexia, muscle and bone pains, conjunctivitis, hepatotoxicity, hyperlipidemia, increased risk of fractures, as well as other problems [4,6,9]. These symptoms should be screened for routinely in patients on supplements, especially in patients known to have elevated levels of retinol, RBP, or both.

Zinc deficiency is also noted in patients with CF and should be considered when vitamin A deficiency is detected. Zinc aids in the synthesis of RBP, so zinc deficiency can cause a decrease in RBP availability, which then decreases the mobility of retinol from the liver to other tissues. Zinc deficiency is difficult to clinically diagnose, as serum zinc values are an unreliable indicator of zinc stores. Some symptoms of zinc deficiency include stunted growth, immune dysfunction, poor wound healing, and low vitamin A. If zinc deficiency is suspected, a dose of 1 mg/kg/day elemental zinc up to 25 mg elemental zinc daily should be initiated [1,7].

16.3 VITAMIN D

Vitamin D, or calciferol, exists in many forms, but the most biologically relevant forms are vitamin D2 (ergocalciferol) and vitamin D3 (cholecalciferol). Dietary forms (D2 and D3) are fat-soluble and are found naturally in beef liver (1 μg/3oz), eggs (1.1 μg/egg), and fatty fish (1.7–11 μg/3oz). In the United States, foods such as milk and orange juice may be fortified with vitamin D up to 2.5 μg/8oz [13,14]. Dietary vitamin D is absorbed in the small intestine by chylomicrons into the lymphatic system and ultimately into the blood stream. Vitamin D3 is also photosynthesized by solar ultraviolet B radiation from the steroid 7-dehydrocholesterol in the skin. However, for children living in northern latitudes, vitamin D3 synthesis in the skin may be nonexistent for 3–6 months out of the year [14]. Regardless of source, dietary or solar, vitamin D must be converted to the biologically active form 25-hydroxyvitamin D (25-OH D) in the liver, which releases the newly hydroxylated form into the blood. In response to parathyroid hormone (PTH) concentrations, the kidneys take up 25-OH D from the blood to synthesize calcitriol (1,25-dihydroxycholecalciferol).

Vitamin D is a complex compound with many functions. Improving and maintaining bone health has historically been considered the primary function, but research continues to identify new benefits of vitamin D and its use in biological functions. Vitamin D increases efficiency of calcium and phosphorus absorption in the small intestines and, with the help of PTH, maintains homeostatic concentrations of blood calcium and phosphorus. NHANES III reported a strong relationship between serum 25-OH D concentrations and pulmonary function [15].

Unfortunately, vitamin D deficiency is commonly seen in the general population as well as people with CF. It has been reported that 10–40% of persons with CF demonstrate deficiency of vitamin D [16]. Deficiency is associated with decreased bone loss, unattained peak bone mass, rickets, and osteoporosis. Children with CF are at particularly high risk for bone disease if they have PI and subsequent malabsorption of vitamins D and K, frequent use of corticosteroids, liver disease, delayed puberty, and nutritional failure [16]. Vitamin D deficiency has also been linked with increased risk of CF related diabetes [17].

Checking serum 25-OH D is recommended to assess vitamin D status in people with CF [4,15,16,18]. It has a half-life of 10 days–3 weeks, meaning levels do not fluctuate quickly. Vitamin D status should be assessed at least annually and ideally at the end of winter when exposure to ultraviolet light is lowest. Serum 25-OH D >75 nmol/L (30 ng/L) is considered sufficient in persons with CF for bone health and disease prevention.

A value of <27.5 nmol/L (11 ng/L) is associated with deficiency and rickets [4]. The CF Foundation recommends rechecking vitamin D status 12 weeks after adjusting vitamin D dosage. Serum 1,25-dihydroxyvitamin D is not recommended for assessment due to its dependence on PTH and fibroblast growth factor. It also has a very short half-life of 4 h, meaning levels fluctuate greatly and may not accurately reflect true vitamin D status [18]. In addition to annual laboratory studies, dual energy X-ray absorptiometry (DEXA) scans should be performed to evaluate skeletal integrity for children >8 years of age with CF who have risk factors for bone disease or are candidates for organ transplantation. Conventional X-ray is not as sensitive as DEXA [16].

The use of either supplemental vitamin D2 or vitamin D3 to improve serum levels continues to be studied [19,20]. In 2012, the CF Foundation recommended treating persons with CF who have serum 25-OH D <75 nmol/L (<30 ng/L) with vitamin D3 (cholecalciferol). The 2012 CF Guidelines recommend 400–10,000 IU vitamin D daily for all patients with CF, depending on their prospective ages, in an effort to achieve and maintain appropriate serum vitamin D levels. If the goal serum vitamin D is not achieved while a person is adherent on maximum dosages, referral to a specialist with expertise in vitamin D therapy is recommended [18]. Since the majority of vitamin D is stored in adipose tissues before hydroxylation, people who are overweight or obese may need more supplementation than their lean counterparts to maintain optimal serum concentrations [13]. The frequency of dosage has been discussed in literature. It has been proposed that a single, high dose of vitamin D3, in conjunction with maintenance therapy, can elevate and maintain goal serum vitamin D [21]; however, further research may be needed to establish safety. The CF Foundation has not found a benefit of bolus versus daily supplementation [18], but financial constraints and adherence must be considered when prescribing therapy.

Toxicity may be seen after the abuse of supplements or in kidney disease. Symptoms of toxicity include hypercalcemia, nephrolithiasis, weakness, fatigue, diarrhea, anorexia, headache, confusion, psychosis, tremor, and hypercalciuria. Vitamin D excess may be defined as >250 nmol/L (100 ng/L), whereas intoxication is >375 nmol/L (150 ng/l) [14].

In summary, maintaining adequate vitamin D is no doubt very important for persons with CF. Vitamin D plays a critical part in bone health, but may also have other significant health benefits. Clinicians should take care to monitor routinely and treat deficiency appropriately to prevent adverse effects. Further research is warranted to continue to search out the most efficient and effective treatment of deficiency.

16.4 VITAMIN E

Vitamin E is a fat-soluble vitamin that functions primarily as an antioxidant by preventing the oxidation of unsaturated fatty acids in the cell membrane. Cells in the lungs, brain, and erythrocytes are particularly vulnerable to oxidation. Chemically, vitamin E is classified into two groups, tocopherols and tocotrienols, with each group containing four forms (α, γ, β, and δ). α-tocopherol is the most biologically active form and is the only form maintained in human plasma [4,13,15,16,22].

The majority of vitamin E found in food contains the tocopherol group. Foods containing vitamin E have varying contents of all forms; however, the liver releases only α-tocopherol into the blood stream [22]. Foods high in polyunsaturated fat, such as nuts and vegetable oil, are relatively good sources of vitamin E. In addition, wheat germ, dairy products, and eggs are also good sources. Like many other vitamins, exposure to light and heat can decrease the vitamin E content of any food [4,13].

Vitamin E deficiency in the healthy population is rare. The risk of deficiency increases in persons with genetic defects to the α-tocopherol transfer protein (α-TTP), protein-calorie malnutrition, and fat malabsorption disorders such as pancreatic insufficiency or hepatobiliary disease. 5–10% of persons with CF may be deficient despite supplementation [16]. A plasma concentration of <5 μg/dL suggests deficiency, which can cause hemolytic anemia, neuromuscular degeneration, retinal deficits, ataxia, strabismus, muscle weakness, dementia, and cardiac arrhythmias [4]. Vitamin E deficiency and malnutrition have also been linked to decreased cognitive scores in children [23].

Unlike vitamin A, vitamin E is not an acute phase reactant, so infection and catabolism do not affect biochemical studies. Serum α-tocopherol is commonly used as a marker of vitamin E status because of its availability. However, blood lipid levels can affect serum tocopherol levels. High lipids can result in falsely high vitamin E levels, whereas low lipids can result in falsely low vitamin E levels. The most reliable method of determining vitamin E status is calculating the tocopherol:total lipid ratio (tocopherol:cholesterol + triglycerides + phospholipids). A ratio <0.6 mg/g indicates deficiency, regardless of what the plasma concentration of α-tocopherol is. The tocopherol:cholesterol ratio may be more readily available and may also correlate well with actual vitamin E status. Vitamin E labs should be checked at least annually [4,16,24,25].

Increasing dietary intake of vitamin E has not correlated with improvement in biochemical studies; rather, pharmacological supplementation is required to improve deficiency [22]. The 2002 CF Guidelines recommend 40–400 IU vitamin E for all patients with CF, depending on their prospective ages [16]. Supplementation may be given individually or as part of CF-specific multivitamins. The use of D-α-tocopheryl polyethylene glycol succinate (TPGS), which enhances micelle formation, can improve absorption of supplementation [4]. Vitamin E can interfere with vitamin K absorption [13], so patients on anticoagulation therapy should be monitored closely when also on supplemental vitamin E.

Toxicity of vitamin E is rare; however, consumption of 200–800 mg of Vitamin E per day from supplements may have adverse effects including nausea, diarrhea, and flatulence [13]. Fatigue and altered coagulation studies may present after chronic intake of >1000 mg per day of vitamin E [4,22]. Despite this, toxicity from natural food sources has not been described.

16.5 VITAMIN K

Vitamin K is a fat-soluble vitamin that is important for coagulation, regulation of calcification, energy metabolism, and inflammation [26]. Its role in regulation of calcification makes it vital in maintaining bone health [26]. Specifically, vitamin K is a cofactor for the enzyme γ-glutamyl carboxylase, which converts glutamic acid residues to γ-carboxyglutamic residues [27,28]. These glutamic acid residues are located on the coagulation factors VII, IX, X, prothrombin, protein C and S, as well as osteocalcin, a protein that is important in the calcification of bone [27,28]. Vitamin K deficiency is also influenced by polymorphisms in apolipoprotein E, which affects clearance of chylomicrons, the main carriers of vitamin K in circulation [29]. There are certain alleles of apolipoprotein E that are associated with higher clearance rates of chylomicrons, making individuals who carry these alleles more susceptible to vitamin K deficiency [29].

There are two forms of naturally occurring vitamin K: phylloquinone (vitamin K1) and menaquinone (vitamin K2). Phylloquinone is found in green leafy vegetables, algae, and vegetables oils including soybean and olive oil [30]. Menaquinone is synthesized by gut bacteria [30,31]. Similar to the other fat soluble vitamins (A, D, and E), vitamin K requires bile acids and pancreatic enzymes for emulsification and micelle formation prior to absorption into enterocytes. Patients with PI are obviously at risk for vitamin K deficiency; however, those with PI in addition to CF associated liver disease (CFALD), small bowel resection, persistent diarrhea, and/or frequent antibiotic use (which decreases the presence of normal gut flora) are at even higher risk [28].

Deficiency in vitamin K results in under carboxylation of coagulation factors, resulting in the most commonly recognized symptoms of easy bruising and bleeding. More often forgotten is the contribution of under carboxylation of osteocalcin, which results in osteopenia and osteoporosis. Vitamin K is more readily taken up by the liver as compared to bone [28]. Therefore, vitamin K deficiency will affect bone health before it will cause a disturbance in coagulation.

This makes measuring accurate vitamin K levels a complicated issue. Serum phylloquinone levels are often inaccurate, as they depend on plasma triglycerides, foods ingested prior to testing, and apolipoprotein E phenotypes [28]. Prothrombin time is considered an insensitive marker of vitamin K deficiency as it does not show deficiency until at least 50% or more of prothrombin is uncarboxylated [32]. However, it is the most readily available and inexpensive method of assessing vitamin K deficiency. PIVKA-II, prothrombin produced in vitamin K absence, and Gluc-OC, the fraction of uncarboxylated osteocalcin to total osteocalcin, are both more sensitivity indicators of vitamin K [32–35]. However, PIVKA-II and Gluc-OC are not readily available and are costly, leading to underutilization of these more sensitive markers in monitoring vitamin K levels. Despite this, the CF Foundation Pediatric Nutrition Consensus Report in 2002, advocates annual PIVKA-II levels in patients with CF and PI [16]. In addition, multiple studies have shown suboptimal levels of PIVKA-II and Gluc-OC in the presence of normal prothrombin times [35–38]. Based on the fact that PIVKA-II reflects vitamin K deficiency as it relates to the liver, whereas Gluc-OC reflects vitamin K deficiency as it relates to bone, both may be needed to understand the complete picture of vitamin K status. In addition, if bone is the first to be affected by vitamin K deficiency and the last to show benefits of repletion, it may be beneficial for practitioners to utilize Gluc-OC in assessing vitamin K deficiency.

There are three forms of synthetic vitamin K [3–5] that can be supplemented in a variety of water-soluble and fat-soluble forms [39,40], including as part of a CF-specific multivitamin, as a single supplement in an oral form, an intramuscular form, or an intravenous form [33,40,41]. Vitamin K tends to be the least individually supplemented fat-soluble vitamin. This is likely because of the fact that most practitioners tend to follow the most readily available, but less accurate, form of measuring vitamin K level, prothrombin time. Vitamin K supplementation has been used routinely to decrease bleeding risk; however, it is not routinely utilized in treatment of osteopenia or osteoporosis. This is despite evidence of its use decreasing fracture risk in patients with osteopenia when utilized in combination with calcium and vitamin D in non-CF patients [42].

Supplementation is also a challenge, as routine supplementation recommendations of 300–500 mcg/day, for all patients with CF and PI seem inadequate, and it is only patients receiving high doses of vitamin K that seem to have adequate levels [16,38]. Naturally occurring forms of vitamin K, phylloquinone (vitamin K1) and menaquinone (vitamin K2), have no known toxicity even when given in extremely high doses [16,28,38]. Menadione, vitamin K3, was previously used to provide intramuscular vitamin K supplementation; however, because of allergic reactions, hemolytic anemia, and cytotoxicity in liver cells, the Federal Drug Administration (FDA) removed the supplement from the market [43]. The most common form of supplementation seems to be in the form of phylloquinone. Current recommendations are based more on expert opinion than prospective studies. The 2002 CF Pediatric Nutrition Consensus Report recommends 300–500 mg/day for all ages [16]. The 2004 CF Adult Nutrition Consensus Report recommends 2.5–5 mg per week with consideration to increase dose while on antibiotics [44], whereas the 2002 CF Trust Nutrition Working Group coming from Europe and the United Kingdom recommends a dose between 1 and 10 mg/day [45].

In conclusion, there is much that remains unknown about the best methods of assessing vitamin K deficiency. What we do know is that vitamin K, previously thought only to be associated with coagulation, plays a vital role in bone health. Vitamin K has no known toxicities in the most commonly supplemented form of phylloquinone and, therefore, practitioners should push the dosages for supplementation to maintain adequate levels for the most sensitive indicators of deficiency, PIVKA-II and Gluc-OC.

16.6 CONCLUSION

Although we know that our patients with CF and PI are at increased risk for fat-soluble vitamin deficiencies, there are many barriers to the diagnosis and management of these deficiencies. Accurate levels of deficiency are often difficult to obtain because of limitations of testing, whether the most sensitive tests are available locally or if tests are costly. Another challenge is that most insurance companies consider vitamin supplementation nonmedicinal and, therefore, do not cover multivitamins. We know that it is best to individually supplement fat-soluble vitamins; however, cost and pill burden make this a daunting task. Most CF-specific multivitamin preparations do not have adequate dosages of any one single vitamin.

Despite these challenges, we must emphasize to patients the importance of fat-soluble vitamin deficiencies and their role in their overall health. We must educate our patients on connections that may seem obvious to us, but not to them. For example, vitamin A plays an extremely important role in vision, vitamin K in addition to vitamin D is essential for bone health, and vitamin E is important in maintaining the integrity of RBC (red blood cell) and preventing hemolysis. We must remind them that in order to optimize supplementation of fat-soluble vitamins, even in their water-soluble form, patients should take their vitamins in the morning with their enzymes. This utilizes the best amount of natural bile acids and pancreatic enzymes, in addition to orally taken pancreatic enzymes. We also need to educate our patients about toxicity, especially with vitamin A, which is more likely than any other fat-soluble vitamin to be implicated in toxicity such as pseudotumor cerebri. In addition, although the CF Foundation recommendations are an annual evaluation and assessment of fat-soluble vitamins, it is important to monitor both very low levels and very high levels more aggressively, at least every 3–4 months. This is essential in order to prevent significant morbidity from fat-soluble vitamin deficiencies and toxicities (Table 16.1).

TABLE 16.1 Fat-soluble Vitamin Dosage Recommendations for CF Patients [15,18]

Vitamins	Age	Recommended Dosage (per day)	Monitoring Recommendations
A	0–12 months	1500 IU	• Annual serum retinol and retinol binding protein (RBP) (not during acute inflammatory state); consider evaluation of retinol:RBP ratio and zinc stores • Every 3–4 months with adjustments in dosage, deficiency states, or high levels • Monitor for toxicity
	1–3 years	5000 IU	
	4–8 years	5000–10,000 IU	
	>8 years	10,000 IU	
D	0–12 months	400–500 IU (max 2000 IU)	• Annual serum 25-OH D • Recheck levels 3 months after dose adjustment • Refer to vitamin D specialist if inadequate levels despite 3 months of maximum dose for age • Ideal time to evaluate for vitamin D deficiency is late fall or winter • DEXA scan to evaluate bone health >8 years of age • Toxicity from supplementation is rare
	1–10 years	800–1000 IU (max 4000 IU)	
	>10 years	800–2000 IU (max 10,000)	
E	0–12 months	40–50 IU	• Annual serum α-tocopherol levels • Consider evaluation of α-tocopherol:total lipid ratio • Every 3–4 months with adjustments in dosage and deficiency states • Toxicity from supplementation is rare
	1–3 years	80–150 IU	
	4–8 years	100–200 IU	
	>8 years	200–400 IU	
K	0–18 years	0.3–0.5 mg	• Annual serum levels of prothrombin time and PIVKA-II levels • Consider Gluc-OC (more of a research tool) • Every 1–2 months with adjustments in dosage • Toxicity from supplementation is rare
	Adult	2.5–5 mg (per week)	

References

[1] Rogers CL. Nutritional management of the adult with cystic fibrosis—part 1. Pract Gastroenterol 2013;113:12–23.

[2] Boeck K, Weren M, Proesmans M, Kerem E. Pancreatitis among patients with cystic fibrosis: correlation with pancreatic status and genotype. Pediatrics 2005;115:e463–9.

[3] Benahmed NA, Manene D, Barbot L, Kapel N. Fecal pancreatic elastase in infants under 2 years of age. Ann Biol Clin 2008; 66(5):549–52.

[4] Sathe MN, Patel AS. Update in pediatrics: focus on fat soluble vitamins. Nutr Clin Pract 2010;25(Suppl. 4):340–6.

[5] Carr SB, McBratney J. The role of vitamins in cystic fibrosis. J R Soc Med 2000;98(Suppl. 38):14–9.

[6] Brei C, Simon A, Krawinkel MB, Naehrlich L. Individualized vitamin A supplementation for patients with cystic fibrosis. Clin Nutr 2013;32:805–10.

[7] Pankau J, Beckett K. Nutrition management in cystic fibrosis: a case study involving enteral nutrition. Support Line 2012;34:12–9.

[8] Bonifant CM, Shevill E, Chang AB. Vitamin A supplementation for cystic fibrosis (review). Cochrane Database Syst Rev 2014, Issue 5. Art. No.: CD006751.

[9] Graham-Maar RC, Schall JI, Stettler N, Zemel BS, Stallings VA. Elevated vitamin A intake and serum retinol in preadolescent children with cystic fibrosis. AM J Clin Nutr 2006;84:174–82.

[10] Ansari EA, Sahni K, Etherington C, Morton A, Conway SP, Moya E, et al. Ocular signs and symptoms and vitamin A status in patients with cystic fibrosis treated with daily vitamin A supplements. Br J Ophthalmol 1999;83:688–91.

[11] James DR, Owen G, Campbell IA, Goodchild MC. Vitamin A absorption in cystic fibrosis: risk of hypervitaminosis A. Gut 1992;33:707–10.

[12] Sommer Alfred. Vitamin A deficiency today: conjunctival xerosis in cystic fibrosis. J R Soc Med 1989;82:1–2.

[13] Gropper SS, Smith JL. Advanced nutrition and human metabolism. 6th ed. Belmont (Ca): Wadsworth; 2013. p.390–407.

[14] Institute of Medicine. Dietary reference intakes for calcium, phosphorus, magnesium, vitamin D, and fluoride. Washington (DC): National Academy Press; 1997. p.250–87.

[15] Maqbool A, Stallings VA. Update on fat-soluble vitamins in cystic fibrosis. Curr Opin Pulm Med 2008;14:574–81.

[16] Borowitz D, Baker RD, Stallings V. Consensus report on nutrition for pediatric patients with cystic fibrosis. J Pediatr Gastroenterol Nutr 2002;35:246–59.

[17] Pincikova T, Nilsson K, Moen IE, Fluge G, Hollsing A, Knudsen PK, et al. Vitamin D deficiency as a risk factor for cystic fibrosis related diabetes in the Scandiavian cystic fibrosis nutritional study. Diabetologia 2011;54(12):3007–15.

[18] Tangpricha V, Kelly A, Stephenson K, Maguiness J, Enders J, Robinson KA, et al. An update on the screening, diagnosis, management, and treatment of vitamin D deficiency in individuals with cystic fibrosis: evidence-based recommendations from the cystic fibrosis foundation. J Clin Endocrin Metab 2012;97(4):1082–93.

[19] Holick MF, Bianncuzzo RM, Chen TC, Klein EK, Young A, Bibuld D, et al. Vitamin D2 is as effective as vitamin D3 in maintaining circulating concentration of 25-hydroxyvitamin D. J Clin Endocrinol Metab 2008;93(3):677–81.

[20] Logan VF, Gray AR, Peddie MC, Harper MJ, Houghton LA. Long-term vitamin D3 supplementation is more effective than vitamin D2 in maintaining serum 25-hydroxyvitamin D status over winter months. Br J Nutr 2012;109:1082–8.

[21] Sheperd D, Belessis Y, Katz T, Morton J, Field P, Jaffe A. Single high-dose oral vitamin D3 (stoss) therapy—a solution to vitamin D deficiency in children with cystic fibrosis? J Cyst Fibros 2012;12:177–82.

[22] Institute of Medicine. Dietary reference intakes for vitamin C, vitamin E, selenium, and carotenoids. Washington (DC): National Academy Press; 1997. p.186–253.

[23] Koscik RL, Farrell PM, Kosorok MR, et al. Cognitive function of children with cystic fibrosis: deleterious effect of early malnutrition. Pediatrics 2004;113:1549–58.

[24] Thurnham DI, Davies JA, Crump BJ, Situnayake RD, David M. The use of different lipids to express serum tocopherol: lipid ratios for the measurement of vitamin E status. Ann Clin Biochem 1986;23:514–20.

[25] Sokol RJ, Heubl JE, Iannaccone ST, Bove KE, Balistreri WF. Vitamin E deficiency with normal serum vitamin E concentrations in children with chronic cholestasis. N Engl J Med 1984;310:1209–12.

[26] Booth SL. Roles for vitamin K beyond coagulation. Annual Rev Nutr 2009;29:89–110.

[27] Kleinman RE, Fracchia MS. Vitamin K and cystic fibrosis: five me a double, please. Am J Clin Nutr 2010;92:469–70.

[28] Conway SP. Vitamin K in cystic fibrosis. J R Soc Med 2004; 97(44):48–51.

[29] Saupe J, Shearer MJ, Kohlmeier M. Phylloquinone transport and its influence on γ-carboxy glutamate residues of osteocalcin in patients on maintenance haemodialysis. Am J Clin Nutr 1993;58:204–8.

[30] Booth SL, Suttie JW. Dietary intake and adequacy of vitamin K1. J Nutr 1998;128:785–8.

[31] Suttie JW. The importance of menaquinones in human nutrition. Annu Rev Nutr 1995;15:399–417.

[32] Suttie JW. Vitamin K and human nutrition. J Am Diet Assoc 1992;92:585–90.

[33] Vermeer C, Jie KS, Knapen MH. Role of vitamin K in bone metabolism. Annu Rev Nutr 1995;15:1–22.

[34] Gundberg CM, Niema SD, Abrams S, Rosen H. Vitamin K status and bone health: an analysis of methods for determination of undercarboxylated osteocalcin. J Clin Endocrinol Metab 1998;83:3258–66.

[35] Conway SP, Wolfe SP, Brownlee KG, White H, Oldroyd B, Truscott JG, et al. Vitamin K status among children with cystic fibrosis and its relationship to bone mineral density and bone turnover. Pediatrics 2005;115:1325–31.

[36] Grey V, Atkinson S, Drury D, Casey L, Ferland G, Gundberg C, et al. Prevalence of low bone mass and deficiencies of vitamins D and K in pediatric patients with cystic fibrosis from 3 Canadian centers. Pediatrics 2008;122:1014–20.

[37] Nicolaidou P, Stavrinadis I, Loukou I, Papdopoulou A, Georgouli H, Dourus K, et al. The effect of vitamin K supplementation on biochemical markers of bone formation in children and adolescents with cystic fibrosis. Eur J Pediatr 2006;165:540–5.

[38] Dougherty KA, Schall JI, Stallings VA. Supoptimal vitamin K status despite supplementation in childrens and young adults with cystic fibrosis. Am J Clin Nutr 2010;92:660–7.

[39] Drury D, Grey VL, Ferland G, Gundberg C, Lands LC. Efficacy of high dose phylloquinone in correcting vitamin K deficiency in cystic fibrosis. J Cystic Fibrosis 2008;7(5):457–9.

[40] Jagannath VA, Fedorowicz Z, Thaker V, Chang AB. Vitamin K supplementation for cystic fibrosis (review). Cochrane Collaboration 2013;4:1–26.

[41] Shearer MJ. Vitamin K. Lancet 1995;345(8944):229–34.

[42] Sato Y, Kanoko T, Satoh K, Iwamoto J. Menatetrenone and vitamin D2 with calcium supplements prevent nonvertebral fracture in elderly women with Alzheimer's disease. Bone 2005;36(1):61–8.

[43] Higdon J. Vitamin K. Linus Pauling Institute, Oregon State University; February 2008. Retrieved 12 April 2008 – look up reference.

[44] Yankaskas JR, Marshall BC, Sufian B, Simon RH, Rodman D, Cystic fibrosis adult care: consensus conference report. Chest 2004;125(1):1S–39S.

[45] Cystic Fibrosis Trust Nutrition Working Group. Nutritional management of cystic fibrosis. London: Cystic Fibrosis Trust; 2002.

SECTION C

VITAMIN DEFICIENCY, ANTIOXIDANTS, AND SUPPLEMENTATION IN CYSTIC FIBROSIS PATIENTS

CHAPTER

17

Vitamin Supplements: A Role in Cystic Fibrosis Patients?

Alison Morton[1], *Sue Wolfe*[2]

[1]Regional Adult Cystic Fibrosis Unit and Department of Nutrition and Dietetics, Leeds Teaching Hospitals NHS Trust, St James's Hospital, Leeds, UK; [2]Regional Paediatric Cystic Fibrosis Unit and Department of Nutrition and Dietetics, Leeds Teaching Hospitals NHS Trust, Leeds Children's Hospital, Leeds General Infirmary, Leeds, UK

17.1 INTRODUCTION

Most patients with cystic fibrosis (CF), particularly those who are pancreatic insufficient, are at risk of fat-soluble vitamin deficiencies. Even when diagnosis is established early by newborn screening, biochemical evidence of deficiencies can occur [1–4].

There are many factors that contribute to vitamin deficiencies, including fat malabsorption and maldigestion as a consequence of pancreatic enzyme insufficiency, bile salt deficiency, altered intestinal transport, and the presence of a mucus barrier lining the intestine. In some, short gut syndrome due to previous bowel resection and CF-related liver disease may also play a role. Deficiencies may be further exacerbated by a poor dietary intake and increased requirements as a consequence of treatment, e.g., chronic antibiotic use, which may affect vitamin K status, and chronic infection, which may increase the requirement for antioxidant vitamins. Finally, poor adherence with prescribed vitamin supplements, which is commonly recognized in patients with CF, may also contribute [5–7].

There have been many advances in the management of CF that have led to improved life expectancy and new and emerging comorbidities. Advances in our understanding of the role and metabolism of fat-soluble vitamins mean that we now need to consider vitamin status in the context of modern disease [8]. In the past, vitamin status was discussed in terms of overt deficiency resulting in the classically recognized deficiency symptoms such as night blindness and xerophthalmia in vitamin A deficiency. Vitamin status may now be considered in the context of health outcomes rather than overt deficiency [8], with levels being described as deficient, insufficient, adequate, optimal, and toxic. It is recognized that subclinical deficiencies defined as low serum, tissue, or airway surface liquid vitamin levels with no visible signs or symptoms of overt deficiency may play an increasing role.

Most pancreatic-insufficient patients will require supplementation with the fat-soluble vitamins A, D, E, and K. Most pancreatic-sufficient patients require supplementation with vitamin D, and some may require additional vitamin A, E, or K. All patients with CF should have plasma fat-soluble vitamin levels checked annually and supplemental doses readjusted according to these measurements [9,10]. Pancreatic-sufficient patients also need to be monitored annually to ensure continuing adequacy of or need for vitamin supplementation.

17.2 VITAMIN A

The term vitamin A refers to a group of fat-soluble compounds known as the retinoids. Preformed vitamin A refers to retinol or the fatty acid ester derivative. This is the active form of the vitamin and in health 70–90% of vitamin A is absorbed in the intestine. Preformed vitamin A is found in liver, fortified margarines, dairy products, oily fish, and fish oils. The carotenoids are precursors of vitamin A (also referred to as provitamin A), and the most common is beta-carotene, which has one-sixth the potency of retinol. Beta-carotene is found in carrots, red

Diet and Exercise in Cystic Fibrosis
http://dx.doi.org/10.1016/B978-0-12-800051-9.00017-1

peppers, tomatoes, green leafy vegetables such as spinach and broccoli, and yellow fruits such as peaches and mangos. Beta-carotene is less efficiently absorbed from the gut, and the conversion rate decreases as oral intake increases, therefore there is less risk of toxicity. Following absorption from the intestine, vitamin A is stored as retinyl esters. From the liver, retinyl esters are exported to the tissues, mostly bound to retinol-binding protein (RBP) and to transthyretin (formerly known as prealbumin) in a 1:1:1 ratio [11].

17.2.1 Role of Vitamin A

Vitamin A plays a role in the eye, skin, respiratory, and immune systems. Biologically, vitamin A is important for dark adaptation and normal vision, proliferation and differentiation of epithelial cells, maintaining integrity of epithelial cells and mucus secreting epithelia, apoptosis, normal organogenesis, and immune competence. Beta-carotene, the precursor of vitamin A, is an antioxidant.

17.2.2 Assessment of Status

Assessment of vitamin A status in CF is not without some inherent difficulties. Serum retinol is most widely used as an indicator of vitamin A status but it is insensitive and remains normal until hepatic stores are almost depleted [12]. There is poor correlation between clinical and biochemical findings and serum retinol levels.

There is a strong correlation between serum retinol and retinol-binding protein levels in CF [13]. Low serum retinol levels may be due to low levels of RBP. Retinol-binding protein is important for the transfer of retinyl esters to the tissues, and synthesis of RBP is responsible for the homeostatic control of serum retinol concentrations. Retinol-binding protein is produced in the liver and is an acute phase reactant [14–16]. Synthesis of RBP decreases during the acute phase response and serum vitamin A levels may be falsely depressed.

These difficulties in interpreting serum levels in patients with CF during acute respiratory exacerbations are well documented [17,18], with levels being low at the beginning, but normal after intravenous antibiotic treatment. In addition, circulating RBP levels may also be low in patients with CF-related liver disease.

Secondary vitamin A deficiency may occur as a consequence of zinc deficiency as this depresses RBP synthesis and hence release of vitamin A from the liver [19].

Serum vitamin A samples need to be processed quickly to avoid degradation or transported in the dark [20]. In addition to serum retinol levels, assessment of vitamin A status in CF would be aided by measurement of retinol-binding protein and plasma zinc, and an assessment of a positive acute phase protein such as C-reactive protein [16,21].

Though not performed in clinical practice, the measurement of retinyl esters obtained through biopsy is considered the gold standard method of assessing vitamin A status as they reflect the liver's reserve of vitamin A [22].

17.2.3 Signs and Symptoms of Deficiency

Maldigestion and malabsorption of dietary fat are the main risk factors for vitamin A deficiency in patients with CF. However, in addition, there is a significant increase in fecal losses of retinol in CF unrelated to the degree of fat lost in the stools. It has been suggested that this may be due to specific retinol-handling defect in the gastrointestinal tract in people with CF unrelated to digestion and absorption of dietary fat [23].

Beta-carotene is an antioxidant, and oxidative stress and increased free radical formation may contribute to a conditional deficiency due to increased needs.

The major consequence of vitamin A deficiency is ocular, with abnormal dark adaptation (night blindness) and conjunctival and corneal xerosis that can lead to blindness. Historically, xerophthalmia and night blindness have been reported in patients with CF [24] and have been presenting features leading to a diagnosis of CF [25,26]. Asymptomatic conjunctival xerosis with associated night blindness or abnormal dark adaptation despite vitamin A supplementation has also been reported [27–29]. Rarely, cases of night blindness in patients with CF continue to be reported in the literature [30].

Keratoconjunctivitis sicca (dry eye) was reported in adult patients receiving the recommended daily supplemental dose of vitamin A [31,32]. None of the patients had biochemical vitamin A deficiency or clinical (ocular) signs of vitamin A deficiency. More recently, with modern vitamin A supplementation, no significant difference was found in vitamin A or retinol-binding protein levels in pancreatic-insufficient children with CF and pancreatic-sufficient controls, and similar levels of retinal function (electroretinogram amplitudes and implicit times) were reported [33]. Taking appropriate vitamin A supplements can prevent the development of the ocular signs of deficiency.

Other cases of overt vitamin A deficiency in CF take the form of case reports of facial nerve palsy [34–36].

The impact of vitamin A deficiency and subclinical vitamin A deficiency on the respiratory tract may be of importance in CF. Vitamin A is important in the maintenance of mucus secreting epithelia, including the respiratory epithelia. Subclinical vitamin A deficiency may impair respiratory epithelium integrity [37]. It leads to a loss of ciliated cells and an increase in mucus secreting goblet-cells, which impairs mucociliary clearance and favors bacterial adherence [38–40]. Bacterial adhesion to nasopharyngeal epithelial cells is increased in vitamin A deficiency and may permit increased colonization [41].

In severe vitamin A deficiency, there is squamous cell metaplasia due to the mucus-secreting ciliated epithelium being replaced by squamous epithelium. Vitamin A also plays a role in immune competence, with suggested mechanisms including T-cell subset depression, cell differentiation, cytokine modulation, and phagocyte stimulation [42–44]. Beta-carotene may help prevent or slow oxidative damage in the lung.

Linking subclinical vitamin A deficiency and lung function is difficult due to the impact of the acute phase response on plasma vitamin A levels and the multifactorial basis of respiratory exacerbations. However, a direct correlation has been shown between increased number of exacerbations and lower vitamin A levels. This correlation was sustained within normal serum levels, across all lung function, and in both pancreatic-sufficient and -insufficient patients [45]. In addition, a significant positive correlation has been reported between serum retinol levels and measures of lung function in patients with CF [46–48].

17.2.4 Vitamin A Requirements

Vitamin A requirements in CF have not been clearly defined.

17.2.5 Vitamin A Supplementation

There is no international consensus on vitamin A supplemental doses for patients with CF. Current recommendations for pancreatic-insufficient patients are based on historical data and vary between countries [9,10,49]. The recommended starting doses in North America are at less than 1 year of age 1500 IU (450 μg) [10,49], 1–3 years of age 5000 IU (1500 μg) [10,49], 4–8 years of age 5000–10,000 IU (1500–3000 μg) [49] and from 8 years of age 10,000 IU (3000 μg) [49]. Whilst in the UK the recommendations are at: less than 1 year of age 4000 IU (1200 μg) and from 1 year of age 4000–10,000 IU (1200–3000 μg) [50], which are largely in agreement with the European recommendations [9].

The 2012 Cochrane review update still reports no randomized controlled trials or quasi-randomized trials that dosage recommendations can be based on [51]. Doses should be guided by individuals' serum vitamin A levels, taking into consideration RBP and zinc levels accepting that requirements will be variable.

Despite only one case report of vitamin A toxicity in a newly diagnosed infant [52], there is currently concern about the potential for toxicity of vitamin A in CF. Higher levels of vitamin A have been reported in children and adults with CF [53,54], especially following lung transplant [55,56]. These studies highlight the importance of regular assessment and adjustment of supplemental doses.

During pregnancy in women with CF, care should be taken to avoid excessive intakes of vitamin A because of the risk of teratogenicity. In practice, supplemental vitamin A (retinol) is usually continued at a dose of less than 10,000 IU/day (3000 μg) and plasma levels are closely monitored [57].

17.3 VITAMIN D

The two main nutritionally significant forms of vitamin D are ergocalciferol (vitamin D2) and colecalciferol (vitamin D3). The dietary intake of vitamin D (mainly ergocalciferol) is relatively low with very few foods being naturally rich sources. Those that are include fatty fish, fish oils, and liver. In addition, some foods are fortified with either ergocalciferol or colecalciferol, e.g., margarine, fat spreads, and breakfast cereals. Most colecalciferol (vitamin D3) is produced photochemically by exposure of 7-dehydrocholesterol present in the skin to sunlight. Specific wavelengths of ultraviolet light are required for this process, and as a consequence seasonal variation is reported in both patients with CF [58] and the general population, with those living at higher latitudes being most at risk of low levels. In addition, the prescription of supplemental vitamin D as either ergocalciferol or colecalciferol is a significant source for patients with CF.

After synthesis in the skin or absorption from food or supplements, vitamin D is transported to the liver where it is hydroxylated to 25 hydroxy vitamin D (25OHD), the major circulating form of the vitamin. Further hydroxylation takes place in the kidney to produce the major active metabolite 1,25 dihydroxy vitamin D ($1,25(OH)_2D$).

17.3.1 Role of Vitamin D

Vitamin D is primarily involved in calcium and phosphate metabolism. The active metabolite, $1,25(OH)_2D$, regulates calcium transport in the proximal tubules of the kidney. It also regulates calcium and phosphate absorption from the small intestine. In addition, $1,25(OH)_2D$ has a crucial role in bone metabolism, stimulating bone resorption and accrual to help maintain bone health and plasma calcium levels. This process is regulated by parathyroid hormone, which is secreted in response to low calcium levels [59,60].

More recently, vitamin D deficiency has been associated with lower lung function and increased respiratory infections in patients with CF [58,61,62]. The specific relationship between vitamin D status and lung function is not clear, but is thought to be due to a role in regulating inflammation, inducing antimicrobial peptides and its action on muscle [61,63–65]. The precise connection between vitamin D status and lung function is unclear at this point, and the question remains as to whether vitamin D deficiency contributes to the etiology of CF lung disease or whether it is simply a manifestation of the disease itself. Vitamin D deficiency has also been linked with diabetes, cardiovascular disease, and some forms of cancer [59,60], which with increasing life expectancy may have relevance in CF.

17.3.2 Assessment of Status

Serum 25OHD levels have a long half-life of approximately 2–3 weeks and therefore because levels do not fluctuate acutely are the most reliable indicator of vitamin D status [66]. Levels should be checked annually and approximately 12 weeks following any change in supplementation regimen, when steady-state levels of serum 25OHD should have been achieved. The assay used to analyze 25OHD levels should measure both vitamin D2 and vitamin D3, as this is essential to accurately determine vitamin D status. It is also important to recognize that inter-assay and inter-laboratory variability have been reported [67,68]. It is uncertain at the present time whether fasting levels are required.

Because there is seasonal variation in serum vitamin D levels, status varies throughout the year. The highest levels occur in the summer months and the lowest in the winter. Therefore the late autumn/winter months may be the best time to assess vitamin D status when it is most likely to be lowest [66,69].

There are also a number of indirect measures to assess vitamin D status, including serum calcium, phosphorus, and parathyroid hormone concentrations. The North American guide to bone health and disease and the North American evidence-based recommendations for vitamin D deficiency recommend against the routine measurement of these and other indirect markers to assess vitamin D status in individuals with CF [66,69].

17.3.3 Signs and Symptoms of Deficiency

Overt clinical deficiency of vitamin D is rare, although low serum 25OHD levels have been widely reported despite supplementation [58,70–73]. Vitamin D deficiency in CF is due to a number of factors, including reduced absorption of dietary and supplemental vitamin D due to pancreatic insufficiency, poor dietary intakes, low body mass index leading to reduced capacity for vitamin D storage in adipose tissue, reduced levels of vitamin D binding protein [74], impaired hepatic hydroxylation of vitamin D, and reduced sunlight exposure due to poor health or photosensitivity with quinolone antibiotics [75].

Reduced bone mineral density (BMD) due to reduced calcification of bones is the main symptom related to vitamin D deficiency. Severe deficiency causes rickets in children and osteomalacia in adults. Overt vitamin D deficiency in patients with CF is rare, but both rickets and osteomalacia have been reported [76–78].

Suboptimal vitamin D status in CF may contribute to the etiology of low BMD and increase risk of low trauma fracture. Reduced BMD is of concern to patients with CF as it is associated with higher fracture rates [75,79–81], kyphosis [79], and pain. These consequences may limit physiotherapy and may also be a contraindication to transplantation [82].

Although low BMD is common in CF [83,84], because the etiology of CF-related low BMD is complex and multifactorial, studies do not show a direct association between BMD and the vitamin D status of patients with CF [83]. However, it is highly likely that vitamin D deficiency contributes to low bone mineral density in CF [75] as it does in the general population [85].

17.3.4 Vitamin D Requirements

There is currently no consensus on desirable vitamin D levels to optimize bone mineralization in the CF population. The North American CF Bone Health Consensus statement [69], the North American evidence-based recommendations from the Cystic Fibrosis Foundation [66], and the UK CF Trust Bone Mineralization Report [86] recommend achieving a minimum plasma 25-hydroxy vitamin D concentration of 75 nmol/L (30 ng/mL). This is inline with recommended level for the healthy population [87–89]. This contrasts with the European CF Bone Mineralization guidelines [90], Lawson Wilkins Pediatric Endocrine Society [91], and the American Academy of Pediatrics [92], who recommend a threshold level of 50 nmol/L (20 ng/mL). The European panel felt there was insignificant evidence regarding the long-term benefits and consequences of levels greater than 50 nmol/L (20 ng/mL) in large groups of patients with CF. Although the safe upper limit for 25OHD levels has not been established, the risk of hypercalcemia increases when levels exceed 250 nmol/L (100 ng/mL) [59]. It is therefore considered that plasma 25OHD should not exceed this level.

17.3.5 Vitamin D Supplementation

Achieving recommended levels can be difficult in patients with CF, even with considerable amounts of supplementation [70,93–95]. Vitamin D3 (colecalciferol) has been shown to be more effective than vitamin D2 (ergocalciferol) at increasing and sustaining serum vitamin D levels [73,96–98]. Vitamin D2 is more rapidly catabolized and therefore its effect may be more transient. In North America, current recommendations for vitamin D supplementation in CF are broad at 400–2000 IU (10–50 μg) vitamin D3 daily for infants under 1 year of age, 800–4000 IU (20–100 μg) for children aged

1–10 years and 800–10,000 IU (20–250 μg) for those over 10 years of age [69]. Recommendations are similar in Europe at 400–5000 IU (10–125 μg) vitamin D3 daily for children over 1 year of age and adults [90]. Patients who struggle to achieve desirable levels despite these recommendations may improve with 50,000 IU (1250 μg) vitamin D weekly for 3 months [98]. More recently, single high-dose (stoss) vitamin D3 therapy has been shown to improve plasma 25OHD levels for up to 12 months [99]. In this study doses of 100,000 IU (2500 μg) and 600,000 IU (15,000 μg) vitamin D3 were given, the dose being determined by plasma levels and age of the patient.

The supplementation regimen chosen should be individualized dependent on serum levels, patient preference, and adherence. Any change in supplementation dose or regimen should be monitored with repeat serum measurements 12 weeks after the change was made. The results should take into account adherence to supplements and seasonal variation. The North American evidence-based recommendations for screening, diagnosis, management, and treatment of vitamin D deficiency are an age-dependent stepwise approach to increase vitamin D intake when optimal vitamin D status is not achieved [66].

A number of other methods to improve vitamin D status have been explored. Ultraviolet radiation and UVB light therapy has been shown to improve serum vitamin D levels in patients with CF [100,101]. However, adherence to this type of therapy can be poor and it may be an unrealistic therapy for many patients to include in already heavy treatment regimens [98]. Given the limited evidence of efficacy and the potential risk of burn in those with increased photosensitivity, this type of therapy cannot be recommended [66]. There have also been two studies examining the efficacy of supplementation with more polar (more water-soluble) forms of vitamin D. In one study, 25OHD was found to improve status [102] and in the other $1,25(OH)_2D$ improved serum PTH levels [103]. The US CF Foundation recommends that polar preparations should be considered if plasma vitamin D levels are unresponsive to conventional therapy. However, they should only be used in consultation with a specialist with expertise in vitamin D therapy [66].

17.4 VITAMIN E

Vitamin E is a generic term for a group of eight lipid soluble compounds synthesized by plants. There are two classes: tocopherols (α, β, γ, δ) and tocotrienols (α, β, γ, δ), which exhibit different levels of activity. Alpha-tocopherol is the major physiological form. It has the highest biological potency and is the standard against which other forms are compared. Most vitamin E in the diet is derived from vegetable oils, particularly corn, soy and sunflower seed oil, margarines and fat spreads depending on the base oil used and fortification, wholegrain bread and cereal, and nuts and seeds.

17.4.1 Role of Vitamin E

Vitamin E (α-tocopherol) acts as a membrane antioxidant closely associated with polyunsaturated fatty acids. The antioxidant properties of vitamin E may help reduce the effects of free radicals. Free radicals are highly reactive molecules and react with oxygen to form reactive oxygen species (ROS). The body forms ROS when it converts food to energy and when the body is also exposed to environmental sources of free radicals, e.g., air pollution. Antioxidants help protect cell membranes from the damaging effects of free radicals, i.e., oxidative damage.

In CF, bacterial infection and chronic inflammation may contribute to increased free radical production. Evidence supporting the occurrence of oxidative stress in CF is now established [104–107].

In addition to its activities as an antioxidant, vitamin E has an immunomodulatory effect [108]. It is also important in maintaining neurological function.

17.4.2 Assessment of Status

Routine monitoring of vitamin E status in patients with CF is recommended at diagnosis and then at least annually. There is no consensus for reporting vitamin E status. Serum or plasma vitamin E levels are most commonly used to assess vitamin E status. Vitamin E circulates in the blood bound to lipoproteins and as a consequence more accurate measurement of status may be assessed using the vitamin E to total lipid ratio, which requires a fasting blood sample. Normal ratios of α-tocopherol to total lipid are >0.6 mg and >0.8 mg α-tocopherol/gram total lipid in children and adults, respectively [20]. In a study carried out in a non-CF population, 47% of low vitamin E levels were normal when re-evaluated using vitamin E to lipid ratio and 58% with elevated plasma vitamin E were normal or low when re-evaluated with lipid ratio. Elevated triglyceride levels in non-fasting specimens were the most common reason for abnormal results when lipid was not considered [109]. In patients with CF, low plasma vitamin E levels may be a reflection of low total lipid levels and the vitamin E to total lipid ratio may actually be within the normal range.

Though not often used in clinical practice, platelet tocopherol is suggested to be a better measure of vitamin E nutritional status because it is more sensitive to vitamin E intake and not dependent on circulating lipid levels [20].

Biomarkers of oxidative stress are elevated in many patients with CF despite normal vitamin E levels. This suggests that vitamin E reference ranges provide incomplete information on the redox status [107].

17.4.3 Signs and Symptoms of Deficiency

The digestive tract requires fat to absorb vitamin E, and consequently fat malabsorption in CF contributes to vitamin E deficiency. Increased oxidative stress in CF may also contribute to increased requirements for vitamin E [110].

Biochemical evidence of vitamin E deficiency, however, occurs early in infants with CF diagnosed by newborn screening, though the prevalence varies [1–4]. Vitamin E status is significantly correlated with pancreatic status, with levels at diagnosis being significantly lower in pancreatic-insufficient patients [4].

Severe vitamin E deficiency is rare in the non-CF population and due to early supplementation, it is now rare in most CF populations. Biochemical deficiency of vitamin E still occurs and has recently been reported as 13.9% in an Australian pediatric CF population [111]. Neurological symptoms of vitamin E deficiency have been described in the past in children and adults with CF including peripheral neuropathy, areflexia, ataxia, pigmentary retinopathy, abnormal somatosensory, and visual evoked potentials [112–114]. More recently, 48.9% of children with CF have been described as having electrophysiological evidence of peripheral neuropathy (predominantly axonal, sensory, and polyneuropathy) [115].

Vitamin E deficiency also causes hemolytic anemia due to decreased erythrocyte survival time and increased red blood cell fragility with susceptibility to hemolysis. Historically, hemolytic anemia has been reported as occurring as early as 6 weeks in three infants with CF [116] and also as a presenting feature of CF [117].

More recently, there has been interest in the role of vitamin E in cognitive function in CF. Patients with prolonged vitamin E deficiency due to delayed diagnosis were found to have a significantly lower Cognitive Skills Index and cognitive factor scores. Newborn screening programs that lead to early diagnosis, prevention of prolonged periods of malnutrition, and reduced duration of vitamin E deficiency are associated with better cognitive function in children with CF [118,119].

Studies suggest that patients with CF have inadequate antioxidant defenses to cope with elevated oxidative stress [110]. There have been reports suggesting reduced levels of vitamin E within the normal range are associated with an increased rate of pulmonary exacerbations [45].

17.4.4 Vitamin E Requirements

The dietary requirement for vitamin E in the non-CF population has been determined from intakes providing plasma α-tocopherol concentrations that inhibit hydrogen peroxide-induced hemolysis or the prevention of lipid peroxidation [120]. The requirement for vitamin E depends on polyunsaturated fatty acid intake, which varies widely [121]. Research is required to address the potential benefits of vitamin E on relevant clinical health outcomes [120].

17.4.5 Vitamin E Supplementation

It is recommended that routine vitamin E supplementation should be started in all pancreatic-insufficient patients at diagnosis and in pancreatic-sufficient patients if low serum vitamin E levels are detected. Current recommendations for supplemental vitamin E doses in pancreatic-insufficient patients are based on historical data and vary between countries [9,10,49,50]. This variability is due to a lack of strong evidence on optimal doses combined with earlier intervention, improved pancreatic enzyme replacement therapy, and the improving clinical condition of patients with CF. It should be recognized that the recommendations were made in the context of preventing clinical and biochemical deficiency and not in the context of vitamin E's antioxidant role. Doses should be adjusted in response to annual measurement of serum levels.

The recommended starting doses in North America are: less than 1 year of age 40–50 IU [10,49], 1–3 years of age 80–150 IU [10,49], 4–8 years of age 100–200 IU [49], and from 8 years of age 200–400 IU [49]. Whilst in the UK the recommendations are less than 1 year of age 15–75 IU, from 1 year of age 75–150 IU and adults 150–300 IU [50] which is largely in agreement with the European recommendations (100–400 IU/day) [9].

Biomarkers of oxidative stress are elevated in many patients with CF despite normal vitamin E levels. This suggests that vitamin E reference ranges provide incomplete information on the redox status [107]. A Cochrane Review of Antioxidant Micronutrients for Lung Disease in Cystic Fibrosis found conflicting evidence regarding the clinical effectiveness of antioxidants and stated that antioxidant supplementation was not recommended beyond routine care, i.e., to prevent vitamin E deficiency [122].

Optimal vitamin E status as conveyed by biomarkers and clinical outcome require further investigation. The impact of changes in supplementation protocols requires close monitoring to determine supplement formulation and best dosing practices. Further studies are required to understand the potential role of vitamin E in the health of patients with CF [8,106].

High plasma vitamin E levels have been reported in pancreatic-insufficient patients with CF [123] and following lung transplantation [55,56]. This emphasizes the need for regular nutritional assessment and surveillance.

Hypervitaminosis E occurs only with very large supplemental doses and may cause bruising and bleeding with increased prothrombin time. This is due to inhibition of vitamin K dependent carboxylase and can be reversed by administering vitamin K. Other symptoms may include fatigue, weakness, and gastrointestinal upset.

17.5 VITAMIN K

Vitamin K is a fat-soluble vitamin that exists in both natural and synthetic forms. Vitamin K1 (phylloquinone) is synthesized by plants and in the diet is present in dark green leafy vegetables such as spinach, broccoli, cabbage, and kale. Some vegetable oils such as rapeseed, soybean, and olive oil are also rich sources. Dairy products, meat, and eggs contain vitamin K1 in smaller amounts. Vitamin K2 (menaquinone) is synthesized by gram-positive bacteria present in the jejunum and ileum. It is unclear how much the bacterial flora contribute to overall vitamin K status [124,125]. In the diet, vitamin K2 is found in fermented foods such as sauerkraut, fermented soybean products such as natto, and fermented cheese (e.g., Gorgonzola, Roquefort, and Parmesan). Vitamin K3 (menadione) and vitamin K4 (menadiol) are synthetic forms of the vitamin. It is recommended that supplementation with vitamin K3 is avoided in humans as it is linked to hemolysis and liver damage in the newborn and also has the potential for mutagenicity [12,121].

17.5.1 Role of Vitamin K

Vitamin K is an essential cofactor in the post-translational carboxylation of glutamyl (Glu) residues to γ carboxyglutamyl (Gla) residues. Undercarboxylated Gla-dependent proteins are functionally inactive. Major Gla-proteins (active) include prothrombin, osteocalcin, and other bone metabolism-related proteins. Hence vitamin K is important for blood coagulation [126–128] and bone health. Interest has also been shown in its role in energy metabolism and inflammation [129], both of which may be of importance in CF.

17.5.2 Assessment of Status

Assessment of vitamin K status is difficult due to a lack of robust measures of vitamin K sufficiency and insufficiency [130,131].

Serum levels of vitamin K are an unreliable measure of status [49,50]. Prothrombin time may be considered an indirect measure; however, it reflects only severe deficiency of the liver and is an insensitive marker of overall vitamin K status. Just 50% of the normal prothrombin concentration produces a normal prothrombin time. As a consequence, prothrombin times are rarely abnormal in CF even if advanced liver disease is present [132]. The most sensitive indicator of vitamin K deficiency is the presence of circulating Gla proteins in their undercarboxylated forms. However, it appears that different vitamin K dependent proteins have different vitamin K requirements. Protein or prothrombin induced by vitamin K absence or antagonism (PIVKA II) is the undercarboxylated form of prothrombin and is more sensitive than prothrombin time to assess vitamin K status of the liver. Undercarboxylated osteocalcin is the most sensitive indicator of vitamin K status of the bone and is the first Gla protein to occur in the undercarboxylated form in deficient states [133]. Unfortunately, PIVKA II and undercarboxylated osteocalcin are not routinely available in many CF centers. They are usually available only as research tools.

17.5.3 Signs and Symptoms of Deficiency

Absorption of vitamin K from the gut is dependent on bile salts and pancreatic lipase secretion, which is stimulated by dietary fat. Risk factors for vitamin K deficiency in CF include fat malabsorption and maldigestion, bile salt deficiency, liver disease, chronic antibiotic use, short gut due to bowel resection, and inadequate dietary intake [134].

Over 50 years ago, bleeding and vitamin K deficiency were reported as an early sign of CF [135,136]. Coagulopathies including hematomas, intracerebral hemorrhage, and severe life-threatening bleeding associated with vitamin K deficiency, although rare, continue to be reported in patients with CF [127,128,137].

Subclinical deficiency is more common and may contribute to CF-related low bone mineral density. Some studies have shown high levels of undercarboxylated osteocalcin, which have been associated with reduced lumbar spine bone mineral content [138], reduced levels of markers of bone mineral accrual [72,133,138], and increased levels of markers of bone turnover [133].

In addition, vitamin K and vitamin K dependent proteins may play a role in energy metabolism and inflammation, which may be of relevance in CF. A recent observational study in the non-CF population has shown that low vitamin K status is inversely associated with circulating measures of inflammation [139]. The mechanism underlying the potential influence of vitamin K on inflammatory cytokine production is unclear, but it has been suggested that vitamin K may suppress inflammation by decreasing expression of genes for individual cytokines. This may be of importance in CF where inflammation and infection contribute to lung damage.

17.5.4 Vitamin K Requirements

Although vitamin K is an essential nutrient, even in the general population, recommendations on requirements have not been established. This is mainly because of the uncertainty of the contribution of vitamin K2 production in the gut.

17.5.5 Vitamin K Supplementation

In the past vitamin K supplementation was given only to patients with overt deficiency characterized by abnormal prothrombin times. With the increasing understanding of the role of vitamin K and the recognition of subclinical deficiency in CF, routine vitamin K supplementation is now recommended for all patients [9,69,86,90].

In the general population, vitamin K intervention studies have reported increased bone mineral density, reduced fracture rates and improved bone strength [129]. However, as with vitamin D, because of the multifactorial nature of low BMD in CF, cause and effect will be difficult to establish.

The correct supplemental dose of vitamin K is yet to be established. A number of studies have been published indicating that despite supplementation with varying doses of vitamin K, subclinical deficiency persists in some patients. In these studies, doses of 5–10mg daily [140], 0.18mg daily for a minimum of 4months [141], <0.25 to >1mg/day mg daily [142], 1–5mg daily for 1month [143], and <150 μg to ≥1000 μg daily [144] have all been used.

Current recommendations for vitamin K supplementation are therefore based on historical data and vary between countries [9,10,49,50,69,86,90]. As there is limited storage capacity for vitamin K and rapid metabolic turnover (within 24h), daily rather than weekly supplementation appears more appropriate. The North American CF Foundation evidence-based guidelines for the management of infants with CF and the North American guide to bone health and disease in CF recommend 300–500 μg daily for all age groups [10,69]. In the UK, the recommended doses are 300 μg/kg body weight daily rounded to the nearest milligram for infants under 2years of age, 5mg daily for children aged 2–7years, and 10mg daily for those greater than 7years of age [86]. The European bone mineralization guidelines advise 0.5–2mg daily for infants and a starting dose of at least 1–10mg daily in children above 1year of age and adults [90]. The Cochrane Review on vitamin K supplementation for CF concluded that evidence from randomized controlled trials on the benefits of vitamin K supplementation for patients with CF is currently weak and limited to two trials of short duration [145]. However, no harm was found from supplementation and because of the potential beneficial effects, the authors concluded that until further evidence is available, the present recommendations should be adhered to [145].

17.6 WATER-SOLUBLE VITAMINS

Small studies that have assessed the water-soluble vitamin status of children and adults with CF have found that water-soluble vitamin status is adequate in the majority. In general, water-soluble vitamins are well absorbed in CF and routine supplementation is not necessary [50]. Many patients may receive additional water-soluble vitamins along with fat-soluble vitamin preparations, in enteral tube feeds or in oral calorie supplements. There is however, an increased risk of deficiency in patients with CF who have advanced lung disease, liver disease, or a poor or restrictive dietary intake, and those undergoing a period of rapid growth.

With the exception of vitamin B12, pancreatic secretions are not involved in the absorption of water-soluble vitamins. However, vitamin B12 deficiency is rare even in those who are pancreatic insufficient [146]. Pancreatic enzyme replacement therapy is very efficient at improving absorption of vitamin B12 [147]. There is an increased risk of deficiency in patients following terminal ileal resection or in those who do not produce intrinsic factor. In these patients vitamin B12 may result in megaloblastic anemia. High serum levels of vitamin B12 have also been reported in people with CF. This finding is possibly associated with higher levels of supplementation in CF-specific vitamin preparations [148].

There have been isolated case reports of thiamine, riboflavin, and pyridoxine deficiency in patients with CF [149,150]. Deficiency may be due to increased requirements, inadequate absorption, or a poor biochemical rate of conversion.

All women with CF planning a pregnancy should be advised to take 400 μg of folate daily for the prevention of neural tube defects. This level of supplementation should be taken until the 12th week of pregnancy [57]. Patients who have diabetes should be advised to take a higher dose of 5mg daily, because the risk of fetal abnormalities is higher in diabetics [151].

Currently the role of antioxidant vitamins is under investigation [107]. Antioxidants may help protect against oxidative lung damage caused by persistent infection. Low levels of vitamin C [152] and impaired status of glutathione and carotenoids [153] have been reported in CF. The Cochrane review examining the effects of antioxidant micronutrients for lung disease in CF concluded that further trials were required before a firm conclusion regarding the need for and effects of antioxidant supplementation can be drawn [122].

References

[1] Sokol RJ, Reardon MC, Accurso FJ, Stall C, Narkewicz M, Abman SH, et al. Fat-soluble-vitamin status during the first year of life in infants with cystic fibrosis identified by screening of newborns. Am J Clin Nutr 1989;50:1064–71.

[2] Feranchak AP, Sontag MK, Wagener JS, Hammond KB, Accurso FJ, Sokol RJ. Prospective, long-term study of fat-soluble vitamin status in children with cystic fibrosis identified by newborn screen. J Pediatr 1999;136:601–10.

[3] Bines JE, Truby HD, Armstrong DS, Carzino R, Grimwood K. Vitamin A and E deficiency and lung disease in infants with cystic fibrosis. J Pediatr Child Health 2005;41:663–8.

[4] Neville LA, Ranganathan SC. Vitamin D in infants with cystic fibrosis diagnosed by newborn screening. J Pediatr Child Health 2009;45:36–41.

[5] Abbott J, Dodd M, Bilton D, Webb AK. Treatment compliance in adults with cystic fibrosis. Thorax 1994;49:15–20.

[6] Borowitz D, Wegman T, Harris M. Preventive care for patients with chronic illness. Multivitamin use in patients with cystic fibrosis. Clin Pediatr 1994;33:720–5.

[7] Modi AC, Lim CS, Yu N, Geller D, Wagner MH, Quittner AL. A multi-method assessment of treatment adherence for children with cystic fibrosis. J Cyst Fibros 2006;5:177–85.

[8] Maqbool A, Stallings VA. Update on fat-soluble vitamins in cystic fibrosis. Curr Opin Pulm Med 2008;14:574–81.

[9] Sinaasappel M, Stern M, Littlewood J, Wolfe S, Steinkamp G, Heijerman HG, et al. Nutrition in patients with cystic fibrosis: a European Consensus. J Cyst Fibros 2002;1:51–75.

[10] Borowitz D, Robinson KA, Rosenfeld M, Davis SD, Sabadosa KA, Spear SL, et al. Cystic Fibrosis Foundation evidence-based guidelines for management of infants with cystic fibrosis. J Pediatr 2009;155:S73–93.

[11] Ross A. Vitamin A and carotenoids. In: Shils M, Shikem M, Ross A, Caballero B, Cousins R, editors. Modern nutrition in health and disease. 10th ed. Baltimore: Lippincott, Williams, and Wilkins; 2006.

[12] Expert Group on Vitamin and Minerals (EVM). Safe upper levels for vitamins and minerals. London: FSA; 2003.

[13] Lindblad A, Diczfalusy U, Hultcrantz R, Thorell A, Strandvik B. Vitamin A concentration in the liver decreases with age in patients with cystic fibrosis. J Pediatr Gastroenterol Nutr 1997;24: 264–70.

[14] Beisel WR. Infection-induced depression of serum retinol-a component of the acute phase response or a consequence? Am J Clin Nutr 1998;68:993–4.

[15] Thurnham DI, McCabe GP, Northrop-Clewes CA, Nestel P. Effects of subclinical infection on plasma retinol concentrations and assessment of prevalence of vitamin A deficiency: meta-analysis. Lancet 2003;362:2052–8.

[16] Duncan A, Talwar D, McMillan DC, Stefanowicz F, O'Reilly D, St J. Quantitative data on the magnitude of the systemic inflammatory response and its effect on micronutrient status based on plasma measurements. Am J Clin Nutr 2012;95:64–71.

[17] Duggan C, Colin AA, Agil A, Higgins L, Rifai N. Vitamin A status in acute exacerbations of cystic fibrosis. Am J Clin Nutr 1996;64:635–9.

[18] Greer RM, Buntain HM, Lewindon PJ, Wainwright CE, Potter JM, Wong JC, et al. Vitamin A levels in patients with CF are influenced by the inflammatory response. J Cyst Fibros 2004;3:143–9.

[19] Christian P, West Jr KP. Interactions between zinc and vitamin A: an update. Am J Clin Nutr 1998;68:435S–41S.

[20] Tietz NW, editor. Clinical guide to laboratory tests. 3rd ed. Philadelphia: W.B. Saunders Company; 1995.

[21] Stephenson CB, Gildengorin G. Serum retinol, the acute phase response, and the apparent misclassification of vitamin A status in the third National Health and Nutrition Examination Survey. Am J Clin Nutr 2000;72:1170–8.

[22] Tanumihardjo SA. Assessing vitamin A status: past, present and future. J Nutr 2004;134:290s–3S.

[23] Ahmed F, Ellis J, Murphy J, Wooton S, Jackson AA. Excessive faecal losses of vitamin A (retinol) in cystic fibrosis. Arch Dis Child 1990;65:589–93.

[24] Petersen RA, Petersen VS, Robb RM. Vitamin A deficiency with xerophthalmia and night blindness in cystic fibrosis. Am J Dis Child 1968;116:662–5.

[25] Lindenmuth KA, Del Monte M, Marino LR. Advanced xerophthalmia as a presenting sign in cystic fibrosis. Ann Ophthalmol 1989;21:189–91.

[26] Joshi D, Dhawan A, Baker AJ, Heneghan MA. An atypical presentation of cystic fibrosis: a case report. J Med Case Rep 2008;2:201–2.

[27] Rayner RJ, Tyrrell JC, Hiller EJ, Marenah C, Neugebauer MA, Vernon SA, et al. Night blindness and conjunctival xerosis caused by vitamin A deficiency in patients with cystic fibrosis. Arch Dis Child 1989;64:1151–6.

[28] Neugebauer MA, Vernon SA, Brimlow G, Tyrrell JC, Hiller EJ, Marenah C. Nyctalopia and conjunctival xerosis indicating vitamin A deficiency in cystic fibrosis. Eye 1989;3:360–4.

[29] Huet F, Semama D, Maingueneau C, Charavel A, Nivelon JL, Vitamin A deficiency and nocturnal vision in teenagers with cystic fibrosis. Eur J Pediatr 1997;156:949–51.

[30] Roddy MF, Greally P, Clancy G, Leen G, Feehan S, Elnazir B. Night blindness in a teenager with cystic fibrosis. Nutr Clin Pract 2011;26:718–21.

[31] Morkeberg JC, Edmund C, Prause JU, Lanng S, Koch K, Michaelson KF. Ocular findings in cystic fibrosis patients receiving vitamin A supplementation. Graefes Arch Clin Exp Ophthalmol 1995;233:709–13.

[32] Ansari EA, Sahini K, Etherington C, Morton A, Conway SP, Moya E, et al. Ocular signs and symptoms and vitamin A status in patients with cystic fibrosis treated with daily vitamin A supplements. Br J Ophthalmol 1999;83:688–91.

[33] Whatham A, Suttle C, Blumenthal C, Allen J, Gaskin K. ERGs in children with pancreatic enzyme insufficient and pancreatic enzyme sufficient cystic fibrosis. Doc Ophthalmol 2009;119:43–50.

[34] Cameron C, Lodes MW, Gershan WM. Facial nerve palsy associated with a low serum vitamin A level in an infant with cystic fibrosis. J Cyst Fibros 2007;6:241–3.

[35] Basu AP, Kumar P, Devlin AM, O'Brien CJ. Cystic fibrosis presenting with bilateral facial palsy. Eur J Paediatr Neurol 2007;11:240–2.

[36] Obeid M, Price J, Sun L, Scantlebury MH, Overby P, Sidhu R, et al. Facial palsy and idiopathic hypertension in twins with cystic fibrosis and hypovitaminosis A. Pediatr Neurol 2011;44:150–2.

[37] McCullough FSW, Northrop-Clewes CA, Thurnham DI. The effect of vitamin A on epithelial integrity. Proc Nutr Soc 1999;58: 289–93.

[38] Biesalski HK, Stofft E. Biochemical, morphological and functional aspects of systemic and local vitamin A deficiency in the respiratory tract. Ann NY Acad Sci 1992;669:325–31.

[39] Biesalski HK, Nohr D. Importance of vitamin-A for lung function and development. Mol Aspects Med 2003;24:431–40.

[40] Biesalski HK, Nohr D. New aspects in vitamin A metabolism: the role of retinyl esters as systemic and local sources for retinol in mucous epithelia. J Nutr 2004;134:3453S–7S.

[41] Chandra RK. Increased bacterial binding to respiratory epithelial cells in vitamin A deficiency. BMJ 1988;297:834–5.

[42] Ross AC. Vitamin A status and relationship to immunity and the antibody response. Proc Soc Exp Biol Med 1992;200:303–20.

[43] Semba RD, West KP, Sommer A, Griffin DE, Ward BJ, Griffin DE, et al. Abnormal T-cell subset proportions in vitamin-A-deficient children. Lancet 1993;8836:5–8.

[44] Stephenson CB. Vitamin A, infection, and immune function. Annu Rev Nutr 2001;21:167–92.

[45] Hakim F, Kerem E, Rivlin J, Bentur L, Stankiewicz H, Bdolach-Abram T, et al. Vitamin A and E and pulmonary exacerbations in patients with cystic fibrosis. J Pediatr Gastroenterol Nutr 2007;45:347–53.

[46] Aird FK, Greene SA, Ogston SA, Macdonald TM, Mukhopadhyay S. Vitamin A and lung function in CF. J Cyst Fibros 2006;5: 129–31.

[47] Low SS, Lee J, Rao P, Reiboldt W. The correlation of serum vitamin A & E levels to lung function values & acute infective exacerbation rates in adult CF patients. Pediatr Pulmonol Suppl 2010;33:427.

[48] Rivas-Crespo MF, González JD, Acuña QMD, Sojo AA, Heredia GS, Diaz MJJ, et al. High serum retinol and lung function in young patients with cystic fibrosis. J Pediatr Gastroenterol Nutr 2013;56:657–62.

[49] Borowitz D, Baker RD, Stallings V. Consensus report on nutrition for pediatric patients with cystic fibrosis. J Pediatr Gastroenterol Nutr 2002;35:246–59.

[50] Cystic Fibrosis Trust. Nutritional management of cystic fibrosis. UK Cystic Fibrosis Trust Nutrition Working Group. Bromley: Cystic Fibrosis Trust; 2002.

[51] Bonifant CM, Shevill E, Chang AB. Vitamin A supplementation for cystic fibrosis. Cochrane Database Syst Rev 2012:CD006751. http://dx.doi.org/10.1002/14651858.CD006751.pub3.

[52] Eid NS, Shoemaker LR, Samiec TD. Vitamin A in cystic fibrosis: case report and review of the literature. J Pediatr Gastroenterol Nutr 1990;10:265–9.

[53] Graham-Maar RC, Schall J, Stettler N, Zemel BS, Stallings VA. Elevated vitamin A intake and serum retinol in preadolescent children with cystic fibrosis. Am J Clin Nutr 2006;84:174–82.

[54] Maqbool A, Graham-Marr RC, Schall J, Zemel B, Stallings VA. Vitamin A intake and elevated serum retinol levels in children and young adults with cystic fibrosis. J Cyst Fibros 2008;7:137–41.

[55] Stephenson A, Brotherwood M, Robert R, Durie P, Verjee Z, Chaparro C, et al. Increased vitamin A and E levels in adult cystic fibrosis patients after lung transplantation. Transplantation 2005;79:613–5.

[56] Ho T, Gupta S, Brotherwood M, Robert R, Cortes D, Verjee Z, et al. Increased serum vitamin A and E levels after lung transplantation. Transplantation 2011;92:601–6.

[57] Edenborough FP, Borgo G, Knoop C, Lannefors L, Mackenzie WE, Madge S, et al. Guidelines for the mamagement of pregnancy in women with cystic fibrosis. J Cyst Fibros 2008;7:S2–32.

[58] Wolfenden LL, Judd SE, Shah R, Sanyal R, Ziegler TR, Tangpricha V. Vitamin D and bone health in adults with cystic fibrosis. Clin Endocrinol 2008;69:374–81.

[59] Holick MF. Medical progress: vitamin D deficiency. N Engl J Med 2007;357:266–81.

[60] Holick MF, Chen TC. Vitamin D deficiency: a worldwide problem with health consequences. Am J Clin Nutr 2008;87:1080S–6S.

[61] Herscovitch K, Dauletbaev N, Lands LC. Vitamin D as an antimicrobial and anti-inflammatory therapy for cystic fibrosis. Paediatr Respir Rev 2014;15:154–62.

[62] McCauley LA, Thomas W, Laguna TA, Regelmann WE, Moran A, Polgreen LE. Vitamin D deficiency is associated with pulmonary exacerbations in children with cystic fibrosis. Ann Am Thorac Soc 2014;11:198–204.

[63] Finklea JD, Grossmann RE, Tangpricha V. Vitamin D and chronic lung disease: a review of molecular mechanisms and clinical studies. Adv Nutr 2011;2:244–53.

[64] Grossmann RE, Zughaier SM, Liu S, Lyles RH, Tangpricha V. Impact of vitamin D supplementation on markers of inflammation in adults with cystic fibrosis hospitalized for a pulmonary exacerbation. Eur J Clin Nutr 2012;66:1072–4.

[65] Bartley J, Garrett J, Grant CC, Camargo Jr CA. Could vitamin D have a potential anti-inflammatory and anti-infective role in bronchiectasis? Curr Infect Dis Rep 2013;15:148–57.

[66] Tangpricha V, Kelly A, Stephenson A, Maguiness K, Enders J, Robinson KA, et al., Cystic Fibrosis Foundation Vitamin D Evidence-Based Review Committee. An update on the screening, diagnosis, management and treatment of vitamin D deficiency in individuals with cystic fibrosis: evidence-based recommendations from the Cystic Fibrosis Foundation. J Clin Endocrinol Metab 2012;97:1082–93.

[67] Binkley N, Krueger D, Lensmeyer G. 25-hydroxyvitamin D measurement, 2009: a review for clinicians. J Clin Densitom 2009;12:417–27.

[68] Carter GD, Berry JL, Gunter E, Jones G, Jones JC, Makin HL, et al. Proficiency testing of 25-hydroxy vitamin D (25-OH) assays. J Steroid Biochem Mol Biol 2010;121:176–9.

[69] Aris RM, Merkel PA, Bachrach LK, Borowitz DS, Boyle MP, Elkin SL, et al. Guide to bone health and disease in cystic fibrosis. J Clin Endocrinol Metab 2005;90:1888–96.

[70] Stephenson A, Brotherwood M, Robert R, Atenafu E, Corey M, Tullis E. Cholecalciferol significantly increases 25-hydroxyvitamin D concentrations in adults with cystic fibrosis. Am J Clin Nutr 2007;85:1307–11.

[71] Conway SP, Oldroyd B, Brownlee KG, Wolfe SP, Truscott JG. A cross-sectional study of bone mineral density in children and adolescents attending a Cystic Fibrosis Centre. J Cyst Fibros 2008;7:469–76.

[72] Grey V, Atkinson S, Dury D, Casey L, Ferland G, Gundberg C, et al. Prevalence of low bone mass and deficiencies of vitamin D and K in pediatric patients with cystic fibrosis from 3 Canadian centers. Pediatrics 2008;122:1014–20.

[73] Green DM, Leonard AR, Paranjape SM, Rosenstein BJ, Zeitlin PL, Mogayzel Jr PJ. Transient effectiveness of vitamin D_2 therapy in pediatric cystic fibrosis patients. J Cyst Fibros 2010;9:143–9.

[74] Speeckaert MM, Wehlou C, Vandewalle S, Taes YE, Robberecht E, Delanghe JR. Vitamin D binding protein, a new nutritional marker in cystic fibrosis patients. Clin Chem Lab Med 2008;46:365–70.

[75] Hall WB, Sparks AA, Aris RM. Vitamin D deficiency in cystic fibrosis. Int J Endocrinol 2010, 218691. http://dx.doi.org/10.1155/2010/218691. Epub January 28, 2010.

[76] Scott J, Elias E, Moult PJ, Barnes S, Wills MR. Rickets in adult cystic fibrosis with myopathy, pancreatic insufficiency and proximal tubular dysfunction. Am J Med 1977;63:488–92.

[77] Friedman HZ, Ingman CB, Favus MJ. Vitamin D metabolism and osteomalacia in cystic fibrosis. Gastroenterology 1985;88:808–13.

[78] Elkin SL, Vedi S, Bord S, Garrahan NJ, Hodson ME, Compston JE. Histomorphometric analysis of bone biopsies from the iliac crest of adults with cystic fibrosis. Am J Respir Crit Care Med 2002;166:1470–4.

[79] Aris RM, Renner JB, Winders AD, Buell HE, Riggs DB, Lester GE, et al. Increased rates of fractures and severe kyphosis: sequelae of living into adulthood with cystic fibrosis. Ann Intern Med 1998;128:186–93.

[80] Stephenson A, Jamal S, Dowdell T, Pearce D, Corey M, Tullis E. Prevalence of vertebral fractures in adults with cystic fibrosis and their relationship to bone mineral density. Chest 2006;130:539–44.

[81] Papaioannou A, Kennedy CC, Freitag A, O'Neill J, Pui M, Ioannidis Get al., Longitudinal analysis of vertebral fracture and BMD in a Canadian cohort of adult cystic fibrosis patients. BMC Musculoskelet Disord 2008;9:125.

[82] Orens JB, Estenne M, Arcasoy S, Conte JV, Corris P, Egan JJ, et al., Pulmonary Scientific Council of the International Society for Heart and Lung Transplantation: International guidelines for the selection of lung transplant candidates: 2006 update – a consensus report from the Pulmonary Scientific Council of the International Society for Heart and Lung Transplantation. J Heart Lung Transplant 2006;25:745–55.

[83] Haworth CS. Impact of cystic fibrosis on bone health. Curr Opin Pulm Med 2010;16:616–22.

[84] Paccou J, Zeboulon N, Combescure C, Gossec L, Cortet B. The prevalence of osteoporosis, osteopenia and fractures among adults with cystic fibrosis: a systematic literature review with meta-analysis. Calcif Tissue Int 2010;86:1–7.

[85] Cranney A, Horsley T, O'Donnell S, Weiler H, Puil L, Ooi D, et al. Effectiveness and safety of vitamin D in relation to bone health. Evid Rep Technol Assess 2007;158:1–235.

[86] Cystic Fibrosis Trust Consensus Document. Bone mineralisation in cystic fibrosis. Bromley: Cystic Fibrosis Trust; 2007.

[87] Bischoff-Ferrari HA, Giovannucci E, Willett WC, Dietrich T, Dawson-Hughes B. Estimation of optimal serum concentrations of 25-hydroxyvitamin D for multiple health outcomes. Am J Clin Nutr 2006;84:18–28.

[88] Dawson-Hughes B, Mithal A, Bonjour JP, Boonen S, Burckhardt P, Fuleihan GE, et al. IOF position statement: vitamin D recommendations for older adults. Osteoporos Int 2010;21:1151–4.

[89] Holick MF, Binkley NC, Bischoff-Ferrari HA, Gordon CM, Hanley DA, Heaney RP, et al. Evaluation, treatment, and prevention of vitamin D deficiency: an endocrine society clinical practice guideline. J Clin Endocrinol Metab 2011;96:1911–30.

[90] Sermet-Gaudelus I, Bianchi ML, Garabedian M, Aris RM, Morton A, Hardin DS, et al. EuroCareCF: European cystic fibrosis bone mineralisation guidelines. J Cyst Fibros 2011;10:S16–23.

[91] Misra M, Pacaud D, Petryk A, Collett-Solberg PF, Kappy M, Drug and Therapeutics Committee of the Lawson Wilkins Pediatric Endocrine Society. Vitamin D deficiency in children and its management: review of current knowledge and recommendations. Pediatrics 2008;122:398–417.

[92] Wagner CL, Greer FR. Prevention of rickets and vitamin D deficiency in infants, children and adolescents. Pediatrics 2008; 122:1142–52.

[93] Boyle MP, Noschese ML, Watts SL, Davis ME, Stenner SE, Lechtzin N. Failure of high-dose ergocalciferol to correct vitamin D deficiency in adults with cystic fibrosis. Am J Respir Crit Care Med 2005;172:212–7.

[94] Rovner AJ, Stallings VA, Schall JI, Leonard MB, Zemel BS. Vitamin D insufficiency in children, adolescents, and young adults with cystic fibrosis despite routine oral supplementation. Am J Clin Nutr 2007;86:1694–9.

[95] Boas SR, Hageman JR, Ho LT, Liveris M. Very high-dose ergocalciferol is effective for correcting vitamin D deficiency in children and young adults with cystic fibrosis. J Cyst Fibros 2009;8:270–2.

[96] Trang HM, Cole DE, Rubin LA, Pierratos A, Siu S, Vieth R. Evidence that vitamin D3 increases serum 25-hydroxyvitamin D more efficiently than does vitamin D2. Am J Clin Nutr 1998;68:854–8.

[97] Armas LA, Hollis BW, Heaney RP. Vitamin D2 is much less effective than vitamin D3 in humans. J Clin Endocrinol Metab 2004;89:5387–91.

[98] Khazai NB, Judd SE, Jeng L, Wolfenden LL, Stecento A, Ziegler TR, et al. Treatment and prevention of vitamin D insufficiency in cystic fibrosis patients: comparative efficacy of ergocalciferol, cholecalciferol and UV light. J Clin Endocrinol Metab 2009;94:2037–43.

[99] Shepherd D, Belessis Y, Katz T, Morton J, Field P, Jaffe A. Single high-dose oral vitamin D3 (stoss) therapy – a solution to vitamin D deficiency in children with cystic fibrosis? J Cyst Fibros 2013;12:177–82.

[100] Gronowitz E, Larko O, Gilljam M, Hollsing A, Lindblad A, Melstrom D, et al. Ultraviolet B radiation improves serum levels of vitamin D in patients with cystic fibrosis. Acta Paediatr 2005;94:547–52.

[101] Chandra P, Wolfenden LL, Ziegler TR, Tian J, Luo M, Stecenko AA, et al. Treatment of vitamin D deficiency with UV light in patients with malabsorption syndromes: a case series. Photodermatol Photoimmunol Photomed 2007;23:179–85.

[102] Enfissi L, Bianchi ML, Galbiati E, Faraifoger S, Arban D, Moretti E, et al. Osteoporosis in CF: calcifediol therapy increases bone mineral density (BMD). Pediatr Pulmonol Suppl 2001;32:334. [Abstract].

[103] Brown SA, Ontjes DA, Lester GE, Lark RK, Hensler MB, Blackwood AD, et al. Short-term calcitriol administration improves calcium homeostasis in adults with cystic fibrosis. Osteopor Int 2003;14:442–9.

[104] Cantin AM, White TB, Cross CE, Forman HJ, Sokol RJ, Borowitz D. Antioxidants in CF. Conclusions from the CF antioxidant workshop, Bethesda, Maryland. Free Radic Biol Med 2007;42: 15–31.

[105] Ntimbane T, Comte B, Mailhot G, Berthiaume Y, Poitout V, Prentki M, et al. Cystic fibrosis-related diabetes: from CFTR dysfunction to oxidative stress. Clin Biochem 2009;30:153–77.

[106] Galli F, Battistoni A, Gambari R, Pompella A, Bragonzi A, Pilolli F, et al. Oxidative stress and antioxidant therapy in cystic fibrosis. Biochim Biophys Acta 2012;1822:690–713.

[107] Lezo A, Biasi F, Massarenti P, Clabrese R, Poli G, Santini B, et al. Oxidative stress in stable cystic fibrosis patients: do we need higher antioxidant plasma levels? J Cyst Fibros 2013;12:35–41.

[108] Salinthone S, Kerns AR, Tsang V, Carr DW. α-tocopherol (vitamin E) stimulates cyclic AMP production in human peripheral mononuclear cells and alters immune function. Mol Immunol 2013;53:173–8.

[109] Winbauer AN, Pingree SS, Nuttall KL. Evaluating serum alpha-tocopherol (vitamin E) in terms of lipid ratio. Ann Clin Lab Sci 1999;29:185–91.

[110] Brown RK, Wyatt H, Price JF, Kelly FJ. Pulmonary dysfunction in cystic fibrosis is associated with oxidative stress. Eur Respir J 1996;9:334–9.

[111] Rana M, Wong-See D, Katz T, Gaskin K, Whitehead B, Jaffe A, et al. Fat-soluble vitamin deficiency in children and adolescents with cystic fibrosis. J Clin Pathol 2014;67(7):605–8.

[112] Willison HJ, Muller DPR, Matthews S, Jones S, Kriss A, Stead RJ, et al. A study of the relationship between neurological function and serum vitamin E concentrations in patients with cystic fibrosis. J Neurol Neurosurg Psychiatry 1985;48:1097–102.

[113] Bye AME, Muller DPR, Wilson J, Wright VM, Mearns MB. Symptomatic vitamin E deficiency in cystic fibrosis. Arch Dis Child 1985; 60:162–4.

[114] Sitrin MD, Lieberman F, Jensen WE, Noronha A, Milburn C, Addington W. Vitamin E deficiency and neurologic disease in adults with cystic fibrosis. Ann Intern Med 1987;107:51–4.

[115] Chakrabarty B, Kabra SK, Gulati S, Toteja GS, Lodha R, Kabra M, et al. Peripheral neuropathy in cystic fibrosis: a prevalence study. J Cyst Fibros 2013;12:754–60.

[116] Wilfond BS, Farrell PM, Laxova A, Mischler E. Severe haemolytic anemia associated with vitamin E deficiency in infants with cystic fibrosis: implications for neonatal screening. Clin Pediatr 1994;33:2–7.

[117] Swann IL, Kendra JR. Anaemia, vitamin E deficiency and failure to thrive in an infant. Clin Lab Haematol 1998;20:61–3.

[118] Koscik RL, Farrell PM, Kosorok MR, Zaremba KM, Laxova A, Lai HC, et al. Cognitive function of children with cystic fibrosis: deleterious effect of early malnutrition. Pediatrics 2004; 113:1549–58.

[119] Koscik RL, Lai HC, Laxova A, Zaremba KM, Kosorok MR, Douglas JA, et al. Preventing early, prolonged vitamin E deficiency: an opportunity for better cognitive outcomes via early diagnosis through neonatal screening. J Pediatr 2005;147:S51–56.

[120] Péter S, Moser U, Pilz S, Eggersdorfer M, Weber P. The challenge of setting appropriate recommendations for vitamin E: consideration of status and functionality to define nutrient requirements. Int J Vitam Nutr Res 2013;83:129–36.

[121] Dietary Reference Values for Food Energy and Nutrients for the United Kingdom. Report of the panel on dietary reference values, committee on medical aspects of food and nutrition policy. London: HMSO; 1991.

[122] Shamseer L, Adams D, Brown N, Johnson JA, Vohra S. Antioxidant micronutrients for lung disease in cystic fibrosis. Cochrane Database of Syst Rev 2010;(12). Art No. CD007020. http://dx.doi.org/10.1002/14651858.CD007020.pub2.

[123] Huang SH, Schall JI, Zemel BS, Stallings VA. Vitamin E status in children with cystic fibrosis and pancreatic insufficiency. J Pediatr 2006;148:556–9.

[124] Conly JM, Stein K, Worobetz L, Rutlege-Harding S. The contribution of vitamin K_2 (menaquinones) produced by the intestinal flora to human nutritional requirements for vitamin K. Am J Gastroenterol 1994;86:915–23.

[125] Paiva SAR, Sepe TE, Booth SL, Camilo ME, O'Brien ME, Davidson KW, et al. Interaction between vitamin K nutriture and bacterial overgrowth in hypochlorhydria induced by omeprazole. Am J Clin Nutr 1998;68:699–704.

[126] Corrigan JJ, Taussig LM, Beckerman R, Wagener JS. Factor II (prothrombin) coagulant activity in immunoreactive protein: detection of vitamin K deficiency and liver disease in patients with CF. J Pediatr 1981;99:254–7.

[127] Hamid B, Khan A. Cerebral hemorrhage as the initial manifestation of cystic fibrosis. J Child Neurol 2007;22:114–5.

[128] McPhail GL. Coagulation disorder as a presentation of cystic fibrosis. J Emerg Med 2010;38:320–2.

[129] Booth SL. Roles for vitamin K beyond coagulation. Annu Rev Nutr 2009;29:89–110.

[130] Mosler K, von Kries R, Vermeer C, Saupe J, Schmitz T, Schuster A. Assessment of vitamin K deficiency in CF-how much sophistication is useful? J Cyst Fibros 2003;2:91–6.

[131] Booth SL, Al Rajabi A. Determinants of vitamin K status in humans. Vitam Horm 2008;78:1–22.

[132] Rashid M, Durie PR, Andrew M, Kalnins D, Shin J, Corey M, et al. Prevalence of vitamin K deficiency in cystic fibrosis. Am J Clin Nutr 1999;70:378–82.

[133] Conway SP, Wolfe SP, Brownlee KG, White H, Oldroyd B, Truscott JG, et al. Vitamin K status among children with cystic fibrosis and its relationship to bone mineral density and bone turnover. Pediatrics 2005;115:1325–31.

[134] Durie PR. Vitamin K and the management of the patient with cystic fibrosis. Can Med Assoc J 1994;151:933–6.

[135] Shwachman H. Therapy of cystic fibrosis of the pancreas. Pediatrics 1960;25:155–63.

[136] Di Sant'Agnese PA, Vidaurreta AM. Cystic fibrosis of the pancreas. JAMA 1960;172:2065–72.

[137] Ngo B, Van Pelt K, Labarque V, Van de Casseye W, Pender J. Late vitamin K deficiency bleeding leading to a diagnosis of cystic fibrosis. Acta Clin Belg 2011;66:142–3.

[138] Fewtrell MS, Benden C, Williams JE, Chomtho S, Ginty F, Nigdikar SV, et al. Undercarboxylated osteocalcin and bone mass in 8-12 year old children with cystic fibrosis. J Cyst Fibros 2008;7:307–12.

[139] Shea MK, Dallal GE, Dawson-Hughes B, Ordovas JM, O'Donnell CJ, Gunberg CM, et al. Vitamin K, circulating cytokines, and bone mineral density in older men and women. Am J Clin Nutr 2008;88:356–63.

[140] De Montalembert M, Lenoir G, Saint-Raymond A, Rey J, Lefrère JJ. Increased PIVKA-II concentrations in patients with cystic fibrosis. J Clin Pathol 1992;45:180–1.

[141] Wilson DC, Rashid M, Durie P, Tsang A, Kalnins D, Andrew M, et al. Treatment of vitamin K deficiency in cystic fibrosis: effectiveness of a daily fat-soluble vitamin combination. J Pediatr 2001;138:851–5.

[142] Van Hoorn JHL, Hendricks JJE, Vermeer C, Forget PP. Vitamin K supplementation in cystic fibrosis. Arch Dis Child 2003;88: 974–5.

[143] Drury D, Grey VL, Ferland G, Gunberg C, Lands LC. Efficacy of high dose phylloquinone in correcting vitamin K deficiency in cystic fibrosis. J Cyst Fibros 2008;7:457–9.

[144] Dougherty KA, Schall JI, Stallings VA. Suboptimal vitamin K status despite supplementation in children and young adults with cystic fibrosis. Am J Clin Nutr 2010;92:660–7.

[145] Jagannath VA, Fedorowicz Z, Thaker V, Chang AB. Vitamin K supplementation for cystic fibrosis. Cochrane Database Syst Rev 2013;30(4):CD008482.

[146] Guéant JL, Champigneulle B, Gaucher P, Nicolas JP. Malabsorption of vitamin B12 in pancreatic insufficiency of the adult and of the child. Pancreas 1990;5:559–67.

[147] Peters SA, Rolles CJ. Vitamin therapy in cystic fibrosis-a review and rationale. J Clin Pharm Ther 1993;18:33–8.

[148] Maqbool A, Schall JI, Mascarenhas MR, Dougherty KA, Stallings VA. Vitamin B12 status in children with cystic fibrosis and pancreatic insufficiency. J Pediatr Gastroenterol Nutr 2014; 58:733–8.

[149] McCabe H. Riboflavin deficiency in cystic fibrosis: three case reports. J Hum Nutr Diet 2001;14:365–70.

[150] McCabe HE, Johnson JK, O'Brien C. B vitamin deficiency in the paediatric cystic fibrosis population. Pediatr Pulmonol 2004;(27): 338–9. [Abstract 427].

[151] National Institute for Health and Clinical Excellence. Diabetes in pregnancy. Management of diabetes and its complications from pre-conception to the post-natal period; 2008. NICE Clinical Guideline 63, London.

[152] Back EI, Frindt C, Nohr D, Frank J, Ziebach R, Stern M, et al. Antioxidant deficiency in cystic fibrosis: when is the right time to take action? Am J Clin Nutr 2004;80:374–84.

[153] Rust P, Eichler I, Renner S, Elmadfa I. Effects of long-term oral beta-carotene supplementation on lipid peroxidation in patients with cystic fibrosis. Int J Vitam Nutr Res 1998;68: 83–7.

CHAPTER

18

Nutritional Strategies to Modulate Inflammation and Oxidative Stress in Patients with Cystic Fibrosis

Izabela Sadowska-Bartosz[1], *Sabina Galiniak*[1], *Grzegorz Bartosz*[1, 2]

[1]Department of Biochemistry and Cell Biology, University of Rzeszów, Rzeszów, Poland;
[2]Department of Molecular Biophysics, University of Łódź, Łódź, Poland

18.1 OXIDATIVE STRESS AND INFLAMMATION IN CYSTIC FIBROSIS

Oxidative stress (OS) was defined originally as "a disturbance in the prooxidant-antioxidant balance in favor of the former" and viewed as a source of oxidative damage to cellular and extracellular molecules by reactive oxygen species (ROS) [1]. A more recent definition takes into account that ROS are not only damaging species, but also play a role in cellular signaling [2,3] and views OS as "disruption of redox control and signaling" [4]. Apparently, both these definitions are compatible since disruption of the redox control leads to uncontrolled oxidative damage [5].

Many studies have documented the occurrence of OS in CF. Increased plasma level of malondialdehyde (MDA), peroxides, and protein carbonyls were revealed in blood plasma of patients with CF, which proves oxidative damage to lipids and proteins [6]. We found elevated MDA, but not peroxides, in the plasma of pediatric patients with CF [7]. Increased levels of chromolipids (lipid adducts of 4-hydroxynonenal (HNE-L) and MDA (MDA-L)) were reported for CF patients [8]. MDA-L were elevated in the majority of patients despite normal plasma vitamin E, A, and C. HNE-L and MDA-L increased with age of the patients, whereas plasma vitamins decreased. The most relevant negative correlation was identified between vitamin C and chromolipids. Moreover, patients with pancreatic insufficiency (PI) showed significantly higher plasma chromolipids, despite no differences in plasma vitamins.

F_2-isoprostanes are believed to be generated by nonenzymatic, free-radical catalyzed peroxidation of arachidonic acid esterified to lipids and then cleaved and released into the circulation by phospholipases. Concentrations of 15-F_{2t}-isoprostane (8-isoprostane, 8-IP), the most known compound belonging to the F_2-isoprostane class, are elevated in blood plasma [9,10], urine [11], and exhaled breath condensate (EBC) [12] in patients with stable CF. Moreover, Lucidi et al. suggested that in patients with CF exhaled 8-isoprostane correlates with functional, clinical, and radiological assessment, and could be a useful marker of disease severity [13].

Chronic bacterial infection of the respiratory tract by microorganisms as *Pseudomonas aeruginosa*, members of the *Burkholderia cepacia* complex or *Staphylococcus aureus*, causes destruction and loss of lung function and is closely related to nutritional status. A major cause of nutritional depletion is the increase in energy expenditure due to inflammation and lung infection [14]. Recently, we reported elevated content carbonyls and advanced glycoxidation end-products (AGEs) in plasma proteins of CF patients chronically infected with *P. aeruginosa* or *S. aureus* [15].

Proteins of the bronchoalveolar lavage fluid (BAL) of CF patients show increased levels of carbonylation [16,17] and halogenation of tyrosine residues [17]. Increased nitrotyrosine was also found in the sputum of CF patients [18].

Increase in the 8-hydroxydeoxyguanosine (8-OHdG) in the urine of CF patients reflects enhanced oxidative damage to DNA and may explain the higher incidence of malignancy compared to normal, healthy age-matched

Diet and Exercise in Cystic Fibrosis
http://dx.doi.org/10.1016/B978-0-12-800051-9.00018-3

controls, although it showed no correlation with the severity of the disease [19].

Increased levels of superoxide dismutase and glutathione reductase found in erythrocytes of the patients most probably reflect a compensatory response to OS [6]. The elevated glutathione *S*-transferase activity found by us in erythrocytes of CF patients can be attributed to the same phenomenon [15].

During acute infection, the balance between oxidants and antioxidants may be further disturbed [20].

CF patients also have altered levels of polyunsaturated fatty acids. The ratio of arachidonic to docosahexaenoic acid was increased in mucosal and submucosal nasal-biopsy specimens [21]. These changes may be due to malabsorption, but the effect of increased lipid peroxidation resulting from oxidative stress cannot be excluded [22].

OS may be due to an increased generation of ROS, an impaired antioxidant barrier, or both. Available data demonstrate that the last possibility is true for CF [23].

One reason for the occurrence of OS in CF may be ascribed to the mutation in the CFTR gene itself. There is a lot of evidence that this protein is involved in glutathione efflux, apart from controlling chloride transport [24]; therefore, its mutation leads to decreased glutathione efflux from cells (see "Antioxidant Deficiencies in CF").

The development of inflammatory and degenerative lesions in target tissues such as lung, pancreas, and liver further exacerbates the shift from normal to abnormal flux of ROS in several organs, thereby leading to develop systemic oxidative stress [25]. A factor that seems of primary importance is the chronic lung infection of CF patients with *P. aeruginosa* and other pathogens. Lung infection is responsible for 90% of the morbidity of patients with CF because of damage of the airways and gradual deterioration of the lung function. One hallmark of the CF airway disease is the persistent inflammatory response consisting in massive recruitment of polymorphonuclear leucocytes (PMNs). Analysis of BAL has shown that the number of PMNs recovered from the lungs of patients with CF is 1000 times higher than that recovered from the lungs of controls. PMNs release leukocyte proteases, myeloperoxidase, and ROS, which are the main mechanisms of lung tissue damage in CF [23,26].

Granulocyte NADPH oxidase was found to be increased in CF; however, this increase did not correlate with bacterial killing or clinical status [27], apparently due to inefficient myeloperoxidase-mediated hypochlorite formation in the phagosomes, resulting from impaired chloride transport [28].

Nrf2 protein plays a critical role in the control of antioxidant response of cells to OS, inducing biosynthesis of such proteins as heme oxygenase 1, γ-glutamylcysteine synthase (the rate limiting enzyme in glutathione synthesis), thioredoxin 1 (the predominant regulator of nuclear redox balance), and catalase [3]. Nrf2 was reported to be dysfunctional in CF air-way epithelial cells, as demonstrated by a significant decrease in the expression of a number of Nrf2 regulated proteins and a significant increase in steady intracellular H_2O_2 in CF versus non-CF cells [29].

Another factor in the genesis of OS in CF is the deficiency of exogenous antioxidants due to malabsorption of dietary antioxidants in the gut, leading to lowered intake and the absorption of fat-soluble antioxidants (vitamin E, carotenoids, coenzyme Q_{10}, some polyunsaturated fatty acids, etc.) and oligoelements (such as Se, Cu, and Zn) that are involved in ROS detoxification by means of enzymatic defenses [25].

The abnormalities in the oxidant/antioxidant balance could be a potential target for a new therapeutic approach. A suboptimal antioxidant protection is believed to represent a main contributor to OS and to the poor control of immuno-inflammatory pathways in CF patients. Observed defects include an impaired, reduced glutathione metabolism and lowered intake and absorption of fat-soluble antioxidants (vitamin E, carotenoids, coenzyme Q_{10}, some polyunsaturated fatty acids, etc.) and oligoelements (such as Se, Cu, and Zn) that are involved in reactive oxygen species (ROS) detoxification by means of enzymatic defenses. Oral supplements and aerosolized formulations of thiols have been used in the antioxidant therapy of this inherited disease, with the main aim of reducing the extent of oxidative lesions and the rate of lung deterioration [25].

18.2 ANTIOXIDANT DEFICIENCIES IN CF

Early studies revealed antioxidant deficiencies in CF. As a result of a decreased production of pancreatic juices, the vast majority (up to 92%) of CF patients are unable to efficiently breakdown and absorb fat, including fat-soluble antioxidants such as vitamins A, E, and β-carotene [22,30]. Decreased level of plasma vitamin E correlated with increased erythrocyte susceptibility to hydrogen peroxide-induced hemolysis and shortened erythrocyte lifespan, other hematological parameters remaining unchanged [31]. Apart from vitamin E, the levels of vitamin A and β-carotene were also found to be decreased in CF patients [9]. Lagrange-Puget et al. (2004) observed that decreased levels of vitamins A, E, and carotenoids were confirmed in the largest study performed on CF patients, including 312 patients. Very low carotenoid levels were found even in young patients with CF, which suggests that these deficiencies appear early during the course of the disease. These early deficiencies may suggest a close relationship with CFTR mutations. In addition, BMI correlated positively with vitamin A and carotenoids, but did not influence plasma concentrations during infection. Thus, nutritional status seems to have less effect on antioxidant and oxidative-marker levels than does FEV1, particularly during bronchial exacerbations [32]. Another study found a deficiency in β-carotene, β-cryptoxanthin and total lycopene, while the concentrations of vitamin C and α-tocopherol were only in the groups over 18 years old [33]. Low levels of vitamin C in patients with CF were also reported [34,35]. Interestingly, vitamin C was reported to control the function of CFTR, inducing the openings of CFTR chloride channels by increasing its probability to be in the open state, the average half-maximal stimulatory constant being $36.5 \pm 2.9\,\mu M$, which corresponds to physiological concentrations [36].

A decreased level of low blood-reduced glutathione (GSH) levels and normal total glutathione were also found [32]. A subsequent study showed a systemic deficiency in the plasma glutathione level in patients with CF, the concentration of GSH being decreased by about 40% (2.7 μM vs. 4.5 μM). CF patients showed a sharp discrepancy in plasma GSH concentrations with a 40% decrease below normal controls in the CF patients [37]. However, the redox potential of plasma glutathione remained unchanged (−139 mV vs. −138 mV) [38]. The mechanism for total plasma GSH concentration differences in CF is not entirely understood, but the CFTR has been implicated in the transport of GSH. In vitro, intracellular glutathione in epithelial cell lines harboring mutated CFTR protein (*delta*F508) was unchanged, but extracellular glutathione was decreased, suggesting a defect in export of this compound, but not in its intracellular metabolism [39]. CFTR shares a structural similarity to ABCC proteins, including multidrug resistant proteins (MRPs), which are believed to export glutathione and/or glutathione-S-conjugates [40,41]. Using current recordings from excised membrane patches, GSH was shown to be permeant in CFTR [42]. Utilization of inhibitors, such as glibenclamide and dinitrostilbene-2,2′-disulfonate (DNDS), prevented GSH efflux. In other studies, inside-out membrane vesicles were used to evaluate CFTR-mediated GSH efflux [43]. Vesicles expressing CFTR demonstrated an increased uptake in radiolabeled GSH ([^{35}S] GSH) compared to vesicles lacking CFTR. In this same report, mutant CFTR (R347D) showed a decreased ability to traffic GSH compared to wild-type. Although these studies suggest GSH export through CFTR, reproducibility of these results and inconsistencies have been problematic for the notion that CFTR is a GSH transporter [38].

Erythrocyte glutathione showed high variability in CF patients and was inversely and significantly correlated to tests of pulmonary function, suggesting a compensatory, but probably inadequate, increase in patients with more severe respiratory deterioration in response to OS [44]. Mitochondrial GSH levels were found to be decreased up to 85% in CFTR-knockout mice, and 43% in human lung epithelial cells deficient in CFTR. A concomitant 29% increase in the oxidation of mitochondrial DNA, and a 30% loss of aconitase activity, a sensitive measure of oxidative stress, confirm the existence of mitochondrial OS in CF lung epithelial cells [45]. Decreases in the glutathione level in the lung lumen are most significant and apparently most important. In the lung airway surface liquid (ASL), GSH concentration is much higher than in blood plasma [37,46]. Total glutathione concentrations in the airway surface liquid have been estimated to be 275–500 μM in healthy subjects [37,47]. In CF patients, total GSH concentrations in ASL are greatly diminished. The total glutathione concentration was found to be only 92 μM, whereas reduced glutathione of about 78 μM, with concentrations of oxidized glutathione (GSSG), was almost unchanged. This results in an increase of the redox potential of the glutathione couple from approximately −175 mV in healthy subjects to −145 mV to −129 mV in CF patients [37], thus creating a much more oxidizing environment. Recently, Kettle et al. (2014) demonstrated that the concentration of glutathione was lower in BAL from children with CF, whereas glutathione sulfonamide, a specific oxidation product of hypochlorous acid, was higher. Oxidized glutathione and glutathione sulfonamide correlated with myeloperoxidase and a biomarker of hypochlorous acid. These authors suggested that therapies targeted against myeloperoxidase may boost antioxidant defense and slow the onset and progression of lung disease in CF [48].

Trolox-equivalent antioxidant capacity of blood plasma, estimated with 2,2′-azino-bis(3-ethylbenzthiazoline-6-sulphonic) (ABTS), was found unchanged in nonhospitalized CF patients (1.40 mmol/L) compared to healthy controls (1.35 mmol/L), but was decreased in patients hospitalized for acute exacerbation (1.09 mmol/L) [49]. However, our study using another method of assay (FRAP) showed a significant reduction in the antioxidant capacity of plasma in CF [7].

Selenium level was also decreased in CF patients [50–52], although the activity of the selenoprotein erythrocyte glutathione peroxidase was reported unchanged [52].

The lowered levels of antioxidants may correlate with the course of the disease. It was reported that reduced serum levels of vitamin A and E are associated with an increased rate of pulmonary exacerbations in CF [53].

Markers of OS (increased levels of 4-HNE and MDA lipid adducts) were observed even in patients with normal levels of vitamins A, C, and E. It suggests that optimal requirements for CF patients, based on antioxidant needs, may be higher than those assumed for the healthy population [8]. Olveira et al. (2013) noted that the MDA content is higher in CF, even despite normal vitamins E and A status. Moreover, supplementation with β-carotene or a mixture of antioxidants can normalize the MDA content [54].

18.3 DIETARY INTERVENTIONS IN CF

Nutrition plays an essential role in the survival and quality of life of CF patients. CF patients have high caloric requirements due to an increased resting energy expenditure (REE), bacterial infection, and malabsorption. REE is higher in CF patients with a more severe phenotype. Lung function and nutritional status are closely correlated, and the severe weight loss can lead to a decrease in lean body mass, with consequences on respiratory muscles [14]. Dorothy Andersen, in the first description of CF in the English language literature, described nutritional intervention for children with the disease and recommended that "*caloric intake be in excess of the usual amount for the age*" [55]. Twenty-five years later, Sproul and Huang reported that there was a significant correlation between growth retardation and the severity of pulmonary involvement and emphasized the importance of early intervention, stating "...gains were greatest when therapy was initiated during infancy" [56]. Recently, Yen et al. found that greater weight at age 4 years is associated with greater height, better pulmonary function, fewer complications of CF, and better survival through the age of 18 years. Furthermore, greater weight-for-age in the peripubertal period is associated, on average, with improved tempo and timing of pubertal height growth [57]. Long-term nutritional management is as integral a part of modern care as pulmonary therapy, and is intimately linked to pulmonary outcomes. Dietetic management is based on the replacement of pancreatic enzymes and fat soluble vitamins (A, D, E, K), together with a high protein diet and high calorie intake (120–125% of the normal recommended daily allowance). The recommended Body Mass Index for adult CF patients is ≥22 kg/m^2 for females and ≥23 kg/m^2 for males. Overnight enteral nutrition via nasogastric tube or gastrostomy provides supplementary nutritional support when BMI is suboptimal [58]. Woestenenk et al. [59] found that both children and adolescents with CF intake less calories than the recommended estimated average requirements. These authors suggested that proper intake of calories due to malnutrition occurred only in childhood. Furthermore, it was confirmed that the children consumed more fat and saturated fatty acids as compared to healthy children, which may lead to cardiovascular diseases in the future. It was shown that protein synthesis in children with CF can speed by 30% in the case of increased protein intake. Surprisingly, the rate of protein products degradation was unchanged [60]. It has been observed that long-term intake the daily mixtures of fatty acids (eicosapentaenoic, docosahexaenoic, linoleic and γ-linolenic acid) at a low dose has positive effect on lung function and inflammation in adult CF patients. The total number of exacerbations after a year supplementation was reduced while lean body mass and spirometry were increased. In addition, supplementation led to improve parameters of oxidation, inflammation (IgG and IgM) and other clinical parameters [61]. 8-isoprostanes, free radical products of lipid peroxidation which is a consequence of oxidative stress, appears to be a prognostic factor in deterioration of lung function in a short time in patients infected with *Burkholderia cenocepacia*. The measured concentration of 8-isoprostane in exhaled breath condensate of 24 patients did not show correlation with the clinical parameters during one- and 3-year study [62]. Diet supplemented with antioxidants may have a positive effect on OS in patients, although we cannot predict exactly what biological impact of intake of antioxidants may have on the body [25].

The early study of Farrell et al. reported normalization of the plasma α-tocopherol concentration by supplementation with 5–10 times the recommended daily allowance of vitamin E in a water-miscible form. This treatment reduced the abnormal oxidant sensitivity of erythrocytes and normalized erythrocyte lifespan [31]. A subsequent study has shown that a 1-year supplementation with 100 mg daily is required to normalize erythrocyte concentration of vitamin E in CF patients, whereas a daily dose of 15 mg is insufficient [63].

Rust et al. examined the effect of long-term oral β-carotene supplementation in patients with CF. Patients of the CF supplementation group received 1 mg β-carotene/kg body weight/day (maximally 50 mg β-carotene/day) for the first 12 weeks; during the following 12 weeks, the dosage was reduced to 10 mg β-carotene/day. At study entry, plasma β-carotene concentrations were significantly lower in CF patients than in controls. In the CF supplementation group, plasma β-carotene concentrations were significantly increased (baseline: 0.08 ± 0.04 μM) at the end of high-dose treatment (12th week; 0.6 ± 0.4 μM), but decreased again during supplementation with 10 mg β-carotene/day to 0.3 ± 0.2 μM. β-Carotene supplementation did not affect plasma concentrations of other carotenoids and retinol, but an increase in plasma α- and γ-tocopherol concentrations was noticed. During high-dose treatment, a significant decrease in the MDA level and a correction of total antioxidative capacity was observed [64].

Wood et al. carried out an 8-week, double-blind, randomized intervention trial, providing one patient group with low-dose (10 mg vitamin E, 500 μg vitamin A) and one group with high-dose (200 mg vitamin E, 300 mg vitamin C, 25 mg β-carotene, 90 μg selenium, 500 μg vitamin A) vitamin supplements. They observed significant changes of clinical indicators after treatment, including increased plasma antioxidant concentrations and a correlation between improved β-carotene status and lung function, as well as between improved selenium status and lung function [65]. Renner et al. reported distinct clinical benefits from high-dose (1 mg/kg body weight/day, maximum 50 mg/day) supplements. Their patients required significantly less antibiotics during the phase of high-dose β-carotene supplementation and showed a decrease in pulmonary exacerbations [66]. Lepage et al. reported that the 2-month supplementation of CF patients with 4.42 mg β-carotene, three times per day, led to the normalization of increased MDA level and increased plasma β-carotene from 0.08 ± 0.03 to $3.99 \pm 0.92\,\mu M$ [67].

Gray et al. supplemented CF patients who were in stable condition were with a whey protein isolate (Immunocal, 10 g twice a day) or casein placebo for 3 months. They found a 46.6% increase in lymphocyte glutathione in the whey protein-treated group, but no effect on lung function [68]. Hyperbaric pressure treatment of whey protein promotes the release of novel peptides for absorption, increases intracellular glutathione in healthy subjects, and reduces in vitro production of interleukin (IL)-8. A pilot open-label study of 1-month dietary supplementation with pressurized whey in CF patients (20 g/day in patients less than 18 years of age and 40 g/day in older patients) showed enhancements in nutritional status, as assessed by body mass index, and children showed improvement in lung function (forced expiratory volume in 1 s) in the children. The majority of patients with an initially elevated CRP showed a decrease [69].

Papas et al. supplemented CF patients with a CF-1 antioxidant preparation containing, per a daily dose of 10 mL, 30 mg of β-carotene, 200 IU of α-tocopherol, 94 mg of γ-tocopherol, 31 mg of other tocopherols, 30 mg of Coenzyme Q, 400 IU of vitamin D_3, and 300 μg of vitamin K_1 for 56 days. They found increased β-carotene, γ-tocopherol, and CoQ_{10} levels, and decreased sputum myeloperoxidase level. Improvements in antioxidant plasma levels were associated with reductions in airway inflammation [70].

Sagel et al. examined the effect of AquADEKs® (Yasoo Health Inc.) softgel containing vitamins A, C, D_3, E, K_1, Coenzyme Q, B_1, B_2, B_6, B_{12}, nicotinamide, folic acid, biotin, calcium panthothenate, zinc, and selenium for 12 weeks. AquADEKs has several theoretical advantages compared with other commercially available CF vitamin preparations. This formulation uses micelle-like particles to enhance absorption of fat-soluble vitamins and micronutrients compared with oil-based products [71]. AquADEKs is a highly absorbable, antioxidant-rich nutritional supplement that is specially formulated to overcome the malabsorption of fat-soluble vitamins. Compared with other CF multivitamin products, such as ADEKs (Axcan Scandipharm Inc., Birmingham, Alabama, USA) or SourceCF (Eurand Pharmaceuticals Inc., Huntsville, Alabama, USA), AquADEKs contains higher amounts of vitamins D and K in water-soluble form to meet current recommendations for supplementation in patients with CF. It also contains important antioxidants such as coenzyme Q_{10}, beta carotene, and γ-tocopherol. This product includes selenium, a cofactor that stimulates endogenous antioxidant enzymes such as glutathione peroxidase [7]. Our study compared various pharmaceutical forms of AquADEKs, such as chewable tablets, capsules, and liquid administered daily for 12 weeks that would reduce oxidative stress and enhance antioxidant status in pediatric patients with cystic fibrosis (CF). Patients were divided into fourgroups: group A received supplementation with vitamins A (3 mg daily), E (200 mg daily), and D_3 (20 μg daily); group B was supplemented with AquADEKs chewable tablets; group C received the recommended amount of AquADEKs capsules; and group D was supplemented with AquADEKs liquid. The level of oxidative stress was determined by the analysis of activities of enzymes neutralizing reactive oxygen species and by the estimated markers of the intensity of free radical processes. There was no difference in the activity of erythrocyte catalase, hydroperoxides level, and sulfhydryl group content in blood plasma between patients with CF and healthy children. The plasma total antioxidant status was decreased in all CF groups compared with the control. The supplementation with either AquADEKs chewable tablets or capsules normalized the malondialdehyde level in plasma. AquADEKs in various pharmaceutical forms normalized the sulfhydryl group content of erythrocytes. The superoxide dismutase activity was increased to near control level in the patients supplemented with either AquADEKs chewable tablets or liquid as compared with the group supplemented with vitamins or with AquADEKs capsules. In conclusion, AquADEKs attenuates selected oxidative stress markers in pediatric patients with CF.

The supplementation resulted in significantly increased plasma β-carotene, coenzyme Q_{10}, and γ-tocopherol concentrations, and decreased proteins induced in vitamin K absence (PIVKA-II) levels, but did not normalize vitamin D and K status [72].

Oudshoorn et al. studied the effect of a nutritional supplement ML1 containing numerous micronutrients, including vitamins A, C, D, E, and Coenzyme Q_{10} for 3 months. They found increased plasma levels of vitamin E and A when compared patients receiving placebo. However, no significant difference between the effect of the ML1 or placebo was observed neither for FEV1, FVC, anthropometry, nor for the parameters for muscle performance [73].

Shamseer et al. performed a meta-analysis of data on the effect of supplementation with vitamin E, vitamin C, β-carotene, and selenium (individually or in combination) comparing to placebo or standard care. According to these authors, only three trials (87 participants) contained data suitable for analysis. Based on two trials, there was no significant improvement in lung function; one trial indicated significant improvement in quality of life favoring control. Based on two trials, selenium-dependent glutathione peroxidase enzyme significantly improved in favor of combined supplementation, and selenium supplementation. All plasma antioxidant levels, except vitamin C, significantly improved with supplementation. The authors concluded that the evidence regarding the clinical effectiveness of antioxidant supplementation in CF is conflicting. Main supplementation studies in CF are summarized in Table 18.1. Based on the evidence, antioxidants appear to decrease quality of life and oxidative stress. Further trials examining clinically important outcomes, and elucidation of a clear biological pathway of oxidative stress in CF, are necessary before a firm conclusion regarding the effects of antioxidants supplementation can be drawn [76].

TABLE 18.1 Nutritional Trials in Cystic Fibrosis (CF)

Rank	Author, Year, Location	Supplementation	Research Group	Results
1	Winklhofer-Roob et al., 1995, Switzerland [23,74]	0.5 mg β-carotene/kg body weight/day for 3 months	34 CF patients (10.8 ± 7.6 year); 42 healthy subjects (31.5 ± 8.0)	Increase in plasma and LDL β-carotene concentration Decrease in plasma MDA concentrations
2	Lepage et al., 1996, Canada [67]	4000 IU vitamin A and 200 mg all-*rac*-α-tocopheryl acetate/day for 72 h 4.42 mg β-carotene three times/day for 2 months*	25 CF patients (9.6 ± 0.8 year); 16 healthy subjects (10.4 year) 12 CF patients (11.5 ± 0.8 year)*	Decrease in plasma MDA concentration* Increase in plasma β-carotene*
3	Rust et al., 2000, Austria [64]	1 mg β-carotene/kg body weight/day (maximum 50 mg/day) for 12 weeks, then 10 mg β-carotene/day for another 12 weeks; placebo group received starch-containing capsules during 24 weeks	24 CF patients (12.8 ± 6.3 year) (CF supplementation or CF placebo group); 14 age-matched healthy subjects	Increase in plasma β-carotene concentration Decrease in plasma MDA-TBA concentration Increase in total antioxidative capacity of plasma
4	Renner et al., 2001, Austria [67]	1 mg β-carotene/kg/day (max 50 mg/day) for 3 months, then 10 mg/day β-carotene in a single dose for another 3 months, placebo group received capsules prepared with starch for 6 month	13 CF patients (12.8 year)—supplementation group, 11 CF patients (10.5 year)—placebo group	Increase in plasma β-carotene concentration Decrease in plasma MDA-TBA concentration Increase in total antioxidative capacity of plasma
5	Grey et al., 2003, Canada [68]	10 g whey protein twice/day for 3 months Placebo group received casein	10 CF patients supplementation group, 11 CF patients placebo group	Increase in lymphocyte GSH
6	Wood et al., 2003, Australia [65]	10 mg vitamin E and 500 μg vitamin A for 8 weeks (low dose-A) 200 mg vitamin E, 300 mg vitamin C, 25 mg β-carotene, 90 μg Se, 500 μg vitamin A for 8 weeks (high dose-B)	46 CF patients (divided into 2 groups—24 subjects low supplementation (A; 10.6 ± 0.7 year) and 22 subjects high supplementation (B; 12.6 ± 0.8 year))	Increase in plasma vitamin E, β-carotene selenium and glutathione peroxidase after high dose of supplementation
7	Innis et al., 2007, Canada [75]	2 g lecithin/day for 14 days 2 g choline/day for 14 days 3 g betaine/day for 14 days	35 CF patients (divided into 3 groups—13 subjects lecithin supplementation, 10.3 ± 1 year; 12 subjects choline supplementation, 10.4 ± 1.1 year; 10 subjects betaine supplementation, 11.6 ± 1.3 year); 15 healthy subjects	Increase in plasma methionine, *S*-adenosylmethionine, *S*-adenosylmethionine: *S*-adenosylhomocysteine and glutathione:glutathione disulfide Decrease in *S*-adenosylhomocysteine
8	Oudshoorn et al., 2007, the Netherlands [73]	100 mL of a liquid micronutrient mixture mL/day for 3 months, placebo group received 100 mL of a liquid contained identical amounts of protein, carbohydrates and fats, lacked the micronutrients	22 CF patients (12.9 ± 2.5 year) divided into 2 groups—supplementation or placebo	Increase in plasma vitamin A and E concentration

TABLE 18.1 Nutritional Trials in Cystic Fibrosis (CF)—cont'd

Rank	Author, Year, Location	Supplementation	Research Group	Results
9	Papas et al., 2008, USA [71]	10 mL of the CF-1 formulation/ day for 56 days (CF-1: β-carotene 30 mg, α-tocopherol 200 IU, γ-tocopherol 94 mg, other tocopherols 31 mg, CoQ_{10} 30 mg, vitamin D_3 400 IU, vitamin K_1 300 μg)	10 CF patients (16.7 ± 4.9 year)	Increase in plasma β-carotene Increase in CoQ_{10} plasma level Increase in γ-tocopherol plasma level Decrease in induced sputum myeloperoxidase
10	Lands et al., 2010, Canada [69]	20 g/day pressurized whey (<18 year) 40 g/day pressurized whey (>18 year)	27 CF patients (9 children, 18 adults)	Decrease in CRP
11	Sagel et al., 2011, USA [72]	AquADEKs® once a day for 12 weeks	17 CF patients—11 with pancreatic insufficient (PI; 15.5 ± 4.6 year); 6 with pancreatic sufficient (PS; 15.0 ± 3.8 year)	Increase in plasma β-carotene, CoQ10, γ-tocopherol (whole CF group) Decrease in PIVKA-II levels (whole CF group) Increase in plasma β-carotene (PI patients) Decrease in PIVKA-II levels (PI patients)
12	Sadowska-Woda et al., 2011, Poland [7]	Chewable tablets, capsule, or liquid AquADEKs® once a day for 12 weeks or vitamin supplementation (3 mg vitamin A/day, 2 mg vitamin E/day, 20 μg D_3/day)	50 CF patients (divided into 4 groups—10 subjects vitamin supplementation 9.6 ± 3.7 year; 15 subjects AquADEKs chewable tablets 9.2 ± 4.65 year; 15 subjects AquADEKs capsules 8.7 ± 3.22 year; 10 subjects AquADEKs pediatric liquid 8.9 ± 3.42 year); 21 healthy subjects (9.6 ± 3.06 year)	Decrease in malondialdehyde after supplementation with chewable tablets or capsule AquADEKs® Increase in sulfhydryl group content of erythrocytes after supplementation Increase in superoxide dismutase activity after supplementation with chewable tablets or liquid AquADEKs®

*$P < 0.05$

References

[1] Cadenas E, Sies H. Oxidative stress: excited oxygen species and enzyme activity. Adv Enzyme Regul 1985;23:217–37.

[2] Bartosz G. Reactive oxygen species: destroyers or messengers? Biochem Pharmacol 2009;77:1303–15.

[3] Forman HJ, Davies KJ, Ursini F. How do nutritional antioxidants really work: nucleophilic tone and para-hormesis versus free radical scavenging in vivo. Free Radic Biol Med 2014;66:24–35.

[4] Jones DP. Redefining oxidative stress. Antioxid Redox Signal 2006;8:1865–79.

[5] Sies H. Role of metabolic H_2O_2 generation: redox signaling and oxidative stress. J Biol Chem 2014;289:8735–41.

[6] Domínguez C, Gartner S, Liñán S, Cobos N, Moreno A. Enhanced oxidative damage in cystic fibrosis patients. Biofactors 1998;8:149–53.

[7] Sadowska-Woda I, Rachel M, Pazdan J, Bieszczad-Bedrejczuk E, Pawliszak K. Nutritional supplement attenuates selected oxidative stress markers in pediatric patients with cystic fibrosis. Nutr Res 2011;31:509–18.

[8] Lezo A, Biasi F, Massarenti P, Calabrese R, Poli G, Santini B, et al. Oxidative stress in stable cystic fibrosis patients: do we need higher antioxidant plasma levels? J Cyst Fibros 2013;12:35–41.

[9] Collins CE, Quaggiotto P, Wood L, O'Loughlin EV, Henry RL, Garg ML. Elevated plasma levels of F2 alpha isoprostane in cystic fibrosis. Lipids 1999;34:551–6.

[10] Wood LG, Fitzgerald DA, Gibson PG, Cooper DM, Collins CE, Garg ML. Oxidative stress in cystic fibrosis: dietary and metabolic factors. J Am Coll Nutr 2001;20:157–65.

[11] Ciabattoni G, Davì G, Collura M, Iapichino L, Pardo F, Ganci A, et al. In vivo lipid peroxidation and platelet activation in cystic fibrosis. Am J Respir Crit Care Med 2000;162: 1195–201.

[12] Montuschi P, Kharitonov SA, Ciabattoni G, Corradi M, van Rensen L, Geddes DM, et al. Exhaled 8-isoprostane as a new noninvasive biomarker of oxidative stress in cystic fibrosis. Thorax 2000;55:205–9.

[13] Lucidi V, Ciabattoni G, Bella S, Barnes PJ, Montuschi P. Exhaled 8-isoprostane and prostaglandin E(2) in patients with stable and unstable cystic fibrosis. Free Radic Biol Med 2008;45: 913–9.

[14] Haack A, Carvalho Garbi Novaes MR. Multidisciplinary care in cystic fibrosis: a clinical-nutrition review. Nutr Hosp 2012;27:362–71.

[15] Sadowska-Bartosz I, Galiniak S, Bartosz G, Rachel M. Oxidative modification of proteins in pediatric cystic fibrosis with bacterial infections. Oxid Med Cell Longevity 2014:389629.

[16] Starosta V, Rietschel E, Paul K, Baumann U, Griese M. Oxidative changes of bronchoalveolar proteins in cystic fibrosis. Chest 2006;129:431–7.

[17] Thomson E, Brennan S, Senthilmohan R, Gangell CL, Chapman AL, Sly PD, et al. Australian Respiratory Early Surveillance Team for Cystic Fibrosis (AREST CF). Identifying peroxidases and their oxidants in the early pathology of cystic fibrosis. Free Radic Biol Med 2010;49:1354–60.

[18] Jones KL, Hegab AH, Hillman BC, Simpson KL, Jinkins PA, Grisham MB, et al. Elevation of nitrotyrosine and nitrate concentrations in cystic fibrosis sputum. Pediatr Pulmonol 2000;30:79–85.

[19] Brown RK, McBurney A, Lunec J, Kelly FJ. Oxidative damage to DNA in patients with cystic fibrosis. Free Radic Biol Med 1995;18:801–6.

[20] Wood LG, Fitzgerald DA, Gibson PG, Cooper DM, Garg ML. Increased plasma fatty acid concentrations after respiratory exacerbations are associated with elevated oxidative stress in cystic fibrosis patients. Am J Clin Nutr 2002;75:668–75.

[21] Freedman SD, Blanco PG, Zaman MM, Shea JC, Ollero M, Hopper IK, et al. Association of cystic fibrosis with abnormalities in fatty acid metabolism. N Engl J Med 2004;350:560–9.

[22] Lands LC. Nutrition in pediatric lung disease. Paediatr Respir Rev 2007;8:305–11. quiz 312.

[23] Winklhofer-Roob BM. Oxygen free radicals and antioxidants in cystic fibrosis: the concept of an oxidant-antioxidant imbalance. Acta Paediatr Suppl 1994;83:49–57.

[24] Gould NS, Min E, Martin RJ, Day BJ. CFTR is the primary known apical glutathione transporter involved in cigarette smoke-induced adaptive responses in the lung. Free Radic Biol Med 2012;52:1201–6.

[25] Galli F, Battistoni A, Gambari R, Pompella A, Bragonzi A, Pilolli F, et al. Working Group on Inflammation in Cystic Fibrosis. Oxidative stress and antioxidant therapy in cystic fibrosis. Biochim Biophys Acta 2012;1822:690–713.

[26] Ciofu O, Riis B, Pressler T, Poulsen HE, Høiby N. Occurrence of hypermutable *Pseudomonas aeruginosa* in cystic fibrosis patients is associated with the oxidative stress caused by chronic lung inflammation. Antimicrob Agents Chemother 2005;49:2276–82.

[27] Berry DH, Brewster MA. Granulocyte NADH oxidase in cystic fibrosis. Ann Allergy 1977;38:316–9.

[28] Painter RG, Bonvillain RW, Valentine VG, Lombard GA, LaPlace SG, Nauseef WM, et al. The role of chloride anion and CFTR in killing of *Pseudomonas aeruginosa* by normal and CF neutrophils. J Leukoc Biol 2008;83:1345–53.

[29] Chen J, Kinter M, Shank S, Cotton C, Kelley TJ, Ziady AG. Dysfunction of Nrf-2 in CF epithelia leads to excess intracellular H_2O_2 and inflammatory cytokine production. PLoS One 2008;3:e3367.

[30] Homnick DN, Cox JH, DeLoof MJ, Ringer TV. Carotenoid levels in normal children and in children with cystic fibrosis. J Pediatr 1993;122:703–7.

[31] Farrell PM, Bieri JG, Fratantoni JF, Wood RE, di Sant'Agnese PA. The occurrence and effects of human vitamin E deficiency. A study in patients with cystic fibrosis. J Clin Invest 1977;60:233–41.

[32] Lagrange-Puget M, Durieu I, Ecochard R, Abbas-Chorfa F, Drai J, Steghens JP, et al. Longitudinal study of oxidative status in 312 cystic fibrosis patients in stable state and during bronchial exacerbation. Pediatr Pulmonol 2004;38:43–9.

[33] Back EI, Frindt C, Nohr D, Frank J, Ziebach R, Stern M, et al. Antioxidant deficiency in cystic fibrosis: when is the right time to take action? Am J Clin Nutr 2004;80:374–84.

[34] Brown LA, Harris FL, Jones DP. Ascorbate deficiency and oxidative stress in the alveolar type II cell. Am J Physiol 1997;273:L782–788.

[35] Winklhofer-Roob BM, Ellemunter H, Frühwirth M, Schlegel-Haueter SE, Khoschsorur G, van't Hof MA, et al. Plasma vitamin C concentrations in patients with cystic fibrosis: evidence of associations with lung inflammation. Am J Clin Nutr 1997;65:1858–66.

[36] Fischer H, Schwarzer C, Illek B. Vitamin C controls the cystic fibrosis transmembrane conductance regulator chloride channel. Proc Natl Acad Sci USA 2004;101:3691–6.

[37] Roum JH, Buhl R, McElvaney NG, Borok Z, Crystal RG. Systemic deficiency of glutathione in cystic fibrosis. J Appl Physiol (1985) 1993;75:2419–24.

[38] Ziady AG, Hansen J. Redox balance in cystic fibrosis. Int J Biochem Cell Biol 2014;52C:113–23.

[39] Gao L, Kim KJ, Yankaskas JR, Forman HJ. Abnormal glutathione transport in cystic fibrosis airway epithelia. Am J Physiol 1999;277:L113–118.

[40] Ballatori N, Hammond CL, Cunningham JB, Krance SM, Marchan R. Molecular mechanisms of reduced glutathione transport: role of the MRP/CFTR/ABCC and OATP/SLC21A families of membrane proteins. Toxicol Appl Pharmacol 2005;204:238–55.

[41] Keppler D, Leier I, Jedlitschky G. Transport of glutathione conjugates and glucuronides by the multidrug resistance proteins MRP1 and MRP2. Biol Chem 1997;378:787–91.

[42] Linsdell P, Hanrahan JW. Glutathione permeability of CFTR. Am J Physiol 1998;275:C323–326.

[43] Kogan I, Ramjeesingh M, Li C, Kidd JF, Wang Y, Leslie EM, et al. CFTR directly mediates nucleotide-regulated glutathione flux. EMBO J 2003;22:1981–9.

[44] Mangione S, Patel DD, Levin BR, Fiel SB. Erythrocytic glutathione in cystic fibrosis. A possible marker of pulmonary dysfunction. Chest 1994;105:1470–3.

[45] Velsor LW, Kariya C, Kachadourian R, Day BJ. Mitochondrial oxidative stress in the lungs of cystic fibrosis transmembrane conductance regulator protein mutant mice. Am J Respir Cell Mol Biol 2006;35:579–86.

[46] Iyer SS, Ramirez AM, Ritzenthaler JD, Torres-Gonzalez E, Roser-Page S, Mora AL, et al. Oxidation of extracellular cysteine/cystine redox state in bleomycin-induced lung fibrosis. Am J Physiol Lung Cell Mol Physiol 2009;296:L37–45.

[47] Moss M, Guidot DM, Wong-Lambertina M, Ten Hoor T, Perez RL, Brown LA. The effects of chronic alcohol abuse on pulmonary glutathione homeostasis. Am J Respir Crit Care Med 2000;161:414–9.

[48] Kettle AJ, Turner R, Gangell CL, Harwood DT, Khalilova IS, Chapman AL, et al. On Behalf of AREST CF. Oxidation contributes to low glutathione in the airways of children with cystic fibrosis. Eur Respir J 2014.

[49] Lands LC, Grey VL, Grenier C. Total plasma antioxidant capacity in cystic fibrosis. Pediatr Pulmonol 2000;29:81–7.

[50] Castillo R, Landon C, Eckhardt K, Morris V, Levander O, Lewiston N. Selenium and vitamin E status in cystic fibrosis. J Pediatr 1981;99:583–5.

[51] Hubbard VS, Barbero G, Chase HP. Selenium and cystic fibrosis. J Pediatr 1980;96:421–2.

[52] Lloyd-Still JD, Ganther HE. Selenium and glutathione peroxidase levels in cystic fibrosis. Pediatrics 1980;65:1010–2.

[53] Hakim F, Kerem E, Rivlin J, Bentur L, Stankiewicz H, Bdolach-Abram T, et al. Vitamins A and E and pulmonary exacerbations in patients with cystic fibrosis. J Pediatr Gastroenterol Nutr 2007;45: 347–53.

[54] Olveira G, Olveira C, Dorado A, García-Fuentes E, Rubio E, Tinahones F, et al. Cellular and plasma oxidative stress biomarkers are raised in adults with bronchiectasis. Clin Nutr 2013;32:112–7.

[55] Andersen DH. Cystic fibrosis of the pancreas and its relation to celiac disease: a clinical and pathologic study. Am J Dis Child 1938;56:344–99.

[56] Sproul A, Huang N. Growth patterns in children with cystic fibrosis. J Pediatr 1964;65:664–76.

[57] Yen EH, Quinton H, Borowitz D. Better nutritional status in early childhood is associated with improved clinical outcomes and survival in patients with cystic fibrosis. J Pediatr 2013;162:530–5. e531.

[58] Garattini E, Bilton D, Cremona G, Hodson M. Adult cystic fibrosis care in the 21st century. Monaldi Arch Chest Dis 2011;75:178–84.

[59] Woestenenk JW, Castelijns SJ, van der Ent CK, Houwen RH. Dietary intake in children and adolescents with cystic fibrosis. Clin Nutr 2013;33:528–32.

[60] Geukers VG, Oudshoorn JH, Taminiau JA, van der Ent CK, Schilte P, Ruiter AF, et al. Short-term protein intake and stimulation of protein synthesis in stunted children with cystic fibrosis. Am J Clin Nutr 2005;81:605–10.

[61] Olveira G, Olveira C, Acosta E, Espíldora F, Garrido-Sánchez L, García-Escobar E, et al. Fatty acid supplements improve respiratory, inflammatory and nutritional parameters in adults with cystic fibrosis. Arch Bronconeumol 2010;46:70–7.

[62] Fila L, Grandcourtová A, Chládek J, Musil J. Oxidative stress in cystic fibrosis patients with *Burkholderia cenocepacia* airway colonization: relation of 8-isoprostane concentration in exhaled breath condensate to lung function decline. Folia Microbiol (Praha) 2014;59:217–22.

[63] Peters SA, Kelly FJ. Vitamin E supplementation in cystic fibrosis. J Pediatr Gastroenterol Nutr 1996;22:341–5.

[64] Rust P, Eichler I, Renner S, Elmadfa I. Long-term oral beta-carotene supplementation in patients with cystic fibrosis - effects on antioxidative status and pulmonary function. Ann Nutr Metab. 2000;44:30–7.

[65] Wood LG, Fitzgerald DA, Lee AK, Garg ML. Improved antioxidant and fatty acid status of patients with cystic fibrosis after antioxidant supplementation is linked to improved lung function. Am J Clin Nutr 2003;77:150–9.

[66] Renner S, Rath R, Rust P, Lehr S, Frischer T, Elmadfa I, et al. Effects of beta-carotene supplementation for six months on clinical and laboratory parameters in patients with cystic fibrosis. Thorax 2001;56:48–52.

[67] Lepage G, Champagne J, Ronco N, Lamarre A, Osberg I, Sokol RJ, et al. Supplementation with carotenoids corrects increased lipid peroxidation in children with cystic fibrosis. Am J Clin Nutr 1996;64:87–93.

[68] Grey V, Mohammed SR, Smountas AA, Bahlool R, Lands LC. Improved glutathione status in young adult patients with cystic fibrosis supplemented with whey protein. J Cyst Fibros 2003;2: 195–8.

[69] Lands LC, Iskandar M, Beaudoin N, Meehan B, Dauletbaev N, Berthiuame Y. Dietary supplementation with pressurized whey in patients with cystic fibrosis. J Med Food 2010;13:77–82.

[70] Papas KA, Sontag MK, Pardee C, Sokol RJ, Sagel SD, Accurso FJ, et al. A pilot study on the safety and efficacy of a novel antioxidant rich formulation in patients with cystic fibrosis. J Cyst Fibros 2008;7:60–7.

[71] Papas K, Kalbfleisch J, Mohon R. Bioavailability of a novel, water-soluble vitamin E formulation in malabsorbing patients. Dig Dis Sci 2007;52:347–52.

[72] Sagel SD, Sontag MK, Anthony MM, Emmett P, Papas KA. Effect of an antioxidant-rich multivitamin supplement in cystic fibrosis. J Cyst Fibros 2011;10:31–6.

[73] Oudshoorn JH, Klijn PH, Hofman Z, Voorbij HA, van der Ent CK, Berger R, et al. Dietary supplementation with multiple micronutrients: no beneficial effects in pediatric cystic fibrosis patients. J Cyst Fibros 2007;6:35–40.

[74] Winklhofer-Roob BM, Puhl H, Khoschsorur G, van't Hof MA, Esterbauer H, Shmerling DH. Enhanced resistance to oxidation of low density lipoproteins and decreased lipid peroxide formation during beta-carotene supplementation in cystic fibrosis. Free Radic Biol Med 1995;18:849–59.

[75] Innis SM, Davidson AG, Melynk S, James SJ. Choline-related supplements improve abnormal plasma methionine-homocysteine metabolites and glutathione status in children with cystic fibrosis. Am J Clin Nutr 2007;85:702–8.

[76] Shamseer L, Adams D, Brown N, Johnson JA, Vohra S. Antioxidant micronutrients for lung disease in cystic fibrosis. Cochrane Database Syst Rev 2010:CD007020.

CHAPTER

19

Vitamin A Supplementation Therapy for Patients with Cystic Fibrosis

Holly M. Offenberger[1], *Ronald Ross Watson*[2]

[1]College of Science, Mel and Enid Zuckerman College of Public Health, University of Arizona, Tucson, AZ, USA; [2]Health Sciences Center, School of Medicine, Mel and Enid Zuckerman College of Public Health, University of Arizona, Tucson, AZ, USA

19.1 OVERVIEW OF VITAMIN A

Vitamin A is the name for a group of fat-soluble retinoids: retinol, retinal, retinoic acid, and retinyl esters. It is involved in reproduction, vision, cellular communication, and immune function. In vision, it is an essential component of the protein rhodopsin, which absorbs light in retinal receptors, and it supports the normal differentiation/functioning of the conjunctival membranes and cornea. Last, vitamin A supports cell growth and differentiation and plays a role in the normal formation and maintenance of the heart, kidney, lungs, and other organs [1].

There are two types of vitamin A that are available in the diet of humans: preformed vitamin A and provitamin A carotenoids. Preformed vitamin A is retinol and its esterified form is retinyl ester. These forms of vitamin A are found in foods from animal sources. Provitamin A carotenoids are β-carotene, α-carotene, and β-cryptoxanthin. These come from plant sources, which the body converts into vitamin A. The most important provitamin A is β-carotene. Both preformed vitamin A and provitamin A carotenoids are metabolized intracellularly into retinal and retinoic acid, which are active forms of vitamin A that are required to complete the biological functions [1].

19.1.1 Foods Rich in Vitamin A

Preformed vitamin A is highest in liver and fish oils. Both milk and eggs also have concentrations of preformed vitamin A as well as slight amounts of provitamin A. The largest dietary concentrations of provitamin A come from leafy green vegetables, orange and yellow vegetables, tomato products, fruits, and some vegetable oils. In an average United States diet the top food sources of provitamin A are carrots, broccoli, cantaloupe, and squash [1].

19.1.2 Vitamin A Supplements

As a supplement, vitamin A is often in the form of retinyl acetate or retinyl palmitate. There are a few supplements that have β-carotene in them in combination with a preformed vitamin A. Multivitamin supplements typically contain approximately 2500–10,000 IU vitamin A. This amount is equivalent to having approximately a half cup of raw carrots (9189 IU) or a half cup of raw sweet red peppers (2332 IU) [1].

19.1.3 Recommended Intakes of Vitamin A

The recommended daily allowance (RDA) for vitamin A is given as micrograms of retinol activity equivalents (RAEs). For those 0–6 months of age the RDA is 400 μg RAEs; those 7–12 months old have an RDA of 500 μg RAEs; those 1–3 years old, 300 μg RAEs; those 4–8 years old, 400 μg RAEs; and those 9–13 years old, 600 μg RAEs. The recommendations become sex-dependent after age 13 for males: the RDA is 900 μg RAEs for those older than 14 years of age. The recommendation for women after age 13 is 700 μg RAEs; however, this differs during pregnancy and lactation. When a female is pregnant the RDA is 750–770 μg RAEs, and when she is lactating the RDA is 1200–1300 μg RAEs [1].

19.1.4 Vitamin A Deficiency in Healthy Individuals

The most common symptom of vitamin A deficiency is xerophthalmia, which causes night blindness, low

Diet and Exercise in Cystic Fibrosis
http://dx.doi.org/10.1016/B978-0-12-800051-9.00019-5

iron status, and Bitot's spots, which are a build-up of keratin debris superficially located in the conjunctiva of the eye. If vitamin A deficiency is prolonged it causes severe visual impairment and possibly complete blindness. Vitamin A deficiency also increases the severity and mortality risk of infections, especially those causing diarrhea and measles [1].

19.2 BIOLOGICAL PROFILE OF VITAMIN A IN CYSTIC FIBROSIS

Patients with cystic fibrosis (CF) produce a thick mucus that obstructs the pancreas, which causes malabsorption and maldigestion of dietary lipids [2]. Vitamin A, a fat-soluble vitamin, is co-absorbed with fat and therefore is difficult for patients with CF to absorb [3]. Low serum concentrations of vitamin A are frequent among patients with CF because of their inability to absorb and digest fats [2]. One study found that 21% of the patients with CF had a vitamin A deficiency [4].

Patients who have CF and low or even undetectable levels of vitamin A experience symptoms such as night blindness [5]. One specific patient, diagnosed with night blindness and an undetectable level of vitamin A, received a regimen of high-dose vitamin A to reverse the symptoms [5]. The treatment was effective, yet the patient's vitamin A concentrations did not return to normal until a year later [5]. Patients with CF with low concentrations of vitamin A also experience more pulmonary exacerbations in a year than patients with high concentrations of vitamin A [6]. A pulmonary exacerbation is due to increased levels of oxidative stress, which create pulmonary damage that contributes to chronic lung disease in patients with CF [6]. Last, vitamin A deficiency in infants with CF is linked to unilateral facial nerve paralysis [7]. One patient, 10 weeks old and previously healthy, was diagnosed with CF and had a vitamin A concentration 10% below the lower normal threshold when they were diagnosed with facial nerve palsy [7].

19.2.1 Effective Dose of Vitamin A Supplements in CF Patients

When vitamin A supplements are prescribed to patients with CF, the aim is to ensure that the dose is high enough to maintain serum concentrations in the normal range yet not too high to induce side effects [2]. One study found that individualized vitamin A supplementation of 0–20,000 IU/day would prevent deficiency and high serum retinol concentrations but could lead to an intake above the tolerable upper threshold [8]. Another study found a more narrow dose of vitamin A supplements between 4000 and 10,000 IU of a fat-soluble preparation [2].

The CF Committee recommends that infants younger than a year old take Poly-Vi-Sol or another liquid supplement [9]. Children aged 2–8 years must take daily one tablet that has 4000 IU of vitamin A in it [9]. Adults and adolescents are required to take a multivitamin preparation (one–two tablets per day) [9].

This high variability in the dosing of vitamin A supplements makes it essential for doctors to monitor serum levels every 3–6 months to ensure patients' vitamin A concentrations are in an optimum range [2]. Another concern when dosing vitamin A supplements for patients with CF is the concern of those who are pregnant. A pregnant woman should take no more than 10,000 IU vitamin A/day because it could cause birth defects [2].

19.3 EFFECTIVENESS OF VITAMIN A TREATMENTS IN PREVENTING NEGATIVE SYMPTOMS

19.3.1 Night Blindness

Vitamin A deficiency most commonly results in night blindness in both healthy patients and those with CF. This has become increasingly important in those with CF for two reasons. The first is the higher prevalence of vitamin A deficiency in patients with CF because of their difficulty in absorbing fat-soluble vitamins. The second is the increasing number of patients reaching the driving age and needing to be able to see properly to drive [10].

One study found 8 of 43 patients with CF had vitamin A deficiencies. These eight patients had abnormal dark adaptation concentration tests and abnormal liver function tests, and three of the eight had conjunctival xerosis. To treat these deficiencies the patients were given orally 100,000–200,000 IU water-miscible vitamin A, and their daily vitamin supplements were increased. In all but one of the patients, the dark adaption tests returned to normal. The other patient required three further doses of water-miscible vitamin A and a daily supplement of 12,000 IU vitamin A before the dark adaptation test was normal. The study found overall that the supplements of vitamin A could return night vision function to normal and that adolescents with CF are at an increased risk for night blindness and conjunctival xerosis if they fail to take daily vitamin supplements [10].

19.3.2 Pulmonary Exacerbation

A causal relationship between pulmonary exacerbations and low concentrations of vitamin A has been shown in patients with CF. In one study, patients who were admitted to a hospital displayed a negative correlation between retinol concentrations and pulmonary exacerbations. The patients who displayed the lowest concentrations of retinol displayed the maximum amount of C-reactive protein contraction. This indicated that the body's response during the acute phase is associated with the depression of retinol concentrations. Overall, it was demonstrated that plasma retinol concentrations are depressed in patients with CF with acute pulmonary exacerbations [11].

In another study, exacerbations were directly correlated with lower vitamin A and E concentrations, even if within the normal range. The study concluded that reduced serum concentrations of both vitamin A and E, even if in the normal range, are associated with exacerbations in patients with CF. However, neither study determined whether supplementation of these vitamins was necessary or how much supplementation would need to be given [6]. To truly improve the pulmonary exacerbations caused by vitamin A deficiency, more research on supplementation needs to be completed.

19.3.3 Unilateral Facial Nerve Paralysis

In the case study of the infant who was brought in with unilateral facial nerve paralysis, there was a strong association between low vitamin A concentrations in patients with CF and facial nerve paralysis. The infant had normal concentrations of all fat-soluble vitamins except vitamin A, which was 10% below the normal range [7]. Another case study presented an infant with bilateral facial palsy. Upon examination, it was found the infant had CF and hypovitaminosis A. The facial palsy was improved after vitamin A supplementation to restore normal vitamin A levels [12]. The vitamin A supplementation to achieve normal concentrations improved the paralysis in both case studies.

References

[1] National Institute of Health. Vitamin A. Retrieved from: http://ods.od.nih.gov/factsheets/VitaminA-HealthProfessional/; 2013.

[2] Sinaasappel M, Stern M, Littlewood J, Wolfe S, Steinkamp G, Heijerman HG, et al. Nutrition in patients with cystic fibrosis: a European consensus. J Cyst Fibros 2002;1(2):51–75.

[3] Bonifant CM, Shevill E, Chang AB. Vitamin A supplementation for cystic fibrosis. Cochrane Database Sys Rev 2012;8:CD006751.

[4] Sokol RJ, Reardon MC, Accurso FJ, Stall C, Narkewicz MR, Abman SH, et al. Fat-soluble vitamins in infants identified by cystic fibrosis in newborn screening. Pediatr Pulmonol 1991;7:52–5.

[5] Roddy MF, Greally P, Clancy G, Leen G, Feehan S, Elnazir B. Night blindness in a teenager with cystic fibrosis. Nutr Clin Pract 2011;26(6):718–21.

[6] Hakim F, Kerem E, Rivlin J, Bentur L, Stankiewicz H, Bdolach-Abram T, et al. Vitamins A and E and pulmonary exacerbations in patients with cystic fibrosis. J Pediatr Gastroenterol Nutr 2007;45(3):347–53.

[7] Cameron C, Lodes MW, Gershan WM. Facial nerve palsy associated with a low serum vitamin A level in an infant with cystic fibrosis. J Cyst Fibros 2007;6(3):241–3.

[8] Brel C, Simon A, Krawinkel MB, Naehrlich L. Individualized vitamin A supplementation for patients with cystic fibrosis. Clin Nutr 2013;32(5):805–10.

[9] Ramsey BW, Farrell PM, Pencharz P, Consensus Committee. Nutritional assessment and management in cystic fibrosis: a consensus report. Am J Clin Nutr 1992;55(1):108–16.

[10] Rayner RJ, Tyrrell JC, Hiller EJ, Marenah C, Neugebauer MA, Vernon SA, et al. Night blindness and conjunctival xerosis caused by vitamin A deficiency in patients with cystic fibrosis. Arch Dis Child 1989;64(1):1151–6.

[11] Duggan C, Colin AA, Agil A, Higgins L, Rifai N. Vitamin A status in acute exacerbations of cystic fibrosis. Am J Clin Nutr 1996;98(6):635–9.

[12] Basu AP, Kumar P, Devlin AM, O'Brien CJ. Cystic fibrosis presenting with bilateral facial palsy. Eur J Pediatr Neurol 2007;11(4):240–2.

The Emergence of Polyphenols in the Potentiation of Treatment Modality in Cystic Fibrosis

Stan Kubow[1], *Manyan Fung*[1], *Noor Naqvi*[1], *Larry C. Lands*[2]

[1]School of Dietetics and Human Nutrition, McGill University, QC, Canada; [2]Montreal Children's Hospital, Pediatric Respiratory Medicine, Montreal, QC, Canada

20.1 INTRODUCTION

Mutations in the gene encoding the cystic fibrosis (CF) transmembrane conductance regulator (CFTR) result in CF and its accompanying morbidities. CFTR is ubiquitously expressed in the plasma membrane of secretory epithelia found in the airways, intestine, pancreas, testis, and exocrine glands as well as in some nonepithelial cell types. The first steps toward gene correction are taking place. Thus, from a practical perspective, the focus of CF therapy is currently about prevention of pulmonary damage, attentiveness to the identification of comorbidities, maintenance of structural integrity in other vital organs, improvement to nutritional status, and enhancement to the overall quality of life. This chapter introduces the topic of nonpharmacological interventions, particularly in the use of polyphenols, organic chemicals that in recent literature have demonstrated their capacity to potentiate the effectiveness of other treatment strategies. Such an alternative approach is worthwhile to consider in the context of CF.

20.2 ROLE OF POLYPHENOLS IN OTHER DISEASES AND POSSIBLY CF

Dietary polyphenols are naturally occurring phytochemicals present in many plant-based foods, such as tea, coffee, wine, legumes, fruits, and vegetables. Polyphenols are composed of several different classes of compounds based on their differing chemical structures and on the location and number of hydroxyl groups. On these bases, polyphenols have been divided into hydroxycinnamic acids, hydroxybenzoic acids, stilbenes, lignans, isoflavones, flavonols, flavones, flavanols, flavanones, anthocyanins, and proanthocyanidins. The various chemical structures and food content of polyphenols have been reviewed comprehensively in excellent reviews [1,2]. Their beneficial effects on human health were based initially on epidemiological studies showing that populations consuming polyphenol-rich food demonstrated a lower incidence of a variety of chronic diseases, including diabetes mellitus, cancer, and respiratory and cardiovascular diseases [3]. Although many prospective cross-sectional and intervention studies have confirmed such associations with dietary consumption of polyphenols, the specific fruits or vegetables or combinations thereof that confer optimal health benefits for specific disease states remain elusive. Among the different dietary sources of polyphenols, however, food staples such as oranges, apples, and potatoes [4] as well as coffee [5] have provided the bulk polyphenols to the diet.

As indicated by chemical assays and other in vitro models, the biological properties of polyphenols are not merely governed by the changes associated with oxidative stress, inflammation, and endothelial function; in vivo bioavailability, biotransformation, and utilization must also be considered to fully understand their functionality. Depending on the degree of polymerization and the glycosylation pattern, a significant fraction of dietary polyphenols persists in the colon to undergo further metabolic transformation (i.e., dehydroxylation and demethylation) by the resident colonic microbiota. Therefore, through examining the diversity of the colonic microbiota using metabolomic methods, one could indirectly correlate/evaluate the antioxidant capacity of polyphenolic compounds and their associated metabolites. The structural

Diet and Exercise in Cystic Fibrosis
http://dx.doi.org/10.1016/B978-0-12-800051-9.00020-1

motif (typically comprised of at least one aromatic ring structure with one or more hydroxyl groups) imparts the antioxidant activity commonly seen in the diverse classes of polyphenols through electron donation to the surrounding free radicals. However, antioxidant capacities of polyphenols based on their capability to donate an electron are limited to the oral cavity and gut lumen because regulatory gut and hepatic mechanisms prevent their access to the systemic circulation, which leads to only low micromolar plasma concentrations of both parent polyphenolics and their microbial metabolites [6]. These compounds are present at insufficient concentrations to exert a significant direct free radical scavenging effect systemically. Despite their low bioavailability, these compounds can act as signaling molecules that can induce endogenous antioxidant defense pathways as well as modulate many intracellular signaling processes associated with oxidative stress and inflammation. In that regard, the health-promoting benefits of polyphenol parent compounds and their microbial metabolites have been related to translocation into the nucleus of nuclear factor erythroid 2-related factor 2 (Nrf2) and induction of Nrf2-antioxidant response element signaling pathway [7], suppression of nuclear factor κB (NF-κB) and activator protein 1 (AP-1) [8], induction of glutathione conjugates [9] and glutathione S-transferase [10], caspase-3 activation [11], c-jun NH_2-terminal kinase [12] and P38 activation [13], and modulation of mitogen-activated protein kinases [14] and the phosphoinositide-3 kinase/Akt protein kinase B pathway [15].

The recent advancement of automated screening instrumentation has helped tremendously in the identification of promising compounds that may be considered as either "correctors" or "potentiators" [1]. Correctors refer specifically to those compounds that can promote the translocation of mutant CFTR to the membrane, whereas potentiators promote the activity of translocated mutant CFTR. In the case when both trafficking and activity are poor (e.g., ΔF508 CFTR) (Box 20.1), a compound with both "corrector" and "potentiator" properties would be ideal. Flavonoids represent the ideal candidates for the consideration in their role as either a "corrector" or "potentiator" in CFTR processing due to the three therapeutic properties they often exhibit in other diseases: anti-inflammatory, antioxidant, and antimicrobial. This is in combination with the fact that these compounds are naturally present in many plant-based foods and are the largest class of polyphenols, with a common structure of diphenylpropanes (consisting of two aromatic rings linked through three carbons, C6-C3-C6) (Table 20.1).

TABLE 20.1 Classification of the Six Major Classes of Dietary Flavonoids Adapted From [22]

Class of Flavonoids	Compound
Flavonols	Quercetin and myricetin
Flavanones	Naringenin and hesperidin
Flavanols (or catechins)	Epicatechin and gallocatechin
Flavones	Apigenin and luteolin
Anthocyanidins	Cyanidin and pelargonidin
Isoflavones	Genistein and daidezin

BOX 20.1

More than 1700 CFTR mutations have now been identified [16], many of which are broadly grouped into six classes according to structural and physiological defects with the CFTR protein [17,18]. Class I to III mutations represent the classic "severe" mutations wherein a nonsense, frame-shift, or amino acid deletion causes no synthesis or processing of the protein, whereas class IV and V mutations represent "mild" mutations wherein changes in amino acids cause reduced synthesis, dysregulation, and altered channel conductance of the protein [19,20]. The latter allows some degree of CFTR function to be retained and is associated with pancreatic sufficiency [21].

Class I	Absence of Synthesis
Class II	ΔF508 (deletion of phenylalanine) is the most common class II mutation worldwide. Abnormal folding of protein → defective protein maturation → premature degradation in the endoplasmic reticulum → very little protein reaches the apical membrane
Class III	G551D is a relatively common mutation, accounting for 3% of CF mutations → regulation of the ion channel is disturbed
Class IV	Shorter polythymidine tracts (5 and 7T as opposed to 9T) → CFTR protein is correctly translocated to the cell membrane but pore region of the channel is affected → defective chloride conductance or channel gating → might or might not show clinical manifestations of CF
Class V	Promoter or splicing abnormality → unstable CFTR mRNA transcript → reduced number → might or might not show clinical manifestations of CF
Class VI	Accelerated turnover from the cell surface → unstable CFTR

20.3 THERAPEUTIC CAPACITY OF POLYPHENOLS TO CFTR FUNCTION

CFTR mutations are classified according to the type of functional alterations observed, with class I to III mutations generally considered as severely defective, many of which are intrinsic to the regulatory functions that CFTR participates in within the overall ion transport systems. On its own, CFTR functions mainly as a chloride channel, but the efflux of other anions such as reduced glutathione is under its regulation as well. On the whole, its role extends well beyond chloride permeability in epithelial cells because CFTR is in proximity to several membrane receptors, ion channels, and the cytoskeleton [18]. Thus, it is not surprising that CFTR can influence other gene products, mainly proteins related to transmembrane ion transport, membrane conductance, regulation of ATP channels, regulation of intracellular vesicle transport, and acidification of intracellular organelles [23–26]. These other proteins can then act as modifiers of the CF phenotype and may help explain the differences in clinical severity and manifestations among patients with the same mutations in CFTR [18]. A notable adverse effect associated with CFTR dysfunction is the electrolyte imbalance that is transforming mucin secretion to become increasingly viscous [27]. The poor solubility of such secretion not only leads to progressive obstruction and fibrosis of organs, but also the poor mucociliary clearance can predispose the luminal environment to recurrent bacterial infection [28–30]. The biological lesions can be seen in the pancreas, liver, gastrointestinal tract, and reproductive organs, secondary to the pulmonary system. Therefore, concurrent comorbidities are common among patients with CF, further complicating disease management and treatment.

Because deletion of Phe508 (ΔF508, a class II mutation) is the most common CFTR mutation, primary therapeutic endpoints lie in the correction of endoplasmic reticulum-to-plasma membrane protein trafficking and efficient potentiation of defective ΔF508-CFTR Cl^- channel gating activity (i.e., enhanced Cl^- conductance). In brief, transepithelial measurements of most compounds are conducted in a bronchial epithelial cell culture model after a 27°C temperature rescue for 24h. In one study, oral ingestion of capsaicin and dihydrocapsaicin, compounds isolated from dried hot red chili peppers, showed activation of CFTR by direct binding and interaction with the CFTR protein [31]. This binding may be attributed to its similarity in structure in the cytoplasmic domains [31] and the ATP regulation properties of CFTR and vanilloid receptor 1 (VR1). Capsaicin binds to the VR1 in the sensory neuron and allows activation of ligand-gated cation channels, leading to the influx of cations [31]. The effect of 100μM capsaicin in the presence of 10μM forskolin (a compound generally used to elevate cAMP levels) is significantly greater than that of forskolin alone. The effects of capsaicin can be compared to that of genistein [32], a tyrosine kinase inhibitor that is known to have a good human safety profile and is a potent stimulator of CFTR channel activity by modulating protein conformation and protein phosphorylation at the R domain, a requirement for CFTR to function as a chloride channel [31]. It has previously been shown to inhibit protein kinase A (PKA) in CFbe41o2 cells, implying that ΔF508-CFTR needs PKA activity for genistein [33]. Both capsaicin and genistein have been shown to affect wild-type and mutant CFTR, and they may be acting at the same binding site at the cytoplasmic side of the membrane, which may lead to potential for competition. At the channel, they both increase the opening rate and decrease the closing rate of the ATP-dependant gating of CFTR [31]. However, they differ in terms of the extent of potentiation because the maximum effect of capsaicin is approximately 60% that of genistein. The effect of genistein potentiation is maximal at the 40–80μM dose [34–36]. Because CFTR channels with lower phosphorylation experience a slower opening rate, the stimulation of intestinal CFTR by oral ingestion of these polyphenols may provide a therapeutic strategy for gastrointestinal disorders in patients with CF. In contrast, the ingestion of capsaicin may cause diarrhea by stimulating ion transport in the intestinal tissue, thereby increasing fluid secretion [31,37].

An in vitro study shows that quercetin, a compound safe for human use even at high doses for 6–7 months, stimulated short-circuit current in Fisher and CFbe41 cells [32,38]. Although it had little effect on cAMP levels and phosphorylation (as shown with genistein [32,36,39,40]) of isolated CFTR R domain, suggesting activation that is not related to classic R domain channel phosphorylation, it had an effect on activation of CFTR-mediated anion transport in respiratory epithelia. This effect was evident both in vitro and in vivo, and despite the fact that there may be a direct effect on the CFTR Cl^- channel, which helps overcome gating defects, the mechanism by which this occurs by quercetin and other flavonoids remains unknown [32]. In terms of activation of CFTR, it can be speculated that quercetin may be providing two synergistic yet different stimuli, independent and dependent on cAMP [32]. Genistein administered in IB3-1 cells pretreated with 4-phenylbutyrate therapy (cell chaperone shown to restore ΔF508-CFTR to processing pathway and cAMP mediated Cl^- conductance) shows increasing CFTR processing and increased Cl^- efflux through immunoblot measurements [41]. In addition, long-term exposure (18h) of 50μM resveratrol has been shown to enhance CFTR trafficking to the cell membrane as revealed by immunofluorescence imagery. However, the relationship of resveratrol to ΔF508-CFTR remains unclear, and the concentration used in the CF pancreatic cell line for this study (and others using similar doses) is unrealistic for validation by human trials [42]. Another polyphenol, naringenin, which is found in citrus fruits, has exhibited inhibition of basal and stimulated Cl^- secretion in the rat and human colon, which may be due to its role in inhibition of basolateral NKCC1 [43]. However, naringenin did not alter cAMP generation. Structurally similar compounds such as eriodyctiol have the ability to inhibit ion transport; yet, apigenin, genistein, and cocoa-related flavonoids, which are structurally similar, may inhibit CFTR at micromolar concentrations.

Another common CF-associated mutation is glycine-to-aspartate missense mutation at the 551 position (class III mutation) [18,34]. Curcumin, a component of the turmeric spice, has been shown to open CFTR channels through a mechanism that is not ATP dependent but is dependent on PKA phosphorylation [34,35]. It can be tolerated in large dosages (45 mg daily per kg of body weight) without evidence of toxicity [44,45]; however, its bioavailability is uncertain [44]. It can crosslink various CFTR polypeptides, including G55-1D-CFTR; however, it has not been shown to have a significant influence on CFTR channel function [34]. In terms of potentiating G551D-CFTR, the effect of curcumin is lower than that of genistein but can increase channel activity synergistically after potentiation mediated by genistein, leading to restoration of the gating effect of G551D-CFTR to up to 50% of the wild-type-CFTR (norm) level [34]. Curcumin has been shown to activate ΔF508-CFTR function with the combination of low-temperature incubation. Expression of ΔF508-CFTR was 25% of that seen with low-temperature incubation alone [44]. However, evidence of activation of ΔF508-CFTR by curcumin alone is not well substantiated [44–48]. Curcumin has also been shown to enhance CFTR expression by downregulation of calreticulin (CRT), a negative regulator of CFTR expression and its function [44,49]. CRT may be a mediator of curcumin's effect on CFTR because curcumin suppressed endogenous CRT mRNA directly, an effect that may be attributable to its ability to block AP-1 binding to its DNA binding sites [44,50] and also block AP-1 at the c-Jun and c-Fos sites [44,51,52]. It is worth noting that most of these compounds perform in a bimodal manner, wherein CFTR activity is enhanced in parallel to a dose-dependent current increase ($<5\,\mu M$), but the reverse is shown at higher concentrations [53]. An indication is that their structural conformation allows them to bind to at least two sites, an activating site and an inhibitory site [54,55]. Hence, their synergistic or antagonistic properties are highly dependent on finding the ideal dosage and the relative position of the mutated residues the channel carries [55], both of which can interfere with the affinity/interaction of the compound in question.

20.4 ANTIMICROBIAL CAPACITY OF POLYPHENOLS

Biofilms are a major cause of concern in CF because the repercussions of their formation contribute to approximately 80% of all infections [56]. A biofilm is produced when microorganisms become attached to biotic surfaces and multiply in the extracellular matrix of polymeric substances comprised of proteins and polysaccharides [57,58]. Biofilms are resistant to antibiotics and disinfectants [57–59], and some may carry resistance genes for penicillin antibiotics [56]. Polyphenolic compounds have shown to have antimicrobial properties and hence may have promising therapeutic effects [56]. Gallic acid, found in tea leaves, oak bark, grape seeds, and various vegetables, has been tested against *Staphylococcus aureus* strains and has been shown to have a higher antibacterial activity than chlorogenic acid [56]. It can also decrease the adhesion capacity of *S. aureus* to polystyrene, thereby attenuating its ability to form biofilms. Furthermore, small phenolics may behave as proline analogs or mimics and thus have the ability to modulate cellular redox response through the proline-linked pentose phosphate pathway [60]. Inhibition of proline dehydrogenase at the plasma membrane level in a prokaryotic cell can inhibit its energy metabolism and hence cause disruption to the process of bacterial growth. Phenolic acids have been shown to have antimicrobial qualities at the range of $100–1000\,\mu g\,mL^{-1}$ [58]. Caffeic acid also plays a role in disrupting membrane function of *S. aureus* by acting as a nucleophile [58] and increasing membrane permeability, depolarizing cell membranes, and reducing respiratory activity, thereby inhibiting the growth of this Gram-positive bacterium [56]. EGCG, a polyphenolic compound found in green tea, has also shown to significantly reduce the metabolic activity to young and mature biofilms; however, this effect was only seen in biofilms produced by *S. maltophilia* and is thus strain specific [60]. Overall, these findings suggest the need for further research in the prevention of biofilm production and treatment of mature biofilms [60].

20.5 ANTI-INFLAMMATORY PROPERTIES OF POLYPHENOLS

A wide variety of inflammation-related pathways are affected by polyphenols, and they are reviewed extensively elsewhere [61]. Depending on their structural nature, polyphenols affect many different inflammatory cascades; however, the most common inhibitory effects on inflammation have been related to their antioxidant action. Inflammatory states are associated with reactive oxygen species–mediated activation of the transcription factors NF-κB and AP-1, both of which are inhibited by a variety of polyphenolic compounds in several tissue types. In addition, polyphenolics can induce changes in nuclear histone acetylation and deacetylation patterns associated with anti-inflammatory effects [62]. Taken together, such evidence supports the concept that polyphenols can exert regulatory effects on inflammatory genes that lead to the ensuing anti-inflammatory response. Several cell culture studies have investigated the anti-inflammatory effects associated with the use of polyphenol-rich extracts of foods. One study examined bergamot (*Citrus bergamia* (Risso)) fruit extracts and their effects on the production of interleukin-8 (IL-8) in CF IB3-1 and CuFi-1 cells. This study found that coumarin and psoralen compounds present in the extract contained the most biologically active constituents, whereby bergapten and citropten were the most active in terms of reducing IL-8 mRNA level in IB3-1 cells treated with TNF-α [63]. Olive mill waste water extract, which contains abundant amounts of hydroxytyrosol, was used in TNF-α exposed bronchial cells. The results of this study showed that the extract inhibits the interactions between NF-κB and DNA, even at high concentrations of 400 ng/μL, thereby reducing IL-8 gene expression. Similarly, the polyphenol epigallocatechin gallate protected respiratory epithelial cells against IL-1β–dependent inflammation and IL-8 gene expression induced via NF-κB [64]. A major limitation of these in vitro studies is the use of parent polyphenol compounds because they undergo extensive degradation to secondary metabolites by gut microbiota as well as intestinal and hepatic first-pass metabolism. Intake of polyphenols such as chlorogenic and ferulic acids results in the appearance of secondary metabolites in human plasma, such as vanillic, isoferulic, phenylpropionic, hippuric, benzoic, and phenylacetic acids [65,66].

20.6 ROLE OF POLYPHENOLS IN CFTR-RELATED COMORBIDITIES

The diagnostic criteria of CF can be challenging because of the wide variability in clinical phenotypes along with varying degrees of disease severity (or lack thereof) and progression. Although disease heterogeneity can be explained by the large number of genetic mutations, the presence of modifier genes and environmental stressors is responsible for driving the clinical outcomes [18]. This enforces the idea that genotype–phenotype correlation cannot reliably predict for one's prognosis with complete certainty. Another concept that is important to understand is the distinction between CFTR-related disorders and CFTR-related comorbidities [67,68]. For example, the organ that appears to be most sensitive to CFTR dysfunction is not the lung, but is in fact the vas deferens, because some individuals with class IV or V mutations (most notably R117H) may exhibit congenital bilateral absence of the vas deferens as their primary or only clinical manifestation of CF [69]. In contrast, CFTR-related comorbidities are life-threatening health issues that emerge in addition to lung conditions, all of which can further complicate treatment and lengthen the choice of a proper modality, because CF is a highly individualized disease.

Polyphenolic compounds are highly diverse in kind with equally diverse properties. These properties allow them to convey a vast array of effects in various parts of the human body. Currently, there is a large gap in the literature pertaining to the direct effect of polyphenols on specific CF-related comorbidities. Hence, Table 10.2 has been formulated to provide an overview of the current research on polyphenolic compounds and their varying effects on diseases that manifest in CF patients with declining prognoses. It must be noted that direct clinical trials on polyphenols and CF-related comorbidities have not been conducted. Thus, the following findings pertaining to polyphenols and various chronic diseases may be of benefit in aiding future CF research.

20.7 CF-RELATED MENTAL IMPAIRMENTS

The principle organs being affected in CF are mainly the respiratory tract and airways of the lungs, but organs in the digestive system (e.g., small intestine, liver, and pancreas) also are typically affected, leading to aberrant nutrient status. Symptoms associated with the myriad of comorbidities can also contribute to nongastrointestinal conditions, such as depression and anxiety, that are often overlooked by medical professionals. Mental health impairment can pose a significant impact on clinical outcomes, treatment adherence, and health-related quality of life. Association of depression and anxiety with declining health outcomes of CF has been a major focus in recent years. Several cross-sectional studies have consistently shown that CF patients with anxiety, depression, or both exhibit worsened lung function, nutritional status, and overall quality of life [102–105]. Specifically, those afflicted with chronic lung conditions and some form of a psychiatric disorder tend to show patterns of worse symptoms and greater functional impairment than those without [106]. Given the current lack of systematic reviews regarding the impact of mental health symptoms on clinical disease outcomes

TABLE 20.2 Principle Comorbidities Associated With Cystic Fibrosis (CF) as Adapted From [70]

Disease Manifestations	Flavonoids as "Potentiators"	Findings
PANCREATIC INSUFFICIENCY		
Fecal fat excretion weight loss/inability to gain weight Fat-soluble vitamin deficiency Flatulence Abdominal discomfort	Resveratrol, effects of EGCG, green tea extract, curcumin	Decreased incidence of acute pancreatitis in rodents [71] Reduction of pancreatic damage: pancreatic macropathology, micropathology, neutrophil infiltrate, trypsin activity, lipoperoxides, and inflammatory cytokines [71] Curcumin reduces serum amylase, TNF-α, IL-6, and bacterial translocation [71]
MALNUTRITION		
Weight loss Nutritional deficiencies	Cocoa, berries, and green tea polyphenols	Increased sense of satiety via potent effect on neuropeptides and neurohormones, hence may be detrimental in situations of malnourishment [72]
CF-RELATED DIABETES MELLITUS		
Abnormal serum glucose levels Impaired glucose tolerance Micro- and macrovascular complications	Quercetin, ferulic acid, catechin, epicatechin, epigallocatechin, epicatechin gallate, isoflavones, tannic acid, glycyrrhizin, chlorogenic acid, saponins	Resveratrol is an antidiabetic agent that modulates SIRT1, thereby improving glucose homeostasis and inhibiting diabetic neuropathy [73–76] Quercetin is protective against oxidative stress and lipid peroxidation [77] Ferulic acid has shown to reduce blood glucose levels [78,79] (−)-Epigallocatechin 3-*O*-gallate regulates glucose homeostasis and decreases renal advanced glycation end products (AGEs) accumulation and proteinuria (common in diabetic nephropathy) [80] Suppression of hyperglycemia, AGEs, and related oxidative stress and cytokine activations by various polyphenols [80] Attenuation of diabetic cataracts, retinopathy, hyperlipidemia, oxidative stress, and increase in insulin sensitivity with green tea polyphenols [81]
HEPATOBILIARY DISEASE		
Fibrosis, cholestasis, chlolithiasis, cirrhosis, portal hypertension, liver failure	Curcumin, quercetin	Natural inducers of heme oxygenase-1, a stress response enzyme induced as a protective mechanism for protection against inflammatory processes, oxidative tissue damage, and induction of liver regeneration [82] Quercetin prevents ethanol toxicity in human hepatocytes through p38 and ERK/Nrf2 transduction pathway [82–84]
RENAL DISEASE		
Acute renal failure Kidney failure Kidney stones	(−)-Epicatechin 3-*O*-gallate and green tea polyphenols	EGCG scavenges peroxynitrite directly, a reactive oxygen and nitrogen metabolite (formed from O_2^- and NO and its decomposition product •OH) that plays a role in cellular damage and renal dysfunction [80,85,86] by instigating antioxidant depletion, alterations in protein structure and function by tyrosine nitration, and oxidative damage [80] Inhibitory effect of EGCG on methylguanidine (strong uremic toxin produced by oxidative reactions) [80] Green tea polyphenols have a positive effect on increased serum creatinine and urinary protein levels and decreased creatinine clearance [80,87,88]
CANCER		
Fever Unexplained weight loss Skin changes Unusual bleeding/discharge Thickening lump or obstruction Pain, fatigue Indigestion or trouble swallowing	Quercetin, catechins, isoflavones, lignans, flavanones, ellagic acid, red wine polyphenols, resveratrol, curcumin, theaflavins, thearubigins	Inhibition and promotion of elimination of procarcinogens and promutagens [81] through variable mechanisms of action [89] Chemopreventative action through estrogenic, antiestrogenic, antiproliferative, apoptotic, anti-oxidative, anti-inflammatory effects [90] Regulation of detoxifaction enzymes (i.e., cytochrome P450 expression) and host immunity [90] such as polyphenolic quinones that pose as substrates for phase II enzymes, hence activating them for their own detoxification and augmenting host immunity [91] Inhibition of cyclooxygenase, hydroperoxide, protein kinase C, Bcl-2 phosphorylation, Akt, focal adhesion kinase, NF-κB, matrix metalloprotease-9, and cell regulators [92]

TABLE 20.2 Principle Comorbidities Associated With Cystic Fibrosis (CF) as Adapted From [70]—cont'd

Disease Manifestations	Flavonoids as "Potentiators"	Findings
GASTROINTESTINAL CONDITIONS		
Distal intestinal obstruction syndrome Epigastric pain, heartburn, dyspepsia	Resveratrol, EGCG/green tea extract, curcumin, quercetin, rutin	Green tea extract effective in preventing and treating intestinal inflammation and injury and increase myeloperoxidase activity in the intestine [71,81,93] Reduction of mortality rates, attenuation of colonic (e.g., diarrhea, bloody stools) and extracolonic (e.g., weight loss) signs of disease, colon macropathology and micropathology (e.g., hyperemia, ulcerations, inflammatory infiltrate, serosal adhesions) [71] Reduction of inflammation, autoimmunity (e.g., colonic myeloperoxidase, NF-κB activity, TNF-α, IL-1β, IL-12, inducible nitric-oxide synthase, IL-10, T cell and neutrophil infiltration [71]
BONE DISEASE		
Low BMD Episodic arthritis Hypertrophic pulmonary osteoarthropathy	Ferulic acid, *p*-coumaric acid, chlorogenic acid, alohexanecarboixylic acid, (−)-epigallocatechin gallate, theaflavin, apigenin, iuteolin, anthocyanin, rutin, genistein, glycitein, lignin, resveratrol, catechin, daizin,	Increase serum osteoblast progenitors/differentiation, bone mass, bone density, and bone mineral mass [94–96] Green tea polyphenols increase serum osteocalcin, bone mineral density (BMD), trabecular volume, strength of femur, trabecular thickness, and bone formation [97] Synergistic effect of phenolic acids in extracts may reverse BMD and calcium loss [98] (phytochemicals—a global perspective of their role in nutrition and health; chapter 22 - polyphenols antioxidants and bone health: a review) Green tea and theaflavin enriched tea plant extract reduces expression of inflammatory mediators and cytokines in animal models [81] (Hirsch JB, Evans D, the state of nutrigenomics, nutraceuticals world; 8(8) (2005)) Supplements rich in variety of polyphenols found to be effective in stimulating osteoblasts for the formation of bone nodules than epicatechin alone, lending to evidence of synergism (phytochemicals—a global perspective of their role in nutrition and health; chapter 22—polyphenols antioxidants and bone health: a review)
COGNITIVE DISORDERS		
Depression Anxiety Insomnia Irritability Decreased compliance with therapeutic regimens	Amentoflavone, apigenin, chlorogenic acid, curcumin, ferulic acid, hesperidin, rutin, quercetin, naringenin, resveratrol, ellagic acid, nobiletin, proanthocyanidin, EGCG	Effect on central noradrenergic, dopaminergic, serotoninergic activity, monoamine oxidase inhibitory action, modulation of brain-derived neurotrophic factor-TrkB-PI3K/Akt pathways, antioxidative effect, and modulation of γ-aminobutyric acid type A receptors [99] EGCG inhibits spontaneous excitatory synaptic transmission in mice therefore having antistress effects without affecting appetite or physical fitness [81,100] Consumption of green tea improves cognitive and psychomotor performance with less effects on quality of sleep at night compared with coffee intake [81,101]

and comorbidities in CF, it is difficult to draw any actual correlations or relationships; thus, this need presents a good opportunity for alternative pharmacotherapy. By definition, depression can be related to a deficient level of serotonin (5-hydroxytryptamine) and noradrenaline, the two selective inhibitors of monoamine oxidase (MAO). MAO is a mitochondrial outer membrane-bound flavoprotein with two isoforms, mainly A and B, and together they regulate levels of neurotransmitters through oxidative deamination [107]. Many flavonoid-based MAO inhibitors have been identified and can potentially be used as antidepressants in a clinical setting. An average dietary intake of flavonoids is about 23 mg/day, of which quercetin contributes to 73% and 76% of the total polyphenol consumption in women and men, respectively [108]. Various reports suggest that quercetin and its associated metabolites can easily pass the through the blood–brain barrier and improve neuronal activity induced by stress [109,110]. Six fractions were extracted from the herbal antidepressant St John's wort (*Hypericum perforatum*) to study the antidepressive mechanism of action in rat brain homogenates in vitro and ex vivo analysis after intraperitoneal application of the extracts to albino rats [111]. This study showed that the fraction rich in quercetin had an MAO-A inhibition of 39% at the concentration of only 10^{-4} mol/L, whereas all the other fractions demonstrated less than 25% inhibition. Similar results were shown for quercetin isolated from St John's wort leaves, with a selective inhibitory activity for MAO-A (IC_{50} = 0.01 μM) and MAO-B (IC_{50} = 20 μM) [112]. A rather interesting study evaluated the antidepressant-like effect of onion powder (*Allium cepa* L., dosage 50 mg/kg, containing 0.4 mg of quercetin glycosides) using a rat behavioral model of depression, the forced swimming test (FST), for a period of 14 days [113]. Daily treatment significantly reduced the immobility time in FST without altering the motor dysfunction, while increasing the dopaminergic activity in the rat hypothalamus by suppressing its increase in the turnover of this neurotransmitter. The results of the study suggest that the active component quercetin exerted antidepressant-like activity in a behavioral model. Oral administration of rutin (5–80 mg/kg) and quercetin (5–40 mg/kg) to adult rats was shown to produce similar electropharmacograms, with dose-dependent decreases of spectral alpha2 and beta1 frequencies within all brain areas, primarily the frontal cortex, hippocampus, striatum, and reticular formation, and with peak effects observed 4 h postadministration [114]. This pattern of changes is comparable to that obtained using 2.5 mg/kg moclobemide after 1 h of observation. Moclobemide is classified as an MAO-A inhibitor. Thus, the functional antidepressant capabilities of rutin and quercetin are indeed present, consistent with previous studies, but the mechanisms of action involved remain unknown, and further clinical testing in patients with symptoms of depression is required.

20.8 CONCLUSION

Despite the accumulation of highly promising research in the past 10 years showing potential beneficial effects of polyphenols on a wide variety of biological processes related to CF, more research is needed to identify the optimal polyphenol or complementary mixture of polyphenols as pharmacological treatment approaches. In that regard, a better understanding of the fundamental information pertaining to polyphenol dose, bioavailability, distribution, excretion, and toxicity is needed before the implementation of clinical trials. Such data are also important in the development of optimal dietary strategies for CF regarding recommendations for intake of polyphenol-rich foods, such as fruits and vegetables and whole grains.

References

[1] Manach C, Scalbert A, Morand C, Remesy C, Jimenez L. Polyphenols: food sources and bioavailability. Am J Clin Nutr 2004;79(5):727–47.

[2] Friedman M. Chemistry, Biochemistry, and dietary role of Potato polyphenols. A review. Journal of Agricultural and Food Chemistry 1997;45(5):1523–40.

[3] Arts IC, Hollman PC. Polyphenols and disease risk in epidemiologic studies. Am J Clin Nutr 2005;81(1 Suppl):317S–25S.

[4] Chun OK, Kim D-O, Smith N, Schroeder D, Han JT, Lee CY. Daily consumption of phenolics and total antioxidant capacity from fruit and vegetables in the American diet. Journal of the Science of Food and Agriculture 2005;85(10):1715–24.

[5] Svilaas A, Sakhi AK, Andersen LF, Svilaas T, Strom EC, Jacobs DR Jr, et al. Intakes of antioxidants in coffee, wine, and vegetables are correlated with plasma carotenoids in humans. J Nutr 2004;134(3):562–7.

[6] Schaffer S, Halliwell B. Do polyphenols enter the brain and does it matter? Some theoretical and practical considerations. Genes Nutr 2012;7(2):99–109.

[7] Tanigawa S, Fujii M, Hou DX. Action of Nrf2 and Keap1 in ARE-mediated NQO1 expression by quercetin. Free Radic Biol Med 2007;42(11):1690–703.

[8] Canali R, Comitato R, Ambra R, Virgili F. Red wine metabolites modulate NF-kappaB, activator protein-1 and cAMP response element-binding proteins in human endothelial cells. Br J Nutr 2010;103(6):807–14.

[9] Moridani MY, Scobie H, Salehi P, O'Brien PJ. Catechin metabolism: glutathione conjugate formation catalyzed by tyrosinase, peroxidase, and cytochrome p450. Chem Res Toxicol 2001;14(7):841–8.

[10] Nishinaka T, Ichijo Y, Ito M, Kimura M, Katsuyama M, Iwata K, et al. Curcumin activates human glutathione S-transferase P1 expression through antioxidant response element. Toxicol Lett 2007;170(3):238–47.

[11] Spencer JP, Schroeter H, Kuhnle G, Srai SK, Tyrrell RM, Hahn U, et al. Epicatechin and its in vivo metabolite, 3′-O-methyl epicatechin, protect human fibroblasts from oxidative-stress-induced cell death involving caspase-3 activation. Biochem J 2001;354(Pt 3):493–500.

[12] Kobuchi H, Roy S, Sen CK, Nguyen HG, Packer L. Quercetin inhibits inducible ICAM-1 expression in human endothelial cells through the JNK pathway. Am J Physiol 1999;277(3 Pt 1):C403–11.

[13] Saeki K, Kobayashi N, Inazawa Y, Zhang H, Nishitoh H, Ichijo H, et al. Oxidation-triggered c-Jun N-terminal kinase (JNK) and p38 mitogen-activated protein (MAP) kinase pathways for apoptosis in human leukaemic cells stimulated by epigallocatechin-3-gallate (EGCG): a distinct pathway from those of chemically induced and receptor-mediated apoptosis. Biochem J 2002;368(Pt 3):705–20.

[14] Kong AN, Yu R, Chen C, Mandlekar S, Primiano T. Signal transduction events elicited by natural products: role of MAPK and caspase pathways in homeostatic response and induction of apoptosis. Arch Pharm Res 2000;23(1):1–16.

[15] Tarahovsky YS. Plant polyphenols in cell-cell interaction and communication. Plant Signal Behav 2008;3(8):609–11.

[16] Collaboration. The cystic fibrosis mutation Database; 2011. http://www.genet.sickkids.on.ca/cftr/.

[17] Bombieri C, Claustres M, De Boeck K, et al. Recommendations for the classification of diseases as CFTR-related disorders. Journal of Cystic Fibrosis 2011;10(Supplement 2(0)):S86–102.

[18] Rowe SM, Miller S, Sorscher EJ. Cystic fibrosis. The New England Journal of Medicine 2005;352(19):1992–2001.

[19] Welsh MJ, Smith AE. Molecular mechanisms of CFTR chloride channel dysfunction in cystic fibrosis. Cell 1993;73(7):1251–4.

[20] Zielenski J, Tsui LC. Cystic fibrosis: genotypic and phenotypic variations. Annu Rev Genet 1995;29:777–807.

[21] Collaboration. Correlation between genotype and phenotype in patients with cystic fibrosis. The Cystic Fibrosis Genotype-Phenotype Consortium. N Engl J Med 1993;329(18):1308–13.

[22] Spencer JP, Abd El Mohsen MM, Minihane AM, Mathers JC. Biomarkers of the intake of dietary polyphenols: strengths, limitations and application in nutrition research. Br J Nutr 2008;99(1):12–22.

[23] Mehta A. CFTR: more than just a chloride channel. Pediatric Pulmonology 2005;39(4):292–8.
[24] Cheng SH, Gregory RJ, Marshall J, Paul S, Souza DW, White GA, et al. Defective intracellular transport and processing of CFTR is the molecular basis of most cystic fibrosis. Cell 1990;63(4):827–34.
[25] Schwiebert EM, Egan ME, Hwang TH, Fulmer SB, Allen SS, Cutting GR, et al. CFTR regulates outwardly rectifying chloride channels through an autocrine mechanism involving ATP. Cell 1995;81(7):1063–73.
[26] Stutts MJ, Canessa CM, Olsen JC, Hamrick M, Cohn JA, Rossier BC, et al. CFTR as a cAMP-Dependent regulator of sodium channels. Science 1995;269(5225):847–50.
[27] Jayaraman S, Joo NS, Reitz B, Wine JJ, Verkman AS. Submucosal gland secretions in airways from cystic fibrosis patients have normal [Na+] and pH but elevated viscosity. Proc Natl Acad Sci U S a 2001;98(14):8119–23.
[28] Shapiro ED, Milmoe GJ, Wald ER, Rodnan JB, Bowen AD. Bacteriology of the maxillary sinuses in patients with cystic fibrosis. Journal of Infectious Diseases 1982;146(5):589–93.
[29] Hansen SK, Rau MH, Johansen HK, Ciofu O, Jelsbak L, Yang L, et al. Evolution and diversification of Pseudomonas aeruginosa in the paranasal sinuses of cystic fibrosis children have implications for chronic lung infection. ISME Journal 2012;6(1):31–45.
[30] Razvi S, Quittell L, Sewall A, Quinton H, Marshall B, Saiman L, et al. Respiratory microbiology of patients with cystic fibrosis in the united states, 1995 to 2005. CHEST Journal 2009;136(6):1554–60.
[31] Ai T, Bompadre SG, Wang X, Hu S, Li M, Hwang TC, et al. Capsaicin potentiates wild-type and mutant cystic fibrosis transmembrane conductance regulator chloride-channel currents. Mol Pharmacol 2004;65(6):1415–26.
[32] Pyle LC, Fulton JC, Sloane PA, Backer K, Mazur M, Prasain J, et al. Activation of the cystic fibrosis transmembrane conductance regulator by the flavonoid quercetin: potential use as a biomarker of DeltaF508 cystic fibrosis transmembrane conductance regulator rescue. Am J Respir Cell Mol Biol 2010;43(5):607–16.
[33] Bebok Z, Collawn JF, Wakefield J, Parker W, Li Y, Varga K, et al. Failure of cAMP agonists to activate rescued deltaF508 CFTR in CFBE41o- airway epithelial monolayers. J Physiol 2005;569(Pt 2): 601–15.
[34] Yu YC, Miki H, Nakamura Y, Hanyuda A, Matsuzaki Y, Abe Y, et al. Curcumin and genistein additively potentiate G551D-CFTR. J Cyst Fibros 2011;10(4):243–52.
[35] Wang W, Bernard K, Li G, Kirk KL. Curcumin opens cystic fibrosis transmembrane conductance regulator channels by a novel mechanism that requires neither ATP binding nor dimerization of the nucleotide-binding domains. J Biol Chem 2007;282(7):4533–44.
[36] Moran O, Galietta LJ, Zegarra-Moran O. Binding site of activators of the cystic fibrosis transmembrane conductance regulator in the nucleotide binding domains. Cell Mol Life Sci 2005;62(4):446–60.
[37] Miller MJ, MacNaughton WK, Zhang XJ, Thompson JH, Charbonnet RM, Bobrowski P, et al. Treatment of gastric ulcers and diarrhea with the Amazonian herbal medicine sangre de grado. Am J Physiol Gastrointest Liver Physiol 2000;279(1):G192–200.
[38] Ferry DR, Smith A, Malkhandi J, Fyfe DW, deTakats PG, Anderson D, et al. Phase I clinical trial of the flavonoid quercetin: pharmacokinetics and evidence for in vivo tyrosine kinase inhibition. Clin Cancer Res 1996;2(4):659–68.
[39] Al-Nakkash L, Hu S, Li M, Hwang TC. A common mechanism for cystic fibrosis transmembrane conductance regulator protein activation by genistein and benzimidazolone analogs. J Pharmacol Exp Ther 2001;296(2):464–72.
[40] Weinreich F, Wood PG, Riordan JR, Nagel G. Direct action of genistein on CFTR. Pflugers Arch 1997;434(4):484–91.
[41] Lim M, McKenzie K, Floyd AD, Kwon E, Zeitlin PL. Modulation of deltaF508 cystic fibrosis transmembrane regulator trafficking and function with 4-phenylbutyrate and flavonoids. Am J Respir Cell Mol Biol 2004;31(3):351–7.
[42] Hamdaoui N, Baudoin-Legros M, Kelly M, Aissat A, Moriceau S, Vieu DL, et al. Resveratrol rescues cAMP-dependent anionic transport in the cystic fibrosis pancreatic cell line CFPAC1. Br J Pharmacol 2011;163(4):876–86.
[43] Collins D, Kopic S, Geibel JP, Hogan AM, Medani M, Baird AW, et al. The flavonone naringenin inhibits chloride secretion in isolated colonic epithelia. Eur J Pharmacol 2011;668(1–2):271–7.
[44] Harada K, Okiyoneda T, Hashimoto Y, Oyokawa K, Nakamura K, Suico MA, et al. Curcumin enhances cystic fibrosis transmembrane regulator expression by down-regulating calreticulin. Biochem Biophys Res Commun 2007;353(2):351–6.
[45] Loo TW, Bartlett MC, Clarke DM. Thapsigargin or curcumin does not promote maturation of processing mutants of the ABC transporters, CFTR, and P-glycoprotein. Biochem Biophys Res Commun 2004;325(2):580–5.
[46] Dragomir A, Bjorstad J, Hjelte L, Roomans GM. Curcumin does not stimulate cAMP-mediated chloride transport in cystic fibrosis airway epithelial cells. Biochem Biophys Res Commun 2004;322(2):447–51.
[47] Song Y, Sonawane ND, Salinas D, Qian L, Pedemonte N, Galietta LJ, et al. Evidence against the rescue of defective DeltaF508-CFTR cellular processing by curcumin in cell culture and mouse models. J Biol Chem 2004;279(39):40629–33.
[48] Grubb BR, Gabriel SE, Mengos A, Gentzsch M, Randell SH, Van Heeckeren AM, et al. SERCA pump inhibitors do not correct biosynthetic arrest of deltaF508 CFTR in cystic fibrosis. Am J Respir Cell Mol Biol 2006;34(3):355–63.
[49] Harada K, Okiyoneda T, Hashimoto Y, Oyokawa K, Nakamura K, Suico MA, et al. Calreticulin negatively regulates the cell surface expression of cystic fibrosis transmembrane conductance regulator. J Biol Chem 2006;281(18):12841–8.
[50] Bierhaus A, Zhang Y, Quehenberger P, Luther T, Haase M, Muller M, et al. The dietary pigment curcumin reduces endothelial tissue factor gene expression by inhibiting binding of AP-1 to the DNA and activation of NF-kappa B. Thromb Haemost 1997;77(4):772–82.
[51] Huang TS, Kuo ML, Lin JK, Hsieh JS. A labile hyperphosphorylated c-Fos protein is induced in mouse fibroblast cells treated with a combination of phorbol ester and anti-tumor promoter curcumin. Cancer Lett 1995;96(1):1–7.
[52] Maheshwari RK, Singh AK, Gaddipati J, Srimal RC. Multiple biological activities of curcumin: a short review. Life Sci 2006;78(18):2081–7.
[53] Illek B, Fischer H. Flavonoids stimulate Cl conductance of human airway epithelium in vitro and in vivo. Am J Physiol 1998;275(5 Pt 1):L902–10.
[54] Lansdell KA, Cai Z, Kidd JF, Sheppard DN. Two mechanisms of genistein inhibition of cystic fibrosis transmembrane conductance regulator Cl- channels expressed in murine cell line. The Journal of Physiology 2000;524(2):317–30.
[55] Zegarra-Moran O, Monteverde M, Galietta LJV, Moran O. Functional analysis of mutations in the Putative binding site for cystic fibrosis transmembrane conductance regulator Potentiators: Interaction between activation and inhibition. Journal of Biological Chemistry 2007;282(12):9098–104.
[56] Luis A, Silva F, Sousa S, Duarte AP, Domingues F. Antistaphylococcal and biofilm inhibitory activities of gallic, caffeic, and chlorogenic acids. Biofouling 2014;30(1):69–79.
[57] Sandasi M, Leonard CM, Viljoen AM. The in vitro antibiofilm activity of selected culinary herbs and medicinal plants against Listeria monocytogenes. Lett Appl Microbiol 2010;50(1): 30–5.

[58] Borges A, Saavedra MJ, Simoes M. The activity of ferulic and gallic acids in biofilm prevention and control of pathogenic bacteria. Biofouling 2012;28(7):755–67.

[59] Raja AF, Ali F, Khan IA, Shawl AS, Arora DS, Shah BA, et al. Antistaphylococcal and biofilm inhibitory activities of acetyl-11-keto-beta-boswellic acid from Boswellia serrata. BMC Microbiol 2011;11(54):1471–2180.

[60] Vidigal PG, Müsken M, Becker KA, Häussler S, Wingender J, Steinmann E, et al. Effects of green tea compound epigallocatechin-3-gallate against Stenotrophomonas maltophilia infection and biofilm. PLoS ONE 2014;9(4):e92876.

[61] Biesalski HK. Polyphenols and inflammation: basic interactions. Curr Opin Clin Nutr Metab Care 2007;10(6):724–8.

[62] Rahman I, Biswas SK, Kirkham PA. Regulation of inflammation and redox signaling by dietary polyphenols. Biochem Pharmacol 2006;72(11):1439–52.

[63] Borgatti M, Mancini I, Bianchi N, Guerrini A, Lampronti I, Rossi D, et al. Bergamot (Citrus bergamia Risso) fruit extracts and identified components alter expression of interleukin 8 gene in cystic fibrosis bronchial epithelial cell lines. BMC Biochem 2011;12(15):1471–2091.

[64] Wheeler DS, Catravas JD, Odoms K, Denenberg A, Malhotra V, Wong HR. Epigallocatechin-3-gallate, a green tea-derived polyphenol, inhibits IL-1 beta-dependent proinflammatory signal transduction in cultured respiratory epithelial cells. J Nutr 2004;134(5):1039–44.

[65] Renouf M, Guy PA, Marmet C, Fraering AL, Longet L, Moulin J, et al. Measurement of caffeic and ferulic acid equivalents in plasma after coffee consumption: small intestine and colon are key sites for coffee metabolism. Mol Nutr Food Res 2010;54(6):760–6.

[66] Rechner AR, Kuhnle G, Bremner P, Hubbard GP, Moore KP, Rice-Evans CA, et al. The metabolic fate of dietary polyphenols in humans. Free Radic Biol Med 2002;33(2):220–35.

[67] Castellani C, Cuppens H, Macek Jr M, Cassiman JJ, Kerem E, Durie P, et al. Consensus on the use and interpretation of cystic fibrosis mutation analysis in clinical practice. J Cyst Fibros 2008;7(3):179–96.

[68] Dequeker E, Stuhrmann M, Morris MA, Casals T, Castellani C, Claustres M, et al. Best practice guidelines for molecular genetic diagnosis of cystic fibrosis and CFTR-related disorders–updated European recommendations. Eur J Hum Genet 2009;17(1):51–65.

[69] Gilljam M, Moltyaner Y, Downey GP, Devlinr R, Durie P, Cantin AM, et al. Airway inflammation and infection in congenital bilateral absence of the vas deferens. Am J Respir Crit Care Med 2004;169(2):174–9.

[70] Sawicki GSTH. Managing treatment complexity in cystic fibrosis: Challenges and Opportunities. PPUL Pediatric Pulmonology 2012;47(6):523–33.

[71] Shapiro H, Singer P, Halpern Z, Bruck R. Polyphenols in the treatment of inflammatory bowel disease and acute pancreatitis. Gut 2007;56(3):426–35.

[72] Panickar KS. Effects of dietary polyphenols on neuroregulatory factors and pathways that mediate food intake and energy regulation in obesity. Mol Nutr Food Res 2013;57(1):34–47.

[73] Pandey KB, Rizvi SI. Plant polyphenols as dietary antioxidants in human health and disease. Oxid Med Cell Longev 2009;2(5):270–8.

[74] Harikumar KB, Aggarwal BB. Resveratrol: a multitargeted agent for age-associated chronic diseases. Cell Cycle 2008;7(8):1020–35.

[75] Milne JC, Lambert PD, Schenk S, Carney DP, Smith JJ, Gagne DJ, et al. Small molecule activators of SIRT1 as therapeutics for the treatment of type 2 diabetes. Nature 2007;450(7170):712–6.

[76] Chen WP, Chi TC, Chuang LM, Su MJ. Resveratrol enhances insulin secretion by blocking K(ATP) and K(V) channels of beta cells. Eur J Pharmacol 2007;568(1–3):269–77.

[77] Rizvi SI, Mishra N. Anti-oxidant effect of quercetin on type 2 Diabetic Erythrocytes. Journal of Food Biochemistry 2009;33(3):404–15.

[78] Barone E, Calabrese V, Mancuso C. Ferulic acid and its therapeutic potential as a hormetin for age-related diseases. Biogerontology 2009;10(2):97–108.

[79] Jung EH, Kim SR, Hwang IK, Ha TY. Hypoglycemic effects of a phenolic acid fraction of rice bran and ferulic acid in C57BL/KsJ-db/db mice. J Agric Food Chem 2007;55(24):9800–4.

[80] Yokozawa T, Noh JS, Park CH. Green Tea polyphenols for the Protection against renal damage Caused by oxidative stress. Evid Based Complement Alternat Med 2012;845917(10):10.

[81] Sajilata MG, Bajaj PR, Singhal RS. Tea polyphenols as Nutraceuticals. Comprehensive Reviews in Food Science and Food Safety 2008;7(3):229–54.

[82] Li Volti G, Sacerdoti D, Di Giacomo C, Barcellona ML, Scacco A, Murabito P, et al. Natural heme oxygenase-1 inducers in hepatobiliary function. World J Gastroenterol 2008;14(40):6122–32.

[83] Patriarca S, Furfaro AL, Cosso L, Maineri E, Balbis E, Domenicotti C, et al. Heme oxygenase 1 expression in rat liver during ageing and ethanol intoxication. Biogerontology 2007;8(3):365–72.

[84] Kluth D, Banning A, Paur I, Blomhoff R, Brigelius-Flohe R. Modulation of pregnane X receptor- and electrophile responsive element-mediated gene expression by dietary polyphenolic compounds. Free Radic Biol Med 2007;42(3):315–25.

[85] Radi R, Beckman JS, Bush KM, Freeman BA. Peroxynitrite-induced membrane lipid peroxidation: the cytotoxic potential of superoxide and nitric oxide. Arch Biochem Biophys 1991;288(2):481–7.

[86] Douki T, Cadet J, Ames BN. An adduct between peroxynitrite and 2′-deoxyguanosine: 4,5-dihydro-5-hydroxy-4-(nitrosooxy)-2′-deoxyguanosine. Chem Res Toxicol 1996;9(1):3–7.

[87] Yokozawa T, Dong E, Nakagawa T, Kashiwagi H, Nakagawa H, Takeuch S, et al. Vitro and in vivo studies on the radical-scavenging activity of Tea. Journal of Agricultural and Food Chemistry 1998;46(6):2143–50.

[88] Yokozawa T, Chung HY, He LQ, Oura H. Effectiveness of green tea tannin on rats with chronic renal failure. Biosci Biotechnol Biochem 1996;60(6):1000–5.

[89] Johnson IT, Williamson G, Musk SR. Anticarcinogenic factors in plant foods: a new class of nutrients? Nutr Res Rev 1994;7(1):175–204.

[90] Garcia-Lafuente A, Guillamon E, Villares A, Rostagno MA, Martinez JA. Flavonoids as anti-inflammatory agents: implications in cancer and cardiovascular disease. Inflamm Res 2009;58(9):537–52.

[91] Talalay P, De Long MJ, Prochaska HJ. Identification of a common chemical signal regulating the induction of enzymes that protect against chemical carcinogenesis. Proc Natl Acad Sci U S a 1988;85(21):8261–5.

[92] Athar M, Back JH, Tang X, Kim KH, Kopelovich L, Bickers DR, et al. Resveratrol: a review of preclinical studies for human cancer prevention. Toxicol Appl Pharmacol 2007;224(3):274–83.

[93] Di Paola R, Mazzon E, Muia C, Crisafulli C, Genovese T, Di Bella P, et al. Green tea polyphenol extract attenuates zymosan-induced non-septic shock in mice. Shock 2006;26(4):402–9.

[94] Folwarczna J, Zych M, Burczyk J, Trzeciak H, Trzeciak HI. Effects of natural phenolic acids on the skeletal system of ovariectomized rats. Planta Med 2009;75(15):1567–72.

[95] Park JA, Ha SK, Kang TH, Oh MS, Cho MH, Lee SY, et al. Protective effect of apigenin on ovariectomy-induced bone loss in rats. Life Sci 2008;82(25–26):1217–23.

[96] Kim TH, Jung JW, Ha BGHong JM, Park EK, Kim HJ, et al. The effects of luteolin on osteoclast differentiation, function in vitro and ovariectomy-induced bone loss. J Nutr Biochem 2011;22(1):8–15.

[97] Shen CL, Cao JJ, Dagda RY, Tenner TE Jr, Chyu MC, Yeh JK. Supplementation with green tea polyphenols improves bone microstructure and quality in aged, orchidectomized rats. Calcif Tissue Int 2011;88(6):455–63.

[98] Arjmandi BH, Johnson CD, Campbell SC, Hooshmand S, Chai SC, Akhter MP. Combining fructooligosaccharide and dried plum has the greatest effect on restoring bone mineral density among select functional foods and bioactive compounds. J Med Food 2010;13(2):312–9.

[99] Pathak L, Agrawal Y, Dhir A. Natural polyphenols in the management of major depression. Expert Opin Investig Drugs 2013;22(7):863–80.

[100] Vignes M, Maurice T, Lante F, Nedjar M, Thethi K, Guiramand J, et al. Anxiolytic properties of green tea polyphenol (-)-epigallocatechin gallate (EGCG). Brain Res 2006;19(1):102–15.

[101] Hindmarch I, Rigney U, Stanley N, Quinlan P, Rycroft J, Lane JA. A naturalistic investigation of the effects of day-long consumption of tea, coffee and water on alertness, sleep onset and sleep quality. Psychopharmacology 2000;149(3):203–16.

[102] Riekert Ka, B.S.J.B.M.P.K.J.A.R.C.S. The association between depression, lung function, and health-related quality of life among adults with cystic fibrosis. Chest 2007;132(1):231–7.

[103] Yohannes AM, Willgoss TG, Fatoye FA, Dip MD, Webb K. Relationship between anxiety, depression, and quality of life in adult patients with cystic fibrosis. Respir Care 2012;57(4):550–6.

[104] Besier T, Goldbeck L. Anxiety and depression in adolescents with CF and their caregivers. J Cyst Fibros 2011;10(6):435–42.

[105] Havermans T, Colpaert K, Dupont LJ. Quality of life in patients with Cystic Fibrosis: association with anxiety and depression. J Cyst Fibros 2008;7(6):581–4.

[106] James AC, James G, Cowdrey FA, Soler A, Choke A. Cognitive behavioural therapy for anxiety disorders in children and adolescents. Cochrane Database Syst Rev 2013;3(6).

[107] Edmondson DE, Mattevi A, Binda C, Li M, Hubalek F. Structure and mechanism of monoamine oxidase. Curr Med Chem 2004;11(15):1983–93.

[108] Sampson L, Rimm E, Hollman PCH, de Vries JHM, Katan MB. Flavonol and flavone intakes in US health Professionals. Journal of the American Dietetic Association 2002;102(10):1414–20.

[109] Paulke A, Noldner M, Schubert-Zsilavecz M, Wurglics M. St. John's wort flavonoids and their metabolites show antidepressant activity and accumulate in brain after multiple oral doses. Pharmazie 2008;63(4):296–302.

[110] Youdim KA, Shukitt-Hale B, Joseph JA. Flavonoids and the brain: interactions at the blood-brain barrier and their physiological effects on the central nervous system. Free Radic Biol Med 2004;37(11):1683–93.

[111] Bladt S, Wagner H. Inhibition of MAO by fractions and constituents of hypericum extract. J Geriatr Psychiatry Neurol 1994;7(1):S57–9.

[112] Chimenti F, Cottiglia F, Bonsignore L, Casu L, Casu M, Floris C, et al. Quercetin as the active principle of Hypericum hircinum exerts a selective inhibitory activity against MAO-A: Extraction, biological analysis, and Computational study. Journal of Natural Products 2006;69(6):945–9.

[113] Sakakibara H, Yoshino S, Kawai Y, Terao J. Antidepressant-like effect of onion (Allium cepa L.) powder in a rat behavioral model of depression. Biosci Biotechnol Biochem 2008;72(1):94–100.

[114] Dimpfel W. Rat electropharmacograms of the flavonoids rutin and quercetin in comparison to those of moclobemide and clinically used reference drugs suggest antidepressive and/or neuroprotective action. Phytomedicine 2009;16(4):287–94.

CHAPTER

21

Chronic Infection with *Pseudomonas aeruginosa* in an Animal Model of Oxidative Stress: Lessons for Patients with Cystic Fibrosis

Oana Ciofu

Department of International Health, Immunology and Microbiology, Costerton Biofilm Center, Faculty of Health Sciences, University of Copenhagen, Denmark

Oxidative stress results from the imbalance between productions of reactive oxygen species (ROS) and antioxidants. In patients with cystic fibrosis (CF) and lung infection, both increased ROS production and a deficiency of antioxidants have a role in the oxidative stress status of these patients.

Like humans, guinea pigs are unable to synthesize vitamin C or L-ascorbic acid (ASC), a potent water-soluble antioxidant present in cells and plasma; therefore, they depend on dietary intake. The guinea pig is one of the rare species that has lost the capability to synthesize vitamin C [1]; thus, it is a good experimental model for studies dealing with the effects of oxidative stress on outcomes of infection.

The leading cause of morbidity and mortality in CF is chronic progressive lung disease, predominantly caused by *Pseudomonas aeruginosa* endobronchial infection [2]. We decided to use a guinea pig model to investigate the role of oxidative stress in the outcome of this type of infection.

21.1 OXIDATIVE STRESS IN CF

21.1.1 Chronic Burden of ROS in the Respiratory Tract of CF Patients

As a consequence of excessively neutrophil-dominated inflammation, there is a chronic burden of ROS in the respiratory tract of CF patients.

Malfunction of the chloride channel cystic fibrosis transmembrane conductance regulator (CFTR) in CF patients leads to decreased volume of the pericilliary fluid in the lower respiratory tract, which in turn leads to impaired mucociliary clearance of inhaled microbes. This impairment of the noninflammatory defense mechanism of the respiratory tract leads to early recruitment of inflammatory defense mechanisms such as polymorphonuclear leukocytes (PMN) and antibodies. Despite the inflammatory response and intensive antibiotic therapy, however, infections caused by *P. aeruginosa* persist and lead to respiratory failure and lung transplantation or death as a result of the ability of *P. aeruginosa* to survive by switching to the biofilm mode of growth. The biofilm mode of growth provides tolerance to inflammatory defense mechanisms and antibiotic treatment [3].

Pulmonary inflammation starts early in infancy in people with CF and is exaggerated in response to pulmonary infection [4]. Polymorphonuclear leukocyte counts in CF airway fluid pre-elevated by bronchoalveolar lavage have been found to be thousands of times higher than in controls [5,6]. A consequence of PMN-dominated inflammation is the release of proteases and ROS, which are believed to be the main modulators of tissue damage in CF [5]. In addition, the ROS might have a role in bacterial adaptation to the lung and persistence, because we have shown that conversion to mucoid phenotype (alginate production) can be caused by ROS-induced mutations in the alginate regulatory system [7], and an association between oxidative stress and bacterial mutability has been demonstrated [8].

There is abundant indirect evidence supporting the role of ROS as mediators of tissue damage in CF [9]. High sputum levels of extracellular myeloperoxidase, a polymorphonuclear neutrophil-derived enzyme that transforms hydrogen peroxide into highly reactive oxygen metabolites, have been detected in CF patients and inversely

Diet and Exercise in Cystic Fibrosis
http://dx.doi.org/10.1016/B978-0-12-800051-9.00021-3

correlated with lung function [6]. Higher concentrations of oxidation products have also been reported in the plasma of CF patients compared with control subjects [6].

Reactive oxygen species, which are the key players in oxidative stress, are thought to cause tissue damage in the lungs by attacking polyunsaturated fatty acids (PUFAs) in cellular membranes found in lung epithelial cells [10]. These PUFAs are one of the main components of dietary fats and are converted to arachidonic acid, a component of phospholipids in cell membranes. It is thought that ROS attack phospholipids (peroxidation) and produce a free radical, which in turn initiates attacks on adjacent arachidonic acid chains, thus compromising cell membrane structure. Free radical damage is propagated until the host defense system counteracts and terminates these actions.

Once in the airways, neutrophils show multiple signs of dysfunction, culminating in their abnormal clearance and necrosis. Sputum neutrophil count and elastase activity are strong correlates to clinical measures of CF lung dysfunction, such as declining functional expiratory volume in 1 s or forced vital capacity, which is consistent with neutrophils having a central role in CF airway destruction [11,12].

Chronic airway infection and high levels of chemoattractants and inflammatory mediators ensure that activated neutrophils continue to accumulate in the airways. In addition to producing chemotactic factors and proinflammatory cytokines, both of which amplify and perpetuate inflammatory cell recruitment, these cells release an array of noxious mediators that may damage the surrounding tissue [13]. Neutrophil elastase is one of these mediators and has a major role in the pathophysiology of chronic inflammation in CF. It directly contributes to tissue damage by degrading structural proteins such as elastin, collagen, and proteoglycans [14], and has many other detrimental biological activities in the CF airways. It is a potent secretagogue and enhances macromolecular secretion from serous gland cells. It also promotes hypertrophy and hyperplasia of the mucus-secreting apparatus and inhibits ciliary beating in vitro. Cystic fibrosis airway neutrophils are the primary source of extracellular actin and DNA, which contribute to mucus hyperviscosity [15].

Taken together, these effects may further impair mucociliary clearance and exacerbate airway obstruction in CF patients. Neutrophil elastase also facilitates the persistence of infection by cleaving immunoglobulins, complement components, and opsonic receptors such as complement receptor (CR)-1 on the surface of phagocytes, and thus has an important impact on opsonophagocytosis. Furthermore, neutrophil elastase in secretions may itself attract more neutrophils into the airway lumen by inducing interleukin-8 (IL-8) production from epithelial cells [13].

21.1.2 Reduced Levels of Antioxidants in CF Patients

The reason for low local antioxidant levels in the epithelial lung fluid is related to the function of the CFTR channel as a major mechanism of glutathione (GSH) efflux into the extracellular milieu of the lung from lung epithelial cells [16]. This efflux is severely compromised in CF, resulting in glutathione system dysfunction. Glutathione is the major intracellular antioxidant. Cystic fibrosis patients experience GSH deficiency locally in the epithelial lining fluid of the lung, and also a systemically extracellular GSH deficiency in plasma [17], a deficiency that predisposes patients with CF to oxidative tissue damage.

A reason for low systemic antioxidant levels is the poor absorption of dietary antioxidants in the gut such as fat-soluble vitamins (vitamin E and beta-carotene).

Vitamin C as a water-soluble vitamin should not pose an appreciable problem in CF; however, plasma vitamin C concentrations have been shown in CF to be inversely related to age and different indices of inflammation [18]. This deficiency is probably mainly related to high consumption as a result of inflammation, either directly as an antioxidant or coupled with the consumption or deficiency of GSH.

Vitamin C and GSH have actions in common and can spare each other; this redundancy reflects the metabolic importance of such antioxidant activity. Both GSH and ASC are present in high concentrations in the epithelial lining fluid and are considered to have a major role in extracellular defense system of the lung.

The interrelationship between GSH and ascorbate has been shown repeatedly in animal models [19]. Both ascorbate and GSH can react with hydrogen peroxide and oxygen free radicals [19]. Glutathione reacts with dehydroascorbate to form ascorbate (Figure 21.1). Depletion of GSH leads to depletion of tissue ascorbate.

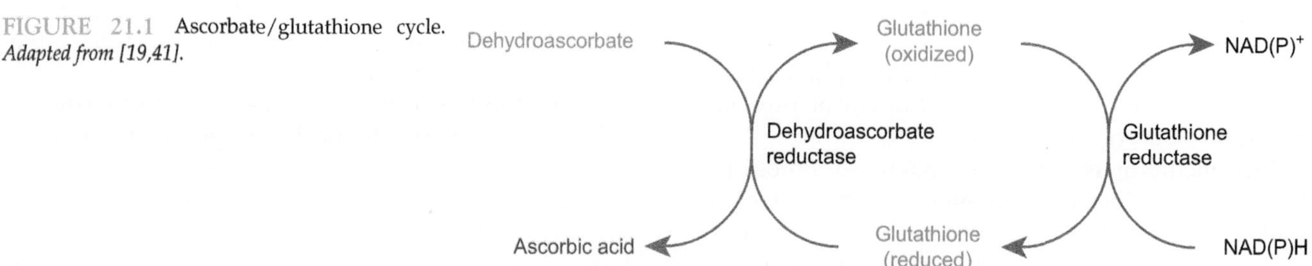

FIGURE 21.1 Ascorbate/glutathione cycle. *Adapted from [19,41].*

It has also been shown that vitamin C has a unique function in CFTR, a cyclic adenosine 3′,5′-cyclic monophosphate–dependent chloride channel that regulates epithelial surface fluid secretion: Cellular vitamin C is a biological regulator of CFTR-mediated chloride secretion in epithelia [20]. An increase in the intracellular ascorbate concentration stimulated the activity of CFTR. Therefore, the cellular pool of vitamin C may serve as a key biological factor for redox-dependent regulation of CFTR activity. This suggests that lack of intracellular levels of vitamin C would impair CFTR function.

Ascorbic acid is easily oxidized by molecular oxygen, and the first oxidation product is dehydroascorbic acid (DHA), which is readily converted back to ASC in vivo by both chemical and enzymatic means. Dehydroascorbic acid has a half-life of only a few minutes at physiological pH [7]. Ascorbic acid (ASC) recycling (i.e., the intracellular regeneration of ASC from its oxidized form DHA) presumably has a key function in maintaining redox homeostasis.

Ascorbate recycling is of major importance particularly in guinea pigs and humans, who do not have the ability to synthesize ASC. Ascorbic acid recycling includes ASC oxidation to DHA, transport of DHA into cells, intracellular reduction of DHA to ASC, and release and accumulation of ASC.

Activated neutrophils produce diffusible oxidants that oxidize extracellular ASC to DHA. Dehydroascorbic acid is at least 10- to 20-fold faster than ascorbate transport and is mediated by glucose transporters. Upon entry intracellularly, DHA is immediately reduced to ASC by glutathione-dependent protein glutaredoxin. After extracellular ASC quenches diffusible oxidants, the oxidative metabolite DHA is recycled intracellularly to increase the intracellular concentration of ASC more than 10-fold, and is available to quench diffusible oxidants.

Ascorbic acid is accumulated by alveolar macrophages and leukocytes. The concentration gradient between leukocytes and plasma is up to 50-fold. Because ascorbic acid is found exclusively in the cytosol, it has been proposed to have a protective effect by reducing ROS that enter the cytosol from the phagolysosome [18]. One of the functions of vitamin C is to minimize tissue damage caused by expulsion of superoxide from activated PMNs during an inflammatory response; therefore, a decrease in ascorbate concentration may result in decreased protection against ROS.

Enhanced rates of oxidation of ascorbate would be expected to favor depletion of vitamin because dehydroascorbate is much less stable than ascorbate and is effectively degraded and excreted. A site of inflammation could therefore be considered a site of vitamin C drainage [21].

This is supported by clinical observations showing a moderate but significant reduction in ascorbate (ASC) levels in CF patients during infection, compared with controls [10]. Treatment of infection with antibiotics showed general improvement in markers of inflammation, oxidative damage, and levels of scavengers; reduced inflammation; and decreased ROS production. Thus, pulmonary rather than nutritional disorders seem essential in imbalance of the oxidant/antioxidant system [10].

21.2 ANIMAL MODELS OF CHRONIC LUNG INFECTION

Animal models of chronic *P. aeruginosa* lung infection have been established in our laboratory in rats and mice [22–25]. To prevent the eradication of lung infection by the immune system, the bacteria are embedded in seaweed alginate beads or native alginate before infecting the lungs [26]. Alginate is the major polysaccharide produced by *P. aeruginosa* when converted to mucoid phenotype. Alginate protects *P. aeruginosa* from the consequences of inflammation by inhibiting activation of complement and decreasing phagocytosis by neutrophils and macrophages [27,28], as well as by scavenging the released free radicals [29]. High amounts of alginate have been measured in the sputum of chronically infected patients [30], and a rise in the antibody response to alginate has been associated with poor prognosis for infection in CF patients. Therefore, it is considered an important virulence factor [31,32].

Embedded in alginate, *P. aeruginosa* form a biofilm in the lung of the animals and persist for 1–2 weeks in the lungs of rodents. It has been shown that the type of inflammation in the lungs of the animals resembles inflammation in the lungs of CF patients with histopathologic features similar to those of chronic *P. aeruginosa* lung infections in humans [26,33].

Early sampling before establishing acquired immune response during experimental biofilm infections in mouse lungs has also demonstrated that accumulation of activated neutrophils in the airways is part of the innate immune response to lung infections with *P. aeruginosa* biofilms. However, PMNs are not able to phagocyte the biofilm-embedded bacteria. The PMNs are lysed by the quorum-sensing regulated rhamnolipid produced in *P. aeruginosa* biofilms [34–38]. Infected CF airways are dominated by endobronchial *P. aeruginosa* growing as biofilms in the shape of dense aggregated bacteria surrounded by numerous neutrophils and few planktonic bacteria, which are readily phagocytosed by the neutrophils [39]. The response by the neutrophils in infected endobronchial secretions in CF resembles the reaction of neutrophils responding to experimental in vivo biofilms with regard to an intense accumulation of neutrophils close to the biofilm, including accelerated oxygen depletion, which is caused by an active respiratory burst where molecular oxygen is reduced to superoxide [40].

21.3 CHRONIC LUNG INFECTION IN ANIMAL MODEL OF OXIDATIVE STRESS

To mimic the situation in CF, we decided to use the chronic lung infection model in guinea pigs exposed to oxidative stress. Oxidative stress was obtained by a diet with low vitamin C content for 2 months before we infected the lung with a *P. aeruginosa* isolate from a CF patient with a chronic lung infection. The *P. aeruginosa* isolate was embedded in alginate beads and an inoculum of 4×10^8 colony forming units (CFU)/mL was instilled in the left lung. Seven days later, the animals were sacrificed and tissues (lung and liver) and blood were pre-elevated for measurement of oxidative stress markers. Bacteriologic and histopathologic analyses of the lungs were also performed.

In our model, the *P. aeruginosa* lung infection led to consumption of the ASC both at the infection site in the lung and at the systemic level (Table 21.1). This was particularly obvious in the lower lung levels of ASC in infected compared with uninfected lungs and in the extremely low antioxidant capacity of the plasma (measured as plasma oxidizability), observed in the infected animals compared with uninfected animals (Figure 21.2). It was remarkable that infection had depleted the antioxidant capacity of the plasma to a higher degree than the ASC deficient diet. The depletion of ASC during infection probably results from the consumption of ASC by activated PMNs at the infection site, because PMNs with an activated respiratory burst consume ASC and infection activates PMNs [21]. Investigation of the dietary effect on PMNs showed that low ASC intake determined a higher oxidative burst of PMNs compared with the ASC-sufficient diet (Figure 21.3). This might be a consequence of ASC depletion inside the PMNs that leads to a lack of the antioxidant GSH [41], which reduces H_2O_2 to water.

Polymorphonuclear leukocytes deficient in glutathione owing to a genetic defect causing glutathione synthetase deficiency (5-oxoprolinuria) accumulated greater amounts of peroxide during phagocytosis but showed

TABLE 21.1 Ascorbic Acid Levels in Plasma (μM), Lung, and Liver (nM/g Tissue) in Animals Receiving Vitamin C–Deficient or –Sufficient Diet

Location	Noninfected Sufficient (n=6)	Noninfected Deficient (n=5)	Infected Sufficient (n=11)	Infected Deficient (n=11)
Plasma	38.0±17.6	2.0±1.2	29.8±23.4	1.2±0.7
Lung	2044±362	523±274*	1646±541*	276±165
Liver	1460±250	209±125	1262±635	138±83

Concentrations are given as means±standard deviation. Significantly lower levels of ASC were measured in the lungs of infected compared to animals. *P=0.03.

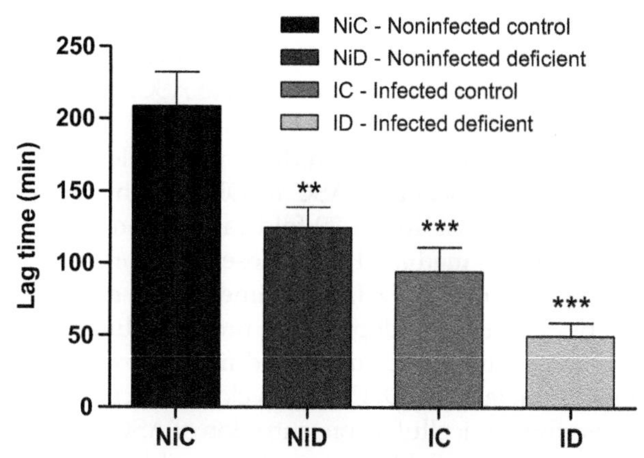

FIGURE 21.2 Resistance to oxidative stress measured as plasma oxidizability lag time in guinea pigs subjected to vitamin C deficiency and/or *P. aeruginosa* infection compared with controls. **P<0.01; ***P<0.001 compared with noninfected control.

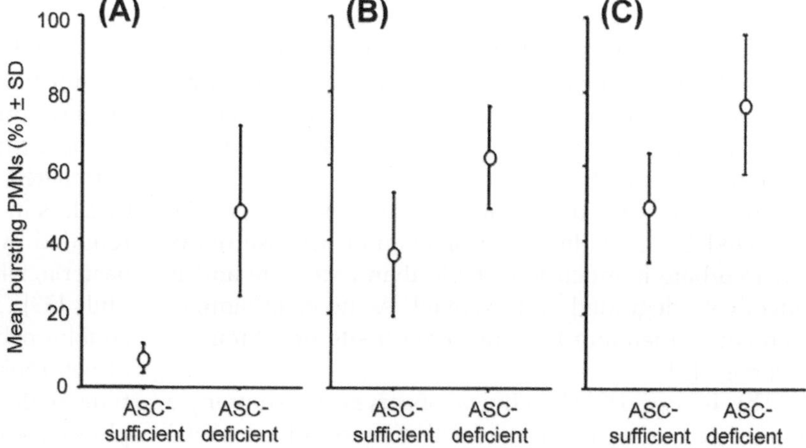

FIGURE 21.3 The respiratory burst in peripheral PMNs from guinea pigs receiving an ASC-sufficient (N=10) diet and ASC-deficient diet (N=10). (A) Spontaneous respiratory burst. (B) Respiratory burst after direct stimulation of *nicotinamide adenine dinucleotide phosphate hydrogen* oxidase by PMA. (C) Respiratory burst during phagocytosis of *Escherichia coli*.

impaired bacterial killing [42]. In accordance, low vitamin C intake has been reported to decrease phagocytic activity in guinea pigs [1].

However, the phagocytic effect of the PMNs is difficult to assess in vivo in this biofilm infection model because the alginate embedding *P. aeruginosa* protects against phagocytosis [43,44], which may explain the similar number of bacteria in the lungs of the two groups of animals after 7 days of infection.

Our results instead suggest that an ASC-deficient diet caused depletion of GSH in PMNs and that the PMNs that accumulate at the infection site of biofilm-growing *P. aeruginosa* caused increased tissue damage as a result of the increased liberation of ROS. The increased respiratory burst of the PMNs in ASC-deficient guinea pigs may have reinforced the inflammatory response to an extent, causing severe lung damage, as demonstrated by the sterile lung injury caused by rabbits treated with a respiratory burst inducer phorbol 12-myristate 13-acetate (PMA) [45].

These data might explain the higher mortality after the *P. aeruginosa* lung infection registered in guinea pigs that received an ASC-deficient diet before infection. In agreement, low ASC intake has been associated with decreased resistance of guinea pigs to various infections [1].

Reactive oxidation species such as H_2O_2 have been implicated in initiating inflammatory response through activation of natural killer-κB, which can induce secretion of IL-8 and tumor necrosis factor-α, leading to increased PMN sequestration [46]. Thus, after infection in ASC-sufficient and -deficient guinea pigs, PMNs have been attracted to the site of lung *P. aeruginosa* biofilm infection, but in ASC-deficient animals, the PMNs liberated a higher amount of ROS, which attracted more PMNs compared with the ASC-sufficient animals. This might explain the larger ratio of PMNs/mononuclear leukocytes (MNs) that surrounded the alginate-embedded *P. aeruginosa* in the lungs of guinea pigs with diet-induced oxidative stress, compared with the infection in animals receiving a normal diet (Figure 21.4).

The PMN-dominated inflammatory response to the biofilm *P. aeruginosa* lung infection seen in the guinea pig model of oxidative stress was similar to the inflammation described in CF patients. This makes the presented guinea pig model an appropriate animal model for the study of anti-inflammatory or antioxidant therapy for lung infection.

In the future, studies on the function of PMNs isolated from infected animals on an ASC-sufficient and -deficient diet will clarify the role of ASC consumption by PMNs during the inflammatory response. The current guinea pig model can be used in the future to relate the type of inflammatory response to the level of oxidative stress during several kinds of *P. aeruginosa* biofilm infections such as chronic wounds [47], or other type of biofilm infections.

21.4 LESSONS FOR CF PATIENTS

Because the main component of the oxidative stress experienced by CF patients is the ROS burden resulting from inflammation, a higher level of antioxidants than normal could theoretically be an advantage, which suggests a possible benefit of supplementary intake of antioxidants. However, the beneficial effect of this type of intervention depends on the concentration of antioxidants achieved at the site of inflammation. Vitamin C homeostasis is tightly regulated by a variety of transport mechanisms, and dose- and concentration-dependent rates of transport have been shown, as well as genetic variations in the active transporters of vitamin C [48]; all of these factors significantly influence vitamin C homeostasis [49].

Pilot studies have shown that it was possible to replete alveolar glutathione after glutathione inhalation therapy, and several clinical trials of GSH or GSH precursor such as *N*-acetyl-cysteine have resulted in improvement in clinically relevant markers in CF [50]. Therefore, both local treatment (inhalations) and systemic administration

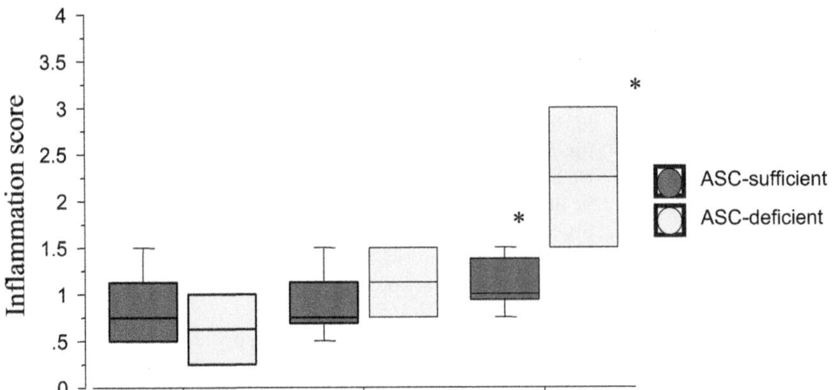

FIGURE 21.4 Type of inflammation, PMN, or mononuclear leukocytes (MN) dominating, assessed by the histopathologic examination of *P. aeruginosa*–infected lungs of guinea pigs receiving an ASC-sufficient diet (N = 6) and an ASC-deficient diet (N = 6). The ratio PMN/MN was significantly higher (P = 0.05) in animals receiving a vitamin C–deficient diet compared with a sufficient diet.

(oral) of GSH or GSH precursor have been proposed to be beneficial for patients with CF.

In a recent multicenter, randomized, clinical trial, GSH supplementation by inhalation for 6 months [51] led to significant improvement in GSH status in the lung, although without sustainable effects on lung function.

A recent Cochrane review on supplementation with thiols (GSH or *N*-acetyl-cysteine) found no evidence to recommend the use of either nebulized or oral thiol derivatives in people with CF. There are few good-quality trials investigating the effect of these medications in CF, and further research is required to investigate the potential role of these medications in improving the outcomes of people with CF [52].

It has previously been shown that a change in the inflammatory response from a PMN- to an MN-dominated response was associated with improved outcome of infection in a mouse model of chronic lung infection [53,54] and that such a switch between the type of inflammatory cell at the site of infection could be beneficial for patients [55].

In the light of the in vitro findings on the role of vitamin C in the function of CFTR [20], a normal level of ASC in epithelial lung cells, where vitamin C is transported via an active transporter, might also be beneficial for normal function of this chloride channel.

The main source of ROS at the infection site is the inflammatory cells; an additional source is the bacterial response to antibiotic stress [56–59] and to the biofilm mode of growth [60]. Increased ROS at the infection site can influence bacterial adaptation. In vitro studies have shown that the addition of antioxidants to biofilm-grown *P. aeruginosa* decreased tolerance of biofilms to tobramycin [61]. Improving the oxidative stress status might have implications not only for the host, but also for bacterial adaptation, by scavenging ROS that have bacterial mutagenic potential.

Many countries in Europe and most in the United States have adopted newborn screening programs for CF. Improved nutritional status is an important clinical outcome in CF and is the most obvious benefit afforded by early diagnosis from newborn screening. Early pulmonary inflammation characterized by free neutrophil elastase and infection with *Staphylococcus aureus* are associated with worse nutritional status in the first few years of life, in the absence of clinically apparent respiratory disease in most children.

Structural changes occur much earlier in life than previously thought [62], and even though those diagnosed by newborn screening appear to have normal lung function during the first 6 months of life, bronchiectasis can be detected shortly after diagnosis [63]. These data are important because they suggest that preventing deterioration in lung structure and lung function requires targeted interventions toward improving respiratory health and nutrition as soon as possible after diagnosis, withmonitoring of outcomes during the preschool years before pulmonary disease has become extensive.

In addition to CF patients, other groups of patients experiencing chronic bacterial infection caused by biofilm-growing microorganisms might have benefit from antioxidant supplementation. Such groups of patients could be patients with chronic obstructive lung disease or chronic otitis media, or chronic wounds. All these patients might benefit from the decreased side effects of inflammation at the infection site, along with anti-biofilm treatment strategies [64].

References

[1] Hemila H. Vitamin C and infectious diseases. In: Paoletti R, Sies H, Bug J, Grossi E, Poli A, editors. Vitamin C. The state of the art in disease prevention sixty years after the nobel prize. Milano,Italy: Springer-Verlag; 1998. pp. 73–85.

[2] Ciofu O, Hansen CR, Hoiby N. Respiratory bacterial infections in cystic fibrosis. Curr Opin Pulm Med 2013;19(3):251–8.

[3] Høiby N. *P. aeruginosa* in cystic fibrosis patients resists host defenses, antibiotics. Microbe 2006;1:571–7.

[4] Armstrong DS, Grimwood K, Carlin JB, Carzino R, Gutierrez JP, Hull J, et al. Lower airway inflammation in infants and young children with cystic fibrosis. Am J Respir Crit Care Med 1997;156(4 Pt 1):1197–204.

[5] Hull J, Vervaart P, Grimwood K, Phelan P. Pulmonary oxidative stress response in young children with cystic fibrosis. Thorax 1997;52(6):557–60.

[6] Brown RK, Kelly FJ. Evidence for increased oxidative damage in patients with cystic fibrosis. Pediatr Res 1994;36(4):487–93.

[7] Mathee K, Ciofu O, Sternberg C, Lindum PW, Campbell JI, Jensen P, et al. Mucoid conversion of *Pseudomonas aeruginosa* by hydrogen peroxide: a mechanism for virulence activation in the cystic fibrosis lung. Microbiology 1999;145(Pt 6):1349–57.

[8] Ciofu O, Riis B, Pressler T, Poulsen HE, Hoiby N. Occurrence of hypermutable *Pseudomonas aeruginosa* in cystic fibrosis patients is associated with the oxidative stress caused by chronic lung inflammation. Antimicrob Agents Chemother 2005;49(6):2276–82.

[9] Brown RK, Kelly FJ. Role of free radicals in the pathogenesis of cystic fibrosis. Thorax 1994;49(8):738–42.

[10] Lagrange-Puget M, Durieu I, Ecochard R, Abbas-Corfa F, Drai J, Steghens JP, et al. Longitudinal study of oxidative status in 312 cystic fibrosis patients in stable state and during bronchial exacerbation. Pediatr Pulmonol 2004;38(1):43–9.

[11] Sagel SD, Kapsner R, Osberg I, Sontag MK, Accurso FJ. Airway inflammation in children with cystic fibrosis and healthy children assessed by sputum induction. Am J Respir Crit Care Med 2001;164(8 Pt 1):1425–31.

[12] Sagel SD, Sontag MK, Wagener JS, Kapsner RK, Osberg I, Accurso FJ. Induced sputum inflammatory measures correlate with lung function in children with cystic fibrosis. J Pediatr 2002;141(6):811–7.

[13] De R,V. Mechanisms and markers of airway inflammation in cystic fibrosis. Eur Respir J 2002;19(2):333–40.

[14] Bruce MC, Poncz L, Klinger JD, Stern RC, Tomashefski Jr JF, Dearborn DG. Biochemical and pathologic evidence for proteolytic destruction of lung connective tissue in cystic fibrosis. Am Rev Respir Dis 1985;132(3):529–35.

[15] Kirchner KK, Wagener JS, Khan TZ, Copenhaver SC, Accurso FJ. Increased DNA levels in bronchoalveolar lavage fluid obtained from infants with cystic fibrosis. Am J Respir Crit Care Med 1996;154(5):1426–9.

[16] Hudson VM. Rethinking cystic fibrosis pathology: the critical role of abnormal reduced glutathione (GSH) transport caused by CFTR mutation. Free Radic Biol Med 2001;30(12):1440–61.

[17] Roum JH, Buhl R, McElvaney NG, Borok Z, Crystal RG. Systemic deficiency of glutathione in cystic fibrosis. J Appl Physiol 1993;75(6):2419–24.

[18] Winklhofer-Roob BM, Ellemunter H, Fruhwirth M, Schlegel-Haueter SE, Khoschsorur G, van't Hof MA, et al. Plasma vitamin C concentrations in patients with cystic fibrosis: evidence of associations with lung inflammation. Am J Clin Nutr 1997;65(6):1858–66.

[19] Meister A. Glutathione-ascorbic acid antioxidant system in animals. J Biol Chem 1994;269(13):9397–400.

[20] Fischer H, Schwarzer C, Illek B. Vitamin C controls the cystic fibrosis transmembrane conductance regulator chloride channel. Proc Natl Acad Sci USA 2004;101(10):3691–6.

[21] Hemila H, Roberts P, Wikstrom M. Activated polymorphonuclear leucocytes consume vitamin C. FEBS Lett December 3, 1984;178(1):25–30.

[22] Pedersen SS. Lung infection with alginate-producing, mucoid *Pseudomonas aeruginosa* in cystic fibrosis. APMIS Suppl 1992;28:1–79.

[23] Johansen HK, Espersen F, Pedersen SS, Hougen HP, Rygaard J, Hoiby N. Chronic *Pseudomonas aeruginosa* lung infection in normal and athymic rats. APMIS 1993;101(3):207–25.

[24] Moser C, Johansen HK, Song Z, Hougen HP, Rygaard J, Hoiby N. Chronic *Pseudomonas aeruginosa* lung infection is more severe in Th2 responding BALB/c mice compared to Th1 responding C3H/HeN mice. APMIS 1997;105(11):838–42.

[25] Ciofu O, Bagge N, Hoiby N. Antibodies against beta-lactamase can improve ceftazidime treatment of lung infection with beta-lactam-resistant *Pseudomonas aeruginosa* in a rat model of chronic lung infection. APMIS 2002;110(12):881–91.

[26] Hoffmann N, Rasmussen TB, Jensen PO, Stub C, Hentzer M, Molin S, et al. Novel mouse model of chronic *Pseudomonas aeruginosa* lung infection mimicking cystic fibrosis. Infect Immun 2005;73(4):2504–14.

[27] Cabral DA, Loh BA, Speert DP. Mucoid *Pseudomonas aeruginosa* resists nonopsonic phagocytosis by human neutrophils and macrophages. Pediatr Res 1987;22(4):429–31.

[28] Meshulam T, Obedeanu N, Merzbach D, Sobel JD. Phagocytosis of mucoid and nonmucoid strains of *Pseudomonas aeruginosa*. Clin Immunol Immunopathol 1984;32(2):151–65.

[29] Simpson JA, Smith SE, Dean RT. Scavenging by alginate of free radicals released by macrophages. Free Radic Biol Med 1989;6(4):347–53.

[30] Pedersen SS, Kharazmi A, Espersen F, Hoiby N. *Pseudomonas aeruginosa* alginate in cystic fibrosis sputum and the inflammatory response. Infect Immun 1990;58(10):3363–8.

[31] Pedersen SS, Hoiby N, Espersen F, Koch C. Role of alginate in infection with mucoid *Pseudomonas aeruginosa* in cystic fibrosis. Thorax 1992;47(1):6–13.

[32] Govan JR, Deretic V. Microbial pathogenesis in cystic fibrosis: mucoid *Pseudomonas aeruginosa* and *Burkholderia cepacia*. Microbiol Rev 1996;60(3):539–74.

[33] Moser C, Hougen HP, Song Z, Rygaard J, Kharazmi A, Hoiby N. Early immune response in susceptible and resistant mice strains with chronic *Pseudomonas aeruginosa* lung infection determines the type of T-helper cell response. APMIS 1999;107(12):1093–100.

[34] Jensen PO, Moser C, Kobayashi O, Hougen HP, Kharazmi A, Hoiby N. Faster activation of polymorphonuclear neutrophils in resistant mice during early innate response to *Pseudomonas aeruginosa* lung infection. Clin Exp Immunol 2004;137(3):478–85.

[35] Jensen PO, Bjarnsholt T, Phipps R, Rasmussen TB, Calum H, Christoffersen L, et al. Rapid necrotic killing of polymorphonuclear leukocytes is caused by quorum-sensing-controlled production of rhamnolipid by *Pseudomonas aeruginosa*. Microbiology 2007;153(Pt 5):1329–38.

[36] Bjarnsholt T, Jensen PO, Burmolle M, Hentzer M, Haagensen JA, Hougen HP, et al. *Pseudomonas aeruginosa* tolerance to tobramycin, hydrogen peroxide and polymorphonuclear leukocytes is quorum-sensing dependent. Microbiology 2005;151(Pt 2):373–83.

[37] Alhede M, Bjarnsholt T, Jensen PO, Phipps RK, Moser C, Christophersen L, et al. *Pseudomonas aeruginosa* recognizes and responds aggressively to the presence of polymorphonuclear leukocytes. Microbiology 2009;155(Pt 11):3500–8.

[38] van GM, Christensen LD, Alhede M, Phipps R, Jensen PO, Christophersen L, et al. Inactivation of the rhlA gene in *Pseudomonas aeruginosa* prevents rhamnolipid production, disabling the protection against polymorphonuclear leukocytes. APMIS 2009;117(7):537–46.

[39] Bjarnsholt T, Jensen PO, Fiandaca MJ, Pedersen J, Hansen CR, Andersen CB, et al. *Pseudomonas aeruginosa* biofilms in the respiratory tract of cystic fibrosis patients. Pediatr Pulmonol 2009;44(6):547–58.

[40] Kolpen M, Hansen CR, Bjarnsholt T, Moser C, Christensen LD, van GM, et al. Polymorphonuclear leucocytes consume oxygen in sputum from chronic *Pseudomonas aeruginosa* pneumonia in cystic fibrosis. Thorax 2010;65(1):57–62.

[41] Washko PW, Wang Y, Levine M. Ascorbic acid recycling in human neutrophils. J Biol Chem 1993;268(21):15531–5.

[42] Spielberg SP, Boxer LA, Oliver JM, Allen JM, Schulman JD. Oxidative damage to neutrophils in glutathione synthetase deficiency. Br J Haematol 1979;42(2):215–23.

[43] Leid JG, Willson CJ, Shirtliff ME, Hassett DJ, Parsek MR, Jeffers AK. The exopolysaccharide alginate protects *Pseudomonas aeruginosa* biofilm bacteria from IFN-gamma-mediated macrophage killing. J Immunol 2005;175(11):7512–8.

[44] Jensen ET, Kharazmi A, Lam K, Costerton JW, Hoiby N. Human polymorphonuclear leukocyte response to *Pseudomonas aeruginosa* grown in biofilms. Infect Immun 1990;58(7):2383–5.

[45] Shasby DM, Vanbenthuysen KM, Tate RM, Shasby SS, McMurtry I, Repine JE. Granulocytes mediate acute edematous lung injury in rabbits and in isolated rabbit lungs perfused with phorbol myristate acetate: role of oxygen radicals. Am Rev Respir Dis 1982;125(4):443–7.

[46] Rahman I, MacNee W. Oxidative stress and regulation of glutathione in lung inflammation. Eur Respir J 2000;16(3):534–54.

[47] Zhao G, Hochwalt PC, Usui ML, Underwood RA, Singh PK, James GA, et al. Delayed wound healing in diabetic (db/db) mice with *Pseudomonas aeruginosa* biofilm challenge: a model for the study of chronic wounds. Wound Repair Regen 2010;18(5):467–77.

[48] Michels AJ, Hagen TM, Frei B. Human genetic variation influences vitamin C homeostasis by altering vitamin C transport and antioxidant enzyme function. Annu Rev Nutr 2013;33:45–70.

[49] Lindblad M, Tveden-Nyborg P, Lykkesfeldt J. Regulation of vitamin C homeostasis during deficiency. Nutrients 2013;5(8):2860–79.

[50] Tirouvanziam R, Conrad CK, Bottiglieri T, Herzenberg LA, Moss RB. High-dose oral N-acetylcysteine, a glutathione prodrug, modulates inflammation in cystic fibrosis. Proc Natl Acad Sci USA 2006;103(12):4628–33.

[51] Griese M, Kappler M, Eismann C, Ballmann M, Junge S, Rietschel E, et al. Inhalation treatment with glutathione in patients with cystic fibrosis. A randomized clinical trial. Am J Respir Crit Care Med 2013;188(1):83–9.

[52] Tam J, Nash EF, Ratjen F, Tullis E, Stephenson A. Nebulized and oral thiol derivatives for pulmonary disease in cystic fibrosis. Cochrane Database Syst Rev 2013;7:CD007168.

[53] Johansen HK, Hougen HP, Cryz Jr SJ, Rygaard J, Hoiby N. Vaccination promotes TH1-like inflammation and survival in chronic *Pseudomonas aeruginosa* pneumonia in rats. Am J Respir Crit Care Med 1995;152(4 Pt 1):1337–46.

[54] Moser C, Jensen PO, Kobayashi O, Hougen HP, Song Z, Rygaard J, et al. Improved outcome of chronic *Pseudomonas aeruginosa* lung infection is associated with induction of a Th1-dominated cytokine response. Clin Exp Immunol 2002;127(2):206–13.
[55] Moser C, Jensen PO, Pressler T, Frederiksen B, Lanng S, Kharazmi A, et al. Serum concentrations of GM-CSF and G-CSF correlate with the Th1/Th2 cytokine response in cystic fibrosis patients with chronic *Pseudomonas aeruginosa* lung infection. APMIS 2005;113(6):400–9.
[56] Dwyer DJ, Kohanski MA, Hayete B, Collins JJ. Gyrase inhibitors induce an oxidative damage cellular death pathway in *Escherichia coli*. Mol Syst Biol 2007;3:91.
[57] Kohanski MA, Dwyer DJ, Hayete B, Lawrence CA, Collins JJ. A common mechanism of cellular death induced by bactericidal antibiotics. Cell 2007;130(5):797–810.
[58] Kohanski MA, DePristo MA, Collins JJ. Sublethal antibiotic treatment leads to multidrug resistance via radical-induced mutagenesis. Mol Cell 2010;37(3):311–20.
[59] Brochmann RP, Toft A, Ciofu O, Briales A, Kolpen M, Hempel C, et al. Bactericidal effect of colistin on planktonic *Pseudomonas aeruginosa* is independent of hydroxyl radical formation. Int J Antimicrob Agents 2014;43(2):140–7.
[60] Driffield K, Miller K, Bostock JM, O'Neill AJ, Chopra I. Increased mutability of *Pseudomonas aeruginosa* in biofilms. J Antimicrob Chemother 2008;61(5):1053–6.
[61] Boles BR, Singh PK. Endogenous oxidative stress produces diversity and adaptability in biofilm communities. Proc Natl Acad Sci USA 2008;105(34):12503–8.
[62] Stick SM, Brennan S, Murray C, Douglas T, von Ungern-Sternberg BS, Garratt LW, et al. Bronchiectasis in infants and preschool children diagnosed with cystic fibrosis after newborn screening. J Pediatr 2009;155(5):623–8.
[63] Sly PD, Brennan S, Gangell C, De KN, Murray C, Mott L, et al. Lung disease at diagnosis in infants with cystic fibrosis detected by newborn screening. Am J Respir Crit Care Med 2009;180(2):146–52.
[64] Bjarnsholt T, Ciofu O, Molin S, Givskov M, Hoiby N. Applying insights from biofilm biology to drug development–can a new approach be developed? Nat Rev Drug Discov 2013;12(10):791–808.

CHAPTER

22

Vitamin K in Cystic Fibrosis

Vanitha Jagannath

Specialist Pediatrician in American Mission Hospital, Manama, Kingdom of Bahrain

22.1 VITAMIN K SUPPLEMENTATION FOR CYSTIC FIBROSIS

Cystic fibrosis (CF) is a multisystem disorder that primarily affects the respiratory and gastrointestinal (GI) systems. It is caused by the homozygous presence of a mutation in the gene encoding the CF transmembrane conductance regulator (CFTR) protein. In the United Kingdom (UK), over 8000 people are affected by CF; in the United States (US), this figure is approximately 30,000. Most are diagnosed by 6 months of age and the median survival has reached the fifth decade of life. It is the most common, life-threatening, autosomal-recessively inherited disease in the white population, with a carrier rate of one in 25 and an incidence of one in 2500 live births; CF is less common in other ethnic groups, approximately one in 46 Hispanics, one in 65 Africans, and one in 90 Asians carry at least one abnormal CFTR gene [1–5].

The CFTR protein is a chloride ion channel important in tissues that produce sweat, digestive juices, and mucus. The dominant symptoms of CF relate to the respiratory and GI systems. In the GI system, liver dysfunction, intestinal obstruction, and exocrine pancreatic insufficiency (PI) are the most common morbidities. Pancreatic insufficiency affects up to 90% of people with CF and causes fat malabsorption. Fat-soluble vitamins (A, D, E, and K) are coabsorbed with fat, and thus deficiency of these vitamins may occur. Hence, vitamin K deficiency is well recognized in patients with CF and PI. Whereas deficiencies may occur from the disease process of CF and from insufficient supplementation, another additional cofactor is the long-term use of certain types of antibiotics. These can put some individuals at additional risk of vitamin K deficiency by altering intestinal flora that produce vitamin K. Long-term effects of bowel resection, an intervention required for intestinal obstruction in some newborns with CF and varying degrees of liver dysfunction, can pose a further additional risk for vitamin K deficiency [6,7].

22.1.1 Sources of Vitamin K

Two forms of vitamin K (K1 and K2) occur naturally, and synthetic forms of the vitamin (K3, K4, and K5) are also available. Naturally occurring vitamin K (K1 [phytonadione]) is found in green vegetables (i.e., kale, collards, spinach, and salads) and a small amount is made in human gut by bacteria (K2 [menaquinones]). The best food sources of vitamin K1 are dark green leafy vegetables, followed by certain vegetable oils (e.g., soybean, canola, and olive oils). Although K2 vitamins represent a minority of overall vitamin K intake, the best common food sources of vitamin K2 are cheese and liver. Vitamin K2 and a fraction of dietary vitamin K1 are converted to menaquinone-4 MK-4, which exerts some unique functions in brain and pancreas. Large daily doses of MK-4 are used as an anti-osteoporotic agent in Japan. Therapeutic vitamin K is available in both water-soluble and -insoluble forms [8]. The enteral forms of vitamin K supplementation are commonly prescribed. These can be in tablet, ampule [9,10], or multivitamin preparations [8]. Vitamin K can also be administered by intramuscular [11] or intravenous injections [12].

22.1.2 Intestinal Absorption and Blood Transport

The pathway whereby dietary vitamin K is absorbed from the intestine follows the same general principles as for the other fats and fat-soluble vitamins. The naturally occurring forms of vitamins K1 and K2 can only be absorbed in the presence of bile salts that serve to emulsify and solubilize the fat components. This solubilization is strongly enhanced by the molecular breakdown of food provided by enzymes secreted into the proximal intestine by the pancreas. The process by which the solubilized vitamin K and fat breakdown products are transported from the intestinal lumen into the intestinal cells results in vitamin K packed in chylomicrons. These

Diet and Exercise in Cystic Fibrosis
http://dx.doi.org/10.1016/B978-0-12-800051-9.00022-5

chylomicrons are secreted into the lymphatic system of capillaries that eventually empty into the systemic blood circulation via a large vessel called the thoracic duct. While flowing through the blood capillaries, the chylomicrons are transformed to smaller particles called chylomicron remnants, which are taken up by the liver. The vitamin K that reaches the liver is then available for the manufacture of blood coagulation proteins.

Synthetic preparations of menadione are usually administered in salt form (e.g., menadione sodium bisulfite). The absorption of menadione in the salt form travels by a different route that does not need bile salts and bypasses the lymphatic pathway.

The leading risk factors for vitamin K deficiency in people with CF are PI and bile salt deficiency [4–7,14]. Poor appetite from chronic illness may also lead to reduced dietary intake. Another specific risk factor is the frequent need for antibiotics to control chronic infection in CF.

22.1.3 Mode of Vitamin K Action

Vitamin K functions as the cofactor of the enzyme vitamin K–dependent carboxylase. This enzyme catalyses the posttranslation formation of gamma-carboxyglutamyl (Gla) residues in specific proteins. Vitamin K–dependent proteins are blood coagulation factors (prothrombin and Factors VII, IX, and X); other plasma proteins (protein C, protein S, and protein Z); two proteins from bone osteocalcin and matrix Gla protein (MGP); and proteins from lung, kidney, spleen, testis, placenta, and other tissues. Blood coagulation requires activation of inactive proenzymes; hence, vitamin K is a vital factor in the synthesis of clotting factors and causes hemostasis in vitamin K–dependent bleeding manifestations. Vitamin K–related carboxylation allows the activation of the bone matrix protein osteocalcin, resulting in osteoblast function and bone formation; vitamin K deficiency impairs this process and thus impairs bone formation. The under-carboxylated forms of MGP resulting from vitamin K deficiency are associated with sites of calcification in the vasculature. This regulation of tissue calcification is important to patients with diabetes in whom vascular calcification is a complication, and hence is relevant in CF, in which the incidence of diabetes is increasing. Proteins C and S, two Gla proteins that are important in maintaining the balance between bleeding and clotting, also have other functions unrelated to coagulation. In its activated form, protein C has a direct anti-inflammatory effect. Protein S maintains the integrity of cellular function [15,16].

The manifestations of vitamin K deficiency can range from mild subclinical identification (e.g., low levels of vitamin K in the blood) to widespread coagulopathy (defect in the body's mechanism for blood clotting). Such manifestations may include mucosal bleeding (e.g., in the nose, GI system, and urine) and subcutaneous bleeding (e.g., oozing from venipuncture sites and susceptibility to bruising). Vitamin K is also involved in calcium binding proteins in bone, and its deficiency is implicated in defective bone remineralization, and thus osteoporosis [13].

22.1.4 Vitamin K Supplementation

Most CF centers routinely administer vitamins A, D, and E as supplements from the neonatal period onward. Vitamin K administration is usually prescribed when clinical deficiencies are detected or after routine investigations. The limited storage capacity and rapid metabolic turnover of vitamin K [17] support the recommendations for daily rather than weekly supplementation of vitamin K [18].

With newer knowledge of the multiplicity of vitamin K-dependent functions in different organs and tissues, the concept has arisen that vitamin K sufficiency should be considered in relation to the site at which these Gla proteins are synthesized. In other words, vitamin K deficiency states may be organ or tissue specific. In part, this concept is based on studies in healthy people showing that higher amounts of vitamin K are required to carboxylate osteocalcin in the bones than amounts needed to carboxylate the coagulation proteins synthesized in the liver.

22.1.5 Assessment of Vitamin K Status

The most common laboratory test for vitamin K deficiency with respect to its coagulation function is the prothrombin time, which in healthy people is usually in the range of 11–13 s. However, it is an insensitive and nonspecific test for diagnosing vitamin K deficiency. During measurement of circulating levels of vitamin K1, low levels of serum vitamin K offer an early indicator of declining body stores. Functional capability can be assessed by measuring the individual Gla proteins, undercarboxylated coagulation Factor II (known as PIVKA-II), and undercarboxylated osteocalcin (ucOC), which reflect the vitamin K status of liver and bone, respectively.

22.1.6 Evidence for Vitamin K Deficiency in CF

The development of an acute bleeding event resulting from severe vitamin K deficiency is fortunately rare in PWCF, but some sporadic cases are reported. The group most at risk for vitamin K deficiency bleeding (VKDB) is young infants below the age of 6 months, particularly if they are breastfed (breast milk has low concentrations of vitamin K compared with milk formulas) and did not receive vitamin K prophylaxis at birth In this age group, VKDB may be the first presenting symptom of CF and bleeding is commonly the serious intracranial type.

Although bleeding is rare, many studies suggest that subclinical vitamin K deficiency is common in CF. A study in the UK showed that 70% of children with CF had a suboptimal vitamin K status based on low serum vitamin K, raised PIVKA-II, or both of these abnormalities. This was consistent with other studies in the US and Canada based on PIVKA-II alone. Another study showed that the carboxylation status of osteocalcin in children with CF was much worse than in healthy children [19].

Vitamin K deficiency is common in CF patients with PI. Supplementation (often oral and occasionally parenteral) appears to be the most immediate measure to address the deficiency, although there is limited consensus on the dose and frequency of supplements for routine or therapeutic use [20]. In 1992, the Consensus Committee of the Cystic Fibrosis Foundation (CFF) suggested a particular dosage [20], but the recommendations were later proven ineffectual by another study [18]. The 2002 American Consensus Committee recommended low-dose supplementation of vitamin K for all ages (0.3–0.5 mg/day) but emphasized that no adverse effects had been reported at any dosage level of vitamin K [9]. Recent recommendations from Europe and the UK have suggested varying dose regimens ranging from 0.3 to 1 mg/day to 10 mg/week [22,23]. Another study suggested that only a higher level of supplementation can normalize vitamin K levels in people with CF [24].

A Cochrane Systematic review evaluated vitamin K supplementation in CF [25]. The objectives of the review were to determine the effects of vitamin supplementation on the morbidities of vitamin deficiency in CF and to determine the optimal dosage for the purpose. a) Type of study: randomized controlled trials and quasi-randomized controlled trials; b) Study population: children and adults with a diagnosis of CF (defined by sweat test, genetic testing, or both); c) Intervention: all preparations at any dose of vitamin K used as a supplement compared with placebo or no supplementation for any duration including studies comparing different doses and regimens; d) Outcomes: primary clinical outcomes related to coagulopathy: (1) time to cessation of bleeding manifestations (symptomatic coagulopathy); (2) time to normalization of subtherapeutic international normalized ratio (asymptomatic or subclinical coagulopathy), bone formation outcome measures; (3) bone mineral density at the spine (L1–L4); (4) total hip (measured by dual-energy X-ray absorptiometry scans with z-score compared with a reference population); (5) reduction in risk of bone fractures; (6) quality of life and secondary outcomes. Nutritional parameters (including z-scores or centiles) (1) weight–height body mass index (BMI); (2) adverse events mild (not requiring intervention) moderate (requiring treatment) severe (life threatening or requiring hospitalization); (3) serum levels such as serum ucOC ucOC/cOC (carboxylated osteocalcin) ratio (UCR); (4) vitamin K–specific laboratory outcomes (i) plasma level of vitamin K1 (measured by high-performance liquid chromatography), and (ii) fluorescence detection proteins induced by the absence of vitamin K or antagonism factor II (PIVKA II) levels, as measured by enzyme-linked immunosorbent assay.

22.1.7 Summary of Cochrane Review

22.1.7.1 Results

Electronic searches retrieved references to nine trials. Two trials were included [26,27] and five were excluded [28–32] because these were not randomized controlled trials. Two trials were awaiting classification until the time for update [33,34]. A search of the Cystic Fibrosis Trials Register in 2012 retrieved one study, which will be assessed for inclusion as soon as it has been published [35].

Two trials reporting on two of the secondary outcomes were included in this review. Although these two trials were assessed as having a moderate risk of bias, and provided limited data, it was considered that their inclusion and the report of their results would provide at least some evidence toward answering the research question. One of the trials was crossover in design but did not include a washout period, and thus the potential risk of bias as a consequence of the carryover of treatment effect could not be ruled out [26]. Although trials with a crossover design were eligible for inclusion in this review, only data from the first intervention period were used. Unfortunately, the trial investigators only provided an analysis across both treatment periods. The second trial included in this review was a dose-ranging, randomized, controlled trial that reported some data for two of secondary outcomes [27].

22.1.7.2 Characteristics of the Trial Setting and Investigators

Both were single-center trials; one was conducted at the Montreal Children's Hospital Cystic Fibrosis Clinic in Canada [27] and the other was carried out at the Cystic Fibrosis Clinic of the Children's National Medical Center, Washington, DC [26]. Although the overall duration of the two trials differed, the active treatment period was the same (i.e., 1 month). Thus, in the crossover trial, participants were allocated to either active intervention or no-treatment control for a 4-week period, and then crossed over for a further 4 weeks but without undergoing a washout period [26]. The care providers in both trials were hospital staff and the assessors of outcomes were the investigators and other health care providers [26,27].

22.1.7.3 *Characteristics of the Participants*

The total sample size was composed of 32 participants between the ages of 8 and 35 years. In the Beker study, the diagnosis of CF was confirmed by the duplicate sweat test [26]. Although Drury did not report the method used to confirm the diagnosis of CF, it is most probable that participants with CF were enrolled in this trial [27]. All of the participants in the Beker trial were PI, as documented by previous fecal fat measurement, and they received replacement therapy of 750–200 U lipase/kg body weight at mealtimes during the course of the trial [26]. The investigators in the Drury trial reported that only PI participants were included, but provided no further details [27]. Participants with liver disease (diagnosed by ultrasound, liver function tests, hepatomegaly, or all) and those who were taking supplemental therapeutic vitamin K to treat coagulopathies at enrollment were excluded from both trials [26,27].

22.1.7.4 *Characteristics of the Interventions*

The active intervention in the Beker trial consisted of 5 mg oral vitamin K1 supplementation per week and the control was no supplementation for 4 weeks; participants then crossed over for a second 4-week period [26]. Compliance with the intervention by participants was verified by the trial coordinator at each visit. Oral antibiotic medications of cephalosporin, sulfamethoxazole, and erythromycin, and concomitant usage of bronchodilators and standard multivitamins and 200–400 IU vitamin E were allowed during the course of the trial. Participants in the Drury trial were randomized to either the orally administered injectable formulation of vitamin K1 phytonadione at 1 mg/day (diluted to 1 mg/mL) or the 5-mg/day dose [27].

22.1.7.5 *Characteristics of the Outcome Measures*

One of the primary outcomes specified in the protocol for the review was considered in either of the included trials [26,27]. Both trials carried out assessments of plasma vitamin K1 levels and serum ucOC levels; these were measured at entry and at the completion of the trial by Drury [27], and at the end of each period by Beker [26]. Dietary intake records to estimate the extent of dietary contribution of vitamin K1 were maintained by participants in one trial [26]. Data on vitamin K1 in foods were analyzed using Nutritionist III software; however, the database did not have complete vitamin K1 data for many foods.

22.1.7.6 *Risk of Bias in Included Studies*

Both included trials were judged as having an unclear risk of bias overall [26,27]. To a certain extent, these assessments were based on inadequate reporting of several criteria that are considered to be important in evaluating methodologic rigor in terms of trial design and conduct.

22.1.7.7 *Effects of Interventions*

22.1.7.7.1 SERUM UNDERCARBOXYLATED OSTEOCALCIN

In the Drury trial, all participants had elevated (>21%) concentrations of ucOC before supplementation, but these levels were reduced after 1 month of vitamin K1 supplementation [27]. The overall level of ucOC decreased from a median of 46.8–29.1%, and ucOC levels decreased and returned to levels within the normal range in three of 13 participants (one in the 5-mg/day group and two in the 1-mg/day group) by the end of the trial. The mean end of trial difference in ucOC between the two intervention groups was −2.20 (95% confidence interval [CI], −14.33 to 9.93).

There was no evidence of a difference between the 5-mg/day and 1-mg/day vitamin K dosages in terms of a statistically significant effect on ucOC levels at the end of the 1-month trial period [27].

In the Beker trial, all participants had increased serum ucOC concentrations at enrollment and before supplementation [26]. It was reported that after supplementation and by the end of the trial, most had successfully achieved normal reference mean levels (21%) of ucOC [26].

22.1.7.7.2 PLASMA LEVEL OF VITAMIN K1

In the Drury trial, baseline vitamin K levels in seven of 14 participants were suboptimal (defined as <0.3 nmol/L) [27]. The article reported that serum vitamin K levels appeared to improve significantly ($P < 0.001$) with supplementation, rising into the normal range in all participants who were below the optimum level. There was no statistically significant difference in effect between the 5-mg/day and 1-mg/day vitamin K doses; mean difference −4.46 (95% CI, −12.65 to 3.73).

In the Beker trial, a substantial number of participants had below-normal serum vitamin K levels at trial entry [26]. Although the mean concentration of plasma vitamin K was higher in the supplemented group, these levels were reportedly brought into the normal range after supplementation with the 5-mg dose in fewer than half the total number of participants [26].

22.1.7.7.3 PROTEINS INDUCED BY VITAMIN K ABSENCE OR PIVKA-II LEVELS

Drury did not consider or report on this outcome [27]. In the Beker trial, PIVKA-II concentrations were elevated before supplementation and almost one-third of participants had PIVKA-II levels within the normal range (≤2 ng/mL) after supplementation [26].

22.1.8 Summary of Main Results

The results from both of the included trials could not be pooled and entered into a meta-analysis; however, both trials reported an increase in serum vitamin K levels and a decrease in ucOC levels that returned to normal after supplementation with oral vitamin K 1 mg/day for 1 month. There appeared to be no significant difference in these outcomes when a dose of 1 mg/day was compared with 5 mg/day in the Drury trial [27]. The PIVKA levels also showed a decrease and a return to normal in the Beker trial after supplementation [26].

22.1.9 Overall Completeness and Applicability of Evidence

The noticeable absence in the included trials of any assessments of important clinical outcomes related to coagulopathy or growth and improvement in quality of life as a result of vitamin K supplementation limits the overall completeness and ultimately the generalizability of the evidence to the wider CF population. Equally, the short duration and follow-up of the included trials does not permit conclusions to be made about the longer-term benefits and any potential harms of vitamin K supplementation.

22.1.10 Quality of the Evidence

The two trials included in this review were underpowered and of short duration, as illustrated by the wide CIs in the comparisons of treatment effect, and reflected the degree of imprecision in these results [26,27]. Some of the practical and methodological difficulties faced by investigators regarding this research question were highlighted in these trials. Key factors that are likely to have had a degree of impact on the quality level of the evidence for the outcomes sought in this review may be the design and implementation of the included trials, and in particular, effective concealment of the allocation sequence and adequate blinding of investigators and outcome assessors.

22.1.11 Agreements and Disagreements with Other Studies or Reviews

Several trials have indicated a degree of support for the requirement of supplementation, in addition to emphasizing the comparative safety of oral supplementation with vitamin K in people with CF [9,21,23,26], but there is no agreement among trials about the dosage of supplementation. However, there is some concern that the adequacy of dosing based on measurement of vitamin K levels may be inaccurate, and that PIVKA levels and ucOC levels may be better indicators of effectiveness of supplementation. Some trials suggest a complementary role for quantified dietary intake of vitamin K in people with CF [24]. This review showed nothing conclusively to agree or disagree with these trials.

22.1.12 Review Authors' Conclusions

22.1.12.1 Implications for Practice

People with CF are at risk of vitamin K deficiency. Routine supplementation along with the other fat-soluble vitamins is common clinical practice; however, there appears to be little consensus about the exact dosage that should be prescribed. Until further evidence is available, supplementation should continue to follow published European Cystic Fibrosis Society (ECFS) and Cystic Fibrosis Foundation (CFF) guidelines [9,22,23] and be guided by monitoring of the PIVKA or ucOC levels at annual clinical reviews.

22.1.12.2 Implications for Research

There is increased understanding of the important effect of vitamin K on gamma-carboxylation of osteocalcin synthesized in bone over and above its well-acknowledged requirement for synthesis of coagulation proteins in the liver. Well-designed, randomized, controlled trials are still needed to determine the impact of vitamin K on outcomes, including clinical measures (e.g., coagulation effects), bone measures (density scans), and biochemical markers, in addition to patient-orientated measures such as quality of life and patient satisfaction in people with CF. Additional issues that need to be considered are the sensitivity and reliability of markers for vitamin K deficiency, as well as the diagnostic accuracy and associated costs of laboratory equipment and tests.

22.1.13 Recommendations [36]

- There is not yet sufficient evidence to recommend universal vitamin K supplementation for patients with CF, but consideration should be given to individuals with low bone mineral density, liver disease, and/or a prolonged prothrombin time [Grade C recommendation].
- When vitamin K supplements are prescribed, they should be given as daily vitamin K1, 10 mg/day for children from 7 years of age upward and adults; a dose 5 mg/day for children 2–7 years of age; and 300 µg/kg/day, rounded to the nearest 1 mg, for babies and infants under 2 years of age [Grade C recommendation].

References

[1] Davis PB. Cystic fibrosis since 1938. Am J Respir Crit Care Med 2006;173(5):475–82.

[2] Goss CH, Rosenfeld M. Update on cystic fibrosis Epidemiology. Curr Opin Pulm Med 2004;10(6):510–4.

[3] Staab D. Cystic fibrosis – therapeutic challenge in cystic fibrosis children. Eur J Endocrinol 2004;151(Suppl. 1):S77–80.

[4] Ratjen F, Döring G. Cystic fibrosis. Lancet 2003;361(9358):681–9.

[5] Bobadilla JL, Macek Jr M, Fine JP, Farrell PM. Cystic fibrosis: a worldwide analysis of CFTR mutations-correlation with incidence data and application to screening. Hum Mutat 2002;19(6): 575–606.

[6] Wagener JS, Headley AA. Cystic fibrosis: current trends in respiratory care. Respiratory Care 2003;48(3):234–45.

[7] Dodge JA, Turck D. Cystic fibrosis: nutritional consequences and management. Best Pract Res Clin Gastroenterol 2006;20(3): 531–46.

[8] Durie PR. Vitamin K and the management of patients with cystic fibrosis. CMAJ 1994;151(7):933–6.

[9] Borowitz D, Baker RD, Stallings V. Consensus report on nutrition for paediatric patients with cystic fibrosis. J Pediatr Gastroenterol Nutr 2002;35:246–59.

[10] UK CF Trust. Bone mineralisation in cystic fibrosis. Report of the UK Cystic Fibrosis Trust Bone Mineralisation Working Group. www.cftrust.org.uk/aboutcf/publications/consensusdoc/Bone-Mineral-Booklet.pdf; [February 2007].

[11] Shearer MJ. Vitamin K. Lancet 1995;345(8944):229–34.

[12] Verghese T, Beverley D. Vitamin K deficient bleeding in cystic fibrosis. Arch Dis Child 2003;88(6):553.

[13] Conway SP, Wolfe SP, Brownlee KG, White H, Oldroyd B, Truscott JG, et al. Vitamin K status among children with cystic fibrosis and its relationship to bone mineral density and bone turnover. Pediatrics 2005;115(5):1325–31.

[14] Fuchs JR, Langer JC. Long-term outcome after neonatal meconium obstruction. Pediatrics 1998;101(4):E7.

[15] Uotila L. The metabolic functions and mechanism of action of vitamin K. Scand J Clin Lab Invest 1990;201(Suppl.):109–17.

[16] Okano T. Vitamin D, K and bone mineral density. Clin Calcium 2005;15(9):1489–94.

[17] Olson RE. Vitamin K. In: Shils ME, Olson JA, Shike M, editors. Modern nutrition in health and disease. 8th ed. Baltimore: Williams and Wilkins; 1994. pp. 343–58.

[18] Beker LT, Ahrens RA, Fink RJ, Sadowski JA, Davidson KW, Sokoll LJ, et al. Abnormal vitamin K status in cystic fibrosis patients. [abstract] Pediatr Pulmonol 1994;18(Suppl. 10):358.

[19] Aris RM, Ontjes DA, Brown SA, Chalermskulrat W, Neuringer I, Lester GE. Carboxylated osteocalcin levels in cystic fibrosis. Am J Respir Crit Care Med 2003;168:1129.

[20] Rashid M, Durie P, Andrew M, Kalnins D, Shin J, Corey M, et al. Prevalence of vitamin K deficiency in cystic fibrosis. Am J Clin Nutr 1999;70(3):378–82.

[21] Ramsey BW, Farrell PM, Pencharz P, the Consensus Committee. Nutritional assessment and management in cystic fibrosis: a consensus report. Am J Clin Nutr 1992;55(1):108–16.

[22] Cystic Fibrosis Trust Nutrition Working Group. Nutritional management of cystic fibrosis. London: Cystic Fibrosis Trust; 2002.

[23] Sinaasappel M, Stern M, Littlewood J, Wolfe S, Steinkamp G, Heijerman HG, et al. Nutrition in patients with cystic fibrosis: a European Consensus. J Cyst Fibros 2002;1(2):51–75.

[24] Dougherty KA, Schall JI, Stallings VA. Suboptimal vitamin K status despite supplementation in children and young adults with cystic fibrosis. Am J Clin Nutr 2010;92(3):660–7.

[25] Jagannath VA, Fedorowicz Z, Thaker V, Chang AB. Vitamin K supplementation for cystic fibrosis. *Cochrane Database of Syst Rev* 2011; (1). http://dx.doi.org/10.1002/14651858.CD008482.pub2. Art. No.: CD008482.

[26] Beker LT, Ahrens RA, Fink RJ, O'Brien ME, Davidson KW, Sokoll LJ, et al. Effect of vitamin K1 supplementation on vitamin K status in cystic fibrosis patients. J Pediatr Gastroenterol Nutr 1997;24(5):512–7.

[27] Drury D, Grey VL, Ferland G, Gundberg C, Lands LC. Efficacy of high dose phylloquinone in correcting vitamin K deficiency in cystic fibrosis. J Cyst Fibros 2008;7(5):457–9.

[28] Cornelissen EA, van Lieburg AF, Motohara K, van Oostrom CG. Vitamin K status in cystic fibrosis. Acta Paediatr 1992;81(9):658–61.

[29] Grey V, Atkinson S, Drury D, Casey L, Ferland G, Gundberg C, et al. Prevalence of low bone mass and deficiencies of vitamins D and K in pediatric patients with cystic fibrosis from 3 Canadian centers. Pediatrics 2008;122(5):1014–20.

[30] Mosler K, von Kries R, Vermeer C, Saupe J, Schmitz T, Schuster A. Assessment of vitamin K deficiency in CF–how much sophistication is useful? J Cyst Fibros 2003;2(2):91–6.

[31] Nicolaidou P, Stavrinadis I, Loukou I, Papadopoulou A, Georgouli H, Douros K, et al. The effect of vitamin K supplementation on biochemical markers of bone formation in children and adolescents with cystic fibrosis. Eur J Pediatr 2006;165(8):540–5.

[32] Wilson DC, Rashid M, Durie PR, Tsang A, Kalnins D, Andrew M, et al. Treatment of vitamin K deficiency in cystic fibrosis: effectiveness of a daily fat-soluble vitamin combination. J Pediatr 2001; 138(6):851–5.

[33] van Hoorn JH, Hendriks JJ, Vermeer C, Forget PP. Vitamin K supplementation in cystic fibrosis. Arch Dis Child 2003;88(11):974–5.

[34] van Hoorn JH, Schurgers JJ, Vermeer C, Escher HC, Hendriks HJ. The effect of vitamin K supplementation on bone status in children with cystic fibrosis. Pediatr Pulmonol 2008;43(S31):421.

[35] Powell M, Kuitert L. Effect of vitamin K supplementation over one year on bone health in adolescents and adults with cystic fibrosis. Pediatr Pulmonol 2010;45(S33):421.

[36] Shekelle PG, Woolf SH, Eccles M, Grimshaw J. Developing clinical guidelines. West J Med 1999 June;170(6):348–51.

APPENDIX 1 [36]

Levels of Evidence and Grades of Recommendations

Evidence Based Health Care—Practice guidelines levels of evidence and grades of recommendations.

Levels of Evidence:

IA Evidence from meta-analysis of randomized controlled trials
IB Evidence from at least one randomized controlled trial
IIA Evidence from at least one controlled study without randomization
IIB Evidence from at least one other type of quasi-experimental study
III Evidence from nonexperimental descriptive studies, such as comparative studies, correlation studies, and case–control studies
IV Evidence from expert committee reports or opinions or clinical experience of respected authorities, or both

Grades of Recommendations:

A Directly based on level I evidence
B Directly based on level II evidence or extrapolated recommendations from level I evidence
C Directly based on level III evidence or extrapolated recommendations from level I or II evidence
D Directly based on level IV evidence or extrapolated recommendations from level I, II, or III evidence

CHAPTER

23

Vitamin K and Cystic Fibrosis

Mary Shannon Fracchia[1, 2], Ronald E. Kleinman[1, 2]

[1]Department of Pediatrics, Massachusetts General Hospital, Boston, MA, USA;
[2]Harvard Medical School, Boston, MA, USA

23.1 INTRODUCTION

Vitamin K, a fat-soluble vitamin, takes its name from the German word *koagulation*, when it was discovered and reported in the German literature in 1929 to prevent hemorrhage in chickens. Vitamin K deficiency can be subclinical or present with symptoms that include full-blown coagulopathy. The typical presentation consists of easy bruising and mucosal bleeding. Newborn babies are at particular risk for vitamin K deficiency because vitamin K in maternal serum does not cross the placenta in amounts needed to support infant requirements in the perinatal period and human milk is a poor source for the nursing infant. Vitamin K prophylaxis is routinely given at birth to prevent hemorrhagic disease of the newborn. Less well known are more recent discoveries of the role of vitamin K in bone formation and the regulation of inflammation, particularly inhibition of arterial calcification. For patients with cystic fibrosis (CF), a chronic and progressive inflammatory disease, vitamin K may play several important roles that remain to be fully explored.

23.2 DIETARY SOURCES

Naturally occurring vitamin K is present in several commonly consumed foods (Table 23.1). Specifically, vitamin K1 (phylloquinone, phytomenadione, phytonadione) is present in plants and is found in green leafy vegetables such as spinach, broccoli, and collard greens. Some vegetable oils, specifically soybean, canola, and cottonseed, also are sources of vitamin K1 [1]. Animal-based foods such as eggs, chicken, certain types of cheese, and butter contain vitamin K2 (menaquinones) [2]. Vitamin K2 is also synthesized by colonic bacteria. However, it is unclear how much intestinal bacteria contribute to total body vitamin K stores because bile salts are not present in the colon but they are needed to absorb vitamin K. Colonic bacteria can also convert vitamin K1 to vitamin K2. Most of the vitamin K directly measured in plasma is phylloquinone whereas more than 90% of the vitamin K stored in the liver consists of menaquinone [3].

23.3 VITAMIN K DIGESTION

Vitamin K is a fat-soluble vitamin; thus, the presence of fat in the diet increases the intestinal absorption of dietary vitamin K. Plasma K1 levels highly correlate with serum triglyceride concentrations [1]. The pancreas secretes lipase, colipase, and phospholipases, enzymes required to hydrolyze triglycerides in advance of absorption. Therefore, those patients with pancreatic insufficiency are at particular risk for vitamin K deficiency. These include patients with CF and with other inherited disorders of pancreatic function such as Schwachman Diamond Syndrome as well as acquired conditions leading to pancreatitis and pancreatic exocrine failure. Newborn infants, patients with advanced liver disease, and those with significant small bowel resection are also at risk for fat malabsorption and fat-soluble vitamin deficiencies. Infants are born with a relatively sterile gut and do not get adequate quantities of vitamin K from placental transfer prepartum or from breast milk. Patients with liver disease have inadequate vitamin K stores and may have cholestasis and fat malabsorption, and those with significant small bowel resection have inadequate intestinal absorptive surface area.

23.4 BIOCHEMISTRY

Biochemically, vitamin K is a group of 2-methyl-1,4-naphthoquinone derivatives. Vitamin K is a cofactor for the enzyme γ-glutamyl carboxylase, which converts

Diet and Exercise in Cystic Fibrosis
http://dx.doi.org/10.1016/B978-0-12-800051-9.00023-7

TABLE 23.1 Dietary Sources of Vitamin K

Vegetables and Herbs	Vitamin K Content, μg/100 g	Oils and Dairy	Vitamin K Content, μg/100 g
Parsley	900	Natto	1000
Kale	817	Soybean	175
Collard greens	440	Canola	130
Spinach	360	Extra virgin olive	80
Brussel sprouts	275	Hard cheeses	76
Watercress	250	Soft cheeses	56
Green leaf lettuce	175	Mayonnaise	40
Cabbage	90		

Median values taken from Refs [1,3–5].

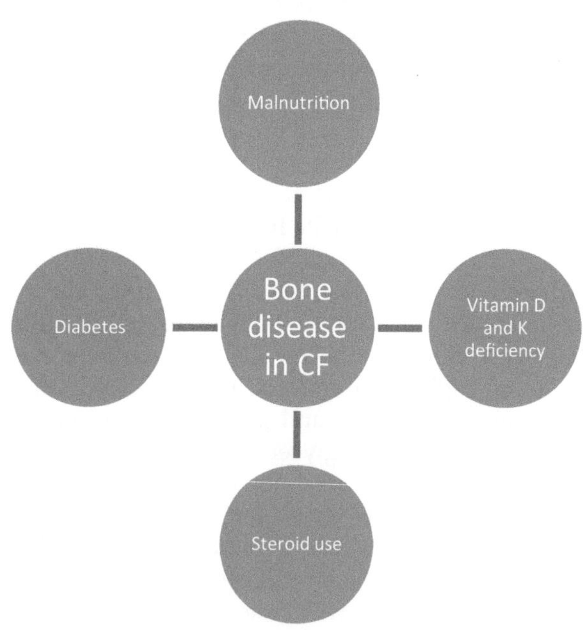

FIGURE 23.1 **Cystic fibrosis (CF) and bone disease.**

glutamic acid (gla) residues to γ-carboxyglutamic residues (Figure 23.1). These gla residues are calcium binding moieties present on prothrombin (as well as clotting factors II, VII, IX, and X and proteins C, S, and Z), the bone proteins osteocalcin and periostin, gla-rich protein, and matrix gla protein. Undercarboxylated prothrombin leads to abnormal clotting and bleeding whereas undercarboxylated osteocalcin (ucOC or Glu-OC) has been associated with osteopenia and osteoporosis. Serum concentrations of ucOC and prothrombin (plasma prothrombin in vitamin K absence II (PIVKAII)) are released from the liver into the blood and are highly sensitive measures of vitamin K status [6]. On the other hand, prothrombin time (PT) is an insensitive marker of vitamin K deficiency because it is only prolonged in advanced deficiency states.

23.5 VITAMIN K AND CF

Vitamin K insufficiency or deficiency is highly prevalent in patients with CF. CF, the most common lethal genetic disease in Caucasians, often presents with pancreatic insufficiency. Over 90% of CF patients are pancreatic insufficient because of an absent or malfunctioning cystic fibrosis transmembrane regulator (CFTR). CFTR is a multifunctional protein located at the apical membrane of epithelia and is mainly involved in modulation of transepithelial ion transport, hydration of epithelial lining fluids, pH regulation, and inflammation [7]. The absence of CFTR in the epithelial lining of the pancreas leads to a protein-rich, viscous exocrine fluid that obstructs the pancreatic ducts as early as the second trimester of gestation. This obstruction leads to autodigestion of the pancreas and therefore pancreatic insufficiency. Patients with pancreatic insufficiency have decreased or absent levels of trypsin, chymotrypsin, amylase, lipase, colipase, and phospholipases, which results in significant maldigestion/malabsorption of dietary macro- and micronutrients. As a result of the lack of those enzymes needed for fatty acid liberation and absorption, patients are at risk for growth failure, fat-soluble vitamin deficiencies (vitamins A, D, E, and K), rectal prolapse, and steatorrhea. In addition, in patients with CF, vitamin K stores are depleted because of low dietary intake, short bowel syndrome in those who have undergone bowel resection as a result of meconium ileus, and the frequent use of antibiotics that diminishes the intraluminal production of vitamin K by gut microorganisms [8].

Although vitamin K ultimately may be shown to play a role in regulating the inflammation that occurs in patients with CF, apart from observations that the vitamin-K-dependent matrix gla protein is a potent inhibitor of arterial calcification [9], this remains a largely unexplored area.

In observational studies, inverse associations have been reported between phylloquinone intake and circulating phylloquinone and bone loss at the hip among the elderly [2]. The decreased bone mass and increased risk of fractures in patients with CF because of inactivity, delayed puberty, increased use of corticosteroids, and persistent systemic inflammation may also be affected by vitamin K status [7]. The mechanism of action of vitamin K on bone mineral density (BMD) is still being explored. The rate of bone turnover is a major determinant of BMD [10]; increased bone turnover leads to decreased bone density. Vitamin K appears to increase osteoblastogenesis and decrease osteoclastogenesis, perhaps through gamma carboxylation of bone-related proteins, including osteocalcin [1].

23.6 DIAGNOSIS OF VITAMIN K DEFICIENCY

As mentioned previously, ucOC (or Glu-OC) and prothrombin (PIVKAII) are serum markers of vitamin K deficiency. It is possible to directly measure serum phylloquinone concentration [8]. Measurements of serum levels of osteocalcin and prothrombin reflect vitamin K function. In healthy, well-nourished individuals, less than 20% of osteocalcin is undercarboxylated. Levels of 20–50% indicate insufficiency, and more than 50% undercarboxylation indicates vitamin K deficiency (Table 23.2).

23.7 VITAMIN K PREPARATIONS

In Europe and North America the only pharmaceutical vitamin K preparation currently available is phylloquinone. Menaquinone-4 is available as a pharmaceutical preparation in Japan [2] and is approved for the prevention and treatment of osteoporosis there. Phylloquinone is available as oral, subcutaneous, intramuscular, and intravenous preparations. The American Academy of Pediatrics recommends a single intramuscular injection of 0.05–1.0 mg of vitamin K for all term infants within an hour of birth [12]. In contrast to repeated oral doses, intramuscular injection does not require parental compliance and protects against early- and late-onset deficiency. Late-onset vitamin K deficiency, seen in those infants that do not receive prophylactic supplemental vitamin K after birth and are exclusively breastfed from birth, occurs between 2 and 12 weeks of age. These infants present with a severe coagulopathy, with a mortality rate of 20%. Intracranial hemorrhage occurs in 50% of these infants [3] (Table 23.3).

23.8 VITAMIN K SUPPLEMENTATION IN CF

There is extensive debate on how much supplemental vitamin K should be provided on a daily basis to patients with CF to achieve sufficiency. Table 23.3 presents the adequate intake of the daily reference intake of vitamin K by age. The 2002 Cystic Fibrosis Foundation guidelines [14] recommend a daily supplemental dose of vitamin K of 300–500 μg in patients with CF. A recent review by Jaganath [15] supports this recommendation. Vitamin K preparations are extremely safe and well tolerated [16].

However, Dougherty et al. [17] examined those factors that might predict vitamin K insufficiency in patients with CF and the daily dose of supplemental vitamin K needed to produce sufficiency in the study population. Subjects between the ages of 8 and 25 years with CF and pancreatic insufficiency and mild-to-moderate lung disease were recruited. Very importantly, a contemporary comparison group of healthy subjects aged 8–21 years was also recruited for this study. Anthropometric measures, body composition, sexual maturation, pulmonary function, and CFTR genotype were obtained. Total osteocalcin and ucOC (%ucOC) were determined in healthy and CF subjects with pancreatic insufficiency. PIVKAII and serum 25-hydroxyvitamin D were also obtained in the CF patients only. Dietary intake was assessed by using 3-day weighed food records. CF patients were divided into three groups according to vitamin K supplementation: <150 μg/day (low), 150–900 μg/day (mid), and >1000 μg/day (high).

Overall, subjects with CF had a higher prevalence of vitamin K deficiency (on the basis of %ucOC) compared with healthy subjects. Of note, most healthy subjects

TABLE 23.2 Biomarkers of Vitamin K Deficiency

Biomarkers	Normal Values	Values in Vitamin K Deficiency
Vitamin K1 (phylloquinone), μg/L	1.1	<1.1
PIVKAII, μg/L	<3	>3
Undercarboxylated osteocalcin, μg/L	<0.6	2.7 ± 1.00
Percentage undercarboxylated osteocalcin, %	<10	24 ± 1
Prothrombin time, s	11–13.5	>13.5

Data taken from Refs [8,11].

TABLE 23.3 The Adequate Intake (AI) of the Daily Reference Intake of Vitamin K by Age

Group	AI
Babies 1–6 months	2 μg/day
Babies 7–12 months	2.5 μg/day
Children 1–3 years	30 μg/day
Children 4–8 years	55 μg/day
Children 9–13 years	60 μg/day
Children 14–18 years	75 μg/day
Women	90 μg/day
Men	120 μg/day

Data taken from Refs [3,4,13].

(63%) were classified as insufficient, although none were deficient. The subjects with CF had median daily dietary intakes of vitamin K below the recommended adequate intake for age and sex of healthy children. Fifty-nine percent of CF subjects also did not meet their recommended daily energy intake. Only those children with CF and pancreatic insufficiency who received >1000 μg/day of vitamin K achieved a status similar to healthy subjects. The dose of supplemental vitamin K predicted vitamin K status for CF patients with pancreatic insufficiency.

23.9 TREATMENT OF VITAMIN K DEFICIENCY

The treatment of patients with vitamin K deficiency is based on the International Normalized Ratio (INR). According to the guidelines put forth by the American College of Chest Physicians [18], patients with a normal INR and vitamin K deficiency should receive a single daily oral dose of 2.5 mg of vitamin K. Those patients with an INR between 1.5 and 1.8 should receive an initial intramuscular dose of 2.5 mg followed by an oral dose of 2.5 mg daily. Patients with an INR greater than 1.8 should receive 2–5 mg intramuscularly once followed by 5 mg orally once daily. Those patients with major bleeding from vitamin K antagonists should receive vitamin K intravenously. With severe obstructive liver disease, intramuscular or intravenous administration should be considered because intraluminal bile salts are needed for optimal absorption of oral vitamin K [19].

23.10 SUMMARY

Patients with CF are at risk of vitamin K deficiency because of an inability to absorb fat because of pancreatic insufficiency. These patients are at risk for clotting disorders and decreased BMD. The most sensitive diagnostic test for vitamin K deficiency is ucOC. PIVKAII is also a sensitive marker of vitamin K deficiency. Supplementation with 0.3–0.5 mg/day of vitamin K is recommended for patients with CF, although larger doses may be more effective in preventing insufficiency.

References

[1] Pearson D. Bone health and osteoporosis. The role of vitamin K and potential antagonism by anticoagulants. Nutr Clin Prac 2007;22:517.

[2] Shea MK, Booth SL. Update on the role of vitamin K on skeletal health. Nutr Rev 2008;66(10):549–57.

[3] Van Winckel M, De Bruyne R, Van De Velde S, Van Biervliet S. Vitamin K, an update for the paediatrician. Eur J Pediatr 2009;168:127–34.

[4] Booth SL, Suttie JW. Dietary intake and adequacy of vitamin K1. J Nutr 1998;128:785–8.

[5] Shearer MJ. Vitamin K. Lancet 1995;345:229–34.

[6] Conway S, Wolfe S, Brownlee K, White H, Holroyd B, Truscott J, et al. Vitamin K status among children with cystic fibrosis and its relationship to bone mineral density and bone turnover. Pediatrics 2005;115:1325–31.

[7] Sermet-Gaudelus I, Castanet M, Retsch-Bogart, Aris RM. Update on cystic fibrosis-related bone disease: a special focus on children. Paediatr Respir Rev 2009;10:134–42.

[8] Rashid M, Durie P, Andrew M, Kalnins D, Shin J, Corey M, et al. Prevalence of vitamin K deficiency in cystic fibrosis. Am J Clin Nutr 1999;70:378–82.

[9] Fodor D, Albu A, Poanta L, Porojan M. Vitamin K and vascular calcifications. Acta Physiol Hug September 2010; 97(3):256–66.

[10] Cashman K. Vitamin K status may be an important determinant of childhood bone health. Nutr Rev 2005;63:284–9.

[11] Mummah-Schnedel LL, Suttie JW. Serum phylloquinone concentration in the adult population. Am J Clin Nutr November 1986;44(5):686–9.

[12] American Academy of Pediatrics, Committee on Nutrition. Controversies concerning vitamin K and the newborn. Pediatrics 2003;112:191–2.

[13] Wilson DC, Rashid M, Durie PR, Tsang A, Kalnins D, Andrew M, et al. Treatment of vitamin K deficiency in cystic fibrosis: effectiveness of a daily fat-soluble vitamin combination. J Pediatr 2001;138:851–5.

[14] Borowitz D, Baker RD, Stallings V. Consensus report on nutrition for paediatric patients with cystic fibrosis. J Pediatr Gastroenterol Nutr 2002;35:246–59.

[15] Jagannath VA, Fedorowicz Z, Thaker V, Chang AB. Vitamin K supplementation for cystic fibrosis. Cochrane Database Syst Rev 2013;4.

[16] Conway SP. Vitamin K in cystic fibrosis. J R Soc Med 2004;97 (Suppl. 44):48–61.

[17] Dougherty KA, Schall JI, Stallings VA. Suboptimal vitamin K status despite supplementation in children and young adults with cystic fibrosis. Am J Clin Nutr 2010;92:660–7.

[18] Hirsh J, Guyatt G, Albers GW, Harrington R, Schünemann HJ. et al. Executive summary: American College of Chest Physicians evidence-based clinical practice guidelines (8th edition). Chest 2008;133(Suppl. 6):71–109.

[19] Weber P. Vitamin K and bone health. Nutrition 2001;17:880–7.

SECTION D

MANAGEMENT OF DIABETES AND CELIAC DISEASE ASSOCIATED WITH CYSTIC FIBROSIS: ROLE OF NUTRITION AND FOOD

CHAPTER

24

Insulin, Body Mass, and Growth in Young Cystic Fibrosis Patients

Shihab Hameed[1, 2], *Charles F. Verge*[1, 2]

[1]Endocrinology Department Sydney Children's Hospital, Randwick, NSW, Australia; [2]School of Women's and Children's Health, University of New South Wales, Kensington, NSW, Australia

24.1 INSULIN SECRETION—NORMAL PHYSIOLOGY

Insulin is synthesized and released from the pancreas by the beta cells of the islets of Langerhans (see Figure 24.1). There are approximately 1 million islets of Langerhans in the healthy human pancreas [1], and larger islets typically contain approximately 2000 beta cells each [1]. Nevertheless, islets make up only 1% of total pancreatic volume [2]. The islets have a complex architecture, with a highly structured arrangement of endocrine and non-endocrine cells, gap junctions, matrix proteins, innervations, and rich vascularizations [3]. It is important to note that in cystic fibrosis (CF), there is a marked reduction in the number of islets [4,5] and the architecture of those islets is severely disrupted [6,7] (as detailed in the next section).

Within the beta cell, a peptide precursor is folded into insulin, then packaged into membrane-bound granules that can be released from the beta cell into adjoining fenestrated vessels. From there, the insulin travels through the portal venous circulation, through the liver, to the systemic circulation.

Insulin is released in a pulsatile manner. In the fasting state, secretion occurs in pulses at a rate of approximately 2 pmol/L/min [3], with pulses occurring approximately every 4 min. However, after a meal, the amplitude of these secretory pulses increases substantially [8].

Thus, insulin secretion is a dynamic process triggered by the availability of metabolic substrates—such as amino and ketoacids, but predominantly glucose [9]. When islets are stimulated by a rapid increase in the circulating glucose concentration, there are two distinct phases of insulin secretion. There is a rapid increase in secretion (the first phase response), lasting approximately 3 min, followed by a more prolonged increase in secretion (the second phase response) lasting several minutes to hours [3]. The first phase response is due to the release of readily available insulin within the secretory granules, whereas the second phase requires maturation of granules prior to secretion [10–12].

Absolute insulin deficiency in humans is rapidly fatal unless insulin treatment is provided [13]. Thus, prior to the discovery of insulin therapy in 1922 [14], type 1 diabetes mellitus caused severe weight loss and hyperglycemia, progressing to diabetic ketoacidosis and death. By contrast, some insulin-secreting ability is preserved in CF and ketoacidosis is extremely rare [15]. Rather, as we will explore in the next section, patients with CF often have delayed and diminished insulin secretion, presenting with postprandial hyperglycemia and gradual catabolism.

24.2 PROGRESSIVE INSULIN DEFICIENCY IS A FEATURE OF CF

CF-related diabetes (CFRD) [16] is classified by the American Diabetes Association as a form of secondary diabetes associated with disease of the exocrine pancreas [17]. Other members of this diagnostic class include diabetes secondary to partial pancreatectomy or through iron deposition from hemochromatosis. These latter disorders result in insulin deficiency due to structural pancreatic disease. Similarly, pancreatic damage in CF

Diet and Exercise in Cystic Fibrosis
http://dx.doi.org/10.1016/B978-0-12-800051-9.00024-9

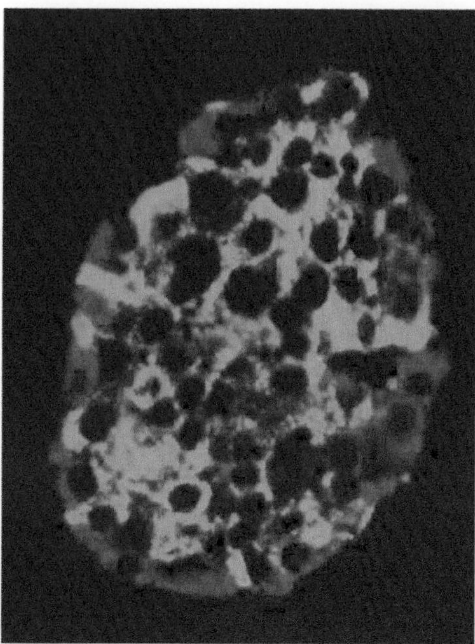

FIGURE 24.1 The human islet of Langerhans has a highly organized and complex architecture. Staining by immunofluorescence demonstrates insulin (green) and glucagon (blue). *Reproduced with permission from Ref. [3].*

is associated with partial insulin deficiency. This section will detail the evidence for this, and discuss the proposed mechanisms of beta cell loss in CF.

24.2.1 Ultrasound and Histopathological Findings in Cystic Fibrosis Demonstrate Beta Cell Loss

From an early age, ultrasound studies of CF patients demonstrate reduced pancreatic size compared to controls [18] as well as abnormal echogenicity [19]. In fact, structural abnormalities are detectable by ultrasound in 75% of CF patients aged less than 5 years, in 95% of those over 5 years, and in 100% of patients older than 5 years who have pancreatic exocrine insufficiency [20]. Furthermore, complete fatty replacement of the pancreas has been demonstrated in children with CF by computed tomography scans [21]. Pancreatic histopathology taken at autopsy of patients with CF shows markedly disrupted islet architecture, characterized by widespread sclerosis and amyloid deposits [6,7], and a reduction in the number of islets [4,5]. Immunohistochemical staining demonstrates reduced beta cell mass in CF, especially among those who have CFRD [4–6,22]. The beta cells are more severely depleted than non-insulin-secreting islet cells, such as alpha cells, which secrete glucagon [22,23]. One explanation is that beta cells are vulnerable to injury by many mechanisms, including hyperglycemia [24]. Hence, islet destruction may cause hyperglycemia, resulting in beta cell loss in the remaining islets, which in turn causes further hyperglycemia in a downward spiral.

24.2.2 Biochemical Findings Confirm Insulin Deficiency Is Common in CF

Insulin deficiency in CF was described in 1969, when it was noted that the oral glucose tolerance test (OGTT) demonstrated reduced insulin secretion in children and adolescents with CF compared to healthy controls [25,26]. In 1977, Lippe and colleagues further described delayed and diminished insulin secretion, with loss of the normal rapid secretory response to a glucose load [23]. Interestingly, they found preservation of glucagon secretion in CF subjects, which were similar in secretion profiles to those of healthy controls. These biochemical findings are consistent with the pattern of preferential beta cell loss seen on histopathology.

Several other groups have also found insulin deficiency in CF [27–31]. We found delayed and impaired insulin secretion in our pediatric cohort, who had a calculated mean beta cell function of only 30% [32]. Lombardo and colleagues followed 14 individual CF subjects with normal fasting glucose for 13 years. They found that over that extensive period of observation, all subjects developed significantly reduced insulin secretion and that insulin secretion was markedly reduced in the four patients who progressed to CFRD diabetes [16,33].

24.2.3 Proposed Mechanisms for Beta Cell Loss in CF

In addition to direct pancreatic tissue destruction from pancreatic duct blockage by thick secretions in CF, other mechanisms leading to beta cell loss have been proposed. These include an intrinsic beta cell defect as a result of the underlying mutations in the CF transmembrane regulator (CFTR) gene [34–36]. However, it is likely that genes other than CFTR also contribute to the development of diabetes in CF. This was demonstrated by Blackman and colleagues in a study of 1366 twins and siblings with CF, which showed that concordance for diabetes was substantially higher in monozygotic twins (0.73) than in siblings or dizygotic twins (0.18) carrying the same CFTR mutation [37]. The same group showed that certain polymorphic alleles of transcription factor 7-like 2 (TCF7L2, which confer increased susceptibility to type 2 diabetes in the non-CF population) also confer increased risk of CFRD [38].

In a recent pilot study, Bellin and colleagues studied the effects of ivacaftor on insulin secretion in five CF patients who carried at least one allele of the relatively rare G551D mutation [39]. When given to patients with this mutation, ivacaftor corrects the underlying CFTR abnormality [40]. They performed intravenous and OGT testing prior to commencement of ivacaftor, and after 1 month of therapy. Interestingly, they found that insulin secretion improved by 66–178% in all subjects, except one with long-standing diabetes. In particular, there was improvement in the acute insulin response. As the CFTR gene is expressed in human beta cells, the authors postulated that CFTR may play a direct role in insulin secretion and that ivacaftor may improve insulin secretion, particularly if commenced at an early age.

In contrast to type 1 diabetes, which is characterized by T-cell mediated autoimmune destruction of islet beta cells, autoimmunity does not appear to be a significant cause of beta cell loss in CF. Autoantibodies to one or more beta cell antigens (insulin, glutamic acid decarboxylase, insulinoma-associated peptide 2, and zinc transporter 8) are detectable in nearly all type 1 diabetes, but are rarely found in patients with CFRD [41,42]. Furthermore, patients with CFRD do not display any increase in the frequency of the human leukocyte antigen (HLA) genotypes that confer susceptibility to type 1 diabetes [42].

24.3 NORMAL PHYSIOLOGY OF INSULIN-MEDIATED GLUCOSE UPTAKE

Glucose is the main fuel for most cells in the body, and uptake of glucose is the rate-limiting step for glucose utilization [43]. To assist in this process, all cells in the human body express one or more members of the glucose transporter family on their cell surface, to allow glucose influx into cells. Of these, the insulin-responsive GLUT-4 transporter is one of the most important, as it is expressed in skeletal muscle, adipose cells, and the heart. Prior to stimulation by insulin, it is found sequestered in intracellular vesicles, and is unable to transport glucose across the plasma membrane. Insulin stimulation promotes translocation of these vesicles to the cell surface and allows cellular glucose uptake by insulin sensitive glucose (GLUT-4) channels [44]. In this way the cell is provided with glucose for adenosine triphosphate (ATP) production and for a wide variety of anabolic reactions. It is also important to note that high circulating glucose levels downregulate GLUT-4 expression in muscle and fat cells [43], resulting in impaired peripheral insulin sensitivity.

24.3.1 Other Factors May Influence the Development of Glycemic Abnormalities through Alterations in Insulin Sensitivity

As mentioned above, hyperglycemia may contribute to insulin insensitivity, thereby worsening glycemia. Additionally, other factors may contribute, including the counter-regulatory hormones growth hormone, cortisol, glucagon, and catecholamines. These hormones oppose the glucose-lowering actions of insulin. Growth hormone is known to cause physiological insulin resistance during puberty [45], and thus may contribute to the higher detection rates of CFRD among adolescents (see Section 24.7.2). This is relevant to the clinical use of growth hormone injection therapy in children with short stature, (including those with CF) as glucose intolerance and diabetes are potential side-effects [46].

Other counter-regulatory hormones, cortisol and catecholamines, act as stress hormones and are elevated during systemic illness. This is particularly important to note in CF during respiratory exacerbations, and during lung transplantation [47]. Bell and colleagues studied 22 adult CF patients during CF exacerbations and found that plasma norepinephrine fell significantly between day 1 and day 8 of therapy for a respiratory exacerbation. Furthermore, the reduction in norepinephrine correlated with a fall in heart rate, suggesting that acute exacerbations temporarily increase counter-regulatory hormone secretion in CF patients. However, in this cohort, the mean plasma insulin and glucagon concentrations remained unchanged before and after treatment of the exacerbation [48]. Inflammatory cytokines such as TNF-alpha may also contribute to worsening glycemic status during systemic inflammation [49].

Exogenous glucocorticoids may be used as part of CF therapy, and these increase glycemic abnormalities by the same mechanism as endogenous cortisol [50,51].

To help detect worsening glycemic status during systemic illness, consensus guidelines recommend screening for CFRD by fasting and 2-h postprandial blood glucose testing during acute pulmonary exacerbations [52]. However, Widger and colleagues recently compared OGTT results from the start of hospital admission with follow-up results performed 4–6 weeks later. They found that the 2-h glucose value was only marginally higher (1.1 mmol/L, 19.8 mg/dL) during the exacerbation [53], with no change to glucose tolerance category in eight out of nine patients.

In our pediatric CF cohort, we avoided periods of respiratory exacerbation when performing annual screening OGTTs to avoid transient deterioration in glycemia due to systemic illness. We found relatively normal insulin sensitivity of 90% during these periods of relative clinical stability [32].

There is also evidence that, in contrast to insulin secretion, insulin sensitivity is largely preserved in CF patients over time. Lombardo and colleagues studied the natural history of glucose tolerance, beta cell function, and peripheral sensitivity in CF patients observed over 13 years. They found a long-term improvement in insulin sensitivity (calculated from the insulin and glucose levels obtained during OGTT), even among those four patients who developed CFRD [33]. These estimates are not as accurate as the results of hyperinsulinemic euglycemic clamp studies, which are a direct measure of peripheral insulin sensitivity [54]. Using clamp studies, Ahmad and colleagues reported markedly increased peripheral insulin sensitivity in CF subjects and an increased rate of metabolic clearance of insulin [55]. However, others have reported either slightly reduced peripheral insulin sensitivity [56] or no change compared to subjects without CFRD [57]. Moran and colleagues studied adult CF subjects to determine both peripheral insulin sensitivity (using hyperinsulinemic clamp studies) as well as hepatic insulin sensitivity (using hyperglycemic clamp studies). They found that exocrine-insufficient CF patients without diabetes appear to compensate for insulin deficiency with increased peripheral sensitivity (leading to increased peripheral glucose utilization) and hepatic insulin resistance (leading to increased hepatic glucose production). They postulated that these changes were metabolic adaptations to increased energy needs. However, once CF subjects had developed diabetes, they found peripheral insulin resistance, concluding that this resulted from extreme insulin depletion. The importance of that finding is that marked insulin depletion is known to cause chronic hyperglycemia, which in turn results in peripheral insulin resistance [56]. The results of a randomized trial to determine if insulin treatment during pulmonary exacerbations improves lung function, body mass index (BMI), and glycemia [58] are expected in 2015.

24.4 IMPLICATIONS OF INSULIN DEFICIENCY ON AIRWAY GLUCOSE AND BACTERIAL COLONIZATION

An important implication for CF of hyperglycemia due to insulin deficiency relates to the findings of Brennan and colleagues. They have shown that airway glucose concentrations are elevated once blood glucose levels exceed a threshold of approximately 8 mmol/L (144 mg/dL), in patients both with [59] and without CF [60]. Furthermore, they demonstrated that such increases in airway glucose concentration promote the growth of bacteria such as *Pseudomonas aeruginosa* and *Staphylococcus* spp. [59]. Garnett and colleagues performed in vitro experiments on human bronchial epithelial monolayers exposed to varying concentrations of glucose and other hallmarks of CF airway surface liquid (such as mucus hyperviscosity and depletion of CFTR-dependent fluid) [61]. They found that elevation of glucose promotes the growth of *P. aeruginosa* on CF airway epithelial monolayers more than non-CF monolayers. Interestingly, they also found that *P. aeruginosa* secretions elicited more glucose flux across CF airway epithelial monolayers than in non-CF monolayers. Elevating glucose levels increased *P. aeruginosa* growth more than any CFTR-dependent effects. These findings suggest that elevated glucose levels in the airway surface liquid may be a major determinant of bacterial infection with *P. aeruginosa* in CF patients. The same group recently demonstrated that in vitro inoculation of human airway epithelial cells with the insulin sensitizer metformin inhibited growth of *S. aureus* and decreased glucose flux, and found that mice infected with *S. aureus* had lower airway bacterial loads if treated with metformin [62]. However, as detailed later in this chapter, insulin (rather than metformin or other oral agents) is the recommended therapy for glycemic abnormalities in CF patients (see Section 24.8.9).

Chotirmall and colleagues published an intriguing paper in 2012, noting that estriol and estradiol induced *P. aeruginosa* strains to convert to mucoid strains in vitro [63]. Estradiol levels in women with CF correlated with infective exacerbations, and the use of oral contraceptive agents was found to be associated with a decreased need for antibiotics. Furthermore, mucoid bacterial strains were more likely to be isolated from patients during their follicular (high estradiol) menstrual phase than in their luteal phase of menstruation. The study did not report glycemic characteristics; however, hormonal variations during the menstrual period are known to modify insulin sensitivity and glycemic outcomes in women without CF [64]. Furthermore, it is known that women with CF are colonized earlier with *P. aeruginosa* and experience earlier conversion to mucoid strains than men [65–67].

This raises questions about potential gender differences in glycemia in CF contributing to lung infections in female CF patients. Coriati and colleagues recently studied the differences in insulin secretion following OGTT according to gender in 230 adult CF patients and compared the results to healthy aged matched controls [68]. They found that both men and women with CF had similarly impaired early insulin secretion and significantly larger glucose excursions post glucose load compared to healthy controls. Additionally, they found that women with CF had the same fasting insulin levels as male CF patients, but had lower peripheral sensitivity than their male counterparts. Thus estrogen-related effects on glycemia and peripheral insulin sensitivity may play a role in lung infections in women.

Further demonstrating the importance of hyperglycemia on respiratory infection, Lanng and colleagues demonstrated in 1994 that insulin treatment reduces sputum culture positivity to respiratory pathogens [69]. Insulin treatment has also been shown to reduce lung infections [70] and pulmonary exacerbations [71]. Recent analysis of the 14,732 patients of the European Cystic Fibrosis Society Patient Registry demonstrated that CFRD was associated with *P. aeruginosa* infection and that this increases the odds of severe lung disease 2.4 fold (95% confidence interval (CI) 2.0–2.7) [72]. Furthermore, infection with *Pseudomonas* spp. and other respiratory pathogens predicts early mortality in CF [73].

Taken together, these findings suggest that frequent postprandial glucose excursions due to insulin deficiency make a significant contribution to the development of lung infections, decline in lung function and early mortality seen in patients with CFRD.

24.4.1 Effects of Diabetes on Lung Function

Patients with type 1 and type 2 diabetes mellitus have reduced lung function, even in the absence of known lung disease such as CF. A meta-analysis of 40 studies of lung function in patients with diabetes mellitus without CF or known lung disease found a pooled mean reduction in forced expiratory volume in 1 s (FEV_1) of 5.1% compared to healthy controls [74]. The mechanism is unclear, but it is possible that hyperglycemia reduces lung elasticity through induced non-enzymatic glycosylation, or microvascular damage [75].

In patients with CF, it is common to note a decline in lung function for several years prior to the diagnosis of CFRD [76,77]. Koch and colleagues analyzed 7566 patients in the European Epidemiologic Registry of Cystic Fibrosis in 2001, identifying a 20% reduction in mean percentage predicted FEV_1 in CF subjects with diabetes compared to those without [78]. A recent analysis of this registry demonstrated that patients with CFRD have a 1.8-fold increased odds ratio (95% CI 1.6–2.2) of severe lung disease compared to patients not affected [72].

Longitudinal studies have also demonstrated the association insulin deficiency with lung function decline. Milla and colleagues studied 152 CF subjects with normal glucose tolerance, impaired glucose tolerance, or CFRD without fasting hyperglycemia. After 4 years follow-up they found that those in the lowest quartile of insulin production at baseline suffered the greatest decline in lung function [79]. An important message from this study is the need to consider the diagnosis of CFRD in any patient with unexplained deterioration in lung function.

24.4.2 Microvascular Disease

In addition to the very important effects on lung infection and respiratory function, microvascular damage from hyperglycemia may also occur, leading to damage of other organs. As with other forms of diabetes mellitus, retinopathy and nephropathy are known to occur in long-standing CFRD, and are found at rates similar to other forms of diabetes [69,80,81]. Nephrotoxic drugs may also contribute to the development of nephropathy in CF [82]. By contrast, macrovascular disease remains very rare in the CF population [83,84]. Given that life expectancy in CF has improved dramatically, screening for microvascular complication is recommended after 5 years of diabetes duration [85–87].

24.5 NORMAL PHYSIOLOGY OF INSULIN ACTION ON GROWTH AND ANABOLISM

Insulin is crucial to growth from embryogenesis throughout adult life [44]. The importance of insulin function on growth in humans is demonstrated by the extreme growth retardation and early mortality seen in the leprechaunism and Rabson-Mendenhall syndromes, which are due to mutations in the insulin receptor gene, causing severe insulin resistance of prenatal onset. At high circulating concentrations, insulin may also bind to the insulin-like growth factor-1 (IGF-1) receptor, which is the main pathway by which growth hormone produces its anabolic effects [44]. IGF-1 and insulin bind to their cell surface receptors on an extracellular domain, which then regulates the activity of an intracellular tyrosine kinase, generating an array of cell- and tissue-specific responses, through phosphorylation of some cytoplasmic proteins and dephosphorylation of others [44]. The actions of insulin can be divided into rapid, intermediate and delayed. Within seconds, insulin rapidly increases transport of glucose, amino acids and potassium into insulin sensitive cells. Over minutes, insulin increases protein synthesis, inhibits protein degradation, and activates glycolytic and glycogen synthase enzymes, whilst inhibiting phosphorylase and gluconeogenic enzymes. Insulin also promotes sustained changes (over hours) by increasing mRNA production for lipogenic and other enzymes.

24.6 IMPLICATIONS OF INSULIN DEFICIENCY ON GROWTH IN CF

The crucial role that insulin plays in growth has several important implications for the management of patients with CF.

24.6.1 Changes in Protein Metabolism Due to Insulin Deficiency Result in Catabolism

Insulin deficiency leads to accelerated protein catabolism as well as diminished protein synthesis. This manifests as poor weight gain and wasting (see Figure 24.3). Protein depletion may also be associated with reduced ability to overcome infections, as seen in patients with protein energy malnutrition [88]. Partial pancreatectomy in animals (resulting in insulin deficiency) is associated with significantly reduced muscle protein synthesis following exercise [89]. This raises the possibility that insulin deficiency may result in impaired muscle gain in CF subjects following exercise. Respiratory muscle strength is, of course, an important factor in effort-dependent respiratory function.

Systemic inflammation results in high circulating TNF-alpha levels. As noted in Section 24.3.1, TNF-alpha can impair insulin action. This effect is thought to be mediated by decreasing the tyrosine kinase activity of the insulin receptor [49]. In this way, systemic illness may contribute to catabolism due to insulin deficiency.

Further adding to these catabolic effects is the loss of glucose through glycosuria. For every gram of glucose excreted in the urine, 4.1 kcal is lost from the body [90], and glycosuria occurs when the renal threshold for glucose reabsorption (approximately 7 mmol/L, 126 mg/dL) is exceeded.

Taken together, these findings raise considerable concerns for CF patients, in whom chronically elevated rates of protein catabolism have been noted [91,92]. Encouragingly, insulin treatment has been shown to successfully reverse protein catabolism in CFRD [93].

24.6.2 Height, Weight, and Body Mass May Be Impaired by Insulin Deficiency

The diagnosis of CFRD is often preceded by declining weight and body mass index (BMI) [32,76,77,94]. Poor weight gain is also associated with declining lung function [95,96]. Analysis of the European Cystic Fibrosis Society Patient Registry demonstrated that patients with a lower BMI experience a six-fold increased odds ratio (95% CI 5.0–7.3) of having severe lung disease (FEV_1 <40% predicted) compared to patients with normal BMI. Patients with CFRD were found to have a 1.8-fold increased odds ratio compared to patients not affected by CFRD. Of note too was the fact that almost 10% of European patients had a BMI ≤2 standard deviation scores (SDS), and this was the strongest predictor of poor FEV_1 [72]. Wasting (or poor weight gain) is also associated with early mortality in CF, and this is independent of lung function [97].

Additionally, in children and adolescents with CFRD, poor linear growth is also common and correlates with the degree of insulin deficiency. Thus poor linear growth in children with CF is an indication to exclude or diagnose CFRD. Cheung and colleagues found that in the 2 years leading to a diagnosis of CFRD (at a mean age of 13.1 years), the mean height velocity was significantly slower than in controls: 4.9 vs. 6.0 cm/year. For the 2 years following diagnosis, height velocity still remained significantly lower at 3.4 vs. 4.4 cm/year [98]. Another important point for consideration of growth in childhood CF is that weight SDS may be a more reliable indicator of appropriate weight gain than BMI, which is calculated by weight in kilograms divided by the square of height in meters [99], and may, therefore, be confounded by short stature. Thus, any child with CF and unexplained decline in weight SDS should be assessed for CFRD.

24.7 SUMMARY OF IMPORTANCE OF ADEQUATE INSULIN SECRETION IN CYSTIC FIBROSIS

Poor weight gain, catabolism, lung infections, and accelerated lung dysfunction have all been associated with insulin deficiency in CF, raising important questions regarding the acceleration of mortality in CF by CFRD and the prevalence of CFRD. In this section, we will explore these aspects.

24.7.1 CFRD is Associated with Early Mortality

The association between CFRD and early mortality has been recognized for many years. The Cystic Fibrosis Foundation Patient Registry Report from 1997, involving

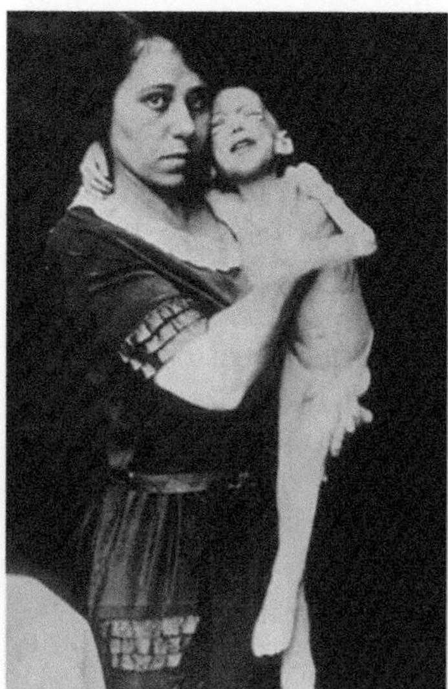

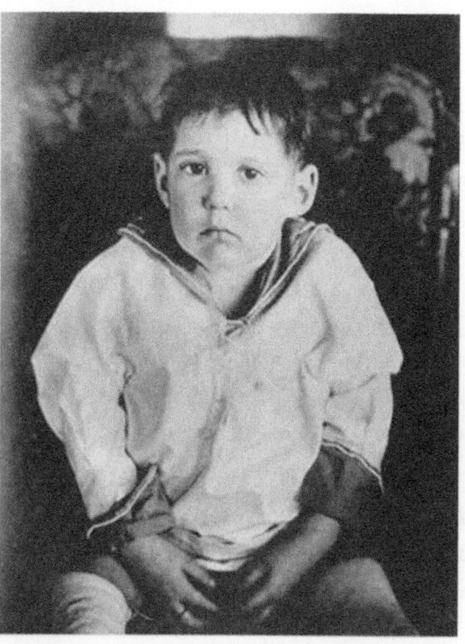

FIGURE 24.3 Highlighting the anabolic effects of insulin, one of the first patients to survive the onset of type 1 diabetes showed dramatic improvement in weight after insulin treatment. *Reproduced with permission from Eli Lilly and Company archives.*

over 21,000 patients, found a six-fold increase in early mortality among subjects with CFRD compared to non-diabetic CF subjects [100]. Finkelstein and colleagues found survival to 30 years of only 25% in patients with CFRD, versus 60% in those without CFRD [101]. CFRD is more common among those with pancreatic exocrine insufficiency, hence genotype severity or inadequate adherence to pancreatic enzyme replacement therapy may have also played a role. Milla and colleagues also demonstrated reduced survival in patients with CFRD, particularly in females over the period 1987–2002. Recently, the same group reported improved survival figures (over the period from 1992 to 2008), with mortality now 3.2–3.8 deaths per 100 patient years for females and males, respectively [102]. These improvements may relate to earlier diagnosis of CFRD and more aggressive treatment with insulin.

24.7.2 Incidence and Prevalence of CFRD

The most common comorbidity in CF is CFRD [100], affecting approximately half of all patients by 30 years of age [103,104]. The prevalence of CFRD increases with age (as expected with progressive beta cell loss), but is nevertheless frequent in children and adolescents. Among patients attending the University of Minnesota, prevalence was 9% among 5–9 year olds and 26% among 10–19 year olds [104]. Other factors associated with increased risk of CFRD include factors associated with more severe genotype, such as pancreatic exocrine insufficiency [105,106] and liver disease [105]. Female sex has also been reported to increase risk [107,108], as has older age [105]. However, it is important to recognize that CFRD can also occur in the very young and has been reported in young infants [109] and even in a neonate [110].

24.8 CURRENT MANAGEMENT PRINCIPLES OF INSULIN THERAPY IN CYSTIC FIBROSIS

We have seen that CFRD is a common complication of CF, and one which is associated with early mortality. In this section, we will review the management principles relating to screening, diagnosis, and management of insulin deficiency and CFRD.

24.8.1 Screening and Diagnosis of CFRD

The classical symptoms of diabetes associated with hyperglycemia (polyuria and polydipsia) are late events in CF. Hence it is important to screen for the development of CFRD and to consider CFRD in patients who present with unexplained deterioration in weight gain or lung function.

24.8.2 Clinical Symptoms

Poor weight gain may be a clue to the diagnosis of CFRD [25,32], as may poor linear growth velocity [98,111] and unexplained lung function decline [32,78,98,112,113]. Only 33% of patients with CFRD have classical diabetes symptoms of polyuria and polydipsia at diagnosis [83]. Hence CFRD will be missed in the majority of patients unless annual screening is performed.

24.8.3 Fasting Blood Glucose and Urinalysis

Although convenient and relatively easy to test for, fasting hyperglycemia (≥7 mmol/L, 126 mg/dL) and persistent glycosuria are late events in CFRD [32,83,102,104,114], making these measures insensitive diagnostic tools [32,115]. Positive results are highly suggestive of diabetes, but should not be used alone or many cases of CFRD will be missed. A category of impaired fasting glucose (defined as fasting glucose level between 6.1 and 6.9 mmol/L, 110–124 mg/dL) has been associated with an increased risk of progression to type 2 diabetes in the general population. Subsequently, a study of 1093 CF patients found that impaired fasting glucose levels in this range also increased the risk of progression to CFRD on average follow-up 3.6 years later. Recently, the American Diabetes Association lowered the threshold for impaired fasting glucose from 6.1 mmol/L to 5.6 mmol/L (100 mg/dL), but Frohnert and colleagues found no increased risk of developing CFRD among patients with this lower threshold [116]. Thus, the clinical significance in CF of impaired fasting glucose levels is incompletely understood.

24.8.4 Glycosylated Hemoglobin

The glycosylated hemoglobin (HbA_{1c}) test is used widely in non-CF patients as it reflects a patient's blood glucose level over the last 2–3 months, with particular weight on the last month. It works by determining the percentage of the 1c component of the hemoglobin A molecule that is glycosylated. It can be performed on a simple finger prick blood sample, and results can be available within minutes using point-of-care machine testing. The use of HbA_{1c} in CF is problematic as it is often spuriously low in CF subjects, a phenomenon that is thought to relate to high red blood cell turnover in CF patients [83]. Thus an elevated HbA_{1c} is highly suggestive of CFRD, but is a late event and HbA_{1c} screening of CF populations will miss about half of all CFRD cases [117]. Furthermore, HbA_{1c} and the measurement of glycosylated fructosamine do not correlate with mean blood glucose readings in CF, and both may therefore be unreliable measures of glycemic control in CF [118].

24.8.5 Oral Glucose Tolerance Test

In 1995, Lanng and colleagues published the results of a 5 year prospective study, and recommended annual OGTTs be performed on all patients with CF aged 10 years or older. In that study, hyperglycemic symptoms, fasting hyperglycemia, and increased levels of HbA_{1C} were unreliable in identifying CFRD [83]. Since then, this recommendation has been endorsed by multiple professional bodies, including the American Diabetes Association and Pediatric Endocrine Society [52], the International Society for Pediatric and Adolescent Diabetes [119], in addition to European consensus guidelines [120]. In Minnesota, Moran and colleagues have performed annual OGTTs at an earlier age (≥6 years) since the 1990s. However, screening practices continue to vary considerably between centers [121,122].

Potential barriers to performing annual OGTTs include their relatively labor-intensive and time-consuming nature, as well as the discomfort they may cause (through venipuncture, the need to fast, or distaste for the oral glucose solution) [122]. This was demonstrated by Rayas and colleagues who recently introduced a clinical care algorithm to ensure OGTTs were ordered appropriately [123]. Although there was an initial increase in OGTT screening rates in their center, the improvement was not sustained into the second year, prompting the authors to suspect issues with patient compliance.

The diagnosis of CFRD by OGTT requires either fasting hyperglycemia (a baseline glucose level of ≥7 mmol/L, 156 mg/dL) or a blood glucose level of >11.1 mmol/L or 200 mg/dL taken 2 h after the glucose load has been consumed (see Table 24.1). However, we recommend intervening samples every 30 min in CF, as hyperglycemia is common at 30, 60 and 90 min. This was well described by Dobson and colleagues who found that the 120-min sample failed to discriminate between CF patients and healthy controls, whereas the intervening samples at 30, 60, and 90 min detected a major glucose peak in the CF group [115] (see Figure 24.2). We found that the peak glucose level occurs earlier than the routinely measured 120-min sample, occurring at 30 min in 18%, at 60 min in 45%, at 90 min in 33%, and at 120 min in only 3% of patients [32].

After performing OGTT with 30 min sampling, Milla and colleagues followed the BMI and lung function in CF patients for 4 years. They found that those with impaired glucose tolerance (7.8–11.0 mmol/L, 140–190 mg/dL at 120 min, see Table 24.1) could not be distinguished from those with CFRD by the rate of subsequent lung function

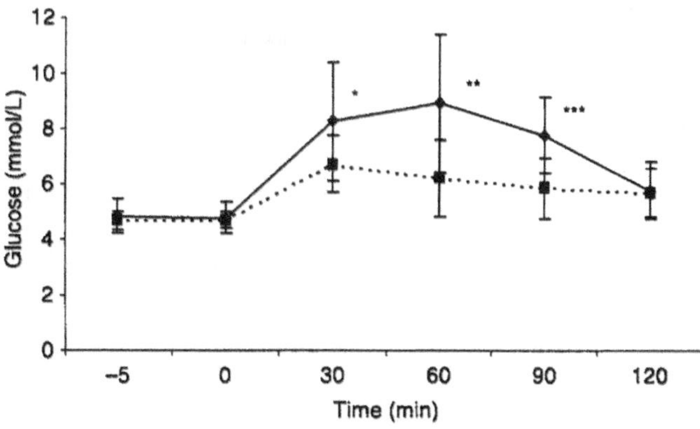

FIGURE 24.2 Plasma glucose values during an oral glucose tolerance test, comparing cystic fibrosis subjects (diamonds) and healthy controls (boxes). * P = 0.01, ** P = 0.0003, *** P = 0.0001. Fasting (−5 and 0 minute) and 120-minute values were similar in the two groups. However, intervening samples collected at 30 minute intervals revealed a significant difference. *Reproduced with permission from Dobson L et al. Diabetic Medicine. 2004;(21): 691–696.*

decline [79]. In contrast, they found that lower baseline insulin secretion (analyzed using area under the curve on OGTT) predicted poorer subsequent lung function. Their findings suggest that the 120-min blood glucose may not be an ideal measure of clinically significant insulin deficiency in CF.

It is important to remember that the OGTT diagnostic cut-offs defining CFRD were adopted from type 2 diabetes, rather than being designed from studies of CF patients. In fact, the cut-offs were created to forecast microvascular complications of diabetes in patients with type 2 diabetes [124], and were not designed to detect decline in weight or lung function. Furthermore, the usual underlying problems in type 2 diabetes (elevated insulin levels and obesity) are very different from those found in CF, where diabetes is a result of insulin deficiency, and preservation of weight gain is of major importance. This has prompted questions regarding the

TABLE 24.1 Categories for the Oral Glucose Tolerance Test (OGTT)

Oral Glucose Tolerance Category	Fasting Blood Glucose Level		120-min Blood Glucose Level	
	mmol/L	mg/dL	mmol/L	mg/dL
Normal glucose tolerance	<5.6	<100	<7.8	<140
Impaired glucose tolerance	<5.6	<100	7.8–11	140–199
CFRD without fasting hyperglycemia	<7.0	<126	≥11.1	≥200
CFRD with fasting hyperglycemia	≥7.0	≥126	≥11.1	≥200

current diagnostic criteria for CFRD and the importance of other glycemic abnormalities [115,125,126].

24.8.6 Catabolic Decline Precedes the Diagnosis of CFRD Using Current Diagnostic Criteria

It has long been noted that weight and lung function decline is present at the time of diagnosis of CFRD and that this decline may precede the diagnosis of CFRD by months to years [28,30,94,127]. Thus, there has been a search for other methods of diagnosing clinically significant glucose abnormalities in CF.

In our pediatric cohort, we compared the results of 30-min sampled OGTT to weight and lung function in the preceding year [32]. We found that the peak blood glucose level (rather than the 0 or 120 min level that are currently used to define diabetes and impaired glucose tolerance in CF) was associated with decline in weight and lung function in the year prior. We used receiver operating characteristic (ROC) analysis to determine the cut-point for peak glucose that was the most sensitive and specific at detecting poor weight gain in the year prior. We found it to be 8.2 mmol/L (148 mg/dL), and it is interesting to note, as mentioned in Section 24.4 of this chapter, that this is very similar to the blood glucose threshold (≥8 mmol/L, 144 mg/dL) above which airway glucose levels are raised in CF patients [59]. Furthermore, blood glucose levels in healthy individuals without CF or diabetes rarely exceed 8.3 mmol/L (150 mg/dL) [128].

24.8.7 An Emerging Diagnostic Role for Continuous Glucose Monitoring

Another method recently used to examine glycemia in CF patients is continuous glucose monitoring (CGM). Originally developed for use in patients with type 1 diabetes, CGM has proved to be useful in CF patients [59,115,129–132]. The CGM device records interstitial fluid glucose levels every 5 min via a subcutaneous probe, thereby providing a detailed assessment of glycemia for several days. Using CGM, patients can consume their usual diet and perform usual activities, whereas OGTT only reflects the glucose profile following exposure to a standard glucose load after overnight fasting. This is important because CGM can detect glucose excursions at any time of day or night, and the carbohydrate load in the CF diet often exceeds the maximal glucose load given during OGTT (1.75 g/kg to a maximum of 75 g). It is therefore possible that in CF patients, glycemia may worsen as the day progresses, following meals with larger carbohydrate portions. Modern CGM devices are wireless, smaller and more discrete than previous models, but still require a subcutaneous sensor probe (which may cause some discomfort during insertion).

Schiaffini and colleagues found that the presence of glucose excursions above 11.1 mmol/L (200 mg/dL) during CGM (or during 30-min sampled OGTT) in CF patients at baseline predicted the development of other OGTT abnormalities, including CFRD, 2.5 years later [133]. Leclercq and colleagues studied CF patients with normal OGTT screening results, and found that those with CGM peaks greater than or equal to 11 mmol/L (198 mg/dL) demonstrated significant lung function impairment and higher rates of colonization with *P. aeruginosa* [134]

In our pediatric cohort, we found that a CGM trace that exceeded 7.8 mmol/L for ≥4.5% of the time reliably detected poor weight gain in the preceding year [32].

There is a pressing need, therefore, to determine whether or not weight and lung function decline in patients with these glycemic markers of early insulin deficiency can be prevented by insulin treatment.

24.8.8 Current Glycemic Management Therapies

Patients with CF should continue with their high energy diet, which has been a cornerstone of CF therapy for many years [135] and which has been shown to be very beneficial [136,137]. Reduction in carbohydrate load or avoidance of high glycemic index foods is not recommended in patients with CFRD as this runs the risk of limiting energy intake and impairing nutritional status.

24.8.9 Oral Agents

In CFRD, the primary problem is insulin deficiency, and agents that stimulate insulin secretion from the remaining beta cells (such as oral sulphonylureas) are not recommended therapy [50,117,138]. Sulphonylureas are known to induce beta cell death by apoptosis in cultured human islets [139]. By contrast, it is known that the early introduction of intensive insulin therapy in newly diagnosed patients with type 1 diabetes preserves insulin secretion, probably by inducing beta cell rest and prolonging the life of the remaining beta cell mass [140]. Another theoretical concern with sulphonylurea therapy is that the sulphonylurea receptor resembles the CFTR [141] and thus it is possible that sulphonylureas may bind to and inhibit CFTR function. The oral agent repaglinide also stimulates insulin release from the beta cell. Repaglinide was used in the cystic fibrosis related diabetes therapy (CFRDT) trial by Moran and colleagues. They found that in contrast to insulin, repaglinide did not result in sustained improvement in BMI in patients with CFRD [16] and found that it was less effective at correcting postprandial glucose than insulin treatment [86,142].

Insulin sensitizers include metformin and the thiazolidinediones, and these agents are also not recommended for use in CFRD. Insulin sensitizers do not address the primary abnormality, namely insulin deficiency. Metformin may cause weight loss and gastrointestinal side-effects. Thiazolidinediones may be associated with osteoporosis, another complication seen in CF [143].

It is known that orally administered glucose results in greater insulin secretion than intravenously administered glucose [144]. The recognition that this is due to the effects of incretin molecules released during meal consumption, which augment insulin secretion, has led to a new class of therapy in diabetes–incretin modifiers. Oral and injectable forms of incretin modifiers are now used in the management of type 2 diabetes mellitus, but have not been studied in CFRD where insulin deficiency and pancreatic destruction are hallmarks. Given these differences it is possible that they may not be efficacious in CF. Furthermore, incretin modifiers are associated with significant weight loss in adult patients with type 2 diabetes [145], and may be associated with nausea and pancreatitis [146], raising significant concerns regarding their suitability for use in CF.

24.8.10 Insulin Therapy in CF Patients

Insulin therapy in CFRD has been shown in multiple studies to improve BMI and lung function [69,77,94]. Pharmacological insulin therapy relies on the injection of human insulin or insulin analogs into the subcutaneous tissues. Injected insulin preparations are designed to be absorbed into the circulation at varying speeds, allowing them to fulfill different therapeutic objectives, For example, rapid-acting insulin analogs such as insulin lispro and insulin aspart (which have an onset of action of 10–15min and peak action of approximately 1–2h) can be given with meals to prevent postprandial hyperglycemia. Whereas, long-acting insulin analogs have minimal peak and are ideal basal insulin replacement. These include glargine (2–3h onset, 24h action) and insulin detemir (2–3h onset and 12–24h duration). Traditionally, insulin therapy in CFRD has consisted of multiple daily injections incorporating a basal insulin as well as premeal rapid-acting insulins.

Modern basal insulin analogs such as insulin detemir have much less variability in absorption than other insulin agents [147,148]. Nevertheless, hypoglycemia is a potential side-effect of all insulin therapies. This may be particularly true of multiple daily injection regimens, which involve a potential risk of hypoglycemia following each injection. Patients with type 1 diabetes undergoing multiple daily injections have a two-to-threefold increase in severe hypoglycemia compared with those on less intensive insulin therapy [149]. Multiple daily injections therefore mandate the need for frequent blood glucose testing, which may cause discomfort and add to treatment burden. In particular, children and adolescents may find it difficult to perform blood glucose testing and inject insulin at lunchtime at school.

Given these considerations, it is encouraging to note that once-daily basal insulin therapy [150] may be of benefit to CF patients, particularly if commenced at an early stage of insulin deficiency. We studied once-daily insulin detemir in 12 patients with early insulin deficiency (defined as peak blood glucose on OGTT ≥8.2mmol/L, without CFRD) and in six patients with established CFRD (aged 7.2–18.1years). They received once-daily insulin detemir (for a median of 0.8years), and we found that weight SDS and lung function decline prior to therapy significantly improved on once-daily insulin alone.

Other uncontrolled studies have also examined the effect of insulin treatment at an earlier stage of insulin deficiency than CFRD. Mozzillo and colleagues gave insulin glargine alone for 1year to 22 children and adolescents (9 with CFRD, 9 with impaired glucose tolerance, and 4 with a peak blood glucose on OGTT of ≥7.8mmol/L and 120-min glucose of <7.8mmol/L) [71]. They found a clinically significant improvement in %FEV_1 of 8.8 and a 42% reduction in the number of lung infection episodes, but BMI-SDS did not improve (except in a subgroup with worse BMI-SDS at baseline). Another study by Bizzarri and colleagues used insulin glargine in six nondiabetic CF children and adults with impaired glucose tolerance [151], finding increased BMI-SDS and a small but clinically significant improvement in FEV_1. There were no episodes of hypoglycemia following once-daily insulin treatment in either of these uncontrolled studies. The CF-IDEA trial (Cystic Fibrosis-Insulin Deficiency, Early Action; clinicaltrials.gov number: NCT01100892 currently under way in Australia) is a randomized controlled trial to determine whether current clinical practice should be altered toward the earlier commencement of insulin treatment in CF, before progression to CFRD.

24.9 FUTURE DIRECTIONS OF INSULIN THERAPIES

In addition to the exciting possibilities of earlier diagnosis and introduction of insulin therapy, other therapeutic advances in the field of endocrinology may also benefit patients with CF in the future. Ultra-long-acting insulin preparations are being developed and are in early trials in non-CF populations. Insulin degludec is an ultra-long-acting insulin that has been shown to be efficacious when given as infrequently as three times per week [152]. Another potential breakthrough that may improve insulin therapy in the future is the closed-loop insulin pump. This will potentially be a major improvement on currently available insulin pumps, which require frequent blood testing and vigilant human oversight for insulin to be administered. This type of standard insulin therapy has been shown to improve glycemia, increase weight, and reduce catabolism in some small CF trials. Experimental "closed-loop" insulin pumps have been developed, but are not yet available for general patient use [153]. In the future this system may make automated insulin adjustments according to results of CGM, thereby minimizing the treatment burden of this therapy for patients requiring insulin including those with CF.

We have seen that adequate insulin secretion is vital for maintaining normoglycemia, and is also critical for growth and anabolism in humans. This chapter has detailed the important implications of this physiology for CF patients, in whom progressive insulin deficiency may result in postprandial hyperglycemia, leading to lung infections and worsening lung function. Insulin deficiency also results in catabolism, leading to further infections, systemic inflammation, worsening pulmonary status, and more insulin deficiency in a deteriorating spiral. Currently available technologies enable us to diagnose insulin deficiency and glucose derangements with great sophistication and much earlier than in the past. Modern insulin therapies also provide greater precision in obtaining treatment outcomes. Future developments may see earlier introduction of insulin replacement, with minimal associated treatment burden, preventing the damage caused by insulin deficiency in CF through hyperglycemia and catabolism.

References

[1] Stefan Y, Orci L, Malaisse-Lagae F, Perrelet A, Patel Y, Unger RH. Quantitation of endocrine cell content in the pancreas of nondiabetic and diabetic humans. Diabetes 1982;31:694–700.

[2] Williams JA, Goldfine ID. The insulin-pancreatic acinar axis. Diabetes 1985;34:980–6.

[3] Meier JJ, Butler PC. Insulin secretion. In: De Groot LJ, Jameson JL, editors. Endocrinology, vol. 1. Philadelphia (PA, USA): Elsevier; 2006. p. 961–73. 3 vols.

[4] Lohr M, Goertchen P, Nizze H, Gould NS, Gould VE, Oberholzer M, et al. Cystic fibrosis associated islet changes may provide a basis for diabetes. An immunocytochemical and morphometrical study. Virchows Arch a Pathol Anat Histopathol 1989;414:179–85.

[5] Soejima K, Landing BH. Pancreatic islets in older patients with cystic fibrosis with and without diabetes mellitus: morphometric and immunocytologic studies. Pediatr Pathol 1986;6:25–46.

[6] Iannucci A, Mukai K, Johnson D, Burke B. Endocrine pancreas in cystic fibrosis: an immunohistochemical study. Hum Pathol 1984;15:278–84.

[7] Couce M, O'Brien TD, Moran A, Roche PC, Butler PC. Diabetes mellitus in cystic fibrosis is characterized by islet amyloidosis. J Clin Endocrinol Metab 1996;81:1267–72.

[8] Ritzel RA, Veldhuis JD, Butler PC. Glucose stimulates pulsatile insulin secretion from human pancreatic islets by increasing secretory burst mass: dose-response relationships. J Clin Endocrinol Metab 2003;88:742–7.

[9] Gabbay KH, Korff J, Schneeberger EE. Vesicular binesis: glucose effect on insulin secretory vesicles. Science 1975;187:177–9.

[10] Hellman B, Sehlin J, Taljedal IB. Calcium and secretion: distinction between two pools of glucose-sensitive calcium in pancreatic islets. Science 1976;194:1421–3.

[11] Gold G, Gishizky ML, Grodsky GM. Evidence that glucose "marks" beta cells resulting in preferential release of newly synthesized insulin. Science 1982;218:56–8.

[12] Daniel S, Noda M, Straub SG, Sharp GW. Identification of the docked granule pool responsible for the first phase of glucose-stimulated insulin secretion. Diabetes 1999;48:1686–90.

[13] Gilliam LK, Lernmark A. Type 1 (Insulin-dependent) diabetes mellitus: etiology, pathogenesis, and natural history. In: De Groot LJ, Jameson JL, editors. Endocrinology 5 edit, vol. 1. Philadelphia (PA, USA): Elsevier; 2006. p. 1073–91. 3 vols.

[14] The discovery of insulin. Diabetes and insulin, vol. 2010. Nobelprize.org. 2009.

[15] Swartz LM, Laffel LM. A teenage girl with cystic fibrosis-related diabetes, diabetic ketoacidosis, and cerebral edema. Pediatr Diabetes 2008;9:426–30.

[16] Moran A, Pekow P, Grover P, Zorn M, Slovis B, Pilewski J, et al. Insulin therapy to improve BMI in cystic fibrosis related diabetes without fasting hyperglycemia: results of the CFRDT trial. Diabetes Care 2009;32:1783–8.

[17] Report of the expert committee on the diagnosis and classification of diabetes mellitus. Diabetes Care 2003;26(Suppl. 1):S5–20.

[18] Swobodnik W, Wolf A, Wechsler JG, Kleihauer E, Ditschuneit H. Ultrasound characteristics of the pancreas in children with cystic fibrosis. J Clin Ultrasound 1985;13:469–74.

[19] Shawker TA, Parks SI, Linzer M, Jones B, Lester LA, Hubbard VS. Amplitude analysis of pancreatic B-scans: a clinical evaluation of cystic fibrosis. Ultrason Imaging 1980;2:55–66.

[20] Wilson-Sharp RC, Irving HC, Brown RC, Chalmers DM, Littlewood JM. Ultrasonography of the pancreas, liver, and biliary system in cystic fibrosis. Arch Dis Child 1984;59:923–6.

[21] Daneman A, Gaskin K, Martin DJ, Cutz E. Pancreatic changes in cystic fibrosis: CT and sonographic appearances. AJR Am J Roentgenol 1983;141:653–5.

[22] Abdul-Karim FW, Dahms BB, Velasco ME, Rodman HM. Islets of Langerhans in adolescents and adults with cystic fibrosis. A quantitative study. Arch Pathol Lab Med 1986;110:602–6.

[23] Lippe BM, Sperling MA, Dooley RR. Pancreatic alpha and beta cell functions in cystic fibrosis. J Pediatr 1977;90:751–5.

[24] McKenzie MD, Jamieson E, Jansen ES, Scott CL, Huang DC, Bouillet P, et al. Glucose induces pancreatic islet cell apoptosis that requires the BH3-only proteins Bim and Puma and multi-BH domain protein Bax. Diabetes 2010;59:644–52.

[25] Handwerger S, Roth J, Gorden P, Di Sant' Agnese P, Carpenter DF, Peter G. Glucose intolerance in cystic fibrosis. N Engl J Med 1969;281:451–61.

[26] Milner AD. Blood glucose and serum insulin levels in children with cystic fibrosis. Arch Dis Child 1969;44:351–5.

[27] Mohan K, Miller H, Dyce P, Grainger R, Hughes R, Vora J, et al. Mechanisms of glucose intolerance in cystic fibrosis. Diabet Med 2009;26:582–8.

[28] Bismuth E, Laborde K, Taupin P, Velho G, Ribault V, Jennane F, et al. Glucose tolerance and insulin secretion, morbidity, and death in patients with cystic fibrosis. J Pediatr 2008;152:540–5, 545. e1.

[29] Lanng S, Thorsteinsson B, Roder ME, Orskov C, Holst JJ, Nerup J, et al. Pancreas and gut hormone responses to oral glucose and intravenous glucagon in cystic fibrosis patients with normal, impaired, and diabetic glucose tolerance. Acta Endocrinol (Copenh) 1993;128:207–14.

[30] Tofe S, Moreno JC, Maiz L, Alonso M, Escobar H, Barrio R. Insulin-secretion abnormalities and clinical deterioration related to impaired glucose tolerance in cystic fibrosis. Eur J Endocrinol 2005;152:241–7.

[31] Elder DA, Wooldridge JL, Dolan LM, D'Alessio DA. Glucose tolerance, insulin secretion, and insulin sensitivity in children and adolescents with cystic fibrosis and no prior history of diabetes. J Pediatr 2007;151:653–8.

[32] Hameed S, Morton JR, Jaffe A, Field PI, Belessis Y, Yoong T, et al. Early glucose abnormalities in cystic fibrosis are preceded by poor weight gain. Diabetes Care 2010;33:221–6.

[33] Lombardo F, De Luca F, Rosano M, Sferlazzas C, Lucanto C, Arrigo T, et al. Natural history of glucose tolerance, beta-cell function and peripheral insulin sensitivity in cystic fibrosis patients with fasting euglycemia. Eur J Endocrinol 2003;149:53–9.

[34] Boom A, Lybaert P, Pollet JF, Jacobs P, Jijakli H, Golstein PE, et al. Expression and localization of cystic fibrosis transmembrane conductance regulator in the rat endocrine pancreas. Endocrine 2007;32:197–205.

[35] Ali BR. Is cystic fibrosis-related diabetes an apoptotic consequence of ER stress in pancreatic cells? Med Hypotheses 2009;72:55–7.

[36] Stalvey MS, Muller C, Schatz DA, Wasserfall CH, Campbell-Thompson ML, Theriaque DW, et al. Cystic fibrosis transmembrane conductance regulator deficiency exacerbates islet cell dysfunction after beta-cell injury. Diabetes 2006;55:1939–45.

[37] Blackman SM, Hsu S, Vanscoy LL, Collaco JM, Ritter SE, Naughton K, et al. Genetic modifiers play a substantial role in diabetes complicating cystic fibrosis. J Clin Endocrinol Metab 2009;94:1302–9.

[38] Blackman SM, Hsu S, Ritter SE, Naughton KM, Wright FA, Drumm ML, et al. A susceptibility gene for type 2 diabetes confers substantial risk for diabetes complicating cystic fibrosis. Diabetologia 2009;52:1858–65.

[39] Bellin MD, Laguna T, Leschyshyn J, Regelmann W, Dunitz J, Billings J, et al. Insulin secretion improves in cystic fibrosis following ivacaftor correction of CFTR: a small pilot study. Pediatr Diabetes 2013;14:417–21.

[40] Eckford PD, Li C, Ramjeesingh M, Bear CE. Cystic fibrosis transmembrane conductance regulator (CFTR) potentiator VX-770 (ivacaftor) opens the defective channel gate of mutant CFTR in a phosphorylation-dependent but ATP-independent manner. J Biol Chem 2012;287:36639–49.

[41] Minicucci L, Cotellessa M, Pittaluga L, Minuto N, d'Annunzio G, Avanzini MA, Lorini R. Beta-cell autoantibodies and diabetes mellitus family history in cystic fibrosis. J Pediatr Endocrinol Metab 2005;18:755–60.

[42] Lanng S, Thorsteinsson B, Pociot F, Marshall MO, Madsen HO, Schwartz M, et al. Diabetes mellitus in cystic fibrosis: genetic and immunological markers. Acta Paediatr 1993;82:150–4.

[43] Klip A, Tsakiridis T, Marette A, Ortiz PA. Regulation of expression of glucose transporters by glucose: a review of studies in vivo and in cell cultures. FASEB J 1994;8:43–53.

[44] White MF. The molecular basis of insulin action. In: DeGroot LJ, Jameson JL, editors. Endocrinology, vol. 1. Philadelphia (Pa, USA): Elsevier; 2006. p. 975–1000. 3 vols.

[45] Amiel SA, Sherwin RS, Simonson DC, Lauritano AA, Tamborlane WV. Impaired insulin action in puberty. A contributing factor to poor glycemic control in adolescents with diabetes. N Engl J Med 1986;315:215–9.

[46] Hardin DS. Growth hormone in cystic fibrosis. J Pediatr Endocrinol Metab 2008;21:727–8.

[47] Navas de Solis MS, Merino Torres JF, Mascarell Martinez I, Pinon Selles F. Lung transplantation and the development of diabetes mellitus in adult patients with cystic fibrosis. Arch Bronconeumol 2007;43:86–91.

[48] Bell SC, Bowerman AM, Nixon LE, Macdonald IA, Elborn JS, Shale DJ. Metabolic and inflammatory responses to pulmonary exacerbation in adults with cystic fibrosis. Eur J Clin Invest 2000;30:553–9.

[49] Hotamisligil GS, Peraldi P, Budavari A, Ellis R, White MF, Spiegelman BM. IRS-1-mediated inhibition of insulin receptor tyrosine kinase activity in TNF-alpha- and obesity-induced insulin resistance. Science 1996;271:665–8.

[50] Moran A, Hardin D, Rodman D, Allen H, Beall R, Borowitz D, et al. Diagnosis, screening and management of cystic fibrosis related diabetes mellitus. A consensus conference report. Diabetes Res Clin Pract 1999;45:61–73.

[51] Moran A. Endocrine complications of cystic fibrosis. Adolesc Med 2002;13:145.

[52] Moran A, Brunzell C, Cohen RC, Katz M, Marshall BC, Onady G, et al. Clinical care guidelines for cystic fibrosis-related diabetes: a position statement of the american diabetes association and a clinical practice guideline of the cystic fibrosis foundation, endorsed by the pediatric endocrine society. Diabetes Care 2010;33:2697–708.

[53] Widger J, Oliver MR, O'Connell M, Cameron FJ, Ranganathan S, Robinson PJ. Glucose tolerance during pulmonary exacerbations in children with cystic fibrosis. PLoS One 2012;7:e44844.

[54] Matsuda M, DeFronzo RA. Insulin sensitivity indices obtained from oral glucose tolerance testing: comparison with the euglycemic insulin clamp. Diabetes Care 1999;22:1462–70.

[55] Ahmad T, Nelson R, Taylor R. Insulin sensitivity and metabolic clearance rate of insulin in cystic fibrosis. Metabolism 1994;43:163–7.

[56] Hardin DS, Leblanc A, Marshall G, Seilheimer DK. Mechanisms of insulin resistance in cystic fibrosis. Am J Physiol Endocrinol Metab 2001;281:E1022–8.

[57] Yung B, Noormohamed FH, Kemp M, Hooper J, Lant AF, Hodson ME. Cystic fibrosis-related diabetes: the role of peripheral insulin resistance and beta-cell dysfunction. Diabet Med 2002;19:221–6.

[58] ClinicalTrials.gov. Cystic fibrosis (CF) exacerbation and insulin treatment; 2010. NCT01149005.

[59] Brennan AL, Gyi KM, Wood DM, Johnson J, Holliman R, Baines DL, et al. Airway glucose concentrations and effect on growth of respiratory pathogens in cystic fibrosis. J Cyst Fibros 2007;6:101–9.

[60] Wood DM, Brennan AL, Philips BJ, Baker EH. Effect of hyperglycaemia on glucose concentration of human nasal secretions. Clin Sci (Lond) 2004;106:527–33.

[61] Garnett JP, Gray MA, Tarran R, Brodlie M, Ward C, Baker EH, et al. Elevated paracellular glucose flux across cystic fibrosis airway epithelial monolayers is an important factor for *Pseudomonas aeruginosa* growth. PLoS One 2013;8:e76283.

[62] Garnett JP, Baker EH, Naik S, Lindsay JA, Knight GM, Gill S, et al. Metformin reduces airway glucose permeability and hyperglycaemia-induced *Staphylococcus aureus* load independently of effects on blood glucose. Thorax 2013;68:835–45.

[63] Chotirmall SH, Smith SG, Gunaratnam C, Cosgrove S, Dimitrov BD, O'Neill SJ, et al. Effect of estrogen on pseudomonas mucoidy and exacerbations in cystic fibrosis. N Engl J Med 2012;366:1978–86.

[64] Valdes CT, Elkind-Hirsch KE. Intravenous glucose tolerance test-derived insulin sensitivity changes during the menstrual cycle. J Clin Endocrinol Metab 1991;72:642–6.

[65] Rosenfeld M, Davis R, FitzSimmons S, Pepe M, Ramsey B. Gender gap in cystic fibrosis mortality. Am J Epidemiol 1997;145:794–803.

[66] Levy H, Kalish LA, Cannon CL, Garcia KC, Gerard C, Goldmann D, et al. Predictors of mucoid *Pseudomonas* colonization in cystic fibrosis patients. Pediatr Pulmonol 2008;43:463–71.

[67] Maselli JH, Sontag MK, Norris JM, MacKenzie T, Wagener JS, Accurso FJ. Risk factors for initial acquisition of *Pseudomonas aeruginosa* in children with cystic fibrosis identified by newborn screening. Pediatr Pulmonol 2003;35:257–62.

[68] Coriati A, Belson L, Ziai S, Haberer E, Gauthier MS, Mailhot G, et al. Impact of sex on insulin secretion in cystic fibrosis. J Clin Endocrinol Metab 2014. jc20132756.
[69] Lanng S, Thorsteinsson B, Nerup J, Koch C. Diabetes mellitus in cystic fibrosis: effect of insulin therapy on lung function and infections. Acta Paediatr 1994;83:849–53.
[70] Franzese A, Spagnuolo MI, Sepe A, Valerio G, Mozzillo E, Raia V. Can glargine reduce the number of lung infections in patients with cystic fibrosis-related diabetes? Diabetes Care 2005;28:2333.
[71] Mozzillo E, Franzese A, Valerio G, Sepe A, De Simone I, Mazzarella G, et al. One-year glargine treatment can improve the course of lung disease in children and adolescents with cystic fibrosis and early glucose derangements. Pediatr Diabetes 2009;10:162–7.
[72] Kerem E, Viviani L, Zolin A, MacNeill S, Hatziagorou E, Ellemunter H, et al. Factors associated with FEV1 decline in cystic fibrosis: analysis of the ECFS patient registry. Eur Respir J 2014;43:125–33.
[73] Courtney JM, Bradley J, McCaughan J, O'Connor TM, Shortt C, Bredin CP, et al. Predictors of mortality in adults with cystic fibrosis. Pediatr Pulmonol 2007;42:525–32.
[74] van den Borst B, Gosker HR, Zeegers MP, Schols AM. Pulmonary function in diabetes: a metaanalysis. Chest 2010;138:393–406.
[75] Vracko R, Thorning D, Huang TW. Basal lamina of alveolar epithelium and capillaries: quantitative changes with aging and in diabetes mellitus. Am Rev Respir Dis 1979;120:973–83.
[76] Lanng S, Thorsteinsson B, Nerup J, Koch C. Influence of the development of diabetes mellitus on clinical status in patients with cystic fibrosis. Eur J Pediatr 1992;151:684–7.
[77] Rolon MA, Benali K, Munck A, Navarro J, Clement A, Tubiana-Rufi N, et al. Cystic fibrosis-related diabetes mellitus: clinical impact of prediabetes and effects of insulin therapy. Acta Paediatr 2001;90:860–7.
[78] Koch C, Rainisio M, Madessani U, Harms HK, Hodson ME, Mastella G, et al. Presence of cystic fibrosis-related diabetes mellitus is tightly linked to poor lung function in patients with cystic fibrosis: data from the European epidemiologic registry of cystic fibrosis. Pediatr Pulmonol 2001;32:343–50.
[79] Milla CE, Warwick WJ, Moran A. Trends in pulmonary function in patients with cystic fibrosis correlate with the degree of glucose intolerance at baseline. Am J Respir Crit Care Med 2000;162:891–5.
[80] Rodman HM, Doershuk CF, Roland JM. The interaction of 2 diseases: diabetes mellitus and cystic fibrosis. Medicine (Baltimore) 1986;65:389–97.
[81] Sullivan MM, Denning CR. Diabetic microangiopathy in patients with cystic fibrosis. Pediatrics 1989;84:642–7.
[82] Andersen HU, Lanng S, Pressler T, Laugesen CS, Mathiesen ER. Cystic fibrosis-related diabetes: the presence of microvascular diabetes complications. Diabetes Care 2006;29:2660–3.
[83] Lanng S, Hansen A, Thorsteinsson B, Nerup J, Koch C. Glucose tolerance in patients with cystic fibrosis: five year prospective study. BMJ 1995;311:655–9.
[84] Dobson L, Sheldon CD, Hattersley AT. Understanding cystic-fibrosis-related diabetes: best thought of as insulin deficiency? J R Soc Med 2004;97(Suppl. 44):26–35.
[85] Schwarzenberg SJ, Thomas W, Olsen TW, Grover T, Walk D, Milla C, et al. Microvascular complications in cystic fibrosis-related diabetes. Diabetes Care 2007;30:1056–61.
[86] O'Riordan SM, Dattani MT, Hindmarsh PC. Cystic fibrosis-related diabetes in childhood. Horm Res Paediatr 2010;73:15–24.
[87] Middleton PG, Wagenaar M, Matson AG, Craig ME, Holmes-Walker DJ, Katz T, et al. Australian standards of care for cystic fibrosis-related diabetes. Respirology 2013.
[88] Dahn M, Kirkpatrick JR, Bouwman D. Sepsis, glucose intolerance, and protein malnutrition: a metabolic paradox. Arch Surg 1980;115:1415–8.
[89] Farrell PA, Fedele MJ, Vary TC, Kimball SR, Jefferson LS. Effects of intensity of acute-resistance exercise on rates of protein synthesis in moderately diabetic rats. J Appl Physiol 1998;85:2291–7.
[90] Barrett KE, Barman S, Boitano S, Brooks HL. Endocrine functions of the pancreas & regulation of carbohydrate metabolism. In: Barrett KE, Barman S, Boitano S, Brooks HL, editors. Ganong's review of medical physiology. 24 ed. New York: McGraw-Hill; 2012.
[91] Miller M, Ward L, Thomas BJ, Cooksley WG, Shepherd RW. Altered body composition and muscle protein degradation in nutritionally growth-retarded children with cystic fibrosis. Am J Clin Nutr 1982;36:492–9.
[92] Holt TL, Ward LC, Francis PJ, Isles A, Cooksley WG, Shepherd RW. Whole body protein turnover in malnourished cystic fibrosis patients and its relationship to pulmonary disease. Am J Clin Nutr 1985;41:1061–6.
[93] Hardin DS, Rice J, Rice M, Rosenblatt R. Use of the insulin pump in treat cystic fibrosis related diabetes. J Cyst Fibros 2009;8:174–8.
[94] Nousia-Arvanitakis S, Galli-Tsinopoulou A, Karamouzis M. Insulin improves clinical status of patients with cystic-fibrosis-related diabetes mellitus. Acta Paediatr 2001;90:515–9.
[95] Konstan MW, Morgan WJ, Butler SM, Pasta DJ, Craib ML, Silva SJ, et al. Risk factors for rate of decline in forced expiratory volume in one second in children and adolescents with cystic fibrosis. J Pediatr 2007;151:134–9.
[96] Lai HJ, Shoff SM, Farrell PM, Wisconsin Cystic Fibrosis Neonatal Screening, G. Recovery of birth weight z score within 2 years of diagnosis is positively associated with pulmonary status at 6 years of age in children with cystic fibrosis. Pediatrics 2009;123:714–22.
[97] Sharma R, Florea VG, Bolger AP, Doehner W, Florea ND, Coats AJ, et al. Wasting as an independent predictor of mortality in patients with cystic fibrosis. Thorax 2001;56:746–50.
[98] Cheung MS, Bridges NA, Prasad SA, Francis J, Carr SB, Suri R, et al. Growth in children with cystic fibrosis-related diabetes. Pediatr Pulmonol 2009;44:1223–5.
[99] Chan SJ,SD. Insulin through the ages: phylogeny of a growth promoting and metabolic regulatory hormone. Am Zool 2000; 40:213–22.
[100] Cystic Fibrosis Foundation Patient Registry. Annual report data 1997. 1998.
[101] Finkelstein SM, Wielinski CL, Elliott GR, Warwick WJ, Barbosa J, Wu SC, et al. Diabetes mellitus associated with cystic fibrosis. J Pediatr 1988;112:373–7.
[102] Moran A, Dunitz J, Nathan B, Saeed A, Holme B, Thomas W. Cystic fibrosis-related diabetes: current trends in prevalence, incidence, and mortality. Diabetes Care 2009;32:1626–31.
[103] Lanng S. Glucose intolerance in cystic fibrosis patients. Paediatr Respir Rev 2001;2:253–9.
[104] Moran A, Doherty L, Wang X, Thomas W. Abnormal glucose metabolism in cystic fibrosis. J Pediatr 1998;133:10–7.
[105] Marshall BC, Butler SM, Stoddard M, Moran AM, Liou TG, Morgan WJ. Epidemiology of cystic fibrosis-related diabetes. J Pediatr 2005;146:681–7.
[106] Geffner ME, Lippe BM, Kaplan SA, Itami RM, Gillard BK, Levin SR, et al. Carbohydrate tolerance in cystic fibrosis is closely linked to pancreatic exocrine function. Pediatr Res 1984;18:1107–11.
[107] Rosenecker J, Eichler I, Kuhn L, Harms HK, von der Hardt H. Genetic determination of diabetes mellitus in patients with cystic fibrosis. Multicenter cystic fibrosis study group. J Pediatr 1995;127:441–3.
[108] Moran A. Abnormal glucose tolerance in CF–when should we offer diabetes treatment? Pediatr Diabetes 2009;10:159–61.
[109] Casas L, Berry DR, Logan K, Copeland KC, Royall JA. Cystic fibrosis related diabetes in an extremely young patient. J Cyst Fibros 2007;6:247–9.

[110] Siahanidou T, Mandyla H, Doudounakis S, Anagnostakis D. Hyperglycaemia and insulinopenia in a neonate with cystic fibrosis. Acta Paediatr 2005;94:1837–40.
[111] Ripa P, Robertson I, Cowley D, Harris M, Masters IB, Cotterill AM. The relationship between insulin secretion, the insulin-like growth factor axis and growth in children with cystic fibrosis. Clin Endocrinol (Oxf) 2002;56:383–9.
[112] Costa M, Potvin S, Hammana I, Malet A, Berthiaume Y, Jeanneret A, et al. Increased glucose excursion in cystic fibrosis and its association with a worse clinical status. J Cyst Fibros 2007;6:376–83.
[113] Rosenecker J, Hofler R, Steinkamp G, Eichler I, Smaczny C, Ballmann M, et al. Diabetes mellitus in patients with cystic fibrosis: the impact of diabetes mellitus on pulmonary function and clinical outcome. Eur J Med Res 2001;6:345–50.
[114] Lanng S, Thorsteinsson B, Erichsen G, Nerup J, Koch C. Glucose tolerance in cystic fibrosis. Arch Dis Child 1991;66:612–6.
[115] Dobson L, Sheldon CD, Hattersley AT. Conventional measures underestimate glycaemia in cystic fibrosis patients. Diabet Med 2004;21:691–6.
[116] Frohnert BI, Ode KL, Moran A, Nathan BM, Laguna T, Holme B, et al. Impaired fasting glucose in cystic fibrosis. Diabetes Care 2010;33:2660–4.
[117] O'Riordan SM, Robinson PD, Donaghue KC, Moran A. Management of cystic fibrosis-related diabetes in children and adolescents. Pediatr Diabetes 2009;10(Suppl. 12):43–50.
[118] Godbout A, Hammana I, Potvin S, Mainville D, Rakel A, Berthiaume Y, et al. No relationship between mean plasma glucose and glycated haemoglobin in patients with cystic fibrosis-related diabetes. Diabetes Metab 2008;34:568–73.
[119] O'Riordan SM, Robinson PD, Donaghue KC, Moran A. Management of cystic fibrosis-related diabetes. Pediatr Diabetes 2008;9:338–44.
[120] Kerem E, Conway S, Elborn S, Heijerman H. Standards of care for patients with cystic fibrosis: a European consensus. J Cyst Fibros 2005;4:7–26.
[121] Allen HF, Gay EC, Klingensmith GJ, Hamman RF. Identification and treatment of cystic fibrosis-related diabetes. A survey of current medical practice in the US. Diabetes Care 1998;21:943–8.
[122] Mohan K, Miller H, Burhan H, Ledson MJ, Walshaw MJ. Management of cystic fibrosis related diabetes: a survey of UK cystic fibrosis centers. Pediatr Pulmonol 2008;43:642–7.
[123] Rayas MS, Willey-Courand DB, Lynch JL, Guajardo JR. Improved screening for cystic fibrosis-related diabetes by an integrated care team using an algorithm. Pediatr Pulmonol 2014.
[124] Bennett PH, Burch TA, Miller M. Diabetes mellitus in American (Pima) Indians. Lancet 1971;2:125–8.
[125] Hameed S, Jaffe A, Verge CF. Cystic fibrosis related diabetes (CFRD)–the end stage of progressive insulin deficiency. Pediatr Pulmonol 2011;46:747–60.
[126] Schmid K, Fink K, Holl RW, Hebestreit H, Ballmann M. Predictors for future cystic fibrosis-related diabetes by oral glucose tolerance test. J Cyst Fibros 2014;13:80–5.
[127] Milla CE, Billings J, Moran A. Diabetes is associated with dramatically decreased survival in female but not male subjects with cystic fibrosis. Diabetes Care 2005;28:2141–4.
[128] Palerm C. Proceedings of the 7th IFAC symposium on modelling and control in biomedical systems; 2009. Aalborg (Denmark).
[129] Dobson L, Sheldon CD, Hattersley AT. Validation of interstitial fluid continuous glucose monitoring in cystic fibrosis. Diabetes Care 2003;26:1940–1.
[130] O'Riordan SM, Hindmarsh P, Hill NR, Matthews DR, George S, Greally P, et al. Validation of continuous glucose monitoring in children and adolescents with cystic fibrosis: a prospective cohort study. Diabetes Care 2009;32:1020–2.
[131] Moreau F, Weiller MA, Rosner V, Weiss L, Hasselmann M, Pinget M, et al. Continuous glucose monitoring in cystic fibrosis patients according to the glucose tolerance. Horm Metab Res 2008;40:502–6.
[132] Jefferies C, Solomon M, Perlman K, Sweezey N, Daneman D. Continuous glucose monitoring in adolescents with cystic fibrosis. J Pediatr 2005;147:396–8.
[133] Schiaffini R, Brufani C, Russo B, Fintini D, Migliaccio A, Pecorelli L, et al. Abnormal glucose tolerance in children with cystic fibrosis: the predictive role of continuous glucose monitoring system. Eur J Endocrinol 2010;162:705–10.
[134] Leclercq A, Gauthier B, Rosner V, Weiss L, Moreau F, Constantinescu AA, et al. Early assessment of glucose abnormalities during continuous glucose monitoring associated with lung function impairment in cystic fibrosis patients. J Cyst Fibros 2013.
[135] Andersen DH. The present diagnosis and therapy of cystic fibrosis of the pancreas. Proc R Soc Med 1949;42:25–32.
[136] Shepherd RW, Holt TL, Thomas BJ, Kay L, Isles A, Francis PJ, et al. Nutritional rehabilitation in cystic fibrosis: controlled studies of effects on nutritional growth retardation, body protein turnover, and course of pulmonary disease. J Pediatr 1986;109:788–94.
[137] Dalzell AM, Shepherd RW, Dean B, Cleghorn GJ, Holt TL, Francis PJ. Nutritional rehabilitation in cystic fibrosis: a 5 year follow-up study. J Pediatr Gastroenterol Nutr 1992;15:141–5.
[138] Rana M, Munns CF, Selvadurai H, Donaghue KC, Craig ME. Cystic fibrosis-related diabetes in children–gaps in the evidence? Nat Rev Endocrinol 2010;6:371–8.
[139] Maedler K, Carr RD, Bosco D, Zuellig RA, Berney T, Donath MY. Sulfonylurea induced beta-cell apoptosis in cultured human islets. J Clin Endocrinol Metab 2005;90:501–6.
[140] Shah SC, Malone JI, Simpson NE. A randomized trial of intensive insulin therapy in newly diagnosed insulin-dependent diabetes mellitus. N Engl J Med 1989;320:550–4.
[141] Gribble FM, Reimann F. Differential selectivity of insulin secretagogues: mechanisms, clinical implications, and drug interactions. J Diabetes Complications. 2003 Mar-Apr;17(Suppl. 2):11–5
[142] Moran A, Phillips J, Milla C. Insulin and glucose excursion following premeal insulin lispro or repaglinide in cystic fibrosis-related diabetes. Diabetes Care 2001;24:1706–10.
[143] McDonough AK, Rosenthal RS, Cao X, Saag KG. The effect of thiazolidinediones on BMD and osteoporosis. Nat Clin Pract Endocrinol Metab 2008;4:507–13.
[144] Porksen N, Munn S, Steers J, Veldhuis JD, Butler PC. Effects of glucose ingestion versus infusion on pulsatile insulin secretion. The incretin effect is achieved by amplification of insulin secretory burst mass. Diabetes 1996;45:1317–23.
[145] Horton ES, Silberman C, Davis KL, Berria R. Weight loss, glycemic control, and changes in cardiovascular biomarkers in patients with type 2 diabetes receiving incretin therapies or insulin in a large cohort database. Diabetes Care 2010;33:1759–65.
[146] Nauck MA. A critical analysis of the clinical use of incretin-based therapies: the benefits by far outweigh the potential risks. Diabetes Care 2013;36:2126–32.
[147] Heise T, Nosek L, Ronn BB, Endahl L, Heinemann L, Kapitza C, et al. Lower within-subject variability of insulin detemir in comparison to NPH insulin and insulin glargine in people with type 1 diabetes. Diabetes 2004;53:1614–20.
[148] Danne T, Datz N, Endahl L, Haahr H, Nestoris C, Westergaard L, et al. Insulin detemir is characterized by a more reproducible pharmacokinetic profile than insulin glargine in children and adolescents with type 1 diabetes: results from a randomized, double-blind, controlled trial. Pediatr Diabetes 2008;9:554–60.
[149] Diabetes Control and Complications Trial Research Group. The effect of intensive treatment of diabetes on the development and progression of long-term complications in insulin-dependent diabetes mellitus. N Engl J Med 1993;329:977–86.

[150] Hameed S, Morton JR, Field PI, Belessis Y, Yoong T, Katz T, et al. Once daily insulin detemir in cystic fibrosis with insulin deficiency. Arch Dis Child 2012;97:464–7.
[151] Bizzarri C, Lucidi V, Ciampalini P, Bella S, Russo B, Cappa M. Clinical effects of early treatment with insulin glargine in patients with cystic fibrosis and impaired glucose tolerance. J Endocrinol Invest 2006;29:RC1–4.
[152] Wakil A, Atkin SL. Efficacy and safety of ultra-long-acting insulin degludec. Ther Adv Endocrinol Metab 2012;3:55–9.
[153] Weinzimer SA, Steil GM, Swan KL, Dziura J, Kurtz N, Tamborlane WV. Fully automated closed-loop insulin delivery versus semiautomated hybrid control in pediatric patients with type 1 diabetes using an artificial pancreas. Diabetes Care 2008;31:934–9.
[154] Falkmer S, El-Salhy M, Titlbach M. Evolution of the neuroendocrine system in vertebrates: a review with particular reference to the phylogeny and postnatal maturation of the islet parenchyma. In: Falkner S, Hakanson R, Sundler F, editors. Evolution and tumour pathology of the neuroendocrine system. Amsterdam: Elsevier; 1984.

CHAPTER

25

Low Glycemic Index Dietary Interventions in Cystic Fibrosis

Ben W.R. Balzer[1, 2], *Fiona S. Atkinson*[3], *Kirstine J. Bell*[3], *Katharine S. Steinbeck*[1, 2]

[1]Academic Department of Adolescent Medicine, The Children's Hospital at Westmead, NSW, Australia; [2]Discipline of Paediatrics and Child Health, Sydney Medical School, The University of Sydney, NSW, Australia; [3]School of Molecular Bioscience, The University of Sydney, NSW, Australia

25.1 INTRODUCTION

This chapter will summarize the evidence for low glycemic index (GI) diets in cystic fibrosis (CF), and highlight how such diets have benefited other groups of patients with prediabetes, with the goal of improving glycemic tolerance to prevent or delay diabetes. Sample dietary options and clinical vignettes will be discussed to demonstrate how a low GI diet could be safely implemented in CF clinical practice.

25.1.1 Glycemic Complications in CF

Cystic fibrosis related diabetes (CFRD) mellitus increases in prevalence with age in the CF population: 20% of those with CF have CFRD at 15 years of age, increasing to 70% by age 30. Glycemic tolerance is assessed using the oral glucose tolerance test (OGTT). Patients have CFRD if their 2 h plasma glucose is greater than or equal to 11.1 mmol/L (200 mg/dL); impaired glucose tolerance (IGT) is diagnosed by 2-h plasma glucose of 7.8–11.1 mmol/L (140–199 mg/dL) [1]. The prediabetic state, IGT, likewise increases dramatically with age: from 20% at 10 years of age to 82% by 30 years [2]. Aberrant glycemic control in CF is thought to be due to both defective insulin secretion (due to pancreatic damage) and insulin resistance [3]. Insulin secretion is impaired in those with exocrine pancreatic insufficiency, with lower peak plasma insulin concentration, and delayed time to achieve this [3]. The role of insulin resistance in early abnormalities of glycemic metabolism is unclear, though it plays a role in CFRD [4]. The role of the traditional diet in CF, which is high in energy, dietary fat, protein, and carbohydrate [5], may also promote insulin resistance.

The diagnosis of CFRD has profound prognostic implications [3]—deterioration in lung function [6], increased infection risk [7], and decreased body mass index (due to protein catabolism) [8–10]. These changes translate to decreased survival: only 25% of patients with CFRD survive beyond the third decade, compared with 60% of those with normal glycemic tolerance [11,12]. Additionally, a prediabetic state of impaired glycemic tolerance has been identified with similar prognostic implications as CFRD when it is diagnosed earlier in life [2]. For example, elevation of plasma glucose at 1 h of an OGTT is associated with worse lung function than those without CFRD [13].

Though improvements in respiratory care, including transplantation, have continued to increase life expectancy in CF, the decline in survival of those with CFRD indicates better therapies and management models are needed. Currently, insulin therapy is the recommended management for CFRD [14], but preventive measures should also be considered, given the profound effects of CFRD on survival. No guidelines currently provide recommendations for prevention of progression of IGT to CFRD. Insulin therapy has been associated with improved outcomes in early glucose derangements; however, its role in IGT is controversial [3], especially when considering insulin is not used to manage IGT in a non-CF population. Additionally, increasing the treatment burden in a group with intensive daily therapy requirements may be unacceptable to some.

Diet and Exercise in Cystic Fibrosis
http://dx.doi.org/10.1016/B978-0-12-800051-9.00025-0

25.1.2 The Glycemic Index

A low GI diet may assist in preventing, or at least delaying, the onset of CFRD. The GI is a measure of carbohydrate quality and classifies carbohydrate foods based on their acute effects on blood glucose levels (see Figure 25.1). The GI was designed to classify carbohydrates based on their physiological response in the body rather than on their chemical structure (i.e., simple sugars and complex carbohydrates). The historical assumption was that complex carbohydrates, such as bread, were slowly digested whereas simple sugars, such as sucrose, were rapidly digested. However, many simple sugars produce smaller fluctuations in blood glucose than many complex carbohydrates. Therefore, the GI refers to a food's ability to increase blood glucose postprandially [15], with higher GI foods causing a greater increase in blood glucose concentrations than lower GI foods.

The GI of a food is a relative measure, with pure glucose having a GI of 100, and all carbohydrate-containing foods assessed relative to that [16]. Foods with a GI of less than 55 are considered to have a "low" GI. Low GI foods tend to produce a slower and smaller rise in blood glucose level, which in turn stimulates a lower insulin secretion. The GI ranks carbohydrates on a gram-for-gram basis. Therefore, the GI value of a given food is not dependent on portion size. The glycemic load (GL) is a mathematical measure of both carbohydrate quality (GI) and quantity (portion size). As will be discussed, a diet rich in low GI foods has proven benefits in non-CF diabetes.

25.2 EVIDENCE FOR EFFECT IN NON-CF POPULATIONS WITH DIABETES

There is a growing body of evidence to support the use of low GI and low GL diets in the management of diabetes. Low GI diets have shown beneficial effects on postprandial glycemia and insulin sensitivity in individuals with Type 1 and Type 2 diabetes mellitus. Carbohydrate is the main component in food that directly causes blood glucose levels to rise, so the quality of the carbohydrate is an important factor in the management of postprandial blood glucose excursions. Chronically elevated glucose and insulin levels increase the risk of both macrovascular and microvascular complications associated with diabetes. Therefore, dietary interventions that can help control blood glucose and insulin excursions may help reduce complications associated with diabetes. While microvascular complications have been described in CFRD, macrovascular have not [3]. Other complications, such as worsened respiratory function, which in part may represent excessive glycation of tissues, may be important to consider.

The evidence for low GI and low GL diets for the treatment of diabetes mellitus was examined in a recent Cochrane Review [17], which showed a clinically significant reduction in glycated hemoglobin A1c (HbA1c) when such diets were prescribed. The authors concluded that low GI diets significantly improve diabetic control in both Type 1 and Type 2 diabetes. They noted that the improvements in HbA1c achieved via a low GI diet are similar to the improvements observed in HbA1c with the use of anti-hyperglycemic medications in newly

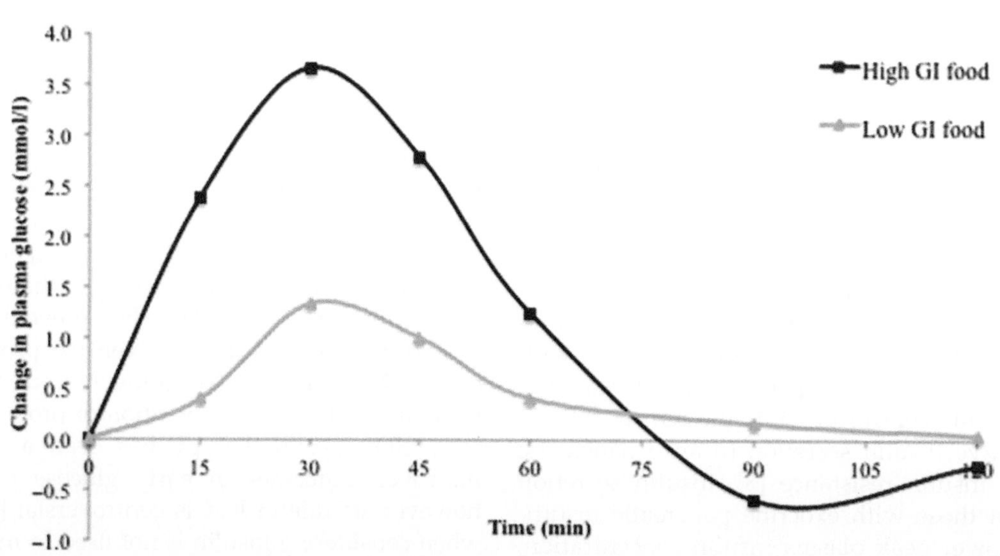

FIGURE 25.1 **Comparison of the effect of high and low GI foods on plasma glucose concentrations over 2h in individuals with normal glucose tolerance.** The low GI food causes lower peak plasma glucose and has a sustained plasma glucose concentration over 2h. In comparison, the high GI food has a large rise in plasma glucose to a higher peak concentration followed by a rapid decline in blood sugar, shown by the negative change in plasma glucose between 80 and 120min.

diagnosed Type 2 diabetes. The low GI dietary interventions did not show an increased risk of hypo- or hyperglycemic episodes. Therefore, the use of low GI diets in the management of diabetes can result in improved glycemic control without an increased risk of acute complications.

One of the studies included in the Cochrane Review was a 12week crossover trial comparing a low versus high GI diet on glycemic control in 16 subjects with well-controlled Type 2 diabetes [18]. The study found that after 12weeks following the dietary interventions the mean HbA1c was 11% lower for the low GI diet compared to the high GI diet, with five of the 16 participants showing a greater than 20% reduction in HbA1c on the low GI diet. Similarly, acute 8h postprandial glucose excursions to standard meals were significantly less on the low GI diet, although mean fasting glucose did not differ between the two diet groups. Weight was maintained throughout the trial period and thus the improvements in glycemic control can be attributed to the dietary intervention rather than an effect of weight loss per se.

A meta-analysis of 14 randomized controlled trials assessing the effect of low GI diets for the management of Type 1 and Type 2 diabetes found that low GI diets produced greater reductions in HbA1c compared to high GI diets [19]. The authors concluded that their analysis showed that using low GI diets had clinically relevant benefits on glycemic control in people with diabetes. Not all individual studies have shown consistent beneficial reductions in HbA1c, potentially due to relatively small sample sizes or the inclusion of subjects with well-controlled diabetes (i.e., HbA1c $<6.2\%$). Additionally, it must be noted that HbA1c is not recommended as a screening test for CFRD [14,20,21]. This means that changes in HbA1c may be difficult to interpret with regard to the glycemic status of the patient with CF.

Low GI diets have also been shown to be better than high cereal fiber diets for the management of Type 2 diabetes. A large randomized, parallel intervention study of 210 individuals with Type 2 diabetes assigned to either a low GI diet or a high cereal fiber diet found that, after 6months, the low GI diet resulted in a 0.5% reduction in HbA1c [22]. The beneficial improvements in HbA1c were observed independently of carbohydrate or fiber intake. This intervention showed that a reduction in the GI of the diet could result in modest but clinically relevant improvements in glycemic control in individuals with Type 2 diabetes treated with anti-hyperglycemic medications.

Low GI diets also have beneficial effects on insulin sensitivity in individuals with diabetes. A crossover intervention conducted in 12 men with Type 2 diabetes compared the effects of low or high GI diets on insulin sensitivity using the gold standard method of a euglycemic hyperinsulinemic clamp. The whole-body peripheral insulin sensitivity was significantly higher in individuals on a low GI diet compared to a high GI diet following a four-week intervention of each diet treatment [23]. Fasting blood glucose levels and chronic glycemic control measured by HbA1c were also significantly lower following 4weeks of a low GI diet compared to the high GI diet. This study shows that low GI diets can result in significant improvements in insulin sensitivity, as well as blood glucose control, in individuals with Type 2 diabetes.

Low GI diets may also help reduce the need for medication use in Type 2 diabetes. A study conducted in 40 poorly controlled Type 2 diabetes patients showed that after 12months, participants following a low GI diet were less likely to increase their hypoglycemic therapy dose or to add new diabetic medications compared to a diet recommended by the American Diabetes Association (ADA) where the GI of the carbohydrates was not considered [24]. This study showed that the potential beneficial effects of low GI diets are not limited strictly to the dietary management of blood glucose levels but may also reduce the need for anti-hyperglycemic medications.

One potential benefit of lower GI carbohydrates is that the smaller glucose fluctuations result in lower compensatory excursions insulin secretion. This may be particularly relevant for CFRD as it is typically associated with insulin insufficiency [14]. Foods that require less insulin secretion may therefore help maintain insulin function by reducing beta-cell stress.

Low GI diets have been shown to help normalize blood glucose levels in individuals with Type 1 and Type 2 diabetes. CFRD is a unique metabolic condition that shows characteristics of both these types of diabetes. Improving glycemic control through dietary interventions may be particularly important for individuals with CFRD to help minimize the need for anti-hyperglycemic medications in a population group that already has a high medication burden. Dietary management of CFRD through the use of low GI diets should also help reduce the risk of diabetic complications.

25.3 EVIDENCE FOR THE USE OF A LOW GI DIET IN CF

There is an assumption that low GI foods increase satiety and help weight loss, both of which would be detrimental to those with CF. Currently, there is a paucity of experimental evidence for the efficacy of a low GI diet in CF. The literature base consists of three papers: two in which GI is not the key focus of the paper [25,26], and the other a systematic review in which the above two studies are summarized, as well as providing the theoretical basis of how low glycemic diets could benefit those with CF [27].

The first original research paper by Skopnik et al. did not use a low GI dietary intervention, but compared the glycemic indices (measured by plasma glucose area under the curve) of two enteral feeds [25]. The lower GI formula (containing lower carbohydrate, higher fat, and protein) resulted in a significantly lower area under the blood glucose curve during the enteral feed challenge when compared to the standard feed formula (solely carbohydrate). The principal finding was that those with CF had lower blood glucose levels using the low GI enteral feed, regardless of their glycemic tolerance status.

The primary focus of Ntimbane's pilot study was tissue oxidative stress in CF in a population of 31 adolescents. A subset of 13 adolescents who had IGT at entry to the study received nutritional recommendations: nine were provided education to avoid high GI foods and simple concentrated sugars, and the other three received no such education [26]. One patient dropped out of the study. Of the nine who received dietary advice, six returned to normal glucose tolerance, one continued to have IGT, and two developed CFRD. The three subjects who did not receive dietary education, when followed up at 12 months, had all developed CFRD [26]. This study remains the best experimental evidence to date regarding the use of a low GI diet in CF, although the sample size was small, there was no randomization and no objective measures of dietary compliance were used. It is the first study to show nutritional recommendations may have a role in delaying progression to CFRD [26], though it has been clinically observed that those with IGT may spontaneously improve glycemic tolerance back to normal.

At present, a low GI diet is recommended for clinical use in CF by the Dietitians Association of Australia [28]. However, this recommendation is based on expert opinion and not randomized controlled trial evidence. In theory, low GI diets may help in the overall management of people with CFRD by minimizing blood glucose and insulin fluctuations, which in turn can help with recovery from infections and maintaining healthy body weight.

25.4 PRACTICAL CONSIDERATIONS OF A LOW GI DIETARY INTERVENTION IN CF

In CF, a dietary intervention requires sufficient energy to meet the demands of an increased metabolic rate, as well as preventing hunger throughout the day. In childhood, the need for increased intake may be met with resistance (especially if unwell), so finding palatable foods for the child with CF is important. That many palatable foods are also higher GI needs to be borne in mind when planning a lower GI dietary intervention. Some foods frequently eaten, such as dairy products, will naturally have a low GI and are suitable choices for people with CF. However, a low GI eating plan does not mean restricting the diet to just those low GI foods. There are four key strategies that can be used to lower the GI of the overall meal within the context of an optimal CF diet.

1. Substitution

Identifying high GI carbohydrate sources within the diet and substituting these for low (or at least lower) GI options will reduce the overall GI of the meal, while still maintaining healthy macronutrient proportions. For example, substituting white bread (high GI) for low GI grain bread or swapping white potato (high GI) for lower GI varieties like sweet potato. See Table 25.1 for other low GI options that can be substituted for high GI options. It is recommended that at least one low GI food be chosen for each meal.

2. Combining with low GI foods

There is no reason to forgo favorite higher GI foods completely. Combining higher GI options with a low GI option will reduce the overall GI of the meal. For example, all dairy foods are low GI, so pouring full cream milk over a bowl of a high GI cereal, such as cornflakes, will create a medium GI meal.

3. Adding protein and/or fat

Foods rich in protein and fat are not only a good source of kilojoules for people with CF, but these also slow the digestion and absorption of carbohydrate foods. Therefore, combining protein and fat with carbohydrate will help lower the overall GI. For example, pairing cheese with high GI crackers, spreading peanut butter on white bread, putting cheese and butter on top of potato, or having a yogurt with a snack bar will all help lower the overall GI.

4. Adding acidic foods

Acidic foods help slow gastric emptying, thereby lowering the GI of the meal. Some foods naturally contain acids, such as sourdough bread, which consequently has a lower GI than white bread. Adding acidic foods like vinegar or lemon or lime juice will also help significantly lower the GI. These foods do not need to be directly applied to the carbohydrate source; as long as these are included in the same meal, the effect will be the same.

Table 25.1 identifies some low and high GI foods within each food category. People with IGT CFRD are encouraged to aim for at least one low GI carbohydrate source at each meal as part of a low GI eating plan.

25.4.1 Sample Meal Plan

The dietary management of IGT and CFRD should involve the standard high-energy, high-protein advice typical of the CF diet along with consideration of carbohydrate quality (GI) and quantity to assist with the management of blood glucose fluctuations. People with CFRD should aim to eat at least three main meals and

TABLE 25.1 Examples of Low and High GI Choices within Common Carbohydrate Food Categories

Food Group	Low GI Choices	Higher GI Choices
Breads, cereals, rice & pasta	*Breads*: Dense whole grain or multigrain breads, fruit breads, multigrain English muffin, and authentic pumpernickel, sourdough rye, and sourdough wheat breads *Cereals*: Unprocessed oats, traditional porridge, muesli, oat bran, wheat bran cereal *Pasta & noodles*: All types *Rice*: Basmati rice, rice noodles *Grains*: Barley, bulgur, quinoa, semolina	*Breads*: White and whole-meal breads, baguette, English muffin, hamburger bun, pita bread, Lebanese bread, flatbread, Turkish bread, naan *Cereals*: Most flaked, puffed, or extruded breakfast cereals, e.g., rice puffs, cornflakes, wheat biscuits *Rice*: White rice, brown rice *Grains*: Couscous
Fruit	Apple (fresh and dried), apricot (fresh and dried), orange, pear, banana, grapes, kiwi fruit, mango, nectarine, peach, plums, strawberries, fruit juices	Watermelon, cantaloupe, pineapple
Vegetables & legumes	Sweet corn, carrot, butternut pumpkin, sweet potato, kidney beans, soybeans, chickpeas, split peas, lentils, baked beans	White potato
Dairy products	Cow's milk, soymilk, yogurt, custard, ice cream, flavored milk, dairy desserts and pudding	
Snack foods & confectionary	*Crackers*: Rye, graham *Biscuits and baked goods*: Digestive biscuits, cakes (made from packet mixture), oatmeal cookies *Confectionary*: Chocolate *Snack foods*: Nuts, trail mix, hummus dip	*Crackers*: Corn thins, rice cakes, rice crackers, water crackers, crispbread *Biscuits*: Arrowroot, shortbread *Confectionary*: Candy *Snack foods*: Potato chips, corn chips

three snacks each day to ensure that sufficient energy is consumed throughout the day, generally 120–150% of recommended dietary intake for energy. This is usually not a difficulty, unless the patient's therapy is complicated by an acute infective exacerbation or severe respiratory failure. For many people with CF, appetite is lower in the morning and increases over the day, with afternoon or evening snacks equivalent in calorie content to main meals. Each meal and snack should include a carbohydrate food with an unrestricted amount of fat. Included in Table 25.2 is a sample meal plan for a person with CFRD. This meal plan provides approximately 13,000 kJ (about 20% protein, 37% fat, and 41% carbohydrate). For a 20-year-old male, with a BMI of 19, such a plan would provide 130% of the estimated energy requirement (10,000 kJ/day).

TABLE 25.2 Sample High-Energy Meal Plan Incorporating Low GI Choices for a Young Man with CFRD

Breakfast:
1 tub of regular plain or fruit-flavored yogurt
½ cup of toasted fruit & nut muesli
½ cup strawberries

Morning snack:
½ cup dried fruit & nut mix (or slice of raisin toast/banana bread + butter)
1 cup orange/apple juice

Lunch:
Sandwich with 2 slices of grain bread, 100 g ham, slice of cheese, lettuce, tomato, 1 T mayonnaise, 1 T mustard.
1 cup of lemonade
Small bag of potato crisps

Afternoon snack:
Large glass of flavored milk or Sustagen sport (Boost energy drink in USA and UK) tetra pack
Peanut butter + crackers

Dinner:
Creamy chicken and vegetable pasta bake with cheese
Water

Supper:
1 banana
200 g vanilla custard

Adapted, with permission, from diet plans written by Christie Graham, CF Dietitian, The Children's Hospital at Westmead, Australia.

25.5 THEORETICAL APPLICATIONS OF A LOW GI DIETARY INTERVENTION IN CF

25.5.1 In Childhood and Adolescence

One of the major goals of management in CF in childhood and adolescence is the maintenance of normal growth and development. These in turn are highly dependent on physical wellbeing and adequate nutrition. Other factors such as the use of glucocorticoids and maturational delay (as a result of chronic illness) may alter growth and development patterns. Adequate nutrition is implicit in general CF guidelines as part of multidisciplinary management [29,30]. Childhood is a time when food tastes develop and where exposure to a wide variety of foods is good practice for the attainment of healthy lifelong nutrition. For many families with a child who has repeated bouts of pulmonary infection and hospitalization, food choice and variety may matter less than simply the urgent need to ingest adequate calories. Often these food choices include higher GI carbohydrates. Thus, high carbohydrate beverages and confectionary (often also high GI) become a regular part of the daily diet, rather than occasional treats. The situation may be exacerbated by the use of nutritional supplements, as discussed later. If these supplements are provided through the intra-gastric route such as percutaneous endoscopic gastrostomy, then the normal oral stimulation to insulin and incretins is diminished, with a resultant increase in hyperglycemia and a delayed but excessive insulin response. It is unknown as to how this situation might contribute to future abnormalities of glucose metabolism, and thus it is not possible to make evidence-based recommendations around the GL of the diet in childhood.

The OGTT and continuous glucose monitoring system (CGMS) in particular have both been used to demonstrate the prevalence of disorders of glucose metabolism in both children and adolescents with CF [31,32]. OGTT remains the suggested screening tool, and the most common abnormality is IGT. CGMS in selected patients often provides evidence of unexpected postprandial hyperglycemia under free-living conditions. Abnormalities in glucose metabolism in CF may be exacerbated by the universal increase in insulin resistance in mid puberty, as a result of increased growth hormone secretion [33]. This situation has not been formally studied in CF, so it is not possible to say whether the insulin resistance of puberty persists to any degree after the growth hormone returns to lower levels after the puberty growth spurt. Given the compromise that already exists for beta-cell function in CF patients with pancreatic insufficiency and if there was any failure to reverse this insulin resistance, then puberty in CF might theoretically be a time to initiate a lower GL eating plan. This change is likely to be difficult in practice, given the general increase in non-adherence to therapies, typical of adolescence [34]. Also, in adolescence, many dietary habits are entrenched and further supported by emerging autonomy where peer preferences and some disposable income also influence food choices.

Numerous CF dietary guidelines exist for childhood and adolescence, and the emphasis has traditionally been on high-fat and high-protein intakes in order to achieve adequate calorie intake with the high calorie density of fat, and adequate protein for growth. Dietary practice is focused on the avoidance of malnutrition, and less emphasis is placed on carbohydrate intake. As an example, in the Australasian nutritional guidelines, it is recommended that carbohydrate intake be as high as required to meet energy requirements, in conjunction with a high fat diet. If CFRD is diagnosed, the guideline states that modification of the distribution of carbohydrate during the day may be required [28].

In the American Clinical Care Guidelines for CFRD [14], the general dietary recommendations are that the need for a high caloric intake should not override sound nutritional guidelines and healthy eating practices, that the diet will likely need an individual approach as compared to Types 1 and 2 diabetes and that carbohydrate counting may be useful. These recommendations are understandably broad given the lack of empirical evidence.

On balance, childhood and adolescence are the age groups where management planning for abnormalities of glucose metabolism needs to begin. This includes screening protocols [3] and the consideration of whether there is sufficient attention given to the types of carbohydrate in the overall diet. There is no empirical evidence that low GI diets in CF prevent progression to IGT or the progression of IGT to CFRD. There is, however, good evidence that children born with CF today will have a life expectancy that demands the consideration of how best to preserve pancreatic islet cell function.

25.5.2 In Adulthood

The progressive nature of pancreatic damage in CF is such that the older the age the greater the chance of the appearance of IGT/CFRD in patients with pancreatic insufficiency. The older a person is, the less adaptable they may be to dietary changes. There is a case to be made for some dietary education and modification prior to the appearance of IGT/CFRD, although there are no data to support this. The presence of insulin resistance and delayed postprandial insulin secretion peaks are a risk for hypoglycemia. While this can occur in childhood and adolescence, the group most commonly affected in practice are young adults, particularly those who perform higher intensity physical activity. Lower GI carbohydrates before and during exercise may be useful. A similar regimen is routine for many elite athletes.

With longer life expectancy and preservation of lung function, many young women with CF are planning pregnancies. If they have pancreatic insufficiency, they may develop diabetes in pregnancy. Low GI diets theoretically may have an advantage by reducing maternal hyperglycemia and fetal hyperinsulinemia, but there are insufficient data to recommend their routine use, even in women without CF [35].

Given the increased longevity of CF, the higher ascertainment of milder cases of CF and the general effectiveness of therapy, clinics are starting to see cases of overweight and obesity. These are generally found in males with milder disease and often without pancreatic insufficiency [36]. Overweight and obesity may also occur with lung transplant recipients prescribed high-dose glucocorticoids. Such patients may find themselves in the unanticipated situation of having to limit calories after a lifetime of *ad libitum* eating. Reduction in dietary fat (and hence caloric density) has a strong empirical base across all age groups [37]. The introduction of lower GL into the diet in the general population has been shown to be useful in the prevention of weight gain, Type 2 diabetes, and cardiovascular disease, through multiple proposed mechanisms, including reduction in appetite stimulation and fuel partitioning [38]. There is yet to be any empirical evidence in CF populations.

25.5.3 Impaired Glucose Tolerance

There is some peripheral empirical evidence available that suggests that low GI diets may retard the progression of IGT to CFRD [26,27], but the evidence base that low GI diets may prevent the transition to diabetes is primarily for Type 2 diabetes mellitus [39]. For those with CF, delaying the onset of diabetes using other interventions recommended for the non-CF community should not be attempted. Such interventions include energy restriction or "diabetic diets", which will not provide sufficient calories for optimal nutrition in CF. Additionally, there is no role for metformin in those with CF and IGT, although it is routinely used in the non-CF patient. This is due to concerns regarding gastrointestinal upset, anorexia, and the risk of lactic acidosis [40].

25.5.4 In those with a Family History of Type 2 Diabetes Mellitus

Type 2 diabetes mellitus has a polygenic inheritance, and an understanding of the patient's family history of this condition should be ascertained as part of the history. First-degree relatives of persons with Type 2 diabetes mellitus already have evidence of insulin resistance [41]. The insulin resistance commonly observed in those with CF may be due to family history as well as chronic infection and glucocorticoid use. Thus, a family history of Type 2 diabetes mellitus may be one situation where a lower GI diet could be recommended in CF, based on empirical evidence [23].

25.5.5 Concerns Regarding the Use of a Low GI Diet in CF

Two potential benefits of low GI diets in non-CF subjects are the ability to enhance satiety, and to facilitate weight loss and help with weight maintenance. These situations are problematic for the individual with CF: increased satiety might prevent adequate energy intake, and weight loss is contraindicated [2].

25.5.5.1 Satiety

A low GI diet may promote satiety by reducing postprandial blood glucose and insulin levels, while simultaneously increasing cholecystokinin levels [42,43]. In the non-CF population, enhancing satiety is advantageous for reducing energy intake, whether as a means to prevent weight gain or enhance weight loss. This would be detrimental in those with CF, who need a higher than estimated dietary intake to maintain their body weight due to intrinsic elevation of metabolic rate, malabsorption, and the increased metabolic requirements of chronic inflammation and infection.

In a study of normal, overweight, and obese children and adolescents, it was observed that lunchtime satiety was significantly increased by a low GI breakfast [44]. Similarly, in a cohort of overweight adult females, lower GI diets were associated with higher subjective fullness and lower desire to eat a fatty meal; however, there was no observed effect on lunchtime *ad libitum* energy intake when compared with a high GI diet [45]. These and other similar studies [46,47] provide evidence of a role for low GI diets in enhancing satiety. However, the potential beneficial effects of low GI diets on satiety are not consistently observed. Some studies in adults, including combinations of high and low protein as well as high and low GI diets, did not find differences in self-reported hunger or satiety after the different meals [48,49]. Therefore, based on the literature, it is difficult to conclude how significant an effect a low GI diet may have on satiety in those with CF. Thus, if a lower GI diet is begun, weight should be carefully monitored and the patient questioned about fullness after meals. It is possible to recommend low GI foods that are less likely to have satiating effects, such as fruit juices rather than whole fruit, which would be able to assist with blood glucose control without negatively impacting on food intake and overall energy intake.

25.5.5.2 Weight Loss

Low GI diets may cause weight loss of their own accord, whether by enhancing satiety or causing fat to be oxidized preferentially to carbohydrate [42]. Therefore the balance between improved glycemic control (to prevent or ameliorate the effects of IGT/CFRD) and weight outcomes must be considered.

Careful analysis of the DiOGenes study, a large, multicenter trial of dietary combinations (low/high-GI and low/high-protein content), found that the low GI diet alone did not cause weight loss [50,51]. The low GI arms of the trial had better glycemic status at the end of the trial and were also more compliant with the intervention. While some other low GI diet trials have shown that greater weight loss is possible using low GI diets compared to high GI diets [52–54], others have not been able to confirm this [45,55]. Isocalorically, a high GI diet is associated with more body fat (including visceral deposits), and greater lipogenic enzyme expression, whereas a low GI diet promotes satiety, decreases postprandial insulin secretion, and maintains insulin sensitivity [42].

Higher fiber content may also play a role in weight loss with lower GI diets, and CF patients do not tolerate high dietary fiber. Therefore, if low GI diets are trialed in CF patients, these should be lower fiber. There should be careful monitoring of weight, bloating, and fullness, and instructions not to reduce intake. If patients cannot consume adequate calories to maintain healthy body weight, then dietary advice needs to be changed.

25.5.6 The Role of Dietary Supplements in CF

It is well understood that maximizing the nutritional status of people with CF is important for their overall health and lung function. However, there is limited evidence in the literature to support the beneficial effects of oral high-energy, high-protein supplements to assist people with CF in achieving their high daily energy targets and improving their nutritional status. In theory, supplements should help improve overall nutritional status; however, studies have not been able to show their effectiveness. A large, multicenter, randomized controlled trial including 102 children with CF conducted over 12 months found that oral energy supplements were not associated with a significant increase in body mass index or improvements in nutritional outcomes compared to standard CF dietary advice alone [56]. The authors found that the group receiving the oral supplements did have an improved overall energy intake, although this improvement did not translate into benefits to their overall nutritional status. A Cochrane Review further found that oral supplements do not confer any additional benefit in moderately malnourished people with CF above and beyond usual dietary advice and monitoring [57]. Despite this, oral supplements are still used regularly as part of managing those people with CF whose nutritional status or dietary intake are not optimal and this is reflected in clinical practice recommendations [28,58]. These supplements may help some people with CFRD reach their energy intake targets, which would ideally have the follow-on benefit of improving weight status and nutritional health.

Oral high-energy, high-protein supplements have a potential dual benefit in the dietary management of CFRD. First, they may help increase overall energy intake and second, many of these products have a low GI value so they can assist with the management of daily blood glucose excursions. Nutritional supplements, such as Sustagen™, Ensure™, and Resource™, are all low GI choices [59] that could be used to enhance the energy intake in people with CFRD whilst assisting with glycemic control.

Enteral feeding has a well-established nutritional benefit in CF; however, its high carbohydrate content and high GL increase the risk of CFRD, with approximately half of enterally fed adolescent subjects progressing to CFRD [60]. Any interventional study would need to determine how best to integrate this crucial therapy into an all-round lower GI approach. Such an approach implies the introduction of a lower GI equivalent—providing the same energy load but producing less burden on beta-cells [25].

25.5.7 Case Studies

The following case studies are examples of how low GI dietary interventions could be implemented in CF.

1. An adolescent with hypoglycemia in the absence of CFRD

Tom is an 18-year-old male, diagnosed with CF at birth. He has pancreatic insufficiency. His FEV_1 is 90% of predicted. He plays soccer competitively and trains twice a week as well as the game on weekends. He also attends the gym for weight training because he finds it hard to put on weight and would like to have bigger muscles. Tom has also heard that physical exercise is good for preventing osteoporosis. Over the last 6 months he has experienced light-headedness and hunger, especially during the soccer game, and describes "the shakes" such that he finds it hard to kick the ball accurately. Eating tends to relieve these symptoms. On closer questioning he has also experienced a similar sensation when he lifts heavy weights. Because he does not like to eat a large meal before sport he tends to rely on carbonated soft drinks and jellybeans for energy—he has been doing this since he was 10 years old. His most recent 75 g OGTT is as shown in Table 25.3.

This OGTT excludes CFRD. There is evidence of an excessive and delayed insulin response. This rapid increase of insulin between 60 and 120 min puts him at risk of later hypoglycemia, particularly if there is exercise that increases both insulin sensitivity and muscle glucose uptake. CF hypoglycemia, which occurs in the absence of medications that are known to cause hypoglycemia, is poorly understood [61]. Tom's high GI food choices have been learned at an early age. In this case, a low GI diet may assist in preventing hypoglycemic episodes by minimizing fluctuations in postprandial glycemia.

2. The recent diagnosis of abnormal glucose metabolism in a young adult

TABLE 25.3 OGTT Results for an Adolescent with CF and Hypoglycemia

	0	60	120 min
Glucose (mmol/L)	3.9 (70 mg/dL)	10.5 (189 mg/dL)	4.7 (85 mg/dL)
Insulin (pmol/L)	45 (7.5 μIU/L)	132 (22 μIU/L)	195 (32.5 μIU/L)

Maya is a 23-year-old female who has not regularly attended the Adult CF clinic after her discharge from pediatric care at the age of 18 years. She is known to have pancreatic insufficiency. She comes to adult clinic complaining of cough, increased sputum, and weight loss. She has lost 4 kg (BMI 18.2 kg/m^2) and has not menstruated for 5 months. Her lung function has deteriorated since her last clinic visit nearly 1 year ago. Among a number of preclinic attendance investigations she has an OGTT, her first one ever. Maya tells you that she has no intention of going on insulin, even if her OGTT reveals that she has diabetes, as she cannot cope with injections. You are concerned that she will have difficulty with any additional management impositions because her adherence to routine pharmacotherapy and home physiotherapy is marginal. Her OGTT is as shown in Table 25.4.

This is IGT. The priority is to improve her lung function by concentrating on improving her adherence to standard best pulmonary practice and an admission is arranged to treat her infection and to re-educate the patient. Until Maya feels physically better she is unlikely to be able to sensibly discuss the management of IGT (which may improve once her chest infection is treated). It would be reasonable to reduce the GL of her diet by choosing lower GI options. CGMS at a later date might also be helpful to determine her glycemic control over the day.

3. A CF patient with a lung transplant

Jack, a 30-year-old male, receives a double lung transplant for CF. He had CFRD prior to transplantation, which was controlled on 20 units of insulin glargine. His HbA1c was 6.0% (42 mmol/mol). He required an insulin infusion in ICU and was discharged post-transplant on 50 units of insulin glargine *mane*. Twelve months posttransplant he is well, has had one episode of rejection requiring pulsed prednisolone, and is now on 7.5 mg prednisolone daily, taken with his morning medications. His HbA1c has risen to 8.0% (64 mmol/mol). Jack is changed to *bd* insulin detemir and rapid-acting insulin with his evening meal, but declines any more preprandial dosing. Jack's afternoon and pre-evening meal blood glucose levels are generally 10–15 mmol/L (180–270 mg/dL). His morning insulin is changed to a pre-mix rapid and insulin isophane so that there is a peak of long-acting insulin mid-afternoon, where the prednisolone effect is maximal. His regular lunch of chicken and vegetable stir-fry with jasmine rice is changed to spaghetti Bolognese and a side salad with vinaigrette dressing.

TABLE 25.4 OGTT Results for a 23-Year-Old Female with CF

	0	60	120 min
Glucose (mmol/L)	5.6 (101 mg/dL)	12.2 (220 mg/dL)	10.9 (196 mg/dL)
Insulin (pmol/L)	60 (10 μIU/L)	122 (20.3 μIU/L)	82 (13.7 μIU/L)

25.6 CONCLUSIONS

Abnormalities in glycemic control (IGT and CFRD) are common complications in CF, with both having important effects on prognosis. While insulin has been established as the standard of care for this form of diabetes mellitus, the role of dietary therapy should be considered, especially in the prediabetic state of IGT. Here, a low GI diet may be beneficial by preventing or delaying progression to diabetes or ameliorating glycemic aberrations. Such effects have been demonstrated in non-CF cohorts. Lower GI diets reduce postprandial insulin requirements and may improve insulin resistance, two aspects of the abnormalities in glycemic metabolism observed in CF. It is hypothesized that the potential to induce weight loss of low GI diets would be ameliorated by the increased caloric (especially dietary fat) intake in CF. This chapter has also addressed practical concerns about applying a lower GI diet to a CF population and provided examples of how these diets could be used as a means to ensure optimal glycemic control in CF.

References

[1] Moran A, Hardin D, Rodman D, Allen HF, Beall RJ, Borowitz D, et al. Diagnosis, screening and management of cystic fibrosis related diabetes mellitus: a consensus conference report. Diabetes Res Clin Pract 1999;45:61–73.

[2] Bismuth E, Laborde K, Taupin P, Velho G, Ribault V, Jennane F, et al. Glucose tolerance and insulin secretion, morbidity, and death in patients with cystic fibrosis. J Pediatr 2008;152:540–5. 545 e541.

[3] Rana M, Munns CF, Selvadurai H, Donaghue KC, Craig ME. Cystic fibrosis-related diabetes in children–gaps in the evidence? Nat Rev Endocrinol 2010;6:371–8.

[4] Kelly A, Moran A. Update on cystic fibrosis-related diabetes. J Cyst Fibros 2013;12:318–31.

[5] Craig M, Twigg S, Donaghue K, Cheung N, Cameron F, Conn J, for the Australian Type 1 Diabetes Guidelines Expert Advisory Group, et al. In: Draft national evidence-based clinical care guidelines for type 1 diabetes in children, adolescents and adults. Canberra: Australian Government Department of Health and Ageing; 2011.

[6] Costa M, Potvin S, Hammana I, Malet A, Berthiaume Y, Jeanneret A, et al. Increased glucose excursion in cystic fibrosis and its association with a worse clinical status. J Cyst Fibros 2007;6:376–83.

[7] Brennan AL, Gyi KM, Wood DM, Johnson J, Holliman R, Baines DL, et al. Airway glucose concentrations and effect on growth of respiratory pathogens in cystic fibrosis. J Cyst Fibros 2007;6:101–9.

[8] Koch C, Rainisio M, Madessani U, Harms HK, Hodson ME, Mastella G, et al. Presence of cystic fibrosis-related diabetes mellitus is tightly linked to poor lung function in patients with cystic fibrosis: data from the European Epidemiologic Registry of Cystic Fibrosis. Pediatr Pulmonol 2001;32:343–50.

[9] Brennan AL, Geddes DM, Gyi KM, Baker EH. Clinical importance of cystic fibrosis-related diabetes. J Cyst Fibros 2004;3:209–22.
[10] Ode KL, Frohnert B, Laguna T, Phillips J, Holme B, Regelmann W, et al. Oral glucose tolerance testing in children with cystic fibrosis. Pediatr Diabetes 2010;11:487–92.
[11] Mohan K, Miller H, Dyce P, Grainger R, Hughes R, Vora J, et al. Mechanisms of glucose intolerance in cystic fibrosis. Diabetic Med 2009;26:582–8.
[12] Finkelstein SM, Wielinski CL, Elliott GR, Warwick WJ, Barbosa J, Wu SC, et al. Diabetes mellitus associated with cystic fibrosis. J Pediatr 1988;112:373–7.
[13] Brodsky J, Dougherty S, Makani R, Rubenstein RC, Kelly A. Elevation of 1-hour plasma glucose during oral glucose tolerance testing is associated with worse pulmonary function in cystic fibrosis. Diabetes Care 2011;34:292–5.
[14] Moran A, Brunzell C, Cohen RC, Katz M, Marshall BC, Onady G, et al. Clinical care guidelines for cystic fibrosis-related diabetes: a position statement of the American Diabetes Association and a clinical practice guideline of the Cystic Fibrosis Foundation, endorsed by the Pediatric Endocrine Society. Diabetes Care 2010;33:2697–708.
[15] Jenkins DJ, Wolever TM, Taylor RH, Barker H, Fielden H, Baldwin JM, et al. Glycemic index of foods: a physiological basis for carbohydrate exchange. Am J Clin Nutr 1981;34:362–6.
[16] Konda R, Kakizaki H, Nakai H, Hayashi Y, Hosokawa S, Kawaguchi S, Reflux Nephrology Forum, J. P. S. G, et al. Urinary concentrations of alpha-1-microglobulin and albumin in patients with reflux nephropathy before and after puberty. Nephron 2002;92:812–6.
[17] Thomas D, Elliott EJ. Low glycaemic index, or low glycaemic load, diets for diabetes mellitus. Cochrane Database Syst Rev 2009:CD006296.
[18] Brand JC, Colagiuri S, Crossman S, Allen A, Roberts DC, Truswell AS. Low-glycemic index foods improve long-term glycemic control in NIDDM. Diabetes Care 1991;14:95–101.
[19] Brand-Miller J, Hayne S, Petocz P, Colagiuri S. Low-glycemic index diets in the management of diabetes: a meta-analysis of randomized controlled trials. Diabetes Care 2003;26:2261–7.
[20] Holl RW, Buck C, Babka C, Wolf A, Thon A. HbA1c is not recommended as a screening test for diabetes in cystic fibrosis. Diabetes Care 2000;23:126.
[21] Hunkert F, Lietz T, Stach B, Kiess W. Potential impact of HbA1c determination on clinical decision making in patients with cystic fibrosis-related diabetes. Diabetes Care 1999;22:1008–10.
[22] Jenkins DJ, Kendall CW, McKeown-Eyssen G, Josse RG, Silverberg J, Booth GL, et al. Effect of a low-glycemic index or a high-cereal fiber diet on type 2 diabetes: a randomized trial. JAMA 2008;300:2742–53.
[23] Rizkalla SW, Taghrid L, Laromiguiere M, Huet D, Boillot J, Rigoir A, et al. Improved plasma glucose control, whole-body glucose utilization, and lipid profile on a low-glycemic index diet in type 2 diabetic men: a randomized controlled trial. Diabetes Care 2004;27:1866–72.
[24] Ma Y, Olendzki BC, Merriam PA, Chiriboga DE, Culver AL, Li W, et al. A randomized clinical trial comparing low-glycemic index versus ADA dietary education among individuals with type 2 diabetes. Nutrition 2008;24:45–56.
[25] Skopnik H, Kentrup H, Kusenbach G, Pfaffle R, Kock R. [Glucose homeostasis in cystic fibrosis. Oral glucose tolerance test in comparison with formula administration]. Monatsschr Kinderheilkd 1993;141:42–7.
[26] Ntimbane T, Krishnamoorthy P, Huot C, Legault L, Jacob SV, Brunet S, et al. Oxidative stress and cystic fibrosis-related diabetes: a pilot study in children. J Cyst Fibros 2008;7:373–84.
[27] Balzer BW, Graham CL, Craig ME, Selvadurai H, Donaghue KC, Brand-Miller JC, et al. Low glycaemic index dietary interventions in youth with cystic fibrosis: a systematic review and discussion of the clinical implications. Nutrients 2012;4:286–96.
[28] Stapleton D, Ash C, King S, editors. Dietitians Association of Australia National Cystic Fibrosis Interest Group. Australasian clinical practice guidelines for nutrition in cystic fibrosis; 2006.
[29] Balfour-Lynn I editor. Clinical guidelines: care of children with cystic fibrosis – Royal Brompton Hospital; 2011.
[30] Fitzgerald DA, editor. Cystic fibrosis standards of care 2008. Australia.
[31] Elder DA, Wooldridge JL, Dolan LM, D'Alessio DA. Glucose tolerance, insulin secretion, and insulin sensitivity in children and adolescents with cystic fibrosis and no prior history of diabetes. J Pediatr 2007;151:653–8.
[32] Franzese A, Valerio G, Buono P, Spagnuolo MI, Sepe A, Mozzillo E, et al. Continuous glucose monitoring system in the screening of early glucose derangements in children and adolescents with cystic fibrosis. J Pediatr Endocrinol 2008;21:109–16.
[33] Caprio S. Insulin: the other anabolic hormone of puberty. Acta Paediatr Suppl 1999;88:84–7.
[34] KyngAs HA, Kroll T, Duffy ME. Compliance in adolescents with chronic diseases: a review. J Adolesc Health 2000;26:379–88.
[35] Louie JC, Brand-Miller JC, Moses RG. Carbohydrates, glycemic index, and pregnancy outcomes in gestational diabetes. Curr Diab Rep 2013;13:6–11.
[36] Stephenson AL, Mannik LA, Walsh S, Brotherwood M, Robert R, Darling PB, et al. Longitudinal trends in nutritional status and the relation between lung function and BMI in cystic fibrosis: a population-based cohort study. Am J Clin Nutr 2013;97:872–7.
[37] Hooper L, Abdelhamid A, Moore HJ, Douthwaite W, Skeaff CM, Summerbell CD. Effect of reducing total fat intake on body weight: systematic review and meta-analysis of randomised controlled trials and cohort studies. BMJ 2012;345:e7666.
[38] Brand-Miller J, McMillan-Price J, Steinbeck K, Caterson I. Dietary glycemic index: health implications. J Am Coll Nutr 2009;28(Suppl.):446S–9S.
[39] Barclay AW, Petocz P, McMillan-Price J, Flood VM, Prvan T, Mitchell P, et al. Glycemic index, glycemic load, and chronic disease risk – a meta-analysis of observational studies. Am J Clin Nutr 2008;87:627–37.
[40] Hardin DS, Moran A. Diabetes mellitus in cystic fibrosis. Endocrinol Metab Clin North Am 1999;28:787–800. ix.
[41] Arslanian SA, Bacha F, Saad R, Gungor N. Family history of type 2 diabetes is associated with decreased insulin sensitivity and an impaired balance between insulin sensitivity and insulin secretion in white youth. Diabetes Care 2005;28:115–9.
[42] Brand-Miller JC, Holt SH, Pawlak DB, McMillan J. Glycemic index and obesity. Am J Clin Nutr 2002;76:281S–5S.
[43] Reynolds RC, Stockmann KS, Atkinson FS, Denyer GS, Brand-Miller JC. Effect of the glycemic index of carbohydrates on day-long (10h) profiles of plasma glucose, insulin, cholecystokinin and ghrelin. Eur J Clin Nutr 2009;63:872–8.
[44] Warren JM, Henry CJ, Simonite V. Low glycemic index breakfasts and reduced food intake in preadolescent children. Pediatrics 2003;112:e414.
[45] Krog-Mikkelsen I, Sloth B, Dimitrov D, Tetens I, Bjorck I, Flint A, et al. A low glycemic index diet does not affect postprandial energy metabolism but decreases postprandial insulinemia and increases fullness ratings in healthy women. J Nutr 2011;141:1679–84.
[46] Pawlak DB, Kushner JA, Ludwig DS. Effects of dietary glycaemic index on adiposity, glucose homoeostasis, and plasma lipids in animals. Lancet 2004;364:778–85.
[47] Bornet FR, Jardy-Gennetier AE, Jacquet N, Stowell J. Glycaemic response to foods: impact on satiety and long-term weight regulation. Appetite 2007;49:535–53.
[48] Makris AP, Borradaile KE, Oliver TL, Cassim NG, Rosenbaum DL, Boden GH, et al. The individual and combined effects of glycemic index and protein on glycemic response, hunger, and energy intake. Obesity 2011;19:2365–73.

[49] Liu AG, Most MM, Brashear MM, Johnson WD, Cefalu WT, Greenway FL. Reducing the glycemic index or carbohydrate content of mixed meals reduces postprandial glycemia and insulinemia over the entire day but does not affect satiety. Diabetes Care 2012;35:1633–7.

[50] Larsen TM, Dalskov SM, van Baak M, Jebb SA, Papadaki A, Pfeiffer AF, et al. Diets with high or low protein content and glycemic index for weight-loss maintenance. N Engl J Med 2010;363: 2102–13.

[51] Papadaki A, Linardakis M, Larsen TM, van Baak MA, Lindroos AK, Pfeiffer AF, et al. The effect of protein and glycemic index on children's body composition: the DiOGenes randomized study. Pediatrics 2010;126:e1143–1152.

[52] Thomas DE, Elliott EJ. The use of low-glycaemic index diets in diabetes control. Br J Nutr 2010;104:797–802.

[53] Thomas DE, Elliott EJ, Baur L. Low glycaemic index or low glycaemic load diets for overweight and obesity. Cochrane Database Syst Rev 2007:CD005105.

[54] Turner-McGrievy GM, Jenkins DJ, Barnard ND, Cohen J, Gloede L, Green AA. Decreases in dietary glycemic index are related to weight loss among individuals following therapeutic diets for type 2 diabetes. J Nutr 2011;141:1469–74.

[55] Giacco R, Parillo M, Rivellese AA, Lasorella G, Giacco A, D'Episcopo L, et al. Long-term dietary treatment with increased amounts of fiber-rich low-glycemic index natural foods improves blood glucose control and reduces the number of hypoglycemic events in type 1 diabetic patients. Diabetes Care 2000;23: 1461–6.

[56] Poustie VJ, Russell JE, Watling RM, Ashby D, Smyth RL, Group CTC. Oral protein energy supplements for children with cystic fibrosis: CALICO multicentre randomised controlled trial. BMJ 2006;332:632–6.

[57] Smyth RL, Walters S. Oral calorie supplements for cystic fibrosis. Cochrane Database Syst Rev 2012;10:CD000406.

[58] Stallings VA, Stark LJ, Robinson KA, Feranchak AP, Quinton H, Clinical Practice Guidelines on Growth and Nutrition Subcommittee, Ad Hoc Working Group. Evidence-based practice recommendations for nutrition-related management of children and adults with cystic fibrosis and pancreatic insufficiency: results of a systematic review. J Am Diet Assoc 2008;108:832–9.

[59] Atkinson FS, Foster-Powell K, Brand-Miller JC. International tables of glycemic index and glycemic load values: 2008. Diabetes Care 2008;31:2281–3.

[60] White H, Pollard K, Etherington C, Clifton I, Morton AM, Owen D, et al. Nutritional decline in cystic fibrosis related diabetes: the effect of intensive nutritional intervention. J Cyst Fibros 2009;8:179–85.

[61] Ruf K, Winkler B, Hebestreit A, Gruber W, Hebestreit H. Risks associated with exercise testing and sports participation in cystic fibrosis. J Cyst Fibros 2010;9:339–45.

For Further Information on the Glycemic Index, Please See:

www.glycemicindex.com.
www.gisymbol.com.

CHAPTER

26

Insulin Resistance in Cystic Fibrosis: Management

David Gonzalez Jiménez[1], *M. Francisco Rivas Crespo*[2], *Carlos Bousoño García*[1]

[1]Pediatric Gastroenterology and Nutrition Unit, Universidad de Oviedo, Hospital Universitario Central de Asturias, Oviedo, Spain; [2]Pediatric Endocrinology, Universidad de Oviedo, Hospital Universitario Central de Asturias, Oviedo, Spain

26.1 INSULIN RESISTANCE IN CYSTIC FIBROSIS: AN OVERVIEW

Glucose homeostasis status is the consequence of the response of the β-cell function (i.e., production of insulin) to organic insulin sensitivity. There is an inverse relationship between the two factors: β-cell function adapts to the prevailing insulin sensitivity, forming a hyperbolic curve in which if the former rises, lower insulin production is required, and when it descends, the insular effort rises. Insulin vigorously controls the balance between these so that insulin sensitivity multiplied by β-cell function is an always-constant product [1]. The main signals for the β-cells to respond to organic requirements are glucose, free fatty acids, autonomic nerves, fat-derived hormones, and the gut hormone glucagon-like peptide-1 [2]. Faced with any increase in blood glucose, β-cells respond with an insulin spike (the ability of insulin to stimulate glucose disposal varies more than sixfold), which balances the situation by inducing hepatic glucose uptake (that will be stored as glycogen) and increasing its peripheral uptake (about 80–85% by skeletal muscle).

In the insulin resistance (IR) state, target cells show a reduced sensitivity to ordinary levels of insulin. That is, insulin achieves a less-than-expected effect from liver and muscle response, breaking the hyperbolic interplay between insulin sensitivity and insulin production. The impairment of hepatic insulin response raises glucose production by the liver, leading to fasting hyperglycemia. Muscular IR is identified by an oral glucose tolerance test (OGTT) as impaired glucose tolerance. Although both impaired hepatic response and muscular IR can lead to diabetes, the latter occurs with earlier onset in patients with cystic fibrosis (CF). In this way, some of these patient may become diabetic, although they maintain normal fasting glucose for a time. Thus, basal glucose determination does not allow an early diagnosis of glucose tolerance disorders in patients with CF. A yearly OGTT is required at least from the age of 10 [3].

Insulin can activate the insulin-like growth factor 1 receptor, although it mainly acts through its specific receptor, the concentration or affinity of which depends on its own concentration and a number of eventual physiological or pathological conditions. Activation of the intrinsic tyrosine kinase receptor starts a complex set of postreceptor signaling pathways within target cells, which lead to a broad spectrum of effects (Figure 26.1). Therefore, IR inconveniences are not confined to glucose metabolism. IR is also responsible for suppressing protein synthesis, lipolysis and proteolysis, and cell growth; it also has a protective function on the vascular endothelium and promotes gene expression [4].

The main determinant of CF-related diabetes (CFRD) is insulin deficiency; however, there is no strict relationship between the degree of pancreatic involvement in patients with CF and advancement of the diabetogenic process. Further, there are insulinopenic patients maintaining a normal glucidic tolerance, whereas some of those with CFRD show hyperinsulinism reactive to IR. Therefore, CFRD is a complex complication conditioned by insufficient insulin secretion, although its occurrence is triggered by other factors, not all of which are well known (to explain its disparate incidence in different areas) but among which IR plays an important role.

The phenomenon of IR is the basic CF anomaly and is compounded by the various pathophysiological components of the disease, among which oxidative stress plays an axial role. (Its main elements are schematically

Diet and Exercise in Cystic Fibrosis
http://dx.doi.org/10.1016/B978-0-12-800051-9.00026-2

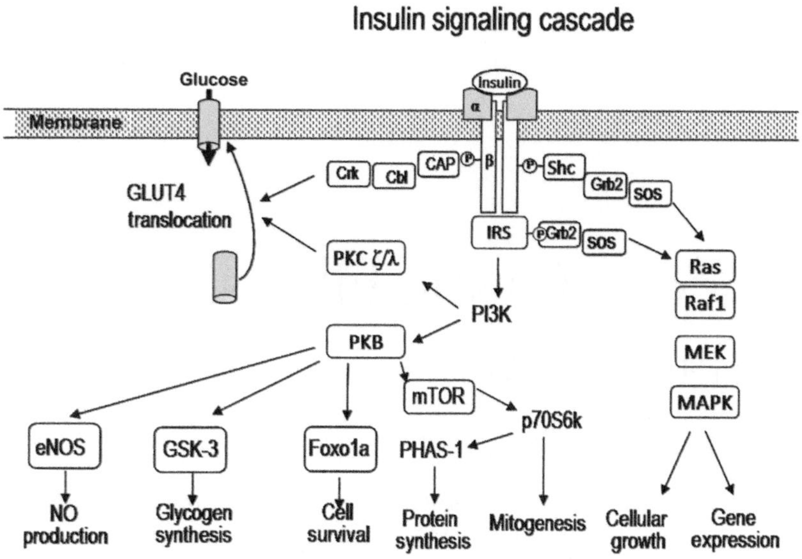

FIGURE 26.1 **Insulin signalin cascade.** Insulin receptor activation induces the phosphorylation of the Tyr residues of several target proteins by means of its intrinsic tyrosine kinase. This causes the activation of the pathways of Cbl/CAP, phosphatidylinositol 3-kinase (PI3K), and mitogen-activated protein kinase (MAPK). Through them, insulin achieves a number of essential organic functions. CAP, Cbl-associated protein; eNOS, endothelial nitric oxide synthase; GLUT4, glucose transporter type 4; GSK-3, glycogen synthase kinase 3; IRS, insulin receptor substrate; MEK, mitogen-activated protein kinase; mTOR, mammalian target of rapamycin; NO, nitric oxide; PKB, protein kinase B; PKC, protein kinase C; SOS, son of sevenless (nucleotide exchange protein).

ordered in Figure 26.2). The endoplasmic reticulum (ER) plays a central role in integrating multiple metabolic signals critical in cellular homeostasis. Together with the Golgi apparatus, the ER is responsible for the synthesis, folding, maturation, storage, and transport of most proteins. Therefore, its ability to adapt and manage diverse metabolic conditions, including adverse ones, is a determinant of cell viability. Under conditions compromising ER function, particularly if folding is disturbed, it evolves the adaptive unfolded protein response to restore the organelle [5]. Patients with CF with an F508del mutation, the most frequent among the white population, have abnormal posttranslational processing of the CF transmembrane regulator (CFTR) protein, which is incorrectly folded (class II mutation). Its accumulation in the ER-Golgi intermediate compartment disrupts ER function ("ER stress"). Further, CFTR-F508del modifies calcium homeostasis because of its interaction with calcium-dependent ER chaperones, reinforcing ER stress [6].

Inflammatory signals interfere with insulin action by means of posttranslational modification of insulin receptor substrate molecules, in particular through serine phosphorylation. This modification, in which c-Jun N-terminal kinase is a central mediator, is universal to diverse forms of IR [7]. The unfolded protein response, which activates c-Jun N-terminal kinase, is closely integrated with stress signaling inflammation [8].

CFRD, one of the most important comorbidities of CF, reinforces and is reinforced by respiratory involvement, which becomes the main determinant of disease prognosis. Patients with CF are especially prone to colonization by *Pseudomonas aeruginosa*, which has a special ability to adapt to conditions of hypoxia of the mucous layer in the airways, developing macrocolonies that are almost impossible to eradicate. This triggers a continuous inflammatory response that includes a high influx of acute inflammatory cells (with a special role for neutrophils, as will be shown later), as well as the release of reactive oxygen species, leading to oxidative stress. This reaction is highly increased during exacerbation, causing severe damage to lung tissues and, therefore, an even higher inflammatory response [6,9].

The CFTR protein operates as an organic anion efflux channel in the cell membrane, which is permeable to Cl^- as well as to other larger organic ions such as reduced glutathione (GSH), the most important intracellular and extracellular antioxidant agent in the body. This is why its concentration in the apical fluid of CFTR-deficient cells (i.e., patients with CF) is markedly deficient (a reduction of 55%) [10]. The consequence is a redox imbalance, which leaves the respiratory apparatus disarmed against oxidative burden.

GSH is a low-molecular-weight tripeptide containing glycine, glutamic acid, and cysteine whose biological activity as a proton donor lies in the sulfhydryl (thiol) group of cysteine. Free radicals can be trapped by this thiol-containing molecule, although its direct antioxidant effect attains several other indirect ways of affecting antioxidant function.

Despite generous supplementation, exogenous antioxidants (retinol, β-carotene, α-tocopherol, ascorbic acid, and ubiquinone) tend to be reduced in the plasma of patients with CF because of their steatorrhea [11]. It should be taken into account that GSH, ascorbic acid, and α-tocopherol are involved in an interdependent regenerating system, where the level of the reduced form of each is dependent on normal levels of the others [9].

Oxidative damage is mainly related to increased production of oxygen or nitrogen free radicals, reduction of antioxidant control, or both. The impairment of insulin action by an excess of free radicals undermines its action, which results in IR and hyperglycemia. The latter causes

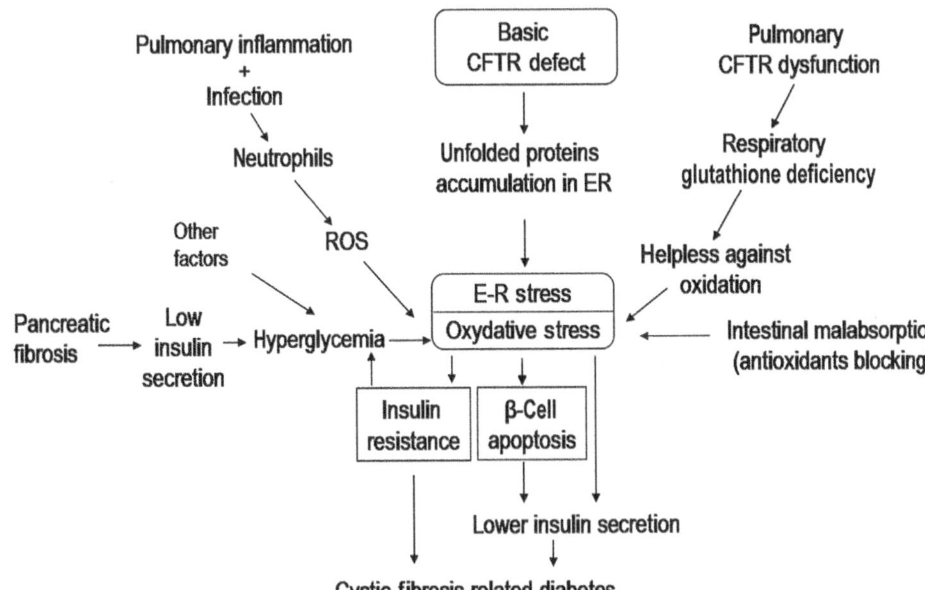

FIGURE 26.2 **Pathophysiological components of insulin resistance in patients with cystic fibrosis.** Oxidative stress plays an axial role (see "Insulin Resistance in Cystic Fibrosis: An Overview" in the text). CFTR, cystic fibrosis transmembrane regulator; ER, endoplasmic reticulum; ROS, reactive oxygen species.

oxidative stress, setting in motion a vicious cycle that depletes antioxidant capacity [12].

The disturbance in mitochondrial function in CF plays a prominent role in the pathophysiology of the disease. Changes in gene expression of mitochondrial proteins, inhibition of oxidative phosphorylation (and increase of mitochondrial DNA damage and disturbance of calcium homeostasis as a consequence of it), oxidative stress (related to the respiratory extracellular GSH deficiency, accumulation of misfolded proteins because of autophagy inhibition, and diminished extracellular superoxide dismutase) allowing chronic inflammation, impairment of innate immunity, and cellular apoptosis are issues participating in the complex CF phenotype and should be included as targets in the therapeutic strategy for these patients [13].

26.2 PATHOPHYSIOLOGY OF INSULIN RESISTANCE

26.2.1 Oxidative Stress

Colonization of the respiratory tract by bacterial pathogens in the mucus of patients with CF leads to a sustained inflammatory response characterized by a massive influx of polymorphonuclear neutrophils and the activation of macrophages, eosinophils, monocytes, and lymphocytes. An integral part of this inflammation is the formation and release of free radicals such as superoxide (O_2^-) and hydrogen ions (OH^-), which can induce oxidative stress. It has been shown that elevated levels of proinflammatory cytokines (interleukin (IL)-1β, IL-6, IL-8), tumor necrosis factor (TNF)-α, and potent chemoattractant neutrophils in bronchoalveolar lavage are involved in the production of pro-oxidants and with tissue damage and apoptosis. Conversely, the production of the immunosuppressive cytokine IL-10, which has anti-inflammatory properties, is reduced or even suppressed [14].

CFTR dysfunction in the pancreas causes exocrine pancreatic insufficiency in almost 90% of patients with CF. This leads to fat malabsorption, which explains the difficulty with gaining or at least maintaining weight, the high incidence of deficits of soluble vitamin antioxidants (vitamins A, E, and D and carotenoids), and deficiencies of essential fatty acids (EFAs). Obviously, the reduced availability of dietary antioxidants may further increase oxidative stress in patients with CF, which apparently plays an important role in the multiorgan pathophysiology of CF.

Consequently, the products of lipid peroxidation, which are markers of oxidative stress, have been detected in exhaled breath condensate, blood, and urine of patients with CF. Thus products of lipid peroxidation are unstable molecules that can reach distant sites and exert various effects, including activation of fibroblast cells in the presence of inflammation, which further increases oxidative stress [15].

In CFRD the damage to exocrine level is conditional of long-term replacement of parenchyma with fibrosis of and damage to Langerhans islets. Basically, it is believed that the obstruction of the pancreatic ducts leads to reduced blood flow in the ischemic tissue, causing pancreatic cell damage and death that is present even from the beginning of life and later leads to overt exocrine deficiency.

Hyperglycemia can cause oxidative stress in many ways, including increased production of mitochondrial free radicals, nonenzymatic glycation of proteins, and glucose auto-oxidation. Furthermore, elevated levels of free fatty acids can also cause oxidative stress due to increased mitochondrial uncoupling and β-oxidation, which ultimately leads to increased formation of free radicals. Oxidative stress can lead to activation of signaling pathways that are sensitive to stress, which in turn may worsen both insulin secretion and IR, impaired glucose tolerance, and, finally, overt diabetes mellitus. Also, compared to other organs such as the lungs or liver, the expression of major endogenous antioxidants such as superoxide dismutase, catalase, and glutathione peroxidase is low in β-cells, and this situation makes them easy targets for free radicals and oxidative stress. Finally, GSH deficiency causes oxidative damage in the pancreas and is associated with the onset of diabetes. CFTR dysfunction, inflammation pathways sensitive to oxidative stress, and the decrease of both GSH and the ratio of GSH to glutathione disulfide (disulfuric glutathione or oxidated glutathione) even in the absence of infection, lead to activation of the nuclear factor κ-light-chain enhancer of activated B cells with a cascade of proinflammatory cytokines such as IL-1β and TNF-α, leading to cell death by the prevailing vicious cycle of chronic infection/inflammation and oxidative stress.

Oxidative stress and inflammation may alter calcium homeostasis by inhibiting the protein that ensures calcium enters the ER for folding and release of insulin. The accumulation of misfolded protein in the ER results in ER stress involved in apoptosis and the death of β-cells.

This IR pathway includes decreased numbers of insulin receptors, reduced insulin binding to receptors, and impaired translocation of glucose transporters. In individuals with CF, insulin-stimulated translocation of the glucose transporter GLUT4 to the plasma membrane is substantially reduced and insulin binding affinity seems to be decreased. IR conditions encountered among patients with CF may result first from the anti-inflammatory therapy and second from increased oxidative stress. It is well known that, among its various effects, corticosteroid administration can cause reduced glucose uptake by skeletal muscle and impaired insulin-mediated suppression of hepatic glucose production [16].

As oxidative stress events are known to maintain inflammatory processes, it is believed that sustained cytokine production leads to the activation of multiple stress-sensitive serine/threonine kinase signaling cascades. These kinases can phosphorylate insulin receptor substrate 1 on specific serine and threonine sites, which lowers the extent of insulin-stimulated tyrosine phosphorylation, thus resulting in diminished translocation of GLUT4 at the membrane surface and lessened insulin action, leading to IR [9,16,17].

26.2.2 Excess Weight and Hypertriglyceridemia

The prevalence of overweight and obesity has been increasing in recent years among patients with CF, as in the general population, placing an additional burden on the endocrine pancreas. The proportion of overweight and obesity in Canadian adults with CF increased from 7.0% before 1990 to 18.4% in the past decade [18]. The 2008–2009 European Cystic Fibrosis Society Patient Registry reports patients with body mass index (BMI) >30 kg/m^2 in many countries in Europe [19,20]. According to the 2011 Australian CF Data Registry, nearly 43% of adult males with CF older than age 30 had a BMI >25 kg/m^2. In fact, nearly half of adult men with CF are overweight or obese [21].

It seems that excess of weight in CF is strongly related to pancreatic status and disease severity, since the vast majority of these patients had pancreatic sufficiency and milder genotypes. They usually carry mutations different than F508del. There also are reported cases of obesity among those with the F508del mutation, although in a smaller proportion [22].

It is well known that BMI is positively associated with forced expiratory volume in 1 s (FEV_1); obese and overweight patients had better lung function than normal and malnourished patients. However, the advantage of FEV_1 is significantly less in the obese group than in the reference group with adequate weight. Beyond a BMI >23 kg/m^2 it is marginal [18].

Overweight pediatric patients with CF also are hyperinsulinemic, as has been recently discovered by our working group [23]. Basal insulin and OGTT- surrogated IR indexes are higher in these patients than in normal and malnourished patients. However, their CFRD prevalence is similar, probably because of their youth, and its evolution will presumably worsen in the coming years.

Regardless of what specific factors have played a role in the increase of overweight and obese subjects, the recommendation of a high-fat, high-calorie diet in CF may now paradoxically have negative consequences in certain patients with CF. It may be disadvantageous for individuals who are pancreatic sufficient and overweight, particularly if they have elevated lipid profiles. Heart disease has been distinctly uncommon in CF to date; however, patients live longer nowadays and complications not previously recognized may arise; risk factors such as excess weight could play a significant role in maintaining health. Therefore, dietary and nutritional support should be recommended to prevent this situation, particularly in pancreatic sufficient patients with mild genotypes.

Insulin plays an important role in lipid metabolism by activating lipoprotein lipase, the key enzyme in adipose tissue triglyceride (TG) storage. Reduction of enzyme activity related to insulin deficiency or IR impairs this process, leading to increased TG levels. Insulin-mediated inhibition of lipolysis in patients with CF tends to be diminished compared to controls, suggesting IR [24]. Hypertriglyceridemia is present in 5–16% of patients with CF, and 4% display hypercholesterolemia [25].

Hypertriglyceridemic patients with CF have glucose levels during OGTT that are similar to those of normotriglyceridemic patients. However, their higher insulin area under the curve suggests IR. Overall, there is a positive association between TG levels and insulin area under the curve in all patients with CF. Hypertriglyceridemic patients represent a subgroup prone to IR-related hyperinsulinemia [26].

26.3 MANAGEMENT OF INSULIN RESISTANCE

26.3.1 Treatment of Oxidative Stress

Malnutrition, infection, and inflammation are the leading causes of weakened antioxidant status, reduced energy intake, increased energy expenditure, and increased oxidative stress in patients with CF. CFRD may further increase calorie losses due to decreased insulin secretion and variable IR, which lead to glycosuria and increased protein catabolism [27]. Therefore, such conditions may contribute to an increased morbidity and mortality in the absence of any attempt to manage energy intake and antioxidant therapy. The importance of nutrition for the quality and length of life in patients with CF has been highlighted in the past, and several correlations have been made between nutritional status and pulmonary function [28].

It is generally accepted that normal growth rates are achieved in most patients with CF who are given access to appropriate nutritional support. Exacerbated lung disease often is followed by anorexia and episodes of vomiting that cause weight loss, which can be worse for survival because they affect respiratory muscle strength, making these ineffective for handling mucus. Because of malabsorption secondary to exocrine pancreatic insufficiency, pancreatic enzyme replacement therapy, a high-energy diet, and essential nutrient supplementation are required to achieve energy demand. For the population with CF, consumption of 120–150% of the Recommended Dietary Allowance for energy intake, including a high proportion of fat (35–40%) and protein (15–20%), is commonly recommended.

Oral pancreatic enzyme supplements such as lipase are mostly required for fat digestion, and patients may need 500–4000 units/g fat ingested per day since they continuously have severe steatorrhea, with fecal energy losses remaining as high as 10–20% of energy intake [29].

Enteral feeding with vegetable oils containing linoleic acid has been shown to improve EFA deficiency in CF. Supplementation with n-3 polyunsaturated fatty acid (PUFA) also has been shown to ameliorate lung function in CF. It is important to note that large amounts of n-3 PUFA can inhibit desaturation and elongation of linoleic acid, resulting in reduced levels of arachidonic acid along with reduced inflammatory processes; these two pathways are in competition for the utilization of a number of common enzymes. However, the benefits of long-term use of n-3 PUFA, as well as the ideal n-6-to-n-3 ratio, remain to be clarified; both groups of fatty acids are oxidizable substrates in cell membranes that are susceptible to oxidative injury and subsequent lung damage [28,29].

Supplementation with vitamin E and β-carotene is effective in the prevention of such damage in patients with CF [30], as the decrease in lipid peroxidation products proves. It is believed that these antioxidants have an important role in maintaining or restoring EFA status by protecting PUFAs from oxidative degradation, as their supplementation increases plasma PUFA levels.

GSH deficiency in the apical cell surface of lung epithelial cells worsens CF as well as several other respiratory diseases [31]. The decline in GSH levels in CF lungs, a phenomenon aggravated by inflammation and infection, has prompted several researchers to propose new therapeutic avenues through the administration of large quantities of GSH [32]. Adequate delivery and a half-life of several hours are required for this antioxidant therapy. Accordingly, in vivo studies have shown an improvement of clinical parameters in CF lung disease using inhaled GSH (600 mg/day) [33]. However, in view of the small number of clinical trials using only a short treatment duration, which limited the resulting enhancement of lung function, further research is needed to determine the extent of the effectiveness of GSH administration on CF symptoms.

The thiol group of cysteine gives GSH a protective capacity against the oxidative action of organic free radicals by acting as an electron donor. GSSG is reconverted to GSH when nicotinamide adenine dinucleotide phosphate yields the necessary electron. The ratio of both forms of molecular glutathione informs of the patient's redox state [34]. GSH also operates as a donor to keep tocopherol and ascorbic acid in their reduced forms [35]. The blood concentration of GSH has not managed to be improved by oral delivery, but the administration of cysteine donor molecules (sulfhydryl groups) enhances antioxidant function, whereas calcitriol elevates serum GSH concentration [36].

26.3.2 Physical Activity

Acute and exhaustive exercise increases oxidative stress, mainly via mitochondrial hydrogen peroxide and superoxide generation and through neutrophil activation, ischemia reperfusion, and catecholamine auto-oxidation. However, regular training eventually provides protection against further exercise-induced oxidative stress by increased endogenous antioxidant production and improved mitochondrial respiratory control. Progressive training and optimized nutritional and antioxidant support are necessary to avoid the negative effects induced by oxidative stress [37].

Several benefits of physical activity have been well documented in CF. Regular aerobic exercise enhances cardiovascular fitness [38], enhances airway clearance, slows the decline in lung function, and improves body mass [39]. Patients who have good aerobic capacity have not only an improved overall prognosis but also an enhanced quality of well-being [40]. Improvements in weight gain (total and fat-free mass), lung function, and leg strength also were described in anaerobic [41] and resistance training programs [42].

Aerobic exercise is effective in improving insulin sensitivity at a variety of intensities and to differing degrees. Significant improvements can be elicited by interval training (high-intensity exercise separated by rest intervals) [43] as well as continuous effort [44]. It seems that this is the most efficient training strategy in improving insulin sensitivity, although further research controlling the volume of training is warranted [45].

While regular participation in physical activity and exercise has been promoted as part of the management of individuals with CF, adherence to exercise in this population can be poor. Significant barriers to regular participation in habitual physical activity include parental and adolescent attitudes, lack of opportunities for participation in sports and exercise, health status and illness or infection, lung function, peripheral skeletal muscle function, poor nutritional status, the vulnerable child syndrome, and the burden of disease.

In summary, exercise training and/or habitual physical activity should become an integral part of the treatment regimen for CF. Optimal training guidelines for patients with CF have yet to be determined, and it is likely that a combination of endurance, anaerobic, and strength exercise and training programs may be the most effective for health benefits and long-term adherence. Clinicians in CF units should encourage children and adults to participate in regular activity to prevent IR and improve glycemic status, among other benefits.

26.3.3 Nutritional Support

26.3.3.1 Macronutrients

26.3.3.1.1 ENERGY

Fat restriction was standard therapy for CF for decades, although in the early 1970s the Toronto CF clinic advocated a high-fat, high-calorie diet with up to 20 or 30 pancreatic enzyme capsules per meal. With the use of data from Toronto, a pivotal study published in 1988 showed that patients in Toronto had a 9-year survival advantage compared with a similar clinic in the United States that prescribed fat restriction. The authors concluded that the improved growth and nutrition seen in Toronto patients contributed to the survival advantage. Since that publication, a high-fat, high-calorie diet has become the standard of care for individuals with CF worldwide to prevent malnutrition. What is less clear is whether there is an advantage of continuing to increase weight over and above the normal range in CF.

Defining the energy needs of patients with CF is a challenge. Individual variables include differences in maldigestion and resultant malabsorption, pulmonary exacerbation, pulmonary function, fat-free mass, sex, pubertal status, genetic mutation, age, and medical complications, including liver disease or CFRD. Daily calorie recommendations provided by various CF societies range from 110% to 200% of those recommended for individuals who do not have CF [46]. Energy requirements vary widely among individuals; therefore, individual estimation of energy requirement is required. A formula that incorporates level of activity, pulmonary function, and degree of malabsorption was included in the Cystic Fibrosis Foundation's nutrition consensus report for use by clinicians in CF centers (Table 26.1).

Overweight and obese patients with CF require special mention. These patients represent a new challenge for clinicians. The current adult and pediatric CF guidelines on nutrition focus on nutritional failure, with no recommendations for the management of individuals who are overweight or obese. They are mostly pancreatic sufficient and have milder genotypes and good lung function. So, the caloric requirements of these patients should be lower than normal and malnourished patients with CF [18].

Although benefit of increasing BMI on lung function is well known, when BMI is $>25\,kg/m^2$ this benefit is small. Moreover, a lack of exercise, high-calorie diet, changing the lipid pattern, hyperinsulinemia associated with overweight and obesity, and increased life expectancy are reasons to believe that these patients may develop macrovascular complications and cardiovascular disease in the future. These new findings suggested that the current diet might not be appropriate for overweight and obese patients with CF. Formulas or indirect calorimetry (if available) can be used as a starting point for calculating energy needs, but gain in weight and height, velocity of weight and height gain, and fat stores may provide a more objective measure of energy balance. The modification of caloric intake should be closely monitored, and in all cases BMI should be above the nutritional objective (50th percentile for children and $22\,kg/m^2$ for female and $23\,kg/m^2$ for male adults).

TABLE 26.1 Energy Requirements According to the US Cystic Fibrosis Foundation

1. Calculate BMR (kcal) from body weight (kg) using World Health Organization equations

Age (Years)	Females	Males
0–3	61.0 wt − 51	60.9 wt − 54
3–10	22.5 wt + 499	22.7 wt + 495
10–18	12.2 wt + 746	17.5 wt + 651
18–30	14.7 wt + 496	15.3 wt + 679

2. Calculate the DEE by multiplying the BMR by activity plus disease coefficients

AC	Disease coefficients	DEE
Confined to bed: BMR × 1.3	FEV_1 >80% predicted: 0	BMR × (AC + 0)
Sedentary: BMR × 1.5	FEV_1 40–79% predicted: 0.2	BMR × (AC + 0.2)
Active: BMR × 1.7	FEV_1 <40% predicted: 0.3–0.5[a]	BMR × (AC + 0.3 to 0.5)

3. Calculate degree of steatorrhea: (100 − stool fat)/fat intake × 100
If a stool collection is not available to determine the fraction of fat intake, an approximate value of 0.85 may be used in the calculation. For PS patients and PI patients with a COA>93% of intake, DER = DEE. For example: for a patient with a COA of 0.78, the factor is 0.93/0.78, or 1.2. If the COA is not known the factor is 1.1.

4. Calculate total DERs = BMR × DEE × (0.93/COA), or 1.1 if not available

Abbreviations: wt, weight; AC, activity coefficient; BMR, basal metabolic rate; COA, coefficient of fat absorption; DEE, daily energy expenditure; DER, daily energy requirement; PS, pancreatic sufficient; PI, pancreatic insufficient.
[a]*May range up to 0.5 with very severe lung disease.*
Adapted from: Ramsey BW, Farrell PM, Pencharz P. Nutritional assessment and management in cystic fibrosis: a consensus report. Am J Clin Nutr 1992; 55:108–16.

26.3.3.1.2 FATS

Nutritional guidelines have traditionally not restricted the quantity and quality of fat intake, and suggested fat should provide 40% of the daily energy needs of individuals with CF. Another practical recommendation for individuals with CF older than the age of five is to consume >100 g fat/day. However, in selected individuals, such as those who are overweight or those who do not require a high-fat diet to achieve adequate energy intake, the recommendations for fat and their subtypes may be different.

Replacing saturated fat with polyunsaturated fat is recommended for the general population to decrease the risk of cardiovascular disease. No similar recommendations exist for people with CF, and little is known about the fat composition of their diet. A recent study of 27 children with CF showed how intakes of saturated fats were well above the reference nutrient intakes (158%), whereas intakes of polyunsaturated fats were consistently below reference nutrient intakes (92%) [47]. The high saturated fatty acids are in part to be expected because a higher intake of fat is required to meet recommendations. However, it is the disproportionate distribution of the fat that is of concern. So we may recommend that patients with CF choose fats in the form of monounsaturated and polyunsaturated fats from foods such us oils, spreads, avocado, nuts, and seeds, especially if they have cardiovascular risk factors.

A peculiar role in the onset and development of IR is played by trans fatty acids, whose major dietary sources are partially hydrogenated vegetable oils and food items that contain them. Indeed, trans fatty acid consumption causes metabolic dysfunction. In particular, it adversely affects circulating lipid levels, triggers systemic inflammation, induces endothelial dysfunction, and, according to some studies, increases visceral adiposity, body weight, and IR [48]. Patients should be advised to minimize consumption of industrially produced trans fatty acids.

26.3.3.1.3 FIBER AND WHOLE CEREAL INTAKES MAY PROTECT AGAINST HYPERINSULINISM

A randomized crossover trial showed benefits of consuming high-fiber rye bread on insulin secretion, suggesting an improvement of β-cell function [49]. Moderate fiber intake (10–30 g/day for adults and 5–10 g plus their age in years in children) in CF also provides overall health benefits for these patients, such us avoiding constipation and abdominal pain.

26.3.3.2 *Micronutrients*

26.3.3.2.1 VITAMINS

Fat-soluble vitamin deficiencies are common among patients with CF, despite routine oral supplementation. These deficiencies, mainly those of vitamins D and E, are associated with poorer clinical status and worse IR and glucose homeostasis. Thus, periodic monitoring and assessments of adherence to treatment should be conducted until appropriate levels are achieved.

Vitamin E. Its predominant forms in diet and plasma are α- and γ-tocopherol, both of which have a high antioxidant ability. In addition, the former regulates the expression of genes involved in a wide range of functions, such as cell cycle regulation, inflammation, and cell adhesion; cell signaling; and lipid uptake [50].

Epidemiologic studies have suggested the positive influence of vitamin E on insulin sensitivity. More than 15 years ago, the improving effect of vitamin E supplementation on the insulin sensitivity , fasting plasma glucose, and glycosylated hemoglobin (HbA_{1c}) of diabetic patients [51] and older, nonobese subjects [52] was reported. The positive association between plasma α-tocopherol concentrations and insulin sensitivity in nonsupplemented patients has been shown more recently [53].

Although for many years it has been thought that vitamin E exerts insulin-sensitizing effects through its antioxidant capacity [54], recent research implicated α-tocopherol in the direct regulation of adiponectin and *PPARγ* gene expression. These two proteins play an important role in insulin sensitivity [55].

Unfortunately, vitamin E deficiency is common in patients with CF, as was proved by the longitudinal monitoring of 127 children, identified as having CF by newborn screening; the ratio of serum α-tocopherol to total serum lipids was low in 6–14% of them over several successive years [56]. The devastating effects of chronic vitamin E deficiency in patients with CF are known [57]. However, to our knowledge, the role of vitamin E on insulin sensitivity in these patients has not yet been satisfactorily studied. Although this effect is suggested by the previously discussed facts, giving another reason to achieve optimal vitamin E levels in these patients, this issue is waiting for the right approach.

Based on the American Consensus Report [58], recommendations for vitamin E supplementation are 40–50 mg (40–50 IU) for those 0–12 months old; 80–150 mg (80–150 IU) for those 1–3 years old; 100–200 mg (100–200 IU) for those 4–8 years old; and 200–400 mg (200–400 IU) for those >8 years old. Water-soluble forms have superior bioavailability compared with fat-soluble formulations [59], and the trend to include multiple vitamin E compounds (γ-tocopherol in addition to or in lieu of α-tocopherol) may be particularly important for patients with CF because of its function as a scavenger of reactive nitrogen species and its synergistic effects with α-tocopherol. Improved antioxidant capacity with γ-tocopherol could potentially decrease oxidant-mediated damage and limit cytokine-mediated neutrophil recruitment [60].

Vitamin D. Impaired absorption of fat-soluble vitamins as well as decreased sunlight exposure and suboptimal intake make vitamin D deficiency a common condition in individuals with CF [61]. Even more, between 50% and 90% of these patients do not achieve the serum calcidiol goal of the CF Foundation (at least 30 ng/mL), despite oral supplementation [62].

Characterized as a regulator of homeostasis of bone and mineral metabolism, it can also provide no skeletal actions through its receptors in many tissues, such as brain, prostate, breast, colon, and pancreas, and immune cells. The discovery of its roles in modulation of the immune response, or regulation of cell proliferation and differentiation, has broadened the view of the physiological role of this molecule. However, there seem to be more aspects that are interesting [63,64].

1,25-Dihydroxyvitamin D plays an important role in glucose homeostasis by improving β-cell function and enhancing insulin sensitivity of the target cells (liver, skeletal muscle, and adipose tissue) [65–67]. In addition, it protects β-cells from detrimental immune attacks, not only directly by its action on β-cells but also indirectly by acting on different immune cells, including inflammatory macrophages, dendritic cells, and a variety of T cells [68].

The study of a large cohort of Scandinavian patients with CF showed that vitamin D insufficiency was a significant risk factor for CFRD [69]. HbA_{1c} values were positively associated with serum calcidiol <12 ng/mL and <20 ng/mL, but this association was only in pediatric patients. However, the Scandinavian Cystic Fibrosis Nutritional Study was not designed for the specific study of diabetes, and they do not have data on C-peptide, fasting plasma glucose, or peripheral insulin sensitivity. Therefore, this study unfortunately cannot address the question of underlying mechanisms behind the associations found.

Several data support the association between IR and serum calcidiol concentration in 90 young patients with CF studied by our working group (publication of these results is pending). Progression through increasing serum calcidiol ranges corresponds with decreasing values of OGTT-surrogated IR indexes. Further, an inverse relationship between serum calcidiol and IR indexes was appreciated.

Worldwide, the currently used vitamin supplementation regimen has not been able to correct vitamin D insufficiency in CF. Recent findings suggested that vitamin D in CF may help to outline future prevention and treatment strategies for IR and CFRD. It should be noted that children with CF have more preserved insulin secretion and greater peripheral insulin sensitivity as a result of a lower degree of inflammation. Thus they have greater potential for the glucose-lowering effect of vitamin D. It seems safe to assume that maintaining serum calcidiol sufficiency from a very early age would protect patients with CF from the pathogenic effect of IR over β-cell destruction.

Routine oral dosing with vitamin D includes 400–500 UI for patients <12 months old; 800–1000 UI from 12 months to 10 years old; and 800–2000 UI for older than 10 years [62]. A change in vitamin D supplementation policy (increase 450%; between 3000 and 4000 UI/day) could prevent vitamin D deficiency but still fail to reach the target level (>30 ng/mL) [70]. In those cases in which the usual supplementation is not sufficient, it is necessary to assess for therapeutic adherence, review pancreatic enzyme replacement therapy, and determine whether it is necessary to increase the vitamin D dose according to evidence-based recommendations from the CF Foundation [62].

Vitamin C. Vitamin C is present in the respiratory lining fluid of human lungs, and local deficit occurs during oxidative stress. Experimental findings confirm that vitamin C induced the openings of CFTR Cl^- channels without a detectable increase in intracellular cyclic AMP levels. Vitamin C instilled into the nasal epithelium of human subjects effectively activates Cl^- transport, too. The pool of vitamin C in the respiratory tract represents a potential nutraceutical and pharmaceutical target for the complementary treatment of oxidative stress in patients with CF [71].

Water-soluble vitamins seem to be well absorbed by patients with CF, but there is documented evidence of poor dietary intake. A supplement of at least 50–100 mg vitamin C/day should be prescribed for patients with an unbalanced diet or if there is evidence of deficiency.

26.3.3.2.2 OTHER ANTIOXIDANTS AND MICRONUTRIENTS

Dietary intake of selenium is inversely related to inflammatory markers such as sialic acid and triacylglycerol [72]. Moreover, a possible role of selenium in the modulation of serum complement 3, which may be an early marker of metabolic syndrome manifestations, also was documented. Consensus regarding supplementation of antioxidants in CF to include selenium is yet to be established. Two studies attempted to demonstrate successfully how the administration of 2.8 μg/kg/day and 90 μg/day of selenium was able to decrease oxidative stress in CF [73].

26.3.3.3 Potential Benefit from Specific Nutritional Components

Besides having a high nutritive value, legumes contain a significant quantity of polyphenolic compounds such as flavonoids, isoflavones, phenolic acids, and lignans. Several soy constituents might have a favorable effect on IR, and some authors propose their therapeutic role in the metabolic syndrome [74]. Not only does soy consumption has interesting benefits on metabolic alterations but more common legumes, such as beans, lentils, and chickpeas, also improve oxidative stress markers [75]. In CF, using soy-based diets as the only dairy intake is not recommended except in cases of intolerance or cow's milk allergy. However, legumes including soy should be recommended three or four times a week for nutritive value and the quality of their compounds.

Nuts have been considered of interest since they provide many bioactive compounds that may benefit metabolic syndrome components. Daily nut consumption reduced insulin and homeostatic model assessment-IR in patients with IR [76]. In patients with CF they are very interesting because they are high-energy foods and a source of unsaturated fatty acids, especially oleic acid (monounsaturated fatty acid) and linoleic acid (PUFA). In addition, they supply significant amounts of tocopherols, squalene, and phytosterols, which are relevant compounds with antioxidant properties. So the consumption of nuts two or more times per week should be recommended in patients with CF.

Flavonoids contain a variable number of hydroxyl phenolic groups. They are located in sources of vegetal origin: fruits, seeds, roots, flowers, tea, or wine. An interesting kind of flavonoid for CF is genistein, which is found principally in soybean and related products. Genistein suppressed lipopolysaccharide-induced TNF-α and IL-6 production in mouse macrophages and blocked the secretion of TNF-α and IL-1 in phytohemaglutinin-stimulated macrophages [77]. Besides anti-inflammatory properties, genistein has been recognized as a CFTR potentiator. Although the exact mechanism of action remains elusive, studies have identified binding sites for genistein on CFTR, whereas other studies have suggested that the drug acts by altering lipid bilayer mechanics [78].

Curcumin, a bright yellow compound derived from the rhizomes of the plant *Curcuma longa*, mitigates inflammatory responses by inhibiting cyclooxygenase-2, lipoxygenase, nuclear factor-κB, inducible nitric oxide synthase, and nitrite oxide production in lipopolysaccharides, macrophages, and natural killer cells [79]. In CF, recent findings confirm that curcumin and genistein additively potentiate G551D-CFTR channels and could potentially restore the activity of some mutations associated with severe CF [80].

References

[1] Kahn SE, Prigeon RL, McCulloch DK, Boyko EJ, Bergman RN, Schwartz MW, et al. Quantification of the relationship between insulin sensitivity and beta-cell function in human subjects. Evidence for a hyperbolic function. Diabetes 1993;42:1663–72.

[2] Ahrén B, Pacini G. Islet adaptation to insulin resistance: mechanisms and implications for intervention. Diabetes Obes Metab 2005;7:2–8.

[3] O'Riordan SM, Robinson PD, Donaghue KC, Moran A, ISPAD Clinical Practice Consensus. Management of cystic fibrosis-related diabetes. Pediatr Diabetes 2008;9(Part I):338–44.

[4] Sesti G. Pathophysiology of insulin resistance. Best Pract Res Clin Endocrinol Metab 2006;20:665–79.

[5] Hotamisligil GS. Inflammation and endoplasmic reticulum stress in obesity and diabetes. Int J Obes (Lond) 2008;32(Suppl. 7):S52–4.

[6] Rottner M, Freyssinet JM, Martínez MC. Mechanisms of the noxious inflammatory cycle in cystic fibrosis. Respir Res 2009;10:23.

[7] Hotamisligil GS. Role of endoplasmic reticulum stress and c-Jun NH2-terminal kinase pathways in inflammation and origin of obesity and diabetes. Diabetes 2005;54(Suppl. 2):S73–8.

[8] Hirosumi J, Tuncman G, Chang L, Gorgun CZ, Uysal KT, Maeda K, et al. A central role for JNK in obesity and insulin resistance. Nature 2002;420:333–6.

[9] Ntimbane T, Comte B, Mailhot G, Berthiaume Y, Poitout V, Prentki M, et al. Cystic fibrosis-related diabetes: from CFTR dysfunction to oxidative stress. Clin Biochem Rev 2009;30:153–77.

[10] Gao L, Kim KJ, Yankaskas JR, Forman HJ. Abnormal glutathione transport in cystic fibrosis airway epithelia. Am J Physiol 1999;277:L113–8.

[11] Maqbool A, Stallings VA. Update on fat-soluble vitamins in cystic fibrosis. Curr Opin Pulm Med 2008;14:574–81.

[12] Ceriello A. Oxidative stress and glycemica regulation. Metabolism 2000;49:27–9.

[13] Valdivieso AG, Santa-Coloma TA. CFTR activity and mitochondrial function. Redox Biol 2013;1:190–202.

[14] Back EI, Frindt C, Nohr D, Frank J, Ziebach R, Stern M, et al. Antioxidant deficiency in cystic fibrosis: when is the right time to take action? Am J Clin Nutr 2004;80:374–84.

[15] Evans JL, Goldfine ID, Maddux BA, Grodsky GM. Are oxidative stress-activated signaling pathways mediators of insulin resistance and beta-cell dysfunction. Diabetes 2003;52:1–8.
[16] Evans JL, Maddux BA, Goldfine ID. The molecular basis for oxidative stress-induced insulin resistance. Antioxid Redox Signal 2005;7:1040–52.
[17] Hardin DS, Leblanc A, Marshall G, Seilheimer DK. Mechanisms of insulin resistance in cystic fibrosis. Am J Physiol Endocrinol Metab 2001;281:E1022–8.
[18] Stephenson AL, Mannik LA, Walsh S, Brotherwood M, Robert R, Darling PB, et al. Longitudinal trends in nutritional status and the relation between lung function and BMI in cystic fibrosis: a population-based cohort study. Am J Clin Nutr 2013;97:872–7.
[19] European Cystic Fibrosis Society. Patient registry. Annual data report (2008–2009 data). Version 03; 2012.
[20] Panagopoulou P, Maria F, Nikolaou A, Nousia-Arvanitakis S. Prevalence of malnutrition and obesity among cystic fibrosis patients. Pediatr Int September 4, 2013.
[21] Cystic fibrosis in Australia 2011. 14th annual report from the Australian Cystic Fibrosis Data Registry; 2012.
[22] Kastner-Cole D, Palmer CN, Ogston SA, Mehta A, Mukhopadhyay S. Overweight and obesity in deltaF508 homozygous cystic fibrosis. J Pediatr 2005;147:402–4.
[23] González D, Bousoño C, Rivas MF, Díaz JJ, Acuña MD, Heredia S, et al. Insulin resistance in overweight cystic fibrosis paediatric patients. An Pediatr (Barc) 2012;76:279–84.
[24] Moran A, Basu R, Milla C, Jensen MD. Insulin regulation of free fatty acid kinetics in adult cystic fibrosis patients with impaired glucose tolerance. Metabolism 2004;53:1467–72.
[25] Figueroa V, Milla C, Parks EJ, Schwarzenberg SJ, Moran A. Abnormal lipid concentrations in cystic fibrosis. Am J Clin Nutr June 2002;75:1005–11.
[26] Ishimo MC, Belson L, Ziai S, Levy E, Berthiaume Y, Coderre L, et al. Hypertriglyceridemia is associated with insulin levels in adult cystic fibrosis patients. J Cyst Fibros 2013;12:271–6.
[27] Brennan AL, Geddes DM, Gyi KM, Baker EH. Clinical importance of cystic fibrosis-related diabetes. J Cyst Fibros 2004;3:209–22.
[28] Luder E, Kattan M, Thornton JC, Koehler KM, Bonforte RJ. Efficacy of a nonrestricted fat diet in patients with cystic fibrosis. Am J Dis Child 1989;143:458–64.
[29] Borowitz DS, Grand RJ, Durie PR. Use of pancreatic enzyme supplements for patients with cystic fibrosis in the context of fibrosing colonopathy. Consensus Committee. J Pediatr 1995;127:681–4.
[30] Rivas-Crespo MF, González Jiménez D, Acuña Quirós MD, Sojo Aguirre A, Heredia González S, Díaz Martín JJ, et al. High serum retinol and lung function in young patients with cystic fibrosis. J Pediatr Gastroenterol Nutr 2013;56:657–62.
[31] Kogan I, Ramjeesingh M, Li C, Kidd JF, Wang Y, Leslie EM, et al. CFTR directly mediates nucleotide-regulated glutathione flux. EMBO J 2003;22:1981–9.
[32] Bishop C, Hudson VM, Hilton SC, Wilde C. A pilot study of the effect of inhaled buffered reduced glutathione on the clinical status of patients with cystic fibrosis. Chest 2005;127:308–17.
[33] Griese M, Ramakers J, Krasselt A, Starosta V, Van Koningsbruggen S, Fischer R, et al. Improvement of alveolar glutathione and lung function but not oxidative state in cystic fibrosis. Am J Respir Crit Care Med 2004;169:822–8.
[34] Pastore A, Piemonte F, Locatelli M, Lo Russo AL, Gaeta LM, Tozzi G, et al. Determination of blood total, reduced, and oxidized glutathione in pediatric subjects. Clin Chem 2003;47:1467–9.
[35] Scholz RW, Graham KS, Gumpricht E, Reddy CC. Mechanism of interaction of vitamin E and glutathione in the protection against membrane lipid peroxidation. Ann NY Acad Sci 1989;570:514–7.
[36] Lands LC, Grey VL, Smountas AA. Effect of supplementation with a cysteine donor on muscular performance. J Appl Physiol (1985) 1999;87:1381–5.
[37] Aguiló A, Tauler P, Fuentespina E, Tur JA, Córdova A, Pons A. Antioxidant response to oxidative stress induced by exhaustive exercise. Physiol Behav 2005;84:1–7.
[38] Moorcroft A, Dodd M, Webb A. Long-term change in exercise capacity, body mass, and pulmonary function in adults with cystic fibrosis. Chest 1997;111:338–43.
[39] Schneiderman-Walker J, Pollock SL, Corey M, Wilkes DD, Canny GJ, Pedder L, et al. A randomized controlled trial of a 3-year home exercise program in cystic fibrosis. J Pediatr 2000;136:304–10.
[40] Orenstein D, Nixon P, Ross E, Kaplan R. The quality of well-being in cystic fibrosis. Chest 1989;95:344–7.
[41] Klijn PH, Oudshoorn A, van der Ent CK, van der Net J, Kimpen JL, Helders PJ. Effects of anaerobic training in children with cystic fibrosis: a randomized controlled study. Chest 2004;125:1299–305.
[42] Orenstein DM, Hovell MF, Mulvihill M, Keating KK, Hofstetter CR, Kelsey S, et al. Strength vs aerobic training in children with cystic fibrosis: a randomized controlled trial. Chest 2004;126:1204–14.
[43] Babraj JA, Vollaard NB, Keast C, Guppy FM, Cottrell G, Timmons JA. Extremely short duration high intensity interval training substantially improves insulin action in young healthy males. BMC Endocr Disord 2009;9:3.
[44] Magkos F, Tsekouras Y, Kavouras SA, Mittendorfer B, Sidossis LS. Improved insulin sensitivity after a single bout of exercise is curvilinearly related to exercise energy expenditure. Clin Sci (Lond) 2008;114:59–64.
[45] Balducci S, Leonetti F, Di Mario U, Fallucca F. Is a long-term aerobic plus resistance training program feasible for and effective on metabolic profiles in type 2 diabetic patients? Diabetes Care 2004;27:841–2.
[46] Stallings VA, Stark LJ, Robinson KA, Feranchak AP, Quinton H, Clinical Practice Guidelines on Growth and Nutrition Subcommittee, Ad Hoc Working Group. Evidence-based practice recommendations for nutrition-related management of children and adults with cystic fibrosis and pancreatic insufficiency: results of a systematic review. J Am Diet Assoc 2008;108:832–9.
[47] Smith C, Winn A, Seddon P, Ranganathan S. A fat lot of good: balance and trends in fat intake in children with cystic fibrosis. J Cyst Fibros 2012;11:154–7.
[48] Micha R, Mozaffarian D. Trans fatty acids: effects on metabolic syndrome, heart disease and diabetes. Nat Rev Endocrinol 2009;5:335–44.
[49] Leinonen K, Liukkonen K, Poutanen K, Uusitupa M, Mykkanen H. Rye bread decreases postprandial insulin response but does not alter glucose response in healthy Finnish subjects. Eur J Clin Nutr 1999;53:262–7e.
[50] Azzi A, Gysin R, Kempná P, Munteanu A, Villacorta L, Visarius T, et al. Regulation of gene expression by alpha-tocopherol. Biol Chem 2004;385:585–91.
[51] Paolisso G, D'Amore A, Giugliano D, Ceriello A, Varricchio M, D'Onofrio F. Pharmacologic doses of vitamin E improve insulin action in healthy subjects and non-insulin-dependent diabetic patients. Am J Clin Nutr 1993;57:650–6.
[52] Paolisso G, Di Maro G, Galzerano D, Cacciapuoti F, Varricchio G, Varricchio M, et al. Pharmacological doses of vitamin E and insulin action in elderly subjects. Am J Clin Nutr 1994;59:1291–6.
[53] Costacou T, Ma B, King IB, Mayer-Davis EJ. Plasma and dietary vitamin E in relation to insulin secretion and sensitivity. Diabetes Obes Metab 2008;10:223–8.
[54] Vinayaga Moorthi R, Bobby Z, Selvaraj N, Sridhar MG. Vitamin E protects the insulin sensitivity and redox balance in rat L6 muscle cells exposed to oxidative stress. Clin Chim Acta 2006;367:132–6.
[55] Landrier JF, Gouranton E, El Yazidi C, Malezet C, Balaguer P, Borel P, et al. Adiponectin expression is induced by vitamin E via a peroxisome proliferator-activated receptor gamma-dependent mechanism. Endocrinology 2009;150:5318–25.

[56] Feranchak AP, Sontag MK, Wagener JS, Hammond KB, Accurso FJ, Sokol RJ. Prospective, long-term study of fat-soluble vitamin status in children with cystic fibrosis identified by newborn screen. J Pediatr 1999;135:601–10.

[57] Koscik RL, Lai HJ, Laxova A, Zaremba KM, Kosorok MR, Douglas JA, et al. Preventing early, prolonged vitamin E deficiency: an opportunity for better cognitive outcomes via early diagnosis through neonatal screening. J Pediatr 2005; 147(3 Suppl.):S51–6.

[58] Borowitz D, Baker RD, Stallings V. Consensus report on nutrition for pediatric patients with cystic fibrosis. J Pediatr Gastroenterol Nutr 2002;35:246–59.

[59] Papas K, Kalbfleisch J, Mohon R. Bioavailability of a novel, water-soluble vitamin E formulation in malabsorbing patients. Dig Dis Sci 2007;52:347–52.

[60] Papas KA, Sontag MK, Pardee C, Sokol RJ, Sagel SD, Accurso FJ, et al. A pilot study on the safety and efficacy of a novel antioxidant rich formulation in patients with cystic fibrosis. J Cyst Fibros 2008;7:60–7.

[61] Hall WB, Sparks AA, Aris RM. Vitamin D deficiency in cystic fibrosis. Int J Endocrinol 2010;2010:218691.

[62] Tangpricha V, Kelly A, Stephenson A, Maguiness K, Enders J, Robinson KA, et al. An update on the screening, diagnosis, management, and treatment of vitamin D deficiency in individuals with cystic fibrosis: evidence-based recommendations from the Cystic Fibrosis Foundation. J Clin Endocrinol Metab 2012;97:1082–93.

[63] Holick MF. Vitamin D: a millenium perspective. J Cell Biochem 2003;88:296–307.

[64] Haussler MR, Whitfield GK, Haussler CA, Hsieh JC, Thompson PD, Selznick SH, et al. The nuclear vitamin D receptor: biological and molecular regulatory properties revealed. J Bone Miner Res 1998;13:325–49.

[65] Chiu KC, Chu A, Go VL, Saad MF. Hypovitaminosis D is associated with insulin resistance and beta cell dysfunction. Am J Clin Nutr 2004;79:820–5.

[66] Deleskog A, Hilding A, Brismar K, Hamsten A, Efendic S, Ostenson CG. Low serum 25-hydroxyvitamin D level predicts progression to type 2 diabetes in individuals with prediabetes but not with normal glucose tolerance. Diabetologia 2012;55:1668–78.

[67] Forouhi NG, Ye Z, Rickard AP, Khaw KT, Luben R, Langenberg C, et al. Circulating 25-hydroxyvitamin D concentration and the risk of type 2 diabetes: results from the European Prospective Investigation into Cancer (EPIC)-Norfolk cohort and updated meta-analysis of prospective studies. Diabetologia 2012;55:2173–82.

[68] Hewison M. Vitamin D and the immune system: new perspectives on an old theme. Endocrinol Metab Clin North Am 2010;39:365–79.

[69] Pincikova T, Nilsson K, Moen IE, Fluge G, Hollsing A, Knudsen PK, et al. Vitamin D deficiency as a risk factor for cystic fibrosis-related diabetes in the Scandinavian Cystic Fibrosis Nutritional Study. Diabetologia 2011;54:3007–15.

[70] Brodlie M, Orchard WA, Reeks GA, Pattman S, McCabe H, O'Brien CJ, et al. Vitamin D in children with cystic fibrosis. Arch Dis Child 2012;97:982–4.

[71] Fischer H, Schwarzer C, Illek B. Vitamin C controls the cystic fibrosis transmembrane conductance regulator chloride channel. Proc Natl Acad Sci USA 2004;101:3691–6.

[72] Zulet MA, Puchau B, Hermsdorff HH, Navarro C, Martinez JA. Dietary selenium intake is negatively associated with serum sialic acid and metabolic syndrome features in healthy young adults. Nutr Res 2009;29:41–8e.

[73] Shamseer L, Adams D, Brown N, Johnson JA, Vohra S. Antioxidant micronutrients for lung disease in cystic fibrosis. Cochrane Database Syst Rev December 8, 2010;12:CD007020.

[74] Merritt JC. Metabolic syndrome: soybean foods and serum lipids. J Natl Med Assoc 2004;96:1032–41.

[75] Crujeiras AB, Parra D, Abete I, Martinez JA. A hypocaloric diet enriched in legumes specifically mitigates lipid peroxidation in obese subjects. Free Radic Res 2007;41:498–506.

[76] Casas-Agustench P, López-Uriarte P, Bulló M, Ros E, Cabré-Vila JJ, Salas-Salvadó J. Effects of one serving of mixed nuts on serum lipids, insulin resistance and inflammatory markers in patients with the metabolic syndrome. Nutr Metab Cardiovasc Dis 2011;21:126–35e.

[77] Kesherwani V, Sodhi A. Involvement of tyrosine kinases and MAP kinases in the production of TNF-alpha and IL-1beta by macrophages in vitro on treatment with phytohemagglutinin. J Interferon Cytokine Res 2007;27:497–505.

[78] Hwang TC, Koeppe II RE, Andersen OS. Genistein can modulate channel function by a phosphorylation-independent mechanism: importance of hydrophobic mismatch and bilayer mechanics. Biochemistry 2003;42:13646–58.

[79] Brouet I, Ohshima H. Curcumin, an anti-tumour promoter and anti-inflammatory agent, inhibits induction of nitric oxide synthase in activated macrophages. Biochem Biophys Res Commun 1995;206:533–40.

[80] Yu YC, Miki H, Nakamura Y, Hanyuda A, Matsuzaki Y, Abe Y, et al. Curcumin and genistein additively potentiate G551D-CFTR. J Cyst Fibros 2011;10:243–52.

CHAPTER

27

Cystic Fibrosis and Celiac Disease

John F. Pohl[1], *Amy Lowichik*[2], *Amy Cantrell*[3]

[1]Department of Pediatric Gastroenterology, Primary Children's Medical Center, University of Utah, Salt Lake City, UT, USA, [2]Division of Pediatric Pathology, Primary Children's Medical Center, Salt Lake City, UT, USA, [3]Division of Pediatric Endocrinology, Scott and White Hospital, Temple, TX, USA

27.1 INTRODUCTION

Cystic fibrosis (CF) is caused by a mutation of the CF transmembrane conductance regulator gene (*CFTR*), which subsequently affects the translated CFTR protein leading to abnormal movement of fluid across the apical region of all epithelial cells. Diagnosis generally is made by sweat chloride measurement or DNA screening for CF mutations [1,2]. Typically, CF is associated with acquired pulmonary disease, although gastrointestinal (GI) disease, including pancreatic insufficiency (PI), meconium ileus in the newborn, malabsorption, steatorrhea, and weight loss are common presentations of the disease process [3]. Therefore, when a CF patient presents with diarrhea, abdominal pain, or weight loss, it may be assumed that the patient is having an exacerbation of GI-associated disease related to CF. Bacterial overgrowth is common in CF due to intestinal dysmotility, chronic antibiotic use, prolonged use of acid suppression medication, and an increased risk of intestinal inflammation [3,4]. Other common causes of malabsorption, such as celiac disease (CD), should be considered in this setting.

CD is an autoimmune disease in which an increased immune response to gliadin (a common protein found in the gluten component of the wheat seed) leads to inflammation primarily of the small intestine, with associated diarrhea, malabsorption, weight loss, and abdominal pain [5]. CD is a T-cell-driven autoimmune process in which the human leukocyte antigens, HLA-DQ2 and HLA-DQ8, of antigen-presenting cells bind to gluten peptides leading to subsequent CD4+ T-cell activation. Interestingly, HLA-DQ2 and HLA-DQ8 bind to negatively charged peptides, and tissue transglutaminase (TTG), a cross-linking calcium-dependent enzyme, can convert glutamine residues of gluten into negatively charged glutamic acid, which has a high affinity for HLA-DQ2 and HLA-DQ8. TTG typically is considered to be intracellular and inactive. In times of tissue injury, however, TTG becomes extracellular and subsequently active causing protein binding and production of T-cell-specific stimulation through gluten peptide exposure. Immunogenic similarities with other peptides, such as those found in barley and rye, also can lead to a similar inflammatory response [6]. The inflammatory response to gluten peptides causes infiltration of intraepithelial lymphocytes (IELs) into the brush border region of the small bowel leading to enterocyte apoptosis, villous atrophy, crypt hyperplasia, and subsequent malabsorption [5]. Evidence suggests that the timing of gluten exposure during infancy, such as in the first 3 months of life or after 7 months of age, may increase the risk of developing CD [7]. An epidemic of CD in Sweden in the 1980s led to an epidemiologic surveillance system, which demonstrated possible risk factors for childhood CD, including delayed gluten exposure until 6 months of age as opposed to 4 months of age, a decrease in breastfeeding rates in infants at 6 months of age, and increased consumption of gluten-containing milk products before 2 years of age [8].

27.2 CLINICAL FEATURES

CD is a common GI disease, especially in Europe, with up to 1 in 99 Finnish children having CD [9]. The incidence of CD in the United States is not as clear, although one study from Denver, Colorado, followed a large group of children with high-risk HLA genotypes and found that the potential risk for developing CD was 0.9% in that pediatric population [10]. Other studies have suggested that CD prevalence in the United States is quite high and similar to many European countries [11].

Diet and Exercise in Cystic Fibrosis
http://dx.doi.org/10.1016/B978-0-12-800051-9.00027-4

Clinical features are consistent with signs and symptoms of malabsorption. Weight loss, abdominal pain, diarrhea, fatigue, and anemia are seen in adults with CD [12]. Interestingly, mutations of the hemochromatosis susceptibility gene appear to be more common in adult CD patients, leading to prevention of iron deficiency anemia [13]. Children with CD often present between 6 and 24 months of age with diarrhea, abdominal pain, vomiting, constipation, failure to thrive, irritability, ascites from a hypoprotein state (kwashiorkor), short stature, or enamel hypoplasia of the permanent teeth [14]. Diarrhea symptoms of CD, in particular, can present in a manner similar to diarrhea-predominant irritable bowel syndrome (IBS) and often can be misdiagnosed as IBS [15]. A pruritic rash specific for CD, dermatitis herpetiformis, can be seen in a blistering pattern throughout all skin regions. Serum testing may demonstrate iron, B12 (cobalamin), and folic acid deficiencies as well as hypoalbuminemia [16]. Calcium and 25(OH) vitamin D3 levels are lower in untreated CD patients, and osteoporosis is a known CD complication, which resolves with the institution of a gluten-free diet [17]. Intestinal malabsorption of calcium, in the setting of CD-related intestinal inflammation, leads to elevated parathyroid hormone levels, subsequently causing cortical-bone loss. Dual-energy X-ray absorptiometry can be used to diagnose osteoporosis, and calcium and Vitamin D supplementation is necessary [18,19].

Other autoimmune diseases can occur concomitantly with CD, and it has been postulated that similar HLA associations may explain why CD patients are at increased risk of type 1 (insulin-dependent) diabetes mellitus, Addison's disease (adrenal insufficiency), autoimmune thyroid disease, and alopecia areata, as well as other autoimmune diseases [19]. Focal white matter lesions leading to seizures, ataxia, or hypotonia have been described in CD as well, suggesting autoimmune-associated demyelination or vasculitis. These central nervous system lesions have not been demonstrated to correlate with gluten-free diet compliance [20]. In particular, patients with type 1 diabetes mellitus have up to a 20-fold increase in the risk of developing CD compared with the general population [21]. In North America, up to 5% of patients with type 1 diabetes mellitus have a concurrent diagnosis of CD [22]. Pediatric patients with autoimmune thyroid disease also may have an increased risk of CD [23]. Various liver diseases are more common in adult CD patients compared with the general population, including hepatitis, primary sclerosing cholangitis, nonalcoholic fatty liver disease, cirrhosis or fibrosis, and primary biliary cirrhosis [24]. Autoimmune hepatitis is associated with pediatric CD, and any patient with an elevation of transaminases of unknown cause should be tested for CD [25]. Finally, patients with Down syndrome (Trisomy 21), Turner syndrome, and Williams syndrome have been identified as high-risk groups for CD [26,27]. These data suggest that improved screening for CD in select patient populations is needed.

27.3 TESTING

Testing for CD can be done by serum antibody testing. Several serum antibody tests have been developed for CD screening, although duodenal biopsy by esophagogastroduodenoscopy (EGD) generally is needed to confirm positive serum antibody results. Antigliadin immunoglobulin A (IgA) antibodies were first used to screen for CD, and this antibody test is useful in the setting of a high-pretest prevalence of CD [11]. Autoantibodies such as antireticulin IgA and antiendomysium IgA also have been used for CD diagnosis, and these tests have good specificity but variable sensitivity [12,28]. TTG is the target antigen for the antiendomysium antibody. Thus, the subsequent discovery of the TTG IgA antibody has demonstrated that TTG IgA is the most sensitive serum antibody test for CD screening, although the TTG IgA, as well as the antiendomysium IgA, may be less sensitive than the antigliadin IgA in children less than 2 years of age [12]. The TTG IgA antibody titer is measured using a standard quantitative enzyme-linked immunoabsorbent assay, and rare false-positive results have been noted in patients with liver disease, IMS, arthritis, and congestive heart failure [28]. IgA deficiency occurs in CD patients at a higher rate than the general population with about 2–3% of CD patients having IgA deficiency, so serum total IgA testing may be needed in clinical scenarios in which a patient clinically has CD but has negative CD antibody testing [12]. Up to 95% of CD patients have HLA-DQ2 and HLA-DQ8 haplotype positivity, and testing for these HLA heterodimers is associated with excellent sensitivity. Such HLA testing may be helpful for patients with negative antibody markers [12,29]. Patients with negative HLA-DQ2 and HLA-DQ8 markers will not have CD. Specificity of this test is quite poor, however, as approximately 40% of the general population has these HLA markers [29,30].

The gold standard for CD diagnosis is use of EGD to obtain multiple small bowel biopsies. Gross endoscopic findings of the duodenum may include a mosaic pattern, nodular mucosa, scalloping, and loss of normal duodenal folds, but biopsies are needed to confirm the diagnosis of CD (Figure 27.1) [31].

Microscopic changes indicative of CD include increased IELs, crypt lengthening, and villous atrophy, which are characterized in an increasing scale of severity developed by Marsh (Table 27.1) [28,32]. Commonly, a modification of the Marsh criteria is needed to better define these duodenal biopsies, as CD is a dynamic process with lesions changing over time (Table 27.2) [33].

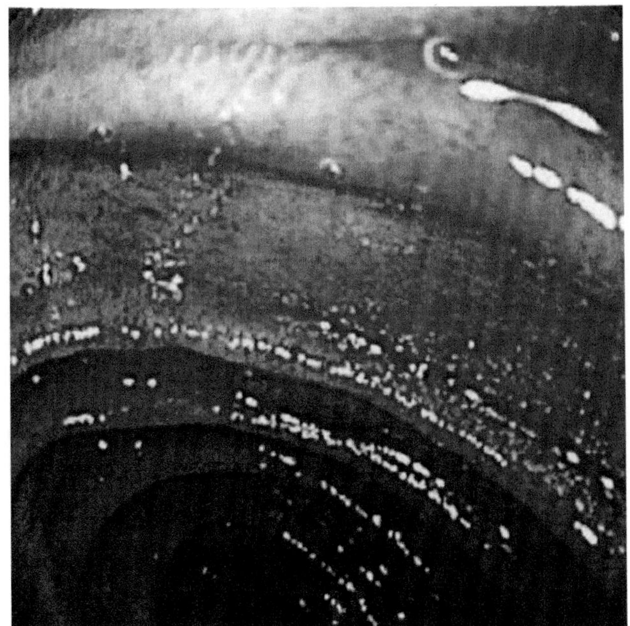

FIGURE 27.1 **Endoscopic view of a patient with CD showing duodenal scalloping.**

TABLE 27.1 Marsh Criteria for Determining Degree of CD Damage Seen in Duodenal Biopsies

Grade	Histology
0	Normal
1	Increased IELs (>25 per 100 enterocytes)
2	Increased IELs, crypt hyperplasia
3a	Increased IELs, crypt hyperplasia, partial villous atrophy
3b	Increased IELs, crypt hyperplasia, subtotal villous atrophy
3c	Increase IELs, crypt hyperplasia, total villous atrophy

Modified from Ref. [32].

Microscopic findings also may be subtle or patchy as patients with asymptomatic or early CD can have increased IELs only at the villous tips compared with the normal population, suggesting that repeat EGD with duodenal biopsy may be needed in some potential CD patients as their disease evolves over time (Figure 27.2) [34].

Patients with early CD may show patchy histologic abnormalities, mainly in the proximal duodenum. Thus, multiple biopsies from different sites in the proximal small bowel are recommended [35]. A very elevated TTG IgA titer or a moderately elevated TTG IgA titer in the setting of a high antiendomysium IgA potentially may lead to the diagnosis of CD in symptomatic adults without duodenal biopsy [36]. Similar findings have been demonstrated in children with a TTG IgA titer greater than seven times the upper limit of normal, and such a scenario may preclude the need for a duodenal biopsy if the antiendomysium antibody titer also is elevated [37]. There is no general consensus, however, that duodenal biopsy can be replaced by serum testing for the final diagnosis of CD. There is also the

TABLE 27.2 Modification of Marsh Criteria

Grade	Histology
0	Normal villous architecture; less than 40 IELs per 100 epithelial cells (ECs)
1	Infiltrative type. Normal villous architecture; more than 40 IELs per 100 ECs. Not indicative of CD and may be present in CD patients not fully treated or in first-degree relatives of patients with CD
2	Hyperplastic type. Normal villous architecture; more than 40 IELs per 100 ECs; crypt hyperplasia. Seen in patients with patchy presentation of CD or in patients with dermatitis herpetiformis
3	Destructive type. Villous atrophy is present. Diagnostic of CD
	3A Mild villous flattening; increase in crypt height; more than 40 IELs per 100 ECs
	3B Moderate villous flattening; increase in crypt height; more than 40 IELs per 100 ECs
	3C Flat or total villous flattening; increase in crypt height; more than 40 IELs per 100 ECs
4	Hypoplastic type. Flat mucosa; normal crypt height; less than 40 IELs per 100 ECs. This is seen in severe malnutrition, as in kwashiorkor

Modified from Ref. [33].

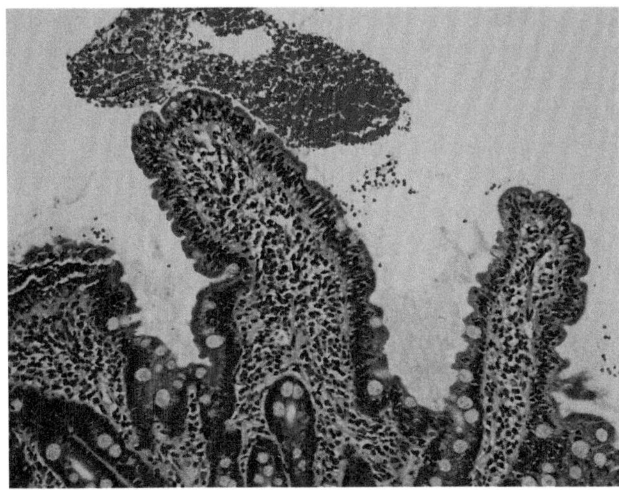

FIGURE 27.2 **Duodenal biopsy of a 7-year-old male with abdominal pain, an elevated TTG IgA, and a family history of CD.** Villi show minimal blunting and IELs are increased, particularly in the upper portion of the villous. Increased IELs with either an even distribution or accentuation at the villous tips suggest possible CD, even in the absence of severe villous atrophy (100×, hematoxylin and eosin stain).

potential of identifying increased TTG IgA deposits in the intestine of patients with possible CD, including those patients with otherwise-normal intestinal mucosa who may develop CD over time, although this testing is strictly experimental at this time and needs further clinical studies [38]. Capsule endoscopy is another potential option for diagnosing equivocal CD, although there are no standard criteria to diagnose CD using this modality [39].

27.4 TREATMENT

The treatment of CD is life-long removal of gluten (wheat) from the diet. Gluten removal is essential for resolution of CD symptoms, including malabsorption [12]. All nutritional deficiency states (iron, vitamin D, vitamin B12, etc.) should be treated as well [29]. CD patients need dietary education from a dietician with CD experience as gluten protein will be found in the prolamin components of wheat (gliadin), barley (hordein), and rye (secalin). Grains and starches that are acceptable for consumption in CD patients include rice (orzenin), corn (zein), hominy, buckwheat, quinoa, sorghum (kafirin), tapioca, teff, dried bean and bean flours, soy, yucca, taro, parsnips, and turnips. Individuals on a gluten-free diet are allowed to have certified gluten-free oats, although oats generally are not recommended as part of the diet of a CD patient because of the risk of contamination with other gluten-containing proteins [40]. The U.S. Food and Drug Administration has developed food-labeling guidelines for "gluten-free" items, which has helped CD patients with food purchasing and preparation [41]. The amount of gluten ingested that can cause histologic damage to the small intestine is small, and it is recommended that gluten intake does not exceed 10 mg per day [42]. Gluten is ubiquitous in everyday use, and patients should be aware that gluten can be present in nonfood products, such as in medication bulking agents [43].

Regular follow-up visits with a medical provider is essential in CD management, as antibody tests, such as the TTG IgA, and duodenal biopsy findings should normalize on a gluten-free diet. All first-degree relatives should have serum antibody testing for CD regardless of symptoms, as up to 5–20% of first-degree relatives of an index case also can have CD [12]. Long-term gluten exposure in the setting of CD can lead to small intestinal lymphoma, typically a T-cell non-Hodgkin lymphoma, which can occur during the sixth decade of life and which further emphasizes the importance of gluten removal [44]. Evidence also suggests that maintaining a gluten-free diet reduces the risk of developing other autoimmune disease. This aspect is especially important as children with CD may have an increased risk of developing other autoimmune diseases over time compared with adults [45].

Dietary compliance is the keystone therapy of CD, and compliance is more of a problem in CD patients with no to minimal symptoms [46]. In particular, adolescents with CD and no symptoms have been shown to have lower compliance rates compared with age-matched CD patients with symptoms [47]. Long-term dietary compliance is improved when a gluten-free diet is made a normal, nonmedicalized, part of a daily diet routine with all family members sharing the same diet as the CD patient [48]. For children in school and young adults in college, open communication about the gluten-free diet with school administrators is needed to allow for dietary compliance [49].

27.5 RELATION TO CYSTIC FIBROSIS

CF and CD have overlapping symptoms, and medical providers should be aware that patients with CF also may have CD if they present with symptoms of malabsorption and malnutrition that do not resolve despite standard CF therapy (Figure 27.3). CF and CD also can present with a diabetes presentation, although CF-related diabetes is a clinically distinct entity from type 1 diabetes mellitus [50]. Testing for exocrine pancreatic insufficiency associated with CF often is done using the fecal elastase-1 test, which is a pancreatic enzyme not degraded by luminal digestion. A low fecal elastase-1 level (usually defined as less than 100 µg/g) is indicative of pancreatic insufficiency. Any GI disease associated with villous atrophy or injury can be associated with a false-positive fecal elastase-1 test. Patients with CD may have a false low-fecal elastase-1 level, and clinical correlation is required in this setting [51,52]. Additionally, CD patients that continue with diarrhea despite continuing a gluten-free diet may have concomitant exocrine pancreatic insufficiency, and such patients often benefit from pancreatic enzyme replacement therapy. CF patients with CD often will have complaints of poor weight gain, diarrhea, steatorrhea, abdominal pain, and irritability. These symptoms may go undiagnosed for a prolonged period of time if CD is not considered, and symptoms will resolve once CD is diagnosed and a gluten-free diet is initiated [53].

The percentage of patients with CD in large CF patient study populations has ranged from 1.2% to 2.13%, which is a higher prevalence of CD compared with the general population [54–56]. Potential theories for the concurrent presentation of both diseases include CF-associated intestinal inflammation causing increased intestinal permeability, upregulation of intestinal TTG in response to oxidative stress and activation of nuclear factor kappa

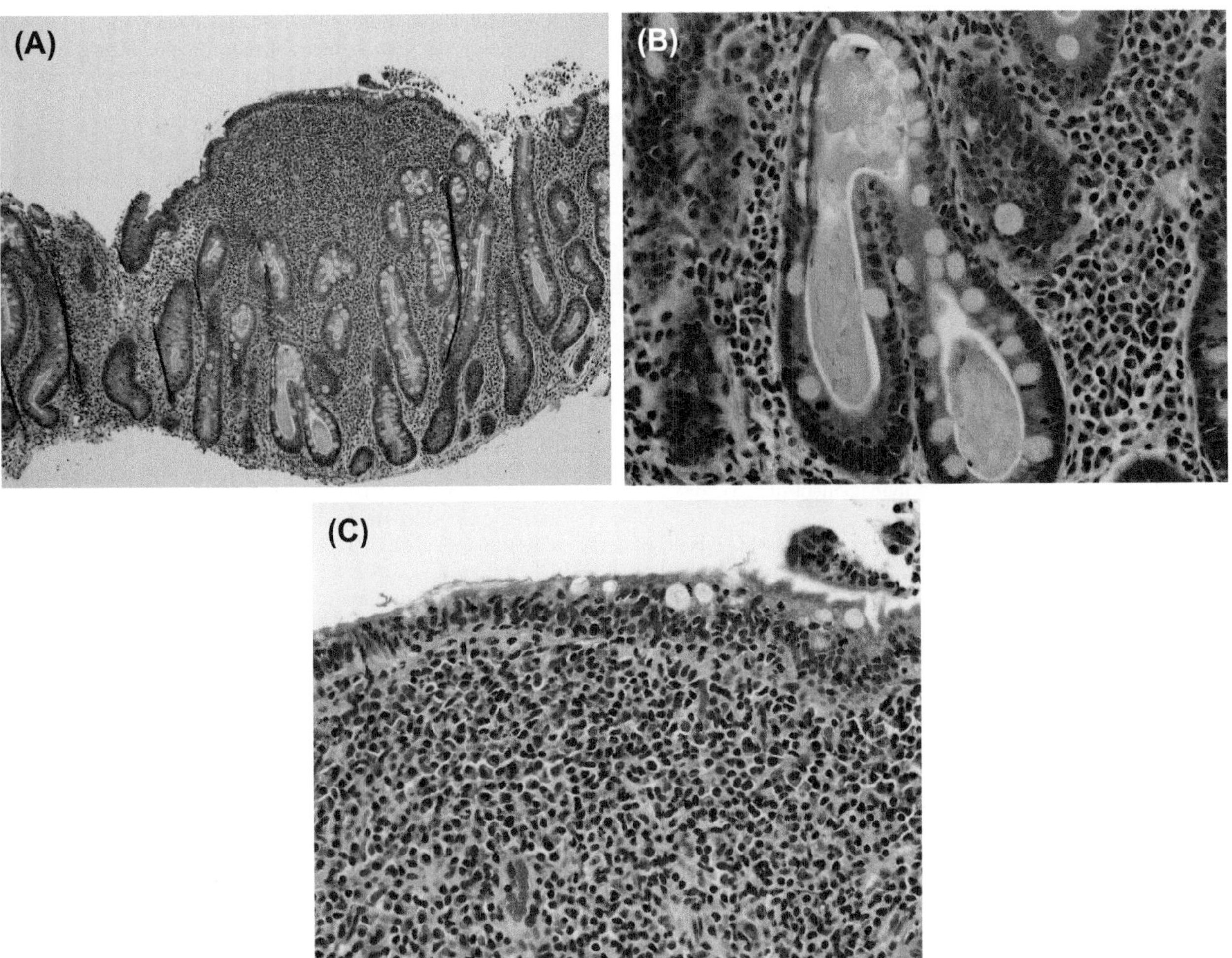

FIGURE 27.3 **Duodenal biopsy of a 10-year-old male with cystic fibrosis and associated pancreatic insufficiency, recent decline in interval weight gain, and an elevated TTG IgA.** (A) Low-power view (10×, hematoxylin and eosin stain) view of duodenum shows marked villous flattening and relatively dense mixed inflammation throughout the mucosa. (B) Higher power view (20×, hematoxylin and eosin stain) shows dense, inspissated mucoid material expanding the crypts, consistent with the patient's CF history. The inspissated material is a nonspecific finding that can be seen with dehydration, renal failure, or pancreatic insufficiency. (C) High-power view (40×, hematoxylin and eosin stain) confirms a marked increase in IELs characteristic of CD.

B, as well as decreased breastfeeding duration in CF patients, which removes a protective factor in the development of CD [3,54]. Avoidance of complementary food before 4 months of age may provide immunologic protection from CD, although this option may not be available in infants with CF requiring increased caloric intake [56]. A genetic understanding of why CF patients develop CD at a higher prevalence than the general population remains unclear.

27.6 CONCLUSION

CD can occur in patients with CF. CD should be considered in any CF patient who presents with symptoms of malabsorption, diarrhea, steatorrhea, abdominal pain, and weight loss despite standard CF therapy. Serum testing, such as the TTG IgA serum test, provides excellent sensitivity and sensitivity; however, EGD with duodenal biopsy is the gold standard to diagnose CD. A gluten-free diet is the lifelong treatment of CD, which typically leads to alleviation of all symptoms.

References

[1] Traeger N, Shi Q, Dozor A. Relationship between sweat chloride, sodium, and age in clinically obtained samples. J Cyst Fibros 2013;13:117–23.

[2] Wagener J, Zemanick E, Sontag M. Newborn screening for cystic fibrosis. Curr Opin Pediatr 2012;24:329–35.

[3] Gelfond D, Borowitz. Gastrointestinal complications of celiac disease. Clin Gastroenterol Hepatol 2013;11:333–42.

[4] Werlin S, Benuri-Silbiger I, Adler S, Goldin E, Zimmerman J, Malka J, et al. Evidence of intestinal inflammation in patients with cystic fibrosis. J Pediatr Gastroenterol Nutr 2010;51:304–8.

[5] Green P, Cellier C. Celiac disease. N Engl J Med 2007;357:1731–43.

[6] Koning F. Celiac disease: caught between a rock and a hard place. Gastroenterology 2005;129:1294–301.

[7] Norris J, Barriga K, Hoffenberg E, Taki I, Miao D, Haas J, et al. Risk of celiac disease autoimmunity and timing of gluten introduction in the diet of infants at increased risk of disease. J Am Med Assoc 2005;293:2343–51.

[8] Olsson C, Hernell O, Hornell A, Lonnberg G, Ivarsson A. Differences in celiac disease risk between Swedish birth cohorts suggests an opportunity for primary prevention. Pediatrics 2008;122:528–34.

[9] Maki M, Mustalahti K, Kokkonen J, Kulmala P, Haapalahti M, Karttunen T, et al. Prevalence of celiac disease among children in Finland. N Engl J Med 2003;348:2517–24.

[10] Hoffenberg E, MacKenzie T, Barriga K, Eisenbarth G, Bao F, Haas J, et al. A prospective study of the incidence of celiac disease of childhood celiac disease. J Pediatr 2003;143:308–14.

[11] Rubio-Tapia A, Ludviqsson J, Brantner T, Murray J, Everhart J. The prevalence of celiac disease in the United States. Am J Gastroenterol 2012;107:1538–44.

[12] Rubio-Tapia A, Hill I, Kelly C, Calderwood A, Murray J. ACG clinical guidelines: diagnosis and management of celiac disease. Am J Gastroenterol 2013;108:656–76.

[13] Butterworth J, Cooper B, Rosenberg W, Purkiss M, Jobson S, Hathaway M, et al. The role of hemochromatosis susceptibility gene mutations in protecting against iron deficiency in celiac disease. Gastroenterology 2002;123:444–9.

[14] Hill I, Dirks M, Liptak G, Colletti R, Fasano A, Guandalini S, et al. Guideline for the diagnosis and treatment of celiac disease in children: recommendations of the North American Society for Pediatric Gastroenterology, Hepatology, and Nutrition. J Pediatr Gastroenterol Nutr 2005;40:1–19.

[15] Sainsbury A, Sanders D, Ford A. Prevalence of irritable bowel syndrome-type symptoms in patients with celiac disease: a meta-analysis. Clin Gastroenterol Hepatol 2013;11:359–65.

[16] American Gastroenterological Association medical position statement: celiac disease. Gastroenterology 2001;120:1522–5.

[17] Zanchi C, Di Leo G, Ronfani L, Martelossi S, Not T, Ventura A. Bone metabolism in celiac disease. J Pediatr 2008;153:262–5.

[18] Scott E, Gaywood I, Scott B. Guidelines for osteoporosis in coeliac disease and inflammatory bowel disease. Gut 2000;46(Suppl. 1): 1–8.

[19] Kaukinen K, Collin P, Mykkanen A, Partanen J, Maki M, Salmi J. Celiac disease and autoimmune endocrinologic disorders. Dig Dis Sci 1999;44:1428–33.

[20] Kieslich M, Errazuriz G, Posselt H, Moeller-Harmann W, Zanella F, Boehles H. Brain white-matter lesions in celiac disease: a prospective study of 75 diet-treated patients. Pediatrics 2001;108:E21.

[21] Barera G, Bonfanti R, Viscardi M, Bazzigaluppi E, Calori G, Meschi F, et al. Occurrence of celiac disease after onset of type 1 diabetes: a 6-year prospective longitudinal study. Pediatrics 2002;109:833–8.

[22] Aktay A, Lee, Kumar V, Parton E, Wyatt D, Werlin S. The prevalence and clinical characteristics of celiac disease in juvenile diabetes in Wisconsin. J Pediatr Gastroenterol Nutr 2001;33:462–5.

[23] Larizza D, Calcaterra V, De Giacomo C, De Silvestri A, Asti M, Badulli C, et al. Celiac disease in children with autoimmune thyroid disease. J Pediatr 2001;139:738–40.

[24] Ludvigsson J, Elfstrom P, Broome U, Ekbom A, Montgomery S. Celiac disease and risk of liver disease: a general population-based study. Clin Gastroenterol Hepatol 2007;5:63–9.

[25] Vajro P, Paolella G, Maggiore G, Giordano G. Pediatric celiac disease, cryptogenic hypertransaminasemia, and autoimmune hepatitis. J Pediatr Gastroenterol Nutr 2013;56:663–70.

[26] Marild K, Stephansson O, Grahnquist L, Cnattinqius S, Soderman G, Ludviqsson J. Down syndrome is associated with elevated risk of celiac disease: a nationwide case-control study. J Pediatr, in press.

[27] Frost A, Band M, Conway G. Serological screening for coeliac disease in adults with Turner's syndrome: prevalence and clinical significance of endomysium antibody positivity. Eur J Endocrinol 2009;160:675–9.

[28] Boige V, Bouhnik Y, Delchier J, Jian R, Matuchansky C, Andre C. Anti-endomysium and anti-reticulin antibodies in adults with celiac disease followed-up in the Paris area. Gastroenterol Clin Biol 1996;20:931–7.

[29] Rostrom A, Murray J, Kagnoff M. American Gastroenterological Association Institute technical review on the diagnosis and management of celiac disease. Gastroenterology 2006;131:1981–2002.

[30] Kaukinen K, Partanen J, Maki M, Collin P. HLA-DQ typing in the diagnosis of celiac disease. Am J Gastroenterol 2002;97:695–9.

[31] Piazzi L, Zancanella L, Chilovi F, Merighi A, De Vitis I, Feliciangeli G, The S.I.E.D. Group, et al. Diagnostic value of endoscopic markers for celiac disease in adults: a multicenter prospective Italian study. Miner Gastroenterol Dietol 2008;54:335–46.

[32] Evans K, Aziz I, Cross S, Sahota G, Hopper A, Hadjivassiliou M, et al. A prospective study of duodenal bulb biopsy in newly diagnosed and established adult celiac disease. Am J Gastroenterol 2011;106:1837–742.

[33] Oberhuber G, Granditsch G, Vogelsang H. The histopathology of coeliac disease: time for a standardization report scheme for pathologists. Eur J Gastroenterol Hepatol 1999;11:1185–94.

[34] Godstein N, Underhill J. Morphologic features suggestive of gluten sensitivity in architecturally normal duodenal biopsy specimens. Am J Clin Pathol 2001;116:63–71.

[35] Tanpowpong P, Broder-Fingert S, Katz A, Camargo C. Predictors of duodenal bulb biopsy performance in the evaluation of coeliac disease in children. J Clin Pathol 2012;65:791–4.

[36] Wakim-Fleming J, Pagadala M, Lemyre M, Lopez R, Kumaravel A, Carey W, et al. Diagnosis of celiac disease in adults based on serology test results, without small-bowel biopsy. Clin Gastroenterol Hepatol 2013;11:511–6.

[37] Alessio M, Tonutti E, Brusca I, Radice A, Linici L, Sonzoqni A, et al, for the Study Group on Autoimmune Diseases of Italian Society of Laboratory Medicine. Correlation between IgA tissue transglutaminase antibody ration and histological findings in celiac disease. J Pediatr Gastroenterol Nutr 2012;55:44–9.

[38] Tosco A, Maqlio M, Paparo F, Rapacciuolo L, Sannino A, Miele E, et al. Immunoglobulin A anti-tissue transglutaminase antibody deposits in the small intestinal mucosa of children with no villous atrophy. J Pediatr Gastroenterol Nutr 2008;47:293–8.

[39] Kurien M, Evans K, Aziz I, Sidhu R, Drew K, Rogers T, et al. Capsule endoscopy in adult celiac disease: a potential role in equivocal cases of celiac disease? Gastrointest Endosc 2013;77:227–32.

[40] Hlywiak K. Hidden sources of gluten. Pract Gastroenterol 2008;8:582–8.

[41] Niewinski M. Advances in celiac disease and gluten-free diet. J Am Diet Assoc 2008;108:661–72.

[42] Akobeng A, Thomas A. Systematic review: tolerable amount of gluten for people with coeliac disease. Aliment Pharmacol Ther 2008;11:1044–52.

[43] Plogsted S. Medications and celiac disease: tips from a pharmacist. Pract Gastroenterol 2007;31:58–65.

[44] Catassi C, Fabiani E, Corrao G, Barbato M, De Renzo A, Gabrielli A, for the Italian Working Group on Coeliac Disease and Non-Hodgkin's Lymphoma, et al. Risk of non-Hodgkin lymphoma in celiac disease. J Am Med Assoc 2002;287:1413–9.

[45] Cosnes J, Cellier C, Viola S, Colombel J, Michaud L, Sarles J, the Groupe D'Etude Et De Recherche Sure La Maladie Coeliaque, et al. Incidence of autoimmune diseases in celiac disease: protective effect of the gluten-free diet. Clin Gastroenterol Hepatol 2008;6:753–8.

[46] Edwards G, Leffler D, Dennis M, Franko D, Blom-Hoffman J, Kelly C. Psychological correlates of gluten-free diet adherence in adults with celiac disease. J Clin Gastroenterol 2009;43:301–6.
[47] Fabiani E, Taccari L, Ratsch I, Di Giuseppe S, Coppa G, Catassi C. Compliance with gluten-free diet in adolescents with screening-detected celiac disease: a 5-year follow-up study. J Pediatr 2000;136:841–3.
[48] Veen M, te Molder H, Gremmen B, van Woerkum C. If you can't eat what you like, like what you can: how children with coeliac disease and their families construct dietary restrictions as a matter of choice. Sociol Health Illn 2013;35:592–609.
[49] Panzer R, Dennis M, Kelly C, Weir D, Leichtner A, Leffler D. Navigating the gluten-free diet in college. J Pediatr Gastroenterol Nutr 2012;55:740–4.
[50] Kelly A, Moran A. Update on cystic fibrosis-related diabetes. J Cyst Fibros 2013;12:318–31.
[51] Walkowiak J, Herziq K. Fecal elastase-1 is decreased in villous atrophy regardless of the underlying disease. Eur J Clin Invest 2001;31:425–30.
[52] Duggan S, Conlon K. A practical guide to the nutritional management of chronic pancreatitis. Pract Gastroenterol 2013;37:24–32.
[53] Leeds J, Hopper A, Hurlsone D, Edwards S, McAlindon M, Lobo A, et al. Is exocrine pancreatic insufficiency in adult coeliac disease a cause of persisting symptoms? Aliment Pharmacol Therapeut 2007;25:265–71.
[54] Fluge G, Olesen H, Gilljam M, Meyer P, Pressler T, Storroston O, et al. Co-morbidity of cystic fibrosis and celiac disease in Scandinavian cystic fibrosis patients. J Cyst Fibros 2009;8:198–202.
[55] Walkowiak J, Blask-Osipa A, Lisowska A, Oralewska B, Pogorzelski A, Cichy W, et al. Cystic fibrosis is a risk factor for celiac disease. Acta Biochim Pol 2010;57:115–8.
[56] Pohl J, Judkins J, Meihls S, Lowichik A, Chatfield B, McDonald C. Cystic fibrosis and celiac disease: both can occur together. Clin Pediatr 2011;50:1153–5.

SECTION E

NUTRITION AND PULMONARY FUNCTION IN CYSTIC FIBROSIS PATIENTS

CHAPTER

28

Probiotic Supplementation and Pulmonary Exacerbations in Patients with Cystic Fibrosis

Tzippora Shalem[1, 2], *Batia Weiss*[1, 2]

[1]Division of Pediatric Gastroenterology and Nutrition, Edmond and Lily Safra Children's Hospital, Tel-Hashomer, Israel; [2]Sackler Faculty of Medicine, Tel-Aviv University, Tel-Aviv, Israel

The hallmark of cystic fibrosis (CF) is recurrent severe and destructive pulmonary inflammation and infection that begins in early childhood and leads to morbidity and mortality due to respiratory failure. The nutritional status of CF patients is a major determinant of pulmonary and survival outcomes. Longitudinal cohort studies in CF report a distinct survival advantage among patients with better nutritional status. More specifically, poor nutritional status is not only strongly linked to poorer lung function but is also an independent risk factor for early death in children with CF. Several factors contribute to impaired nutritional status in CF. These include malabsorption, recurrent sinopulmonary infections, increased energy expenditure, and suboptimal intake. The malabsorptive state in CF is likely multifactorial. The primary cause of malabsorption is due to maldigestion from pancreatic exocrine insufficiency. However, children with CF can continue to have malabsorption despite pancreatic enzyme replacement therapy (PERT) administration. It has been previously suggested that the presence of an acidic intestinal milieu that impairs enzyme activity contributes to the failure of PERT in nutrient assimilation in CF. More recently, intestinal inflammation has been hypothesized as another contributing factor.

28.1 INTESTINAL INFLAMMATION

Although the exact underlying mechanism for inflammation is unknown, several pathogenic mechanisms have been proposed:

1. The defective cystic fibrosis transmembrane receptor (CFTR) results in viscid, inspissated intestinal mucus secretions that, together with the acidic milieu and altered glycosylation of mucins, may predispose to alterations in the balance and/or composition of the gut flora, resulting in mucosal inflammation [1].
2. Quantitative polymerase chain reaction of bacterial 16S RNA in CF mice revealed a more than 40-fold increase in bacterial load in the small intestine. Intestinal dysmotility was also observed in CF mice and was attributed to the further development of dysbiosis and bacterial overgrowth [2].
3. Secondary factors such as intestinal resection (e.g., due to meconium ileus) or early and/or repeated antibiotic exposure to treat sinopulmonary infection(s) may also influence gut microbial ecology and cause inflammation [1].
4. It has also been suggested that inflammation is induced by an increased and abnormal antigenic load, which is a result of the inadequate digestive enzymes present in the small intestines [1].
5. Studies show that the intestinal barrier function is severely compromised in CF [3].

Multiple studies aiming to investigate intestinal inflammation in CF patients have been performed:

1. Smyth et al. reported measurements of intestinal inflammatory proteins from whole-gut lavage of 21 children with pancreatic-insufficient CF and 12 controls. This study demonstrated increased production of inflammatory biomarkers including albumin, immunoglobulin (Ig)G, IgM, interleukin (IL)-1β, IL-8, neutrophil, elastase, and eosinophilic cationic protein in children with CF [4].
2. Bruzzese and colleagues used fecal calprotectin (S100A8/S100A9) and rectal nitric oxide production for measuring intestinal inflammation. Mean (standard deviation) fecal calprotectin levels were

Diet and Exercise in Cystic Fibrosis
http://dx.doi.org/10.1016/B978-0-12-800051-9.00028-6

significantly higher in 30 CF children with pancreatic insufficiency compared with healthy controls (219 (94) μg/g vs. 46 (31) μg/g). Twenty CF patients and 20 controls had rectal nitric oxide production measurements by the rectal dialysis bag technique. Likewise, nitric oxide levels were significantly higher in children with CF than controls [5].

3. Raia et al. compared duodenal mucosal specimens (obtained endoscopically) from 14 pancreatic insufficiency CF patients, 20 healthy controls, and 4 non-CF patients with pancreatic insufficiency due to chronic pancreatitis. An increased mononuclear cell infiltrate that expressed intercellular adhesion molecule-1, IL-2 receptor alpha (CD25), IL-2, and interferon-γ was observed in the lamina propria of specimens from CF children. Two CF patients who had commenced PERT before the endoscopy had evidence of intestinal inflammation. Small intestinal inflammation was not observed in subjects with non-CF pancreatic insufficiency-chronic pancreatitis, suggesting that pancreatic insufficiency itself is unlikely to be the cause of intestinal inflammation [6].
4. Steven et al. examined the small intestine of patients with CF without overt evidence of gastrointestinal disease using capsule endoscopy (CE). Forty-two patients with CF (10–36 years) were included, and 29 had pancreatic insufficiency. Review of the CE videos showed that most patients had varying degrees of diffuse areas of inflammatory findings in the small bowel, including edema, erythema, mucosal breaks, and frank ulcerations [7].

On the basis of these studies, there is clear evidence of intestinal inflammation in CF, particularly those with pancreatic insufficiency. Understanding that patients with CF also have a gastrointestinal inflammatory process, combined with the frequent use of antibiotics and their effect on gastrointestinal microbiota, led to the thought of using probiotics to treat patients with CF. The use of probiotics might improve their quality of life and reduce morbidity and antibiotic use.

28.2 PROBIOTICS

The human large intestine is a densely populated microbial ecosystem. Most of the bacteria are obligate anaerobes. Some gut bacteria are beneficial to health whereas others may be harmful. The main potentially health-enhancing bacteria are the bifidobacteria and lactobacilli, both of which belong to the lactic acid bacteria group and are being used as probiotics. Probiotics are defined as "live microorganisms which confer a health benefit on the host when administered in adequate amounts." The mechanism of action of probiotic microorganisms can be explained by enhancement of the nonspecific and specific immune response of the host, production of antimicrobial substances, and competition with pathogens for binding sites. Probiotics may protect from gastrointestinal diseases by acting on intestinal permeability, as witnessed by their capacity to improve intestinal barrier function. However, it is becoming increasingly clear that several effects induced by probiotics are related to modifications of the immune response, which may explain the clinical effects of probiotics observed in nonintestinal and systemic diseases. Several probiotic strains have been investigated in several clinical conditions. The best characterized probiotic is *Lactobacillus rhamnosus* strain GG (LGG), for which several clinical effects have been identified. Recent data showed that LGG affects the expression of genes involved in immune response and inflammation [8]. Probiotics are gaining more and more interest as alternatives for antibiotics or anti-inflammatory drugs. They have been widely tested, in animal and human studies, for their beneficial actions in the prevention or treatment of a broad spectrum of gastrointestinal and extraintestinal disorders.

1. Probiotics improve intestinal microbial balance, reducing the duration and the severity of acute infectious diarrhea. The European Society for Pediatric Gastroenterology, Hepatology and Nutrition and the European Society of Pediatric Infectious Diseases guidelines make a strong recommendation for the use of probiotics for the management of acute gastroenteritis, particularly those with documented efficacy such as LGG, *Lactobacillus reuteri*, and *Saccharomyces boulardii* [9].
2. Probiotics reduced antibiotic-associated diarrhea and diarrhea with *Clostridium difficile* infection [10].
3. Probiotics are frequently used by patients with inflammatory bowel disease; however, literature data are conflicting related to their importance. In mild to moderate ulcerative colitis, probiotics can be used effectively in induction and maintaining remission and prevention of pouchitis. Thus far, there is not sufficient evidence to support the use of probiotics in daily clinical practice in Crohn's disease [11].
4. The administration of probiotics that colonize the vaginal tract can be important in maintaining a normal urogenital health and to prevent or treat infections [12].
5. It has been observed that allergic children have a different microbiota composition than healthy infants, and this assumption supports the rationale for modulating the gut microbiota. It has been demonstrated in cow's milk-allergic infants that supplementation of hydrolyzed formula with *Bifidiobacterium lactis* Bb-12 or LGG led to an earlier recovery than standard treatment alone. In atopic

dermatitis children, a combination of *Lactobacillus* strains was found to significantly reduce the clinical scoring of atopic dermatitis [13].

6. A significantly lower incidence of respiratory tract infection (RTIs) was detected in infants receiving prebiotics or probiotics compared with those receiving placebo. In addition, the incidence of rhinovirus-induced episodes, which comprised 80% of all RTI episodes, was found to be significantly lower in the prebiotic and probiotic groups compared with the placebo group. No differences emerged among the study groups in rhinovirus RNA load during infections, duration of rhinovirus RNA shedding, duration or severity of rhinovirus infections, or occurrence of rhinovirus RNA in asymptomatic infants [14].

Probiotics are reported to have roles in various diseases. The preceding examples deal mainly with situations related to the digestive system and lungs in non-CF patients.

28.2.1 Probiotics for CF Patients

Respiratory and gastrointestinal tract symptoms are the main causes of morbidity and reduced quality of life in CF patients. Recurrent respiratory exacerbations result in constant exposure to large-spectrum antibiotics. CF patients have increased intestinal permeability, abnormal microflora, and dysregulated innate immune mediators. Because probiotics have beneficial effects on those mechanisms, it may be assumed that probiotics may be beneficial for CF patients.

Few studies evaluating probiotic treatment in CF patients are available in the literature.

Di Nardo et al. conducted a prospective, randomized, double-blind, placebo-controlled study enrolling 61 CF patients with mild to moderate lung disease. Patients were randomly assigned to receive *L. reuteri* (LR; 30 patients) in five drops per day (10 colony-forming units) or placebo (31 patients) for 6 months. Pulmonary exacerbations were significantly reduced in the LR group compared with the placebo group. Likewise, the number of upper RTIs was significantly reduced in the LR group compared with the placebo. The two groups did not statistically differ in the mean number and duration of hospitalizations for pulmonary exacerbations and gastrointestinal infections. There was no significant statistical difference in the mean delta value of forced expiratory volume in 1 s, fecal calprotectin concentration, and tested cytokines (tumor necrosis factor-α and IL-8) between the two groups [15].

Bruzzese and colleagues treated 10 CF children with *Lactobacillus casei* strain GG once a day for 4 weeks and remeasured fecal calprotectin and rectal nitric oxide levels. Of note, both biomarkers of calprotectin and nitric oxide were significantly reduced in 8 and 5 of the 10 children, respectively, suggesting an association between intestinal inflammation and modifications in intestinal microflora [5].

Weiss et al. conducted a prospective pilot study of 10 CF patients with mild to moderate lung disease and *Pseudomonas aeruginosa* colonization to determine the effect of a mixed probiotic preparation on pulmonary exacerbations and inflammatory characteristics of the sputum. Pulmonary function tests (PFTs), sputum cultures with semiquantitative bacterial analysis, and sputum neutrophil count and IL-8 levels were compared to pretreatment and post-treatment values. The rate of pulmonary exacerbations was compared to that 2 years before the study. After 6 months of treatment, the exacerbation rate was significantly reduced in comparison with the previous 2 years and to 6 months post-treatment ($P=0.002$). PFTs did not change at the end of treatment and during 6 months post-treatment. No changes in sputum bacteria, neutrophil count, and IL-8 levels were observed. The authors concluded that probiotics reduce the pulmonary exacerbation rate in patients with CF [16].

28.3 SUMMARY

CF is a prototype of chronic inflammatory diseases characterized by a dysregulation of the first-line host defense to the environment. CF patients suffer recurring illness and poor nutritional status. These factors influence each other, the quality of life, and the prognosis of patients.

As we have shown, probiotics have a role in improving immune function and lowering the incidence of various diseases. Here we have chosen to show the effect of probiotics in non-CF diseases involving the gastrointestinal tract and airways. The beneficial effect of probiotics in these diseases suggests that CF patients will benefit from treatment with probiotics.

Few studies have been conducted on the effectiveness of probiotics in patients with CF. These studies have shown some benefit in reducing exacerbations of the disease. However, data are scarce, and further prospective randomized controlled trials are needed to determine the effect probiotics in CF patients.

References

[1] Lee JM, Leach ST, Katz T, Day AS, Jaffe A, Ooi CY. Update of faecal markers of inflammation in children with cystic fibrosis. Mediators Inflamm 2012;25:1–6.

[2] De Lisle RC. Altered transit and bacterial overgrowth in the cystic fibrosis mouse small intestine. Am J Physiol Gastrointest Liver Physiol 2007;293:G104–11.

[3] Hallberg K, Grzegorczyk A, Larson G, Strandvik B. Intestinal permeability in cystic fibrosis in relation to genotype. J Pediatr Gastroenterol Nutr 1997;25:290–5.

[4] Smyth RL, Croft NM, O'Hea U, Marshall TG, Ferguson A. Intestinal inflammation in cystic fibrosis. Arch Dis Child 2000;82:394–9.
[5] Bruzzese E, Raia V, Gaudiello G, Polito G, Buccigrossi V, Formicola V, et al. Intestinal inflammation is a frequent feature of cystic fibrosis and is reduced by probiotic administration. Aliment Pharmacol Ther 2004;20:813–9.
[6] Raia V, Maiuri L, de Ritis G, de Vizia B, Vacca L, Conte R, et al. Evidence of chronic inflammation in morphologically normal small intestine of cystic fibrosis patients. Pediatr Res 2000;47:344–50.
[7] Werlin SL, Benuri-Silbiger I, Kerem E, Adler SN, Goldin E, Zimmerman J, et al. Evidence of intestinal inflammation in patients with cystic fibrosis. J Pediatr Gastroenterol Nutr 2010;51:304–8.
[8] Macfarlane GT, Cummings JH. Probiotics, infection and immunity. Curr Opin Infect Dis 2002;15:501–6.
[9] Ciccarelli S, Stolfi I, Caramia G. Management strategies in the treatment of neonatal and pediatric gastroenteritis. Infect Drug Resist 2013;6:131–61.
[10] Goldenberg JZ, Ma SS, Saxton JD, Martzen MR, Vandvik PO, Thorlund K, et al. Review: probiotics prevent *C. difficile*–associated diarrhea in patients using antibiotics. Ann Intern Med 2013;159:JC7.
[11] Müllner K, Miheller P, Herszényi L, Tulassay Z. Probiotics in the management of Crohn's disease and ulcerative colitis. Curr Pharm Des 2013.
[12] Borges S, Silva J, Teixeira P. The role of lactobacilli and probiotics in maintaining vaginal health. Arch Gynecol Obstet 2014;289: 479–89.
[13] Vitaliti G, Pavone P, Spataro G, Falsaperla R. The immunomodulatory effect of probiotics beyond atopy: an update. J Asthma 2014;21:320–32.
[14] Luoto R, Ruuskanen O, Waris M, Kalliomäki M, Salminen S, Isolauri E. Prebiotic and probiotic supplementation prevents rhinovirus infections in preterm infants: a randomized, placebo-controlled trial. J Allergy Clin Immunol 2013;133:405–13.
[15] Di Nardo G, Oliva S, Menichella A, Pistelli R, Biase RV, Patriarchi F, et al. Randomized clinical trial: *Lactobacillus reuteri* ATCC55730 in cystic fibrosis. J Pediatr Gastroenterol Nutr 2014;58:81–6.
[16] Weiss B, Bujanover Y, Yahav Y, Vilozni D, Fireman E, Efrati O. Probiotic supplementation affects pulmonary exacerbations in patients with cystic fibrosis: a pilot study. Pediatr Pulmonol 2010;45:536–40.

CHAPTER

29

Buteyko: Better Breathing = Better Health

Russell Stark[1,2], *Jennifer Stark*[1,2]

[1]Buteyko Institute of Breathing and Health (Inc), Fitzroy, VIC, Australia; [2]Buteyko Breathing Educators Association, IN, USA

29.1 INTRODUCTION

In a condition such as cystic fibrosis (CF), in which a lack of adequate breathing is the primary cause of morbidity and mortality [1,2], it may seem odd that the Buteyko method, which usually encourages hypoventilation for short periods of time, could be helpful. However, Buteyko techniques are not encouraged in the final stage of the disease but rather earlier on, when their main benefits of improved symptom control and sense of well-being are considered most valuable.

In the 1950s, Ukrainian doctor, Konstantin Buteyko, realized the importance that breathing plays in maintaining good health. He also recognized the concept of plasticity and theorized that as the respiratory center could become accustomed to increased minute volume, the opposite must also be true. Following this train of thought, he developed the Buteyko method, which aims to restore a healthy breathing pattern by having students follow an education program that consists of individually tailored breathing exercises and strategies.

In the first study of the Buteyko method conducted in the western world, the asthmatic subjects were initially breathing on average 14L of air per minute while resting. It was noted in the 3-month study that the average reduction in minute volume was 31%. This reduction was accompanied by an average 96% reduction in bronchodilators within the first 6 weeks and an average 49% reduction in preventive medications by the end of the study. There was an average 71% reduction in symptoms, and these improvements led to significant advancement in quality of life measurements [3].

This study and others like it suggest that if children with CF followed even the most basic points of the Buteyko method, which are to breathe through the nose virtually all of the time and to use the diaphragm, their quality of life seems likely to improve and the expected progression of the disease may be altered or delayed. However, it should be noted that the Buteyko method becomes a lifestyle with the chronically ill to better manage their symptoms rather than a short-lived program.

When CF is discussed in conventional medical literature, the following flow chart seems typical of current thinking in regard to the respiratory system:

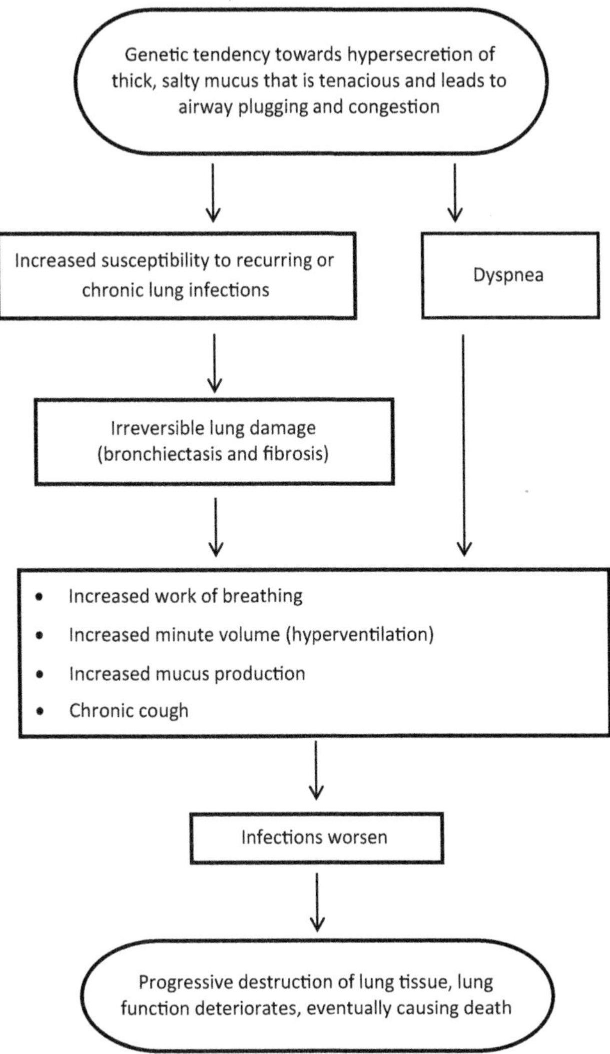

Diet and Exercise in Cystic Fibrosis
http://dx.doi.org/10.1016/B978-0-12-800051-9.00029-8

It is also well documented that CF affects epithelial cell function in the respiratory system. Although this includes the upper airways, and sufferers tend to have chronic nasal infections, polyps, and blocked or running sinuses that result in cough and chronic mouth-breathing [4,5], most literature primarily discusses only the lungs [2,6,7].

The Buteyko approach to CF differs slightly from the medical model because Buteyko instructors consider mouth-breathing to be detrimental to the lungs. This is because the mouth is the beginning of the digestive tract; although it can be used for breathing, it is an inferior substitute for the nose, particularly with the physical problems of CF. Therefore, the Buteyko flow chart looks a little different from the conventional view.

Although there is no clinical evidence of the effectiveness of using the Buteyko method with CF, there is

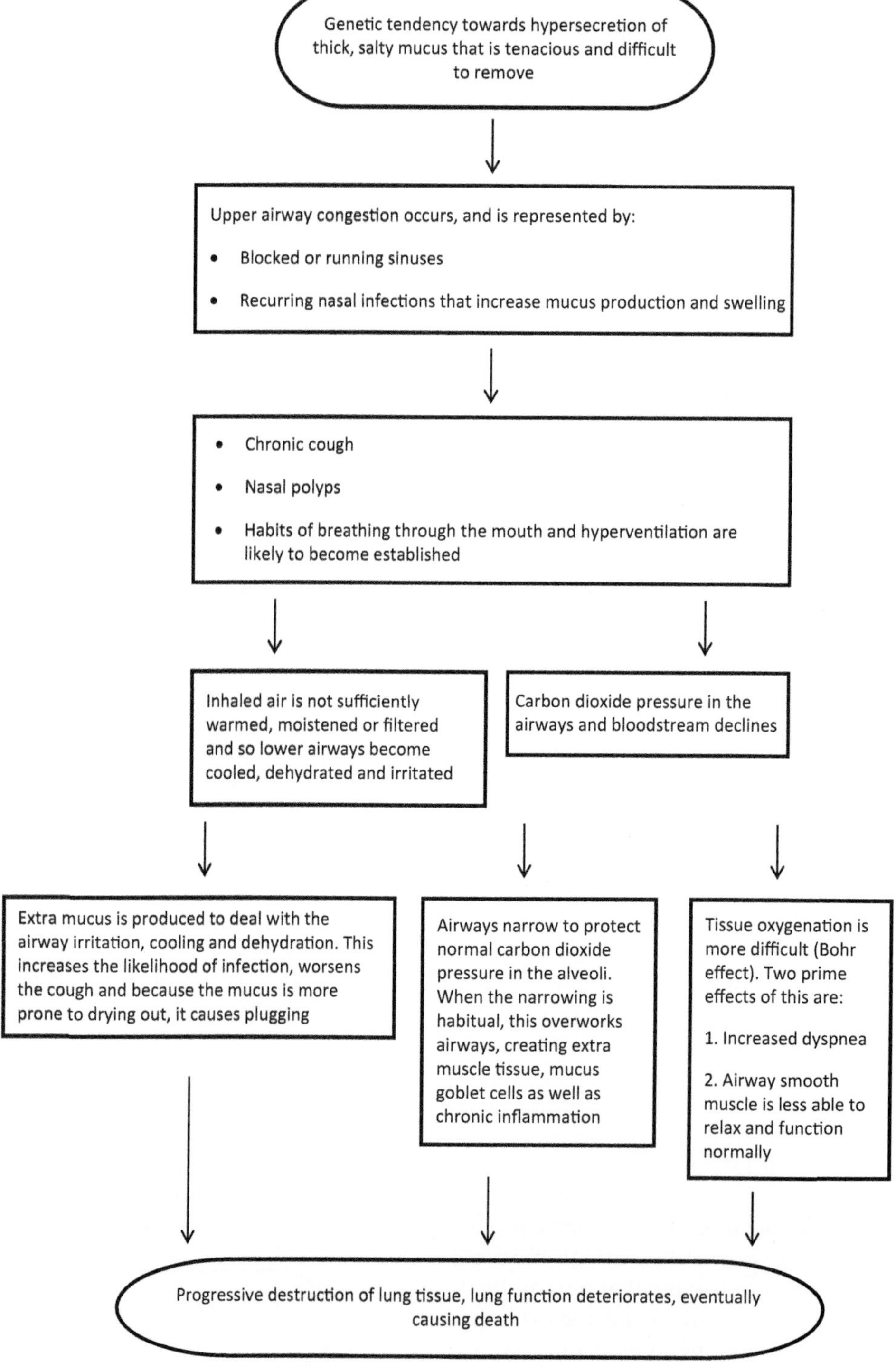

evidence that breathing exercises that change the way people breathe are useful adjuncts to medical treatments. For example, Delk (1993) showed that using biofeedback to encourage diaphragmatic breathing provided the following remarkable results in patients with CF [8]:

- Forced expiratory volume in 1 s (FEV1) improved by 32%
- Forced vital capacity increased by 29%
- Forced expiratory flow 25–75% increased by 38%

Similar to Buteyko, part of a technique called "autogenic drainage" involves small-volume breathing, which is said to "mobilize peripheral mucus," enhancing its clearance. There are also many other particular breathing techniques, such as huffing, that assist people around the world to better manage their lung problems [9].

29.2 BREATHING

To understand why Buteyko instructors consider breathing correctly so important, we must first look at what it does.

The most basic purpose of breathing is to meet the demands of metabolism by maintaining proper concentrations in arterial blood of oxygen and carbon dioxide. Although different levels of metabolism have differing oxygen requirements and produce different volumes of carbon dioxide, breathing automatically changes on a breath-by-breath basis in such a precise way that the pressures virtually never change.

Common knowledge stresses how vitally important sufficient oxygen is and how without it, a person will die. However, it must also be considered that without sufficient carbon dioxide, death will also occur. Furthermore because of hemoglobin's capacity to carry large volumes of oxygen, the constancy of carbon dioxide is actually more important at any given time. This is why the body micromanages it, creating short apneas or breathing stimulation as required to maintain a constant 40 mmHg, which keeps blood pH steady at 7.35–7.45.

29.2.1 The Breathing Paradox

When a person is tired, sleeping restores their sense of well-being, and when thirsty, drinking satisfies the thirst, but breathing has something of a counterintuitive aspect to it. This is because when feeling breathless, breathing vigorously frequently worsens the situation, leading to a vicious circle in which the more you breathe, the more you feel the need to breathe.

When at rest, people seldom feel short of breath while they breathe quietly and at a reasonable pace because only approximately one tenth of total lung capacity is required for sufficient gas exchange while resting [10]. However, when a fast or vigorous breathing pattern is applied, the person can quickly feel airway restriction and shortness of breath even when sitting still.

This is because breathing excessively has the effect of driving off extra carbon dioxide from the airways and consequently the bloodstream, causing hypocapnia and respiratory alkalosis. Christian Bohr explained over 100 years ago that this reduces oxygen delivery to tissues (Bohr effect) [11]. This concept is important for everyone, but especially in the management of chronic lung disease, because

- Hypocapnia strongly increases smooth muscle tension whereas hypercapnia is a highly effective bronchodilator that works directly on the action of airway smooth muscles independent of airway pH [12]. This is demonstrated by the fact that an atmosphere of 8% carbon dioxide provided a 42% dilation of rat airways [13].
- Hypocapnia destabilizes mast cells, increasing degranulation and allergic response [14].
- Hypocapnia increases the release of histamines, with consequential inflammation [15].
- Carbon dioxide is effective at killing bacteria, molds, and yeasts [16], which is why in many cases permissive hypercapnia allows speedier recovery in intensive care and is useful in the treatment of infection in acute lung injury or acute respiratory distress syndrome [17,18]. This property of carbon dioxide was used in the 1920s by Henderson and Haggard, who developed a tent that had an atmosphere of 8% carbon dioxide and was used successfully to treat pneumonia [19].

When breathing becomes excessive for any length of time, it creates difficulties for the respiratory system.

29.2.2 What is Hyperventilation?

Hyperventilation is clinically defined as "abnormally increased pulmonary ventilation, resulting in reduction of carbon dioxide tension, which, if prolonged, may lead to alkalosis" [20]. Hyperventilation is generally achieved by a combination of

- Breathing faster than is required for the task at hand
- Breathing through the mouth instead of the nose
- Upper-chest breathing instead of diaphragmatic
- Poor posture
- Repeated forceful breathing, such as yawning, sighing, gasping, or coughing

When these types of breathing patterns become habitual, the breathing also frequently becomes noisy and erratic.

Hyperventilation does not have to be excessive to be detrimental in the long run. If a healthy adult is asked to take a breath every 3 s instead of at the normal 5-s interval, the person perhaps breathes 8 L of air each minute instead of the usual 4–6 L. Although barely perceptible, breathing like this will lower carbon dioxide and water vapor in the lungs and cause many more irritants and bacteria to be inhaled within a short space of time.

Not many people recognize breathing 20 times a minute as hyperventilation because generally only acute hyperventilation, which is represented by noisy, rapid breathing, is noticed. Acute hyperventilation has a massive effect in the short term; therefore, it cannot be maintained for long. However, the respiratory center has plasticity and will adapt to a pattern of low-grade hyperventilation that can continue for a lifetime. Dr Buteyko called this habit of breathing slightly more than is required for metabolism "hidden hyperventilation." He perceived that a mismatch of ventilation and metabolic requirement can lead to chronic washing out of carbon dioxide with devastating effects on good health in the long term.

29.2.3 What Causes Hyperventilation?

Hyperventilation is caused by any type of stress that is placed on the body and, in the case of a child with CF, the stressors are likely to be a mixture of biological or physical, environmental and psychological.

- CF puts the body under physical stress by the respiratory effort and discomfort as well as the associated digestive problems it produces [21].
- Environmental factors such as high air humidity, temperature, or pollution typically play a large role in the outcome of this condition [22,23].
- Anxiety and depression are common factors of CF [24]. Children are quick to tune into fear from their loved ones in regard to treatments and anticipation of an early death. Hospital stays are likely to cause separation anxiety, apprehension, and loneliness, and the passing of friends who shared hospital stays also undoubtedly adds extra stress.

Therefore, chronic stress plays a large role in the life of people with CF, and this encourages hidden hyperventilation.

29.2.4 What Does Hyperventilation Do?

Because of diffusion and normal lung structure, low-grade hyperventilation has little effect on the pressure of oxygen; however, it lowers carbon dioxide concentration in the alveoli, affecting the rest of the body in the following ways:

- Hemoglobin has a strong affinity to oxygen when carbon dioxide is low, causing oxygen levels to appear normal, but paradoxically tissue oxygenation is not occurring at the required rate (Bohr effect).
- Although several chemical and physical reflexes regulate breathing, the most important of these is keeping arterial carbon dioxide constant because this directly controls blood pH. Once the normal pressure is deviated from for any length of time and compensation is made to protect the pH of the blood, the plasticity of the respiratory center allows the automatic breathing rhythm to change, maintaining the new lower pressure, although it is not healthy.

While it may seem strange that the body will sabotage itself in this way, it could be compared to how the brain sometimes allows blood pressure or sugar levels to become habitually higher than usual, although these malfunctions are not healthy either.

Because of hyperventilation's effect on the oxygenation and pH of the blood, it can negatively affect the whole body. The following list of the most common symptoms of chronic hyperventilation has been compiled from researchers Lum (1977), Magarian (1982), Bradley (2006), and Timmons (1994) [25–28]:

Shortness of breath; chest tightness; extra-sensitive airways; excessive production of mucus resulting in frequent sneezing, coughing, nose-blowing, repeated throat clearing, and long-term blocked or running sinus; frequent yawning and sighing; light-headedness, dizziness, unsteadiness, and feeling "spaced out"; poor concentration; numbness, tingling and coldness, especially in the hands, feet, and face; heartbeat that is irregular, pounding, or racing; degrees of anxiety, tension, phobias, depression, irritability, and apprehension; dry mouth; digestive disorders including abdominal bloating, belching, flatulence, diarrhea, and constipation; tiredness, general weakness, and chronic exhaustion; poor sleep patterns; chest pain (not heart-related); pale or itchy skin; sweaty palms; sore muscles and cramps; tremor; and headaches.

29.2.5 Could Hyperventilation Remodel Airways?

The observation that breathing is more vigorous and faster than usual during an asthma attack, and when bronchodilators are taken the breathing usually calms down, led to the common belief that a prime cause of hyperventilation is airway restriction.

Although the production of carbon dioxide is at the cellular level, regulation of its pressure is controlled by lung ventilation. Dr Buteyko became interested in the fact that even hidden hyperventilation kept carbon dioxide concentrations low in the airways. He postulated that airways narrow to protect stores of carbon dioxide in the alveoli; therefore, hyperventilation was actually a prime cause of airway narrowing rather than the other way around. He determined that although there is no doubt that more work is required to move the same volume of air through a narrowed breathing tube, if the breathing was not excessive for the level of activity in the first place, then the airway would be less likely to narrow.

Dr Buteyko further hypothesized that when chronic hyperventilation is present, airways chronically narrow in an effort to maintain eucapnia. Because nasal-breathing is natural, the narrowing usually commences in the nose, and if this is unsuccessful in trapping sufficient carbon dioxide, the lower airways then become involved. His theory appears to be supported by the fact that the upper airways are greatly affected in those with CF, with chronic sinusitis the norm and the upper lobes of the lungs being the worst affected by epithelial thickening or bronchiectasis [29,30].

This idea was novel in the 1950s, but today there is ample research to support it. For instance, in 1968 McFadden and Lyon showed that in the case of asthma, hyperventilation is present and hypocapnia the norm when the person is asymptomatic. This continues until the airways become so restricted that the FEV1 lung function test result is only approximately 25% of the person's predicted normal reading. Once the FEV1 reading becomes even lower, carbon dioxide in the bloodstream rises rapidly and oxygen, which until this point has been normal, drops [31]. When hypoxic, the body cannot function properly, making it virtually impossible for airway smooth muscles to relax even when the person takes heavy doses of bronchodilators.

Although there is some debate over whether the likelihood of developing asthma may be predicted by the testing of lung function in babies [32], there seems to be no clear sign of lung abnormalities at birth, but hyperventilation is present in asthma and over time asthmatic airways become abnormal [33] with the following changes:

- The extent of smooth muscle wrapping the bronchioles is approximately 50–80% greater than in someone without asthma [34].
- Mucus-producing cells are much larger and more numerous, greatly increasing their ability to produce mucus [35].
- Mast cells are approximately 5-fold the usual number [36].
- Airways become "remodeled" and some so deformed or plugged with mucus that gas exchange is greatly impaired [37].

This same idea of hyperventilation being present and possibly causing airway damage seems to be supported in the case of CF, in which it has been noted that lung architecture is normal at birth [30]. However, the breathing pattern quickly becomes abnormal because babies and young children with CF tend to have a significantly increased respiratory rate—at least six extra breaths per minute with a lower tidal volume than their healthy peers [38]. Breathing ordinarily uses only 2–3% of the body's total expenditure of energy when resting, but when breathing is fast with a small volume, the work of breathing increases and this perpetuates the need for increased breathing [39].

Presumably because it seems illogical that breathing less may improve oxygenation and dyspnea, treatments of forced or lengthy exhalations and deliberate coughing, along with the regular use of bronchodilating drugs that encourage deeper and more forceful breathing, are conventionally used by those with damaged lungs. In actual fact, when a level of breathing is maintained that is suitable for the present metabolic requirement, then carbon dioxide concentration is more likely to remain in the normal range, improving oxygenation.

29.2.6 Hyperventilation Frequently Leads to Chronic Mouth-Breathing

Once the habit of hyperventilation is established, nasal breathing becomes more difficult to maintain, making breathing through the mouth more likely. Even without the additional problem of hyperventilation, breathing through the mouth is linked to the following problems:

- Nasal swelling and blockage, with increased mucus production, infections, polyps, and coughing [40]. Apart from the obvious difficulty in breathing that this causes, another factor is that sinuses that are filled with mucus produce less nitric oxide, which has a protective property in sterilizing air, relaxing smooth muscles, and improving mucus clearance [41,42]. Nitric oxide is typically low in children with CF [43].
- The lower airways also become swollen and produce more mucus, increasing the need to cough, reducing room for airflow, and causing the plugging of tiny airways [44,45].
- According to the Buteyko theory, smooth muscles tighten as they attempt to reduce the loss of carbon dioxide from the alveoli [46,47].
- Histamine levels increase, compounding the problem and leading to swelling, mucus production, and further spasm of airway smooth muscles [15].
- Direct inhalation of cool air and pollen/irritants/bacteria into the throat and lungs because filtration by the nose is bypassed [48].
- Polyps and enlarged tonsil and adenoid swelling [49].
- Chronic cough [50].
- Exercise-induced bronchial spasm [51].
- Increased work of breathing [52].

- A dry mouth and tendency toward dehydration, which is demonstrated by the increased need to drink water by people who habitually breathe through their mouth.
- A long, narrow face, crooked teeth, high arched palate, difficulty in swallowing, headache, gum disease, and tooth decay [53,54].
- Poor sleep, snoring, and sleep apnea, which cause impaired daytime cognitive function [49].
- Digestive problems of belching, bloating, and flatulence [55].
- Forward head posture that may lead to neck or back pain or stiffness, cervical joint damage, and kyphosis (In 1994, Henderson and Specter found that 77% of females and 36% of males with CF who were over the age of 15 years had a kyphosis greater than 40°) [56].

However, it is important to remember that hyperventilation is more likely to occur when the mouth is used to breathe through because mouth-breathing tends to involve the accessory muscles of breathing rather than the diaphragm. These muscles encourage larger volumes of air to be inhaled, and when this is coupled with a habit of fast breathing, minute volume greatly increases. Hyperventilation, and especially when accompanied by breathing through the mouth, may result in more damage to the CF patient's already compromised lungs through the following processes:

- Additional bacteria are inhaled and the lungs fight back with smooth muscle spasm, swelling, and increased mucus and histamine production, which is attempting to at least prevent bacteria from going further into the lungs. This process causes structural damage, such as scarring and the loss of cilia.
- Rapid breathing and airway obstruction may lead to hyperinflation of the lungs, which increases the work of breathing and may result in a loss of elasticity and damage to the alveoli because of excessive air pressure [57,58].
- Because water vapor is exhaled in every breath, hyperventilation increases the tendency toward dehydration. Because hydration is already a problem in the CF disease process, this could worsen symptoms [59].
- A high influx of cool, dry air is inhaled, possibly triggering local irritation and leading to increased swelling in an effort to keep the irritants out of the airways [60]. Therefore, chronic hyperventilation may be a reason for the increased goblet cell numbers and mucus secretions found in CF [35].
- Mouth-breathing does not impose sufficient resistance to airflow, and this can lead to poor lung ventilation or atelectasis [61].
- The associated chronic cough causes a marked decrease in the quality of life, and coughing worsens when breathing increases [62], which could be for any of the above-mentioned reasons or because of the upper respiratory tract infections or sinus problems that are frequently associated with mouth-breathing [63]. Repeated vigorous coughing causes the airways to become stretched, corrugated, and floppy, which encourages pooling of mucus where bacteria easily multiply. Therefore, lung function and structure may deteriorate further [64].

29.2.7 The "One Airway, One Disease" Concept

Until recent times upper and lower airway diseases were considered to be completely separate entities, but there has been a shift in this thinking, which is commonly known as the "one airway, one disease" concept [65]. Although this may not appear as true for CF as it is for allergic asthma and rhinitis for instance, it is still the airway epithelium that is mostly affected in CF, and essentially the same type of tissue extends from the start of the nasal passages down to the alveoli. This means that if hyperventilation and mouth-breathing can damage the lower airway epithelium, it could potentially also damage the upper airways. For example, sinus infections damage nasal tissue, leading to polyps [66], and Fernald (1990) confirms that "Mouth breathing and anterior open bite have been associated with chronic nasal and sinus obstruction often seen in patients with CF" [67].

It has been determined that it is more difficult to eliminate nasal infections than lung infections and that nasal infections can initiate lung infections [5,68], making it seem even more important to consider the one airway, one disease concept in regard to CF, which is what the Buteyko method aims to do.

29.3 THE BUTEYKO METHOD

The Buteyko method approaches breathing difficulties in a two-pronged attack:

1. The Buteyko program encourages the acceptance of the counterintuitive aspect of breathing and suggests that when people experience breathing difficulties, they should try breathing a little less air for a short period of time or to put manageable pauses in their breathing. If this is not successful, then prescribed treatments are used in the normal way.
2. To support and actively encourage nasal and diaphragmatic breathing. When healthy babies are born, they almost exclusively breathe this way, and when observing other species in the animal kingdom, it is noted that this continues life-long. However, humans can form unhealthy breathing habits.

Although breathing through the mouth is useful in some situations, such as when a healthy person is under great physical pressure, filtering, moistening, and warming of the air are not completed effectively; therefore, the function of airways that are already sensitive, irritated, or damaged become even more impaired.

Buteyko principles also acknowledge that mucus is a protective coating secreted by the body and not an excretion to be eradicated. It is noted by pharmacists and the medical fraternity from around the world that when mucus production is suppressed by the use of drugs for any length of time, mucus-producing cells go into overdrive in an effort to restore this necessary covering. Therefore, going to extreme lengths to remove mucus is not part of the Buteyko method.

Part of the program involves deliberately breathing to a pattern that is unique to each person but follows the same basic principles of making each breath have three separate components: an inhalation, an exhalation, and what Dr Buteyko called the "automatic pause." The rhythm changes frequently as the person continues with the program, but the goal is always the same: for the respiratory center in the brain to adapt to a more normal pattern so that the work of breathing reduces and breathing becomes more efficient, allowing for improved oxygenation.

The Buteyko exercises are usually practiced for a few minutes several times a day for several weeks, and each time they are practiced small changes to the automatic breathing rhythm occur. Because the respiratory center has plasticity, it can be trained to change the speed, depth, and rhythm of breathing so that these changes continue to a certain extent even when the person is not concentrating on their breathing [69], such as when watching television or sleeping. Because the Buteyko exercises are practiced so regularly, over time the change in breathing from the original initially dysfunctional pattern becomes more pronounced and the healthier, reestablished rhythm becomes more permanent.

In addition to the breathing exercises, one of the strategies is for Buteyko students to become more mindful of the way they breathe while doing ordinary daily activities, which assists the flexible breathing center to maintain the improved patterns. For example, if students are unable to maintain an easy breathing pattern while lifting weights or walking, they are instructed to slow down their pace or to lift lighter weights.

Asthma responds exceptionally well to Buteyko, and the program is scaled down as symptoms reduce. Should a person relapse, the program is resumed until once again symptoms no longer exist. However, with chronic CF, a portion of the program needs to be continued daily to maintain and reinforce the achieved progress and possibly slow lung deterioration.

Checking a chronic condition is vital, and the Buteyko method provides an accurate way of doing this. When people are ill, heartbeat and breathing rates increase; therefore, students of the Buteyko method use the pulse and their ability to comfortably suspend the breathing to check for early warning of deterioration in health. This encourages prompt medical intervention when the person is less stable and in a more critical condition.

Anyone considering the Buteyko method should remember that the enlargement of dead air space in those with CF necessitates an increase in minute volume; therefore, students should proceed carefully and take rests from the exercises as required. It is also important not to change breathing patterns too quickly because this could result in rebound hyperventilation, making symptoms worse.

29.3.1 Physical Exercise

Most of the carbon dioxide in a person's body is present because of muscle movement. Therefore, it comes as no surprise that regular physical exercise is an integral part of the Buteyko method. Students are encouraged to exercise at any intensity they are comfortable with providing they do not breathe through their mouth. The only time that mouth-breathing is permitted is when swimming or in competitions. Once it becomes difficult to breathe through the nose, or the person becomes abnormally short of breath, the level of activity is reduced. Students are taught to prepare their breathing for exercise and to switch off hyperventilation afterward so that they have less dyspnea during and after exercise.

Research shows that physical exercise is important in improving the health and well-being of people with CF [70]; therefore, following the above approach would be useful even if the person does not have access to a Buteyko teacher to coach them on the remainder of the program.

29.3.2 Dietary Modifications

Hyperventilation has a huge effect on the digestive system. For example, hydrochloric acid production automatically drops when the person is in a state of chronic hypocapnia [71], which may affect nutrition and digestion by

- Stimulating the appetite, leading to a habit of overeating as the body tries to obtain a balanced diet
- Creating a lack of energy or stamina because the person is malnourished
- Leading toward food intolerances as food passes from the stomach into the intestines in a state that is less digested than usual
- Causing the person to become irritable and have disturbed sleep if the calorific intake is insufficient

When the lack of hydrochloric acid is combined with increasing histamine levels and smooth muscle contraction that are also affected by hypocapnia, it is not surprising that the Buteyko method involves guidelines on the way people eat. Although the Buteyko method does not usually include dietary restrictions, students taking part in the program are generally requested to make some dietary modifications, of which the three most important points are

1. Eat when hungry rather than by the clock.
2. Do not aim to empty the plate, but instead stop eating when you have consumed sufficient food.
3. Drink water when you are thirsty until you feel satisfied, which should result in urine being pale yellow and not dark.

Babies and young children naturally follow these principles until they become indoctrinated into eating to a timetable rather than by hunger and by a growing awareness that it is socially acceptable to eat the served portion.

In addition to these primary guidelines, it is recommended that when ill or under stress, the following foods are to be avoided as much as possible:

- Easily digested proteins (e.g., seafood, animal stock, nuts, cottage cheese, ice cream, and yogurt)
- Caffeine (e.g., chocolate, coffee, tea, and cocoa)
- Milk of any kind
- Alcohol
- Fruit

Dr Buteyko recommended that when a person is sick, they should drink water that has been made from melted ice, even melted ice cubes from your freezer are preferable to ordinary tap water.

It is not recommended that people eat anything immediately before going to sleep, but at any other time the following list of foods are considered suitable for the person with a breathing problem:

- Fresh vegetables, including onions
- Proteins that are not easily digested (e.g., lamb, beef, and pork)
- Whole grains (e.g., brown rice)
- Spices such as garlic, black pepper, mustard, chili, and horseradish
- Unprocessed fully mineralized sea salt

Perhaps one of the most controversial aspects of the method for the general population is Dr Buteyko's recommendation to use unprocessed, fully mineralized sea salt on a daily basis; however, with CF this suggestion is far more commonplace. Ordinary table salt is usually made from 100% sodium chloride, or perhaps a little less when fillers and drying agents are added. On the other hand, unprocessed sea salt contains only approximately 83% sodium chloride and the remaining 17% is a mixture of 80 or more trace elements and minerals, such as magnesium and potassium.

Dr Buteyko instructed people to use this type of salt in cooking and on food and as a "medicine" when feeling off-color. His "prescription" was to add ¼ to ½ teaspoon of unprocessed sea salt to a cup of hot water, stir to dissolve, and sip until finished. Repeat as required unless the salty water tastes unpleasant. It is important to note that anyone on a salt-restricted diet is instructed to check with their physician before following Dr Buteyko's recommendation.

29.3.3 Unblocking the Nose

The habit of nasal-breathing is natural, but because nasal congestion, infection, and polyps are so common in CF sufferers, and mouth-breathing can lead to lung infection [72], the Buteyko nose-clearing exercise, which is outlined as follows, is used frequently at the commencement of learning the Buteyko method:

- Sit down.
- If possible, exhale through the nose without using any force (i.e., emptying the lungs is not the objective, but instead a relaxed exhalation is required).
- If the nose is blocked, then exhale gently through the mouth and close it immediately.
- Without inhaling, pinch the nostrils closed with one hand and keep the lips together so that you cannot breathe in or out.
- Being careful that you do not hurt your neck, nod your head until you need to breathe.
- When you need to breathe, keep the mouth closed, and release the nostrils.
- Inhale softly through the nose and remember to keep the mouth closed as you continue to breathe.

This exercise can be used whenever the nose feels congested, and if the nose still feels too congested to easily breathe through after completing this process, it can be repeated up to 5 times.

Although breathing repeatedly through the mouth tends to make the nose block, when the nose is used continuously for breathing, it becomes easier to breathe through and when nose-breathing is maintained, hyperventilation is less and the entire body will benefit.

29.3.4 Closed-Mouth Cough

Coughing involves drawing large volumes of cool, dry, and dusty air through the mouth directly into the lungs before it is exploded out again. Forceful or chronic cough can cause collapse of central airways, bronchospasm, and fatigue as well as impairing the natural movement of mucus and damaging the airways [73]. Evidence also shows that the cough reflex has plasticity and can become excessive and inappropriate [74]. Therefore, those learning the Buteyko method are encouraged to cough with a closed mouth.

It is of course impossible to always control a cough, and when a coughing fit occurs, the person is encouraged to put a short pause (2–5 s) in the breathing as soon as possible before inhaling through the nose again. By doing this, the previously uncontrolled cough generally subsides quickly and chronic cough usually becomes a thing of the past when the Buteyko program is continued. This may be because

- A higher pressure of carbon dioxide depresses the cough reflex [75], and the Buteyko method may facilitate this [3].
- Airways dilate slightly when Buteyko breathing techniques are correctly applied, allowing healthy cilia to perform their natural function with less impediment.
- Mucus production reduces when the Buteyko method is applied correctly.
- The most common cause of cough is postnasal drip, and the Buteyko method effectively deals with this.

29.3.5 Basic "Horse-Rider" Exercise

Breathing repeatedly with the accessory muscles increases the work of breathing and makes it harder to exhale, which may cause hyperinflation along with the feeling of needing to inhale before the previous exhalation is completed. Therefore, Buteyko students are routinely reminded to relax the shoulders so that less oxygen is required to breathe and personal comfort is increased. The "horse-rider" stance detailed below helps in this regard:

- Sit on the front of a straight-backed chair with the legs long, stretched out in front of the chair or bent with the feet tucked under the seat. Either way, the thighs should slope slightly toward the floor as they do when sitting on a horse.
- Look straight ahead with the neck in a neutral position.
- Lengthen the spine.
- Relax and drop the shoulders.
- Relax the abdomen and notice that it expands slightly during each inhalation.
- Sit like this for a few minutes.

29.3.6 Buteyko is Not Just about Breathing

In addition to the previously mentioned exercises and strategies, there are others that make up the Buteyko method to provide symptom relief of chest tightness, shortness of breath, panic, and excessive mucus production and that change the respiratory pattern on a more permanent basis. In short, the Buteyko method goes further than providing simple breathing exercises. Instead, it is a total educational approach to good breathing that is delivered in a manner that empowers people because they have fewer symptoms, less dependency on medications, and a clearer idea of when to seek medical advice, making them feel more in control of their condition.

When the individual Buteyko program is followed and the student is guided by their teacher, making sure that the person is not trying too hard or doing the exercises incorrectly, hyperventilation begins to subside, reducing stress on the body. After each Buteyko session, the following immediate effects are frequently commented on by students:

- The airways (in the nose and lungs) seem to open spontaneously, allowing easier breathing. Exhaling or the taking of a full breath, both of which were previously more difficult to do, become easier.
- A natural and effortless expulsion of excess mucus from the nasal passages and the lower airways begins to occur. This becomes more pronounced as practice improves their techniques.
- The urge to take repeated deep, forceful breaths lessens somewhat and a regular rhythm is more likely.
- An increase in saliva production and sometimes an increase in sweating along with a sensation of warmth in the hands, face, and feet [76].

The following long-term improvements are commonly noted:

- Less shortness of breath and increased stamina during physical exercise.
- In the case of asthma, chronic obstructive pulmonary disease, and sinus problems, there is significantly less need for dilating or preventive medicines because symptoms greatly diminish.
- Less constipation and improved digestion.
- Sounder sleep with reports from bed partners that breathing is quieter and more regular.
- Feeling calmer and less anxious with a clearer mind.
- Skeletal muscles tend to relax and become less sore [77].

29.4 CONCLUSION

Although hypercapnia becomes the norm when lungs are severely damaged, it is Buteyko theory that this is more likely to occur when episodes of hypocapnia are found in the earlier stages of CF as a result of the habitual low-grade hyperventilation that begins at a very early age. It is also Buteyko theory that the body will not deal effectively with episodes of hypercapnia while carbon dioxide remains chronically and subclinically low in the upper airways because

- Hyperventilation always lowers carbon dioxide in the airways and bloodstream, causing airway narrowing, hypocapnia, and hypoxia.
- Furthermore, hyperventilation cools, dehydrates, and irritates the airways, increasing mucus production. As overbreathing continues, secretions dry out, becoming stickier and harder to shift through the normal process.
- Postnasal drip and tenacious mucus cause coughing that exhausts the person and damages airways.

These dangers to the body create a vicious circle until eventually the person lacks sufficient oxygen and has hypercapnia, both of which further stimulate breathing, creating a deeper vicious circle.

By reducing hyperventilation as much as possible through the Buteyko method in the early stages of the disease, and by at least teaching children with CF to breathe correctly and to use nasal-breathing as a guide as to how much physical activity is possible at any given time, much can be done toward improving the quality of life and slowing the process of lung deterioration.

References

[1] Meyer KC, Sharma A, Rosenthal NS, Peterson K, Brennan L. Regional variability of lung inflammation in cystic fibrosis. Am J Resp Crit Care Med 1997;156:1536–40.

[2] Lumb AB, editor. Nunn's applied respiratory physiology. London: Reed; 2000. p. 535.

[3] Bowler SD, Green A, Mitchell CA. Buteyko breathing techniques in asthma: a blinded randomised controlled trial. Med J Aust 1998;169:575–8.

[4] Mainz JG, Koitschev A. Pathogenesis and management of nasal polyposis in cystic fibrosis. Curr Allergy Asthma Rep 2012;12(2):163–74.

[5] Davidson TM, Murphy C, Mitchell M, Smith C, Light M. Management of chronic sinusitis in cystic fibrosis. Laryngoscope 1995;105(4 Pt 1):354–8.

[6] Hough A, editor. Physiotherapy in respiratory care. Cheltenham: Stanley Thornes; 1997. p. 70.

[7] Pryor JA, Webber BA, editors. Respiratory and cardiac problems. London: Churchill Livingston; 1998. p. 469.

[8] Delk KK, Gevirtz R, Hicks DA, Carden F, Rucker R. The effects of biofeedback assisted breathing retraining on lung functions in patients with cystic fibrosis. Chest 1993;105:23–8.

[9] Pryor JA, Webber BA, editors. Respiratory and cardiac problems. London: Churchill Livingston; 1998. pp. 140–68.

[10] Coope R, editor. Diseases of the chest. Edinburgh: E and S Livingstone Ltd; 1948. p. 177.

[11] Lumb AB, editor. Nunn's applied respiratory physiology. London: Reed; 2000. p. 267.

[12] El Mays TY, Saifeddine M, Choudhury P, Hollenberg MD, Green FH. Carbon dioxide enhances substance P-induced epithelium-dependent bronchial smooth muscle relaxation in Sprague-Dawley rats. Can J Physiol Pharmacol 2011;89(7):513–20.

[13] El Mays TY. The effect of carbon dioxide on airway tone [thesis]; 2007. [Personal correspondence with El Mays TY].

[14] Strider JW, Masterson CG, Durham PL. Treatment of mast cells with carbon dioxide suppresses degranulation via a novel mechanism involving repression of increased intracellular calcium levels. Allergy 2011;66(3):341–50.

[15] Kontos HA, Richardson DW, Raper AJ, Zubair-ul-Hassan, Patterson JL. Mechanisms of action of hypocapnic alkalosis on limb blood vessels in man and dog. Am J Physiol 1972;223:1296–307.

[16] Haas GJ, Prescott Jr HE, Dudley E, Dik R, Hintlian C, Kean L. Inactivation of microorganisms by carbon dioxide under pressure. J Food Safety 1989;9(4):253–65.

[17] Laffey JG, Kavanagh BP. Carbon dioxide and the critically ill – too little of a good thing? Lancet 1999;354(9186):1283–6.

[18] Curley G, Hayes M, Laffey JG. Can 'permissive' hypercapnia modulate the severity of sepsis-induced ALI/ARDS? Crit Care 2011;15:212.

[19] Henderson Y. Reasons for the use of carbon dioxide with oxygen in the treatment of pneumonia. N Engl J Med 1932;206:151–5.

[20] Dorland's medical dictionary for health consumers. Philadelphia: Elsevier (Saunders); 2012. p. 898.

[21] American Lung Association [Internet]. Available from: http://www.lung.org/lung-disease/cystic-fibrosis/; 2006 [accessed 28.12.13].

[22] Collaco JM, McGready J, Green DM, Naughton KM, Watson CP, Shields T, et al. Effect of temperature on cystic fibrosis lung disease and infections: a replicated cohort study. PLoS One 2011;6(11):e27784.

[23] Goeminne PC, Kiciński M, Vermeulen F, Fierens F, De Boeck K, Nemery B, et al. Impact of air pollution on cystic fibrosis pulmonary exacerbations: a case-crossover analysis. Chest 2013;143(4):946–54.

[24] Cruz I, Marcie KK, Quittner AL, Schechter MS. Anxiety and depression in cystic fibrosis. Semin Respir Crit Care Med 2009;30(5):569–78.

[25] Lum LC. Breathing exercises in the treatment of hyperventilation and chronic anxiety states. Chest Heart Stroke J 1977;2:1.

[26] Magarian GJ. Hyperventilation syndromes: infrequently recognized common expressions of anxiety and stress. Medicine 1982;61(4):219–36. Williams and Wilkins.

[27] Bradley D, editor. Hyperventilation syndrome. Auckland: Random House; 2006. pp. 16–7.

[28] Timmons BH. In: Timmons BH, Ley R, editors. Behavioral and psychological approaches to breathing disorders. New York: Plenum; 1994. p. 4.

[29] Nemec SF, Bankier AA, Eisenberg RL. Upper lobe–predominant diseases of the lung. Am J Roentgenol 2013;200(3):W222–37.

[30] Meyer KC, Sharma A, Rosenthal NS, Peterson K, Brennan L. Regional variability of lung inflammation in cystic fibrosis. Am J Respir Crit Care Med 1997;156(5):1536–40.

[31] McFadden ER, Lyons HA. Arterial-blood gas tension in asthma. N Engl J Med 1968;278:1027–32.

[32] Håland G, Lødrup Carlsen KC, Sekhar Devulapalli C, Sandvik L, Cheng Munthe-Kaas M, Pettersen M, et al. Reduced lung function at birth and the risk of asthma at 10 Years of age. N Engl J Med 2006;355:1682–9.

[33] Elias JA, Zhu Z, Chupp G, Homer RJ. Airway remodeling in asthma. J Clin Invest 1999;104(8):1001–6.

[34] Woodruff PG, Dolganov GM, Ferrando RE, Donnelly S, Hays SR, Solberg OD, et al. Hyperplasia of smooth muscle in mild to moderate asthma without changes in cell size or gene expression. Am J Respir Crit Care Med 2004;169(9).

[35] Lumb AB, editor. Nunn's applied respiratory physiology. London: Reed; 2000. pp. 23–4.

[36] Asthma medications in action. In: MacLennon L, editor. Asthma action. Wellington: The Asthma Foundation of New Zealand, Allen and Hanburys; 1995. pp. 1–3.

[37] Bergeron C, Al-Ramli W, Hamid O. Remodeling in asthma. Proc Am Thorac Soc 2009;6(3):301–5.

[38] Lum S, Gustafsson P, Ljungberg H, Hülskamp G, Bush A, Carr SB, et al. Early detection of cystic fibrosis lung disease: multiple-breath washout versus raised volume tests. Thorax 2007;62:341–7.

[39] Naifeh K. In: Timmons BH, Ley R, editors. Behavioral and psychological approaches to breathing disorders. New York: Plenum; 1994. pp. 24–5.

[40] Naifeh K. In: Timmons BH, Ley R, editors. Behavioral and psychological approaches to breathing disorders. New York: Plenum; 1994. p. 31.

[41] Bradley D, Clifton-Smith T, editors. Breathing works for asthma. Auckland: Tandem Press; 2002. p. 25.

[42] Lumb AB, editor. Nunn's applied respiratory physiology. London: Reed; 2000. pp. 71–315.

[43] Lundberg JO, Nordvall SL, Weitzberg E, Kollberg H, Alving K. Exhaled nitric oxide in paediatric asthma and cystic fibrosis. Arch Dis Child 1996;75:323–6.

[44] Guyton AC, editor. Human physiology and mechanisms of disease. Philadelphia: WB Saunders; 1982. p. 58.

[45] O'Cain CF, Hensley MJ, McFadden Jr ER, Ingram Jr RH. Pattern and mechanism of airway response to hypocapnia in normal subjects. J Appl Physiol 1979;47(1):8–12.

[46] Ameisen PJ, editor. Every breath you take. Sydney: New Holland; 1997. pp. 10, 20, 100, 116.

[47] Stark RJ, Stark JJ, editors. The carbon dioxide syndrome. Brisbane: Buteyko Works; 2002. pp. 95–102.

[48] Henriksen AH, Sue-Chu M, Lingaas Holmen T, Langhammer A, Bjermer L. Exhaled and nasal NO levels in allergic rhinitis: relation to sensitization, pollen season and bronchial hyperresponsiveness. Eur Respir J 1999;13:301–6.

[49] Nayak AS, Schenkel E. Desloratadine reduces nasal congestion in patients with intermittent allergic rhinitis. Allergy 2001; 56(11):1077–80.

[50] Royal Berkshire NHS Foundation Trust [Internet]. Available from: http://www.royalberkshire.nhs.uk/pdf/Chronic_cough_dec10.pdf; [accessed 09.02.14].

[51] Bradley D, Clifton-Smith T, editors. Breathing works for asthma. Auckland: Tandem Press; 2002. p. 74.

[52] Chaitow L, Bradley D, Gilbert C, editors. Multidisciplinary approaches to breathing pattern disorders. London: Churchill Livingstone; 2002. p. 117.

[53] John Flutter Dental Pty. Ltd. [Internet]. Available from: http://www.jfdental.com/pdf/article-malocclusion.pdf; 2006 [accessed 09.02.14].

[54] Cheng MC, Enlow DH, Papsidero M, Broadbent Jr BH, Oyen O, Sabat M. Developmental effects of impaired breathing in the face of the growing child. Angle Orthod 1988;58:309–20.

[55] Flutter J. The negative effect of mouth breathing on the body and development of the child. Int J Orthod 2006;17(2):31–7.

[56] Henderson RC, Specter BB. Kyphosis and fractures in children and young adults with cystic fibrosis. J Pediatr 1994;125:208–12.

[57] Grippi MA, editor. Pulmonary physiology. Philadelphia: JB Lippincott Co; 1995. p. 99.

[58] Hough A, editor. Physiotherapy in respiratory care. Cheltenham: Stanley Thornes; 1997. p. 55.

[59] Shaw KN, Spandorfer PR. In: Fleisher GR, Ludwig S, editors. Textbook of pediatric emergency medicine. Philadelphia: Lippincott Williams & Wilkins; 2010. p. 207.

[60] Davis MS, Freed AN. Repeated hyperventilation causes peripheral airways inflammation, hyperreactivity, and impaired bronchodilation in dogs. Am J Respir Crit Care Med 2001;164(5): 785–9.

[61] Barelli P. In: Timmons BH, Ley R, editors. Behavioral and psychological approaches to breathing disorders. New York: Plenum; 1994. p. 50.

[62] Tatar M, Karcolova D, Pecova R, Kollarik M, Plevkova J, Brozmanova M. Experimental modulation of the cough reflex. Eur Respir Rev 2002;12(85):264–9.

[63] Korpas J, Shannon R, Widdicombe JG. Chairmen's summary. Eur Respir Rev 2002;12(85):278–82.

[64] Morice AH. The epidemiology of chronic cough. Eur Respir Rev 2002;12(85):222–5.

[65] Grossman J. One airway, one disease. Chest 1997;111(Suppl. 2): 11S–6S.

[66] Pawankar R. Nasal polyposis: an update. Curr Opin Allergy Clin Immunol 2003;3(1):1–6.

[67] Fernald GW, Roberts MW, Boat TF. Cystic fibrosis: a current review. Am Acad Pediatr Dent 1990;12(2).

[68] Aanæs K. Bacterial sinusitis can be a focus for initial lung colonization and chronic lung infection in patients with cystic fibrosis. J Cyst Fibros 2013;12(Suppl. 2):S1–20.

[69] Mitchell GS, Johnson SM. Plasticity in respiratory motor control. Invited review: neuroplasticity in respiratory motor control. J Appl Physiol 2003;94:358–74.

[70] Nixon PA, Orenstein DM, Kelsey SF, Doershuk CF. The prognostic value of exercise testing in patients with cystic fibrosis. N Engl J Med 1992;327:1785–8.

[71] Lum LC. Hyperventilation: the tip and the iceberg. J Psychosom Res 1975;19:375–83.

[72] Guyton AC, editor. Human physiology and mechanisms of disease. Philadelphia: WB Saunders; 1982. p. 301.

[73] Hough A, editor. Physiotherapy in respiratory care. 2nd ed. Cheltenham: Stanley Thornes; 1997. pp. 74–139.

[74] Mazzone SB, Canning BJ. Plasticity of the cough reflex. Eur Respir Rev 2002;12(85):236–42.

[75] Nishino T, Hiraga K, Honda Y. Inhibitory effects of CO_2 on airway defensive reflexes in enflurane-anesthetized humans. J Appl Physiol 1989;66(6):2642–6.

[76] Stark RJ, Stark JJ. Personal feedback from face-to-face teaching the Buteyko method to approximately 10,000 students by the Starks; 2014.

[77] Stark RJ, Stark JJ. Data is taken from student questionnaires who have attended Buteyko classes conducted by the Starks; 2014.

CHAPTER

30

Lactoferrin and Cystic Fibrosis Airway Infection

Piera Valenti[1], Angela Catizone[2], Alessandra Frioni[1], Francesca Berlutti[1]

[1]Department of Public Health and Infectious Diseases, Sapienza University of Rome, Rome, Italy; [2]Department of Anatomy, Histology, Forensic Medicine, and Orthopedics, Sapienza University of Rome, Rome, Italy

30.1 INTRODUCTION

Airway infections in subjects with cystic fibrosis (CF) are the most difficult diseases to treat and represent a real challenge in the management of patients with CF, affecting their quality of life.

Bacterial infection in the airway occurs early in the life of CF patients [1]. The transmembrane conductance regulator (CFTR) dysfunction affecting the chloride transport and hydration of airway impairs mucociliary clearance, favoring the establishment of *Pseudomonas* infections and the strong inflammatory status [2,3]. *Pseudomonas aeruginosa* is the most frequent bacterial pathogen detected in approximately 50% of patients with the disease overall and almost 80% of all adult patients [4]. Airway infections correlate with the onset of symptomatic lung disease, excessive airway inflammation, and eventual loss of pulmonary function. Early *Pseudomonas* infection may be intermittent/transient and sustained by planktonic non-mucoid strains, whereas chronic infections are caused by *Pseudomonas* mucoid strains grown in biofilm lifestyle [1].

Moreover, several lines of evidence suggest that CFTR dysfunction affects not only chloride but also iron homeostasis and components of innate immunity [1,5,6]. In particular, high iron concentrations in airway secretions have been found and clinical studies suggest that iron homeostasis is aberrant in the lung of CF patients [7–9]. Excess iron can derive from dysfunction of cellular iron homeostasis in CF bronchial epithelium [10], dysregulation of systemic iron homeostasis, inflammation, and infection [6,11]. As a consequence, CF airway secretions contain micromolar concentrations of iron much higher than those present in healthy subjects (10^{-18} M). The high concentration of this micronutrient favors the multiplication of inhaled bacteria, thus increasing the inherent susceptibility of CF patients to bacterial infection. Moreover, the iron overload in CF airway secretions induces *P. aeruginosa* to adopt the biofilm lifestyle, favoring bacterial persistence and chronic infection and worsening the severity of the disease [12–14].

Lactoferrin (Lf), a cationic glycoprotein able to chelate two Fe^{3+} ions per molecule with high affinity is one of the most important glycoproteins of natural immunity [15–17] present in airway secretions [18]. Recently, Lf has emerged as an important regulator of iron and inflammatory homeostasis, exerting multiple functions both dependent and independent on its iron-withholding ability.

Herein, the role of iron and Lf in CF airway infection is revised.

30.2 IRON HOMEOSTASIS IN HEALTHY AND CF AIRWAYS

Iron, an essential element for cell growth and proliferation, is a component of fundamental processes, such as DNA replication and energy production. However, iron can also be toxic when present in excess because of its capacity to donate electrons to oxygen, thus causing the generation of reactive oxygen species (ROS), such as superoxide anions and hydroxyl radicals [19]. ROS are known to cause tissue injury and organ failure by damaging several cellular components, including DNA, proteins, and membrane lipids. This dichotomy of iron, able to gain and lose electrons, needs tight controls at both cellular and systemic levels. The iron uptake and recycling essentially performed by enterocytes and circulating macrophages ensure the correct balance of iron. The iron absorption, storage, and export are strongly regulated because mammals lack a pathway of iron excretion [19,20].

Diet and Exercise in Cystic Fibrosis
http://dx.doi.org/10.1016/B978-0-12-800051-9.00030-4

The correct balance of iron, defined as iron homeostasis, involves a physiological ratio between iron in tissues/secretions and blood, thus avoiding its delocalization as iron accumulation in tissues/secretions and iron deficiency in blood [21]. At the cellular level, iron homeostasis ensures that the amount of iron absorbed by cells is appropriate to their requirements to avoid the deficiency or overload of this important microelement. Generally, iron is acquired via the binding of iron-saturated transferrin (Tf) [Tf-Fe(III)] to the Tf receptors (TfR1 and TfR2), with subsequent endocytosis. In the endosome, the low pH strips the iron from the receptor–ligand complex: the TfR1-Tf is cycled back to the plasma membrane for the next cycle of iron uptake, and the released ferric ion is reduced by a ferrireductase to ferrous ions and transported across the endosomal membrane into cytoplasm via divalent metal transporter 1 (DMT1) [22].

In the cytoplasm, iron is stored by ferritin, a protein composed of 24 subunits endowed with ferroxidase activity, able to sequester 4500 ferric iron atoms per molecule, [23] or constitutes the cytosolic iron pool, whose nature and function remain to be fully characterized [24].

Cellular iron export is ensured by ferroportin (Fpn), a protein with a molecular weight of 67 kDa and 12 putative transmembrane domains [25]. Fpn, the only known cellular iron exporter from cells into blood, has been found in all cell types involved in iron export to circulation, including enterocytes, hepatocytes, placental cells, and macrophages; they recycle 20 mg/day of endogenous iron from lysed erythrocytes [26]. The basolateral localization of Fpn protein in enterocytes, hepatocytes, and placental cells is relevant, attesting to its role in exogenous dietary iron export to blood [27]. Fpn acts in partnership with a ferroxidase (i.e., hephaestin in enterocytes and ceruloplasmin in macrophages) that oxidizes exported ferrous iron, facilitating its binding to Tf [28,29].

Iron export by Fpn is regulated by hepcidin, another pivotal component of systemic iron metabolism. Hepcidin is a peptide hormone, containing 25 amino acids synthesized by hepatocytes, as a precursor containing 84 amino acids, and secreted in blood and urine [30,31]. Hepcidin regulates the export of iron through the binding and degradation of Fpn [32,33]. Fpn degradation hinders iron export, thus enhancing the cytosolic iron storage. Hepcidin synthesis is suppressed by iron deficiency in serum and anemia, whereas it is upregulated in response to iron overload and inflammation [32]. Of interest, hepcidin synthesis is significantly induced by an inflammatory status. In particular, interleukin 6 (IL-6) induces transcription of the hepcidin gene in hepatocytes, thus inhibiting iron release from enterocytes and macrophages [32]. The hepatic release of hepcidin can also be induced by IL-1α and IL-1β [34]. Independent of hepcidin synthesis, high levels of serum IL-6 and intracellular iron deficiency appear to downregulate Fpn mRNA expression, thus blocking iron flow into plasma and increasing iron overload inside cells [35,36].

Iron homeostasis in the normal airways is complex and remains poorly understood [37]. Airways are composed of several cell types that obtain their basic iron supplies from the plasma. In addition, airway cells are exposed to active iron from inhaled air [38] that can damage host tissues. Therefore, airway cells must also be able to either store or export any excess iron to reduce the ROS formation and to prevent toxic intracellular iron overload. Airway cells adopt several strategies to reduce the oxidative stress as DMT1-mediated iron uptake in conjunction with stimulated Ft synthesis [39,40] and the upregulation of Fpn expression [40a,41]. The expression of iron homeostasis–related genes in airway epithelial cells is regulated differently from enterocytes, as shown in Table 30.1 [38,42]. Unlike enterocytes, Fpn seems to be localized mainly in the apical membrane of airway epithelial cells and overexpressed in iron overloaded cells, thus supporting the hypothesis of the tissue-specific role for this protein in lung iron detoxification rather than iron export to blood [38,41].

In CF airways, a relation between the dysfunction of CFTR and the abnormalities in iron transport across the airway epithelial cells in CF subjects has been hypothesized [43]. As already reported in the introduction, high iron concentrations in airway secretions suggest that iron homeostasis is aberrant in the airways of CF patients [7–9]. Excess iron can derive from dysregulation of intracellular iron homeostasis of CF bronchial epithelial cells [6,10]. The altered expression of several proteins required for correct handling of iron has been found. In particular, the increased expression of Ft, the iron importer DMT1, and the exporter Fpn in lung tissue of CF patients both in early and end-stage lung diseases has been described [40a]. Human airway epithelial cells express hepcidin, and the hepcidin expression is upregulated in response to interferon-γ. However, different from enterocytes and macrophages, hepcidin does not influence cellular iron export, being unable to decrease Fpn protein [44].

In CF airway secretions, iron is represented by a pool of ferrous and ferric iron. Interestingly, ferrous iron,

TABLE 30.1 Response of Airway Epithelial Cells vs. Enterocytes to Iron

	Airway Epithelial Cells		Enterocytes	
Protein	Iron Deficiency	Iron Overload	Iron Deficiency	Iron Overload
Dcytb	↓	↑	↑	↓
DMT1	↓	↑	↑	↓
Fpn1	↓	↑	↑	↓

abundant (>35 μM) for severely sick patients, correlates with disease severity and, over time, it comes to dominate the iron pool [45]. Iron concentration in airway secretions negatively correlates with lung functionality, as measured by forced expiratory volume in 1 second percentage, while it positively correlates with cell injury and disease severity [7,13]. As a consequence, CF airway secretions contain micromolar concentrations of iron (up to >100 μM), significantly higher compared with those in healthy humans (10^{-18} M).

30.3 CF AIRWAY INFECTIONS

CF airways are essentially normal at birth but later become obstructed with mucus plugs. The CF airway is inherently prone to infection, and respiratory tract infections begin in early life [46]. Clinically apparent infections are generally sustained with the typical pathogens *Staphylococcus aureus*, *P. aeruginosa*, or both, even in infants at a young age [1]. Airway infections correlate with the onset of symptomatic lung disease, excessive airway inflammation, and eventual loss of pulmonary function [1].

However, the microbiota of CF airways is complex, and it changes during the life. Studies on the composition of the microbiota of CF airways show a progressive loss of microbiota diversity over time. In particular, young patients are mostly colonized by *Pasteurellaceae*, *Mycobacteriaceae*, *Actinomycetaceae*, and anaerobes, whereas the adult CF airway microbiota is nearly stable and largely composed of *Pseudomonas* and *Burkholderia* [47]. Different microbial communities may colonize the same patient, and the colonization probably is sequential, rather than simultaneous [48]. When *Pseudomonas* and *Burkholderia* co-infections are established, they are responsible for severe nosocomial infections rarely eradicated by antibiotic therapy [49]. Recently, it has been shown that *Burkholderia cenocepacia* positively influenced *P. aeruginosa* biofilm development by increasing the biomass. Interestingly, co-infection experiments in the mouse model reveal that *P. aeruginosa* does not change its ability to establish chronic infection in the presence of *B. cenocepacia*, but the co-infection increases the host inflammatory response [50].

The microbiota composition does not change significantly in patients during an exacerbation or when clinically stable. Moreover, the antibiotic treatment during exacerbation does not affect significantly the microbiota, suggesting that exacerbations may represent intrapulmonary spread of microorganisms rather than a change in microbial community composition [51].

Early *Pseudomonas* infection may be intermittent/transient and sustained by planktonic non-mucoid strains. By adulthood, the chronic airway infections are essentially sustained by *P. aeruginosa* becoming the dominant pathogen in 80% of cases [52]. Chronic infection is characterized by mucoid strains that live in a structured microbial community immersed in an exopolymeric matrix (i.e., in biofilm lifestyle) [53,54]. *Pseudomonas aeruginosa* infection is related to an increased rate of lung function decline, morbidity, and mortality [55].

The iron overload in CF airway secretions strongly favors the multiplication of inhaled *P. aeruginosa*, thus increasing the inherent susceptibility of CF patients to infection [8,9,56,57]. Iron availability favors *Pseudomonas* multiplication and further stimulates bacteria to adopt the biofilm lifestyle [10,58]. The relevance of iron availability is also demonstrated by the observation that several synthetic chelators prevent *Pseudomonas* biofilm development and that such ability is reversed by iron addition to bacterial cultures [59]. In CF airways, *P. aeruginosa* expresses genes associated with siderophore-mediated ferric iron uptake, heme transport systems, and ferrous ion uptake, indicating the key requirement for iron of *P. aeruginosa* [60]. Finally, iron contributes to the structural integrity of the biofilm by cross-linking exopolysaccharide strands [61].

Bacteria in biofilm are more resistant to antibiotics than the planktonic counterparts; consequently, the eradication of *Pseudomonas* biofilm is rarely possible [62]. Although the use of aerosolized antibiotics improves the outcomes of CF patients, it does not reduce *P. aeruginosa* load [9,10]. Therefore, the iron overload in CF airways and the *Pseudomonas* biofilm lifestyle correlate with *P. aeruginosa* persistence in the CF lung [9,57]. *Pseudomonas aeruginosa* biofilm is able to invade and persist inside CF airway epithelial cells, [10,15] thus evading the antibacterial activity of antibiotics.

30.4 INFLAMMATION IN CF AIRWAYS

Inflammation is a hallmark of CF airway disease. Increased airway inflammation in CF patients compared with healthy subjects is shown in the absence of detectable infection, [63] as demonstrated by a high level of IL-8 and accumulation of neutrophils [64–67]. Interestingly, human fetal CF lung grafts develop progressive intraluminal inflammation before any infection [68]. The involvement of the genetic CFTR defect in the inflammatory process is suggested by the perturbation in intracellular signaling. In particular, the dysregulation of several intracellular signaling pathways, including inhibitor of κB and nuclear factor kB, results in excessive production of nuclear factor κB–dependent cytokines, such as IL-1, tumor necrosis factor (TNF), IL-6, and IL-8 [69]. Recently, the over-production of IL-8 has been related to an intrinsic CFTR mutation-dependent mechanism [70]. Furthermore, the different localization of some toll-like receptors

(TLRs) as TLR2 and TLR5 on CF with respect to healthy epithelial cells correlates with increased inflammatory responses to bacterial molecules [71]. For example, the increased availability of TLR2 at the apical surfaces of CF epithelial cells is consistent with the increased pro-inflammatory responses seen in CF airways and suggests a selective participation of TLRs in the airway mucosa [72]. These defects may self-initiate, or at least enhance, the pulmonary inflammatory status in CF patients.

As previously reported, the dysregulation of iron homeostasis that can precede the introduction of microbes into the respiratory tract contributes to an increased production of ROS [73,74]. The CFTR dysfunction, affecting the chloride transport and hydration of airway, impairs mucociliary clearance and favors the establishment of bacterial infections and the strong inflammatory status [2,3]. The presence of bacterial lipopolysaccharide (LPS), derived from the lysis of Gram-negative bacterial cells, is associated with a further increase of inflammation related to both the TLR localization and the altered pathway of the lysosomal degradation of the LPS-TLR4 complex [72,75]. Airway infection with bacteria in biofilm and especially *Pseudomonas* worsens the inflammatory phenotype. Bacterial biofilm is able to increase the inflammation in CF airways. Pyocyanin, a phenazine pigment produced by all *P. aeruginosa* strains, particularly those growing in biofilms, activates pro-inflammatory signaling [76] and inhibits the activity of the antioxidant glutathione [71]. The expression of more inflammatory LPS by *P. aeruginosa* and of lipooligosaccharides by *B. cenocepacia* and the presence of extracellular DNA are able to enhance the already strong inflammatory status by the activation of TLR-mediated pro-inflammatory signaling [77–79]. As already reported, biofilm is able to invade and persist inside CF airway epithelial cells, which activates a robust inflammatory response [10,80]. In particular, high levels of IL-1β, positively related to iron availability, [8] are produced by CF epithelial cells in response to bacterial infection [80].

In this scenario, a dangerous vicious circle involving dysregulation of iron homeostasis, bacterial infection, and inflammation damage is established.

30.5 ANTIMICROBIALS OF THE HUMAN AIRWAY: LACTOFERRIN, A COMPONENT OF INNATE IMMUNITY

In human airways, several antimicrobial factors produced by serous cells of airway glands are found. The most abundant airway antimicrobial factors are lysozyme, secretory leukocyte protease inhibitor, and Lf [18].

Lf is an iron-binding cationic glycoprotein of molecular weight 80 kDa, belonging to the transferrin family. Lf is able to reversibly chelate two Fe(III) per molecule with high affinity (*Kd* ~10–20 M) and retains ferric iron until pH values as low as 3.0 [81]. In the presence of Lf, the concentration of free iron in body fluids as airway secretions cannot exceed 10^{-18} M, thus preventing the precipitation of this metal as insoluble hydroxides, hindering formation of ROS, and inhibiting microbial growth [12,58,82]. Lf is highly conserved among human, bovine (Figure 30.1(A) and (B)), mouse, and porcine species [83]. As apparent from the three-dimensional structure of human Lf and bovine Lf, the molecules are folded into two homologous lobes. The two lobes, connected by a peptide that forms a three-turn α-helix, are further divided into two domains for each lobe (N1 and N2, C1 and C2). Each lobe binds one Fe(III) ion in a deep cleft between two domains. The iron-saturated form of Lf is closed and much more resistant to proteolytic enzymes than the unsaturated form [84] (Figure 30.2). Lf is expressed and secreted by glandular epithelial cells and by neutrophils. Lf is found in human colostrum, human mature milk and tears, and at low levels in most exocrine secretions [85]. Recently, the levels of human Lf recovered in the main different body secretions have been revised by Alexander et al. (2012) [86]. The concentration of Lf synthesized by exocrine glands in several secretions and by neutrophils is summarized in Table 30.2.

Lf functions are both dependent and independent on its iron-withholding ability [15–17]. The Lf function first described was the antibacterial activity related to its ability in sequestering iron necessary for microbial survival and growth [15].This Lf activity is bacteriostatic, because the addition of ferric iron restores bacterial multiplication [87]. Successively, multiple antimicrobial activities

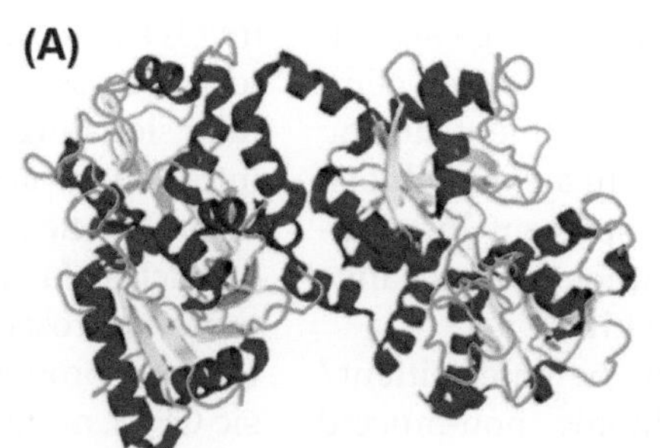

FIGURE 30.1 Human (A) and bovine lactoferrin (B) crystal structures. (For color version of this figure, the reader is referred to the online version of this book.) *From PDB* http://www.rcsb.org/pdb/home/home.do.

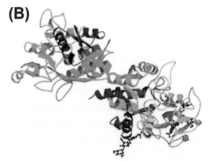

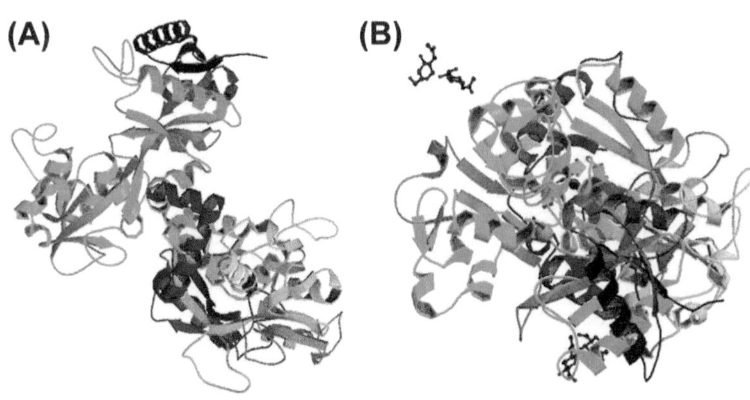

FIGURE 30.2 Crystal structures of unsaturated (A) and saturated (B) human lactoferrin. (For color version of this figure, the reader is referred to the online version of this book.) *From PDB* http://www.rcsb.org/pdb/home/home.do.

TABLE 30.2 Concentrations (μg/mL) of Lactoferrin

Tissue	Lf Concentration (μg/mL)
Human colostrum	3000–8000
Human milk	1000–4800
Nasal fluid	3.1 [18]
Saliva	4–20
Vaginal fluid	0.9
Intestinal fluid	<0.1
Bronchoalveolar lavage fluid	0.2–1.9 [18,85]
Lung alveoli	Absent
Neutrophils (10^6 cells)	3 [100]

independent from Lf iron-binding function were demonstrated. The Lf bactericidal activity is related to a direct interaction with LPS of Gram-negative or lipoteichoic acid of Gram-positive bacteria [88]. In particular, the binding of Lf with lipid A of LPS induces the release of LPS and the bacterial lysis [89].

Among the Lf antibacterial activities, the modulation of biofilm development has been more recently described. Lf modulates biofilm development in different ways according to the bacterial species and the relative habitat. In particular, in saliva, iron-unsaturated Lf induces the cariogenic *Streptococcus mutans* to form aggregates and biofilm, whereas iron-saturated Lf decreases *S. mutans* aggregation and biofilm development [90]. Similarly, the periodontopathogen *Aggregatibacter actinomycetemcomitans* adopts the biofilm lifestyle in an iron-limiting condition mediated by the iron-withholding property of Lf. The iron limitation upregulates the biofilm gene expression of *A. actinomycetemcomitans*, contributing to biofilm formation [91]. On the contrary, *P. aeruginosa* and *B. cenocepacia* differently respond to iron limitation mediated by iron-unsaturated Lf. These bacterial species adopt the planktonic instead of the biofilm lifestyle when grown in the presence of iron-unsaturated Lf [12,58].

Lf interferes with the bacterial adhesion process. The ability of microbes to adhere, colonize, and form biofilm on both abiotic surfaces and host cells is a crucial step in the development and persistence of infections. Both Gram-positive and Gram-negative bacteria possess specific adhesins that mediate the adhesion process on abiotic surfaces or epithelial host cells. The anti-adhesive activity of Lf was first demonstrated on *S. mutans* adhesion to hydroxyapatite [92]. The anti-adhesive property of Lf does not rely on its iron-binding property because the residues 473–538 of the Lf C-lobe are involved in this function, residues not involved in iron-chelating property [93]. However, the contrasting results on the mechanisms involved in Lf anti-adhesive function reported in the literature are related to the different nature of abiotic surfaces and the several bacterial adhesion mechanisms [15,94]. Bacterial adhesins mediate the adhesion process to epithelial host cells. Lf is able to bind to bacterial surfaces as LPS and pili as well as to host cells (glycosaminoglycans and heparan sulfate) [15,95]. Therefore, Lf anti-adhesive function seems to be related to its interaction with bacterial or host cell or both. Although the anti-adhesive property of Lf has been shown for several decades, the actual mechanism remains to be fully understood.

Several mucosal pathogenic bacteria are capable of entering into host cells. Bacterial invasion of nonprofessional phagocytes, such as epithelial cells, requires that both bacteria and cells are viable. The fate of intracellular bacteria depends on specific bacterial mechanisms. Inside the host cells, bacteria are localized within the endosome or escape into cytoplasm. In any case, the intracellular lifestyle ensures the bacteria a protective niche to replicate, persist, and avoid the host defenses. Lf interferes with the bacterial invasion process of host cells, probably through its binding to bacterial and/or cell surfaces [15,95–97]. This ability is not related to the Lf iron-chelating function because apo-saturated, iron unsaturated, and iron-saturated Lf exerts an inhibiting activity against the microbial internalization. In contrast to inhibition of bacterial adhesion, Lf binding to glycosaminoglycans of host cells seems crucial in inhibiting bacterial internalization [15,94,96]. However, the Lf ability to interfere with invasion ability of facultative intracellular bacteria is dependent on the bacterial species. For example, Hyvönen, [97] using a bovine mammary epithelial cell model, has shown that Lf has a limited effect on the adhesion and invasion of coagulase-negative *Staphylococcus* strains, but, interestingly, significantly decreases intracellular replication rates. Antifungal and antiprotozoal activities of Lf have been demonstrated using in vitro models [15]. Recently, the Lf antimalaria infection has been reported (Spaccapelo *et al.*, personal communication, XI International Conference on Lactoferrin, Rome, Italy, 2013).

In addition to the antimicrobial functions, Lf exerts several functions in innate immunity processes. Lf acts in synergy with antimicrobial molecules of the airway secretions as lysozyme and secretory leukocyte protease inhibitor (SLPI). The triple combination of lysozyme, SLPI, and Lf shows even greater synergy [98]. The concentration of the antimicrobial molecules is modulated locally by inflammation. Inflammatory stimuli induce the differentiation of respiratory epithelial cells into secretory cell types [99]. Moreover, inflammatory stimuli as IL-8 also recruit neutrophils containing large amounts of antimicrobial factors in secondary granules, including Lf [100].

Lf contributes to the protective activities of neutrophils. Neutrophils recruited in the infection and inflammation sites release several molecules from secondary granules. Among these molecules, apo-Lf is secreted and, consequently, its level is increased in biological fluids. Lf exerts a multiple role in the antimicrobial activity of neutrophils by its antibacterial function. Moreover, Lf chelates free iron accumulated in inflamed tissues, thus inhibiting the ROS production and the related cell damage; in the phagocytosis process of neutrophils, Lf is released into phagosomes and contributes to the intracellular killing; finally, Lf is secreted together with neutrophil DNA in the so-called neutrophil extracellular traps (NETs). Bacteria are entrapped in the NETs, where they are killed by Lf [101].

Finally, even if the mechanism of action is not fully elucidated, Lf demonstrates anti-inflammatory activity, thus contributing to mucosal protection from inflammation-related damage [102–104].

The anti-inflammatory activity of Lf is unrelated to its iron-withholding function. Lf is a potent scavenger of LPS. The Lf-LPS binding allows the reduction of the LPS toxicity, thus preventing the activation of the LPS-mediated pro-inflammatory pathway cascade and related tissue damage, and the favoritism of the Lf interaction with cell-surface receptors, thus favoring Lf activity [17,105].

The Lf protective activity against LPS-induced damage in mice has been recently demonstrated. In these experiments, Lf treatment markedly attenuated lung edema, alveolar hemorrhage, and inflammatory cell infiltration. Moreover, a decrease of TNF-α (pro-inflammatory cytokine) and an increase of IL-10 (anti-inflammatory cytokine) were recorded in the bronchoalveolar lavage fluid of Lf-treated mice [106]. In mice and piglets, Lf administration can prevent not only microbial infections, but also septic shock [107]. Interestingly, Lf not only reduces the expression of pro-inflammatory cytokines as IL-1α, IL-6, and TNFα, and granulocyte-macrophage colony-stimulating factor [108,109], but also upregulates that of anti-inflammatory cytokines. In particular, the Lf oral administration in rats with colitis increases IL-4 and IL-10 [110].

The anti-inflammatory activity of Lf has been demonstrated in human cell lines. Human intestinal epithelial cells infected with facultative intracellular pathogens respond to bacterial injury, activating a strong inflammatory response. The addition of Lf at the moment of infection modulates the inflammatory response of infected cells, dramatically downregulating pro-inflammatory cytokine gene expression and synthesis. Interestingly, Lf does not influence uninfected cells, suggesting a specific function in modulating inflammation during bacterial infections [102]. The observation that Lf reaches the nucleus of human intestinal epithelial cells leads to the hypothesis of a direct involvement of Lf in the regulation of gene expression [111,112]. Similar anti-inflammatory activity has been shown also in different human host cell models. In mouse and human T cells treated with staphylococcal pyrogenic enterotoxin B (SEB), Lf is able to attenuate SEB-induced proliferation and IL-2 production [104].

Recently, the anti-inflammatory function of Lf has been also confirmed in clinical studies, in which Lf oral administration decreases the levels of serum IL-6 [113,114,114a].

Even if the mechanism of action is not fully elucidated, Lf ability to downregulate expression of pro-inflammatory cytokines might represent an important natural mechanism regulating epithelial cell responses to pathogenic bacteria or toxins, limiting cell damage and spreading of infections. Overall, Lf represents the most relevant protein symbolizing a brick in the wall of natural non-immune defenses of human mucosal fluids against microbial infections [15].

30.6 LACTOFERRIN AND CF AIRWAYS

As previously described, in the CF airways, a dangerous vicious circle involving dysregulation of iron homeostasis, bacterial infection, and inflammation damage is established. In particular, the high levels of pro-inflammatory cytokines, including IL-8, recruit neutrophils that synthesize and secrete Lf. Among the CF subjects, Lf increased with age and correlated positively with neutrophil counts but not with bacterial load [41].

Unfortunately, the Lf functions as well as the synergistic and additive killing activity with the antimicrobial factors of the human airway surface liquid may be inhibited by high ionic strength of CF airway secretions [18,98]. Moreover, in the presence of *P. aeruginosa* infection, the CF lung displays relatively low levels of Lf. This reduction may be due partly to proteolytic degradation by the high concentrations of bacterial proteases or the cell proteases as metalloproteases and cathepsins derived from cell lysis present in the CF airways [115,116]. Concerning cathepsins, even if Lf is resistant and inhibits the activity of cathepsin G [117,118], it may be sensitive to cathepsins B and S found in CF airway secretions [119]. In addition, protease activity may increase the ability of *P. aeruginosa* pyoverdine to acquire iron from Lf that, undergoing to proteolysis, provides a potential supply of iron [120–122].

Notwithstanding the criticisms previously highlighted, Lf may be considered a potential important factor of human defense against lung infections in CF patients. It has been shown that the expression of Lf receptor on respiratory tract epithelial cells increased after exposure to iron. This observation is consistent with the Lf function in diminishing the iron overload by chelating ferric iron. Consequently, Lf participates in decreasing the oxidative stress presented to the lower respiratory tract by complexing catalytically active iron [123].

Even if concentrated mucus in CF airways limits the diffusion of the chemical mediators of quorum sensing, thus favoring biofilm development [124], Lf may exert its anti-biofilm activity against *Pseudomonas* strains [12,58]. Interestingly, even if it has been shown that the anti-biofilm property of Lf is related to its iron-chelating property [12,125], conflicting lines of evidence have been reported. The observation that the Lf demonstrates enhanced anti-biofilm effects as iron supplementation increased suggests that the anti-biofilm inhibition activity of Lf occurs through more complex mechanisms than the sole iron chelation property [59]. This observation is of interest considering that in CF airway secretions, the iron overload can saturate Lf and that only iron-saturated Lf can be found in CF airway secretions. Moreover, Lf is able to reduce also with preformed biofilm, suggesting that Lf may have a destructive effect on developed biofilm [125,126].

Lf ability in inhibiting the host cell invasion by some facultative intracellular process has been shown also in human epithelial airway cell models infected with CF airway bacterial pathogens as *Burkholderia* and *Pseudomonas* [96]. Lf binds to *Burkholderia* cable pili, leading to reduced bacterial adhesion efficiency to mucins. Because cable pili are implicated in mediating the bacterial interactions with mucins and epithelial cells, Lf binding to these structures could play an important role in neutralizing bacterial infection in CF patients [95]. Even if the addition of Lf to host cells just before the infection has little influence on adhesion efficiency of *B. cenocepacia* or slightly increases that of *P. aeruginosa*, Lf significantly inhibits bacterial invasion efficiencies [96]. Interestingly, the anti-invasive function of Lf is not related to its iron-binding capability because this function is exerted both by apo- and holo-Lf [96].

Finally, Lf demonstrates anti-inflammatory activity also in the infected CF airway model [80]. Likely to human intestinal epithelial cells [102], the anti-inflammatory activity of Lf is shown only in infected human CF airway epithelial cells. In fact, Lf added to uninfected CF airway cells does not induce any significant change of inflammatory gene expression. On the contrary, Lf regulates inflammatory response of CF epithelial cells infected with *B. cenocepacia* by regulating both mRNA and protein synthesis of several pro- and anti-inflammatory cytokines. Particularly relevant is the observation that Lf decreases IL-1β and increases IL-11 production [80]. Because IL-1β over-production is associated with the malfunction of the CFTR channel in murine macrophages in response to *B. cenocepacia* infection [127], a similar mechanism may be also hypothesized in human CF epithelial airway cells.

Because Lf changed the expression of inflammatory genes of CF infected cells, but not that of CF uninfected cells, it should be argued that the Lf modulates only genes regulated in response to bacterial invasion [80]. This hypothesis is supported by the observation that Lf is absorbed by CF airway epithelial cells and reaches the nucleus within 3h, [80] as previously shown in enteric epithelial cells [21,112]. Interestingly, Lf reaches the nucleus of the CF airway epithelial cells both uninfected and infected with *B. cenocepacia* strains, but modulates the inflammatory gene expression of infected cells only [80]. Similar nuclear localization of Lf is shown in CF airway cells infected with *P. aeruginosa* (Figure 30.3). Recently, by using a more complex CF epithelium model infected with *P. aeruginosa*, the anti-invasive and anti-inflammatory abilities of Lf have been evaluated. Even if the adhesion efficiency was not affected by Lf, this molecule reduces significantly the survival of intracellular bacteria and the inflammatory response of infected CF epithelium [127a].

30.7 FUTURE APPROACHES IN THE THERAPY OF CF AIRWAY INFECTIONS

The treatment of CF lung infection is essentially based on long-term administration of antibiotics [5,128]. However, high antibiotic concentrations are difficult to achieve and maintain in airway secretions and bacteria are repeatedly exposed to sub-inhibitory levels, thus developing resistance. Indeed, sub-inhibitory doses of antibiotics increase the mutation frequency in *P. aeruginosa*, facilitating adaptation of bacteria to CF lung [128a] and positively influencing biofilm formation [129–132]. Although *P. aeruginosa* infection of CF airways may be eradicated if treatment is started early, [133] no antibiotic is able to

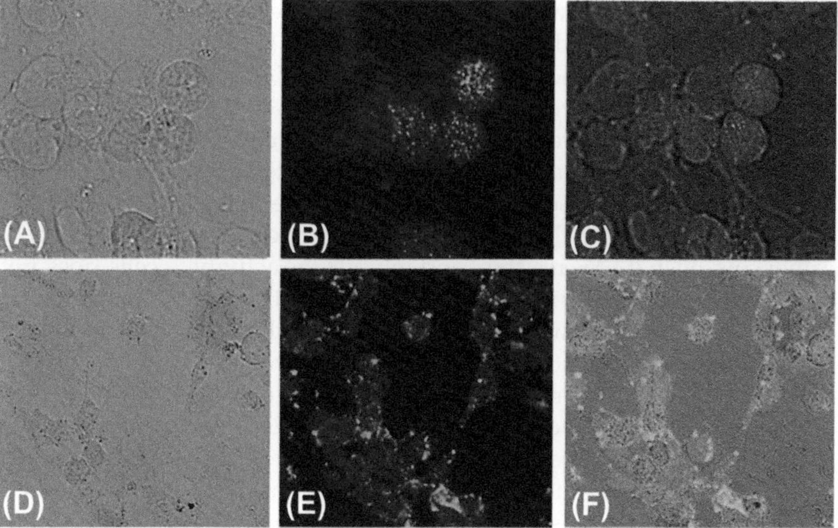

FIGURE 30.3 Laser scanner confocal microscopy images of the optical sections recovered to the nucleus level of CF human bronchial cells (IB3-1 cell line) uninfected or infected with *Pseudomonas aeruginosa*. Bright field microscopy images of uninfected (A) or infected (D) cells; green fluorescent bovine Lf (bLf) recovered in uninfected (B) or infected (E) cells; merged images showing bLf localization in uninfected (C, merged image of A and B) or infected (F, merged image of D and E) cells. (For color version of this figure, the reader is referred to the online version of this book.)

eradicate a chronic *P. aeruginosa* biofilm infection and, at the moment, there is no such agent on the horizon [134].

The emergence of increasingly resistant bacteria and the infections sustained by biofilm make the treatment of CF airway infection a great medical challenge [135]. Therefore, there is a pressing need to develop new therapeutic approaches to cure CF airway infections.

Various strategies, including the inhibition of quorum sensing, efflux pumps, and lectins, and the use of bacteriophages, immunization, and immunotherapy, have been proposed [136]. A novel strategy based on the use of high-affinity iron chelators that may compete with infecting bacteria for available iron is emerging. The rationale is based on the *Pseudomonas* absolute requirement for iron for infection success. Several approaches have been proposed to restrict iron availability based on the use of mimetic iron element, synthetic, and natural iron chelators [14,137,138]. By this point of view, Lf may represent a good candidate to reduce iron availability.

In addition to the previously mentioned functions and abilities, Lf may act in synergy with antibiotics. Although only in vitro experiments have been performed, the Lf ability to increase the antibacterial activity of rifampicin and doxycycline against both planktonic and biofilm cultures of *B. cepacia* and *P. aeruginosa* [139,140] may represent a future strategy for counteracting *P. aeruginosa* infection.

30.8 CONCLUDING REMARKS

The complexity in the relationships between the dysfunctions of CFTR, infection, and inflammation and the difficulties in treating CF airway infections justify all efforts in developing new therapeutics. In this respect, Lf, the multifunction glycoprotein of innate immunity, possesses several properties (iron chelating, antibacterial, and anti-inflammatory) that may characterize it as a potential, future molecule for the CF airway infection treatment. Interestingly, the bovine milk derivative Lf, generally recognized as a safe molecule by the US Food and Drug Administration and available in large quantities, is being used in clinical trials to identify medical applications (http://clinicaltrials.gov). It is attractive that, in the future, the bovine Lf could be evaluated using in vivo models and clinical trials involving CF patients.

Acknowledgments

We thank the Italian Cystic Fibrosis Research Foundation, which supported the research on Lf and CF airway infections over the years. In particular, we thank Delegazione FFC di Palermo e di Vittoria Ragusa Catania 2, who adopted the research project FFC#13/2013 to FB.

References

[1] Cohen TS, Prince A. Cystic fibrosis: a mucosal immunodeficiency syndrome. Nat Med 2012;18:509–19.

[2] Knowles MR, Robinson JM, Wood RE, Pue CA, Mentz WM, Wager GC, et al. Ion composition of airway surface liquid of patients with cystic fibrosis as compared with normal and disease-control subjects. J Clin Invest 1997;100:2588–95.

[3] Matsui H, Grubb BR, Tarran R, Randell SH, Gatzy JT, Davis CW, et al. Evidence for periciliary liquid layer depletion, not abnormal ion composition, in the pathogenesis of cystic fibrosis airways disease. Cell 1998;95:1005–15.

[4] Cystic Fibrosis Foundation. US cystic fibrosis foundation annual registry report 2009. Cystic Fibrosis Foundation; 2010.

[5] O'Sullivan BP, Flume P. The clinical approach to lung disease in patients with cystic fibrosis. Semin Respir Crit Care Med 2009;30:505–13.

[6] Wang G. State-dependent regulation of cystic fibrosis transmembrane conductance regulator (CFTR) gating by a high affinity Fe3+ bridge between the regulatory domain and cytoplasmic loop 3. J Biol Chem 2010;285:40438–47.

[7] Reid DW, Withers NJ, Francis L, Wilson JW, Kotsimbos TC. Iron deficiency in cystic fibrosis: relationship to lung disease severity and chronic *Pseudomonas aeruginosa* infection. Chest 2002;121:48–54.

[8] Reid DW, Lam QT, Schneider H, Walters EH. Airway iron and iron-regulatory cytokines in cystic fibrosis. Eur Respir J 2004;24:286–91.

[9] Reid DW, Carroll V, O'May C, Champion A, Kirov SM. Increased airway iron as a potential factor in the persistence of *Pseudomonas aeruginosa* infection in cystic fibrosis. Eur Respir J 2007;30:286–92.

[10] Moreau-Marquis S, Bomberger JM, Anderson GG, Swiatecka-Urban A, Ye S, O'Toole GA, et al. The DeltaF508-CFTR mutation results in increased biofilm formation by *Pseudomonas aeruginosa* by increasing iron availability. Am J Physiol Lung Cell Mol Physiol 2008;295:25–37.

[11] Wessling-Resnick M. Iron homeostasis and the inflammatory response. Annu Rev Nutr 2010;30:105–22.

[12] Singh PK, Parsek MR, Greenberg EP, Welsh MJ. A component of innate immunity prevents bacterial biofilm development. Nature 2002;417(6888):552–5.

[13] Smith DJ, Anderson GJ, Bell SC, Reid DW. Elevated metal concentrations in the CF airway correlate with cellular injury and disease severity. J Cyst Fibros 2013;1993(13). 00223-3.

[14] Smith DJ, Lamont IL, Anderson GJ, Reid DW. Targeting iron uptake to control *Pseudomonas aeruginosa* infections in cystic fibrosis. Source the Prince Charles Hospital, Brisbane. Eur Respir J 2013;42:1723–36.

[15] Valenti P, Antonini G. Lactoferrin: an important host defence against microbial and viral attack. Cell Mol Life Sci 2005;62:2576–87.

[16] Berlutti F, Pantanella F, Natalizi T, Frioni A, Paesano R, Polimeni A, et al. Antiviral properties of Lf- a natural immunity molecule. Molecules 2011;16:6992–7018.

[17] Latorre D, Berlutti F, Valenti P, Gessani S, Puddu P. LF immunomodulatory strategies: mastering bacterial endotoxin. Biochem Cell Biol 2012;90:269–78.

[18] Travis SM, Conway BAD, Zabner J, Smith JJ, Anderson NN, Singh PK, et al. Activity of abundant antimicrobials of the human airway. Am J Cell Mol Biol 1999;20:872–9.

[19] Andrews NC. Iron metabolism: iron deficiency and iron overload. Annu Rev Genomics Hum Genet 2000;1:75–98.

[20] Frazer DM, Anderson GJ. The orchestration of body iron intake: how and where do enterocytes receive their cues? Blood Cells Mol Dis 2003;30:288–97.

[21] Paesano R, Natalizi T, Berlutti F, Valenti P. Body iron delocalization: the serious drawback in iron disorders in both developing and developed countries. Pathogens and global health. Ann Trop Med Parasitol 2012;106:200–16.

[22] Mackenzie B, Garrick MD. Iron imports: II. Iron uptake at the apical membrane in the intestine. Am J Physiol Gastrointest Liver Physiol 2005;289:G981–6.

[23] Theil EC. Ferritin: at the crossroads of iron and oxygen metabolism. J Nutr 2003;133:1549S–53S.

[24] Hider RC, Kong X. Iron speciation in the cytosol: an overview. Dalton Trans 2013;42:3220–9.

[25] De Domenico I, Ward DM, Musci G, Kaplan J. Evidence for the multimeric structure of ferroportin. Blood 2007;109:2205–9.

[26] Donovan A, Lima CA, Pinkus JL, Zon LI, Robine S, Andrews NC. The iron exporter ferroportin/Slc40a1 is essential for iron homeostasis. Cell Metab 2005;1:191–200.

[27] Núñez MT, Tapia V, Rojas A, Aguirre P, Gómez F, Nualart F. Iron supply determines apical/basolateral membrane distribution of intestinal iron transporters DMT1 and ferroportin 1. Am J Physiol Cell Physiol 2010;298:C477–85.

[28] Knutson MD, Oukka M, Koss LM, Aydemir F, Wessling-Resnick M. Iron release from macrophages after erythrophagocytosis is upregulated by ferroportin 1 overexpression and down-regulated by hepcidin. Proc Natl Acad Sci USA 2005;102:1324–8.

[29] Ganz T. Cellular iron: ferroportin is the only way out. Cell Metab 2005;1:155–7.

[30] Ganz T. Hepcidin—a regulator of intestinal iron absorption and iron recycling by macrophages. Best Pract Res Clin Haematol 2005;18:171–82.

[31] Schranz M, Bakry R, Creus M, Bonn G, Vogel W, Zoller H. Activation and inactivation of the iron hormone hepcidin: biochemical characterization of prohepcidin cleavage and sequential degradation to N-terminally truncated hepcidin isoforms. Blood Cells Mol Dis 2009;43:169–79.

[32] Nemeth E, Rivera S, Gabayan V, Keller C, Taudorf S, Pedersen BK, et al. IL-6 mediates hypoferremia of inflammation by inducing the synthesis of the iron regulatory hormone hepcidin. J Clin Invest 2004;113:1271–6.

[33] Qiao B, Sugianto P, Fung E, Del-Castillo-Rueda A, Moran-Jimenez MJ, Ganz T, et al. Hepcidin-induced endocytosis of ferroportin is dependent on ferroportin ubiquitination. Cell Metab 2012;15:918–24.

[34] Lee P, Peng H, Gelbart T, Wang L, Beutler E. Regulation of hepcidin transcription by interleukin-1 and interleukin-6. Proc Natl Acad Sci USA 2005;102:1906–10.

[35] Weinstein DA, Roy CN, Fleming MD, Loda MF, Wolfsdorf JI, Andrews NC. Inappropriate expression of hepcidin is associated with iron refractory anemia: implications for the anemia of chronic disease. Blood 2002;100:3776–81.

[36] Ludwiczek S, Aigner E, Theurl I, Weiss G. Cytokine-mediated regulation of iron transport in human monocytic cells. Blood 2003;101:4148–54.

[37] Reid DW, Anderson GJ, Lamont IL. Role of lung iron in determining the bacterial and host struggle in cystic fibrosis. Am J Physiol Lung Cell Mol Physiol 2009;297:L795–802.

[38] Ghio AJ. Disruption of iron homeostasis and lung disease. Biochim Biophys Acta 2009;1790:731–9.

[39] Turi JL, Yang F, Garrick MD, Piantadosi CA, Ghio AJ. The iron cycle and oxidative stress in the lung. Free Radical Biol Med 2004;36:850–7.

[40] Ghio AJ, Piantadosi CA, Wang X, Dailey LA, Stonehuerner JD, Madden MC, et al. Divalent metal transporter-1 decreases metal-related injury in the lung. Am J Physiol Lung Cell Mol Physiol 2005;289:L460–7.

[40a] Kang GS, Li Q, Chen H, Costa M. Effect of metal ions on HIF-1alpha and Fe homeostasis in human A549 cells. Mutat Res 2006;610:48–55.

[41] Sagel SD, Sontag MK, Accurso FJ. Relationship between antimicrobial proteins and airway inflammation and infection in cystic fibrosis. Pediatr Pulmonol 2009;44:402–9.

[42] Yang F, Wang X, Haile DJ, Piantadosi CA, Ghio AJ. Iron increases expression of iron-export protein MTP1 in lung cells. Am J Physiol Lung Cell Mol Physiol 2002;283:L932–9.

[43] O'Sullivan BP, Freedman SD. Cystic fibrosis. Lancet 2009;373:1891–904.

[44] Frazier MD, Mamo LB, Ghio AJ, Turi JL. Hepcidin expression in human airway epithelial cells is regulated by interferon-γ. Respir Res 2011;12:100.

[45] Hunter RC, Asfour F, Dingemans J, Osuna BL, Samad T, Malfroot A, et al. Ferrous iron is a significant component of bioavailable iron in cystic fibrosis airways. MBio 2013;20:4.

[46] Ranganathan SC, Parsons F, Gangell C, Brennan S, Stick SM, Sly PD. Evolution of pulmonary inflammation and nutritional status in infants and young children with cystic fibrosis. Thorax 2011;66:408–13.

[47] Cox MJ, Allgaier M, Taylor B, Baek MS, Huang YJ, Daly RA, et al. Airway microbiota and pathogen abundance in age-stratified cystic fibrosis patients. PLoS One 2010;5:e11044.

[48] Harrison F. Microbial ecology of the cystic fibrosis lung. Microbiology 2007;153:917–23.

[49] Eberl L, Tümmler B. *Pseudomonas aeruginosa* and *Burkholderia cepacia* in cystic fibrosis: genome evolution, interactions and adaptation. Int J Med Microbiol 2004;294:123–31.

[50] Bragonzi A, Farulla I, Paroni M, Twomey KB, Pirone L, Lorè NI, et al. Modelling co-infection of the cystic fibrosis lung by *Pseudomonas aeruginosa* and *Burkholderia cenocepacia* reveals influences on biofilm formation and host response. PLoS One 2012;7:e52330.

[51] Fodor AA, Klem ER, Gilpin DF, Elborn JS, Boucher RC, Tunney MM, et al. The adult cystic fibrosis airway microbiota is stable over time and infection type, and highly resilient to antibiotic treatment of exacerbations. PLoS One 2012;7:e45001.

[52] Tunney MM, Field TR, Moriarty TF, Patrick S, Doering G, Muhlebach MS, et al. Detection of anaerobic bacteria in high numbers in sputum from patients with cystic fibrosis. Am J Respir Crit Care Med 2008;177:995–1001.

[53] Singh PK, Schaefer AL, Parsek MR, Moninger TO, Welsh MJ, Greenberg EP. Quorum-sensing signals indicate that cystic fibrosis lungs are infected with bacterial biofilms. Nature 2000;407:762–4.

[54] Lee TW, Brownlee KG, Conway SP, Denton M, Littlewood JM. Evaluation of a new definition for chronic *Pseudomonas aeruginosa* infection in cystic fibrosis patients. J Cyst Fibros 2003;2:29–34.

[55] Emerson GG, Herndon CN, Sreih AG. Thrombotic complications after intravenous immunoglobulin therapy in two patients. Pharmacotherapy 2002;22:1638–41.

[56] Stites SW, Plautz MW, Bailey K, O'Brien-Ladner AR, Wesselius LJ. Increased concentrations of iron and isoferritins in the lower respiratory tract of patients with stable cystic fibrosis. Am J Respir Crit Care Med 1999;160:796–801.

[57] Gifford AH, Miller SD, Jackson BP, Hampton TH, O'Toole GA, Stanton BA, et al. Iron and CF-related anemia: expanding clinical and biochemical relationships. Pediatr Pulmonol 2011;46:160–5.

[58] Berlutti F, Morea C, Battistoni A, Sarli S, Cipriani P, Superti F, et al. Iron availability influences aggregation, biofilm, adhesion and invasion of *Pseudomonas aeruginosa* and *Burkholderia cenocepacia*. Int J Immunopathol Pharmacol 2005;18:661–70.

[59] O'May CY, Sanderson K, Roddam LF, Kirov SM, Reid DW. Iron-binding compounds impair *Pseudomonas aeruginosa* biofilm formation, especially under anaerobic conditions. J Med Microb 2009;58:765–73.

[60] Konings AF, Martin LW, Sharples KJ, Roddam LF, Latham R, Reid DW, et al. *Pseudomonas aeruginosa* uses multiple pathways to acquire iron during chronic infection in cystic fibrosis lungs. Infect Immun 2013;81:2697–704.

[61] Chen X, Stewart PS. Role of electrostatic interactions in cohesion of bacterial biofilms. Appl Microbiol Biotechnol 2002;59(6):718–20.

[62] Ballmann M, Smyth A, Geller DE. Therapeutic approaches to chronic cystic fibrosis respiratory infections with available, emerging aerosolized antibiotics. Respir Med 2011;105:S2–8.

[63] Dakin CJ, Numa AH, Wang H, Morton JR, Vertzyas CC, Henry RL. Inflammation, infection and pulmonary function in infants and young children with cystic fibrosis. Am J Respir Crit Care Med 2002;165:904–10.

[64] Verhaeghe C, Delbecque K, de Leval L, Oury C, Bours V. Early inflammation in the airways of a cystic fibrosis foetus. J Cyst Fibros 2007;6:304–8.

[65] van Heeckeren AM, Schluchter MD, Drumm ML, Davis PB. Role of cftr genotype in the response to chronic *Pseudomonas aeruginosa* lung infection in mice. Am J Physiol Lung Cell Mol Physiol 2004;287:L944–52.

[66] Weber AJ, Soong G, Bryan R, Saba S, Prince A. Activation of NF-κB in airway epithelial cells is dependent on CFTR trafficking and Cl– channel function. Am J Physiol Lung Cell Mol Physiol 2001;281:L71–8.

[67] Al Alam D, Deslee G, Tournois C, Lamkhioued B, Lebargy F, Merten M, et al. Impaired interleukin-8 chemokine secretion by Staphylococcus aureus–activated epithelium and T-cell chemotaxis in cystic fibrosis. Am J Respir Cell Mol Biol 2010;42:644–50.

[68] Tirouvanziam R, de Bentzmann S, Hubeau C, Hinnrasky J, Jacquot J, Peault B, et al. Inflammation and infection in naive human cystic fibrosis airway grafts. Am J Respir Cell Mol Biol 2000;23:121–7.

[69] Nichols D, Chmiel J, Berger M. Chronic inflammation in the cystic fibrosis lung: alterations in inter- and intracellular signalling. Clin Rev Allergy Immunol 2007;34:146–62.

[70] Tsuchiya M, Kumar P, Bhattacharyya S, Chattoraj S, Srivastava M, Pollard HB, et al. Differential regulation of inflammation by inflammatory mediators in cystic fibrosis lung epithelial cells. J Interferon Cytokine Res 2013;33:121–9.

[71] Greene CM, McElvaney N,G. TLR-induced inflammation in cystic fibrosis and non-cystic fibrosis airway epithelial cells. J Immunol 2005;174:1638–46.

[72] Muir A, Soong G, Sokol S, Reddy B, Gomez MI, van Heeckeren A, et al. Toll-like receptors in normal and cystic fibrosis airway epithelial cells. Am J Respir Cell Mol Biol 2004;30:777–83.

[73] Khan TZ, Wagener JS, Bost T, Martinez J, Accurso FJ, Riches DW. Early pulmonary inflammation in infants with cystic fibrosis. Am J Respir Crit Care Med 1995;151:1075–82.

[74] Tirouvanziam R, Khazaal I, Peault B. Primary inflammation in human cystic fibrosis small airways. Am J Physiol Lung Cell Mol Physiol 2002;283:L445–51.

[75] Kelly C, Canning P, Buchanan PJ, Williams MT, Brown V, Gruenert DC, et al. Toll-like receptor 4 is not targeted to the lysosome in cystic fibrosis airway epithelial cells. Am J Physiol Lung Cell Mol Physiol 2013;304:L371–82.

[76] Look DC, Stoll LL, Romig SA, Humlicek A, Britigan BE, Denning GM. Pyocyanin and its precursor phenazine-1-carboxylic acid increase IL-8 and intercellular adhesion molecule-1 expression in human airway epithelial cells by oxidant-dependent mechanisms. J Immunol 2005;175:4017–23.

[77] Bamford S, Ryley H, Jackson SK. Highly purified lipopolysaccharides from *Burkholderia cepacia* complex clinical isolates induce inflammatory cytokine responses via TLR4-mediated MAPK signalling pathways and activation of NFkB. Cell Microb 2007;9:532–43.

[78] Ciornei CD, Novikov A, Beloin C, Fitting C, Caroff M, Ghigo JM, et al. Biofilm-forming *Pseudomonas aeruginosa* bacteria undergo lipopolysaccharide structural modifications and induce enhanced inflammatory cytokine response in human monocytes. Innate Immun 2010;16:288–301.

[79] Fuxman Bass JI, Russo DM, Gabelloni ML, Geffner JR, Giordano M, Catalano M, et al. Extracellular DNA: a major proinflammatory component of *Pseudomonas aeruginosa* biofilms. J Immunol 2010;184:6386–95.

[80] Valenti P, Catizone A, Pantanella F, Frioni A, Natalizi T, Tendini M, et al. Lactoferrin decreases inflammatory response by cystic fibrosis bronchial cells invaded with *Burkholderia cenocepacia* iron-modulated biofilm. Int J Immunopathol Pharmacol 2011;24:1057–68.

[81] Baker EN, Rumball SV, Anderson BF. Transferrins: insights into structure and function from studies on lactoferrin. Trends Biochem Sci 1987;12:350–3.

[82] Schaible UE, Kaufmann HE. Iron and microbial infection. Nat Rev Microbiol 2004;2:946–53. Erratum in Nat. Rev. Microbiol (2005)3, 268.

[83] Baker EN, Baker HM. Molecular structure, binding properties and dynamics of Lf. Cell Mol Life Sci 2005;62:2531–9.

[84] Grossmann JG, Neu M, Pantos E, Schwab FJ, Evans RW, Townes-Andrews E, et al. X-ray solution scattering reveals conformational changes upon iron uptake in Lf, serum and ovotransferrins. J Mol Biol 1992;225:811–9.

[85] Alexander DB, Iigo M, Yamauchi K, Suzui M, Tsuda H. Lf: an alternative view of its role in human biological fluids. Biochem Cell Biol 2012;90:279–306.

[86] Thompson AB, Bohling T, Payvandi F, Rennard SI. Lower respiratory tract Lf and lysozyme arise primarily in the airways and are elevated in association with chronic bronchitis. J Lab Clin Med February 1990;115(2):148–58.

[87] Weinberg ED. The development of awareness of iron-withholding defense. Perspect Biol Med 1993;36:215–21.

[88] Arnold RR, Cole MF, Mcghee JR. A bactericidal effect for human Lf. Science 1977;197:263–5.

[89] Appelmelk BJ, Vandenbroucke-Grauls CMJE. Lipopolysaccharide lewis antigens. In: Mobley HLT, Mendz GL, Hazell SL, editors. *Helicobacter pylori*: physiology and genetics. Washington (DC): ASM Press; 2001. [Chapter 35].

[90] Berlutti F, Ajello M, Bosso P, Morea C, Petrucca A, Antonini G, et al. Both lactoferrin and iron influence aggregation and biofilm formation in Streptococcus mutans. Biometals 2004;17:271–8.

[91] Amarasinghe JJ, Scannapieco FA, Haase EM. Transcriptional and translational analysis of biofilm determinants of *Aggregatibacter actinomycetemcomitans* in response to environmental perturbation. Infect Immun 2009;77:2896–907.

[92] Visca P, Berlutti F, Vittorioso P, Dalmastri C, Thaller MC, Valenti P. Growth and adsorption of *Streptococcus mutans* 6715-13 to hydroxyapatite in the presence of Lf. Med Microbiol Immunol 1989;178:69–79.

[93] Oho T, Mitoma M, Koga T. Functional domain of bovine milk lactoferrin which inhibits the adherence of Streptococcus *mutans* cells to a salivary film. Infect Immun 2002;70:5279–82.

[94] Superti F, Berlutti F, Paesano R, Valenti P. Structure and activity of lactoferrin, a multifunctional protective agent for human health. In: Fuchs H, editor. "Iron metabolism and disease", chapter 8. Transworld Research Network; 2008.

[95] Ammendolia MG, Bertuccini L, Iosi F, Minelli F, Berlutti F, Valenti P, et al. Bovine lactoferrin interacts with cable pili of *Burkholderia cenocepacia*. Biometals 2010;23:531–42.

[96] Berlutti F, Superti F, Nicoletti M, Morea C, Frioni A, Ammendolia MG, et al. Bovine lactoferrin inhibits the efficiency of invasion of respiratory A549 cells of different iron-regulated morphological forms of *Pseudomonas aeruginosa* and *Burkholderia cenocepacia*. Int J Immunopathol Pharmacol 2008;21:51–9.

[97] Hyvönen P, Käyhkö S, Taponen S, von Wright A, Pyörälä S. Effect of bovine Lf on the internalization of coagulase-negative staphylococci into bovine mammary epithelial cells under in-vitro conditions. J Dairy Res 2009;76:144–51.

[98] Singh PK, Pradeep K, Tack Brian F, McCray Jr Paul B, Welsh Michael J. Synergistic and additive killing by antimicrobial factors found in human airway surface liquid. Am J Physiol Lung Cell Mol Physiol 2000;279:L799–805.

[99] Ganz T. Antimicrobial polypeptides in host defense of the respiratory tract. J Clin Invest 2002;109:693–7.

[100] Masson PL, Heremans JF, Schonne E. Lactoferrin, an iron-binding protein in neutrophilic leukocytes. J Exp Med 1969;130:643–58.

[101] Vogel HJ. Lactoferrin, a bird's eye view. Biochem Cell Biol 2012;90:233–44.

[102] Berlutti F, Schippa S, Morea C, Sarli S, Perfetto B, Donnarumma G, et al. Lactoferrin downregulates pro-inflammatory cytokines upexpressed in intestinal epithelial cells infected with invasive or non invasive *Escherichia coli* strains. Biochem Cell Biol 2006;84:351–7.

[103] Legrand D, Elass E, Carpentier M, Mazurier J. Lactoferrin: a modulator of immune and inflammatory responses. Cell Mol Life Sci 2005;62:2549–59.

[104] Hayworth JL, Kasper KJ, Leon-Ponte M, Herfst CA, Yue D, Brintnell WC, et al. Attenuation of massive cytokine response to the staphylococcal enterotoxin B superantigen by the innate immunomodulatory protein lactoferrin. Clin Exp Immunol 2009;157:60–70.

[105] Miyazawa K, Mantel C, Lu L, Morrison DC, Broxmeyer HE. Lactoferrin-lipopolysaccharide interactions. Effect on lactoferrin binding to monocyte/macrophage-differentiated HL-60 cells. J Immunol 1991;146:723–9.

[106] Li XJ, Liu DP, Chen HL, Pan XH, Kong QY, Pang QF. Lactoferrin protects against lipopolysaccharide-induced acute lung injury in mice. Int Immunopharm 2012;12:460–4.

[107] Zagulski T, Lipiński P, Zagulska A, Broniek S, Jarzabek Z. Lactoferrin can protect mice against a lethal dose of *Escherichia coli* in experimental infection in vivo. Br J Exp Pathol 1989;70:697–704.

[108] Broxmeyer HE, Williams DE, Hangoc G, Cooper S, Gentile P, Shen RN, et al. The opposing actions in vivo on murine myelopoiesis of purified preparations of lactoferrin and the colony stimulating factors. Blood Cells 1987;13:31–48.

[109] Kruzel ML, Harari Y, Mailman D, Actor JK, Zimecki M. Differential effects of prophylactic, concurrent and therapeutic lactoferrin treatment on LPS-induced inflammatory responses in mice. Clin Exp Immunol 2002;130:25–31.

[110] Togawa J, Nagase H, Tanaka K, Inamori M, Nakajima A, Ueno N, et al. Oral administration of lactoferrin reduces colitis in rats via modulation of the immune system and correction of cytokine imbalance. J Gastroenterol Hepatol 2002;17:1291–8.

[111] Jiang R, Lopez V, Kelleher SL, Lönnerdal B. Apo- and hololactoferrin are both internalized by lactoferrin receptor via clathrin-mediated endocytosis but differentially affect ERK-signaling and cell proliferation in Caco-2 cells. J Cell Physiol 2011;226:3022–31.

[112] Suzuki YA, Wong H, Ashida KY, Schryvers AB, Lönnerdal B. The N1 domain of human lactoferrin is required for internalization by Caco-2 cells and targeting to the nucleus. Biochem 2008;47:10915–20.

[113] Paesano R, Berlutti F, Pietropaoli M, Goolsbee W, Pacifici E, Valenti P. Lf efficacy versus ferrous sulfate in curing iron disorders in pregnant and non-pregnant women. Int J Immunopathol Pharmacol 2010;23:577–87.

[114] Paesano R, Pietropaoli M, Berlutti F, Valenti P. Bovine Lf in preventing preterm delivery associated with sterile inflammation. Biochem Cell Biol 2012;90:468–75.

[114a]Paesano P, Pacifici E, Benedetti S, Berlutti F, Frioni A, Polimeni A,Valenti P. Safety and efficacy of lactoferrin versus ferrous sulphate in curing iron deficiency and iron deficiency anaemia in hereditary thrombophilia pregnant women: an interventional study. Biometals 2014; Mar 4.[Epub ahead of print].

[115] Rogan MP, Taggart CC, Greene CM, Murphy PG, O'Neill SJ, McElvaney NG. Loss of microbicidal activity and increased formation of biofilm due to decreased lactoferrin activity in patients with cystic fibrosis. J Infect Dis 2004;190:1245–53.

[116] Ratjen F, Hartog CM, Paul K, Wermelt J, Braun J. Matrix metalloproteases in BAL fluid of patients with cystic fibrosis and their modulation by treatment with dornasealpha. Thorax 2002;57:930–4.

[117] He S, McEuen AR, Blewett SA, Li P, Buckley MG, Leufkens P, et al. The inhibition of mast cell activation by neutrophil Lf: uptake by mast cells and interaction with tryptase, chymase and cathepsin G. Biochem Pharmacol 2003;65:1007–15.

[118] Komine K, Kuroishi T, Ozawa A, Komine Y, Minami T, Shimauchi H, et al. Cleaved inflammatory Lf peptides in parotid saliva of periodontitis patients. Mol Immunol 2007;44:1498–508.

[119] Martin SL, Moffitt KL, McDowell A, Greenan C, Bright-Thomas RJ, Jones AM, et al. Association of airway cathepsin B and S with inflammation in cystic fibrosis. Pediatr Pulmonol 2010;45:860–8.

[120] Döring G, Dauner HM. Clearance of *Pseudomonas aeruginosa* in different rat lung models. Am Rev Respir Dis 1988;138:1249–53.

[121] Britigan BE, Hayek MB, Doebbeling BN, Fick Jr RB. Transferrin and lactoferrin undergo proteolytic cleavage in the *Pseudomonas aeruginosa*-infected lungs of patients with cystic fibrosis. Infect Immun 1993;61:5049–55.

[122] Wolz C, Hohloch K, Ocaktan A, Poole K, Evans RW, Rochel N, et al. Iron release from transferrin by pyoverdin and elastase from *Pseudomonas aeruginosa*. Infect Immun 1994;62:4021–7.

[123] Ghio AJ, Carter JD, Dailey LA, Devlin RB, Samet JM. Respiratory epithelial cells demonstrate lactoferrin receptors that increase after metal exposure. Am J Physiol 1999;276:L933–40.

[124] Matsui H, Wagner VE, Hill DB, Schwab UE, Rogers TD, Button B, et al. A physical linkage between cystic fibrosis airway surface dehydration and *Pseudomonas aeruginosa* biofilms. Proc Natl Acad Sci USA 2006;103:18131–6.

[125] Ammons MC, Ward LS, James GA. Anti-biofilm efficacy of a lactoferrin/xylitol wound hydrogel used in combination with silver wound dressings. Int Wound J 2011;8:268–73.

[126] Kamiya H, Ehara T, Matsumoto T. Inhibitory effects of lactoferrin on biofilm formation in clinical isolates of *Pseudomonas aeruginosa*. J Infect Chemother 2012;18:47–52.

[127] Kotrange S, Kopp B, Akhter A, Abdelaziz D, Abu Khweek A, Caution K, et al. *Burkholderia cenocepacia* O polysaccharide chain contributes to caspase-1-dependent IL-1beta production in macrophages. J Leukoc Biol 2011;89:481–8.

[127a]Frioni A, Conte MP, Cutone A, Longhi C, Musci G, di Patti MC, Natalizi T, Marazzato M, Lepanto MS, Puddu P, Paesano R,Valenti P, Berlutti F. Lactoferrin differently modulates the inflammatory response in epithelial models mimicking human inflammatory and infectious diseases. Biometals 2014; Apr 26. [Epub ahead of print].

[128] Döring G, Flume P, Heijerman H, Elborn JS, Group Consensus Study. Treatment of lung infection in patients with cystic fibrosis: current and future strategies. J Cyst Fibros 2012;11:461–79.

[128a]Nair, C.G, Chao, C., Ryall B., and Williams H.D. Sub-lethal concentrations of antibiotics increase mutation frequency in the cystic fibrosis pathogen Pseudomonas aeruginosa. Lett Appl Microbiol 2013;56:149–154.

[129] Rachid S, Ohlsen K, Witte W, Hacker J, Ziebuhr W. Effect of subinhibitory antibiotic concentrations on polysaccharide intercellular adhesin expression in biofilm forming *Staphylococcus epidermidis*. Antimicrob Agents Chemother 2000;44:3357–63.

[130] Majtán J, Majtánová L, Xu M, Majtán V. In vitro effect of subinhibitory concentrations of antibiotics on biofilm formation by clinical strains of *Salmonella enterica* serovar Typhimurium isolated in Slovakia. J Appl Microbiol 2008;104:1294–301.

[131] Cummins J, Reen FJ, Baysse C, Mooij MJ, O'Gara F. Subinhibitory concentrations of the cationic antimicrobial peptide colistin induce the pseudomonas quinolone signal in *Pseudomonas aeruginosa*. Microbiology 2009;155:2826–37.

[132] Shafreen RMB, Srinivasan S, Manisankar P, Pandian SK. Biofilm formation by Streptococcus pyogenes: modulation of exopolysaccharide by fluoroquinolone derivatives. J Biosci Bioeng 2011;112:345–50.

[133] Langton-Hewer SC, Smyth AR. Antibiotic strategies for eradicating *Pseudomonas aeruginosa* in people with cystic fibrosis. Cochrane Database Syst Rev 2009;4:CD004197.

[134] Talbot GH, Bradley J, Edwards Jr JE, Gilbert D, Scheld M, Bartlett JG, et al. Bad bugs need drugs: an update on the development pipeline from the antimicrobial availability task force of the infectious diseases society of America. Clin Infect Dis 2006;42:657–68.

[135] Hurley MN, Forrester DL, Smyth AR. Antibiotic adjuvant therapy for pulmonary infection in cystic fibrosis. Cochrane Database Syst Rev 2013;5(6):CD008037.

[136] Hurley MN, Ca´mara M, Smyth AR. Novel approaches to the treatment of *Pseudomonas aeruginosa* infections in cystic fibrosis. Eur Respir J 2012;40(4):1014–23.

[137] Reid DW, O'May C, Kirov SM, Roddam L, Lamont IL, Sanderson K. Iron chelation directed against biofilms as an adjunct to conventional antibiotics. Am J Physiol Lung Cell Mol Physiol May 2009;296(5):L857–8.

[138] Reid DW, O'May C, Roddam LF, Lamont IL. Chelated iron as an anti-*Pseudomonas aeruginosa* biofilm therapeutic strategy. J Appl Microbiol 2009;106(3):1058.

[139] Caraher EM, Gumulapurapu K, Taggart CC, Murphy P, McClean S, Callaghan M. The effect of recombinant human Lf on growth and the antibiotic susceptibility of the cystic fibrosis pathogen *Burkholderia cepacia* complex when cultured planktonically or as biofilms. J Antimicrob Chemother 2007;60:546–54.

[140] Alkawash M, Head M, Alshami I, Soothill JS. The effect of human Lf on the MICs of doxycycline and rifampicin for *Burkholderia cepacia* and *Pseudomonas aeruginosa* strains. J Antimicrob Chemother 1999;44:385–7.

CHAPTER

31

Cystic Fibrosis–Related Diabetes: Lung Function and Nutritional Status

María Martín-Frías, Raquel Barrio

Diabetes Pediatric Unit, Ramón y Cajal Hospital, Alcalá University, Madrid, Spain

31.1 INTRODUCTION

Cystic fibrosis (CF) is a monogenic disease caused by a defect in the CF transmembrane regulator protein (CFTR). It is located in many organs, including the pancreas and the lungs. In the pancreas, this defect causes both impaired pancreatic exocrine function (responsible for the synthesis and secretion of enzymes necessary for the absorption of nutrients) and endocrine function (producing several hormones, including insulin). This endocrine defect leads to the progressive destruction of pancreatic ß-cells, resulting in a progressive decline in insulin secretion that ultimately leads to progressive alteration of carbohydrate metabolism, and finally resulting in CF-related diabetes (CFRD).

Cystic fibrosis–related diabetes occurs mainly in patients with the presence of mutations in the CFTR gene associated with severe disease and exocrine pancreatic insufficiency. Since the beginning of the 1990s, progress made in the treatment of CF has notably improved the life expectancy of patients. Because of improvements in both nutritional and pulmonary care, this longer life expectancy achieved in CF patients has led to an increase in the prevalence of CFRD, which has become the leading comorbidity in this disorder [1]. The diagnosis of CFRD is usually preceded by a long period of abnormal glucose metabolism that is associated with a decline in lung function and aggravates the nutritional deficiencies of CF patients [1–4]. The association with increased morbidity of CFRD has emphasized the need for accurate monitoring of glycemia in all CF patients.

Cystic fibrosis–related diabetes is usually diagnosed at the end of the second decade of life, with a mean age at diagnosis of 21 years [5,6]; it is present in 2% of children (≥6 years of age), 19% of adolescents, and 40–50% of adults with CF [1]. A rise in annual incidence is reported depending on the age of patients: 5% in patients ≥10 years of age and 9.3% in those ≥20 years of age [7]. There is a further increase in the prevalence of CFRD according to the age of patient cohorts. Differences in prevalence data between studies result from different ages and population monitoring, especially according to the time and methods used as screening [8].

The most important risk factors for the development of CFRD are age, female gender, presence of severe mutations of the CFTR gene, exocrine pancreatic insufficiency, severe lung disease, liver dysfunction, need of corticosteroids treatment, family history of diabetes, and prior organ transplantation [9]. Most patients with CF have exocrine pancreatic insufficiency to varying degrees, but not all of them develop CFRD, so there must be additional factors involved determining individual etiopathologic risk. Furthermore, some patients develop CFRD without exocrine pancreatic insufficiency.

Cystic fibrosis–related diabetes is a unique type of diabetes with clinical features of type 1 and 2 diabetes mellitus, but whereas it shares characteristics of both, significant differences also exist that require specific approaches, diagnoses, and management. Different factors affect CF carbohydrate metabolism in these patients, because of the state of pulmonary inflammation and infection, nutritional status, increased energy expenditure, glucagon deficiency, and gastrointestinal disorders (including malabsorption, altered gastric emptying, impaired intestinal motility, and liver disease).

Other types of diabetes mellitus are also present in patients with CF, such as posttransplant diabetes, gestational diabetes, and autoimmune type 1 diabetes, but they are considerably less frequent. Therefore, in this chapter, we will focus on CFRD.

Diet and Exercise in Cystic Fibrosis
http://dx.doi.org/10.1016/B978-0-12-800051-9.00031-6

31.2 CYSTIC FIBROSIS–RELATED DIABETES: PHYSIOPATHOLOGY

The pathogenesis of abnormal carbohydrate metabolism in patients with CF is multifactorial; genetic and environmental factors influence its development [10]. Delayed insulin secretion characterizes the oral glucose tolerance test (OGTT) in CF patients, even in the absence of CFRD or other glycemic abnormalities [11–13]. Basal insulin secretion is initially partially preserved.

It is thought that the primary alteration of CFRD is a destruction of pancreatic islets, resulting in gradual deterioration of the function of the pancreatic ß-cell and, finally, a deficiency of insulin. Abnormal chloride channel function in CF results in thick viscous secretions as a result of alteration of the electrolyte composition. These secretions initially produce obstructive damage to the exocrine pancreas.

Progressive fibrosis and fatty infiltration destroy pancreatic islets. Islet cell destruction is linked to exocrine pancreas lesions and is not cell-selective. Studies in CF patients identified decreased insulin and glucagon secretion in response to different tests [12].

However, not all findings are explained by fibrosis of the pancreatic islets. The presence of amyloid has been also identified in the islets of patients with CFRD, similar to that found in patients with type 2 diabetes mellitus, whose pathogenesis is unclear. It appears that the accumulation of amyloid is toxic to the pancreatic ß-cells, which collaborate in their destruction. Mutation of the CFTR protein alters intracellular pH and could facilitate the local accumulation of amyloid. Amyloid was detected in patients with CFRD, but not in those with CF without diabetes.

Insulin deficiency is the most important defect in CFRD. On the other hand, peripheral and hepatic insulin resistance has also been detected in CFRD patients [14,15]. Insulin resistance worsens with glucocorticoid therapy and acute exacerbation or chronic severe lung disease [16].

Insulin is an anabolic hormone and its deficiency increases protein catabolism. Deterioration of lung function in patients with CF and disturbances in carbohydrate metabolism coexist [4,17], and the severity of lung deterioration correlates with the severity of glucose metabolism disturbance. This emphasizes the need for early diagnosis and treatment of CFRD.

Moreover, in patients with diabetes mellitus without CF lung disturbances have been found, both at functional and histologic levels; it suggests a negative effect of hyperglycemia on the lung [18]. Major alterations in lung function are described in patients with diabetes mellitus and poor metabolic control (higher levels of glycated hemoglobin [HbA_{1c}]).

31.2.1 Genetics

The etiology of CFRD is complex. It occurs mostly in patients with severe mutations of the CFTR gene, but there is not a complete correlation [19]. The expression of this gene in pancreatic tissue is well known, with a high expression in pancreatic islets, even more than other areas of the pancreas. The most common mutation described is the F508del. That there are patients with the same genotype and large phenotypic spectrum suggests the influence of other factors in the development of disturbances of the carbohydrate metabolism in CF, including modifier genes and environmental factors.

31.2.2 Inflammation and Oxidative Stress

In CF, sustained chronic inflammation and frequent bacterial infections result in an increase of free radicals. These free radicals generate an oxidative stress situation. Defects in CFTR gene could alter the transport and homeostasis of glutathione. Furthermore, intestinal malabsorption limits the uptake of endogenous antioxidant vitamins. All of these situations disturb the balance between pro-oxidant and antioxidant factors and promote oxidative stress [20,21]. CF patients are particularly sensitive to oxidative stress because of their high production of oxidants and impaired protection against them. Pancreatic ß-cell has a high rate of protein synthesis that makes it particularly susceptible to stress in the endoplasmic reticulum [20].

Hyperglycemia itself causes mitochondrial oxidative stress, nonenzymatic glycosylation of proteins, and autooxidation of glucose. Oxidative stress may also activate some stress-sensitive signaling pathways, and they may worsen insulin secretion and insulin action.

31.2.3 Other Factors

Other studies analyzed the role of fat malabsorption, accelerated gastric emptying, and incretins (glucagon like peptide and glucose dependent insulinotropic polypeptide) in the development of disturbances in the carbohydrate metabolism of CF patients [22]. The fat acts as a stimulator of the secretion of incretins. In CF, despite pancreatic enzyme supplementation, there is poor digestion of lipids from food. The digestion of fat is essential for slow gastric emptying and to stimulate the release of incretins. Incretins secreted by intestinal K and L cells have an important role in postprandial insulin secretion.

In summary, disruption of endocrine pancreas function in patients with CF is progressive, and does not become evident until it has significantly decreased insulin production [23]. Thus, initially the basal insulin secretion is preserved, but there is progressive altered response of insulin to stimuli, with a delay in the peak of insulin secretion, even in CF patients without an alteration in carbohydrate metabolism. First, postprandial hyperglycemia will develop, and then, fasting hyperglycemia.

31.3 CYSTIC FIBROSIS–RELATED DIABETES: SCREENING AND DIAGNOSIS

Screening for abnormal glucose tolerance in the CF population is recommended because this disturbance is usually insidious. Classical diabetes symptoms (polyuria and polydipsia) are rarely present. On the other hand, nonspecific symptoms, such as unexplained pulmonary function decline, failure to gain weight or poor growth, and delay in growth and/or puberty, may be the only symptoms to suggest the onset of CFRD [17]. The clinical expression of CFRD is influenced by malnutrition, infections, steroid treatment, increased energy expenditure, malabsorption, glucagon deficiency, hepatic dysfunction, and puberty, among other factors. These factors fluctuate; consequently, hydrocarbon tolerance could oscillate over time in patients with CF [24] (Figure 31.1).

Diagnosis is based on fasting blood glucose levels >125 mg/dL (6.94 mmol/L) or symptomatic diabetes for random glucose levels >200 mg/dL (11.11 mmol/L); or glycated hemoglobin levels of at least 6.5%.

Most CF patients with glucose disturbance do not have fasting hyperglycemia. Therefore, the United States Cystic Fibrosis Foundation (CFF) and the International Society of Pediatric and Adolescent Diabetes (ISPAD) recommend annual screening for CFRD through OGTT (1.75 g/kg glucose, maximum dose 75 g) starting at the age of 10 [3,25]. Tests should be performed during a period of stable baseline health, with at least 6–12 weeks free of CF decompensations and without corticosteroid treatment. Oral glucose tolerance test should also be performed before an organ transplant and during pregnancy.

Based on the response to OGTT, normal glucose tolerance, impaired glucose tolerance, CFRD with and without fasting hyperglycemia, and indeterminate glycemic alteration are defined (Table 31.1).

There are special situations in CF patients in which the diagnosis of CFRD should be considered [3] (Table 31.2).

In non-CF patients, a new category of indeterminate glycemia has been defined as 1-hour plasma glucose >155 mg/dL (>8.6 mmol/L) [26]. This glucose alteration is associated with early atherosclerosis and is a risk factor for type 2 diabetes mellitus. However, an association between the earliest alterations in carbohydrate metabolism and clinical status decline has not yet well defined in CF [27].

Regarding the time of initial screening, the presence of abnormal carbohydrate metabolism in children between 6 and 9 years of age has been associated with the risk of developing early CFRD [28]. In our experience, glucose disturbances are present in the first decade of age in CF patients [13]. This study included 19 prepubertal children; the mean age at the moment of any abnormal glucoses tolerance was 8.6 years (range, 6.4–11.1 years); this supports the usefulness of earlier glycemic screening than is currently recommended.

31.3.1 Other Screening Tools

- Glucose measurements

  In CF patients, it is recommended to measure fasting and postprandial glucose during hospitalization for acute illness [3,12]. If hyperglycemia is present for >2 days (fasting glycemia ≥126 mg/dL [≥7 mmol/L] and/or postprandial glucose ≥200 mg/dL [≥11.1 mmol/L]), a diagnosis of CFRD is consistent.

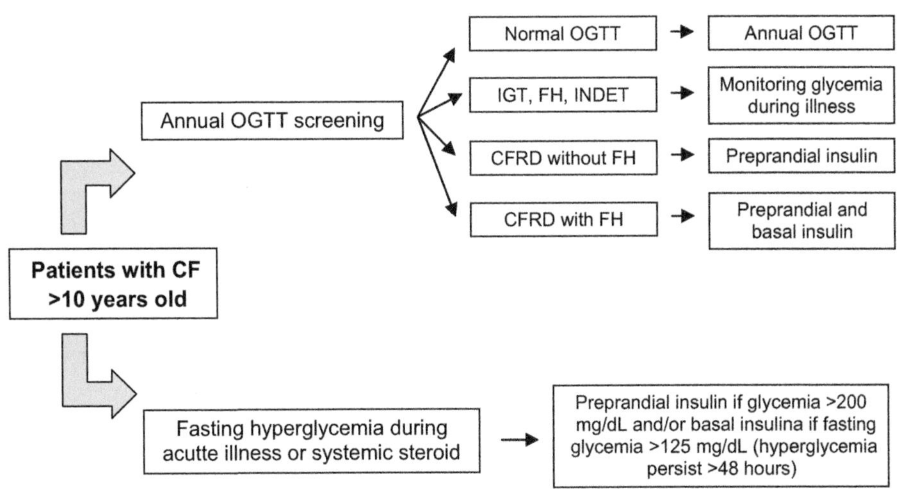

FIGURE 31.1 **Screening and treatment in patients with CFRD.** OGTT: oral glucose tolerance test, IGT: impaired glucose tolerance, FH: fasting hyperglycemia, INDET: indeterminate glycemia, CFRD: cystic fibrosis-related disease.

TABLE 31.1 Diagnosis of Disturbances in Carbohydrate Metabolism in CF

Normal glucose tolerance	1-h plasma glucose <200 mg/dL (<11.1 mmol/L) 2-h plasma glucose <140 mg/dL (<7.8 mmol/L)
Impaired glucose tolerance	2-h plasma glucose ≥140 and <200 mg/dL (≥7.8 and <11.1 mmol/L)
CFRD without fasting hyperglycemia	Fasting glycemia <126 mg/dL (<7 mmol/L) 2-h plasma glucose ≥200 mg/dL (≥11.1 mmol/L)
CFRD with fasting hyperglycemia	Fasting glycemia ≥126 mg/dL (≥7 mmol/L) 2-hour plasma glucose ≥200 mg/dL (≥11.1 mmol/L)
Indeterminate glycemia	Plasma glucose ≥200 mg/dL between 30 and 90 min (≥11.1 mmol/L) 2-h plasma glucose <140 mg/dL (<7.8 mmol/L)

TABLE 31.2 Diagnoses of CFRD in Special Situations

Ambulatory patient	Annual OGTT Diagnostic base on: • Fasting glycemia ≥126 mg/dL (two times) • 2-hour glycemia ≥200 mg/dL (two times) • Sporadic glycemia ≥200 mg/dL with symptoms • HbA_{1c} ≥6.5% (two times)
Continuous enteral nutrition	Diagnostic based on: Glycemia in the middle or postprandial ≥200 mg/dL It must be confirmed two different nights.
Pregnancy (OGTT 75 g)	Diagnostic based on: • Fasting glycemia ≥92 mg/dL • 1-h glycemia ≥180 mg/dL • 2-h glycemia ≥153 mg/dL
Acute illness or systemic steroid	Diagnostic based on maintain hyperglycemia >48 h: • Fasting glycemia ≥126 mg/dL • 2-h glycemia ≥200 mg/dL All studies made with capillary glycemia must be confirmed by laboratory study.

- Glycosylated hemoglobin

  If using HbA_{1c} as the only tool for screening, the diagnosis of CFRD for most patients could be missed [1]. Glycosylated hemoglobin is normal in about 70% of CFRD patients with an abnormal OGTT [7]. A normal HbA_{1c} does not exclude CFRD, but HbA_{1c} > 6.5% is consistent with diabetes [29].
- Continuous glucose monitoring system

  To date, OGTT is the reference method for the screening of CFRD, but the definition of diabetes based on the 2-hour postload plasma glucose level may not be the most accurate method for the early detection of glucose tolerance abnormalities in CF. A long prediabetic phase of abnormal glucose tolerance is described in subjects with CF since childhood. The continuous glucose monitoring system (CGMS) has been described as a useful tool for the early detection of hyperglycemia in CF patients. The continuous glucose monitoring system measures interstitial glucose, which is then translated into blood glucose. It has been validated in patients with CF [30]. This method is receiving increasing attention [6,29,31–38]. The continuous glucose monitoring system could be a useful tool to predict glucose metabolism derangements in CF patients; it reveals more glucose metabolism abnormalities than does OGTT in patients with unexplained altered general status. A systematic review on screening for CFRD concluded that the best screening test may be the CGMS, but further evidence is required [39].

31.4 CYSTIC FIBROSIS–RELATED DIABETES: TREATMENT

The CFF guidelines recommend insulin as the treatment of choice of CFRD. Diabetes education is an important tool for managing CFRD. Monitoring of these patients should be performed within a multidisciplinary team with expertise in CFRD and smooth communication between the CF and Diabetic units (Figure 31.1).

The goals of treatment are:

- to maintain adequate nutritional status
- to normalize blood glucose levels with insulin, maximizing its anabolic effects
- to prevent hypoglycemia
- to perform daily strict self-monitoring
- to adapt to the lifestyle of the patient with flexibility
- to ensure adequate psychological, social, and emotional adjustments.

31.4.1 Diabetic Education

The team monitoring and controlling care of these patients should be multidisciplinary and be formed of a diabetologist, an educator on diabetes, a dietitian, a psychologist, and other specialists required for the proper control and treatment of CF [3]. Patients with CFRD should receive continuous education about diabetes and 24-hour support through telephone contact. Patients should be informed that CFRD is a common complication of CF and that its handling differs from other types of diabetes mellitus.

Education regarding the etiology and implications of CFRD, insulin therapy, blood glucose monitoring, treatment of hypoglycemia/hyperglycemia, and the effects of food intake, stress, illness and physical activity on glycemic control is crucial for success in managing CFRD [29] in a population undertaking multiple medical therapies. Therefore, the introduction of this additional treatment should be done carefully to promote acceptance by patients and to maintain a stable emotional situation. It should be emphasized to patients that treatment for diabetes will improve their overall clinical status.

31.4.2 Nutritional Treatment

Nutritional therapy is a fundamental part of CFRD management. People with CF need high-caloric diets to maintain enough muscle mass to face breathing difficulties resulting from lung damage and to maintain an ideal body weight.

To CF patients, a high-caloric, high-protein, high-fat, high-salt diet is advised to help achieve and maintain a healthy body weight. The diagnosis of CFRD does not change the general dietary recommendations for CF. However, in CF, limiting sugared soft drinks is recommended, as is eating more complex carbohydrates rather than simple sugars. This recommendation is useful to decrease blood glucose excursions [29]. Glucose excursions also improve with the use of pancreatic enzymes [22]. The use of nonnutritive sweeteners should be moderated and controlled and not exceed the recommended intake of 2.5 mg/day.

A dietitian with special knowledge about CF and diabetes should teach these patients and continuously reassess the diet upon each outpatient visit. Frequent concomitant diseases in CF patients need corresponding dietary adjustments.

31.4.3 Insulin Treatment

The CFF recommends both short-term and long-acting insulin therapy when CFRD has been diagnosed. In general, basal/bolus coverage is the most physiologic insulin regimen.

The treatment of CFRD should be started promptly after diagnosis to achieve an improvement in nutritional status and lung function [40]. Insulin, a potent anabolic hormone, is the recommended treatment for CFRD, but its use in earlier stages of insulin deficiency is not established. It is still not precisely defined when insulin therapy should start, possibly because of difficulties in detecting early but clinically relevant abnormalities in blood glucose metabolism among CF patients. Before the 2008 ISPAD guidelines [16] and 2010 CFF guidelines [3], active treatment of CFRD without fasting hyperglycemia was recommended only in the presence of symptoms.

Basal/bolus coverage with exogenous insulin is achieved by providing basal insulin, such as detemir or glargine, once or twice a day, with the addition of

TABLE 31.3 Action Characteristics of Different Types of Insulin

Insulin		Onset Action	Duration Action	Maximum Action
Rapid-acting insulin	Lispro	10–15 min	3–4 h	30–90 min
	Aspart	10–15 min	3–4 h	30–90 min
	Glulisine	10–15 min	3–4 h	30–90 min
Long-acting insulin	Glargine	2–4 h	<24 h	–
	Detemir	2–4 h	12–20 h	–

short-acting insulin analogs (lispro, aspart or glulisine) before meals (Table 31.3).

Insulin regimens should adapt to each patient [16]. Some patients who have impaired glucose tolerance or CFRD without fasting hyperglycemia may achieve euglycemia exclusively with the use of basal insulin (0.1 U/kg/day) as early therapy. In general, insulin is administered in the morning. A second dose may be required in the evening. However, in case postprandial hyperglycemia persists, the use of rapid insulin analogs before meals may be necessary. Basal insulin given as the only type of insulin may also be used to improve weight gain or pulmonary function.

Patients with CFRD with fasting hyperglycemia generally require both basal and rapid-acting insulin analogs.

When people with CFRD are not acutely ill, they generally require 1 U insulin for every 12–15 g of carbohydrates in meals. Calculation of an individual's sensitivity factor is necessary per patient when estimating insulin requirements.

Nighttime enteral feedings are best managed by a combination of intermediate-acting plus short-acting insulin therapy immediately before commencing enteral feeding. During an acute illness or treatment with steroids, insulin requirements increase two- to fourfold, but decrease once the situation has resolved (approximately 4–6 weeks). In pregnant women with CFRD, daily self-monitoring of capillary blood glucose (minimum 4–6 determinations/day) is important to adjust the insulin dose and achieve adequate metabolic control, monitoring weight gain [41].

Treatment with continued subcutaneous insulin infusion (CSII) offers the advantage of excellent control over blood sugars without the need to give multiple daily shots. This allows great dietary and physical activity flexibility. People with CRFD generally require only 20–30% of their total insulin dose as a basal infusion. The rest should be divided as bolus doses to be given at meals and snacks [42]. Few studies have determined the efficacy and tolerance of the CSII via an insulin pump for treatment of CFRD. In all of them, the results were favorable, with no concomitant problems related to the treatment [42], and with an improvement in diabetes control and quality of life [43] and in nutritional status [44].

Conventional dosing (with four or more insulin injections per day) may carry the risk of hypoglycemia. However, uncontrolled trials suggest that a daily injection of intermediate or long-acting insulin improves weight and lung function, with minimal hypoglycemic risk in CFRD and also in early insulin deficiency [38]. It is plausible that insulin may be of greater benefit to respiratory function when given before the diagnosis of CFRD, after which structural lung disease may be irreversible. It is also plausible that early insulin treatment may prolong the lifespan of the remaining insulin-secreting β-cells [38]. In addiction, early use of insulin therapy might improve the weight gain and lung function of CF patients, including those with normal OGTT results [24]. The role of treating prediabetes with basal insulin has been also investigated. In this sense, insulin glargine therapy had been demonstrated to be safe [45] and to prevent lung disease progression in patients with CF and early glucose derangements [46]. On the other hand, insulin detemir therapy improved weight and lung function in a cohort of 12 patients with CF [47]. Randomized controlled trials are needed to determine whether current clinical practice should be altered toward the earlier onset of insulin in CF individuals.

31.4.4 Self-Monitoring of Blood Glucose and Adjustments

For handling and adjusting CFRD treatment daily, the patient must perform several daily capillary blood controls (before each meal and 2h later (needed to adjust the insulin bolus), and fasting and midnight controls (needed to adjust the basal insulin)). The frequency of self-monitoring of capillary blood glucose should be individualized according to each patient's metabolic control and the presence of concurrent situations. Glycemic targets to achieve adequate glycemic control should also be individualized for each patient. Usually, a fasting or before-meal glycemia between 80 and 120mg/dL (provided that at least 3h has passed since the last dose of rapid-acting insulin) and 2-h blood glucose postprandial between 100 and 140mg/dL are recommended. Nocturnal blood glucose must be >90mg/dL [3].

To adjust the intake and glycemic control, the following are used:

- The sensitivity factor (SF) determines the sensitivity to insulin in each patient. It reports the expected glucose decrease, in milligrams per deciliter after administering an extra dose of rapid-acting insulin. It allows correct hyperglycemia values, taking into account the sensitivity factor, the value of hyperglycemia, and the glycemic goal. The calculation is made with the following formulas:
  SF (mg/dL): 1800/total daily insulin dose
  Current glycemia − glycemic target/SF = extra insulin dose (IU)
- The ratio or insulin dose per each carbohydrate ration tells us the amount of rapid-acting insulin needed to metabolize one ration of carbohydrates (10–15g). There are formulas to calculate and approximate the right dose, but it is preferable to assess the dose of insulin that is adequate to achieve a normal capillary glucose level in each intake individually, starting from a normal glycemia. It is useful because it allows the dose of insulin to be adapted to different amounts of carbohydrate [48].

31.4.5 Oral Agents for Treatment of CFRD

The CF Consensus Conference on Diabetes did not recommend oral hypoglycemic agents to be used in the treatment of CFRD [49] owing to the lack of sufficient published research studies on the issue. No demonstrated advantage has been established for using oral hypoglycemic agents over insulin, and some of these agents may have negative side effects for the CF population.

Sulfonylureas are at risk for causing hypoglycemia, so they are not indicated in CF. Metformin decreases insulin resistance, but it is not indicated because of digestive side effects, abdominal pain, nausea, and diarrhea. Acarbose, which reduces postprandial glycemia, may produce anorexia, abdominal distension, and diarrhea, and is not being recommended for CF. Troglitazone have recently been associated with osteoporosis, which limits their use in this group of patients. Oral insulin secretagogues such as repaglinide increase endogenous insulin secretion but are less effective than rapid-acting insulin, analogous to controlling hyperglycemic excursions after eating [50].

Some oral medications may address specific metabolic changes and be beneficial to patients. Agents that potentiate insulin action, especially agents with additional anti-inflammatory action, should be further investigated to determine whether there are clinical advantages to adding these medications to insulin as adjuvant therapy.

31.4.6 Exercise

Regular physical exercise is beneficial to improving glycemic control. Moreover, physical activity improves lung function and has a beneficial psychological effect, helping the patient increase wellbeing. Furthermore, it is beneficial in the long term for the lipid profile, blood pressure, and cardiac function, and thereby controls cardiovascular risk factors.

Moderate aerobic exercise is advised in CF, and it has to be individualized according to each patient's physical capacity and specific situation. Before exercise, patients should adjust treatment: They have to self-monitor capillary glucose and control carbohydrate intake and the dosage of insulin [51].

31.4.7 Complications: Hypoglycemia

Hypoglycemia can occur in the setting of treatment for CFRD; it is the main acute complication and is defined as glucose levels <70 mg/dL in CFRD. It is caused by an excess of insulin in a specific situation, inadequate food intake or absorption, or uncontrolled excessive exercise. There is increased risk of hypoglycemia in the case of poor nutritional status and/or an excess of energy needs in specific situations. In patients with CFRD, although there is an exaggerated response of catecholamine, during hypoglycemic events there might be a defective glucagon response owing to impaired α-pancreatic cells. This requires specific education for the early and intensive control of hypoglycemia [52].

Patients and families must be educated about the symptoms, treatment, and, most important, prevention of hypoglycemia. If hypoglycemia occurs without an associated altered level of consciousness, treatment should consist of the intake of 10–15 g carbohydrates with a high proportion of simple sugars, such as juice or sugar, with subsequent intake of 10 g complex carbohydrates once the resolution of hypoglycemia is achieved. If hypoglycemia is associated with an altered level of consciousness, glucagon must be administered to the patient intramuscularly or subcutaneously.

31.5 CYSTIC FIBROSIS–RELATED DIABETES: COMORBIDITY AND COMPLICATIONS

Microvascular complications are not the most relevant clinical complications; a decline in lung function and/or weight loss is more common. Both events are interrelated and may proceed up to 2–4 years [53] before the diagnosis of CFRD involving major morbidity and mortality. In CF, there is an important relationship between protein catabolism and malnutrition. The important anabolic effect of insulin is well known; the nutritional impact of insulin deficit could be more important than the glucose effect in these patients.

Intermittent postprandial hyperglycemia does not seem to result in an increase of microvascular complications, but the nutritional consequences of insulin deficiency could disturb the course of the disease. It has been demonstrated that treatment with analogs of rapid-acting insulin before a meal changes protein catabolism, weight loss, and the decline in pulmonary function to before the complications [54].

31.5.1 Morbidity/Mortality

In CF patients, worsening lung function and weight loss are predictors of early mortality. Both situations can be present in patients with CFRD several years before diagnosis. In this regard, the association between the presence of CFRD and increased mortality in CF has been demonstrated [55]. Thus, from the 1980s until the current time, studies have shown lower survival of patients with CFRD compared with CF patients without diabetes (25% versus 60%) [56,57]. However, in recent years, annual screening has allowed early diagnosis and initiation of more intensive treatment, and has reduced mortality in this group of patients [1].

In CF, bacterial superinfections have great clinical relevance. The presence of hyperglycemia increases glucose concentration in the exhaled air of patients with CF, a situation that favors bacterial growth. Postprandial hyperglycemia itself causes increased oxidative stress; it will also promote infection. These exacerbations produce overall worsening of the patient with CF. Therefore, it is essential to maintain adequate glycemic control of patients with CF and glucose abnormalities to prevent these complications.

31.5.2 Pulmonary Function

The lung is the main target organ of CFRD. In patients with CF, mortality is mainly caused by respiratory failure and not by vascular complications [7]. Cystic fibrosis–related diabetes is associated with a progressive decline in lung function years before diagnosis in most cases [56]. Hyperglycemia in patients with diabetes appears to have a direct negative effect on pulmonary impairment by increasing oxidative stress. Moreover, loss of the anabolic effect of insulin deficit produces a situation of protein catabolism in the lung with consequent functional impairment [58]. This deterioration of lung function correlates with the degree of insulin deficiency [59]. When continuous interstitial glucose monitoring is performed in CF patients, poorer lung function in those with hyperglycemic excursions along monitoring are detected, whether they were not patients diagnosed with CFRD following international criteria defined [53].

In this sense, improvement and/or stabilization of pulmonary function is clearly demonstrated in patients with CF and CFRD after starting treatment with insulin [60,61]. We reported the long-term impact of insulin treatment of CFRD on pulmonary function in a male patient with CF since the diagnosis of diabetes. The significant and sustained improvement in pulmonary function allowed his withdrawal from the lung transplantation program 4 months later; 8 years later, he no longer met criteria for lung transplantation [62].

31.5.3 Nutritional Status

In addition to lung function alterations, the presence of CFRD is associated with a progressive deterioration in nutritional status. This is reflected by a decrease in body mass index in the years before the diagnosis of CFRD [15,57,63]. Parallel to lung function, a study with continuous monitoring of interstitial glucose showed worse nutritional status in patients with a larger number of hyperglycemic excursions, even without CFRD [53]. This situation is explained by the state of protein catabolism in patients with varying degrees of insulin deficiency.

Studies have also shown improvement in nutritional status for both adults and children when patients start insulin replacement therapy. This advantage occurs in patients with CFRD and in those with prediabetes related to CF [45–47,64].

Although CFRD is different from type 1 and type 2 diabetes mellitus, the development of diabetes-induced complications resulting from uncontrolled hyperglycemia is similar to other types of diabetes mellitus. These complications can include the following:

31.5.4 Microvascular Complications: Retinopathy, Nephropathy, and Neuropathy

The rise in life expectancy for CF patients and those with CFRD has favored that chronic microvascular complications begin to appear get to appear in this group of patients [65]. Microvascular complications result from prolonged periods of hyperglycemia, which inflict pathogenic changes within small blood vessels. Prevalence rates of 10–23% of retinopathy, 4–21% of nephropathy, and 2.9–17% of neuropathy have been described, and their presence is influenced by both metabolic control and the duration of diabetes [66–68]. Manifestations of gastrointestinal neuropathy, alterations in bowel habits, and delayed gastric emptying are described as well. Furthermore, patients with CF have a high risk of developing renal disease owing to neurotoxicity of the treatments received.

The presence of microvascular complications in patients with CFRD is lower than in those with diabetes mellitus type 1 and type 2. This is influenced by the lower life expectancy of patients with CF with a lower duration of diabetes, the presence of pancreatic reserve to maintain some insulin secretion, and the absence of other risk cardiovascular factors.

Screening for these complications should be performed initially at the time of diagnosis, because diabetes can remain silent for years. Thereafter, screening of complications must be performed annually by studying the fundus of the eye, renal excretion of albumin, and renal function, and detecting neuropathy by assessing tendon reflexes and vibratory sensation.

31.5.5 Macrovascular Complications

In patients with CFRD an increased risk of developing macrovascular complications [69] is not found. The presence of hyperlipidemia, obesity, and/or hypertension or toxic habits such as smoking is unusual. Few studies have identified obesity and hypertension in patients with CF [70–72]. Although life expectancy has increased, to date, patients do not survive long enough to develop cardiovascular disease.

31.5.6 Psychosocial Impact

In patients with a diagnosis of CFRD associated with the start of a new therapy, timing is critical. For the underlying disease, patients with CF have a strict regimen of treatment and sometimes certain limitations. There are few data in the literature regarding the psychosocial impact of this new clinical situation. Collins and Reynolds [73] detected a wide range of reactions, from rejection to shock, and they gave recommendations for the management of patients. A multidisciplinary medical team should help patients to accept this second chronic disease associated with CF. It requires specific educational programs, for both patients and family members, to facilitate the progressive adaptation to this new change [29]. A flexible approach to each individual patient should be made. Insulin treatment should be adapted to the patient, and the patient should not be forced to adapt to a fixed regimen of diet and insulin dose.

31.6 PERSPECTIVES AND CONCLUSIONS

The survival rate of people with CF extends well into adulthood; as consequence, secondary disease processes such as CFRD are increasingly being recognized. This entity is the major comorbidity in CF, with a sevenfold increase in mortality [7]. Therefore, it is essential to continue to research the mechanisms underlying the defect of pancreatic endocrine function in patients with CF.

Earlier diagnosis and treatment of glucose abnormalities are associated with an overall clinical improvement in patients with CF, for both nutritional and pulmonary function. To date, the diagnosis of glucose abnormalities in CF is based on the determination of fasting glucose and 2 hours after the completion of OGTT, according to international guidelines. A study by Phillips et al. [74] postulated the use of a lower

OGTT glucose intake (50 g) and a shorter-duration test (60 minutes) as initial screening to identify patients at higher risk; a multicenter study is being conducted with the same approach.

Another field of study for a better understanding of pancreatic function in CF patients is the definition of early alterations in carbohydrate metabolism and its clinical significance in this specific group of patients [29].

As for treatment, further studies to better define the role of early insulin therapy on residual pancreatic function, lung function, and nutritional status, and studies related to new therapies such as oral agents are needed. Multicenter, randomized trials are still required to best assess the effectiveness of therapeutic options in the control of CFRD. Until we know these results, widespread early treatment with insulin in pre-diabetes phases of patients with CF [31] should not be recommended.

References

[1] Moran A, Dunitz J, Nathan B, Saeed A, Holme B, Thomas W. Cystic fibrosis-related diabetes: current trends in prevalence, incidence, and mortality. Diabetes Care 2009;32:1626–31.
[2] Mohan K, Miller H, Dyce P, Grainger R, Hughes R, Vora J, et al. Mechanisms of glucose intolerance in cystic fibrosis. Diabet Med 2009;26:582–8.
[3] Moran A, Brunzell C, Cohen RC, Katz M, Marshall BC, Onady G, et al. CFRD Guidelines Committee. Clinical care guidelines for cystic fibrosis-related diabetes: a position statement of the American Diabetes Association and a clinical practice guideline of the Cystic Fibrosis Foundation, endorsed by the Pediatric Endocrine Society. Diabetes Care 2010;33:2697–708.
[4] Stecenko AA, Moran A. Update on cystic fibrosis-related diabetes. Curr Opin Pulm Med 2010;16:611–5.
[5] O'Riordan SM, Dattani MT, Hindmarsh PC. Cystic fibrosis-related diabetes in childhood. Horm Res Paediatr 2010;73:15–24.
[6] Van der Berg JMW, Kouwenberg JM, heijerman HGM. Demographics of glucose metabolism in cystic fibrosis. J Cyst Fibros 2009;8:276–9.
[7] Lanng S, Hansen A, Thorsteinsson B, Nerup J, Koch C. Glucose tolerance in patients with cystic fibrosis: five year prospective study. BMJ 1995;311:655–9.
[8] Lek N, Acerini CL. Cystic fibrosis related diabetes mellitus - diagnostic and management challenges. Curr Diabetes Rev 2010;6:9–16.
[9] Adler AI, Shine BS, Chamnan P, Haworth CS, Bilton D. Genetic determinants and epidemiology of cystic fibrosis-related diabetes: results from a British cohort of children and adults. Diabetes Care September 2008;31(9):1789–94.
[10] Rana M, Munns CF, Selvadurai H, Donaghue KC, Craig ME. Cystic fibrosis-related diabetes in children. Gaps in the evidencie? Nat Rev Endocrinol 2010;6:371–8.
[11] Mohan V, Alagappan V, Snehalatha C, Ramachandran A, Thiruvengadam KV, Viswanathan M. Insulin and C-peptide responses to glucose load in cystic fibrosis. Diabete Metab 1985;11:376–9.
[12] Moran A, Diem P, Klein DJ, Levitt MD, Robertson RP. Pancreatic endocrine function in cystic fibrosis. J Pediatr 1991;118:715–23.
[13] Martín-Frías M, Lamas Ferreiro A, Enes Romero P, Cano Gutiérrez B, Barrio Castellanos R. Abnormal glucose tolerance in prepubertal patients with cystic fibrosis. An Pediatr (Barc) 2012;77:339–43.
[14] Hardin DS, LeBlanc A, Para L, Seilheimer DK. Hepatic insulin resistance and defects in substrate utilization in cystic fibrosis. Diabetes 1999;48:1082–7.
[15] Tofé S, Moreno JC, Máiz L, Alonso M, Escobar H, Barrio R. Insulin-secretion abnormalities and clinical deterioration related to impaired glucose tolerance in cystic fibrosis. Eur J Endocrinol 2005;152:241–7.
[16] O'Riordan SM, Robinson PD, Donaghue KC, Moran A. ISPAD clinical practice consensus. Management of cystic fibrosis-related diabetes. Pediatr Diabetes 2008;9:338–44.
[17] Moran A. Cystic fibrosis-related diabetes: an approach to diagnosis and management. Pediatr Diabetes 2000;1:41–8.
[18] Pitocco D, Fuso L, Conte EG, Zaccardi F, Condoluci C, Scavone G, et al. The diabetic lung–a new target organ? Rev Diabet Stud 2012;9:23–35.
[19] Cystic Fibrosis Fundation. Cystic Fibrosis Fundation patient registry: 2007 annual data report to the centres directors. Bethesda: MD; 2008. 1–24.
[20] Ntimbane T, Comte B, Mailhot G, Berthiaume Y, Poitout V, Prentki M. Cystic fibrosis-related diabetes: from CFTR dysfunction to oxidative stress. Clin Biochem Rev 2009;30:153–77.
[21] Ali BR. Is cystic fibrosis-related diabetes an apoptotic consequence of ER stress in pancreatic cell? Med Hypotheses 2009;72:55–7.
[22] Kuo P, Stevens JE, Russo A, Maddox A, Wishart JM, Jones KL. Gastric emptying, incretin hormone secretion and postprandial glycemia in cystic fibrosis. Effects of pancreatic enzyme supplementation. J Clin Endocrinol Metab 2011;96:E851–5.
[23] Battezzati A, Mari A, Zazzeron L, Alicandro G, Claut L, Battezzati M. Identification of insulina secretory defects and insulin resistance turing oral glucose tolerante test in a cohorte of cystic fibrosis patients. Eur J Endocrinol 2011;165:69–76.
[24] Frohnert BI, Ode KL, Moran A, Nathan BM, Laguna T, Holme B. Impaired fasting glucose in cystic fibrosis. Diabetes Care 2010;33:2660–4.
[25] O'Riordan SM, Robinson PD, Donaghue KC, Moran A. Management of cystic fibrosis-related diabetes in children and adolescents. Pediatr Diabetes 2009;10:43–50.
[26] Succurro E, Marini MA, Arturi F, Grembiale A, Lugarà M, Andreozzi F, et al. Elevated one-hour post-load plasma glucose levels identifies subjects with normal glucose tolerance but early carotid atherosclerosis. Atherosclerosis 2009;207:245–9.
[27] Brodsky J, Dougherty S, Makani R, Rubenstein RC, Kelly A. Elevation of 1-hour plasma glucose during oral glucose tolerance testing is associated with worse pulmonary function in cystic fibrosis. Diabetes Care 2011;34:292–5.
[28] Ode KL, Frohnert B, Laguna T, Philips J, Holme B, Regelmann W. Oral glucose tolerance testing in children with cystic fibrosis. Pediatr Diabetes 2010;11:487–92.
[29] Kelly A, Moran A. Update on cystic fibrosis-related diabetes. J Cyst Fibros July 2013;12(4):318–31.
[30] Dobson L, Sheldon CD, Hattersley AT. Validation of interstitial fluid continuous glucose monitoring in cystic fibrosis. Diabetes Care 2003;26:1940–1.
[31] Dobson L, Sheldon CD, Hattersley AT. Conventional measures underestimate glycaemia in cystic fibrosis patients. Diabet Med 2004;21:691–6.
[32] Jefferies C, Solomon M, Perlman K, Sweezey N, Daneman D. Continuous glucose monitoring in adolescents with cystic fibrosis. J Pediatr 2005;147:396–8.
[33] Moreau F, Weiller MA, Rosner V, Weiss L, Hasselmann M, Pinget M, et al. Continuous glucose monitoring in cystic fibrosis patients according to the glucose tolerance. Horm Metab Res 2008;40:502–6.

[34] Franzese A, Valerio G, Buono P, Spagnuolo MI, Sepe A, Mozzillo E, et al. Continuous glucose monitoring system in the screening of early glucose derangements in children and adolescents with cystic fibrosis. J Pediatr Endocrinol Metab 2008;21:109–16.
[35] Martín-Frías M, Lamas Ferreiro A, Colino Alcol E, Alvarez Gómez MA, Yelmo Valverde R, Barrio Castellanos R. Continuous glucose monitoring system in the screening of glucose disorders in cystic fibrosis. An Pediatr (Barc) 2009;70:120–5.
[36] Khammar A, Stremler N, Dubus JC, Gross G, Sarles J, Reynaud R. Value of continuous glucose monitoring in screening for diabetes in cystic fibrosis. Arch Pediatr 2009;16:1540–6.
[37] Schiaffini R, Brufani C, Russo B, Fintini D, Migliaccio A, Pecorelli L, et al. Abnormal glucose tolerance in children with cystic fibrosis: the predictive role of continuous glucose monitoring system. Eur J Endocrinol 2010;162:705–10.
[38] Hameed S, Jaffé A, Verge CF. Cystic fibrosis related diabetes (CFRD)–the end stage of progressive insulin deficiency. Pediatr Pulmonol 2011;46:747–60.
[39] Waugh N, Royle P, Craigie I, Ho V, Pandit L, Ewings P, et al. Screening for cystic fibrosis-related diabetes: a systematic review. Health Technol Assess 2012;16:1–179.
[40] Hardin DS. Pharmacotherapy of diabetes in cystic fibrosis patients. Expert Opin Pharmacother 2010;11:771–8.
[41] Lau EM, Moriarty C, Ogle R, Bye PT. Pregnancy and cystic fibrosis. Paediatr Respir Ver 2010;11:90–4.
[42] Hardin DS, Rice J, Rice M, Rossenblatt R. Use of insulin pupm in treat cystic fibrosis related diabetes. J Cyst Fibros 2009;8:174–8.
[43] Klupa T, Małecki M, Katra B, Cyganek K, Skupień J, Kostyk E, et al. Use of sensor-augmented insulin pump in patient with diabetes and cystic fibrosis: evidence for improvement in metabolic control. Diabetes Technol Ther 2008;10:46–9.
[44] Sulli N, Bertasi S, Zullo S, Shashaj B. Use of continuous subcutaneous insulin infusion in patients with cystic fibrosis related diabetes: three case reports. J Cyst Fibros 2007;6:237–40.
[45] Bizzarri C, Lucidi V, Ciampalini P, Bella S, Russo B, Cappa M. Clinical effects of early treatment with insulin glargine in patients with cystic fibrosis and impaired glucose tolerance. J Endocrinol Invest 2006;29:RC1–4.
[46] Mozzillo E, Franzese A, Valerio G, Sepe A, De Simone I, Mazzarella G, et al. One-year glargine treatment can improve the course of lung disease in children and adolescents with cystic fibrosis and early glucose derangements. Pediatr Diabetes May 2009;10:162–7.
[47] Hameed S, Morton JR, Field PI, Belessis Y, Yoong T, Katz T, et al. Once daily insulin detemir in cystic fibrosis with insulin deficiency. Arch Dis Child 2012;97:464–7.
[48] Barrio R, Martín-Frías M. Alteración hidrocarbonada en la fibrosis quística. Rev Esp Pediatr 2010;66:211–7.
[49] Onady GM, Stolfi A. Insulin and oral agents for managing cystic fibrosis-related diabetes. Cochrane Database Syst Rev 2013;7. CD004730.
[50] Moran A, Philips, Milla C. Insulin and glucose excursion following premeal insulin lispro or replaginide in cystic fibrosis-related diabetes. Diabetes Care 2001;24:1706–10.
[51] Nathan BM, Laguna T, Moran A. Recent trends in cystic fibrosis-related diabetes. Curr Opin Endocrinol Diabetes Obes 2010;17:335–41.
[52] American Diabetes Association Standards of medical Care in Diabetes. Diabetes Care 2011;34(Suppl. 1):S11–61.
[53] Hammed SH, Morton JR, Jaffé A, Field PI, Belessis Y, Toong T. Early glucose abnormalities in cystic fibrosis are preceded by poor weight gain. Diabetes Care 2010;33:221–6.
[54] Mohan K, Israel KL, Miller H, Grainger R, Ledson MJ, Walshaw MJ. Long effect of insulin treatment in cystic fibrosis-related diabetes. Respiration 2008;76:181–6.
[55] Moran A, Hardin D, Rodman D, Allen HF, Beall JR, Borowitz D. Diagnosis, screening and management of cystic fibrosis related diabetes mellitus: a consensus conference report. Diabetes Res Clin Pract 1999;45:61–70.
[56] Finkelstein SM, Wielinski CL, Elliott GR, Warwick WJ, Barbosa J, Wu SC. Diabetes mellitus associated with cystic fibrosis. J Pediatr 1988;112:373–7.
[57] Marshall BC, Butler SM, Stoddard M, Moran AM, Liou TG, Morgan WJ. Epidemiology of cystic fibrosis-related diabetes. J Pediatr 2005;146:681–7.
[58] Moran A, Milla C, Ducret R, Nair KS. Protein metabolism in clinically stable adult cystic fibrosis patient with abnormal glucose tolerance. Diabetes 2001;50:1336–43.
[59] Milla CE, Warwick WJ, Moran A. Trends in pulmonary function in patients with cystic fibrosis correlate with the degree of glucose intolerance at baseline. Am J Respire Care Med 2000;162:891–5.
[60] Lanng S, Thorsteinsson B, Nerup J, Koch C. Diabetes mellitus in cystic fibrosis: effect of insulin therapy on lung function and infections. Acta Paediatr 1994;83:849–53.
[61] Dobson L, Hattersley AT, Tiley S, Elworthy S, Oades PJ, Sheldon CD. Clinical improvement in cystic fibrosis with early insulin treatment. Arch Dis Child 2002;87:430–1.
[62] Martín-Frías M, Máiz L, Carcavilla A, Barrio R. Long-term benefits in lung function and nutritional status of strict metabolic control of cystic fibrosis-related diabetes. Arch Bronconeumol 2011;47:531–4.
[63] Lanng S, Thorsteinsson B, Nerup J, Koch C. Influence of the development of diabetes mellitus on clinical status in patients with cystic fibrosis. Eur J Pediatr 1992;151:684–7.
[64] Moran A, Pekow P, Grover P, Zorn M, Slovis B, Pilewski J. Insulin therapy to improve BMI in cystic fibrosis-related diabetes without casting hyperglycemia: results of the cystic fibrosis related diabetes therapy trial. Diabetes Care 2009;32:1783–8.
[65] Sullivan MM, Denning CR. Diabetes microangiopathy in patients with cystic fibrosis. Pediatrics 1989;84:642–7.
[66] Andersen HU, Lanng S, Pressler T, Laugesen CS, Mathiesen ER. Cystic fibrosis-related diabetes: the presence of microvascular diabetes complications. Diabetes Care 2006;29:2660–3.
[67] Dobson L, Stride A, Bingham C, Elworthy S, Sheldon CD, Hattersley AT. Microalbuminuria as a screening tool in cystic fibrosis-related diabetes. Pediatr Pulmonol 2005;39:103–7.
[68] Schwarzenberg ST, Thomas W, Olsen TW, Grover T, Walk D, Milla C. Microvascular complications in cystic fibrosis-related diabetes. Diabetes Care 2007;30:1056–61.
[69] Costa M, Potvin S, Berthiaume Y, Gauthier L, Jeanneret A, Lavoie A. Diabetes: a major co-morbidity of cystic fibrosis. Diabetes Metab 2005;31:221–32.
[70] Kastner-Cole D, Palmer CN, Ogston SA, Mehta A, Mukhopadhyay S. Overweight and obesity in deltaF508 homozygius cystic fibrosis. J Pediatr 2005;147:402–4.
[71] Yahiaoui Y, Jablonski M, Hubert D, Mosnier-Pudar H, Noel LH, Stern M. Renal involvement in cystic fibrosis: diseases spectrum end clinical relevance. Clin J Am Soc Nephrol 2009;4:291–8.
[72] Coderre L, Fadaina C, Belson L, Belisle V, Ziai S, Maillhot G. LDL-cholesterol and insulin are independently associated with body mass index in adult cystic fibrosis patients. J Cyst Fibros 2012;11:393–7.
[73] Collins S, Reynolds F. How do adults with cystic fibrosis cope following a diagnosis of diabetes? J Adv Nurs 2008;64:478–87.
[74] Phillips LS, Ziemer DC, Kolm P, Weintraub WS, Vaccarino V, Rhee MK. Glucose challange test screening for prediabetes and undiagnosed diabetes. Diabetologia 2009;52:1798–807.

SECTION F

EXERCISE AND BEHAVIOR IN MANAGEMENT OF CYSTIC FIBROSIS

CHAPTER

32

Exercise Testing in CF, the What and How

Larry C. Lands[1], *Helge Hebestreit*[2]

[1]Pediatric Respiratory Medicine, McGill University, Montreal, QC, Canada; [2]Universitaets-Kinderklinik, Josef-Schneider-Strasse 2, Wuerzburg, Germany

Exercise testing has many roles to play in the management of CF patients. It can be used to assess physical limitations and the factors contributing to limitation (see Chapter 34), evaluate the risk for exercise participation, and aid in the development of a personalized exercise program and assessment of progress during such a program. Furthermore, the results of exercise testing have important prognostic value for mortality and help in the assessment before lung transplantation. There are both laboratory and field tests available, with the gold standard for testing being a progressive exercise test with measured gas exchange. Exercise testing is recommended in several cystic fibrosis (CF) communities [1,2].

32.1 UTILITY OF EXERCISE TESTING

The first result from a progressive exercise test is the determination of exercise ability. This is important as it is the standard measure of fitness for an adult. Fitness is associated with quality of life [3], and as CF patients now are living longer, it is important that they also live better. As for healthy adults, exercise testing has been shown to be prognostic for the risk of subsequent mortality. Testing is also useful in children; however, fitness in children includes various other aspects, such as coordination and power [4] that are not measured during a progressive test. However, these other key outcomes can be evaluated quantitatively by other means.

32.1.1 Assessment of the Factors Limiting Exercise Ability

Standard exercise testing is symptom-limited testing. In other words, the person stops performing the test when they are no longer able to continue, or they experience adverse reactions that warrant termination of the test. However, various factors contribute to exercise limitation. These are covered in depth in a subsequent chapter, but do need to be recognized. Three main organs contribute to exercise limitation, namely, the lungs, heart, and peripheral skeletal muscle. All may be adversely affected in CF.

Because lung function can be significantly impaired by CF, it is reasonable to assume that this is the major contributor. However, typically lung function can only explain about 35% of exercise limitation [5]. Ascribing exercise capacity by simply measuring resting lung function is akin to evaluating how a car performs by looking at it in the showroom; it really needs to be taken out on the road to assess its ability. Lung function can contribute in various ways to exercise limitation, including expiratory airflow limitation, leading to dynamic hyperinflation and dyspnea; excess dead space ventilation, resulting in excessive ventilatory demands; oxygen desaturation and pulmonary hypertension, leading to poor oxygen delivery; and cardiopulmonary interaction, which can also impair oxygen delivery.

Peripheral skeletal muscle function also significantly contributes to exercise limitation [5,6]. The primary determinant of muscular function is muscle mass [7], and CF patients may be particularly prone to peripheral muscle wasting. However, mutations in CFTR, the gene responsible for CF, may also impair muscle energetics [8], as may chronic inflammation [9]. Furthermore, deconditioning due to inactivity also contributes to decreased skeletal and cardiac muscle function [6]. Medical therapy, such as systemic corticosteroids [10], can result in decreased muscular force. Alternatively, growth hormone may improve function [11].

Diet and Exercise in Cystic Fibrosis
http://dx.doi.org/10.1016/B978-0-12-800051-9.00032-8

Cardiac function may be impaired [12] due to malnutrition, causing cardiac muscle wasting, pulmonary hypertension, and/or cardiopulmonary interaction. This latter factor comes into play with advanced pulmonary disease, resulting in the wide pleural pressure swings necessary for ventilation.

32.1.2 Assessing Risk for Activity Participation

In some instances, patients must limit their participation in physical activity due to adverse effects of exercise. Exercise testing can unmask a need for supplemental oxygen during exertion. Various lung function thresholds place patients more at risk. For example, oxygen desaturation is more likely with a forced expiratory volume in 1-second (FEV_1) below 70% [13,14], or a diffusing capacity below 65%. However, direct exercise testing is the only way to determine exercise oxygen desaturation. Such testing can then be used to make an exercise oxygen prescription. The oxygen is not used in this instance to prolong life, but to enable the patient to exercise for longer periods, and thus gain greater benefit. Cardiac arrhythmias [13], as well as ischemia, can also be assessed during exercise testing. Again, this may not be evident from resting examinations. Exercise testing in patients with significant disease is typically recommended before commencing an exercise training program.

32.1.3 Exercise Program Planning

With a test of maximal exercise capacity that includes gas exchange, an evaluation of the limiting factors can be made and a program designed that focuses on improving the limiting factors, while being safe and tolerable. Because cycle ergometry is done with the subjected seated, the results need to be translated into tolerable ranges for heart rate and oxygen saturation while performing other activities. Exercise testing results to prescribe targets for activities other than cycling may be somewhat easier when the testing is done with a treadmill. Still, exercise testing is able to evaluate the degree of deconditioning. This will help to design training programs to work on the modifiable factors, principally muscular function, although some patients may be helped with bronchodilators, if these reduce airtrapping. Exercise testing can also assess progress made during an exercise program [15–17]. Furthermore, exercise testing may also be useful to demonstrate to patients that they can safely exercise and thus facilitate their participation in activities.

32.1.4 Prognosis

There are numerous studies linking exercise capacity with survival in CF [9,18–23]. Maximal exercise capacity seems to be a more important prognostic function than the traditional measure of FEV_1. Other work suggests that exercise testing can assist with projecting future declines in lung function [14]. Exercise testing has also been used to help determine the urgency for lung transplantation [24,25].

So, who should be tested? The consensus of CF caregivers and exercise specialists is that CF patients 10 years of age or older should be tested. This can be particularly helpful in providing prognostic information and counseling about exercise participation. Patients with new exercise-related symptoms should also be tested. The frequency of testing is debatable. Although yearly testing is often desired, cost and timing constraints make this difficult.

32.2 EXERCISE TESTS

As described above, exercise testing can be used for many purposes. There are a variety of tests, both laboratory-based and field tests, that can be used, depending on the reasons for testing (Table 32.1). The gold standard for assessing aerobic exercise capacity is an incremental test to measure peak oxygen uptake on either a cycle ergometer or a treadmill [26–28]. Typical protocols will use stepwise increases every minute with steps of sufficient size to complete the test within 8–12 min [27,29].

With the addition of gas exchange measurements, accurate oxygen consumption, ventilation, and circulatory aspects can be assessed. This information is vital for assessing the factors leading to exercise limitation, as discussed previously and in the subsequent chapter.

There are also field-based tests. Examples include the 6-min walk test, shuttle/run tests, and step tests. These tests have the advantage of portability and being inexpensive. However, they generally require more space (e.g., a 30–50 m long corridor for the 6-min walk test). Furthermore, they provide only limited information concerning true exercise capacity, the causative limiting factors, or potential adverse responses, such as oxygen desaturation. In addition, these tests might not sufficiently exert the subject, especially when lung disease is mild.

There are many outcomes that can be measured. The essential measurements are oxygen saturation before, during, and after exercise, along with heart rate, either combined with the oximeter, but preferably by electrocardiography. The measurement of gas exchange is also preferred.

32.3 THE GODFREY PROTOCOL

The preferred exercise test uses 1 min stepwise increments on an electronically braked cycle ergometer until volitional fatigue. An electronically braked

TABLE 32.1 Exercise Test Selection

Test Indication	Preferred Test	Alternative Tests
Routine and assessment of symptoms	Progressive exercise test with a cycle ergometry (Godfrey protocol) with pulse oximetry and gas exchange measurements	1. Godfrey protocol with pulse oximetry 2. Treadmill exercise (Bruce protocol) with pulse oximetry and gas exchange measurements 3. Treadmill exercise (Bruce protocol) with pulse oximetry
Assessment for lung transplantation	Godfrey protocol with pulse oximetry with/without gas exchange measurements	1. Treadmill exercise (Bruce protocol) with pulse oximetry with/without gas exchange measurements 2. Six minute walk test with pulse oximetry
Training program planning	Godfrey protocol with pulse oximetry and gas exchange measurements	1. Godfrey protocol with pulse oximetry with/without gas exchange measurements 2. Treadmill exercise (Bruce protocol) with pulse oximetry with/without gas exchange measurements 3. Incremental maximal field test

cycle ergometer maintains the workload constant, over a range of pedaling frequencies. Using a cycle controlled with a friction strap results in highly variable work rate, due to mild variations in pedaling frequency. Typically, the subject is asked to pedal at 60 rpm. This so-called Godfrey protocol [30] selects the step size based on patient size and lung function [31]. The goal is to complete the exercise test within 8–12 min. This allows for testing of both smaller or younger patients, as well as adults. Gas exchange measurements and heart rate are typically derived from the last 15 s of each increment. The test is stopped when the patient can no longer maintain the pedaling cadence, demonstrates oxygen desaturation <80%, or feels uncomfortable (Conducting an Exercise Test). The test has been shown to be reproducible in CF patients [32]. The large number of studies using this protocol has led to the development of well-accepted prediction equations that have been validated in CF patients (Table 32.2). Although many of these are linear equations combining factors, scaling may be best suited to comparing between individuals of different size and age [33].

32.4 THE BRUCE PROTOCOL

The Bruce protocol [35] is the standard treadmill protocol [36–40]. The test has been used to compare maximal oxygen consumption to a variety of clinical parameters, such as lung function and fat-free mass [7,41], and clinical scores in CF patients [41,42]. Each increment in the standard Bruce protocol lasts 3 min at a specific percentage grade and speed. A modified protocol has two additional smaller increments. Typically, oxygen consumption is measured directly using equipment to measure gas exchange. Alternatively, it can be estimated from exercise time [43,44] or estimated metabolic cost. There are equations available to estimate work output based upon the percentage grade and speed of the treadmill, time, subject body mass, and the force of gravity. The test has been demonstrated to be reproducible in healthy individuals [45,46]. Reference values are available for children [39,40,45,47] and adults [35,45]. Results improve with respiratory muscle training or treatment of acute pulmonary exacerbations in patients with CF [42,48] and also reflect changes over time in lung function and fat-free mass in children [7].

32.5 FIELD TESTS

There are several tests that can be used without the expense of exercise equipment. These tests provide some information about exercise capacity and tolerance that can serve as a practical guide.

32.5.1 Six-Minute Walk Test

This is generally a submaximal test for all except those with significantly advanced disease. It is commonly used during the assessment for lung transplantation, and after transplantation [21,24]. This is a self-paced test performed over a 30–50 m course. The distance covered over the 6 min, including pauses for rest is recorded. Typically, two tests are done and the best is retained. The subject is not permitted to run, and there is standardized vocal encouragement. A straight corridor is required, typically marked off every 5 m. The subject is often accompanied by the tester who will monitor and record oxygen saturation and heart rate, and also the time to recover baseline values after the test. Reproducibility is good [49–51], and reference values are available [52–54]. Although typically the distance walked is recorded, the estimated work of walking seems to correlate better with maximal oxygen consumption in children with CF [55].

TABLE 32.2 Prediction Equations for Godfrey Protocol

Outcome Variable	Population Studied	Prediction Equation	Reference
VO_2 peak (L/min)	Not stated	F: VO_2 peak = 3.08806 × height (m) − 2.877 M: VO_2 peak = 4.4955 × height (m) − 4.640	Orenstein [30]
Wpeak (W)	117 healthy females and males, aged 6–16 years	F: Wpeak = 2.38 × height (cm) − 238 M: Wpeak = 2.87 × height (cm) − 291	Godfrey et al. [29]
Wpeak (W)	Adults (20–70 years), number not stated	F: Wpeak = 290 × height $(m)^{1.78}$ × $age^{-0.46}$ M: Wpeak = 417 × height $(m)^{1.78}$ × $age^{-0.46}$	Modified from Jones [33]
VO_2 peak (L/min)	Adults (20–70 years), number not stated	F: VO_2 peak = 3.55 × height $(m)^{1.88}$ × $age^{-0.49}$ M: VO_2 peak = 5.14 × height $(m)^{1.88}$ × $age^{-0.49}$	Jones [33]
VO_2 peak (mL/min)	140 adolescent girls and 223 adolescent boys with CF, aged 14.8 ± 1.7 years	VO_2 peak = 216.3–138.7 × sex + 11.5 × Wpeak Sex: male = 0; female = 1	Werkman et al. [34]
VO_2 peak (mL/min)	92 patients with CF, aged 12–42 years	VO_2 peak = 7.908 × Wpeak (W) − 247.5 × sex + 10.6377 × weight (kg) + 33.4995 × resting oxygen saturation (%) + 5.3415 × HRpeak (bpm) − 3716.5	Hebestreit (personal communication)

VO_2 peak: maximal oxygen consumption; Wpeak: maximal work rate; HRpeak: maximal heart rate; bpm: beats per minute.

32.5.2 Incremental Shuttle Tests

Both a 20m [56] and 10m shuttle [57] test have been developed, and adapted and validated in CF patients [58–60]. The test has the subject moving between two cones, with the time to travel being progressively shortened. A recorded program is used that makes a beep at set times, with the subject attempting to cover the distance before the subsequent beep. The distance covered during the test, or the number of laps, is recorded. Both tests have been shown to be reproducible [58,61]. Reference data are available [62–64]. The estimation of maximal oxygen consumption from these results has limited accuracy.

32.5.3 Three-Minute Step Test

This test requires a commercially available exercise step. A metronome is used to keep a steady pace, and the subject is asked to step on and off the step at a rate of 30 steps per minute. Thus, there is no progression in stepping rate. Oxygen saturation, heart rate, and breathlessness are recorded. This test is used more for determining desaturation, than performance. It has been used in assessing children for lung transplantation [65]. However, individuals with mild-to-moderate disease will typically not demonstrate oxygen desaturation (>4%, [70]). The test is reproducible [66].

32.6 CONDUCTING AN EXERCISE TEST

The physician ordering the exercise test should review the medical history and conduct a physical examination, including vital signs, to assess whether testing can proceed in a secure manner. The subject needs to be prepared before the test, including a thorough explanation of the testing procedures. A light meal should be consumed not less than 2h before the test, and no caffeine should be consumed the day of the test. The subject should refrain from strenuous exercise the day before testing. The subject needs to be appropriately dressed to perform vigorous exercise. Routine medication, including bronchodilators, should be used as prescribed and airway clearance should be done before testing.

The staff, equipment, and infrastructure need to be in place to conduct exercise tests. The staff need to be appropriately trained to recognize and intervene in case of adverse events occurring during exercise testing [26–28]. A physician should be available, and actually present if there is increased risk, such as poor pulmonary function or a history of exercise-associated arrhythmias. The equipment needs to be size appropriate, so that the subject can comfortably cycle.

There are recommended monitoring procedures for a 3min period before the test and for at least 2min after the test, until there is a return to baseline values. These measurements include oxygen saturation and heart rate. Typically, gas exchange is recorded before exertion to establish a baseline. Many centers will monitor blood pressure throughout. Many centers will also record perceived exertion during testing, typically using the Borg scale of perceived exertion [67,68].

The test is typically stopped when the subject is no longer able to maintain the pedaling cadence. There are also reasons to end the test prematurely. These include an oxygen saturation <80% with accompanying symptoms, other signs of respiratory failure, chest pain suggestive of pneumothorax or angina, hemoptysis, sudden pallor or feeling dizzy or faint or significant decreases in systolic blood pressure from rest, or severe elevations in systolic or diastolic blood pressure.

32.7 INTERPRETING AN EXERCISE TEST

Exercise testing provides a wealth of data. Maximal oxygen consumption and work capacity along with maximal heart rate and minimal oxygen saturation are the fundamental results. Electrocardiogram changes can be seen, but these are uncommon in CF. If perceived exertiona and dyspnea scores are recorded, this provides further information about what the subject perceives as limiting his or her ability. With gas exchange measurements, maximal respiratory exchange ratio (the ratio of carbon dioxide production to oxygen consumption), maximal minute ventilation, along with maximal respiratory rate, tidal volume, and deadspace to tidal volume ratio are important. Further insight into factors contributing to exercise limitation can be gained from the slope of the relation between minute ventilation and carbon dioxide production; the ratio of ventilation to oxygen consumption or carbon dioxide production at peak exercise; and the ventilatory threshold, or the point at which ventilation increases at a greater rate than carbon dioxide production. The ventilatory threshold reflects the point at which blood lactate begins to rise, a measure of physical fitness. Some centers will assess inspiratory capacity during exercise. Because total lung capacity does not change during exercise, a decrease in inspiratory capacity means that the end-expiratory lung volume has increased; in other words, there is dynamic hyperinflation. Dynamic hyperinflation significantly increases the elastic work of breathing, and the subject senses this increased effort requirement, that is, the subject experiences dyspnea.

The fundamental questions to be answered when interpreting an exercise test are whether the test truly represents a maximal effort; and if so, was the response to exercise abnormal? If the result is abnormal, then there is the intriguing question of which factors are limiting exercise ability. To assess whether the test was maximal, it is important that the staff witnessing the test record why the test was stopped, and what was the subject feeling at that time. Because this is a symptom-limited test, the subject should appear to have made a maximal effort. Then, there are several questions to be asked of the data: Was there a plateau in maximal oxygen consumption? Classically, as maximal exercise capacity is approached, there should be more reliance on glycolytic muscle fibers that can produce force without significantly increasing oxygen consumption. Therefore, there would be a plateau in the oxygen consumption even though the external work rate continued to increase. Trained athletes will certainly demonstrate this, but most subjects are not trained or ready to push themselves to that limit, so this plateau is not often achieved. Other factors to assess as to maximal effort include whether the maximal achieved oxygen consumption or workload exceed predicted values, whether the maximal heart rate approached predicted values, whether the maximal ventilation used at least 60–70% of the ventilatory reserve (typically estimated at 35–40 times the FEV_1), and whether the respiratory exchange ratio exceeded 1.05.

Once it is decided that a maximal effort was made, then it is important to assess whether the responses to exercise were normal. First the maximal exercise capacity and/or oxygen consumption results need to be compared to predicted values, typically using sex-specific age and height-based predicted equations. If these are low, then exercise limitation is present. It is important to assess whether there is significant oxygen desaturation (>4% or oxygen saturation <90% at peak exercise [13,69]). Maximal minute ventilation is typically 60–70% of maximal ventilatory capacity. When it exceed 85%, then it is likely that there is a ventilatory limitation to exercise. Of course, aerobic training can decrease the ventilatory demands of exercise, by reducing carbon dioxide and lactate production. Typically, the ratio of deadspace to tidal volume decreases from a resting value of 33% to less than 20% at maximal exercise. If this ratio does not decrease, or actually increases, then again there is significant pulmonary pathology contributing to ventilatory limitation.

The contribution of decreased muscular capacity can often be inferred from excessive ventilation despite a low deadspace-to-tidal volume ratio, and an early ventilatory threshold. Muscular capacity can be assessed separately, for example, by assessing strength or short-term power or work capacity [5].

Oxygen desaturation during exercise represents cardiac or pulmonary impairment. Recent development of novel diffusion tests using nitric oxide can assess diffusing capacity and capillary blood volume during exercise to help differentiate cardiac disease from pulmonary disease [70].

Exercise testing provides important information about functional capacity. It permits a discussion of the factors limiting exercise capacity in an individual and can be used for development of a personalized therapeutic plan to improve functional capacity. Laboratory-based testing, including cycle ergometry using the Godfrey protocol, or treadmill testing using a Bruce protocol, provides the maximum amount of information. If these tests cannot been conducted, then some information can be gained from field tests such as the 6 min walk test, shuttle tests, or step tests. These latter tests provide more a measure of exercise tolerance and the need for supplemental oxygen than detailed analysis of the factors limiting exercise ability.

References

[1] Gruber W, Hebestreit A, Hebestreit H. Arbeitskreis Sport des Mukoviszidose e.V. Leitfaden Sport bei Mukoviszidose. Bonn: Mukoviszidose e.V; 2004. p. 24–32.

[2] CF Trust. Standards of care and good clinical practice for the physiotherapy management of cystic fibrosis. 2nd ed. June 2011. http://www.cftrust.org.uk/aboutcf/publications/consensusdoc/Physio_standards_of_care.pdf.

[3] Hebestreit H, Schmid K, Kieser S, Junge S, Ballmann M, Roth K, et al. Quality of life is associated with physical activity and fitness in cystic fibrosis. BMC Pulm Med February 27, 2014;14(1):26.

[4] Malina RM, Katzmarzyk PT. Physical activity and fitness in an international growth standard for preadolescent and adolescent children. Food Nutr Bull 2006;27(4 Suppl. Growth Standard):S295–313.

[5] Lands LC, Heigenhauser GJF, Jones NL. Analysis of factors limiting maximal exercise performance in cystic fibrosis. Clin Sci 1992;83:391–7.

[6] Hebestreit H, Kieser S, Rüdiger S, Schenk T, Junge S, Hebestreit A, et al. Physical activity is independently related to aerobic capacity in cystic fibrosis. Eur Respir J 2006;28:734–9.

[7] Klijn PH, van der Net J, Kimpen JL, Helders PJ, van der Ent CK. Longitudinal determinants of peak aerobic performance in children with cystic fibrosis. Chest 2003;124:2215–9.

[8] Selvadurai HC, Blimkie CJ, Myers N, Mellis CM, Cooper PJ, van Asperen PP. Randomized controlled study of in-hospital exercise training programs in children with cystic fibrosis. Pediatr Pulmonol 2002;33(3):194–200.

[9] van de Weert-van Leeuwen PB, Slieker MB, Hulzebos HJ, Kruitwagen CLJJ, van der Ent CK, Arets HGM. Chronic infection and inflammation affect exercise capacity in cystic fibrosis. Eur Respir J 2012;39(4):893–9.

[10] Barry SC, Gallagher CG. Corticosteroids and skeletal muscle function in cystic fibrosis. J Appl Physiol 2003;95:1379–84.

[11] Hütler M, Schnabel D, Staab D, Tacke A, Wahn U, Böning D, et al. The effect of growth hormone on exercise tolerance in children with cystic fibrosis. Med Sci Sports Exerc 2002;34:567–72.

[12] Pianosi P, Pelech A. Stroke volume during exercise in cystic fibrosis. Am J Respir Crit Care Med March 1996;153(3):1105–9.

[13] Ruf K, Hebestreit H. Exercise-induced hypoxemia and cardiac arrhythmia in cystic fibrosis. J Cyst Fibros 2009;8:83–90.

[14] Holland AE, Rasekaba T, Wilson JW, Button BM. Desaturation during the 3-minute step test predicts impaired 12-month outcomes in adult patients with cystic fibrosis. Respir Care 2011;56:1137–42.

[15] Schneiderman-Walker J, Pollock SL, Corey M, Wilkes DD, Canny GJ, Pedder L, et al. A randomized controlled trial of a 3-year home exercise program in cystic fibrosis. J Pediatr 2000;136:304–10.

[16] Hebestreit H, Kieser S, Junge S, Ballmann M, Hebestreit A, Schindler C, et al. Long-term effects of a partially supervised conditioning programme in cystic fibrosis. Eur Respir J 2010; 35:578–83.

[17] Urquhart D, Sell Z, Dhouieb E, et al. Effects of a supervised, outpatient exercise and physiotherapy programme in children with cystic fibrosis. Pediatr Pulmonol 2012;47(12):1235–41.

[18] Nixon PA, Orenstein DM, Kelsey SF, Doershuk CF. The prognostic value of exercise testing in patients with cystic fibrosis. N Engl J Med 1992;327:1785–8.

[19] Moorcroft AJ, Dodd ME, Webb AK. Exercise testing and prognosis in adult cystic fibrosis. Thorax 1997;52:291–3.

[20] Pianosi P, Leblanc J, Almudevar A. Peak oxygen uptake and mortality in children with cystic fibrosis. Thorax 2005;60:50–4.

[21] Kadikar A, Maurer J, Kesten S. The six-minute walk test: a guide to assessment for lung transplantation. J Heart Lung Transplant 1997;16:313–9.

[22] Aurora P, Wade A, Whitmore P, Whitehead B. A model for predicting life expectancy of children with cystic fibrosis. Eur Respir J December 2000;16(6):1056–60.

[23] Rüter K, Staab D, Magdorf K, Bisson S, Wahn U, Paul K. The 12-min walk test as an assessment criterion for lung transplantation in subjects with cystic fibrosis. J Cyst Fibros March 2003;2(1):8–13.

[24] Radtke T, Faro A, Wong J, Boehler A, Benden C. Exercise testing in pediatric lung transplant candidates with cystic fibrosis. Pediatr Transplant 2011;15:294–9.

[25] Tantisira KG, Systrom DM, Ginns LC. An elevated breathing reserve index at the lactate threshold is a predictor of mortality in patients with cystic fibrosis awaiting lung transplantation. Am J Respir Crit Care Med 2002;165:1629–33.

[26] American Thoracic Society; American College of Chest Physicians. ATS/ACCP statement on cardiopulmonary exercise testing. Am J Respir Crit Care Med January 15, 2003;167(2):211–77.

[27] Balady GJ, Arena R, Sietsema K, Myers J, Coke L, Fletcher GF, et al. American Heart Association Exercise, Cardiac Rehabilitation, and Prevention Committee of the Council on Clinical Cardiology; Council on Epidemiology and Prevention; Council on Peripheral Vascular Disease; Interdisciplinary Council on Quality of Care and Outcomes Research. Clinician's guide to cardiopulmonary exercise testing in adults: a scientific statement from the American Heart Association. Circulation July 13, 2010;122(2):191–225.

[28] ERS Task Force, Palange P, Ward SA, Carlsen KH, Casaburi R, Gallagher CG, Gosselink R, et al. Recommendations on the use of exercise testing in clinical practice. Eur Respir J January 2007;29(1):185–209.

[29] Paridon SM, Alpert BS, Boas SR, Cabrera ME, Caldarera LL, Daniels SR, et al. American Heart Association Council on Cardiovascular Disease in the Young, Committee on Atherosclerosis, Hypertension, and Obesity in Youth. Clinical stress testing in the pediatric age group: a statement from the American Heart Association Council on Cardiovascular Disease in the Young, Committee on Atherosclerosis, Hypertension, and Obesity in Youth. Circulation April 18, 2006;113(15):1905–20.

[30] Godfrey S, Davies CT, Wozniak E, Barnes CA. Cardio-respiratory response to exercise in normal children. Clin Sci May 1971;40(5):419–31.

[31] Orenstein DM. Assessment of exercise pulmonary function. In: Rowland TW, editor. Pediatric laboratory exercise testing. Clinical guidelines. Champaign/IL: Human Kinetics Publishers; 1993. p. 141–63.

[32] McKone EF, Barry SC, FitzGerald MX, Gallagher CG. Reproducibility of maximal exercise ergometer testing in patients with cystic fibrosis. Chest 1999;116:363–8.

[33] Jones NL. Interpretation of Stage 1 exercise test results. In: Clinical exercise testing. 4th ed. Philadelphia: WB Saunders; 1997. p. 124–49.

[34] Werkman MS, Hulzebos EH, Helders PJ, Arets BG, Takken T. Estimating peak oxygen uptake in adolescents with cystic fibrosis. Arch Dis Child January 2014;99(1):21–5.

[35] Bruce RA, Kusumi F, Hosmer D. Maximal oxygen intake and nomographic assessment of functional aerobic impairment in cardiovascular disease. Am Heart J 1973;85(4):546–62.

[36] Stuart Jr RJ, Ellestad MH. National survey of exercise stress testing facilities. Chest 1980;77(1):94–7.

[37] Chang RK, Gurvitz M, Rodriguez S, Hong E, Klitzner TS. Current practice of exercise stress testing among pediatric cardiology and pulmonology centers in the United States. Pediatr Cardiol January–February 2006;27(1):110–6.

[38] American College of Sports Medicine, Thompson WR, et al. ACSM's guidelines for exercise testing and prescription. Philadelphia: Lippincott Williams & Wilkins; 2010.

[39] van der Cammen-van Zijp MH, Ijsselstijn H, Takken T, Willemsen SP, Tibboel D, Stam HJ, et al. Exercise testing of pre-school children using the Bruce treadmill protocol: new reference values. Eur J Appl Physiol 2010;108(2):393–9.
[40] van der Cammen-van Zijp MH, van den Berg-Emons RJ, Willemsen SP, Stam HJ, Tibboel D, Usselstijn H, et al. Exercise capacity in Dutch children: new reference values for the Bruce treadmill protocol. Scand J Med Sci Sports 2010;20(1):e130–6.
[41] Pouliou E, Nanas S, Papamichalopoulos A, Kyprianou T, Perpati G, Mavrou I, et al. Prolonged oxygen kinetics during early recovery from maximal exercise in adult patients with cystic fibrosis. Chest 2001;119(4):1073–8.
[42] Robinson PD, Cooper P, Van Asperen P, Fitzgerald D, Selvadurai H. Using index of ventilation to assess response to treatment for acute pulmonary exacerbation in children with cystic fibrosis. Pediatr Pulmonol 2009;44(8):733–42.
[43] Foster C, Jackson AS, Pollock ML, Taylor MM, Hare J, Sennett SM, et al. Generalized equations for predicting functional capacity from treadmill performance. Am Heart J 1984;107:1229–34.
[44] Pollock ML, Foster C, Schmidt D, Hellman C, Linnerud AC, Ward A. Comparative analysis of physiologic responses to three different maximal graded exercise test protocols in healthy women. Am Heart J 1982;103:363–73.
[45] Cumming GR, Everatt D, et al. Bruce treadmill test in children: normal values in a clinic population. Am J Cardiol 1978;41(1):69–75.
[46] Fielding RA, Frontera WR, Hughes VA, Fisher EC, Evans WJ. The reproducibility of the Bruce protocol exercise test for the determination of aerobic capacity in older women. Med Sci Sports Exer 1997;29(8):1109–13.
[47] Wessel HU, Srasburger JF, Mitchell BM. New standards for the Bruce treadmill protocol in children and adolescents. Ped Exer Sci 2001;13:392–401.
[48] Sawyer EH, Clanton TL. Improved pulmonary function and exercise tolerance with inspiratory muscle conditioning in children with cystic fibrosis. Chest 1993;104(5):1490–7.
[49] Beriault K, Carpentier AC, Gagnon C, et al. Reproducibility of the six-minute walk test in obese adults. Int J Sports Med 2009;30:725–7.
[50] Li AM, Yin J, Yu CC, Tsang T, So HK, Wong E, et al. The six-minute walk test in healthy children: reliability and validity. Eur Respir J 2005;24:1057–60.
[51] Morinder G, Mattsson E, Sollander C, Marcus C, Larsson UE. Six-minute walk test in obese children and adolescents: reproducibility and validity. Physiother Res Int 2009;14:91–104.
[52] Enright PL, Sherrill DL. Reference equations for the six-minute walk in healthy adults. Am J Respir Crit Care Med 1998;158:1384–7.
[53] Geiger R, Strasak A, Treml B, Gasser K, Kleinsasser A, Fischer V, et al. Six-minute walk test in children and adolescents. J Pediatr 2007;150:395–9.
[54] Lammers AE, Hislop AA, Flynn Y, Haworth SG. The 6-min walk test: normal values for children of 4–11 years of age. Arch Dis Child 2008;93:464–8.
[55] Lesser DJ, Fleming MM, Maher CA, Kim SB, Woo MS, Keens TG. Does the 6-min walk test correlate with the exercise stress test in children? Pediatr Pulmonol 2010;45:135–40.
[56] Leger LA, Lambert J. A maximal multistage 20 m shuttle run test to predict VO2 max. Eur J Appl Physiol 1982;49:1–5.
[57] Singh SJ, Morgan MD, Scott S, Walters D, Hardman AE. Development of a shuttle walking test of disability in patients with chronic airways obstruction. Thorax 1992;47:1019–24.
[58] Selvadurai HC, Cooper PJ, Meyers N, et al. Validation of shuttle tests in children with cystic fibrosis. Pediatr Pulmonol 2003;35:133–8.
[59] Bradley J, Howard J, Wallace E, Elborn S. Validity of a modified shuttle test in adult cystic fibrosis. Thorax 1999;54:437–9.
[60] Elkins MR, Dentice RL, Bye PT. Validation of the MST-25: an extension of the modified shuttle test (MST). J Cyst Fibros 2009; 8(Suppl. 2):S70.
[61] Bradley J, Howard J, Wallace E, Elborn S. Reliability, repeatability and sensitivity of the modified shuttle test in adult cystic fibrosis. Chest 2000;117:1666–71.
[62] Ortega FB, Artero EG, Ruiz JR, España-Romero V, Jiménez-Pavón D, Vicente-Rodriguez G, et al. Physical fitness levels among European adolescents: the HELENA study. Br J Sports Med 2011;45:20–9.
[63] Silva G, Aires L, Mota J, Oliveira J, Ribeiro JC. Normative and criterion-related standards for shuttle run performance in youth. Pediatr Exer Sci 2012;24:157–69.
[64] Probst VS, Hernandes NA, Teixeira DC, Felcar JM, Mesquita RB, Gonçalves CG, et al. Reference values for the incremental shuttle walking test. Respir Med February 2012;106(2):243–8 (Epub 2011 Aug 23).
[65] Aurora P, Prasad SA, Balfour-Lynn IM, Slade G, Whitehead B, Dinwiddie R. Exercise tolerance in children with cystic fibrosis undergoing lung transplantation assessment. Eur Respir J 2001;18:293–7.
[66] Balfour-Lynn IM, Prasad SA, Laverty A, Whitehead BF, Dinwiddie R. A step in the right direction: assessing exercise tolerance in cystic fibrosis. Pediatr Pulmonol 1998;25:278–84.
[67] Borg E, Kaijser L. A comparison between three rating scales for perceived exertion and two different work tests. Scand J Med Sci Sports 2006;16:57–69.
[68] Mahler DA. The measurement of dyspnea during exercise in patients with lung disease. Chest 1992;101(Suppl.):242S–7S.
[69] Legrand R, Ahmaidi S, Moalla W, Chocquet D, Marles A, Prieur F, et al. O2 arterial de-saturation in endurance athletes increases muscle deoxygenation. Med Sci Sports Exerc 2005;37(5):782–8.
[70] Zavorsky GS, Quiron KB, Massarelli PS, Lands LC. The relationship between single-breath diffusion capacity of the lung for nitric oxide and carbon monoxide during various exercise intensities. Chest March 2004;125(3):1019–27.

Further Reading

[1] Hulzebos HJ, Werkman MS, van Brussel M, Takken T. Towards an individualized protocol for workload increments in cardiopulmonary exercise testing in children and adolescents with cystic fibrosis. J Cyst Fibros December 2012;11(6):550–4.
[2] Narang I, Pike S, Rosenthal M, Balfour-Lynn IM, Bush A. Three-minute step test to assess exercise capacity in children with cystic fibrosis with mild lung disease. Pediatr Pulmonol 2003;35:108–13.

CHAPTER

33

Mechanisms of Exercise Limitation in Cystic Fibrosis: A Literature Update of Involved Mechanisms

H.J. Hulzebos[1], *M.S. Werkman*[1], *B.C. Bongers*[1], *H.G.M. Arets*[2], *T. Takken*[1]

[1]**Child Development & Exercise Center, Wilhelmina Children's Hospital, University Medical Center Utrecht, Utrecht, The Netherlands;** [2]**Department of Pediatric Respiratory Medicine, Cystic Fibrosis Center, University Medical Center Utrecht, Utrecht, The Netherlands**

33.1 INTRODUCTION

Nearly two decades ago, Nixon et al. [1] reported a significant association between the exercise capacity of young Cystic Fibrosis (CF) patients and survival over 8 years [1]. Nowadays, physical training to increase or maintain exercise capacity is implemented in the usual care package offered to most patients with CF. In addition, clinicians encourage patients with CF to perform physical exercise to develop age-appropriate fitness and to maintain physical fitness in order to preserve or enhance: (1) exercise capacity, (2) muscular endurance and strength, (3) normally developed and retained bone mineral density, (4) a good posture, and (5) maintain or improve mobility of the chest wall (*International Physiotherapy Group for Cystic Fibrosis, 2009*).

An immediate aim for the young patient is to maintain a similar level of exercise in comparison with peers and friends. This is likely to influence self esteem and the type of everyday life activities. The aim of rehabilitating dysfunction is to strive to regain what has been lost (*International Physiotherapy Group for Cystic Fibrosis, 2009*).

Nevertheless, exercise capacity in patients with CF is limited, which seems to have a multifactorial cause [2,3]. If there is a possible relationship between CF genotype and some measures of exercise capacity, the mechanisms remain to be determined [3,4]. It seems that there is an interrelationship between lung function, muscle mass, energy expenditure, respiratory and/or skeletal muscle function, and exercise capacity in patients with CF [5]. The pathophysiology of reduced lung function and reduced muscle mass are known to be the most important factors leading to exercise limitation [6–8].

The exact mechanisms leading to exercise limitation in patients with CF are still a question of debate. The objective of this literature review was to give an overview of which cardiorespiratory and metabolic determinants are known to play a role in the limited exercise capacity in patients with CF. Progressive insight in the possible cardiorespiratory and metabolic limiting factors might be helpful (1) to understand the physiological mechanisms, and (2) for providing appropriate therapeutic interventions such as exercise training in patients with CF.

33.2 METHODS

33.2.1 Study Identification

We searched Medline, EMBASE, and CINAHL for studies about limiting factors in exercise capacity in patients with CF. We used no restriction in time period. The search strategy included the terms "Cystic Fibrosis" AND "Exercise Capacity" OR "exercise tolerance" OR "exercise performance" AND "limiting factor" OR limitation. The databases were searched for the terms in title, abstract or both. Titles and abstracts of search results

Diet and Exercise in Cystic Fibrosis
http://dx.doi.org/10.1016/B978-0-12-800051-9.00033-X

were screened for eligibility. The search strategy and search results are available in supplemental appendix I.

33.2.2 Study Inclusion

Studies were eligible for inclusion if they: (1) were available as full article (no posters or congress abstracts were included), and/or (2) reported effects of interventions on exercise capacity in patients with CF, and/or (3) reported associations between exercise capacity and possible limiting variables in patients with CF.

We used no design and methodological quality threshold; the language restriction was English.

33.2.3 Data Extraction

One author (MW) selected potentially eligible studies for inclusion by abstract and full articles. Reference lists from selected studies were screened for further eligible studies meeting the inclusion criteria.

33.3 RESULTS

The search retrieved 38 articles of which 18 met all the inclusion criteria. Screening the reference lists of these articles revealed an extra 49 articles meeting the inclusion criteria articles of various methodological qualities and study characteristics. No meta-analysis could be performed due to the heterogeneity in interventions, outcomes, and associations described in the included studies (Figure 33.1).

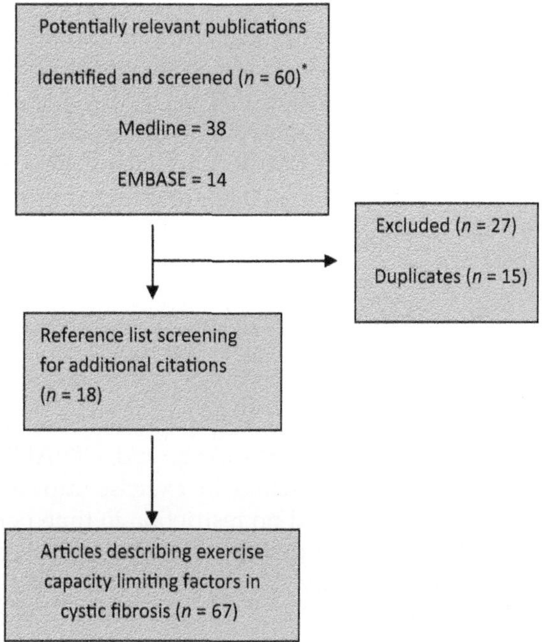

FIGURE 33.1 Result literature search.

33.4 VENTILATORY PARAMETERS

33.4.1 Ventilatory Constraints

Whether (mechanical) ventilatory constraints contribute to limited exercise capacity has traditionally been evaluated by the ventilatory reserve, which reflects the relationship of peak minute ventilation (VE_{peak}) at maximal exercise to the estimated maximal voluntary ventilation ($MVV = 35 \times FEV_1$ (L/min)) [9]. When the minute ventilation (VE) exceeds the arbitrary border of 70% of MVV, a ventilatory limited exercise capacity is suggested [9]. Furthermore, exercise dyspnea, assessed by the Borg scale, which is closely related to the level of ventilation as expressed by the VE/MVV ratio, was found to have an influence on exercise performance [10]. However, ventilatory demand and ventilatory capacity are dependent on multiple factors as lung function, anatomical dead space ventilation, respiratory muscle function, and ventilatory control [9]. Furthermore, Moorcroft et al. [11] found that VE/MVV exceeded the 70% border only in severe patients with CF ($FEV_1 < 40\%$ of predicted), suggesting that ventilatory factors only contribute to exercise limitation in severe disease state [11,12]. Furthermore, medicinal effective bronchodilation in ventilatory-limited patients with moderate CF (FEV_1 $58 \pm 17\%$ of predicted) [13] and antibiotic therapy [14] showed no effect on exercise capacity, suggesting that the subjects in the studies were not truly ventilatory limited or that the primary determinant of ventilatory limitation may not be bronchoconstriction alone [13,14]. Additionally, increasing deadpsace (VD) during exercise to volitional exhaustion in mild patients with CF (FEV_1 $76 \pm 8\%$ of predicted) did not induce changes in cardiopulmonary exercise parameters or subjective measures of exhaustion. This suggests that these mild patients still have an adequate ventilatory reserve to overcome added VD and implies that mild patients with CF are not primarily ventilatory limited [15]. Additionally, a group of patients with CF followed longitudinally had an annual decline (2.7% of predicted) in lung function (FEV_1) and exercise capacity (decrease of VO_{2peak} 0.162 mL/min kg per month) only when FEV_1 fell below 80% of predicted [16]. This indicates that ventilatory constraints play a role in limiting exercise capacity in a more progressive disease state.

Although lung function at rest, as determined by the FEV_1 or inspiratory capacity, was a longitudinal and cross-sectional determinant of exercise capacity [6,8,10,16–18], it seems that the presence of static hyperinflation (ratio between residual volume and total lung capacity (RV/TLC) > 30% after bronchodilator) in adolescents with CF by itself does not strongly influence ventilatory constraints during exercise. This could suggest that static hyperinflation is only a slightly stronger predictor of exercise capacity than the FEV_1 (% of predicted), which only reflects the degree of airflow obstruction and does not account for ventilatory mechanisms at maximal exercise as dynamic hyperinflation [19]. Thin and coworkers [20] have shown that wasted ventilation depends on a higher VD and on the ventilatory pattern during exercise, specifically, a high breathing frequency and a low tidal volume [20].

Furthermore, in contrast to other work [21], lung function independent measurements such as severity of bronchiectasis, sacculations, and abscesses are shown to be independent predictors of exercise capacity in patients with CF. Some authors have noted stronger correlations between exercise capacity and CT findings than between exercise capacity and FEV_1 or body mass index [22,23].

33.4.2 Exercise-Induced Hypoxemia

Progressive lung disease in CF, which involves thick, dehydrated, mucus-impairing airway mucociliary clearance, predisposes the patient to recurrent bronchial infections, inflammation, and airway obstruction [24,25]. As a consequence, lung disease in CF develops from bronchiolitis to bronchitis, and eventually to bronchiectasis [26]. During exercise in the presence of severely impaired pulmonary function (FEV_1 % of predicted $31.1 \pm 12.4\%$), an increased physiological VD and arterio-venous shunting results in ventilation-perfusion mismatching, contributing to the development of hypoxemia [27]. Additionally, CF patients are reported to have a reduced alveolar membrane diffusion capacity (DLCO) at rest [28,29], and also a limited, exercise-induced increase in DLCO [29]. During exercise, pulmonary blood flow increases, which is not adequately met by an increased DLCO in the study of [29], leading to a drop in O_2 saturation. The authors suggested that this limitation in increasing DLCO in the alveoli is the consequence of a reduction in alveolar ventilation during exercise.

A Cochrane review about O_2 therapy for patients with CF pointed to evidence of modest enhancement of exercise capacity and duration with O_2 supplementation, especially in participants with more advanced lung disease [30,31]. O_2 supplementation was accompanied by a lower VE_{peak} and HR_{peak}, suggesting that decreasing or preventing exercise induced hypoxemia might prevent the occurrence of cardiac or ventilatory constraints as the primary limiting factors [31]. Further, the major findings of a study by McKone et al. [32] in moderate to severe adult patients with CF indicate that stressing the respiratory system with added dead space impairs exercise capacity (exercise duration to voluntary exhaustion) with no change in VE_{peak} and the peripheral measured O_2 saturation, suggesting that exercise was limited by the ventilation reaching its maximal capacity. Supplemental O_2 with added dead space caused a small improvement in exercise capacity with an increase in VE_{peak}. These results suggest that arterial hypoxemia is a limiting factor during maximal exercise in adult patients with CF by decreasing O_2 availability to exercising muscles or inducing the sensation of dyspnea associated with arterial hypoxemia [32]. Furthermore, hypoxemia is a partial explanation for the observed slowed oxygen uptake kinetics in the skeletal muscle in patients with CF [33].

33.5 MUSCLE FUNCTION

33.5.1 Muscle Weakness

Compared with healthy controls, reduced peripheral muscle strength has been found in patients with CF [34–39], which was found to be significantly correlated with BMI and FEV_1 [34,35]. Moreover, peripheral muscle strength was even lower when corrected for fat free mass (FFM) and was found to be of contractile origin [39]. However, although the systemic inflammation in patients with CF is suggested to be related with reduced muscle force [37], it does not seem to be an independent predictor of respiratory and limb muscle strength [40].

Whether inspiratory muscle weakness is present or not in CF patients, remains controversial. Inspiratory muscle strength, as reflected by PI_{max}, is found to be relatively well preserved [41] or even higher compared with healthy peers [38] in stable adult patients with CF, although there is a relationship (r .370; $p<0.05$) between the loss of inspiratory muscle work capacity and FFM. On the other hand, other studies found lower PI_{max} values in patients with CF [42,43]. Furthermore, loss of FFM [41] and hyperinflation [44] are associated with the loss of diaphragm muscle strength. The higher PI_{max} values were ascribed to a conditioning effect on the inspiratory muscles, as WOB is increased in patients with CF [38]. In addition, inspiratory muscle endurance may be reduced in CF patients and is strongly related to exercise dyspnea. However, inspiratory muscle endurance limitation was independent of nutritional status, ventilatory obstructive defect, pulmonary hyperinflation, inspiratory muscle strength, or maximal exercise capacity [45].

Hence, as it is still questionable whether inspiratory muscle training can improve exercise capacity [46], the role of inspiratory muscle weakness alone as a limiting factor in exercise capacity in patients with CF remains unclear. On the contrary, unloading the inspiratory muscles by overnight, noninvasive ventilation in hypercapnic patients with CF showed improvement in exercise capacity compared with placebo-controls, suggesting that a nocturnal reduction in the WOB might lead to improved exercise capacity during the day [47].

This finding might still indicate a role of the inspiratory muscles in limiting exercise capacity in patients with CF.

33.5.2 CF Specific Mitochondrial Dysfunction

Compared with healthy controls, study findings indicate that the efficiency of oxidative work performance of skeletal muscle in patients with CF is reduced by 19–25% [48]. A decrease in mitochondrial function secondary to clinical or nutritional factors may be the explanation for this finding [48]. However, until 2010, the CFTR had not been shown to be expressed in human skeletal muscles. Recently, the expression of CFTR has been demonstrated in human skeletal muscle, and its localization in the sarcotubular network [49,50]. Additionally, using ^{31}Phosphorus magnetic resonance spectroscopy, recent literature showed a slower phosphocreatine (PCr) recovery time after 90 s of intense exercise in patients with CF and in patients with primary cilliary dyskinesia [51]. This points towards a possible intrinsic abnormality in mitochondrial oxidative metabolism; however, currently there is no firm evidence available [33,48,52–54]. Furthermore, as similar exercise physiology has been found between patients with CF and non-CF bronchiectasis [54], it is still questionable if the possible mitochondrial impairment may be CF specific or suggestive of a nonspecific effect of chronic systemic inflammation, as present in patients with CF and primary cilliary dyskinesia [51]. In conclusion, it remains unclear whether an intrinsic abnormality in muscle energy metabolism is present in CF [3,54], or whether the exercise physiology of CF skeletal muscles is hampered due to impaired O_2 delivery to these muscles [34,52].

33.5.2.1 Oxidative versus Glycolytic Energy Metabolism

Boas et al. [55] investigated aerobic and anaerobic exercise capacity in children with CF ($n=25$, FEV_1 92.5±17.1% of predicted), children with asthma ($n=22$, FEV_1 100.3±17.7% of predicted), and healthy controls ($n=23$, FEV_1 110.1±8.3% of predicted). They found a similar aerobic and anaerobic exercise capacity among the three groups; however, children with CF used a lower percentage of their VO_{2peak} during each phase of anaerobic exercise testing. They applied mathematical modeling on the exercise data in order to clarify this result, which, compared to children with asthma and healthy controls, suggests the preferential use of the phosphocreatine/adenosine triphosphate (PCr/ATP) and glycolytic energy systems compared with oxidative pathways. In addition, Klijn et al. [6,8] reported a higher anaerobic power output normalized for FFM in patients with CF and moderate lung disease ($n=19$, FEV_1 62.9±14.2% of predicted) than in patients with CF and mild lung disease ($n=20$, FEV_1 99.2±10.6% of predicted). Their results indicate that with progressive lung disease, there is a shift from oxidative to glycolytic energy metabolism during exercise. In children with asthma, it was suggested that reduced aerobic capacity might be compensated for by a maintained or even enhanced anaerobic capacity [56], leading to enhanced CO_2 production during exercise. This phenomenon could explain the higher respiratory exchange ratios (VCO_2/VO_2) at rest and during submaximal exercise in patients with CF [3,48,52].

33.5.3 Nutritional Status

CF has detrimental effects on the patient's nutritional status and thereby induces malnutrition. Malnutrition can lead to the loss off body fat and FFM with muscle mass as the main part of it [57]. For instance, diaphragmatic performance declines as nutritional status, evaluated on the basis of BMI, decreases [44]. Indeed, peak anaerobic capacity [7,58], maximum work capacity [59,60] and, to a lesser extend, aerobic capacity [6,8,59] are found to be related with FFM. Additionally, low FFM is associated with the observed lower peak heart rate [57] and stroke volume (SV) [61] at maximal exercise in patients with CF, which is explained by the lower muscle mass performing the work leading to a decreased cardiovascular load [42] and output [61].

33.6 CARDIAC CONSTRAINTS

More than 20 years ago, a decreased SV was found in malnourished patients with CF [61], which might have been caused by the occurrence of both right and left ventricular dysfunction during stress, without clinical signs or symptoms [62]. However, this impaired SV was indirectly measured using the Fick equation, and thus could not provide distinction in involvement of right ventricular (RV) or left ventricular (LV) dysfunction [61].

A postmortem study showed evidence of RV hypertrophy of 70% in children with CF [63]. Florea et al. [64] and Ionescu et al. [65] confirmed the presence of significant RV systolic and diastolic dysfunction in clinical-stable and nonclinical-stable patients with CF. This RV dysfunction may be caused by pulmonary hypertension, secondary to chronic hypoxemia [66,67], by the chronic inflammation as present in patients with CF [65] or by ventilatory mechanics as airflow limitation, leading to increased intrathoracic pressure [66]. Contrary to what was previously suggested, a decreased SV during exercise could not be ascribed to hyperinflation of the thorax [54]. The increase in the intrathoracic pressure caused by the thoracic hyperinflation would limit the extent to which the Frank-Starling mechanism could be recruited to maintain SV in the face of an increased RV afterload

[66]. This lower SV could not be compensated for by an increase in heart rate, resulting in impaired blood flow to the lungs [54].

Overall, secondary RV enlargement develops in a proportion of patients with CF via pulmonary hypertension and pulmonary vascular remodeling. The exact prevalence of subclinical RV dysfunction in the population is unknown, but prognosis is poor once RV failure is evident [68]. Although involvement of LV dysfunction in patients with CF remains subject to debate [64,69,70], there is evidence for involvement of LV dysfunction [68]. This LV dysfunction could be caused be a regional LV myocardial perfusion deficit due to hemodynamic changes secondary to pulmonary hypertension and RV hypertrophy. A dysfunctional LV seems not to be of major importance in mild disease state, but, next to other previously mentioned factors, may become clinical evident in more progressive disease state in patients with CF, where a LV dysfunction could limit cardiac output during exercise [71].

33.7 CONCLUDING REMARKS

In patients with mild-to-moderate disease, nonpulmonary factors, as muscle mass and muscle function, predominate in limiting exercise capacity [11,12]. In more severe patients with CF ($FEV_1 < 40\%$ of predicted), ventilatory (mechanical) constraints and hypoxemia become more important determinants. However, in any state of progression of CF, none of these factors are the main limiting factor, suggesting that other factors, such as a possible CF specific muscle defect, and/or systematic inflammation, independent of the severity or progression of CF, are attributing to exercise limitation. This might result in specific, unique individual combinations of factors that limit exercise capacity in separate patients with CF.

33.8 CLINICAL IMPLICATIONS

The unique, individual combinations of factors that limit exercise capacity in patients with CF would implicate that exercise training in each patient should focus on different goals and indications. These indications are not only interindividual dependent, but are also dependent on disease progression and exercise induced limitations. These distinctive, interindividual characteristics require detailed cardiopulmonary exercise testing prior to the initiation of exercise training in order to provide the patients with CF with safe training recommendations [72].

In a less severe, mild to moderate disease state of CF, when nonpulmonary factors predominate in limiting exercise capacity, the focus of exercise training should emphasize prevention of the deterioration of lung function by focusing on optimizing chest mobility and airway clearance techniques [73]. Furthermore, general training programs should focus on peripheral muscle function according to the general ACSM Guidelines for exercise testing and prescription. (*ACSM Guidelines for exercise testing and prescription.*)

When ventilatory limitations become predominant, besides optimizing chest mobility and airway clearance techniques, the focus should be decreasing WOB by inspiratory muscle training [74]. Additionally, local peripheral muscle oxidative capacity could be stabilized or even improved by more intermittent, local, peripheral muscle training as high-intensity interval training (HIT), with less burden on the ventilatory system [75,76].

As stated in the Cochrane review about exercise training for cystic fibrosis, the benefits obtained from physical training may be influenced by the type of training. Further research is needed to understand the (physiological) benefits of exercise programs in people with cystic fibrosis, and the relative benefits of the addition of aerobic versus anaerobic versus a combination of both types of physical training to the care of people with cystic fibrosis [77]. Overall, there should be no such form of "one size fits all principle" in patients with CF, and tailored care should be the policy in the domain of exercise training.

Acknowledgments

This study was funded by an unconditional research grant (DO-IT) from the Committee on Physiotherapy Research of the Royal Dutch Society for Physiotherapy (Wetenschappelijk College Fysiotherapie, Koninklijk Nederlands Genootschap voor Fysiotherapie (KNGF)).

References

[1] Nixon PA, Orenstein DM, Kelsey SF, Doershuk CF. The prognostic value of exercise testing in patients with cystic fibrosis. N Eng J Med 1992;327(25):1785–8.

[2] Ferrazza AM, Martolini D, Valli G, Palange P. Cardiopulmonary exercise testing in the functional and prognostic evaluation of patients with pulmonary diseases. Respiration 2009;77(1):3–17.

[3] Selvadurai HC, McKay KO, Blimkie CJ, Cooper PJ, Mellis CM, Van Asperen PP. The relationship between genotype and exercise tolerance in children with cystic fibrosis. Am J Respir Crit Care Med 2002;165:762–5.

[4] Kaplan TA, Moccia-Loos G, Rabin M, McKey RM. Lack of effect of ΔF508 mutation on aerobic capacity in patients with cystic fibrosis. Clin J Sport Med 1996;6(4):226–31.

[5] Schöni MH, Casaulta-Aebischer C. Nutrition and lung function in cystic fibrosis patients: review. Clin Nutr 2000;19(2):79–85.

[6] Klijn PH, Terheggen-Lagro SW, Van der Ent CK, Van der Net J, Kimpen JL, Helders PJM. Anaerobic exercise in pediatric cystic fibrosis. Pediatr Pulmonol 2003;36:223–9.

[7] Shah AR, Gozal D, Keens TG. Determinants of aerobic and anaerobic exercise performance in cystic fibrosis. Am J Respir Crit Care Med 1998;157:1145–50.

[8] Klijn PHC, Net van der J, Kimpen JL, Helders PJM, Ent van der CK. Longitudinal determinants of peak aerobic performance in children with cystic fibrosis. Chest 2003;124(6):2215–9.

[9] American Thoracic Society/American College of Chest Physicians (ATS/ACCP). ATS/ACCP statement on cardiopulmonary exercise testing. Am J Respir Crit Care Med 2003;167(2):211–77.

[10] De Jong W, Van der Schans CP, Mannes GPM, Van Aalderen WMC, Grevink RG, Koëter GH. Relationship between dyspnoea, pulmonary function and exercise capacity in patients with cystic fibrosis. Respir Med 1997;91:41–6.

[11] Moorcroft AJ, Dodd ME, Morris J, Webb AK. Symptoms, lactate and exercise limitation at peak cycle ergometry in adults with cystic fibrosis. Eur Respir J 2005;25:1050–6.

[12] Regnis JA, Donnelly PM, Robinson M, Alison JA, Bye PTP. Ventilatory mechanics at rest and during exercise in patients with cystic fibrosis. Am J Respir Crit Care Med 1996;154:1418–25.

[13] Serisier DJ, Coates AD, Bowler SD. Effect of albuterol on maximal exercise capacity in cystic fibrosis. Chest 2007;131:1181–7.

[14] Cox NS, McKay KO, Follett JM, Alison JA. Home IV antibiotic therapy and exercise capacity in children with CF: a case series. Cardiopulm Phys Ther 2011;22(1):16–9.

[15] Dodd JD, Barry SC, Gallagher CG. Respiratory factors do not limit maximal symptom-limited exercise in patients with mild cystic fibrosis lung disease. Respir Physiol Neurobiol 2006;152:176–85.

[16] Pianosi P, LeBlanc J, Almudevar A. Relationship between FEV_1 and peak oxygen uptake in children with cystic fibrosis. Pediatr Pulmonol 2005;40:324–9.

[17] Ziegler B, Rovedder PM, Lukrafka JL, Oliveira CL, Menna-Barreto SS, Dalcin P de T. Submaximal exercise capacity in adolescent and adult patients with CF. J Bras Pneumol 2007;33(3):263–9.

[18] Perpati G, Nanas S, Pouliou E, Dionyssopoulou V, Stefanatou E, Armeniakou E, et al. Resting respiratory variables and exercise capacity in adult patients with cystic fibrosis. Respir Med 2010;104:1444–9.

[19] Werkman MS, Hulzebos HJ, Arets HGM, Van der Net J, Helders PJM, Takken T. Is static hyperinflation a limiting factor during exercise in adolescents with cystic fibrosis? Pediatr Pulmonol 2011;46(2):119–24.

[20] Thin AG, Dodd JD, Gallagher CG, Fitzgerald MX, Mcloughlin P. Effect of respiratory rate on airway deadspace ventilation during exercise in cystic fibrosis. Respir Med 2004;98:1063–70.

[21] Edwards EA, Narang I, Hansell DM, Rosenthal M, Bush A. HRCT lung abnormalities are not a surrogate for exercise limitation in bronchiectasis. Eur Respir J 2004;24:538–44.

[22] Coates AL, Boyce P, Shaw DG, Godfrey S, Mearns M. Relationship between the chest radiograph, regional lung function studies, exercise tolerance, and clinical condition in cystic fibrosis. Arch Dis Child 1981;56:106–11.

[23] Dodd JD, Barry SC, Barry RBM, Gallagher CG, Skehan SJ, Masterson JB. Thin-section CT in patients with cystic fibrosis: correlation with peak exercise capacity and body mass index. Radiology 2006;240(1):236–45.

[24] Elizur A, Cannon CL, Ferkol TW. Airway inflammation in cystic fibrosis. Chest 2008;133:489–95.

[25] Wheatley CM, Wilkins BW, Snyder EM. Exercise is medicine in cystic fibrosis. Exerc Sport Sci Rev 2011;39(3):155–60.

[26] Gershman AJ, Mehta AC, Infeld M, Budev MM. Cystic fibrosis in adults: an overview for the internist. Clev Clin J Med 2006;73(12):1065–74.

[27] Coffey MJ, FitzGerald MX, McNicholas WT. Comparison of oxygen desaturation during sleep and exercise in patients with cystic fibrosis. Chest 1991;100:659–62.

[28] Godfrey S, Mearns M. Pulmonary function and response to exercise in cystic fibrosis. Arch Dis Child 1971;46:144–51.

[29] Wheatley CM, Foxx-Lupo WT, Cassuto NA, Wong EC, Daines CL, Morgan WJ, et al. Impaired lung diffusing capacity for nitric oxide and alveolar-capillary membrane conductance results in oxygen desaturation during exercise in patients with cystic fibrosis. J Cyst Fibros 2011;10:45–53.

[30] Elphick HE, Mallory G. Oxygen therapy for cystic fibrosis. Cochrane Database Syst Rev 2009;1.

[31] Marcus CL, Bader D, Stabile MW, Wang CI, Osher AB, Keens TG. Supplemental oxygen and exercise performance in patients with cystic fibrosis with severe pulmonary disease. Chest 1992;101: 52–7.

[32] McKone EF, Barry SC, Fitzgerald MX, Gallagher CG. Role of arterial hypoxemia and pulmonary mechanics in exercise limitation in adults with cystic fibrosis. J Appl Physiol 2005;99:1012–8.

[33] Hebestreit H, Hebestreit A, Trusen A, Hughson RL. Oxygen uptake kinetics are slowed in cystic fibrosis. Med Sci Sports Exerc 2005;37:10–7.

[34] Meer de K, Gulmans VAM, Laag van der J. Peripheral muscle weakness and exercise capacity in children with cystic fibrosis. Am J Respir Crit Care Med 1999;159(3):748–54.

[35] Hussey J, Gormley J, Leen G, Greally P. Peripheral muscle strength in young males with cystic fibrosis. J Cyst Fibros 2002;1(3):116–21.

[36] Sahlberg ME, Svantesson U, Magnusson Thomas EML, Strandvik B. Muscular strength and function in patients with cystic fibrosis. Chest 2005;127(5):1587–92.

[37] Troosters T, Langer D, Vrijsen B, Segers J, Wouters K, Janssens W, et al. Skeletal muscle weakness, exercise tolerance and physical activity in adults with cystic fibrosis. Eur Respir J 2009;33:99–106.

[38] Dunnink MA, Doeleman WR, Trappenburg JCA, de Vries WR. Respiratory muscle strength in stable adolescent patients with cystic fibrosis. J Cyst Fibros 2009;8:31–6.

[39] Vallier JM, Gruet M, Mely L, Pensini M, Brisswalter J. Neuromuscular fatigue after maximal exercise in patients with cystic fibrosis. J Electromyogr Kines 2011;21(2):242–8.

[40] Dufresne V, Knoop C, Van Muylem A, Malfroot A, Lamotte M, Opdekamp C, et al. Effect of systemic inflammation on inspiratory and limb muscle strength and bulk in cystic fibrosis. Am J Respir Crit Care Med 2009;180:153–8.

[41] Enright S, Chatham K, Ionescu AA, Unnithan VB, Shale DJ. The influence of body composition on respiratory muscle, lung function and diaphragm thickness in adults with cystic fibrosis. J Cyst Fibros 2007;6:384–90.

[42] Lands LC, Heigenhauser GJ, Jones NL. Analysis of factors limiting maximal exercise performance in cystic fibrosis. Clin Sci 1992;83(4):391–7.

[43] Keochkerian D, Chliff M, Delanaud S, Gauthier R, Maingourd Y, Ahmaidi S. Timing and driving components of the breathing strategy in children with cystic fibrosis during exercise. Pediatr Pulmonol 2005;40:449–56.

[44] Hart N, Tounian P, Clément A, Boulé M, Polkey MI, Lofaso F, et al. Nutritional status is an important predictor of diaphragm strength in young patients with cystic fibrosis. Am J Clin Nutr 2004;80:1201–6.

[45] Leroy S, Perez T, Neviere R, Aguilaniou B, Wallaert B. Determinants of dyspnea and alveolar hypoventilation during exercise in cystic fibrosis: impact of inspiratory muscle endurance. J Cyst Fibros 2011;10(3):159–65.

[46] Reid WD, Geddes EL, O'Brien K, Brooks D, Crowe J. Effects of inspiratory muscle training in cystic fibrosis: a systematic review. Clin Rehabil 2008;22:1003–13.

[47] Young AC, Wilson JW, Kotsimbos TC, Naughton MT. Randomised placebo-controlled trial of non-invasive ventilation for hypercapnia in cystic fibrosis. Thorax 2008;63:72–7.

[48] Meer de K, Jeneson JAL, Gulmans VAM, Laag van der J, Berger R. Efficiency of oxidative work performance of skeletal muscle in patients with cystic fibrosis. Thorax 1995;50(9):980–3.

[49] Lamhonwah AM, Bear CE, Huan LJ, Chiaw PK, Ackerley CA, Tein I. Cystic fibrosis transmembrane conductance regulator in human muscle dysfunction causes abnormal metabolic recovery in exercise. Ann Neurol 2010;67:802–8.

[50] Divangahi M, Balghi H, Danialou G, Comtois AS, Demoule A, Ernest S, et al. Lack of CFTR in skeletal muscle predisposes to muscle wasting and diaphragm muscle pump failure in cystic fibrosis mice. PLoS Genet 2009;5(7):1–14.

[51] Wells GD, Wilkes DL, Schneiderman JE, Rayner T, Elmi M, Selvadurai H, et al. Skeletal muscle metabolism in cystic fibrosis and primary ciliary dyskinesia. Pediatr Res 2011;69(1):40–5.

[52] Moser C, Tirakitsoontorn P, Nussbaum E, Newcomb R, Cooper DM. Muscle size and cardiorespiratory response to exercise in cystic fibrosis. Am J Respir Crit Care Med 2000;162(5):1823–7.

[53] Hjeltnes N, Stanghelle JK, Skyberg D. Pulmonary function and oxygen uptake during exercise in 16 year old boys with cystic fibrosis. Acta Paediatr Scand 1984;73(4):548–53.

[54] Rosenthal M, Narang I, Edwards L, Bush A. Non-invasive assessment of exercise performance in children with cystic fibrosis (CF) and non-cystic fibrosis bronchiectasis: is there a CF specific muscle defect? Pediatr Pulmonol 2009;44(3):222–30.

[55] Boas SR, Danduran MJ, McColley SA. Energy metabolism during anaerobic exercise in children with cystic fibrosis and asthma. Med Sci Sports Exerc 1999;31(9):1242–9.

[56] Varray A, Mercier J, Ramonatxo M, Préfaut C. L'exercise physique maximal chez l'enfant astmatique: limitation aérobie et compensation anaérobie? Sci Sport 1989;4:199–207.

[57] Lands LC, Heigenhauser GJ, Jones NL. Cardiac output determination during progressive exercise in cystic fibrosis. Chest 1992;102:1118–23.

[58] Boas SR, Joswiak ML, Nixon PA, Fulton JA, Orenstein DM. Factors limiting anaerobic performance in adolescent males with cystic fibrosis. Med Sci Sports Exerc 1996;28(3):291–8.

[59] Hütler M, Schnabel D, Staab D, Tacke A, Wahn U, Böning D, et al. Effect of growth hormone on exercise tolerance in children with cystic fibrosis. Med Sci Sports Exerc 2002;34(4):567–72.

[60] Gulmans V, de Meer K, Brackel HJL, Helders PJM. Maximal work capacity in relation to nutritional status in children with cystic fibrosis. Eur Respir J 1997;10:2014–7.

[61] Marcotte JE, Canny GJ, Grisdale R, Desmond K, Corey M, Zinman R, et al. Effects of nutritional status on exercise performance in advanced cystic fibrosis. Chest 1986;90:375–9.

[62] Benson LN, Newth CJL, Desouza M, Lobraico R, Kartodihardjo W, Corkey C, et al. Radionuclide assessment of right and left ventricular function during bicycle exercise in young patients with cystic fibrosis. Am Rev Respir Dis 1984;130:987–92.

[63] Royce SW. Cor pulmonale in infancy and early childhood: report on 34 patients, with special reference to the occurrence of pulmonary heart disease in cystic fibrosis of the pancreas. Pediatrics 1951;8:255–74.

[64] Florea VG, Florea ND, Sharma R, Coats AJS, Gibson DG, Hodson ME, et al. Right ventricular dysfunction in adult severe cystic fibrosis. Chest 2000;118:1063–8.

[65] Ionescu AA, Ionescu A, Payne N, Obieta-Fresnedo I, Fraser AG, Shale DJ. Subclinical right ventricular dysfunction in cystic fibrosis: a study using tissue Doppler echocardiography. Am J Respir Crit Care Med 2001;163:1212–8.

[66] Hortop J, Desmond KJ, Coates AL. The mechanical effects of expiratory airflow limitation on cardiac performance in cystic fibrosis. Am Rev Respir Dis 1988;137:132–7.

[67] Coates AL, Desmond K, Asher MI, Hortop J, Beaudry PH. The effect of digoxin on exercise capacity and exercising cardiac function in cystic fibrosis. Chest 1982;82:543–7.

[68] Bright-Thomas RJ, Webb AK. The heart in cystic fibrosis. J R Soc Med 2002;95(Suppl. 41):2–10.

[69] Koelling TM, Dec GW, Ginns LC, Semigran MJ. Left ventricular diastolic function in patients with advanced cystic fibrosis. Chest 2003;123:1488–94.

[70] Ionescu AA, Chatham K, Davies CA, Nixon LS, Enright S, Shale DJ. Inspiratory muscle function and body composition in cystic fibrosis. Am J Respir Crit Care Med 1998;158(4):1271–6.

[71] De Wolf D, Franken P, Piepsz A, Dab I. Left ventricular perfusion deficit in patients with cystic fibrosis. Pediatr Pulmonol 1998;25: 93–8.

[72] Radtke T, Stevens D, Benden C, Williams CA. Clinical exercise testing in children and adolescents with cystic fibrosis. Pediatr Phys Ther 2009;21(3):275–81.

[73] Robinson KA, McKoy N, Saldanha I, Odelola OA. Active cycle of breathing technique for cystic fibrosis. Cochrane Database Syst Rev 2010;11:CD007862.

[74] Houston BW, Mills N, Solis-Moya A. Inspiratory muscle training for cystic fibrosis. Cochrane Database Syst Rev 2008;4: CD006112.

[75] Williams CA, Benden C, Stevens D, Radtke T. Exercise training in children and adolescents with cystic fibrosis: theory into practice. Int J Pediatr 2010;2010:ID 670640.

[76] Hulzebos HJ, Snieder H, Van der Net J, Helders PJ, Takken T. High-intensity interval training in an adolescent with cystic fibrosis: a physiological perspective. Physiother Theory Pract 2011;27(3):231–7.

[77] Bradley J, Moran F. Physical training for cystic fibrosis. Cochrane Database Syst Rev 2008;23(1):CD002768.

Further Reading

Massin MM, Leclercq-Foucart J, Sacre J-P. Gas exchange and heart rate kinetics during binary sequence exercise in cystic fibrosis. Med Sci Monit 2000;6:55–62.

Kusenbach G, Wieching R, Barker M, Hoffmann U, Essfeld D. Effects of hyperoxia on oxygen uptake kinetics in cystic fibrosis patients as determined by pseudo-random binary sequence exercise. Eur J Appl Physiol 1999;79:192–6.

Webb AK, Dodd ME, Moorcroft J. Exercise and cystic fibrosis. J R Soc Med 1995;88(Suppl. 25):30–6.

Coates AL, Canny G, Zinman R, Grisdale R, Desmond K, Roumeliotis D, et al. The effects of chronic airflow limitation, increased dead space, and the pattern of ventilation on gas exchange during maximal exercise in advanced cystic fibrosis. Am Rev Respir Dis 1988;138: 1524–31.

Dempsey JA, Amann M, Romer LM, Miller JD. Respiratory system determinants of peripheral fatigue and endurance performance. Med Sci Sports Exerc 2008;40:457–61.

CHAPTER

34

Physical Activity Assessment and Impact

Nancy Alarie[1], *Lisa Kent*[2, 3]

[1]Physiotherapy Department, Montreal Children's Hospital, McGill University Health Centre, Montreal, QC, Canada; [2]Centre for Health and Rehabilitation Technology (CHaRT), Institute of Nursing and Health Research, University of Ulster, Northern Ireland, UK; [3]Department of Respiratory Medicine, Belfast Health and Social Care Trust, Belfast, Northern Ireland, UK

34.1 INTRODUCTION

The promotion of physical activity is now a global priority for the general population. According to the World Health Organization (WHO) Global Recommendations on Physical Activity for Health [1], improving levels of physical activity is thought to improve cardiorespiratory and muscular fitness, bone health, and cardiovascular and metabolic health biomarkers and reduce symptoms of anxiety and depression. These outcomes are of importance in people with cystic fibrosis (CF) as they are in the general population. Indeed, aerobic fitness, which is linked to physical activity levels, has been shown to be a predictor of lung function and survival in patients with CF [2,3]. The WHO Global Recommendations on Physical Activity for Health provide recommendations for children aged 5–17 years, adults 18–64 years and adults 65 years old and above. However, they do not provide recommendations for children younger than age 5 years. Regular physical activity should be encouraged in all our patients from as early in life as possible. This may mean targeting the family and not just the individual with CF, particularly in our pediatric patients. There is a strong argument that healthy physical activity behaviors adopted in childhood have a positive effect on health in adulthood; "the child is the father of the man."

For the purposes of this chapter, physical activity is defined as "any bodily movement produced by skeletal muscles that require energy expenditure" [1].

People with CF are living longer and with this many are participating in life to a greater extent than before; for example, they are taking on challenging careers and becoming parents. That's not to say that the WHO guidelines for physical activity can be applied directly to CF. In CF, we see a wide range in severities of disease and physical capacities. Some research shows people with CF spend less time than their healthy peers in moderate to vigorous intensity activity and overall physical activity declines with age, similar to healthy populations [4,5]. Physical deconditioning from reduced activity and lung disease leads to lactic acidosis at relatively low workloads, which increases ventilatory drive and the sensation of shortness of breath [6]. Physical activity advice for individuals needs to be tailored, taking into consideration age, lifestyle, stage of lung disease, nutritional status, burden of therapy, physical capacity, and interests.

In this chapter, we will discuss the following:

1. Tools for measuring physical activity in CF
2. Normative data for physical activity in CF
3. Methods of analyzing and reporting physical activity data
4. Evidence for impact of physical activity on health of patients with CF

34.2 TOOLS FOR MEASURING PHYSICAL ACTIVITY

Physical activity can be thought of as a complex "construct" of behaviors and physiological responses, both of which can be measured in many ways. The behaviors can be directly measured by video analysis or direct observation that, although is the most direct way, has many obvious practical disadvantages. Surrogates for measuring physical activity behaviors include objective measurement using motion sensors and subjective measurement of patient perception of their physical activity behaviors. The advantages and disadvantages to both are summarized in Table 34.1.

We can also measure the determinants of physical activity behavior. For example self-efficacy, decisional balance (i.e., how a person weighs up the pros and

Diet and Exercise in Cystic Fibrosis
http://dx.doi.org/10.1016/B978-0-12-800051-9.00034-1

TABLE 34.1 Physical Activity Assessment Tools

Advantages	Disadvantages
OBJECTIVE METHODS (ACCELEROMETERS, ACTIVITY MONITORS, PEDOMETERS)	
Can be measured in a free-living environment	Not a direct measurement of physical activity behavior
Ability to combine with physiological measurements such as heart rate (NB., however, this is influenced by fitness)	Cannot capture all physical activity • e.g., if placed on the lower limb, it does not capture upper limb movements • e.g., water-based activity such as swimming • e.g., flexibility/stretching exercise
Ability to calibrate to the individual	
Sedentary timer can be accurately quantified (with the exception of pedometers)	Continuing evolution of devices and the number of devices on the market make comparisons difficult
Pedometers are inexpensive and can be used as motivational feedback for patients	Energy expenditure measurement is problematic in CF because the algorithms are developed in healthy populations
More complex activity monitors provide complex information on body position, quantity and intensity of activity, temporal patterns	Cannot differentiate between types of activities • e.g., weight-bearing versus non–weight-bearing • e.g., movements against resistance such as lifting free weights
	The more complex devices can be expensive and require expertise for analysis and data interpretation
SUBJECTIVE METHODS (QUESTIONNAIRES, DIARIES)	
Inexpensive	Not a direct measurement of physical activity behavior
Easy to administer	
Questionnaires and diaries can be used to generate discussion around physical activity	A level of understanding is required of the individual
Diaries can be used as motivational tools	Questionnaires rely on recall
Diaries can be used as aids to interpret activity monitor data (e.g., identify non-wear time and additional activities such as swimming)	Diaries can be burdensome
	Diaries may have a reactivity effect

cons of physical activity) and processes of change (i.e., the strategies used by individuals to progress through different stages of change). For example, Marcus et al. provide questionnaires for measuring these constructs that, although not tested specifically in CF, have been shown to have acceptable internal consistency and reliability in other populations [7–9]. In addition to these, the CF clinician may be interested in monitoring the patient's perceived symptoms during physical activity (e.g., shortness of breath). It should be noted that exercise tests do not measure physical activity behavior but have their use in assessing the consequences of physical activity behaviors. In theory, healthier physical activity patterns will have a beneficial effect on aerobic capacity and muscular strength and endurance.

34.2.1 Tools Evaluated in Cystic Fibrosis

A systematic review was conducted by the European CF Society Working Group for Exercise that focused on evidence of clinimetric properties of physical activity assessment tools. For definitions of clinimetric properties and their relevance to physical activity assessment, the reader is also referred to Bradley [10]. The Working Group's systematic review found more information for adults with CF than children or adolescents. Information presented in this chapter has been facilitated by the work of the Working Group.

34.2.2 Assessing Physical Activity Levels Using Physical Activity Questionnaires and Diaries

Regular physical activity and exercise are known to contribute positively to the health and well-being of CF patients. They can contribute to a slower rate of decline in lung function, lead to improved nutritional status, and improved quality of life [11–13]. High aerobic fitness is associated with improved survival [14]. There are several methods available to assess exercise tolerance (cardiopulmonary exercise testing, 6-Minute Walk Test, and shuttle tests), but what about the assessment of a patient's level of physical activity? Supervised exercise interventions are expensive and not easily implemented in regular patient care. Furthermore, long-term adherence to supervised programs is often low [14,15].

Strategies aimed at monitoring and increasing habitual physical activities may be more effective in maintaining the health and quality of life of CF patients [16]. It follows that access to reliable and valid tools that measure activity levels in the CF population is essential. Despite the recognized importance of physical activity, the measurement of the amount, intensity, and duration of physical activity is not routinely measured in the clinical setting [17]. This may be partially explained by the fact that there are very few validated tools that can reliably, easily, and quickly assess levels of physical activity in the busy clinic setting [17]. Accelerometry and motion sensors can provide a more objective measure of levels of physical activity; however, they are labor-intensive, expensive, and time-consuming. An accurate measurement of physical activity levels is fundamental to the treatment of and the clinical monitoring of our interventions.

Subjective measures of physical activity such as questionnaires and diaries are typically inexpensive, easy to administer, and quick to perform. They can be administered via interviewer format, self-report, or proxy report, and in some instances be performed over the phone. Completion rates are high when the questionnaires are administered during a clinic visit. There are numerous questionnaires that claim to measure physical activity levels. The Habitual Activity Estimation Scale (HAES) [18], the Seven Day Physical Activity Recall (7D-PAR) [19a], and the International Physical Activity Questionnaire (IPAQ) [20] have been used in healthy children and adults [18–20] and have also been used in patients with CF by Schneiderman-Walker, Wells, and Ruf [12,16,17]. In patients with CF, a validation against more objective measures has been performed for the HAES by Wells [16].

34.2.2.1 Habitual Activity Estimation Scale

The Habitual Activity Estimation Scale (HAES) was originally created to measure the habitual activity of children with chronic diseases in a clinical setting [18]. It was designed to distinguish between weight-bearing and non–weight-bearing activities. The HAES has been used in leukemia, epilepsy, type 1 diabetes, and CF. It can be administered by interview for younger children and interviewer administered or supervised for adolescents and adults. The HAES questionnaire is a retrospective two-week recall of one typical weekday (Tuesday, Wednesday, or Thursday) and one typical Saturday. The percentage of time spent in each of the four time periods is reported: waking up to breakfast, breakfast to lunch, lunch to supper, and supper to bedtime. Four levels of activity are reported for each time period: inactive, somewhat inactive, somewhat active, and very active. The use of wakeup, bed, and meal times is used to calculate the total number of hours per day spent in each of the activity categories. Total activity is reported as the sum of somewhat active and very active categories for each of the days as reported by Schneiderman-Walker [12]. The HAES combines easily recognizable segments of time, distinct activity categories, and allows for the estimation of intensity and duration of physical activity from postinfancy to adulthood. Results can be expressed in hours per day spent in each activity category or in percentage of awake time spent in each activity category for one weekday and one Saturday.

Wells and colleagues in 2008 were the first group to report on the reliability and validity of the HAES in a group of CF patients [16]. They compared habitual physical activity scores form the HAES questionnaire with scores from an accelerometer activity monitor and an activity diary in adolescent and adults with CF. The participants were studies over two consecutive weeks. The results from the first week were compared with the results of the second week for each of the instruments to establish reliability and results were compared between instruments to establish validity. Interclass coefficient analysis suggested significant reliability from week 1 and week 2 for the HAES questionnaire, three-day activity diary, and the accelerometer. Reliability was also assessed for the time of day, weekday and weekend periods, and intensity levels. Significant reliability estimates were obtained for all three instruments in all categories except for accelerometer activity counts in the somewhat active category. They reported significant positive correlations between activity levels estimated by the HAES and diary and HAES and accelerometer. Significant correlations were also demonstrated between activity measures when assessed for the time of day, weekday, and weekend days. The significance held true for the different intensity levels, although the HAES had slightly lower agreement with accelerometer data than the activity diary. Their findings suggest that the HAES is a generally reliable and valid instrument that can be used to assess different activity levels in patients with CF. These results should be interpreted carefully because this analysis was performed on a relatively small group of participants.

Standard operating procedures have recently been developed by the authors of the HAES in an attempt to standardize its administration and reinforce the value of the information obtained from the questionnaire. These standardized administration procedures are available by communicating with the author, Dr John Hay, at jhay@brocku.ca.

The advantages of the HAES include its reasonable completion time, low cost, and an acceptable level of difficulty; the interviewer administered/supervised format leads to excellent completion rates; it directly measures activity and inactivity and therefore there is more confidence in describing the habitual pattern of the child, content, and construct validity [16]; can be used in the clinical and research settings; the data can be easily and quickly analyzed and can be used to provide immediate feedback; and the availability of standard operating procedures.

The disadvantages of the HAES include: it requires some basic mathematical skills and language proficiency for completion (the person must state the percentage of time spent in each activity level in the given time block), it may not be useful in detecting small changes over a period of time, energy expenditure cannot be derived from activity intensity, and it provides only an estimation (this is the tradeoff between utility and accuracy). The sensitivity of the HAES to changes in activity levels remains to be established as discussed by Wells et al. [16].

34.2.2.2 *Seven-Day Physical Activity Recall*

The Seven-Day Physical Activity Recall (7D-PAR) as described by Sallis et al. [19a], is an interviewer-administered recall of physical activity at various intensity levels over the previous seven days. The 7D-PAR can also be administered by phone or self-administered. Weekday and weekend activities are scored separately and the questionnaire requires approximately 10–15 min to administer. Intensity categories are based on metabolic equivalents (METs) that are multiples of resting metabolic rate. Participants are asked to report the time spent in sleep (1 MET) and at moderate (3.0–4.9 METs), hard (5.0–6.9 METs), and very hard (7.0 or more METs) activities. The time spent in light activities is derived by subtraction. The 7D-PAR is designed to elicit reports of activities of at least 10–15 min duration. This measure is used frequently in studies of adults and has demonstrated test retest reliability by Sallis et al. [19b] and validity in relation to electronic monitoring by Taylor and colleagues [21]. Sallis and colleagues published in 1993 on the reliability and validity of the 7D-PAR in healthy children 10–16 years of age and found that it can provide reasonably reliable and valid reports of physical activity levels [19a]. In children, the reliability of recall was found to be higher when the interval between interviews was shorter, and the pattern of correlation suggested that reports of very hard activities were well validated and less valid in the reporting of less vigorous activities. Ruf and colleagues in 2012 were the first to report on the validity of the 7D-PAR in a group of CF patients [17]. They were able to demonstrate that the level of physical activity derived from the 7D-PAR showed significant correlation to the physical activity measured by accelerometry.

The advantages of the 7D-PAR include: it is inexpensive and requires only 10–15 min to complete, the intensity categories are based on METs, and a detailed interviewer manual is available at: http://sallis.ucsd.edu/Documents/Measures_documents/7daypar_protocol.pdf.

The disadvantages of the 7D-PAR include: an activity can only be scored if it is carried out for more than 15 min. The reliability is influenced by the delay of recall, and both reliability and validity improve with age. The questionnaire should only be used to describe physical activity within a population.

34.2.2.3 *International Physical Activity Questionnaire*

This questionnaire is a self-report measure of physical activity in adults. The purpose of the IPAQ is to provide a set of well-developed instruments that can be used internationally to obtain comparable estimates of physical activity. IPAQ assesses physical activity undertaken across a comprehensive set of domains including: (1) leisure time physical activity, (2) domestic and gardening (yard) activities, (3) work-related physical activity, and (4) transport-related physical activity. Only bouts of activity of 10 min duration or more are recorded.

The IPAQ short form asks about three specific types of activity undertaken in the four domains introduced previously. The specific types of activity that are assessed are walking, moderate-intensity activities, and vigorous-intensity activities. A total score for the IPAQ Short form can be expressed as MET-min per week: MET level × minutes of activity/day × days per week. The authors used the Ainsworth and colleagues' [22] Compendium of physical activities: an update of activity codes and MET intensities to establish an average MET score for each type of activity. Domain specific scores cannot be estimated. The short version is suitable for use in national and regional surveillance systems.

The long IPAQ form was designed to provide separate domain specific scores for walking, moderate-intensity, and vigorous-intensity activity within each of the work, transportation, domestic chores and gardening (yard), and leisure-time domains. Scoring of the long form is expressed and calculated in the same manner as the short form; however, domain subscores and activity intensity subscore can be computed and reported. The long version provides more detailed information often required in research work or for evaluation purposes.

Preliminary work to validate the IPAQ in CF adults was performed by Button and colleagues [23]. Adult CF patients wore an accelerometer for seven days and then completed the IPAQ questionnaire. They found that the correlation of the IPAQ estimate of energy expenditure with accelerometer derived energy expenditure was moderate with a trend toward higher energy expenditure measured by the IPAQ versus accelerometer. At this time, this is the only work performed to attempt to validate the IPAQ in CF adults.

The advantages of the IPAQ are that a short and a long form are available, it can be completed as a self-report or over the phone, and it provides an estimate of energy expenditure. A complete set of "Guidelines for Data Processing and Analysis of the International Physical Activity Questionnaire (IPAQ)–Short and Long Forms" is available at http://www.ipaq.ki.se/scoring.pdf.

The disadvantages of the IPAQ are that it was designed for a predetermined age range (18–69 years of age) and testing in a younger or older group is not recommended by the authors. It was not designed for interventional studies. Work has shown that it appears to underestimate physical activity with low-level activities and overestimate in higher intensity activities. There is a lack of evidence that it is sensitive to change.

34.2.2.3.1 DIARIES

Detailed self report of one's physical activity on a daily basis can be recorded using a written or electronic diary. The 24-h period is typically broken down in to 15-min segments and individuals record their activity for each period. In diaries, individuals are asked to record their activity from a predefined list which is coded. The list of activities is typically grouped according to their MET value. The intensity of the activity (low, moderate, or vigorous) is also recorded. Diaries may provide more details on the type, duration, and intensity of physical activities but are quite burdensome for participants to complete.

34.2.2.3.1.1 BOUCHARD'S 3-DAY PHYSICAL ACTIVITY DIARY Bouchard and colleagues described a self-administered 3-day activity record for the estimation of energy expenditure with 2 days of any day of the week, with the third day being a Saturday or a Sunday. Each day is divided into 96 periods of 15 min [24]. The participant enters a categorical value ranging from 1 to 9 (where level 1 is described as sleeping to level 9 that describes high-intensity activities), corresponding to the dominant activity for that period. The energy expenditure calculated for each of the nine categories is used to calculate the total energy expenditure of the participant. The diary was designed to be used by older children and adults. An association with physiological measures has been demonstrated.

34.2.2.3.1.2 BRATTEBY 7-DAY PHYSICAL ACTIVITY DIARY The Bratteby 7-Day Physical Activity Diary is a modified version of the Bouchard's 3-Day Physical Activity Diary. The modifications made by Bratteby and colleagues in 1997 affect the energy cost attributed to each of the nine activity levels [25]. The authors based these energy costs on work published by James & Schofield, Ainsworth, and Haggarty [26–28]. Activities are categorized into nine levels according to their energy costs, which represent a multiple of their resting basal metabolic rate (1 sleep, 2 sitting, 3 standing, 4 walking inside, 5 walking outside, and 6 to 9 represent low-, moderate-, high-, and very high-intensity activity). Each day is divided into 96 periods of 15 min; for each 15-min period, participants are asked to enter the categorical value of the predominant activity in that time period. The activities are converted to METs to provide an estimate of energy expenditure. When Bratteby [25] and colleagues compared the use of the diary against doubly labeled water, they found a close correlation but the variation was very large.

The advantages of diaries include: they provide detailed and comprehensive information about the physical activity undertaken in a day across all domains, bouts of physical activity can be quantified, and pattern of activities can be identified. The method of recording activities is prospective; therefore, it does not rely on recall and memory. Activity diaries provide the best subjective method to estimate of energy expenditure and they can be administered at a low cost.

Disadvantages of diaries include: diaries impose a considerable burden on the participant, The level of compliance required makes the method unsuitable for younger children (younger than 10 years), the data processing is reasonably complex and time-consuming, undertaking a physical activity diary may influence an individual's behavior (i.e., a reactivity effect in which the act of monitoring activity causes an individual to increase his or her activity; this may be especially problematic in intervention studies), the estimate of energy expenditure from physical activity will not be as accurate as by an objective measure such as accelerometers or combined motion sensor.

34.2.3 Assessing Physical Activity Using Objective Tools

34.2.3.1 *Activity Monitors*

The European Cystic Fibrosis Society (ECFS) Working Group's current position on activity monitor devices is that the SenseWear and ActiGraph represent informed choices for activity monitoring in CF. The SenseWear is a device worn on the upper limb and incorporates data from a triaxial accelerometer with skin temperature, heat flux, and galvanic skin response sensors. The ActiGraph is a triaxial accelerometer worn on a belt at the mid axillary line of the hip. Both the SenseWear and ActiGraph claim to measure energy expenditure, step counts, time spent in different intensities and sedentary time.

34.2.3.2 *SenseWear*

There are as yet no data on reliability of the SenseWear in CF. Data on convergent validity in adults suggest a good correlation in energy expenditure with indirect calorimetry during flat and incline walking. Despite this, there is evidence that the SenseWear overestimates energy expenditure in walking on the flat and underestimated step counts on both flat and incline walking [29]. This study, however, was conducted in a laboratory during short duration activities, and further research should investigate if the correlation extends to the complex activities of the free-living environment. Troosters provide some data on SenseWear's ability to discriminate between adults with CF and healthy adults in step count, time spent in moderate activity, and time spent in vigorous activity [4]. Troosters et al. (2009) and Garcia et al. (2011), provide some information on the SenseWear's ability to show a relationship between physical activity and clinical outcome measures and hence supports the validity of this device [4,30]. These studies are discussed in more detail in the Section 34.5. Responsiveness to treatment with intravenous antibiotics has also been shown for the SenseWear. Wieboldt demonstrated a response in energy expenditure between admission and discharge and between admission and one month follow-up [31]. This study also showed a response in step count between admission and one month follow-up and between discharge and one month follow-up.

34.2.3.3 *ActiGraph*

The ActiGraph has been demonstrated to be reliable over two time points in children through to adults [16,17]. Convergent validity of the ActiGraph has been investigated by comparison to questionnaires and diaries [16,17]. The evidence of validity gleaned from these studies relies on the validity of the questionnaires and it can be seen that data are not consistent between activity categories or between types of questionnaire or diary. There is evidence of discriminate validity of the ActiGraph from two studies [11,17]. These studies show that the ActiGraph can detect differences in physical activity between different severities of disease in prepubescents and pubescents, and between genders in children through to adults. Hebestreit [32] demonstrates that physical activity measured using an ActiGraph correlates with aerobic capacity, thus adding to the evidence of validity. There is also some evidence that the ActiGraph is responsive to a program of physical conditioning in children through to adults [33].

34.2.3.4 *Pedometers*

Pedometers are designed to detect vertical movement at the hip and so measure the number of steps and provide an estimate of distance walked. They cannot provide information on the temporal pattern of physical activity or the time spent in different activity at different intensity.

The pedometer with the most available data in CF is the DigiWalker and hence the ECFS Working Group state that this represents the most informed choice of pedometer. Unfortunately there is no data on reliability of the DigiWalker however there is information from one study on validity and responsiveness Quon [34] found a significant correlation between step count and some important domains of respiratory symptoms diary and FEV_1 (%predicted) in adolescents and adult (see Section 34.5). The study also found evidence of responsiveness in step rate and step count to intravenous antibiotic treatment in adolescents and adults.

34.3 NORMATIVE DATA FOR PHYSICAL ACTIVITY IN CF

As is the case with all clinical outcome measures, normative data must be specific to the tool being used to measure physical activity and the local population. The implication is that centers would need to collect their own normative data or at least compare to those collected using the same methods in a similar population. In addition, the cut-offs for different activity intensities measured by objective motion sensors will not be appropriate for all patients. Patients with more severe disease, or those who are deconditioned, will experience a greater physiological response to physical activities. This exposes a problem when we define physical activity intensities in terms of shortness of breath, heart rate and perspiration. Movements that represent moderate activity in some patients, may be vigorous to others.

34.4 METHODS OF ANALYZING AND REPORTING PHYSICAL ACTIVITY DATA

It is well recognized that characteristics of physical activity behavior include frequency, duration, intensity, type and total amount. There are also novel data emerging that highlight the importance of breaking up periods of sedentary time and the link between health and the complexity of physical activity patterns [35]. There is a wide variation in the types of data reported in studies investigating physical activity in CF. The ECFS Working Group for Exercise recommend that all research using activity monitors should report, at the least, step counts, time spent physically active (divided by both intensity and time of day), time spent sedentary, and energy expenditure.

Each motion sensor comes with proprietary software and instruction manuals. In saying that, there is still a wide variation in how the monitors are applied to individuals, and how the data are processed and interpreted highlighting the need for centers to develop and share standard operating procedures. It is only through a standardized approach to physical activity measurement that we will be able to make comparisons between interventions and devices and pool physical activity information from different centers. Items that should be considered for standardization include: number of days assessed and whether they include weekend days, duration of wear time, methods of positioning device, procedures for "cleaning" the raw data, methods for averaging the data (e.g., minute-by-minute), and thresholds used to classify the intensity of activity.

In children, physical activity patterns differ to those of adults. Children tend to perform short bursts of higher intensity activity lasting only seconds, interspersed with periods of lower intensity activity [36]. Children's physical activity data are therefore of a greater complexity. This complexity may be lost if the data are averaged over time periods, for example, minute by minute. Different algorithms for classifying intensity of activity would need to be developed to account for the differences in efficiency of body movements depending on the stage of maturity.

34.5 EVIDENCE FOR IMPACT OF PHYSICAL ACTIVITY ON HEALTH OF PATIENTS WITH CF

We know that physical training has a positive effect on lung function in patients with CF [37]. In addition, the combination of physical activity and airway clearance has been shown to be of benefit in CF [38]. However, it is important that patients do not substitute physical activity or exercise for airway clearance.

There is a growing evidence base to support that healthy physical activity behaviors have a positive impact on people with CF. Troosters [4] investigated the relationship between physical activity and aerobic capacity and muscular strength. This study showed a positive relationship between time spent in either moderate intensity activity or vigorous intensity activity and both quadriceps muscle force and VO_{2peak}. Step count in this study also showed a positive relationship with VO_{2peak}. Hebestreit, Ruf, Baker and Wideman, and Selvadurai also found a positive relationship between healthy physical activity behaviors and aerobic capacity [11,17,32,39]. Garcia investigated the relationship between physical activity and bone health in CF [30]. They were able to demonstrate a positive relationship between time spent in low- and moderate-intensity activity and Z-scores for bone mineral density in the lumbar column, neck of femur, and hip. Quon [34] found a significant correlation between step rate/count and some domains of respiratory symptoms diary (difficulty breathing, cough, chest tightness, tired, worried, cranky, frustrated, missed school/work, reduced usual activity, well) in adolescents and adults, indicating that it is a valid tool. Step rate was also found to be significantly correlated to forced expiratory volume in 1 s (FEV_1; %predicted) in adolescents and adults. This reiterates and earlier study by Schneiderman [12] that showed that activity levels were linked to lung function. In addition to this cross-sectional data, Schneiderman [40] found that increases in physical activity were associated with a slower rate of decline in FEV1. This was the first study to show the longitudinal effect of improving physical activity behavior on lung function.

The mounting evidence demonstrates that healthy physical activity behavior is a key component of effective management of CF. This supports the need to measure physical activity and to discuss it with our patients.

34.6 DISCUSSION

Higher levels of physical activity in CF are recognized as desirable, and accurate measurement of the level of physical activity of CF patients is critical in developing intervention programs and assessing their effectiveness. However, there is already a great burden of care placed upon many people with CF to maintain health and be able to participate in society. If physical activity is viewed as yet another therapy, prescribed by the CF team, this could be detrimental to participation. It is better to view physical activity as part of normal life. Clinicians should take care to strike a balance between highlighting the benefits of an active lifestyle versus "medicalizing" physical activity. For some patients, the measurement of their physical activity in the CF clinic may be motivating; however, for some it may have the opposite effect. Any discussion we have with our patients should highlight their abilities and not disabilities and encourage them to participate as much as possible with their healthy peers. We need further research to guide us as to the most important elements of physical activity behavior to measure in the busy clinic: objective versus subjective measurement of patterns, or the determinants of physical activity such as stage of change, self-efficacy, and decisional balance. Until then, the clinician's choice of measurement should be based on the goals to be achieved, and restricted to those measures that have been shown to be reliable and validated against more stringent measures.

References

[1] WHO Global Recommendations on Physical Activity for Health. NLM classification:QT255; 2010. ISBN:978 92 4159 9979.

[2] Pianosi P, Le Blanc J, Almudevar A. Relationship between FEV1 and peak oxygen uptake in children with cystic fibrosis. Pediatr Pulmonol 2005;40:324–9.

[3] Pianosi P, Le Blanc J, Almudevar A. Peak oxygen uptake and mortality in children with cystic fibrosis. Thorax 2005;60:50–4.

[4] Troosters T, Langer D, Vrijsen B, Segers J, Wouters K, Janssens W, et al. Skeletal muscle weakness, exercise tolerance and physical activity in adults with cystic fibrosis. Eur Respir J 2009;33:99–106.

[5] Britto MT, Garrett JM, Konrad TR, Majure JM, Leigh MW. Comparison of physical activity in adolescents with cystic fibrosis versus age-matched controls. Pediatr Pulmonol 2000;30:86–91.

[6] Jolley CJ, Moxham J. A physiological model of patient reported breathlessness during daily activities in COPD. Eur Respir Rev 2009;18:66–79.

[7] Marcus BH, Rakowski W, Rossi JS. Assessing motivational readiness and decision making for exercise. Health Psychol 1992;11:257–61.

[8] Marcus BH, Selby VC, Niaura RS, Rossi JS. Self-efficacy and the stages of exercise behaviour change. Res Q Exerc Sport 1992;63:60–6.

[9] Marcus BH, Rossi JS, Selby VC, Niaura RS, Abrams DB. The stages and processes of exercise adoption and maintenance in a worksite sample. Health Psychol 1992;11:386–95.

[10] Bradley JM, Kent L, Elborn JS, O'Neill B. Motion sensors for monitoring physical activity in cystic fibrosis: what is the next step? Phys Ther Rev 2010;15(3):197–203.

[11] Selvadurai HC, Blimkie CJ, Cooper PJ, Mellis CM, VanAsperen PP. Gender differences in habitual activity in children with cystic fibrosis. Arch Dis Child 2004;89(10):928–33.

[12] Schneiderman-Walker J, Wilkes DL, Strug I, Lands LC, Pollock SL, Selvadurai HC, et al. Sex differences in habitual activity and lung function decline in children with cystic fibrosis. J Pediatr 2005;147(3):321–6.

[13] Schneiderman-Walker J, Pollock SL, Corey M, Wilkes DD, Canny GJ, Pedder L, et al. A randomized controlled trial of a three year home exercise program in cystic fibrosis. J Pediatr 2000;136:304–10.

[14] Nixon PA, Orenstein DM, Kelsey SF, Doershuk CF. The prognostic value of exercise testing in patients with cystic fibrosis. N Engl J Med 1992;327(25):1785–8.

[15] Gulmans VAM, de Meer K, Brackel HJL, Faber JAJ, Berger R, Helders PJM. Outpatient exercise training in children with cystic fibrosis: physiological effects, perceived competence, and acceptability. Pediatr Pulmonol 1999;28:39–46.

[16] Wells GD, Wilkes DL, Schneiderman-Walker J, Elmi M, Tullis E, Lands LC, et al. Reliability and validity of the habitual activity estimation scale (HAES) in patients with cystic fibrosis. Pediatr Pulmonol 2008;43:345–53.

[17] Ruf K, Fehn S, Bachmann M, Möller A, Roth K, Kriemler S, et al. Validation of activity questionnaires in patients with cystic fibrosis by accelerometry and cycle ergometry. BMC Med Res Methodol 2012;12:43.

[18] Hay J. Development and testing of the habitual activity estimation scale. In: Armstrong N, editor. 2nd ed. Children and exercise XIX, vol. 2006. Exeter: Singer Press; 1997. pp. 125–9.

[19] [a] Sallis JF, Buono MJ, Roby JJ, Micale FG, Nelson JA. Seven-day recall and other physical activity self-report in children and adolescents. Med Sci Sports Exerc 1993;25(1):99–108.
[b] Sallis JF, Haskell WL, Wood PD. Physical activity assessment methodology in the Five-City Project. Am J Epidemiol 1985;121:91–106.

[20] Craig CL, Marshall AL, Sjostrom M. International physical activity questionnaire: 12-country reliability and validity. Med Sci Sports Exerc 2003;35:1381–95.

[21] Taylor CB, Coffey T, Berra K, Iaffaldano R, Casey K, Haskell WL. Seven-day activity and self-report compared to a direct measure of physical activity. Am J Epidemiol 1984;120:818–24.

[22] Ainsworth BE, Haskell WL, Whitt MC, Irwin ML, Swartz AM, Strath SJ, et al. Compendium of physical activities: an update of activity codes and MET intensities. Med Sci Sports Exerc 2000;32:S498–504.

[23] Button BM, Rasekaba T, Wilson JW, Holland A. Validation of the international physical activity questionnaire in adults with cystic fibrosis. Pediatr Pulmol 2010;45(S33):382. [Abstract #453].

[24] Bouchard C, Tremblay A, Leblanc C, Lortie G, Savard R, Thériault G. A method to assess energy expenditure in children and adults. Am J Clin Nutr 1983;37:461–7.

[25] Bratteby LE, Sandhagen BO, Fan H, Samuelson GA. 7-day activity diary for assessment of daily energy expenditure validated by the doubly labelled water method in adolescents. Eur J Clin Nutr 1997;51:585–91.

[26] James WPT, Schoefield EC. In: Human energy requirements. A manual for planners and nutritionists. Oxford: Oxford University Press; 1990.

[27] Ainsworth BE, Haskell WL, Leon AS, Jacobs Jr DR, Montoye HJ, Sallis JF, et al. Compendium of physical activities: classification of energy costs of human physical activities. Med Sci Sports Exerc 1993;25:71–80.

[28] Haggarty P, McNeill G, Abu Manneh MK, Davidson L, Milne E, Duncan G, et al. The influence of exercise on the energy requirements of adult males in the UK. Br J Nutr 1994;72:799–813.

[29] Dywer TJ, Alison JA, McKeough ZJ, Elkins MR, Bye PTP. Evaluation of the SenseWear activity monitor during exercise in cystic fibrosis and in health. Respir Med 2009;103:1511–7.

[30] Garcia ST, Giraldez Sanchez MA, Cejudo P, Quintana Gallego E, Dapena J, Garcia Jiminez R, et al. Bone health, daily physical activity and exercise tolerance in patients with cystic fibrosis. Chest 2011;140(2):475–81.

[31] Weiboldt J, Atallah L, Kelly JL, Shrikrishna D, Gyi KM, Lo B, et al. Effect of acute exacerbation on skeletal muscle strength and physical activity in cystic fibrosis. J Cystic Fibrosis 2012;11(3):209–15.

[32] Hebestriet H, Kieser S, Rudiger S, Schenk T, Junge S, Hebestreit A, et al. Physical activity is independently related to aerobic capacity in cystic fibrosis. Eur Respir J 2006;28:734–9.

[33] Hebestriet H, Kieser S, Junge S, Ballmann M, Hebestreit A, Schindler C, et al. Long-term effects of a partially supervised conditioning programme in cystic fibrosis. Eur Respir J 2010;35:578–83.

[34] Quon BS, Partick DL, Edwards TC, Aitken ML, Gibson RL, Genatossio A, et al. Feasibility of using pedometers to measure daily step counts in cystic fibrosis and an assessment of its responsiveness to changes in health state. J Cystic Fibrosis 2012;11:216–22.

[35] Burton C, Knoop H, Popovic N, Sharpe M, Bleijenberg G. Reduced complexity of activity patterns in patients with chronic fatigue syndrome: a case control study. BioPsychoSoc Med 2009;3:7.

[36] Berman N, Bailey, Barstow TJ, Cooper DM. Spectral and bout detection analysis of physical activity patterns in healthy, prepubertal boys and girls. Am J Human Biol 1998;10:289–97.

[37] Bradley JM, Moran F. Physical training for cystic fibrosis. Cochrane Database Syst Rev 2008;(1). http://dx.doi.org/10.1002/14651858.CD002768.pub2. Art. No.: CD002768.

[38] McIlwaine M. Chest physical therapy, breathing techniques and exercise in children with CF. Paediatr Respir Rev 2007;8(1):8–16.

[39] Baker CF, Wideman L. Attitudes towards physical activity in adolescents with cystic fibrosis: sex differences after training: a pilot study. J Pediatr Nurs 2006;21(3):197–210.

[40] Schneiderman JE, Wilkes DL, Atenafu EG, Nguyen T, Wells GD, Alerie N, et al. Longitudinal relationship between physical activity and lung health in patients with cystic fibrosis. Eur Respir J 2014;43:817–23.

CHAPTER

35

Motivating Physical Activity: Skills and Strategies for Behavior Change

Heather Chambliss

Physical activity is a behavior that is a complex behavior determined by multiple factors. Individuals' physical activity—or sedentary—choices are influenced by an interaction of personal history, health and medical status, sociocultural influences, psychological and motivational factors, and environmental considerations. These factors shape one's goals, needs, preferences, barriers, and habits to ultimately affect behavior. Because there are multiple influences on individual physical activity behavior, and these influences are not static, behavioral skill-building is a key component in any program aimed at promoting long-term lifestyle change.

35.1 BEYOND A PRESCRIPTION

For health and fitness professionals, it is easy and perhaps most comfortable to fall into the role of simply providing an exercise prescription to patients, communicating what patients "should" be doing to improve their health and manage disease symptoms. However, the likelihood of patients following a general prescription to exercise is only likely to happen in individuals who are highly motivated and only for the time in which they can readily follow the prescribed plan of action. Once motivation wanes, or barriers arise, adherence will decrease and patients will return to sedentary behaviors.

Consistent research shows that cystic fibrosis (CF) patients are receptive to physical activity and receive numerous health benefits from exercise [1–3]. However, long-term adherence is difficult, and not all patients will have access to structured physical activity or supervised exercise or know how to incorporate exercise into their lives. Therefore, it is important to consider different approaches to increasing physical activity among individuals with CF.

35.2 A'S OF PHYSICAL ACTIVITY COUNSELING: A FRAMEWORK FOR BEHAVIOR CHANGE

The A's framework has been used for counseling in health-care settings for various behavioral targets, including smoking cessation [4] and weight loss [5,6]. The five A's model has been used in health-care settings to promote physical activity with the following steps: Assess, Advise, Agree, Assist, and Arrange [7–9].

35.2.1 A's of Physical Activity Counseling: Assess

Medical Status: The first step in coaching physical activity behavior change for CF patients is assessment of medical history and current health status. Priority should be given to CF history and status, but other medical factors unrelated to CF should also be considered, including mental health and musculoskeletal complaints. Physical fitness, including cardiorespiratory fitness and muscular strength and endurance, should be assessed as a basis for the exercise prescription. Nutritional status should also be assessed as relevant to energy expenditure recommendations.

Physical Activity Level: The next step in assessment is determining current physical activity level as well as past physical activity history. Comprehensive assessment follows a FITTE model: Frequency, Intensity, Time (duration), Type (mode), and Energy Expenditure [10]. In addition, it is important to include assessment of sedentary activity as well as activities of daily living, particularly if patients are not engaged in more structured physical activity or exercise.

Along with objective measures of physical activity level, professionals should also consider readiness for change. The transtheoretical model classifies individuals as being in one of five stages for any given

Diet and Exercise in Cystic Fibrosis
http://dx.doi.org/10.1016/B978-0-12-800051-9.00035-3

behavior: precontemplation (not thinking about change), contemplation (thinking about change but not currently active), preparation (doing some activity but not meeting behavioral target), action (meeting behavioral target but for less than 6 months), and maintenance (active for more than 6 months) [11]. On the basis of an individual's stage, physical activity advice can be appropriately tailored to match the patient's behavioral readiness and move individuals toward progressive stages (Figure 35.1) [12–14].

35.2.2 A's of Physical Activity Counseling: Advise

The next step of physical activity counseling is to provide brief, clear, structured advice considering patient needs (medical and self-stated goals), current behavior and stage of readiness for change, patient self-efficacy, and accepted physical activity guidelines. In particular, CF patients should be counseled on the general and disease-specific benefits of physical activity that can help provide motivation for behavior change. Patients will differ in their understanding and interest in detailed information, but it is important that there is a degree of understanding of the "Whys" for physical activity. However, health practitioners may not routinely give specific information regarding exercise recommendations even when physical activity is addressed [9]. This may be due to time limitations or lack of knowledge and tools for physical activity counseling [8].

Physical activity recommendations for healthy adult populations propose that "most adults engage in moderate-intensity cardiorespiratory exercise training for ≥30 min/day on ≥5 days/week for a total of ≥150 min/week, vigorous-intensity cardiorespiratory exercise training for ≥20 min/day on ≥3 days/week (≥75 min/week), or a combination of moderate- and vigorous-intensity exercise to achieve a total energy expenditure of ≥500–1000 MET·min/week. On 2–3 days/week, adults should also perform resistance exercises for each of the major muscle groups as well as neuromotor exercise involving balance, agility, and coordination" [15]. This position stand is consistent with the Physical Activity Guidelines for Americans [16]. Physical activity guidelines for children emphasize the importance of daily physical activity with a goal of 60 or more minutes a day for healthy youth [16]. On 3 days each week, children and adolescents should incorporate vigorous-intensity activity as well as bone and muscle strengthening activities. These guidelines recognize the importance of individual tailoring based on medical needs, fitness level, and self-stated goals. As part of the recommendations, patients can accumulate physical activity in bouts of at least 10 min in duration, and benefits can be obtained at less than the recommended dose [15,16].

Because CF patients will vary in physical activity habits, fitness level, and health status, it is important to consider patient self-efficacy when advising on physical activity and exercise [17,18]. For patients struggling with disease management or sedentary individuals who have been unsuccessful in adopting exercise, physical activity may seem overwhelming. To build confidence, advice may include education on the different options for physical activity, including lifestyle activity, transportation, recreation, and occupational and traditional sports and fitness opportunities. A Physical Activity Pyramid

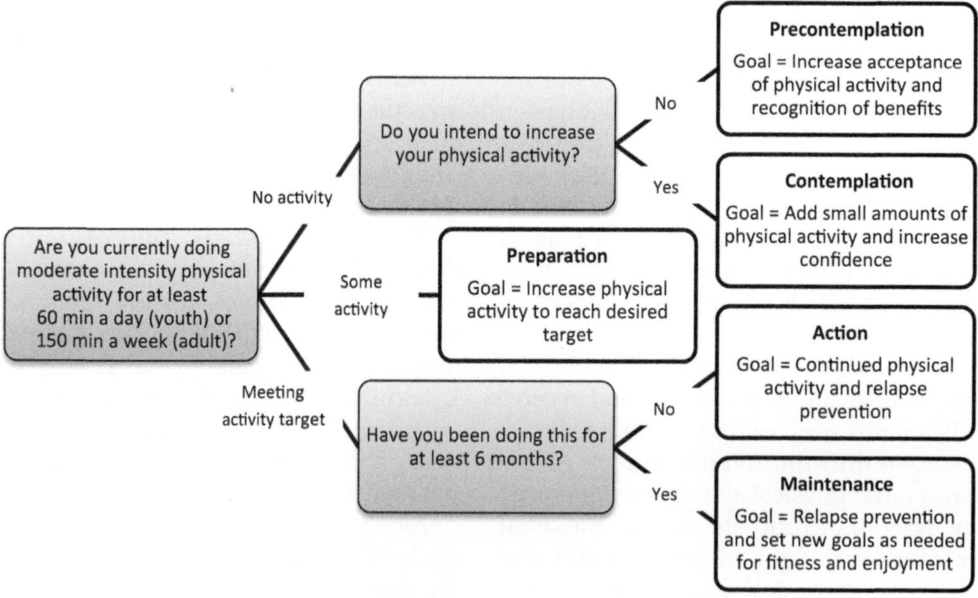

FIGURE 35.1 Assessing stage of change.

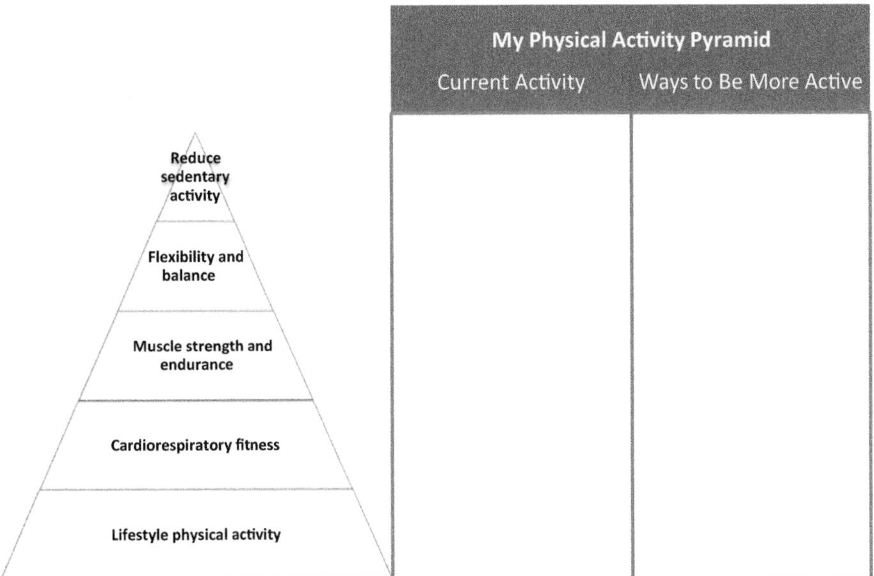

FIGURE 35.2 Physical Activity Pyramid worksheet.

model (Figure 35.2) can be a useful tool for advising patients on physical activity recommendations and assisting patients in establishing physical activity goals.

35.2.3 A's of Physical Activity Counseling: Assist

The assist step of the A's framework provides the development of an action plan for physical activity and application of behavior skill-building. In terms of successful behavior change, this step is critical, but it may not be routinely addressed in many health-care settings [5,9]. Behaviorally based physical activity counseling revolves around the provision of support and assistance for various behavioral skills, including self-monitoring, goal-setting, problem-solving, social support, and relapse prevention.

Self-monitoring is the skill of keeping records of specific goals and behaviors. Ideally, a health professional provides feedback at subsequent visits, but individuals can also independently use the skill of self-monitoring. Self-monitoring serves several purposes: (1) it helps professionals and patients understand current behavior patterns, (2) it provides objective data for goal-setting and tracking progress toward goals, (3) it establishes prompts and reminders for the behavior itself, and (4) it helps uncover barriers to change.

Self-monitoring physical activity can take many forms. The act of self-monitoring and feedback can be an intervention itself, with or without additional components. Various physical activity monitors can be used, such as step counters or pedometers, body monitors, and GPS location devices. Logging systems range from written records with low-tech paper/pencil methods to mobile apps that integrate with body monitoring devices.

In deciding which self-monitoring method to use, these questions may help match.

- What is the patient's age and familiarity with technology?
- What is the patient's physical activity goal (e.g., reducing sedentary activity, walking, increasing structured exercise)?
- How much time is the patient willing to spend monitoring?
- What resources does the patient have access to or is willing to purchase?

Goal-setting and rewarding of goals is critical to successful behavior change [19]. Establishing concrete short-term goals for behaviors—also known as process goals—is as important, or even more important, than setting long-term outcome goals. Another way to think of the goal-setting process is using health outcome goals to develop an action plan consisting of process or behavior goals.

Health professionals should keep in mind the following guidelines for assisting patient goal-setting:

- Consider patient characteristics (needs, medical limitations, preferences, etc.) and best practices for physical activity prescription.
- Choose goals that are appropriate for the individual's readiness to change.
- Include feedback and reinforcement in the planning process to motivate and facilitate progress.

Many people set goals on a regular basis, but few people naturally set goals that guide long-term behavior change. Common goal-setting problems include failing to set behavioral goals, setting goals that are too general, not adjusting goals as needed, setting too many goals at one time, and not evaluating progress toward goals.

Working with patients to develop concrete, objective goals that are challenging yet achievable is an important role of the health professional. The SMART framework can help patients develop the skill of effective goal-setting [10,18]. Effective goals should meet the following SMART-SE characteristics:

- Specific—concrete and detailed; focused on one element
- Measurable—allows progress to be tracked and evaluated
- Action-oriented and adjustable—centers around a behavior or uses a behavior plan and adjustable in case the plan is not working or circumstances arise
- Realistic—challenging yet achievable
- Time bound—includes a deadline that provides guidance for the action plan and evaluation of progress
- Self-directed—patients should be the primary author of the goal
- Evidence-based—patients should be educated so that goals produce desired outcomes and follow best practices rather than popular trends or myths

It is important to include patients, even younger children, in the goal-setting process. This empowers the patient, leading to greater self-efficacy, independence, and sense of control [20]. Goal-setting should be seen as a collaborative process, with the ultimate responsibility being that of the patient (see Table 35.1).

Decisional balance is the skill of identifying and evaluating pros and cons for adopting a behavior such as physical activity [12,13,18]. It is important to identify personal benefits for physical activity and understand the individual barriers to change. Most people can name several pros and cons, and people will have many common; however, the importance that each person places on certain benefits and barriers will differ. Having patients make a list of personal benefits and barriers is an important step in developing a physical activity action plan.

TABLE 35.1 Good vs. Better Goals: Developing SMART-SE Goals

Good	Better
I am going to start exercising next month.	Starting March 1, I will walk 3 days a week on Monday, Wednesday, and Friday for 30 min.
I want to build muscle.	I will do free weights and kettle bell exercises at the gym with Donnie on Tuesday and Thursday afternoons after school. I will plan out and track my routine using a workout log and the My Fitness Pal app.
We are going to be more active as a family.	We will plan an active family outing every Saturday and have a family screen time challenge to limit "fun" screen time to less than 2 h a day.

In the case of CF patients, further exploration of health-related and general categories can help patients better understand how pros and cons may influence their behaviors. CF patients have reported various perceived benefits of physical activity, including improved aerobic and muscular fitness, better lung function, sputum clearance, weight management, increased energy, social interaction and feelings of accomplishment, confidence and normalcy, and enjoyment [1,21–23]. However, numerous barriers to exercise are also reported, many of which are consistent with the general population, such as lack of time, boredom, fatigue, competing activities, and the "normal" physical discomfort associated with exercise. Reported barriers more specific to CF include coughing and shortness of breath, fluctuations in health status, illness, nutritional status, burden of CF treatment regimens, and overprotection [1,21–24].

Prioritizing benefits and barriers in terms of personal meaning and importance assists in identifying factors that can be used for motivation (benefits/pros) and that need to be addressed through problem-solving (barriers/cons).

Step 1: Write in down. Have the patient develop a list of benefits and barriers. In this case, the patient may be a teenage male who is active during the winter with skiing but is inconsistent during other times (Table 35.2).
Step 2: Prioritize. Learn more about what is really going on. In this example, the patient may know that exercise is important for lung function, but what may be most motivating is to build muscle and increase fitness for social and confidence purposes. You might also learn that boredom is the big barrier. He likes skiing, but he does not like to exercise at a gym without a buddy because it is not fun.
Step 3: Problem solve. The more you know about what is really going on, the better you can address barriers

TABLE 35.2 Decisional Balance and Problem-Solving Activity

My Benefits	My Barriers
• Build muscle	• Do not have a gym membership
• Help breathing	• No one to exercise with
• Increased fitness for skiing	• Don't have time/school activities and homework
• Reduce stress	• Get bored

and develop an appropriate and motivating physical activity plan that will maximize motivation and minimize barriers.

Problem-solving can be effectively accomplished by using the steps of the IDEA method: (1) Identify the barrier, (2) Develop a list of potential solutions, (3) Evaluate the solutions, and (4) Act and Assess how well the plan worked [12,14]. As discussed in the concept of decisional balance, the first step in problem-solving is to identify personal barriers. In working with patients on the skill of practical problem-solving, priority should be given to barriers that are currently exerting the greatest influence on behavior and those barriers that the patient has some degree of control over. However, it is also important to acknowledge that there are significant barriers that cannot be "solved," and the solution may instead involve how one reacts to the barrier.

Each barrier should be identified as specifically as possible. Once the barrier has been defined, patients should develop a list of potential solutions through a brainstorming process. The key to this step is for the patient to be nonjudgmental and resist evaluating the solutions. Once potential solutions are exhausted, then the next step is to evaluate each solution for potential effectiveness. The temptation for a health professional may be to "suggest" solutions to a patient; however, this often results in "yes-butting" ("Yes, but here is why that won't work"). When the patient is empowered to develop his or her own solutions, the plan is more likely to be perceived as acceptable and achievable. The final step is to act on the selected solution and assess how well the plan worked. Successful problem-solvers do not get frustrated when the first solution does not work but view an unsuccessful plan as a learning experience, and because a list of solutions was developed, there are other options to try. These problem-solving steps can be learned at an early age. Systematic problem-solving empowers individuals to have control, which builds self-efficacy and can be implemented across situations and behavioral targets.

Research continues to demonstrate the importance of social support in successful behavior change [25,26]. Because people have different motivational and logistical needs for physical activity, matching social support to each individual is important. It is also important to consider the developmental needs of younger patients when planning physical activity so that appropriate social support is included. Family physical activity and role-modeling may be particularly important for children so that they do not feel that exercise is a punishment and yet another negative consequence and burden of CF [24].

Categories of support include knowledge, encouragement, participation, and facilitation. Knowledge support includes those sources that provide education about fitness, exercise options, and the interaction of medical needs with physical activity. Encouragement support revolves around the provision of motivation, having cheerleaders that motivate patients to be active when it is difficult and celebrate successes, big or small. Participatory support includes workout buddies and teammates, those individuals that do physical activity alongside the patient and increase the enjoyment and accountability of exercise. Finally, practical support assists with logistical needs such as transportation, childcare, and other practical tasks [10].

In developing and applying the skill of social support, patients should first consider what type of help they need to facilitate physical activity and then identify people who can help. Sources of social support include family, friends, health-care providers, and fitness professionals. As children progress into adolescence, developmental changes and desire for autonomy may lead to less reliance on parental influence and greater emphasis on peer relationships [20,23].

For some people, online sources of social support are highly effective, and people may feel more comfortable with social media support such as Facebook, Twitter, Instagram, and interactive websites and groups. Caution should certainly be taken when recommending these sources for younger patients, but interactive technology can be a helpful resource for CF patients, allowing the creation of a virtual social support network of other individuals with CF.

It is also helpful to identify potential sources of negative social support (intentional or unintentional saboteurs) and discuss skills to deal with them. There may be people that pose barriers to physical activity out of attempts to protect the patient from illness and injury, or the behavior change may be a disruption to the existing relationship. Given the potential schedule conflicts and transportation needs, families must embrace physical activity as something that takes priority in the family system.

In using social support, patients may need to develop communication skills to ask for what is needed and then positively reinforce supportive behavior so that the source feels valued and appreciated. Professionals can help patients build social support by working through the following questions: What do I need help with? Who can help me? What keeps me from asking for help? What is in it for them? [14].

Stimulus control, the skill of recognizing and manipulating cues for choices that affect physical activity, has been shown to help people change exercise behavior [27,28]. Habits happen; they are mindless behaviors that occur automatically. To develop new habits, individuals must first recognize mental, physical, and social cues for their usual choices. Prompts include environmental, emotional, social, and learned cues that influence behavior, often outside of conscious awareness. Therefore, it is important to first recognize the cues and the responses they produce. Individuals can then remove or limit cues that prompt physical inactivity and establish cues for more active behavior (see Table 35.3).

Once patients have started to be more physically active, relapse prevention should be addressed before lapses happen. All individuals experience times when healthy behaviors lapse, and the behavioral skills are critical to getting back on track. For CF patients, relapse prevention is especially important because it is expected that disease exacerbation will impede physical activity and exercise routines. Other high-risk situations may include travel, holidays, periods of stress, or school/work demands.

The key to relapse prevention is to plan ahead so that the patient is not surprised when a lapse occurs. Planning ahead involves identifying likely triggers for lapses and developing an appropriate action plan using behavioral skills such as goal-setting, time management, and social support. In addition, cognitive restructuring is important to avoid unhelpful thought patterns such as all-or-none or polarized thinking; catastrophizing, in which a small lapse is blown out of proportion; and focusing on the negative while minimizing the positive [10,14].

During high-risk times, adjusting goals and focusing on what the patient can do helps to maintain healthy behavior and a more productive outlook [18]. Health professionals can help in the relapse prevention process by reviewing physical activity history using a life timeline approach. By looking at past exercise and sedentary behavior and associated triggers and situations, patients and professionals can learn about individual patterns and what worked and did not work in getting back on track.

A recent review summarizing lessons learned from physical activity interventions in overweight youth can be applied across populations. Practical recommendations include the following: (1) consider the setting, including physical and sociocultural environment; (2) match the fitness professional to the patient; (3) make physical activity fun; (4) promote parental support; (5) consider patient characteristics for tailoring interventions; (6) set realistic goals; (7) use reminders; (8) involve a multidisciplinary team; (9) problem-solve barriers; and (10) communicate purpose and meaning [29].

TABLE 35.3 Example List of Triggers for Physical Activity and Sedentary Behavior

My Active Triggers	My Inactivity Triggers
• Use pedometer app	• Phone (games, texting)
• Wear running shoes	• Couch and TV when I get home
• Put workouts on calendar	

35.2.4 A's of Physical Activity Counseling: Arrange

The final step for health professionals in physical activity coaching is to arrange follow-up for short-term and long-term progress and provide feedback on behavior change. This is often a neglected step in behavioral counseling in health-care settings [5,9]. For medical practitioners, this can include a formal or informal exercise prescription as well as intentional inclusion of physical activity counseling in the medical plan. Between medical visits, practitioners can arrange for interim support, including telephone, e-mail, online group discussion, newsletters, etc. Other methods of follow-up support include appropriate referrals to fitness professionals and community resources. Documentation of patient contact regarding physical activity should include health and physical activity data as well as information regarding behavioral skills, goals, and action plans. This allows for ongoing coaching and personalized feedback. Figure 35.3 is an exercise prescription template from *Exercise Is Medicine*, which includes referral to a certified fitness professional [30]. Fitness professionals and coaches may not have previous experience working with CF patients; therefore, ongoing communication is essential.

An important role of all health professionals is that of "coach" to patients. These ABCs of coaching can be helpful reminders to promote successful behavior change:

- *A—Accountability*: Enhance patient responsibility and provide feedback on progress.
- *B—Belief*: Communicate confidence in patients and let them know that their "team" believes in them.
- *C—Confidence*: Build patient self-efficacy by promoting behavioral skill-building and facilitating goal-setting.
- *D—Direction*: Provide resources and assist patients in developing individually tailored, patient-centered action plans.
- *E—Encouragement*: Provide motivation and positive feedback through the ups and downs of behavior change.

These ABCs can remind professionals that physical activity behavior change is about more than an exercise prescription and telling patients what to do. The behavioral skills discussed in this chapter can be considered as tools in the coaching toolbox so that

EXERCISE PRESCRIPTION & REFERRAL FORM

PATIENT'S NAME:________________________ DOB:__________ DATE:__________

HEALTH CARE PROVIDER'S NAME:____________________ SIGNATURE: ______________

PHYSICAL ACTIVITY RECOMMENDATIONS

Type of physical activity:	Aerobic	Strength
Number of days per week:		
Minutes per day:		
Total minutes per week*:		

*PHYSICAL ACTIVITY GUIDELINES
Adults aged 18-64 with no chronic conditions: Minimum of 150 minutes of moderate physical activity a week (for example, 30 minutes per day, five days a week) *and* muscle-strengthening activities on two or more days a week (2008 Physical Activity Guidelines for Americans).
For more information, visit www.acsm.org/physicalactivity.

REFERRAL TO HEALTH & FITNESS PROFESSIONAL

Name: ______________________

Phone: ______________________

Address: ______________________

____

Web Site: ______________________

Follow-up Appointment Date: ______________

Notes: ______________________

____

____

____

FIGURE 35.3 Exercise prescription template.

providers can partner with patients and provide individuals with behavioral tools for long-term lifestyle change.

35.3 PHYSICAL ACTIVITY RESOURCES FOR PATIENTS AND PRACTITIONERS

Various evidence-based resources for physical activity promotion are available to health professionals to assist in the development and implementation of physical activity interventions. Some examples of recommended resources are summarized in Table 35.4.

Recent years have given rise to numerous technology-assisted methods for increasing physical activity, including web-based interventions, body monitoring devices, and mobile apps and communication. Web-based physical activity interventions typically involve some degree of tailoring based on consumer input of data, including self-stated biometric, medical, and goal information. Feedback may include immediate generation of content or use a stepped format through periodic e-mails. Programs are also available to allow health professionals to mediate the delivery of content. Considerations for web-based interventions include the credibility of the sponsoring organization, degree of individual tailoring and participant interaction, frequency of content updates, matching the intervention to patient needs and interests, availability of additional components (i.e., print, texts, telephone, apps) and inclusion of professional support. In addition, the frequency of log ins and patient use is expected to decline over time; therefore, long-term use of web-based interventions may be somewhat limited and more research is needed to identify approaches to enhance engagement and adherence [31].

As previously discussed, self-monitoring is one of the most important behavioral skills for behavior change. However, the tedious nature of tracking behavior can make long-term self-monitoring problematic. Advances in technology have allowed more self-monitoring options, including mobile apps and body monitors, which many patients may find more acceptable. Technology in self-monitoring also facilitates sharing of information with professionals through computer generated reports or automated synchronization of information. In weight loss research, use of remote feedback increased self-monitoring adherence [32].

Many mobile apps and devices for physical activity use self-monitoring as the central feature using self-entered exercise data, heart rate, accelerometry, and/or GPS data with automated feedback features. Various low-cost mobile apps are available for smartphone and tablet use. Wearable activity monitoring devices range in price from approximately $20 for a standard pedometer to $200 and up for high-end heart rate, GPS, or accelerometer monitors. Monitoring applications and devices are useful for goal-setting, tracking progress, and acquiring feedback. Some systems allow for personalized programing and professional supervision. Considerations include the specific behavior

TABLE 35.4 Print and Online Resources for Health Professionals and Patients

Title	Description	Source
ACSM's Behavioral Aspects of Physical Activity and Exercise	Book geared toward health professionals and students focusing on physical activity behavior change; includes resources for practitioners and consumers	Published by Lippincott Williams & Wilkins
ACSM's Keys to Exercise Success	Tools for assessing factors that influence exercise behavior, including benefits, goals, and action plans; also includes links to programs and information	http://www.myexerciseplan.com/assessment/
ACSM's Resource Manual for Guidelines for Exercise Testing and Prescription	Book geared toward health professionals and exercise science students providing a comprehensive overview of exercise science topics	Published by Lippincott Williams & Wilkins
Active Living Every Day	Book geared toward consumers that promotes the learning of behavioral skills for physical activity in sequential sessions	Published by Human Kinetics
CDC Youth Physical Activity Guidelines Toolkit	Toolkit for practitioners working to promote physical activity among youth and adolescents in schools, communities, and families	http://www.cdc.gov/healthyyouth/physicalactivity/guidelines.htm
Compendium of Physical Activities	Website that provided updated MET codes for quantification of energy costs across a comprehensive list of physical activities	https://sites.google.com/site/compendiumofphysicalactivities/
Exercise Is Medicine	Resources provided in partnership with the *American College of Sports Medicine* to promote exercise prescription in health-care settings and support physical activity across various settings	http://exerciseismedicine.org/
Motivating People to Be Physically Active	Book for health professionals to assist in the planning, development, implementation, and evaluation of physical activity programs	Published by Human Kinetics
National Diabetes Prevention Program	Research-based curriculum for delivery of a lifestyle intervention for nutrition and physical activity behavior change; includes sessions and facilitation guide	http://www.cdc.gov/diabetes/prevention/recognition/curriculum.htm
Physical Activity Guidelines for Americans	Comprehensive recommendations for physical activity promotion for ages 6 and older; includes guides, fact sheets, and reports	http://www.health.gov/paguidelines/
We Can! Ways to Enhance Children's Activity and Nutrition	Tools for health professionals and consumer resources for healthy eating, physical activity, and reducing screen time for kids and families	http://www.nhlbi.nih.gov/health/public/heart/obesity/wecan/

ASCM, American College of Sports Medicine; CDC, Centers for Disease Control.

monitored, the cost of the device, the accuracy of data, long-term wearability, and patient acceptability. A recent study comparing traditional intervention and technology-based intervention using a multisensor armband for weight loss found that the technology approach was more cost-effective and successful in producing lifestyle change [33]. These approaches warrant further research in other patient populations such as CF.

Monitoring is a central feature of most mobile apps and devices for physical activity. However, technology can also be used to guide workouts, enhance enjoyment, connect patients to people with similar goals, and enhance accountability (Table 35.5). Mobile applications are available that guide participants through specific exercises using audio and video prompts. Unlike the exercise videos of the past, apps allow users to customize duration and type of exercises with additional features such as music, prompts, video tutorials, and integration with social media. Apps and gaming devices are also available to enhance the enjoyment of physical activity through games, collection of points, and competition with friends. Evidence suggests that enjoyment of physical activity through music, distraction, or games may promote adherence [15,34].

Although the research is limited in CF patient populations, evidence suggests that exercise gaming devices (e.g., Xbox Kinect™, Wii™) may be useful in promoting exercise for children [35] and adults [36]. Initial research also supports the potential use of multisensor armbands

TABLE 35.5 Features to Consider when Selecting Physical Activity Technology

- Customization (individual body data, goals)
- Tracking of physical activity information
  - Frequency
  - Duration
  - Distance
  - Calories
  - Steps
  - Points
  - Heart rate
- Prompts/reminders
- Feedback and report generation
- Integration with social media
- Competition with self or other users
- Integration with music or other devices
- Educational component/tutorials
- Physical activity planning/workout customization

in individuals with CF [37]; however, more information is needed regarding how the devices could serve as an intervention to increase physical activity in this population.

Although technology is ever-changing, these methods can be evaluated based on key elements recommended for physical activity behavior change. Questions for evaluating technology for personal and professional use include the following:

- Do physical activity and exercise recommendations follow evidence-based guidelines?
- Is the device accurate and reliable?
- Does the technology include behavioral skill-building elements?
- Is the technology appropriate for user abilities, preferences, and budget?
- What professional support does the patient need, and does the device support sharing and feedback?
- Will the device keep user interest and be a long-term resource?

Examples of technology for physical activity promotion are given in Table 35.6. Health professionals should make an effort to stay informed on new developments in technology and e-health so that they can assist patients in selecting tools that will best meet their physical activity needs and goals. Additional research is needed on the use of different methods to promote physical activity in patients with CF. Relatively few exercise and physical activity interventions have been conducted to examine long-term effectiveness [38]. However, Internet and telemedicine interventions as have been implemented in other populations warrant consideration [39]. Regardless of method, technology selection should be patient focused, with primary consideration given to physical activity goal, current technology use, patient interest, and motivational support.

TABLE 35.5 Examples of Technology-Assisted Resources for Physical Activity

Tool	Examples	Features
Exercise and nutrition tracking applications	My Fitness Pal SparkPeople Livestrong	Exercise and food databases; customizable weight goals; Internet synchronization with additional online features
GPS Walk/Run applications	RunKeeper MapMyRun Nike+	Real-time information on time, distance, etc.; training plans and maps; exercise prompts and tracking
Exercise training applications	Nike Training Club Fitness Buddy Obstacles XRT Pocket Yoga	Photo and video instruction on specific exercises; training plans and exercise routines customized to fitness goals
Motivational applications	Charity Miles Zombies, Run! Endomondo TempoRun Fitocracy GymPact Fleetly	Social media integration; customizable challenges and competition with self or others; entertainment features to enhance workouts
Wearable devices	Basis B1 Band BodyMedia Fit Core Fitbit devices Jawbone UP MOTOACTV Nike + Fuelband Heart rate monitors GPS watches	Worn on clothing, wrist, or arm; accelerometer tracks physical activity and synchs data to app or computer; may include additional features including tracking energy expenditure and sleep quality
Exergaming devices	Nintendo Wii Xbox Kinect PlayStation Move Dance Dance Revolution	Commercial and home-use video-gaming devices lead users through activity and detect movement to award points based on task completion

Note: These are examples only. This is not an exhaustive list, and no product endorsement is implied.

References

[1] Wilkes DL, Schneiderman JE, Nguyen T, Heale L, Moola F, Ratjen F, et al. Exercise and physical activity in children with cystic fibrosis. Paediatr Respir Rev September 2009;10(3):105–9.

[2] Hebestreit H, Kieser S, Junge S, Ballmann M, Hebestreit A, Schindler C, et al. Long-term effects of a partially supervised conditioning programme in cystic fibrosis. Eur Respir J March 2010;35(3):578–83.

[3] Schmidt AM, Jacobsen U, Bregnballe V, Olesen HV, Ingemann-Hansen T, Thastum M, et al. Exercise and quality of life in patients with cystic fibrosis: a 12-week intervention study. Physiother Theory Pract November 2011;27(8):548–56.

[4] Fiore MC, Jaén CR, Baker TB, Bailey WC, Benowitz NL, Curry SJ, et al. Treating tobacco use and dependence: 2008 update. Washington, DC: US Department of Health and Human Services; 2008. Available from: www.ahrq.gov/clinic/tobacco/treating_tobacco_use08.pdf. [accessed 12.01.14].

[5] Alexander SC, Cox ME, Boling Turer CL, Lyna P, Østbye T, Tulsky JA, et al. Do the five A's work when physicians counsel about weight loss? Fam Med March 2011;43(3):179–84.
[6] Vallis M, Piccinini-Vallis H, Sharma AM, Freedhoff Y. Clinical review: modified 5 As: minimal intervention for obesity counseling in primary care. Can Fam Physician January 2013;59(1):27–31.
[7] Joy EA. Practical approaches to office-based physical activity promotion for children and adolescents. Curr Sports Med Rep Nov-Dec 2008;7(6):367–72.
[8] Meriwether RA, Lee JA, Lafleur AS, Wiseman P. Physical activity counseling. Am Fam Physician April 15, 2008;77(8):1129–36.
[9] Carroll JK, Antognoli E, Flocke SA. Evaluation of physical activity counseling in primary care using direct observation of the 5As. Ann Fam Med 2011;9(5):416–22.
[10] ACSM. ACSM's resource manual for guidelines for exercise testing and prescription. 7th ed. Lippincott Williams & Wilkins; 2013.
[11] Prochaska JO, DiClemente CC. The stages and processes of self-change in smoking: towards an integrative model of change. J Consult Clin Psychol 1983;51:390–5.
[12] Marcus B, Forsyth L. Motivating people to be physically active. Champaign, IL: Human Kinetics; 2003. 220 p.
[13] Pekmezi D, Barbera B, Marcus BH. Using the transtheoretical model to promote physical activity. ACSM Health Fit J 2010;14(4):8–13.
[14] Blair SN, Dunn AL, Marcus BH, Carpenter RA, Jaret P. Active living every day. 2nd ed. Champaign, IL: Human Kinetics; 2011.
[15] Garber CE, Blissmer B, Deschenes MR, Franklin BA, Lamonte MJ, Lee IM, American College of Sports Medicine. Quantity and quality of exercise for developing and maintaining cardiorespiratory, musculoskeletal, and neuromotor fitness in apparently healthy adults: guidance for prescribing exercise. Med Sci Sports Exerc 2011;43:1334–59. Retrieved from: http://journals.lww.com/acsm-msse/pages/default.aspx.
[16] U.S. Department of Health & Human Services. Physical activity guidelines for Americans; 2008. http://www.health.gov/paguidelines/.
[17] Bandura A. Self-efficacy: the exercise of control. New York, NY: Freeman; 1997.
[18] American College of Sports Medicine. In: Nigg CR, editor. ACSM's behavioral aspects of physical activity and exercise. Philadelphia: Wolters Kluwer Health/Lippincott Williams & Wilkins; 2013.
[19] Nothwehr F, Yang J. Goal setting frequency and the use of behavioral strategies related to diet and physical activity. Health Educ Res August 2007;22(4):532–8.
[20] Ernst MM, Johnson MC, Stark LJ. Developmental and psychosocial issues in cystic fibrosis. Child Adolesc Psychiatr Clin N Am April 2010;19(2):263–83.
[21] Baker CF, Wideman L. Attitudes toward physical activity in adolescents with cystic fibrosis: sex differences after training: a pilot study. J Pediatr Nurs June 2006;21(3):197–210.
[22] White D, Stiller K, Haensel N. Adherence of adult cystic fibrosis patients with airway clearance and exercise regimens. J Cyst Fibros May 2007;6(3):163–70.
[23] Swisher AK, Erickson M. Perceptions of physical activity in a group of adolescents with cystic fibrosis. Cardiopulm Phys Ther J December 2008;19(4):107–13.
[24] Happ MB, Hoffman LA, DiVirgilio D, Higgins LW, Orenstein DM. Parent and child perceptions of a self-regulated, home-based exercise program for children with cystic fibrosis. Nurs Res September–October 2013;62(5):305–14.
[25] Sallis JF, Grossman RM, Pinski RB, Patterson TL, Nader PR. The development of scales to measure social support for diet and exercise behaviors. Prev Med November 1987;16(6):825–36.
[26] Treiber FA, Baranowski T, Braden DS, Strong WB, Levy M, Knox W. Social support for exercise: relationship to physical activity in young adults. Prev Med November 1991;20(6):737–50.
[27] Epstein LH, Paluch RA, Kilanowski CK, Raynor HA. The effect of reinforcement or stimulus control to reduce sedentary behavior in the treatment of pediatric obesity. Health Psychol July 2004;23(4):371–80.
[28] Lowther M, Mutrie N, Scott EM. Identifying key processes of exercise behaviour change associated with movement through the stages of exercise behaviour change. J Health Psychol March 2007;12(2):261–72.
[29] Alberga AS, Medd ER, Adamo KB, Goldfield GS, Prud'homme D, Kenny GP, et al. Top 10 practical lessons learned from physical activity interventions in overweight and obese children and adolescents. Appl Physiol Nutr Metab March 2013;38(3):249–58.
[30] Exercise Is Medicine. Exercise is medicine health care providers' action guide. Retrieved from: http://exerciseismedicine.org/documents/HCPActionGuide.pdf.
[31] Vandelanotte C, Spathonis KM, Eakin EG, Owen N. Website-delivered physical activity interventions a review of the literature. Am J Prev Med July 2007;33(1):54–64.
[32] Burke LE, Styn MA, Sereika SM, Conroy MB, Ye L, Glanz K, et al. Using mHealth technology to enhance self-monitoring for weight loss: a randomized trial. Am J Prev Med July 2012;43(1):20–6.
[33] Archer E, Groessl EJ, Sui X, McClain AC, Wilcox S, Hand GA, et al. An economic analysis of traditional and technology-based approaches to weight loss. Am J Prev Med August 2012;43(2):176–82.
[34] Papandonatos GD, Williams DM, Jennings EG, Napolitano MA, Bock BC, Dunsiger S, et al. Mediators of physical activity behavior change: findings from a 12-month randomized controlled trial. Health Psychol July 2012;31(4):512–20.
[35] O'Donovan C, Greally P, Canny G, McNally P, Hussey J. Active video games as an exercise tool for children with cystic fibrosis. J Cyst Fibros November 1, 2013. pii: S1569-1993(13) 00165-173.
[36] Holmes H, Wood J, Jenkins S, Winship P, Lunt D, Bostock S, et al. Xbox Kinect™ represents high intensity exercise for adults with cystic fibrosis. J Cyst Fibros December 2013;12(6):604–8.
[37] Cox NS, Alison JA, Button BM, Wilson JW, Morton JM, Dowman LM, et al. Validation of a multi-sensor armband during free-living activity in adults with cystic fibrosis. J Cyst Fibros December 26, 2013. pii: S1569-1993(13) 00225-7.
[38] Cox NS, Alison JA, Holland AE. Interventions for promoting physical activity in people with cystic fibrosis. Cochrane Database Syst Rev December 13, 2013;12:CD009448.
[39] van Sluijs EM, McMinn AM, Griffin SJ. Effectiveness of interventions to promote physical activity in children and adolescents: systematic review of controlled trials. BMJ October 6, 2007;335(7622):703.

CHAPTER

36

Diet, Food, Nutrition, and Exercise in Cystic Fibrosis

Andrea Kench[1], Hiran Selvadurai[2]

[1]Department of Nutrition & Dietetics and Respiratory Medicine, The Children's Hospital Westmead, Westmead, NSW, Australia; [2]Respiratory Medicine, The Children's Hospital Westmead, Westmead, NSW, Australia

36.1 NUTRITION CONSIDERATIONS FOR EXERCISE

Optimizing nutrition status is an important part of therapy for the cystic fibrosis (CF) patient and can significantly affect patient outcomes. The link between body mass index (BMI) and lung function forced expiratory volume in 1 s (FEV_1) is well documented [1–4], and since the introduction of pancreatic enzyme replacement therapy, a high-fat diet has been the backbone of the CF diet. With advances in therapies and improved life expectancy, nutrition advice is becoming increasingly tailored to individual needs. There is now evidence to support dietary modification for the management of multiple comorbidities, including cystic fibrosis related diabetes (CFRD), liver disease (CFLD), and transplantation. Although less common, weight management strategies for the overweight CF individual with raised lipid profiles are also becoming an issue and a consideration once thought unlikely. The role of nutrition in exercise for CF is an area of growing interest. It has been shown that declining nutrition status is associated with reduced activity levels and exercise capacity [5–8]. Optimizing nutrition from a young age and throughout childhood will not only affect overall CF outcomes but may also offset the decline in physical activity with age. Despite this, evidence to support nutrition recommendations for exercise in CF is limited. Further research is still required to first fully understand metabolism and substrate utilization for this population. This chapter will present the current evidence and best available recommendations for optimizing nutrition for CF patients when exercising.

36.1.1 Energy Requirements with Exercise

Determining energy requirements for exercise in healthy individuals can be difficult. In the practical setting, health professionals usually rely on predictive equations to give a rough estimate of total energy expenditure (EE). Multiple factors including the duration, frequency, and intensity of exercise as well as an individual's age, body size, genetics, and fat-free mass can affect metabolism, substrate utilization, and total EE [9]. An imbalance in energy intake (EI) vs. expenditure can result in altered substrate utilization. A poor EI is usually associated with a lack of available carbohydrate, the initial fuel source for aerobic exercise. This results in the preferential metabolism of fat and lean tissue. The breakdown of lean tissue and muscle mass not only has detrimental effects on exercise and performance but can also compromise immune, endocrine, and musculoskeletal function [10].

EE in CF has been studied with often conflicting results. Some studies have found resting energy expenditure (REE) to not be elevated in the CF population [11–13], but most report REE to be elevated [14–18]. Despite this, there is strong evidence to support that a rise in REE is correlated with worsening lung disease [19]. In the practical setting, EE is difficult to calculate for the CF individual because the predictive equations used for the healthy population are not validated in CF. In the past, an estimate of 120–150% of the recommended daily intake for age and gender was recognized as a starting point to account for increased expenditure and losses via malabsorption [20–22]. It has now been recognized that individual variation in estimated energy requirements (EERs) is larger than

Diet and Exercise in Cystic Fibrosis
http://dx.doi.org/10.1016/B978-0-12-800051-9.00036-5

first thought. As per the 2005 Cystic Fibrosis Foundation Clinical Practice Recommendations, EI ranging from 110% to 200% of that for the healthy population is shown to result in weight gain for children and adults with CF [23]. Regardless of the method used to estimate EER in CF, health professionals are encouraged to recognize these figures as crude estimates only. A thorough nutritional assessment combined with experience and clinical judgment remains an important consideration.

The literature describing the effect of exercise on EE in CF is similar to that of REE in that the results are varied. Some studies have found EE to be elevated during exercise with CF [17,24], but most report EE to be unchanged when compared with the healthy population [11,16,18,25]. Postexercise, EE has been shown to be elevated in CF [11,18]. This is thought to be a compensatory response to the ventilation-perfusion disparity in which higher ventilatory requirements result in an increased work of breathing postexercise [18]. As with the healthy population, it is difficult to endorse a one-size-fits-all recommendation for EI with exercise in CF. The effect of the type of exercise and individual characteristics (e.g., age, body size, genetics, and fat-free mass) as well as disease-specific genotype and phenotype considerations may all affect total EE with exercise.

36.1.2 Protein Requirements with Exercise

Protein requirements are also difficult to estimate, especially for the active individual. A person's age, sex, EI, and carbohydrate availability as well as the duration, intensity, and type of exercise will all affect an individual's protein metabolism [9]. In practice, it is often recommended that protein requirements exceed the current recommended daily allowance of 0.8g/kg body weight and 10–35% of total EI for healthy individuals over the age of 18years [9]. This is especially the case for those participating in endurance or resistance training. Despite this, there is limited evidence to support the theory that protein intake should be increased with any form of exercise.

Protein metabolism in CF has been found to be altered [26]. Malabsorption, increased nitrogen losses via feces and sputum, and altered protein metabolism all contribute to increased protein requirements for CF individuals. Protein catabolism in CF also increases with significant systemic inflammation, pulmonary disease, and nutrient deficits [26]. This is particularly the case with protein and energy deficits or in severe cases in which protein energy malnutrition is present. The result of protein and energy deficits and a catabolic state will ultimately affect growth for the CF individual and likely have a negative effect on lean body mass. In addition to this, it has been found that deficits in protein intake and EI result in reduced exercise tolerance and pulmonary muscle function [27].

Protein intake in the CF population generally exceeds the recommended daily intake for the healthy population that is based on age and gender. The intake of protein has also been found to increase as total daily EI increases [28,29]. Despite the lack of evidence to support specific protein recommendations for the CF population, it has been generally accepted that protein should contribute to approximately 15% of total EI for the CF individual [26]. As with most areas of CF sports nutrition, protein-specific requirements for exercise are yet to be determined. This does not take away from the importance of encouraging adequate nutrition and optimal body composition, specifically lean body mass, for the CF person participating in regular exercise.

36.1.3 Substrate Utilization with Exercise

Substrate utilization during exercise has been studied extensively in the healthy population with a focus on athletes, training, and performance. Although the results of these studies are beyond the scope of this chapter, the summary of specific macronutrient recommendations for exercise, as outlined in Table 36.1, was developed with the underlying knowledge of substrate utilization during exercise.

Substrate utilization in CF is an area of growing interest and research. To date, studies looking at substrate utilization focus on aerobic exercise and most use respiratory quotient (RQ) data to determine carbohydrate and fat utilization. The results of early studies indicate that during and after maximal and submaximal exercise, there is no difference in the RQ between CF patients and their matched controls [11,18]. This is suggestive that carbohydrate and therefore fat metabolism is unchanged during exercise for the CF individual. Spicher et al. [16], by analyzing the RQ, demonstrated that substrate utilization may be altered in CF. In this study, the RQ was elevated at rest and during exercise, suggesting that carbohydrate oxidation contributed to a greater percentage of EE in CF versus the controls [16]. Recently, Nguyen et al. studied whole-body oxidation rates of fat and carbohydrate during prolonged submaximal exercise in CF children [25]. To date, the results of this study provide the most comprehensive insight into substrate utilization in the CF population. Whole-body fat and carbohydrate oxidation and blood plasma fatty acid levels were analyzed for six CF boys and matched controls during two 30-min bouts of submaximal exercise at regular intervals. Substrate utilization was determined using gas exchange and oxidation calculations. Similar to Spicher, Nguyen found that substrate utilization is altered in CF [25]. The main results are summarized in the following subsections [25].

The rate (Figure 36.1) and total amount of fat oxidized were significantly lower in CF.

TABLE 36.1 Summary of the American College of Sports Medicine, American Dietetic Association, and Dietitians of Canada's Joint Position Statement for Nutrition and Athletic Performance [9]

	Recommended Dietary Allowance	Acceptable Macronutrient Density Range	Role and Considerations
Carbohydrate	6–10 g/kg		Blood glucose maintenance with exercise Muscle glycogen replacement Mid-exercise recommendations: • 30–60 g/h for blood glucose maintenance Postexercise recommendations: • 1–1.5 g/kg in the first 30 min and every 2 h for 4–6 h to replace glycogen stores Inconclusive evidence to support the recommendation of low glycemic index carbohydrates
Protein	0.8 g/kg	10–35% energy intake	Limited evidence to support that athletes require greater than the recommended dietary allowance; however, in practice, athletes are being recommended the following: • Endurance training: 1.2–1.4 g/kg • Strength training: 1.2–1.7 g/kg It is important to recommend that adequate energy from carbohydrate is provided to ensure that additional protein and amino acids are reserved for protein synthesis and are not oxidized as a fuel source. When possible, natural protein sources from diet alone are encouraged over protein supplements.
Fat		20–30% energy intake	A breakdown of 10% saturated, 10% polyunsaturated, and 10% monounsaturated fat is recommended to ensure an adequate intake of all essential fatty acids.

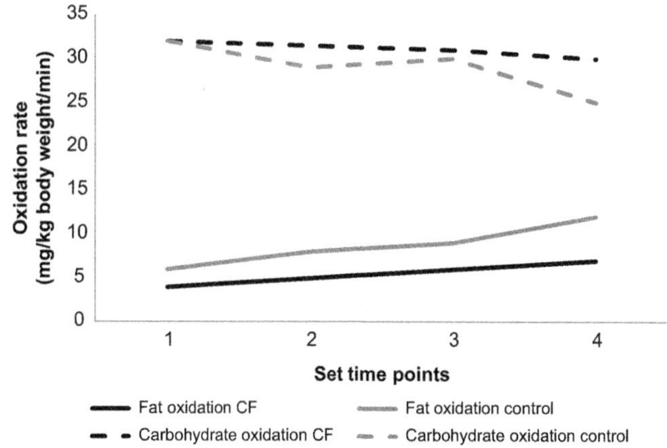

FIGURE 36.1 Fat and carbohydrate oxidation in cystic fibrosis (CF) patients vs. controls adapted from Nguyen et al. [25]. Each time point represents 6 min of exercise. Points 1 and 3 were taken between the 12th and 18th minutes and points 2 and 4 were taken between the 23rd and 29th minutes of each 30-min bout of exercise.

Fat as a substrate contributing to total EE was also significantly lower (Figure 36.2).

Carbohydrate oxidation rates remained unchanged in CF when compared with the controls that experienced the expected decline in carbohydrate oxidation over time (Figure 36.1). There was also no significant difference between the average rate and total carbohydrate oxidation between the two groups. Carbohydrate as a substrate contributing to total EE was significantly higher in CF (Figure 36.2).

The CF cohort exhibited lower plasma free fatty acid (FFA) concentrations during exercise.

Nguyen et al. concluded that CF children have altered fat metabolism that may be affected by the inability to readily mobilize FFA. To date, there are no other published studies investigating the pathophysiology of altered substrate utilization in CF.

Given the relative paucity of good evidence, it is difficult to provide the specific recommendations for exercise and sports nutrition for subjects with CF. However, with what we know to date, the strongest argument would be to support what is already recommended around carbohydrate consumption and exercise in the healthy population: Adequate carbohydrate should be consumed before and during exercise to ensure blood glucose maintenance, muscle glycogen replacement, and adequate substrate for fuel metabolism and amino acid preservation [9]. This is particularly important in CF, in which it is has been

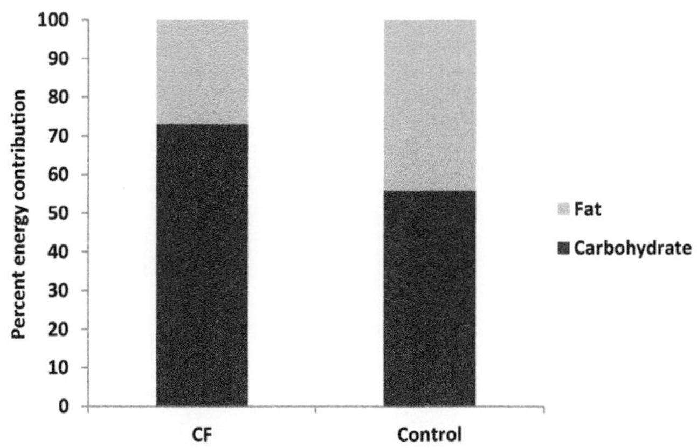

FIGURE 36.2 Percentage of energy expenditure contribution of fat and carbohydrates in children with cystic fibrosis (CF) vs. controls. *Adapted from Ref. [25].*

TABLE 36.2 Summary of the Micronutrient Considerations for Nutrition and Athletic Performance as per the American College of Sports Medicine Position Stand: Nutrition and Athletic Performance [9]

	Role and Considerations
B vitamins	Thiamin, riboflavin, niacin, pyridoxine, pantothenic acid, and biotin: • Play an important role in energy production with exercise Folate and B12: • Involved in red blood cell production and tissue repair These vitamins can be at increased risk for the vegetarian or female athlete, especially those with disordered eating.
Antioxidants • Vitamin C and E • β-carotene • Selenium	Assist in the prevention of oxidative damage to cell membranes. These vitamins and trace elements are at greatest risk for people following a low-fat diet and those that have a poor intake of fruit, vegetables, and whole grains or have a restricted energy intake (EI).
Vitamin D	Plays an important role in bone health because it is involved in calcium absorption and the regulation of serum calcium and phosphorus levels.
Calcium	Involved primarily in the maintenance of blood calcium levels and bone health. It is also plays a role in the regulation of muscle contraction, nerve conduction, and blood clotting. Vitamin D and calcium together play a role in the maintenance of bone mineral density. Low bone mineral density is of particular concern for the female athlete. This is especially the case if EI is low and there is an inadequate intake of calcium-rich foods.
Iron	Primarily involved in the production of hemoglobin and myoglobin and the oxygen-carrying capacity of red blood cells. Deficiencies can result in anemia, impaired muscle function, and reduced work capacity. Iron deficiency in the athlete is most often a result of inadequate EI. The vegetarian diet; periods of rapid growth; and increased losses with donation, menstruation, sweat, urine, and feces should also be considered.
Zinc	Involved primarily in the growth, building, and repair of muscle, but it also plays a role in energy production and immune function. Deficiency is most commonly associated with vegetarian diets low in animal protein.
Magnesium	Plays a role in cellular metabolism. This includes glycolysis and fat and protein metabolism. It has also been linked to the regulation of membrane stability as well as neuromuscular, cardiovascular, immune, and hormonal functions.

shown that carbohydrate, as a substrate, contributes to a greater percentage of total EE and that postexercise energy requirements are elevated [16,25].

36.1.4 Micronutrient Considerations

Vitamins and minerals play an important role in several of the body's physiological processes. These include energy production, hemoglobin synthesis, immune function, prevention against oxidative damage, and the maintenance of bone health [9]. Regular exercise has been shown to increase the body's turnover and loss of some micronutrients [9]. As a result, micronutrient requirements for the repair, building, and maintenance of lean body tissue may be elevated [9]. The most common nutrients at risk for athletes are summarized in Table 36.2 [9].

When adequate EI and dietary variety is unable to be sustained, a multivitamin is recommended [9].

The effect of exercise on micronutrient status in CF has not been studied. Fat-soluble vitamin (FSV) supplementation is the most common form of micronutrient supplementation in CF, especially for pancreatic-insufficient patients. Malabsorption, inadequate intake, liver disease, progressive lung disease, poor compliance with pancreatic enzyme replacement therapy, and vitamin supplementation as well as previous bowel resections can all affect the level of supplementation required to maintain FSV serum levels within the normal range [22]. Other nutrients of increasing interest in CF, as with exercise, include iron, calcium, zinc, and magnesium. Further information regarding these nutrients in CF is available in national CF nutrition guidelines and consensus documents [21–23].

36.1.5 Dietary Sports Supplements

The use of dietary supplements and ergogenic aids in exercise performance is an ever evolving and often lucrative market. To date, very few ergogenic aids have been shown to significantly improve performance [30]. The most commonly used and available dietary supplements and ergogenic aids are listed in Table 36.3. The A–D grouping is according to the Australian Institute of Sport (AIS) group classification system [31]. More detailed information regarding each supplement, including supporting literature, can be found at www.ausport.gov.au/ais/nutrition/supplements/classification_test.

At present, there are no studies looking at the effect of ergogenic aids on exercise performance in CF. As a result, the use of any supplement and ergogenic aid in CF should be done with caution and after consultation with the treating physician. Considerations for use of these products in CF, as per the AIS grouping, are listed in Table 36.4.

36.1.6 Protein Supplements

The effects of protein supplementation on the maintenance, repair, and synthesis of skeletal muscle remain an area of interest in sports nutrition. Studies looking at the effects of protein supplements originally focused on the use of specific amino acids [9]. Today the focus has shifted to recommending higher quality proteins, including whey, casein, and soy, with food sources being recommended over commercial protein powders [9]. Regardless of the protein source, for maximal skeletal muscle benefits, it is recommended that protein be consumed with adequate carbohydrate and around the time of training [32].

Protein powders and amino acid supplements should be used with caution in CF because they are a potential source of contamination with illegal and banned substances. Patients with CFLD should first discuss any significant increase in protein intake with the treating gastroenterologist.

36.1.7 Key Recommendations and Practical Considerations

With limited research specific to nutrition in exercise and CF, it is likely to be some time before evidence-based guidelines for this area will be possible. Until then, it would be reasonable to recommend that advice remains in line with the available pediatric and adult evidence-based general CF nutrition guidelines. Aspects of sports nutrition can then be applied after careful consideration.

The following key recommendations include practical advice for how nutrition and exercise considerations can be applied to the CF population.

36.1.7.1 The Nutrition Assessment

Always complete a thorough nutrition assessment before advising the CF patient on dietary modifications for exercise. Special considerations include

- The effect of CF-related comorbidities (see chapter on CFRD)
- Hydration (See Section 36.3)

36.1.7.2 Energy Requirements

Consider the effect of exercise and disease severity when estimating energy requirements.

1. Choose a predictive equation to calculate the resting metabolic rate (RMR).
   The Harris–Benedict equation [33] (Table 36.5) is most often used for the healthy adult population when considering exercise requirements [9].
   Most recently, the Institute of Medicine (IOM) has released the IOM equation [34] (Table 36.6) to replace the Harris–Benedict equation. Its use in the clinical and sporting environment is likely to increase; however, weight and height must be available for this calculation.
   The Schofield equation [35] (Table 36.7) is often used in CF.
2. Include a physical activity level factor
   Harris–Benedict [33] (Table 36.5), IOM [34] (Table 36.6), or Schofield [35] (Table 36.7).
3. Include a disease factor
   Multiply the RMR by 1.1–2.0 depending on disease severity to account for the 110–200% increased losses and expenditure in CF [23].

TABLE 36.3 Australian Institute of Sport Supplement Group Classification System [31]

Group A Supported for Use in Specific Situations	
Sports drinks	Refer to section on hydration for more details
Sports confectionary and gels	A concentrated, compact, and convenient source of carbohydrate (65–70% in gels and 75–90% in confectionary) that may contain "active ingredients" including caffeine. They provide an excellent source of fuel and are easily digested.
Liquid meals	Usually powders that can be dissolved in water or milk, but ready-to-drink varieties are also available. They are typically high in carbohydrate, low in fat, and contain moderate amounts of protein as well as being fortified with vitamins and minerals.
Whey protein	
Sports bars	Primarily used as a concentrated form of carbohydrate to fuel athletes before, during, or after exercise. Also contain variable amounts of protein (5–30g/serving) and micronutrients.
Calcium supplement	Refer to micronutrient considerations (Table 36.2).
Iron supplement	Refer to micronutrient considerations (Table 36.2).
Probiotics	*Lactobacillus acidophilis* and *Bifidobacterium bifidum* are the most commonly used and commercially available strains. An area of emerging sports nutrition research, but the general benefits include improved gastrointestinal and immune health and allergy prevention.
Multivitamin mineral	These are not found to be performance enhancing unless used to correct an existing deficiency. It is recommended that supplementation is considered if a deficiency exists or dietary quality is poor.
Vitamin D	Emerging evidence to support that strength, power, reaction time, and balance may be enhanced if provided with vitamin D supplementation to correct deficiency.
Electrolyte replacement	Used to replace losses via sweat during exercise and are a good alternative to sports drinks.
Caffeine	Has been shown to primarily affect the central nervous system (fatigue perception) but is also linked to adrenaline stimulation, fat mobilization, and muscle contractility. The diuretic effect of caffeine-containing beverages is minimal. Inconsistent evidence to support use as a performance enhancer and is no longer banned by the World Anti-Doping Agency.
Creatine	Plays a role in the ATP-creatine phosphate pathway, in which ATP is the primary fuel source for high-intensity exercise of up to 10-s duration. Most beneficial for resistance and high-intensity interval training.
Bicarbonate	Most commonly used to act as a buffer in the blood because it plays a role in the maintenance of pH and electrolyte gradients between intracellular and extracellular environments. Used for anaerobic, high-intensity exercise to prevent muscle fatigue.

Group B Deserving of Further Research	Group C No Meaningful Proof of Beneficial Effects	Group D Banned or at High Risk of Contamination
β-alanine	Ribose	*Stimulants*
Beetroot juice/nitrate	Coenzyme Q10	Ephedrine
Antioxidants C and E	Vitamins outside of Group A use	Strychnine
Carnitine	Ginseng	Sibutramine
HMB	Other herbals (coryceps, rhodiola, and rosea)	Methylhexanamine
Fish oils	Glucosamine	Other herbal stimulants
Probiotics for immune support	Chromium picolinate	*Prohormones/hormone boosters*
Other polyphenols as antioxidants and anti-inflammatories	Oxygenated waters	Androstenedione
	MCT oils	19-norandrostenione/Ol
	ZMA	Other prohormones
	Inosine	*Tribulus terrestris* and other testosterone boosters
	Pyruvate	DHEA

More information regarding each product can be found at http://www.ausport.gov.au/ais/nutrition/supplements/group_a.
ATP, adenosine triphosphate; HMB, hydroxymethylbutyrate; MCT, medium-chain triglycerides; ZMA, zinc monomethionine aspartate and magnesium aspartate; DHEA, dehydroepiandrosterone.

TABLE 36.4 Australian Institute of Sport Supplement Group Classification System [31] with Potential Considerations for People with CF

	Group A	Groups B–D
Considerations	Known to be safe for use and appropriate for recommendation in CF: • Sports drinks and electrolyte replacement • Sports bars • Calcium, iron, vitamin D, and multivitamin supplements • Probiotics	The potential risk of using these products for the exercising CF patient include, but may not be limited to, medication interactions and harmful effects to the liver.
	Should be used with caution in CF: • Sports gels and confectionary–CFRD considerations • Caffeine—not advisable for use in children • Creatine—no studies to support use in CF • Bicarbonate	Different regulatory laws regarding health and therapeutic claims exist among countries. With most products being available online, the ease of accessing dietary supplements and ergogenic aids has increased. Consumers need to be aware that not all packaging claims are scientifically proven.
		Several of these substances are banned by the World Anti-Doping Agency and are not permitted for use in competitive sport.
		There is significant risk that these products may be contaminated with other products not listed as active ingredients. This is again of particular concern for patients with CFLD.
		Some patients may not be forthcoming and willing to disclose their use. Adolescent and young adult males who regularly train in the gym setting are most likely to show interest in these products. Although most protein powders are usually safe, preworkouts are of particular concern because they often contain multiple supplements listed in these categories.

CF, cystic fibrosis; CFRD, cystic fibrosis related diabetes; CFLD, cystic fibrosis related liver disease.

TABLE 36.5 Harris–Benedict Equation and Physical Activity Levels [33]

HARRIS BENEDICT EQUATION

Males: $RMR = 66.47 + (13.75 \times wt) + (5 \times ht) - (6.76 \times age)$

Females: $RMR = 65.51 + (9.56 \times wt) + (1.85 \times ht) - (4.68 \times age)$

HARRIS PHYSICAL ACTIVITY LEVELS

Activity Level	Activity Factor	Definition
Sedentary	1.2	No exercise—inactive.
Mild activity	1.375	Minimum of 20 min exercise 1–3 days each week or the maintenance of a busy lifestyle including walking for long periods of time.
Moderate activity	1.55	Minimum of 30–60 min of intense exercise 3–4 times each week. This also includes labor intense occupations.
Heavy activity	1.7	≥60 min of intense exercise 5–7 times per week. This also includes those working in labor-intensive jobs.
Strenuous or very heavy exercise	1.9	Extremely active. This is most often athletes with demanding training schedules (multiple sessions daily). Some demanding jobs including shovel coal would also fall in this category.

TABLE 36.6 Institute of Medicine Equation and Activity Levels [34]

IOM EQUATION	
0–2 years	EER = (89 × wt) − 100
Boys 3–18 years	EER = (88.5 − (61.9 × age) + PA × [(26.7 × wt) + (903 × ht)]
Girls 3–18 years	EER = (135.3 − (30.8 × age) + PA × [(10 × wt) + (934 × ht)]
Men >18 years	EER = (662 − (9.53 × age) + PA × [(15.91 × wt) + (539.6 × ht)]
Women >18 years	EER = (354 − (6.91 × age) + PA × [(9.36 × wt) + (726 × ht)]

IOM PHYSICAL ACTIVITY LEVELS

Activity Level	Male Children	Female Children	Male Adults	Female Adults	Definition
Sedentary	1.0	1.0	1.0	1.0	Light physical activity required for independent living.
Low active	1.13	1.16	1.11	1.12	30 min of moderate—vigorous exercise daily.
Active	1.26	1.31	1.25	1.27	60 min of moderate—vigorous exercise daily.
Very active	1.42	1.56	1.48	1.45	≥60 min of moderate—vigorous exercise daily.

TABLE 36.7 Schofield Equation and Activity Levels [35]

SCHOFIELD EQUATION			
Male 0–3 years	(0.249 × wt) − 0.127	Female 0–3 years	(0.244 × wt) − 0.130
Male 3–10 years	(0.095 × wt) + 2.11	Female 3–10 years	(0.085 × wt) + 2.033
Male 10–18 years	(0.074 × wt) + 2.754	Female 10–18 years	(0.056 × wt) + 2.898
Male 18–30 years	(0.063 × wt) + 2.896	Female 18–30 years	(0.062 × wt) + 2.036
Male 30–60 years	(0.048 × wt) + 3.653	Female 30–60 years	(0.034 × wt) + 3.538
Male 60+	(0.049 × wt) + 2.459	Female 60+	(0.038 × wt) + 2.75

SCHOFIELD ACTIVITY FACTORS

Activity Level	Males	Females	
Bed rest	1.2	1.2	Inactive.
Sedentary	1.3	1.3	Very physically inactive both at work and in leisure. Little to no exercise.
Lightly active	1.6	1.5	20 min of intense exercise 1–2 times per week or the daily routine includes some walking.
Moderately active	1.7	1.6	20–45 min of intense exercise 3–4 times per week or a job that contains a significant amount of walking or intensity.
Very active	2.1	1.9	≥60 min of intense exercise 5–7 times per week. This also includes those working in labor-intensive jobs.
Extremely active	2.4	2.2	Very-high intensity exercise on most days of the week. This usually applies to athletes with multiple daily training sessions.

36.1.7.3 Protein Requirements

1. Calculate protein requirements on the basis of the following recommendations for the healthy population [36]:
   Infants: 1.5 g/kg/day
   1–3 years: 0.95 g/kg/day
   4–14 years: 0.85 g/kg/day
   Adults: 0.8 g/kg/day
2. Determine if the calculated protein requirements fall within the acceptable macronutrient density range:
   Healthy population: 10–35% EI [36]
   CF population: minimum 15% EI [26]

Given the importance of adequate protein with exercise, it would not be unreasonable to advise 15–35% EI from protein for all CF people engaging in regular exercise.

36.1.7.4 Energy Balance

- Discourage patients and parents from limiting participation in sports and exercise because of the fear of burning too much energy. Parents often hesitate at the idea of encouraging too much physical activity for their child with CF because, if not balanced with the appropriate diet, it can result in energy deficits and weight loss.
- From a young age, a team approach to promote the health benefits of exercise in CF as well as the role of nutrition may help reduce parental anxiety.

36.1.7.5 Macronutrient Recommendations

- Further research on substrate utilization during exercise in CF is required before providing set macronutrient targets.
- Table 36.8 provides practical considerations regarding macronutrient distribution and meal/snack ideas for exercise and CF.

36.1.7.6 Micronutrient Recommendations

- Many of the same micronutrients of concern with exercise are also at risk in the CF population, and it would be advisable to ensure adequate monitoring is completed for the CF patient involved in regular exercise.
  - Aim to monitor micronutrients at risk at a minimum of each year.
- Supplementation should be according to biochemical profile and clinical status.
- Consider the following when interpreting biochemical results:
  - Adherence to vitamin supplementation
  - Clinical status—acute illness and inflammation can affect some serum biochemical markers. This is particularly relevant for vitamin A and iron studies.
  - Seasonal variation—particularly with vitamin D
- As per the healthy population, a multivitamin should only be required if a balanced diet with adequate energy is not able to be maintained.

36.1.7.7 The Type of Exercise

- Consider the effect of different types of exercise on body composition.
- Resistance (strength) training has been shown to improve total and fat-free mass when compared with aerobic training in CF [38].

36.1.7.8 Dietary Supplements and Ergogenic Aids

- Most dietary supplements and ergogenic aids should be used with caution and in consultation with the medical team.

36.1.7.9 Children and Sport

36.1.7.9.1 ENERGY

- Aim to maintain a positive energy balance to allow for periods of rapid growth during childhood and adolescence.
- The effects of chronic negative energy balance include short stature, delayed puberty, irregular menstrual cycle, poor bone health, and an increased injury risk [30].

36.1.7.9.2 MEALS AND SNACKS

- Encourage regular meals and snacks with good dietary variety. Fussy eating strategies should be discussed and used from a young age.
- Encourage a protein food source with each meal and snack because children and adolescents have increased protein requirements for growth [30].
- Carbohydrates remain an important fuel source for children and should not be limited, but it is important to encourage good dental hygiene for the prevention of dental caries.

36.1.7.9.3 VITAMINS AND MINERALS

- Calcium, iron, and zinc intake is often inadequate for children and adolescents, especially girls [30], and should be monitored closely in the pediatric CF population.

36.1.7.9.4 PERFORMANCE-ENHANCING SUBSTANCES

- There is no evidence to support the use of performance-enhancing substances in the pediatric population (<18 years of age) [30].

TABLE 36.8 Practical Macronutrient and Food Considerations for People in CF Who Engage in Regular Exercise

	Considerations
MACRONUTRIENTS	
Carbohydrate	*General CF recommendations:* • Relatively high intake is required to help meet energy requirements [22] (see chapter on CFRD for more details). *During times of exercise:* • Consider low glycemic index meal/snack options, especially for patients with CFRD (see chapter on CFRD for more details). • Avoid low-carbohydrate diets with exercise
Protein	*General CF recommendations:* • Requirements are thought to be elevated in CF because of increased losses with malabsorption, nitrogen losses in the feces and sputum, and a potential altered protein metabolism [22,26]. • Should account for ~15% of total daily EI [26]. • In CF, protein intake usually increases as total daily EI increases [37]. *During times of exercise:* • Ensure adequate energy and carbohydrates are consumed with increased protein for best results. • Aim for natural protein food sources when possible. • Consider increasing protein intake up to 35% of total EI to align with the upper recommended range for adults. Note that 10–35% of energy from protein is recommended for healthy individuals over 18 years of age [9].
Fat	*General CF recommendations:* • Unrestricted intake (unless overweight). • Should account for ~40% of total daily EI [22]. • Aim for >100 g/day for patients older than 5 years [22]. • Although not yet used by all CF centers, annual monitoring of lipid profiles is becoming increasingly common. *During times of exercise:* • A high-fat meal should only be discouraged immediately before exercise because it can slow gastric emptying and may consequently result in abdominal discomfort with exercise.
MEALS AND SNACKS	
Pre-exercise	As per advice for the general population, the pre-exercise meal should be as follows [9]: • Relatively low fat and fiber to promote gastric emptying • High in carbohydrate for blood glucose maintenance • Moderate in protein • A familiar food that is well tolerated • Combined with adequate fluid to maintain hydration Liquid meals would be an appropriate pre-exercise snack for the CF patient because they meet the pre-exercise meal recommendations and are food source familiar to many CF patients. Sports bars are another good option for use in CF.
During exercise	As per the general population, any exercise lasting longer than 1 h should include a carbohydrate fuel source [9]. There is no evidence to support that a CF patient requires carbohydrates for exercise <1 h duration; however, hydration is a priority (see Section 36.3). Sports drinks and oral rehydration solutions are highly recommended for use in CF, especially for the active patient.
Postexercise	As per advice for the general population, the postexercise meal should aim to replace all losses [9]: • High in energy with adequate carbohydrates to replace muscle glycogen stores • High protein to provide amino acids for muscle repair and growth • Adequate fluid and electrolytes to replace losses (see Section 36.3) Liquid meals are again a good and well-balanced postexercise snack for the busy CF patient. We are always aiming to replace losses without adding to the current burden of disease and expecting patients to eat a significantly larger quantity of food. Other options include sports bars and encouraging high-energy, high-fat fortification of familiar and preferred meals/snacks. For those patients interested in strength training to improve muscle bulk and therefore high-protein options, a protein shake would be appropriate for use in the postexercise period after discussion with the CF team.

CF, cystic fibrosis; CFRD, cystic fibrosis related diabetes; EI, energy intake.

- Muscle-building supplements are popular among adolescent males. When possible, natural protein food sources should be encouraged for use in CF, and any additional supplement should be discussed with the treating physician before use.

36.1.7.9.5 INFLUENCES ON FOOD CHOICES

- Factors effecting food choices for adolescents should be considered before dietary changes. These include hunger, food cravings, timing of meals/ snacks, convenience, media, cost, body image, habits, mood, health benefits, parental effect, and the effect of peers [30].

36.1.7.9.6 HYDRATION

- See Section 36.3.

36.2 HYDRATION CONSIDERATIONS FOR EXERCISE

As with most areas of sports nutrition, evidence-based hydration guidelines for the exercising CF population are limited. With minimal research in the field, it is primarily the application of what we know in the healthy population that continues to drive recommendations for hydration in CF.

Water is an essential part of the human body and significantly affects hydration because it is involved in the maintenance of blood volume and body temperature regulation and it allows muscle contractions to take place. The type of activity, environment, and clothing and the individual characteristics of size, genetics, heat acclimatization, training level, and age will also significantly influence hydration [39]. Dehydration is generally classified as more than 2% body mass loss from water [39], but signs and symptoms of dehydration have been reported with mild dehydration as low as 1% body mass loss [40]. These include physiological symptoms (increased heart rate and core body temperature) as well as impaired cognition and alertness [40]. Significant changes to an individual's water intake can also alter cellular volume and affect the cellular functions of metabolism, hormone release, excitation, and cell proliferation or death [41].

Hydration and the effect of a person's hydration status on exercise have been studied extensively in the healthy population. Many national and international sporting bodies including the International Olympic Committee [42], the American College of Sports Medicine (ACSM) [39], and AIS [43] have released consensus documents and position statements highlighting specific hydration considerations and recommendations. Table 36.9 provides a brief summary of the ACSM hydration recommendations, including their evidence statement hierarchy.

Fluid and electrolyte requirements during exercise largely depend on an individual's sweat rate and sweat electrolyte losses. It is difficult to give a global recommendation for fluid and electrolyte replacement because the duration and intensity of exercise, environmental conditions, clothing, and individual characteristics of body weight, genetic predisposition, heat acclimatization, and metabolic efficiency all affect fluid and electrolyte losses [39]. In CF, it is postulated that dysregulation of the cystic fibrosis transmembrane conductance regulator (CFTR) alters the transport of sodium chloride across the cell membrane, which impairs hydration.

Table 36.10 summarizes key points to consider when applying general hydration recommendations to the CF population.

The risk and effect of dehydration during exercise for the CF individual has been highlighted in several case reports over the past two decades. In 1991, a 24-year-old infantryman was diagnosed with CF after having collapsed with hyponatremia when training in warm climates [47]. He presented with exercise-associated hyponatremia (EAH) and reported that in comparison to his colleagues, he would sweat more, consumed more water, and often noted salt crusts on his skin while training [47]. A similar case of a 48-year-old male who diagnosed with CF after a 30 year history of profuse sweating, muscle cramps, and salt crusts during exercise was published in 2007 [48]. Salt tablets were used to relieve muscle cramps and nausea before diagnosis for this individual [48]. Most recently, in 2012, a case of an adolescent with known CF presented with a case of hyponatremia-associated rhabdomyolysis after a summer football training session [49]. The authors of this case concluded that research is needed regarding the appropriate amount and composition of oral rehydration fluids in exercising individuals with CF because the physiology encountered in these patients provides a unique challenge to maintaining electrolyte balance and stimulation of thirst [49].

36.2.1 Key Recommendations

Although more research is required in this area to establish evidence-based practice guidelines, it would not be unreasonable to recommend the following

TABLE 36.9 Summary of the ACSM Position Stand for Exercise and Fluid Replacement

	ACSM Evidence Statement[a] [39]
Fluid and electrolyte requirements during exercise	*Evidence category A:* Water and electrolyte losses differ considerably between individuals and are affected by the type of activity. Sustained exercise in warm weather can result in substantial water and electrolyte losses. If the losses are not replaced, then dehydration will occur.
Fluid and electrolyte requirements postexercise	*Evidence category A:* It is recommended that sweat electrolyte losses should be replaced fully.
Hydration assessment	*Evidence category A:* An individual's fluid replacement requirements for a specific exercise and environmental condition can be determined by body weight changes pre- and postexercise because percentage change in body weight is reflective of sweat losses during exercise. *Evidence category B:* Urine and body weight measurements can be used by individuals to monitor their hydration status.
Hydration and performance	*Evidence category A:* Dehydration is classified as >2% body weight loss and can be associated with reduced aerobic exercise performance (especially in warm-hot weather).
Other health outcomes: 1. Hyponatremia 2. Muscle cramps	*Evidence category A:* Exercise-associated hyponatremia is rare but can occur if fluid consumption exceeds the rate of sweating. *Evidence category B:* Exercise-associated hyponatremia is more likely if sweat sodium losses are large and body mass (and total body water) is small. *Evidence category C:* Skeletal muscle cramps can be a result of dehydration, sodium depletion, and muscle fatigue. Muscle cramps are more likely in those individuals who sweat more and have large sweat sodium losses.
Dietary considerations	*Evidence category A:* The consumption of a meal (especially sodium-containing foods) before exercise can help retain water and promote euhydration. It is also important that all sweat electrolyte losses are fully replaced. *Evidence category B:* Caffeine has a limited effect on hydration status and daily urine output. The consumption of alcohol can hinder rehydration because it increases urine output.

ACSM, American College of Sports Medicine.

[a]*Evidence statement hierarchy: A, recommendation is based on consistent and good quality experimental evidence (morbidity, mortality, exercise and cognitive performance, physiologic responses); B, recommendation is based on inconsistent or limited quality experimental evidence; C, recommendation is based on consensus, usual practice, opinion, disease oriented evidence, case series or studies of diagnosis, treatment, prevention, or screening, or extrapolations from quasi experimental research.*

to individuals with CF who participate in regular exercise:

1. *Monitor hydration status:* This is most important for CF individuals exercising for extended periods of time (especially in the heat).

   Urine color:
   a. A practical and convenient indicator of hydration (not scientifically validated)
   b. Aim for urine that is pale yellow in color (dark urine is indicative of dehydration)

   Weight change:
   a. Monitoring weight pre- and postexercise will give a rough estimation of lost fluids and provide a starting point for rehydration. One-kilogram weight loss approximately equates to a 1-l fluid deficit.
   b. Fluid and electrolyte losses continue postexercise via sweat and urine. Aim to replace 125–150% of the fluid deficit over a period of 2–6h postexercise [64].
2. *Encourage sodium-containing fluids (and foods):* Sports drinks and oral rehydration solutions are beneficial to the exercising CF population. Compares the sodium content of sports drinks commonly consumed with exercise. Note that sugar-free alternatives may be considered for the CFRD population.
3. *Do not rely on thirst:* Significant fluid losses have occurred before the onset of the thirst mechanism.
4. *Look out for signs of dehydration:* Early signs of dehydration can include fatigue and lethargy, headache, nausea, muscle cramps, concentrated and dark urine, and flushed skin. If nauseous, then encourage frequent small volumes.
5. *Special considerations for children:* Children and particularly those with CF have an attenuated thirst drive and do not stop for fluids as regularly as adults. They should always be encouraged to drink regularly.
   a. Offer flavored, cooled, and sodium-containing drinks in an attempt to improve voluntary rehydration.
   b. When combined with sports drinks, salty snacks will help bring the sodium intake closer to more than 50mmol/L and stimulate thirst.
   c. Allow children plenty of time to rest, stay cool, refuel, rehydrate, and ultimately prevent dehydration.

TABLE 36.10 Hydration Considerations for the Healthy and CF Population [39]

	Healthy Population	CF Population
Hydration assessment	Daily water balance depends on the net difference between water gain (liquid or food and the metabolic production of water) and loss (respiratory, gastrointestinal, renal, and sweat losses) [44]. Sweating is the main source of water loss during exercise [39]. There is no gold standard for hydration assessment of athletes. The most practical and commonly used protocols include the urine specific gravity assessment and fluid balance assessment [43].	No CF-specific recommendations exist. It is important to remember the considerable burden of disease that CF individuals already endure before implementing hydration assessment strategies. How would the commonly used protocols in the healthy population affect the quality of life of a CF individual?
Sweating rate, electrolyte losses, and EAH	*Sweating rates:* • Individual sweating rates are thought to vary from 0.5 to 2.0 l/h with sodium losses of 35 mEq/L (10–70 mEq/L) [39]. *Electrolyte losses:* • The loss of electrolytes in sweat depends on the volume of sweat lost and concentration of sweat electrolytes [39]. • As the rate of sweating increases, the concentration of sodium and chloride lost in sweat also increases [39]. *EAH:* • EAH is defined as a serum sodium concentration <135 mmol/L [45] and most commonly results from the overconsumption of fluids at a rate that exceeds the sweating rate [46].	*Sweating rates:* • CF individuals are reported to sweat more than their non-CF counterparts [47–49]. *Electrolyte losses and EAH:* • CFTR dysfunction results in the excessive loss of sodium and chloride via sweat gland ducts and places CF individuals at a greater risk of dehydration, hyponatremia, and hypochloridemia [50,51]. • The sodium concentration of sweat for CF individuals is usually 3–5 times that of the healthy population [52]. • The excretion of sweat that is nearly isotonic to plasma increases the CF individual's risk of dehydration and hyponatremia during any prolonged exercise (especially in the heat) [47].
Fluid and electrolyte replacement	*Fluid replacement:* • Sufficient fluid should be consumed during exercise to limit dehydration to <2% of body mass loss [42]. ◦ Athletes should not drink so much that they gain weight during exercise [42]. If body mass loss has occurred, then water should be consumed in a quantity greater than those in the losses [53]. *Evidence to support sodium supplementation with exercise:* • Sodium should be included in fluids during exercise when sweat losses are high (3–4 g) and/or exercise is >2 h in duration [42,54]. • All fluid and electrolyte losses should be fully replaced for optimal rehydration [39,55]. • Sodium stores be replaced for euhydration to be restored and maintained [56]. There is no evidence to support sodium supplementation in the prevention of EAH in any athlete other than for CF individuals [57].	No evidence-based practice guidelines are available for fluid and salt supplementation in the CF population. With knowledge of the physiology behind hydration in the CF individual, it would not be unreasonable to recommend electrolyte replacement for all exercising CF individuals regardless of the environmental conditions, duration, or intensity of exercise [52,58]. Further research is required to determine optimal electrolyte replacement doses and strategies.

(Continued)

TABLE 36.10 Hydration Considerations for the Healthy and CF Population [39] Cont'd

	Healthy Population	CF Population
Thirst drive	Thirst is a physiological response stimulated by the rise in blood osmolality with plasma water loss [59]. Stimulation of the thirst drive occurs in the healthy population at 1–2% body mass water loss [60]. Involuntary dehydration is largely based on perceived thirst and is common after strenuous exercise when ad libitum water intake is less than the sweat output [61].	CF children have been observed to drink less during exercise. This was originally hypothesized to be a result of a reduced hyperosmotic trigger attenuating the thirst drive [51,62]. More recently, it has been shown that despite high sweat sodium concentrations, perceived thirst is unchanged during exercise for CF participants. The greater plasma volume loss associated with exercise-induced dehydration is thought to act as a compensatory response triggering thirst in the absence of a strong hyperosmotic signal [58,62].
The use of sports drinks	The use of sports drinks and their role in hydration with exercise remains controversial. *Composition:* • 4–8% carbohydrate, 10–30 mmol/L sodium, and 3–5 mmol/L potassium [9,39]. • Designed to provide a rapid delivery of fluid and fuel during and after exercise. • Some sports drinks containing protein (2%) are marketed to be superior; however, there is limited evidence to support their use for performance enhancement or recovery. *Evidence to support use:* • The sodium concentration increases the thirst drive and voluntary intake of fluid (when compared with water). It is also thought to improve fluid retention postexercise by reducing urine losses. • Carbohydrate content <8% increases gastric emptying and the rapid absorption via the small intestine. • A source of fuel (in addition to fluid). • Reduced immune stress. • Voluntary rehydration improves in children with beverages that are flavored, cooled, and have a sodium content >18 mmol/L [63]. *Concerns associated with use:* 1. Carbohydrate replacement has not been shown to improve performance in exercise <1 h in duration. 2. Dental decay and concerns regarding regular consumption of high-sugar drinks.	Children: • CF children drink less during exercise (most likely result of the decreased hyperosmotic trigger that would in most circumstances attenuate the thirst drive) [51]. • The sodium content of fluid must be >50 mmol/L to promote voluntary rehydration in the pediatric CF population. This is above the sodium content of most sports drinks [62]. Powdered supplements can be beneficial because concentrations can be altered.

CF, cystic fibrosis; EAH, exercise-associated hyponatraemia; CFTR, cystic fibrosis transmembrane conductance regulator.

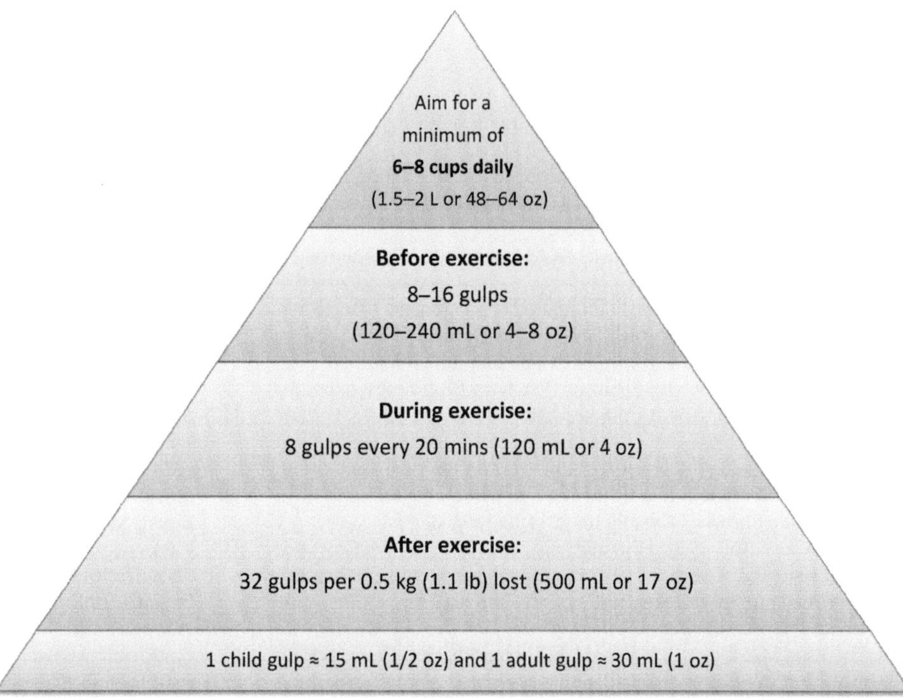

FIGURE 36.3 Practical rehydration guide for patients and their families. *Adapted from Gatorade ©.*

6. *Make a conscious effort to hydrate before, during, and after exercise:* Figure 36.3 outlines a practical rehydration guide for patients and their families.
7. *Consider the environmental effect:* Hot and humid environments increase the risk of dehydration. If ambient temperatures exceed body temperature and the humidity is high, then it becomes difficult for the body to dissipate heat [9]. Dehydration can still occur in cold environments, especially if sweating with insulated clothing [9].

References

[1] McPhail GL, Acton JD, Fenchel MC, Amin RS, Seid M. Improvements in lung function outcomes in children with cystic fibrosis are associated with better nutrition, fewer chronic *Pseudomonas aeruginosa* infections, and dornase alfa use. J Pediatr 2008;153(6):752–7.

[2] Konstan MW, Butler SM, Wohl MEB, Stoddard M, Matousek R, Wagener JS, et al. Growth and nutritional indexes in early life predict pulmonary function in cystic fibrosis. J Pediatr 2003; 142(6):624–30.

[3] Corey M, McLaughlin FJ, Williams M, Levison H. A comparison of survival, growth, and pulmonary function in patients with cystic fibrosis in Boston and Toronto. J Clin Epidemiol 1988;41(6):583–91.

[4] Zemel BS, Jawad AF, FitzSimmons S, Stallings VA. Longitudinal relationship among growth, nutritional status, and pulmonary function in children with cystic fibrosis: analysis of the Cystic Fibrosis Foundation National CF Patient Registry. J Pediatr 2000;137(3):374–80.

[5] Boucher GP, Lands LC, Hay JA, Hornby L. Activity levels and the relationship to lung function and nutritional status in children with cystic fibrosis. Am J Phys Med Rehab 1997;76(4):311–5.

[6] Selvadurai HC, Blimkie CJ, Cooper PJ, Mellis CM, Van Asperen PP. Gender differences in habitual activity in children with cystic fibrosis. Arch Dis Child 2004;89(10):928–33.

[7] Marcotte JE, Grisdale RK, Levison H, Coates AL, Canny GJ. Multiple factors limit exercise capacity in cystic fibrosis. Pediatr Pulmonol 1986;2(5):274–81.

[8] Shah AR, Gozal D, Keens TG. Determinants of aerobic and anaerobic exercise performance in cystic fibrosis. Am J Respir Crit Care Med 1998;157(4 Pt 1):1145–50.

[9] American Dietetic A, Dietitians of C, American College of Sports M, Rodriguez NR, Di Marco NM, Langley S. American College of Sports Medicine position stand. Nutrition and athletic performance. Med Sci Sports Exerc 2009;41(3):709–31.

[10] Burke LM, Loucks AB, Broad N. Energy and carbohydrate for training and recovery. J Sports Sci 2006;24(7):675–85.

[11] Wideman L, Baker CF, Brown PK, Consitt LA, Ambrosius WT, Schechter MS. Substrate utilization during and after exercise in mild cystic fibrosis. Med Sci Sports Exerc 2009;41(2):270–8.

[12] Johnson MR, Ferkol TW, Shepherd RW. Energy cost of activity and exercise in children and adolescents with cystic fibrosis. J Cyst Fibros 2006;5(1):53–8.

[13] Marín VB, Velandia S, Hunter B, Gattas V, Fielbaum O, Herrera O, et al. Energy expenditure, nutrition status, and body composition in children with cystic fibrosis. Nutrition 2004;20(2):181–6.

[14] Allen JR, McCauley JC, Selby AM, Waters DL, Gruca MA, Baur LA, et al. Differences in resting energy expenditure between male and female children with cystic fibrosis. J Pediatr 2003;142(1):15–9.

[15] Shepherd RW, Greer RM, McNaughton SA, Wotton M, Cleghorn GJ. Energy expenditure and the body cell mass in cystic fibrosis. Nutrition 2001;17(1):22–5.

[16] Spicher V, Roulet M, Schutz Y. Assessment of total energy expenditure in free-living patients with cystic fibrosis. J Pediatr 1991;118(6):865–72.

[17] Richards ML, Davies PS, Bell SC. Energy cost of physical activity in cystic fibrosis. Eur J Clin Nutr 2001;55(8):690–7.

[18] Ward SA, Tomezsko JL, Holsclaw DS, Paolone AM. Energy expenditure and substrate utilization in adults with cystic fibrosis and diabetes mellitus. Am J Clin Nutr 1999;69(5):913–9.

[19] Dorlochter L, Roksund O, Helgheim V, Rosendahl K, Fluge G. Resting energy expenditure and lung disease in cystic fibrosis. J Cyst Fibros 2002;1(3):131–6.

[20] Borowitz D, Baker RD, Stallings V. Consensus report on nutrition for pediatric patients with cystic fibrosis. J Pediatr Gastroenterol Nutr 2002;35(3):246–59.

[21] Sinaasappel M, Stern M, Littlewood J, Wolfe S, Steinkamp G, Heijerman HG, et al. Nutrition in patients with cystic fibrosis: a European consensus. J Cyst Fibros 2002;1(2):51–75.

[22] Dietetics Association of Australia and Australian National Cystic Fibrosis Interest Group. Australasian clinical practice guidelines for nutrition in cystic fibrosis; 2006.

[23] Stallings VA, Stark LJ, Robinson KA, Feranchak AP, Quinton H, Clinical Practice Guidelines on G, et al. Evidence-based practice recommendations for nutrition-related management of children and adults with cystic fibrosis and pancreatic insufficiency: results of a systematic review. J Am Diet Assoc 2008;108(5):832–9.

[24] Grunow JE, Azcue MP, Berall G, Pencharz PB. Energy expenditure in cystic fibrosis during activities of daily living. J Pediatr 1993;122(2):243–6.

[25] Nguyen T, Obeid J, Baker JM, Takken T, Pedder L, Parise G, et al. Reduced fat oxidation rates during submaximal exercise in boys with cystic fibrosis. J Cyst Fibros 2014;13(1):92–8.

[26] Kien CL, Zipf WB, Horswill CA, Denne SC, McCoy KS, O'Dorisio TM. Effects of feeding on protein turnover in healthy children and in children with cystic fibrosis. Am J Clin Nutr 1996;64(4):608–14.

[27] Thomson MA, Quirk P, Swanson CE, Thomas BJ, Holt TL, Francis PJ, et al. Nutritional growth retardation is associated with defective lung growth in cystic fibrosis: a preventable determinant of progressive pulmonary dysfunction. Nutrition 1995;11(4):350–4.

[28] Powers SW, Patton SR. A comparison of nutrient intake between infants and toddlers with and without cystic fibrosis. J Am Diet Assoc 2003;103(12):1620–5.

[29] White H, Morton AM, Peckham DG, Conway SP. Dietary intakes in adult patients with cystic fibrosis–do they achieve guidelines? J Cyst Fibros 2004;3(1):1–7.

[30] Burke LDV. Clinical sports nutrition. 3rd ed. Australia: McGraw-Hill; 2006.

[31] AIS Supplement Group Classification System. Australian Institute of Sport: Australian Sports Commission. Available at: www.ausport.gov.au/ais/nutrition/supplements/classification_test.

[32] Tipton KD, Rasmussen BB, Miller SL, Wolf SE, Owens-Stovall SK, Petrini BE, et al. Timing of amino acid-carbohydrate ingestion alters anabolic response of muscle to resistance exercise. Am J Physiol Endocrinol Metab 2001;281(2):E197–206.

[33] Harris JA, Benedict FG. A biometric study of basal metabolism in man. Publication No. 279 ed. Washington DC: Carnegie Institute of Washington; 1919.

[34] Brooks GBN, Rand W, Flatt J, Caballero B. Chronicle of the Institute of Medicine physical activity recommendation: how a physical recommendation came to be among dietary recommendations. Am J Clin Nutr 2004;79:921S–30S.

[35] Schofield WN. Predicting basal metabolic rate, new standards and review of previous work. Hum Nutr Clin Nutr 1985;39(Suppl. 1):5–41.

[36] Dietary reference intakes for energy, carbohydrate, fiber, fat, fatty acids, cholesterol, protein, and amino acids (macronutrients). The National Academies Press; 2005.

[37] Erdman SH. Nutritional imperatives in cystic fibrosis therapy. Pediatr Ann 1999;28(2):129–36.

[38] Selvadurai HC, Blimkie CJ, Meyers N, Mellis CM, Cooper PJ, Van Asperen PP. Randomized controlled study of in-hospital exercise training programs in children with cystic fibrosis. Pediatr Pulmonol 2002;33(3):194–200.

[39] Sawka MN, Burke LM, Eichner ER, Maughan RJ, Montain SJ, Stachenfeld NS. American College of Sports Medicine position stand. Exercise and fluid replacement. Med Sci Sports Exerc 2007;39(2):377–90.

[40] Maughan RJ. Impact of mild dehydration on wellness and on exercise performance. Eur J Clin Nutr 2003;57(Suppl. 2):S19–23.

[41] Lang F, Busch GL, Ritter M, Volkl H, Waldegger S, Gulbins E, et al. Functional significance of cell volume regulatory mechanisms. Physiol Rev 1998;78(1):247–306.

[42] International Olympic Committee. Consensus statement on sports nutrition; 2010.

[43] Cox G, Burke L, Nick P, Cort M. AIS hydration assessment protocol. Australian Sports Commission; 2006.

[44] Medicine Io. Dietary reference intakes for water, potassium, sodium, chloride, and sulfate. Washington DC: National Academy Press; 2005. p. 73–185.

[45] Rosner MH. Exercise-associated hyponatremia. Semin Nephrol 2009;29(3):271–81.

[46] Montain SJ, Cheuvront SN, Sawka MN. Exercise associated hyponatraemia: quantitative analysis to understand the aetiology. Br J Sports Med 2006;40(2):98–105. discussion 98.

[47] Smith HR, Dhatt GS, Melia WM, Dickinson JG. Cystic fibrosis presenting as hyponatraemic heat exhaustion. BMJ (Clinical Research ed) 1995;310(6979):579–80.

[48] Morton A. An unusual cause of exercise-induced hyponatremia. Emerg Med Australas 2007;19(4):377–8.

[49] Kaskavage J, Sklansky D. Hyponatremia-associated rhabdomyolysis following exercise in an adolescent with cystic fibrosis. Pediatrics 2012;130(1):e220–3.

[50] Wheatley CM, Wilkins BW, Snyder EM. Exercise is medicine in cystic fibrosis. Exerc & Sports Sci Rev 2011;39(3):155–60.

[51] Bar-Or O, Blimkie CJ, Hay JA, MacDougall JD, Ward DS, Wilson WM. Voluntary dehydration and heat intolerance in cystic fibrosis. Lancet 1992;339(8795):696–9.

[52] Emrich HM, Stoll E, Friolet B, Colombo JP, Richterich R, Rossi E. Sweat composition in relation to rate of sweating in patients with cystic fibrosis of the pancreas. Pediatr Res 1968;2(6):464–78.

[53] Shirreffs SM, Armstrong LE, Cheuvront SN. Fluid and electrolyte needs for preparation and recovery from training and competition. J Sports Sci 2004;22(1):57–63.

[54] Coyle EF. Fluid and fuel intake during exercise. J Sports Sci 2004;22(1):39–55.

[55] Australian Institute of Sports. Electrolyte replacement supplements; 2006. Available at: www.ais.org.au/nutrition2006.

[56] Shirreffs SM, Taylor AJ, Leiper JB, Maughan RJ. Post-exercise rehydration in man: effects of volume consumed and drink sodium content. Med Sci Sports Exerc 1996;28(10):1260–71.

[57] Lewis DP, Hoffman MD, Stuempfle KJ, Owen BE, Rogers IR, Verbalis JG, et al. The need for salt: does a relationship exist between cystic fibrosis and exercise-associated hyponatremia? J Strength Cond Res 2013.

[58] Brown MB, McCarty NA, Millard-Stafford M. High-sweat Na^+ in cystic fibrosis and healthy individuals does not diminish thirst during exercise in the heat. Am J Physiol Regul Integr Comp Physiol 2011;301(4):R1177–85.

[59] Greenleaf JE. Problem: thirst, drinking behavior, and involuntary dehydration. Med Sci Sports Exerc 1992;24(6):645–56.

[60] Adolph EF, editor. Physiology of man in the desert. New York: Interscience; 1947.

[61] Maresh CM, Gabaree-Boulant CL, Armstrong LE, Judelson DA, Hoffman JR, Castellani JW, et al. Effect of hydration status on thirst, drinking, and related hormonal responses during low-intensity exercise in the heat. J Appl Physiol 2004;97(1):39–44.

[62] Kriemler S, Wilk B, Schurer W, Wilson WM, Bar-Or O. Preventing dehydration in children with cystic fibrosis who exercise in the heat. Med Sci Sports Exerc 1999;31(6):774–9.

[63] Rivera-Brown AM, Gutierrez R, Gutierrez JC, Frontera WR, Bar-Or O. Drink composition, voluntary drinking, and fluid balance in exercising, trained, heat-acclimatized boys. J Appl Physiol (1985) 1999;86(1):78–84.

[64] Sports Dietitians Australia. Fluids in sport; 2011. Available at: www.sportsdietitians.com.au.

CHAPTER

37

Personalizing Exercise and Physical Activity Prescriptions

Matthew Nippins[1,2]

[1]Northeastern University, Boston, MA, USA; [2]Massachusetts General Hospital, Wang Ambulatory Care Center, Boston, MA, USA

37.1 INTRODUCTION

Enhancing physical activity and promoting exercise in persons with cystic fibrosis is an essential component to treatment of the disease. Physical activity and exercise promote improved quality of life, aerobic capacity, strength, posture, body image, and appetite; delay the onset of osteoporosis; and enhance airway clearance [1–5]. In addition, persons with cystic fibrosis report that regular exercise and activity enhance their ability to perform functional activities and activities of daily living. Individualizing the activity and exercise prescription enhances the prospects for the success of and adherence to the prescribed regimen (see Chapter 35).

37.2 COMPONENTS OF A COMPREHENSIVE EXERCISE PRESCRIPTION

The foundation of an individualized exercise prescription is an exercise test. An exercise prescription should be adapted from the results of these tests whenever possible. In cases of patients with more severely advanced disease, the baseline information can be obtained through functional activities with a vital sign assessment (see Chapter 32).

A comprehensive exercise prescription is broken up into several key parts: habitual activity, aerobic exercise, resistance training, and flexibility exercises. Each of these parts can further be classified with a specific mode, intensity, duration, frequency, and exercise progression. Mode is the type of exercise prescribed, such as walking or bicycling. Intensity determines how hard the person is working during the exercise. This may be obtained with objective measurements such as heart rate (HR) or oxygen saturation in patients with gas exchange impairments. Subjective scales such as the rating of perceived exertion (RPE) or dyspnea on exertion (DOE) scales are beneficial objective measurements for prescribing and monitoring intensity [6,6a]. Duration is the amount of time the activity is performed. Frequency describes the number of times the activity is performed per day and the number of days per week. Lastly, progression is often the most overlooked portion of exercise prescription. Progression is the final component of exercise prescription and provides necessary information for how the activity is safely advanced over time. The progression strives to prevent a plateau in gains from the prescribed activity. Progression can be achieved by increasing the duration, frequency, or intensity of the activity. We will look at each of these aspects of physical activity and exercise prescription with the adult and pediatric cystic fibrosis population.

37.3 INDIVIDUALIZING PHYSICAL ACTIVITY AND EXERCISE PRESCRIPTION FOR THE ADULT WITH CYSTIC FIBROSIS

37.3.1 Physical Activity

Increased physical activity levels in adults with cystic fibrosis have been associated with higher quality of life scores, higher VO_2 max, and increased muscle strength as well as lower rates of lung function decline in young adults [7–9]. Much like healthy adults, adults with cystic fibrosis are advised to achieve at least 150min of moderate- to vigorous-intensity physical activity per week [10,11]. Moderate intensity is defined as 11–13 of 20 on the RPE scale or 50–85% of the maximum HR defined by exercise testing or

Diet and Exercise in Cystic Fibrosis
http://dx.doi.org/10.1016/B978-0-12-800051-9.00037-7

maximal predicted HR if those data are not attained. Vigorous intensity is defined as 13–15 of 20 on the RPE scale or 70–85% of maximum HR [11]. In persons with an identified gas exchange impairment from exercise testing, it is important to educate the patient to maintain an oxygen saturation of greater than or equal to 90% throughout the activity with or without supplemental oxygen.

The specific recommendations should be completely customized into the patient's everyday life to enhance adherence. For example, 30 min per day, 5 days per week would more than meet the total week's requirements, but the patient may not be able to perform the duration during the day all at once. The 30 min may be further broken into intervals of no less than 10 min. In addition, the health-care provider must take into consideration that some physical activity is better than no physical activity and that periods of activity of less than 10 min may be required, especially at the initiation of a physical activity program. Longer time periods may be performed on fewer days, but the days per week should not be less than 3 in any instance. Less than 3 days per week leads to diminished gains and possibly lower levels of adherence.

The mode of physical activity chosen should be one that includes multiple large muscle groups such as walking, bicycling, swimming, jogging, or sport-related activities. It is essential to choose the mode with the assistance of the patient to determine what the patient most enjoys. This may include activities that they have enjoyed in the past or may be currently interested in. This improves adherence to the activity and thus maximizes the benefit. Modes also must take into consideration musculoskeletal or other impairments that may preclude a patient from certain modes. In addition, choosing modes that are accessible and available to the individual patient is important as well as taking into account potential financial costs and time commitment, above and beyond the activity time, of the mode. The mode should be enjoyable and not unduly strain the person's time or financial resources.

If time is a significant patient consideration, 10 min of walking can be easily achieved via enhancing certain functional activities. This could include a 10-min walk 1–2 times during the work day or adding 10 min before or after work by lengthening the walk from the person's transportation. The health-care professional educating the person on physical activity should strive to break down any perceived time barriers in creative and innovate ways such as this (see Chapter 35).

37.3.2 Aerobic Exercise

The benefits of aerobic exercise in adults with cystic fibrosis have been extensively researched and include airway clearance enhancement, slower rate of lung function decline, increased aerobic capacity, decreased frequency of pulmonary exacerbations, and increased survival rates [1,12–14].

Aerobic activities are continuously performed by large muscle groups for greater than or equal to 15 min. This is different in the sense that physical activity may be broken down into 10-min intervals or less at a time whereas aerobic activities are typically of longer duration. Aerobic activity is included in the 150 min of total activity per week when considering a comprehensive exercise prescription.

Multiple modes of exercise can be considered when prescribing aerobic capacity in adults. These are similar to those considered in the physical activity prescription from the prior section (walking, bicycling, swimming, jogging, or sport-related activities). Again, as with physical activity, care should be taken by the health-care professional to choose the mode that best fits the patient's situation and needs. For persons with severely impaired aerobic capacity, modified positioning with aerobic training may be beneficial. Activities in which the upper extremities are braced, such as holding onto the arm supports of a treadmill, or seated aerobic activities, such as recumbent biking, may help those with advanced disease to maximize ventilatory muscle use, allowing for longer duration or higher intensities achieved with training. Health-care professionals should also be aware of choosing an appropriate mode of aerobic training that does not exacerbate musculoskeletal impairments that the person may have. Lower-impact activities may be required with this particular population, especially with aging, because of the recent increase in expected lifespan.

The duration of aerobic activity will vary based on the persons exercise test. Durations of at least 15–20 min are needed to achieve the positive benefits of aerobic exercise listed previously, and they may be up to 60 min or more in length. In general, duration of aerobic exercise should be emphasized before advancing intensity to achieve maximal benefits. In patients with poor aerobic capacity, interval training may be required to achieve aerobic benefits. This is done with several shorter duration intervals with rest breaks in between intervals. These intervals are increased, or the rest periods decreased, as the person's aerobic capacity improves, eventually leading to continuous activity of greater than 15 min. The frequency of aerobic training should be 3–5 times per week, with lower frequency to begin training or with patients performing shorter duration or interval training.

Intensities for prescribing aerobic exercise in adults are also similar to the guidelines previously mentioned for physical activity. Moderate intensities on the RPE scale [11–13] (or 50–85% of maximum HR) to vigorous intensities on the RPE scale [13–15] (or 70–85% of maximum HR) have been shown to achieve the benefits of aerobic activities. Lower intensities may be required for patients with more severely impaired aerobic capacity. Oxygen saturation should always be maintained at or above 90% during aerobic activity. The DOE scale may also be used as a measure of intensity for persons who have a primary limitation of dyspnea discovered on exercise testing. In general, DOE scores of 2–4 are adequate for aerobic training, but perceived dyspnea may vary significantly from person to person. The DOE scale is an excellent adjunct to measuring intensity in most cases.

In addition to these duration and intensity considerations, warm-up and cool-down phases are critical in aerobic training. For a warm-up, 3–5 min at lower intensities (40–60% of maximal HR or RPE of 8–10) is sufficient to increase peripheral blood flow, increase muscle range of motion, and facilitate oxygen transport, thus decreasing the risk of adverse effects from the activity. Lower intensities during the last 3–5 min that bring the HR and RPE down into the previously stated ranges help mobilize waste products from muscle metabolism, decrease venous pooling in the distal extremities, and decrease the severity or presence of muscle soreness after activity. Ideally, the warm-up and cool-down time should be in addition to the duration of training duration stated previously. Longer warm-up and cool-down times, up to 10 min, may be required for persons with significant deconditioning or patients with comorbid cardiac or vascular conditions.

Care should also be taken when exercising persons with cystic fibrosis-related diabetes mellitus (CFRD). Monitoring of blood glucose levels before and after exercise reduces the risk of developing hypoglycemia with exercise. It may also encourage independent monitoring practices, which may improve blood glucose levels and adherence to diabetes regimens over time. Having access to a high-carbohydrate snack during exercise is another consideration for enhancing safety with activity in this population.

37.3.3 Resistance Training

Adults with cystic fibrosis may present with peripheral skeletal muscle weakness [15]. Peripheral skeletal muscle weakness seems to be associated with a decrease in maximal workload regardless of pulmonary function and nutritional status [15]. The cystic fibrosis transmembrane conductance regulator (CFTR) is known to be present in the skeletal muscles and affects calcium and ATP-mediated actin-myosin movement, which may contribute to skeletal muscle weakness in patients with cystic fibrosis [16]. In addition, corticosteroid use in persons with cystic fibrosis along with inactivity and deconditioning may further impair peripheral muscle strength. This skeletal muscle weakness has been objectively observed in grip strength, knee extension, and elbow flexors when compared with healthy controls [17–19]. Studies have also shown that resistance training in adults with cystic fibrosis increases muscle strength, positively affects pulmonary function, and improves the ease of everyday functional activities. [14,20]

Prescribing resistance exercise for adults with cystic fibrosis is performed with a formal strength assessment. This is done in various ways that include manual muscle testing, hand-held dynamometry, or more commonly the assessment of a maximal load in one repetition (1RM max). Resistance training should be performed two or more times per week and should include all major muscle groups (chest, back, core, upper extremities, and lower extremities) with preference to activities using multiple muscle groups (cdc.org). Resistance may be achieved simply with the person's body weight, resistance bands, weights, or resistance provided by various machines. As noted in the aerobic exercise and physical activity sections, choosing activities and exercises that the person has access to is key to adherence to this part of the exercise prescription. Intensity for resistance training should be at least 70% of 1RM max or an intensity that allows the person to perform 1–3 sets of 8–15 repetitions with moderate exertion. Moderate exertion is identified by muscle fatigue and decreased ROM with subsequent repetitions. Patients with more severely impaired strength may benefit from lower intensities at 40–60% of 1RM max with higher repetitions of 15–25. Progression of the program can easily be achieved by increasing resistance or the number of repetitions and sets. Resistance training may become a larger part of a person's overall exercise prescription in persons with advanced disease and with severe exercise limitations.

37.3.4 Flexibility Activities

There is a lack of evidence for prescribing flexibility activities for adults with cystic fibrosis. Clinical experience has shown that adults with cystic fibrosis may have impaired flexibility, which may be related to inactivity. The American College of Sports Medicine (ACSM) flexibility activity guidelines for adults in the general population should serve as a guide to prescribing flexibility activities. ACSM guidelines call for flexibility exercises to be performed at least 2 times per week and holding the stretch for at least 30 s to the point of tightness or minimal discomfort. These exercises should not be performed with "cold" muscles, and ideally they should be performed after physical activity, aerobic training, or resistance exercises [21].

General stretching of the major muscle groups of the upper and lower extremities, such as the hamstrings, gastrocnemius, soleus, rectus femoris, iliopsoas, pectoralis major, and pectoralis minor, as well as the many rotators and flexors of the trunk should be considered.

Specific attention should be paid to the stretching of the anterior chest wall because increased use of the accessory muscles of breathing, typically seen with adults that have moderate to severe pulmonary impairments, can cause hypertrophy and shorten those anterior muscle groups. This leads to postural impairments seen with protracted shoulder girdle and increased thoracic kyphosis (See Figure W). Impairments such as these may lead to further limitations in ventilation because of poor chest wall mobility and compliance.

Activities such as yoga can be very beneficial to the adult with cystic fibrosis because this enhances flexibility and emphasizes breathing control with activity. Positions that put the head and trunk in downward positions may need to be avoided in those patients with a history of gastrointestinal reflux. The health-care professional should assess the flexibility and posture of the adult with cystic fibrosis and tailor flexibility activities accordingly.

37.3.5 Putting All of the Exercise Pieces Together for the Adult with Cystic Fibrosis

Separately, the components of a comprehensive exercise program seem daunting, especially in a population in which time constraints of everyday adult life collide with burden-of-care issues from multiple medical treatments. The health-care provider should strive to resolve these barriers because the evidence is clear that regular aerobic exercise and physical activity slow the decline of lung function and improve the quality of life in adults with cystic fibrosis along with the multiple other aforementioned benefits.

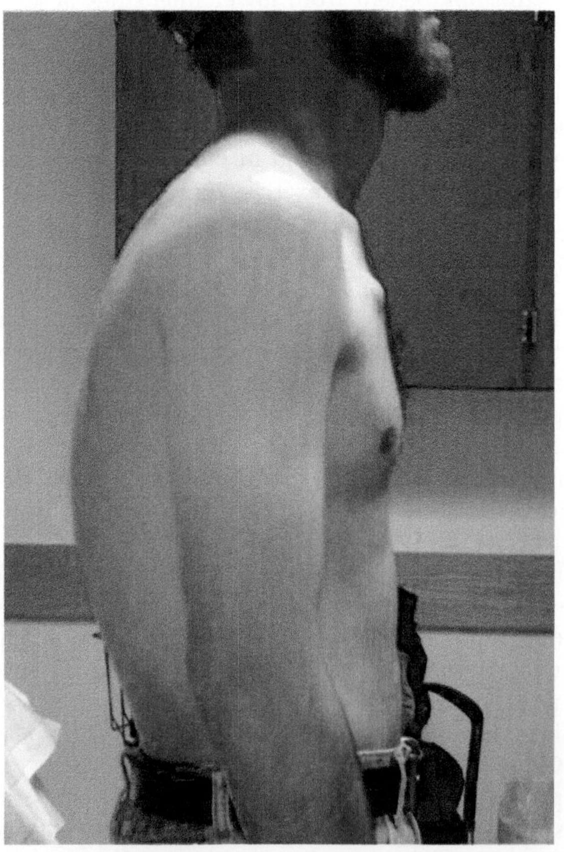

FIGURE 37.1 Lateral view of a 37 year old male with cystic fibrosis displaying protracted shoulder girdle and increased thoracic kyphosis. *Photo credit: Matthew Nippins.*

The key aspects of exercise can be combined in several ways to produce these desired benefits. Achieving the 150 min of recommended physical activity can be a combination of slowly increasing the person's current activity level and adding aspects to the program that may not have been present in their current regimen, such as strengthening and flexibility activities. Having a health-care professional take a comprehensive exercise and activity assessment that includes exercise testing is the first step toward this goal. Starting slowly from the person's baseline and adding just minutes per week in duration or increasing frequency 1 time per week moves the person toward the physical activity goal at a reasonable pace and decreases the risk of injury and poor adherence. Health-care providers must realize that changes are unlikely to be made overnight in persons with cystic fibrosis, and a longer term view should be considered and the patient educated repeatedly to the potential benefits.

37.4 INDIVIDUALIZING PHYSICAL ACTIVITY AND EXERCISE PRESCRIPTION IN THE CHILD WITH CYSTIC FIBROSIS

37.4.1 Physical Activity and Aerobic Exercise

The evidence for an increased physical activity in children with cystic fibrosis is better defined than that of the adult with cystic fibrosis. Higher level of physical activity in children with cystic fibrosis is associated with increased quality of life and a slower rate of lung function decline [8,22]. Lower physical activity levels in the child with cystic fibrosis are associated with steeper declines of lung function [8]. Studies demonstrate that in addition to physical activity benefits, aerobic exercise in children with cystic fibrosis improves airway clearance, slows the rate of lung function decline, and increases aerobic capacity [13,23,24].

Children with cystic fibrosis should perform at least 60 min of moderate to vigorous physical activity daily. Much like the adult population with cystic fibrosis recommendation for physical activity, the recommendation for children with cystic fibrosis is based on the current recommendation to the general population. For teenagers with cystic fibrosis, the adult definition of moderate and vigorous intensity are consistent (11–13 on the RPE scale or 50–85% of maximum HR defined by exercise testing or maximal predicted HR if those data are not available and vigorous intensity would be defined as 13–15 on the RPE scale or 70–85% of maximum HR). With younger children, intensity may be measured on a simple 0–10 scale, such as the OMNI scale, with 0 being sitting and 10 being maximal activity. Moderate intensity would be 5–6 on the scale and vigorous activity would be 7–8 [25]. The duration of activity may be dispersed throughout the day, especially with the younger population.

There are special considerations when choosing a mode of physical activity with children. It should be varied and fun should be maximized when contemplating the choice. The activity should be one that involves multiple large muscle groups such as walking, running, bicycling, and swimming, including sport and active game activities. Creativity, flexibility, and exposure of the child to different types of physical activities are critical when attempting to increase physical activity in children. Various modes should also be considered to achieve this. The health-care provider must also consider developmentally appropriate activities for the child's age, especially in younger children. Modes of physical activity should include weight-bearing exercises such as jumping rope, running, hopping, and sport-specific activities to encourage development of appropriate bone density (see Chapter 36).

The choice of physical activity may ultimately be largely influenced by the activities of their peers, siblings, and parents. Choosing an activity that gets the child involved with others facilitates adherence to physical activity (see Chapter 35). In addition, there is strong evidence in the general population that the child's level of physical activity is influenced by the activity level of their immediate family and friends [26]. Time should be taken to educate the patient's family on the importance of physical activity not only with their child but as a healthy lifestyle choice for the entire family. Emphasizing the importance of physical activity at a young age establishes healthy lifestyle choices early in life.

Prescribing aerobic exercise in children with cystic fibrosis is tied closely to physical activity. With a recommendation of 60 min of daily activity for healthy children, most of the time should be aerobic in nature. From the research on aerobic training in children with cystic fibrosis, a minimum of 15 min and up to 45 min of training 3–5 times per week was shown to provide the abovementioned benefits [13,20,24,27]. There is no difference in choosing the mode or the recommended intensity for aerobic activity in children with cystic fibrosis as compared with prescribing physical activity. In children with significantly impaired aerobic capacity, lower intensities or intervals of exercise to achieve the desired duration may be required. Oxygen saturation should be monitored during exercise to avoid desaturation below 90%. Special considerations also need to be made in children with CFRD. Blood glucose testing before and after exercise is important to avoid episodes of hypoglycemia with exercise.

37.4.2 Resistance Training

Children with cystic fibrosis may present with peripheral skeletal muscle weakness [28]. Much like the adult with cystic fibrosis, peripheral skeletal muscle weakness in children is associated with a decrease in maximal achievable workload [28]. In addition, corticosteroid use, inactivity, and deconditioning may further impair peripheral muscle strength. This skeletal muscle weakness has been objectively observed in handgrip, knee extension, knee extension, hip flexion, and ankle dorsiflexors [17,29]. Resistance training in children has been shown to increase muscle strength, lung function, fat-free muscle mass, and total body weight [24,27].

Resistance training in children may be included in normal play activities or in addition to these activities. For example, swinging on bars, tug of war, and rope or tree climbing are excellent ways to incorporate resistance training into play activities to enhance peripheral muscle strength. Other modes using body weight, bands, or weights as resistance can also be used in children with cystic fibrosis. Unlike resistance training for adults, resistance training in children counts toward the physical activity recommendation of 60 min per day. Resistance training should be performed at least 3 times or more per week and include all major muscle groups (see 37.3.3 Resistance Training above) [25]. Intensities of resistance vary in the literature from 50% to 70% of 1RM max with one to three sets of 8–15 repetitions. The American Academy of Pediatrics also recommends avoiding maximal resistance exercise until children achieve skeletal maturity [30]. Strength should be assessed before prescribing resistance exercise and intensity adjusted accordingly.

37.4.3 Flexibility Exercise

There is little to no evidence for prescribing flexibility exercises in children with cystic fibrosis. Flexibility and posture should be assessed by the health-care provider on an individual case basis. Inactivity may lead to decreased range of motion and flexibility of the peripheral muscles. Yoga may be used to promote flexibility and breathing control. There are several yoga programs that have specific classes tailored for children. This may be beneficial to children with cystic fibrosis. Case reports exist that reveal that postural abnormalities of the trunk occur in children with cystic fibrosis as early as the onset of puberty [31]. This may be seen with children with moderate to severe pulmonary impairments because of an increased use of accessory muscles of breathing on a developing child [31]. Stretching the anterior, lateral, and rotation of the chest wall is valuable to combat these impairments before becoming fixed impairments in adulthood.

37.4.4 Putting All of the Exercise Pieces Together for the Child with Cystic Fibrosis

Evidence exists that physical activity and aerobic exercise slow the rate of lung function decline and improve quality of life for children with cystic fibrosis. It is important to start building healthy lifestyles at a young age so as to take advantage of the benefits of exercise over a longer period of the patient's life. Simple mathematics reveals a slowing the rate of lung function decline by just 0.5–1% over a prolonged portion of the patient's lifespan adds up to saving significant amounts of lung function in addition to multiple other benefits.

Prescribing a comprehensive exercise program that adds up to the recommended 60min of activity per day should be primarily aerobic; however, resistance training or flexibility activities can contribute to a portion of that time. Strengthening can be added with simple playground activities or more formal strengthening for patients using more specific exercises along with weights or bands. For younger children, a 20-min walk or bike trip to and from the playground in addition to 20min of routine playground activities would add up to a well-rounded exercise and physical activity program. This can also be accomplished easily with several combinations of varied play and sports activities that are interesting and fun for the child.

Getting the parents, family, and peers physically active is also of vital importance to increase adherence to physical activity and exercise. This sets a positive example for the child when choosing healthy lifestyle choices over time. Careful evaluation of the patient's current physical activity, aerobic capacity, strength, and flexibility helps to guide prescription in the child with cystic fibrosis. This sets the starting point to adding physical activity, aerobic exercise, resistance training, and flexibility over time with the goal of being more active no matter whether the baseline is no activity or already at the recommended activity level.

References

[1] Nixon P, Orenstein D, Kelsey S, Doershuck C. The prognostic value of exercise testing in patients with cystic fibrosis. N Engl J Med 1992;327:1785–8.

[2] Bradley JM, Moran F. Physical training for cystic fibrosis. Cochrane Database Syst Rev 2008;1.

[3] O'Neil S, Leahy F, Pasterkamp H. The effects of chronic hyperinflation, nutritional status, and posture on respiratory muscle strength in cystic fibrosis. Am Rev Respir Dis 1983;128:1051–4.

[4] Peebles AD. Physiotherapy. (Chapter 5). In: Hill CM, editor. Practical guidelines for cystic fibrosis care. first ed. London: Churchill Livingstone; 1998.

[5] Wolman RL. Osteoporosis and exercise. In: McLatchie G, Harris M, King J, Williams C, editors. ABC of sports medicine. 4th ed. BMJ Publishing; 1999.

[6] Borg G, Linderholm H. Perceived exertion and pulse rate during graded exercise in various age groups. Acta Med Scand Suppl 1967;472:194–206.

[6a] Burdon JGW, Juniper EF, Killian KJ, Hargrave FE, Campbell EJM. The perception of breathlessness in asthma. Am Rev Respir Dis 1982;126:825–8.

[7] Hebestreit H, Kieser S, Rüdiger S, Schenk T, Junge S, Hebestreit A, et al. Physical activity is independently related to aerobic capacity in cystic fibrosis. Eur Respir J 2006;28:734–9.

[8] Schneiderman JE1, Wilkes DL, Atenafu EG, Nguyen T, Wells GD, Alarie N, et al. Longitudinal relationship between physical activity and lung health in patients with cystic fibrosis. Eur Respir J 2013:817–23.

[9] de Jong W, Kaptein AA, van der Schans CP, Mannes GPM, van Aalderen WMC, Grevink RG, et al. Quality of life in patients with cystic fibrosis. Pediatr Pulmonol 1997;23:95–100.

[10] Cerny F, LaMonte M. Learn about exercise and CF; December 26, 2012. Retrieved from: www.cff.org.

[11] Centers for Disease Control. 2008 Physical activity guidelines for Americans; 2008. Retireved from: www.cdc.gov.

[12] O'Neill PA, Dodds M, Phillips B, Poole J, Webb AK. Regular exercise and reduction of breathlessness in patients with cystic fibrosis. Br J Dis Chest 1987;81(1):62–9.

[13] Schneiderman-Walker J, Pollock SL, Corey M, Wilkes D, Canny G, Pedder L. A randomised controlled trial of a 3-year home exercise program in cystic fibrosis. J Pediatr 2000;136(3):304–10.

[14] Moorcroft AJ, Dodd ME, Morris J, Webb AK. Individualised unsupervised exercise training in adults with cystic fibrosis: a 1 year randomised controlled trial. Thorax 2004;59(12):1074–80.

[15] Troosters T1, Langer D, Vrijsen B, Segers J, Wouters K, Janssens W, et al.Skeletal muscle weakness, exercise tolerance and physical activity in adults with cystic fibrosis. Eur Respir J 2009;33:99–106.

[16] Lamhonwah A-M, Bear CE, Huan LJ, Chiaw PK, Ackerley CA, Tein I. Cystic fibrosis transmembrane conductance regulator in human muscle: dysfunction causes abnormal metabolic recovery in exercise. Ann Neurol 2010;67:802–8.

[17] Sahlberg ME, Svantesson U, Thomas EM, Strandvik B. Muscular strength and function in patients with cystic fibrosis. Chest 2005;127(5):1587–92.

[18] Elkin SL, Williams L, Moore M. Relationship of skeletal muscle mass, muscle strength and bone mineral density in adults with cystic fibrosis. Clin Sci (Lond) 2000;14:309–14.

[19] de Jong W, van Aalderen WMC, Kraan J, Koeter GH. Skeletal muscle strength in patients with cystic fibrosis. Physiother Theory Pract 2001;17(1):23–8.

[20] Kriemler S, Kieser S, Junge S, Ballmann M, Hebestreit A, Schindler C, et al. Effect of supervised training on FEV1 in cystic fibrosis. J Cyst Fibros 2013;12:714–20.

[21] Garber CE, Blissmer B, Deschenes MR, Franklin BA, Lamonte MJ, Lee IM, American College of Sports Medicine, et al. American College of Sports Medicine position stand. Quantity and quality of exercise for developing and maintaining cardiorespiratory, musculoskeletal, and neuromotor fitness in apparently healthy adults: guidance for prescribing exercise. Med Sci Sports Exerc 2011;43(7):1334–59.

[22] Hebestreit H, Schmid K, Kieser S, Junge S, Ballmann M, Roth K, et al. Quality of life is associated with physical activity and fitness in cystic fibrosis. BMC Pulm Med 2014;14:26.

[23] Dwyer TJ, Alison JA, McKeough ZJ, Daviskas E, Bye PT. Effects of exercise on respiratory flow and sputum properties in patients with cystic fibrosis. Chest 2011;139(4):870–7.

[24] Selvadurai HC, Blimkie CJ, Meyers N, Mellis CM, Cooper PJ, Van Asperen PP. Randomized controlled study of in hospital exercise training programs in children with cystic fibrosis. Pediatr Pulmonol 2002;33(3):194–200.

[25] Centers for Disease Control. Aerobic, bone, muscle and joint strengthening-what counts? 2011. Retireved from: www.cdc.gov.

[26] Fogelholm M, Nuutinen O, Pasanen M, Myöhänen E, Säätelä T. Parent–child relationship of physical activity patterns and obesity. Int J Obes Relat Metab Disord 1999;23(12):1262–7.

[27] Orenstein DM, Hovell MF, Mulvihill M, Keating KK, Hofstetter CR, Kelsey S, et al. Strength vs aerobic training in children with cystic fibrosis: a randomised controlled trial. Chest 2004;126(4):1204–14.

[28] de Meer K, Gulmans VA, van Der Laag J. Peripheral muscle weakness and exercise capacity in children with cystic fibrosis. Am J Respir Crit Care Med 1999;159(3):748–54.

[29] Swisher AK, Baer L, Moffett K, Yeater R. The influence of lean body mass and leg muscle strength on 6-minute walk test performance in children with cystic fibrosis. Cardiopul Phys Ther 2005; 16(3):5–9.

[30] Council on Sports Medicine and Fitness. Strength training by children and adolescents. Pediatrics 2008;121(4):835–40.

[31] Massery M. Musculoskeletal and neuromuscular interventions: a physical approach to cystic fibrosis. J R Soc Med 2005;98(Suppl. 45):55–66.

SECTION G

FAT AND LIPID METABOLISM IN CYSTIC FIBROSIS

CHAPTER

38

The Pancreatic Duct Ligated Pig as a Model for Patients Suffering from Exocrine Pancreatic Insufficiency—Studies of Vitamin A and E Status

A. Mößeler, T. Schwarzmaier, M. Höltershinken, J. Kamphues

Institute of Animal Nutrition, University of Veterinary Medicine Hanover, Foundation, Hanover, Germany

38.1 INTRODUCTION

Exocrine pancreatic insufficiency (EPI) results in a maldigestion and malabsorption of nutrients, especially of fat [1]. Therefore, patients with EPI are at risk of developing a deficiency of fat-soluble vitamins [2–5]. Children with cystic fibrosis (CF) commonly (about 90%) suffer from EPI [6,7], so the supply of fat-soluble vitamins is of special interest. Because of their manifold functions as antioxidants and coenzymes and in neurodevelopment, bone development, and coagulation [8,9], fat-soluble vitamins are essential. The importance of sufficient vitamin supply during adolescence is beyond dispute, even in healthy children, but it becomes much more important in children suffering from CF because a deficiency in vitamins or antioxidants, respectively, may contribute to adverse health effects [5] and increase the risk of epithelium damage and impair defense against infection [8,10]. Even a mild vitamin A deficiency impairs the cilia-mediated cleaning of the lung because of a decrease of ciliated cells [11]. Although today requirements for fat-soluble vitamins in children are precisely defined and knowledge about the bioavailability of different vitamin compounds has improved, information about the situation in patients with EPI or other diseases associated with malabsorption are less detailed [12].

Two different factors contribute to the impaired vitamin status in patients with EPI, and both of them are difficult to quantify. On the one hand there is an impaired absorption of fat-soluble vitamins due to reduced fat absorption [8]. Absorption is influenced by composition of the diet, the residual function of the exocrine pancreas, the type and dosage of enzyme supplementation, resection of parts of the small intestine (if applicable), and the dosage and galenic preparation of vitamin supplements. A low level of circulating retinol-binding protein [13] and impaired enterohepatic circulation of bile acids in patients with CF are other factors that might be relevant; both factors may contribute to the malabsorption of fat-soluble vitamins from the small intestine [14]. On the other hand, it can be conjectured that there is a higher need for fat-soluble vitamins in addition to reduced absorption because chronic inflammatory processes and bacterial infections are known to increase the requirements of antioxidants—especially vitamin E [10,15].

Previous studies [16,17] showed reduced serum concentrations of vitamin A and E common in patients suffering from EPI. Deficiencies of vitamin A, D, and E were detected in very young children (<2 months old) [18]. Levels of α-tocopherol decreased with age in patients with CF [19], and the highest prevalence of vitamin E deficiency was found in adult patients. Although clinical manifestations of deficiencies of fat-soluble vitamins are no longer common nowadays because of pancreatic enzyme replacement therapy (PERT) and routine vitamin supplementation [20], low levels of fat-soluble vitamins (in particular A, E, and K) in the blood are common in patients with CF [21].

As a consequence, it is generally recommended that fat-soluble vitamins be supplemented [12,18,21,22] and vitamin status be checked on a regular basis (at least yearly [7,12]). As the antioxidant depletion in patients

Diet and Exercise in Cystic Fibrosis
http://dx.doi.org/10.1016/B978-0-12-800051-9.00038-9

with CF is a progressive process, vitamin status should be monitored carefully and supplementation should be considered before severe deficits occur [19]. Optimization of vitamin and oxidant status of patients with CF is an important clinical goal [4] because it might influence lung function and health outcomes [5,23,24]. In some studies, the antioxidant status in patients with CF was suboptimal despite the fact that supplements had been prescribed to all patients; this raises the question as to the underlying reasons. An unsatisfactory compliance rate, as well as insufficient bioavailability of vitamin E from the supplements used or even an incorrect vitamin dosage, have been discussed as reasons for this [18,19].

38.2 ADVANTAGES OF AN ANIMAL MODEL TO STUDY EFFECTS OF EPI ON VITAMIN STATUS

The manifold factors (age of patients, dosage and type of enzyme product used for PERT, composition of diet, vitamin intake, vitamin preparations used, compliance, nutritional status, and health status in general) complicate investigations of vitamin status in human patients with EPI. In experimental studies (animal trials) different factors such as genetics, age, and especially nutrition (diet composition and daily intake) can be completely standardized. Furthermore, there is no risk of low compliance regarding treatment with enzymes or vitamin additives in an animal model.

38.2.1 The Pancreatic Duct Ligated Pig as a Model for EPI in Humans

The pancreatic duct ligated (PL) (mini)pig is an established model for studying EPI in humans [25–28], and it also is used for in vivo testing of the efficacy of pancreatic enzymes [26,29,30]. Ligation of the ductus pancreaticus accessorius results in a complete loss of exocrine pancreatic function without affecting endocrine function [31,32].

Although the PL pig is an established model for studying EPI in adults [25,26,29,30], studies of growing individuals are limited [26,33–35]. Because children and adolescents have much higher nutrient requirements, young piglets with experimentally induced EPI seem to be a good model for children suffering from EPI. The much higher growth rate of the piglets allows effects to be studied in a relatively short period of time: within 6 months a crossbred piglet grows from about 1.5 kg (birth weight) to 120 kg (sexually mature), and therefore the whole adolescent period can be covered within this period (see Table 38.1).

A corresponding study was conducted with young piglets with experimentally induced EPI to generate data regarding retinol and α-tocopherol concentrations in serum and liver tissue under standardized feeding conditions, standardized PERT, and a defined vitamin supply via supplements.

TABLE 38.1 Survey of the Body Weight of Children and Piglets at Birth and Age 6 months and Age at Onset of Puberty

	Child	Piglet (Crossbred)
Body weight at birth (kg)	3.5	1.5
Body weight at age 6 months (kg)	8	120
Age at onset of puberty (months)	120–140	4–5

The effect of PERT was studied as well as the effect of different application forms of the fat-soluble vitamins (oral versus parenteral) to generate more detailed information regarding the requirements of juveniles with EPI in comparison to healthy controls and to optimize vitamin supply. Measurement of serum concentrations of retinol and α-tocopherol allow an estimate of the supply situation in humans and animals. In addition, in the animal model there is the possibility of determining vitamin concentrations in the liver tissue (predominant resource) at the end of the trial to achieve a more differentiated look on vitamin resources.

In this context of fat-soluble vitamins in patients with EPI, two studies were conducted using the PL pig as a model for EPI in humans (study 1 used adult minipigs; study 2 used juvenile crossbred pigs). The project was approved by the ethics committee on animal welfare of the Hannover District Government in accordance with German legislation on animal welfare.

38.3 EFFECTS OF EPI ON SERUM α-TOCOPHEROL CONCENTRATIONS IN ADULT MINIPIGS, STUDY 1

The first study was performed on adult female minipigs (Ellegaard, Dalmose, Denmark) to evaluate the effects of EPI and PERT on serum α-tocopherol concentrations in adults at maintenance (without increased requirements due to growth, pregnancy, or lactation).

38.3.1 Materials and Methods

The control group ($n=4$; body weight, 40.2±2.45 kg) had an intact pancreas, whereas in the other animals (PL group; $n=6$; body weight, 45.3±2.87 kg) the pancreatic duct (ductus pancreaticus accessorius) was ligated at least 4 months before starting the trial, according to the method described by Tabeling [36]. Pancreatic enzyme status was confirmed by measurement of fecal chymotrypsin activity (Chymotrypsin Testkit; Boehringer, Mannheim, Germany) and only pigs with a value <0.8 IU chymotrypsin/g feces were defined as PL pigs.

The animals were fed two meals per day (233 g dry matter/meal, given at 7:00 and 19:00) of a complete diet rich in fat (349 g crude fat/kg dry matter) containing 150 mg α-tocopherol/kg diet (recommendations for healthy non-lactating pigs according to the Committee for Requirement Standards of the Society of Nutrition Physiology [37]: 15 mg vitamin E/kg diet). The α-tocopherol concentration in serum was measured twice in the group of PL pigs: after receiving PERT (Creon) continuously (336.000 IE lipase/meal; 4280 IE lipase/g fat) for at least 3 weeks and additionally after a period of 15 days without PERT. Serum samples (use of serum blood test tube; S-Monovette; Sarstedt, Nümbrecht, Germany) were taken from the jugular vein 90 and 210 min after feeding. Chromium oxide was added to the diet (2.5 g/kg diet) to calculate the coefficient of fat absorption via the marker method using a total feces collection for 5 days. All samples (diet, serum, feces) were stored at −20 °C until analysis. The diet and frozen feces samples were freeze dried for at least 3 days and homogenized by a knife mill (Grindomix; Retsch, Haan, Germany) for laboratory use; Weender analysis [38] was performed to determine contents of dry matter (drying for at least 8 h at 103 °C) and crude fat (Soxhlet extraction after acid hydrolysis). Serum α-tocopherol was measured using high-performance liquid chromatography [39,40].

38.3.2 Results

Pancreatic duct ligation had marked effects on the digestibility of nutrients, with the coefficient of fat absorption being mostly affected (controls, 97.4%; PL group, 29.2%). The substitution of pancreatic enzymes resulted in a considerable increase of the coefficient of fat absorption of up to 78.5%.

There was no significant effect of time when blood samples were taken (90 or 210 min postprandial) on serum α-tocopherol concentrations. Data given here represent the values measured 210 min after feed intake. Control pigs had a serum α-tocopherol concentration (2.95 ± 0.42) within the reference range (1–5 mg/L [41]). Experimentally induced EPI resulted in a significantly reduced serum α-tocopherol concentration (0.49 ± 0.082 mg/L), although duration of withdrawal from the enzymes was only 15 days for the PL pigs. After PERT for 3 weeks, a significant increase in serum α-tocopherol concentrations—up to 1.20 ± 0.280 mg/L—was observed in PL pigs (see Figure 38.1).

38.3.3 Conclusions

The results of this study indicate that PERT in adult pigs with experimentally induced EPI is sufficient to

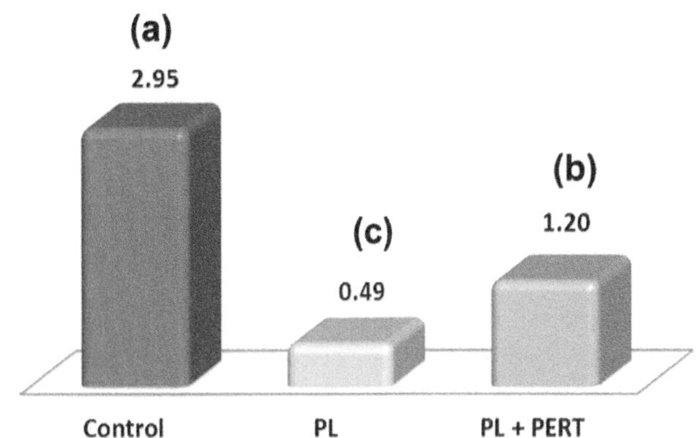

FIGURE 38.1 Serum α-tocopherol concentrations (milligrams/liter) in healthy control pigs and pancreatic duct ligated pigs without (PL) or with pancreatic enzyme replacement therapy (PL + PERT) for at least 15 days; blood was taken 210 min after last feed intake. Different letters indicate significant differences ($p < 0.05$).

maintain serum α-tocopherol concentrations above the lower reference value (>1 mg/L) when they are fed a diet rich in vitamin E (150 mg/kg). Neither coefficient of fat absorption nor α-tocopherol concentrations in serum reached the values of the control pigs. The finding that serum α-tocopherol concentrations were markedly reduced and below the reference range when the animals did not receive any PERT for 15 days illustrates that reserves of α-tocopherol were low or nearly absent. This indicates that the combination of PERT and a diet containing high levels of vitamin E (10-fold higher than requirements) is suitable to prevent deficiency in the animals; however, it does not allow them to build up longer-lasting reserves. Taking into account that these adult animals had only the nutritional requirement for maintenance (no situation with higher nutritional needs, such as growth, pregnancy, or lactation) and serum concentrations decreased below the lower limit within 15 days without PERT, the need for closer study of patients with EPI with increased requirements becomes apparent.

38.4 EFFECTS OF EPI ON RETINOL AND α-TOCOPHEROL CONCENTRATIONS IN SERUM AND LIVER TISSUE OF JUVENILE PIGS, STUDY 2

The second study aimed to test the effects of EPI on serum concentrations of retinol and α-tocopherol after experimental induction of EPI in young pigs—this was used as a model for children suffering from EPI. All PL pigs received pancreatic enzymes (orally via their feed) to imitate the situation in human patients, and the diet used contained vitamin concentrations exceeding the recommendations for healthy pigs as described in the first study. One group of PL pigs (PL+0) did not receive any vitamin supplements in addition to the basal diet to prove the necessity of vitamin supplementation in juvenile pigs receiving PERT. Furthermore, to improve vitamin status and maintain vitamin levels within the physiological range in growing pigs with EPI, we tested which application form of fat-soluble vitamins (oral versus parenteral) is best.

38.4.1 Materials and Methods

The pancreatic duct was ligated in 12 pigs aged 8 weeks (PL group) and an additional four pigs were sham operated at the same age and served as controls. After EPI was confirmed by negative chymotrypsin test in feces (<0.8 IU/g using the Chymotrypsin Testkit; Boehringer) 14 days after the operation, the PL pigs were supplemented with the porcine multienzyme product Creon (19.8 g = 1,048,727 IU lipase/kg feed, which is equivalent to 10,182 IU lipase/g fat). The enzyme product (without capsules) was mixed with the diet to ensure parallel intake of the diet and the enzyme product.

The pigs were housed individually and fed a diet in liquid form (25% dry matter). To make sure that no extracorporal digestion processes took place, the feed was offered restrictively several times per day to ensure immediate ingestion (<10 min). The diet consisted mainly of wheat, barley, and soybean meal, with skim milk powder, fish meal, linseed and soybean oils, as well as isolated amino acids and a mineral and vitamin premix to fulfill the requirements of pigs of that age. Chemical composition of the diet (per kilogram of dry matter) was 407 g starch, 210 g crude protein, and 103 g crude fat. Levels of vitamins A and E were boosted (13,393 IU vitamin A and 122 mg vitamin E/kg dry matter) to exceed recommendations [37] by a factor of 6 (vitamin A) and 9 (vitamin E) (see Table 38.2). During the first part of the trial, all animals were fed identical amounts of the diet, whereas from day 44 after the operation until the end of the trial, the diet was given ad libitum (in that phase the diet was offered in a dry form to make sure that there was no extracorporal hydrolysis of nutrients by activity of the added pancreatic enzyme product).

After EPI was confirmed, the PL pigs were split into three groups (each $n = 4$).

38.4.1.1 *PL + Oral*

These pigs were fed the basal diet described earlier plus a vitamin concentrate (see Table 38.3; 18,000 IU vitamin A/mL, 100 IU vitamin D/mL, and 120 mg vitamin E/mL) that was added to the diet (5 mL/kg diet). The vitamin supply via this additive resulted in additional oral vitamin A and E supplementation about 40 times higher than the requirement of healthy pigs (see Table 38.3). The vitamin concentrate contained two emulsifiers (E484 and E1520 as a co-emulsifier) and butylated hydroxyanisole and butylated hydroxytoluene as antioxidants. This vitamin supplementation was provided by a company specializing in feed additives for farm animals and was designed to supply vitamins to poultry via drinking water. This preparation was chosen since it was expected to be highly bioavailable; it forms small micelles, resulting in a stable emulsion due to the efficient emulsifier (E484 and E1520).

38.4.1.2 *PL + IM*

The pigs in this group were supplemented with an intramuscular injection of 0.735 mL Ursovit AD$_3$EC wässrig pro inj. (Serumwerk, Bernburg, Germany) per kilogram body weight, resulting in 5250 IU vitamin A, 525 IU vitamin D, 3.15 mg vitamin E, and 10.5 mg vitamin C per kilogram body weight per week IM (aqueous) once per week. To improve the vitamin E supply, 700 mg vitamin E/animal/week IM (oily) were applied, resulting in additional parenteral vitamin A and E supplementation about five times higher than the requirement.

The parenteral supplementation was chosen to overcome effects of impaired absorption. However, there were difficulties regarding the availability of appropriate vitamin preparations because there was no product on the market containing the fat-soluble vitamins in a proper ratio. For vitamin E there were only commercial products that were combined either with other fat-soluble vitamins (vitamin A and/or vitamin D) or with selenium. Therefore, the vitamin E product for this trial had to be produced by a pharmacist. For practical reasons, the vitamin E dosage was not boosted with increasing age and bodyweight of the pigs but was kept constant during the whole trial (one prefilled syringe of 1 mL contained 700 mg vitamin E).

TABLE 38.2 Recommended Concentrations of Vitamin A (Retinol) and Vitamin E (α-Tocopherol) in Diets for Healthy Growing Pigs [37] and of the Diet Used in the Study

	Recommendations	Experimental Diet
Vitamin A (IU/kg)	2,200	13,393
Vitamin E (mg/kg)	15	122

TABLE 38.3 Chemical Composition of the Liquid Vitamin Preparation Used for Oral Substitution in the PL + Oral Group (per Milliliter) and Dose per Kilogram Diet (5 mL Vitamin Supplement/kg Diet)

	Preparation per mL	Extra Supply per kg Diet
Vitamin A (retinyl palmitate)	18,000	90,000
Vitamin D$_3$ (cholecalciferol)	100	500
Vitamin E (α-tocopherol acetate)	120	600

Further compounds per kilogram of vitamin additive: 100 mL propyl gallate, 240 mg butylated hydroxytoluene (antioxidant), 240 mg butylated hydroxyanisole (antioxidant), 3000 mg sodium citrate, and 1000 mg potassium sorbate; E484 was used as an emulsifier and E1520 was used as a co-emulsifier.

38.4.1.3 PL + 0

This group received PERT like all the other PL pigs but no extra vitamins besides the vitamin content of the basal diet (see Tables 38.2 and 38.4).

Blood samples were taken from the jugular vein using S-Monovetten blood test tubes (Sarstedt) every second week (weeks 0, 2, 4, 6, and 8 after operation) to get further insights regarding the kinetics of vitamin status after experimentally induced EPI. For blood sampling, animals were anaesthetized for a short time with Ursotamin (Serumwerk, Bernburg, Germany; dosage, 0.023 mg ketamine hydrochloride/kg body weight) and Stresnil (Janssen-Cilag GmbH, Neuss, Germany; dosage, 2 mg azaperone/kg body weight). On the day of dissection (week 8) the animals had unlimited access to feed until anesthesia by injection (as described above); after blood sampling the animals were killed by intracardial application of T61 (MSD Animal Health GmbH, Unterschleissheim, Germany; dosage of 4000 mg embrutamide, 1000 mg mebenzonium iodide, and 100 mg tetracaine hydrochloride per animal). Liver samples were taken during dissection to quantify retinol and α-tocopherol concentrations and to estimate risk of vitamin toxicosis. Vitamins in serum and liver tissue were analyzed simultaneously using high-performance liquid chromatography [39,40].

38.4.2 Results

38.4.2.1 Effect of Experimentally Induced EPI and Vitamin Supply on Body Weight and Body Length of Growing Pigs

Pancreatic duct ligation in piglets caused a significant reduction of body weight (see Table 38.5), but there was no effect of whether or not the animals received additional vitamin supply (parenteral or oral). Body length (measured from nose to tail with a tape measure) did not differ between all groups.

38.4.2.2 Effect of Experimentally Induced EPI on Crude Fat Absorption

The digestibility of fat was determined in all pigs 2 weeks (PL pigs did not receive PERT or extra vitamin supply at that time) and 7 weeks after surgery (PL pigs received PERT at that time).

Experimentally induced EPI resulted in a massive reduction of the coefficient of fat absorption (controls: 81.0 ± 2.28%; PL pigs: 25.9 ± 16.5%). PERT resulted in a distinct increase in the coefficient of fat absorption measured at week 7 after surgery (PL + IM group: 60.8 ± 3.04%; PL + oral group: 57.1 ± 5.44%; PL + 0 group: 54.5 ± 8.81%), but levels of controls (85.0 ± 2.14%) were not reached. Vitamin supply (oral or via intramuscular injection) had no effect on the coefficient of fat absorption.

38.4.2.3 Effects of Parenteral and Oral Vitamin Supply on Serum Retinol and α-Tocopherol Concentrations in Crossbred Piglets with Experimentally Induced EPI under Conditions of PERT

38.4.2.3.1 RETINOL CONCENTRATION IN SERUM

Most retinol concentrations in serum varied within the reference range (0.24–0.48 mg/L [41]) or slightly below. There was a directional effect of neither time after surgery nor treatment group on retinol concentrations in serum ($P > 0.05$). There were only two exceptions: the mean retinol concentration in serum of group PL + 0 was lower at week 4, and the mean values in the PL + IM group 8 weeks after

TABLE 38.4 Vitamin A and Vitamin E Supply in Control Pigs and Different Treatment Groups of Pancreatic Duct Ligated (PL) Pigs Applied via Diet (All Animals; Calculated by Taking into Account the Feed Intake) by Extra Vitamin Supply via Feed Additive (PL + Oral) or Intramuscular Injection (PL + IM)

	Daily Vitamin A Supply (IU/kg body weight)				Daily Vitamin E Supply (mg/kg body weight)			
	Controls	PL + Oral	PL + IM	PL + 0	Controls	PL + Oral	PL + IM	PL + 0
Diet	~470				~4.3			
Extra supply	0	~2160	~750	0	0	~14,4	~3,1	0

TABLE 38.5 Body Weight (kilograms) of Juvenile Pigs at the Beginning (7 weeks of Age) and at the End of the Trial (16 weeks of Age) as well as Body Length (cm) at the Age of 16 weeks

	Controls	PL + IM	PL + Oral	PL + 0
Start (age 7 weeks)	16.0 ± 2.22^a	15.8 ± 0.885^a	15.8 ± 2.60^a	16.8 ± 0.780^a
End (age 16 weeks)	57.7 ± 4.17^a	50.8 ± 2.79^b	50.4 ± 3.54^b	51.7 ± 1.85^b
Body length (age 16 weeks)	117 ± 7.19^a	113 ± 0.957^a	116 ± 2.63^a	113 ± 2.22^a

EPI (exocrine pancreatic insufficiency) was experimentally induced at the age of 8 weeks.
Different superscripts indicate significant differences ($p < 0.05$) between different groups within one row. IM, intramuscular; PL, pancreatic duct ligated.

the operation were above the reference value and higher than in controls ($P<0.05$). The higher values at week 8 in all groups are supposed to have resulted from the ad libitum feeding during the last weeks of life and the fact that on the day of dissection the animals had ad libitum access to feed until anesthesia before dissection, whereas feed was withdrawn for at least 6h on days of blood sampling in the weeks before dissection (Figure 38.2).

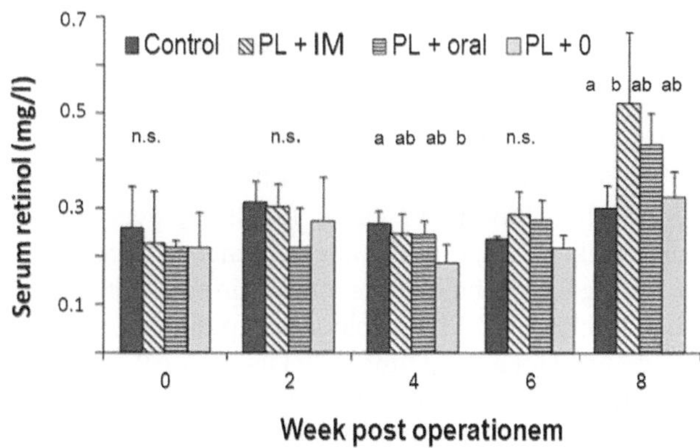

FIGURE 38.2 Serum retinol concentrations (milligrams/liter) of healthy control pigs and pancreatic duct ligated (PL) pigs with (PL+IM, PL+oral) or without (PL+0) vitamin supplementation. Different letters indicate significant differences ($p<0.05$) between different groups within one point of time. n.s., not significant.

38.4.2.3.2 α-TOCOPHEROL CONCENTRATION IN SERUM

In all PL-pigs, serum α-tocopherol concentrations decreased within 2weeks after the operation ($P<0.05$). Since the control pigs underwent a sham operation, this result seems to be more closely related to EPI than to the surgery itself. After starting PERT in the third week after surgery, serum α-tocopherol concentrations increased and were within the reference range (1–5mg/L). The extra supply of vitamins via a special feed additive (PL+oral group) increased serum concentrations ($P<0.05$) and the concentrations of controls were reached, whereas the PL+IM and PL+0 groups showed lower concentrations ($P<0.05$) but were still within the reference range. However, the serum levels decreased in the latter two groups within the following 4weeks and dropped below the reference range. Serum α-tocopherol concentrations were higher in the PL+oral group compared to the PL+0 and PL+IM groups ($P<0.05$) at all times after week 4 (see Figure 38.3).

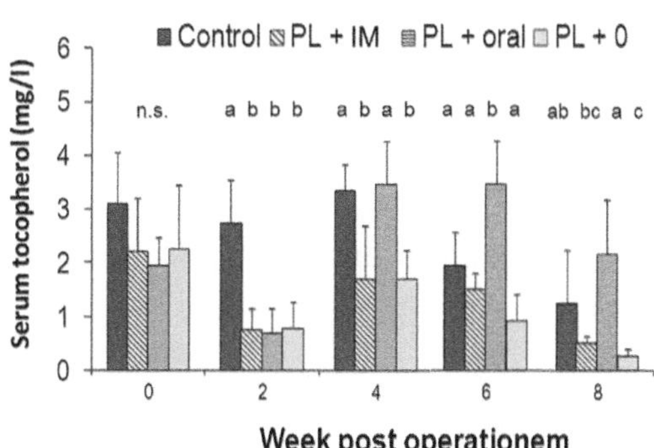

FIGURE 38.3 Serum α-tocopherol concentrations (milligrams/liter) of healthy control pigs and pancreatic duct ligated (PL) pigs with (PL+IM, PL+oral) or without (PL+0) vitamin supplementation. Different letters indicate significant differences ($p<0.05$) between different groups within one point of time. n.s., not significant.

38.4.2.4 *Effects of Parenteral and Oral Vitamin Supply on Retinol and α-Tocopherol Concentrations in Liver Tissue of Crossbred Piglets with Experimentally Induced EPI under Conditions of PERT*

38.4.2.4.1 RETINOL

Retinol concentrations in the liver samples taken during autopsy varied within the reference range (200–400mg retinol/kg dm [41]) in the control, PL+oral, and PL+IM groups, but there was a very high variation in the latter group (in two animals the values were around the upper limit of the reference range [429 and 389mg/kg dry matter], whereas in the other two animals the retinol concentrations were below the lower limit [196 and 139mg/kgdm]). In the PL+0 group the value was lower than that of the PL+oral and control groups ($P<0.05$) and below the lower reference value, whereas the PL+oral group reached higher concentrations than controls ($P<0.05$; see Figure 38.4).

38.4.2.4.2 α-TOCOPHEROL

Regarding α-tocopherol concentrations in liver tissue, the control and PL+oral groups reached levels within the reference range. Concentrations of the PL+IM and PL+0 groups were significantly lower (see Figure 38.5) and below the reference range (15–50mg/kgdm [41]).

38.4.3 Conclusions

In spite of a high dietary vitamin supply and PERT, PL pigs without extra vitamin supply showed reduced α-tocopherol concentrations in serum as well as reduced concentrations of retinol and α-tocopherol in liver tissue, although serum concentrations of retinol were still comparable to that of controls. Vitamin supply via intramuscular injection resulted in a high individual variation of retinol concentrations in the liver but failed to maintain normal concentrations of α-tocopherol in liver tissue (values were not different from those of the PL+0 group). The additional oral supply of high doses of vitamins A and E in combination with a highly efficient emulsifier was adequate to restore α-tocopherol concentrations in serum and to retain retinol and α-tocopherol concentrations in liver tissue within reference values. Retinol concentration in liver tissue distinctly increased above those of controls, indicating that the dosage used in this study was too high. Clinical symptoms of vitamin deficiency or oversupply were clinically apparent in none of the pigs.

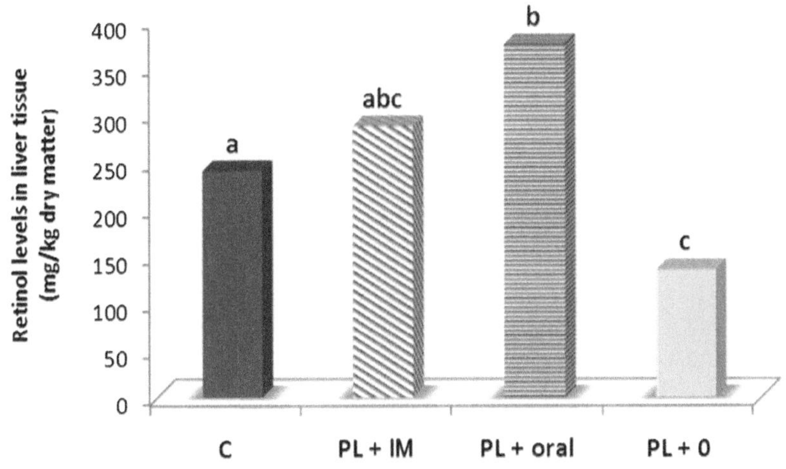

FIGURE 38.4 Retinol concentrations in liver tissue (milligrams/kilogram of dry matter) of healthy control pigs (C) and pancreatic duct ligated (PL) pigs with (PL+IM; PL+oral) or without (PL+0) vitamin supplementation 8 weeks after surgery. Different letters indicate significant differences ($p<0.05$) between treatment groups.

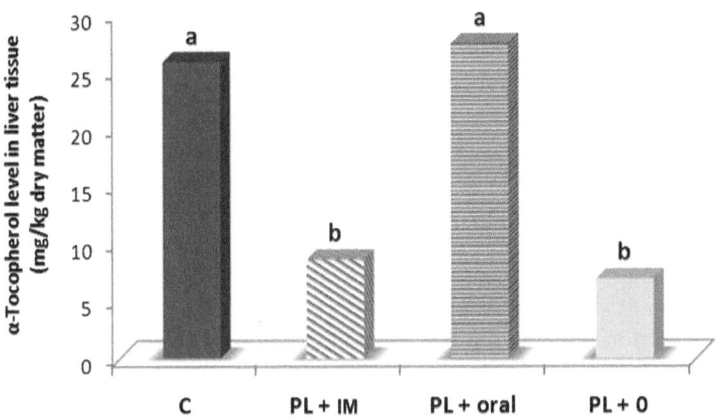

FIGURE 38.5 α-Tocopherol concentrations in liver tissue (milligrams/kilogram of dry matter) of healthy control (C) pigs and pancreatic duct ligated (PL) pigs with (PL+IM, PL+oral) or without (PL+0) vitamin supplementation 8 weeks after surgery. Different letters indicate significant differences ($p<0.05$) between treatment groups.

38.5 DISCUSSION

38.5.1 The PL Pig as a Model for Adults: The PL Piglet as a Model for Children—Do We Really Need a Differentiation?

The depletion of retinol and α-tocopherol in piglets with experimentally induced EPI reflects well the situation in human patients with CF in general: maldigestion and malabsorption of fat often results in malabsorption and a shortage of fat-soluble vitamins [8,16,17]. The possibility of complete standardization of nutritional supply as well as the possibility of using samples of liver tissue for vitamin analysis is a great advantage of animal trials because the concentrations of vitamins in liver tissue are the best indicators of vitamin A and E supply [42,43].

In the PL pig there was an absorption of the fat-soluble vitamins from the diet when animals were treated with PERT, but this was sufficient only in adult pigs and not in the juvenile piglets, as indicated by the reduced vitamin concentrations in the PL+0 group. The fact that in the adult PL minipigs the serum α-tocopherol concentrations dropped quickly when PERT was stopped emphasizes the exhaustion of resources and the impaired absorption of α-tocopherol from the diet without PERT, although the diet contained 10-fold higher α-tocopherol concentrations than what is recommended for maintenance of healthy adult pigs, and the animals had no increased need (no body weight gain, no pregnancy, no lactation, no infection).

Although the growing PL piglets also received PERT and large amounts of fat-soluble vitamins A and E in the diet (almost comparable to those used for the adult PL pigs), they were not able to maintain serum concentrations within the reference ranges without extra vitamin supplementation. This finding underlines the differences between adult and juvenile individuals and the need to use appropriate animal models to imitate the situation in children.

38.5.2 Informative Value of Liver Samples

The possibility of investigating liver samples in the animal model allows to get a more detailed look at vitamin status since blood concentrations stay relatively constant until liver stores are largely exhausted or filled up [43]. Serum retinol concentrations are therefore nonsensitive indicators in the case of subtoxicity or toxicity [44]. If only data for serum concentrations of retinol had been available in study 2 (it is standard in human patients), no differences between the different treatment groups would have become apparent because serum levels still varied within in a physiological range in all PL pigs. Therefore, serum levels alone would have led to an overestimation of vitamin status in the PL+0 group and the oversupply of retinol in the PL+oral group would not have been detected without knowledge of retinol concentration in the liver.

Since in human patients analysis of liver tissue is not possible, any under- or oversupply of vitamin A might be concealed. As a consequence, the frequency of routine blood analysis should be chosen carefully to detect under- or overdosing of vitamin supplements. This seems to be of special importance because low serum concentrations in patients with CF are often subclinical [20] but are nevertheless important [21] because diverse negative effects of vitamin A deficiency can occur before typical symptoms of eye lesions (xerophthalmia) are established. The increased binding of bacteria to epithelial cells of the respiratory tract in the case of vitamin A deficiency [45] is an aspect of crucial relevance for patients with CF. After exhaustion of the hepatic retinol resources the serum concentrations decrease quickly. Therefore, the recommended yearly interval of monitoring the status of retinol in the blood might be too long to detect vitamin deficiencies under some circumstances.

Liver tissue can also be a good indicator of vitamin E in the muscle cells [43]; therefore, a reduced vitamin E concentration in muscle tissue can be assumed in pigs in the PL+IM and PL+0 groups, although no clinical symptoms were obvious. In human patients with CF, buccal mucosa cells seem to be an adequate matrix to measure vitamin E status because its vitamin content responds to dietary intake [19]. The use of this easily accessible tissue as a marker for vitamin supply is an interesting option that should be pursued further in future since the information value of serum concentrations is limited.

The status of fat-soluble vitamins should be checked a few months after substitution and whenever changes in the treatment of malabsorption occur or values are abnormal; routine checks are recommended on a yearly basis [12,22].

38.5.3 Bioavailability of Vitamins after Parenteral Vitamin Supply in Piglets

Another interesting and unexpected result of study 2 is the large variation of vitamin levels in the group of animals supplemented by IM injection. Although injection was suggested to result in more homogenous values because of an exclusion of the effects of reduced intestinal absorption in the case of malabsorption of fat, absorption from the place of injection seems to be highly variable, even when the injection was given in the same manner in all animals. Taking into account the negative effects of intramuscular application of fat-soluble vitamins (pain, reaction of surrounding tissue) and the highly variable effects on serum concentrations, this form of application should not be used for extra vitamin supply despite the theoretical advantage that there is no need for absorption from the gut. Whether other vitamin preparations would have a better availability after intramuscular application can only be speculated. However, positive effects of intramuscular injections of vitamin E in two patients with decreased intraluminal concentrations of bile salts were found [14].

38.5.4 Dosages of Vitamin Supplements

The dosage used for oral substitution in study 2 was much higher than recommendations for healthy individuals. It was chosen to prove the potential of this type of vitamin supplementation in principle. Further studies are needed to generate data for orientation as to the dosage required to maintain normal levels; because this was an orientative study, the vitamin concentrations chosen were relatively high to test the suitability of this model in principle.

The high retinol concentration in the liver of the pigs in the PL+oral group indicates that the dosage was too high. Animals did not show any clinical signs of hypervitaminosis or toxicosis, but increasing serum concentrations during the last 2 weeks of the trial became apparent even though the supplementation period was relatively short (6 weeks). Therefore, the therapeutic dosage should be much lower than that used in this pilot study. The necessity to adapt the dosage is beyond dispute; however, this also clearly demonstrates how fast an increase in vitamin concentrations can occur when supplements with high bioavailability are administered in high doses. Nowadays almost everyone involved in taking care of patients with CF is aware of the risk of vitamin deficiency, and there seems to be an increasing risk of overdosing. Today, highly concentrated vitamin preparations are commercially available, and water-soluble preparations in particular pose a higher risk of overdose or toxicosis in patients because they are absorbed much easier [46–48]. Vitamin concentrations should be monitored regularly [7,49] because the vitamin requirement to maintain serum concentrations within the reference ranges is highly variable in patients with CF [48]. Once again, the restraint of measurement of serum levels must be mentioned; concentrations will only rise in the case of a saturation of liver storage. Hypervitaminosis A is associated with a higher risk of bone disease and liver damage [44] and should therefore be avoided. Vitamin E is stated to be safe across a broad range of doses [50], and high levels might be protective for patients with CF [15,51]. On the other hand, high serum concentrations of vitamin E were found in children with CF when taking typical supplementation of vitamins [52].

To conclude, the PL pig (piglet) turns out to be an appropriate model with which to study the effects of PERT and efficacy of different kinds of vitamin supplements in patients (children) suffering from EPI.

Acknowledgments

The authors thank Miavit GmbH, Essen, Germany, for providing the vitamin additive for oral substitution and to Abbott Laboratories (GmbH), Germany, for providing the digestive enzyme product (Creon) for enzyme replacement therapy.

The authors also thank Dr Johanna Grunemann and Petra Beckmann for support regarding the surgical intervention necessary for this trial.

References

[1] Layer P, Keller J. Pancreatic enzymes: secretion and luminal nutrient digestion in health and disease. J Clin Gastroenterol 1999;28:3–10.

[2] Farrell PM, Bieri JG, Fratononi JF, Wood RE, di SantÀgnese PA. The occurrence and effects of human vitamin E deficiency – a study in patients with cystic fibrosis. J Clin Invest 1977;60:233–41.

[3] Huet F, Semama D, Maingueneau C, Charavel A, Nivelon JL. Vitamin A deficiency and nocturnal vision in teenagers with cystic fibrosis. Eur J Pediatr 1997;156:949–51.

[4] Sagel SD, Sontag MK, Anthony MM, Emmett P, Papas KA. Effect of an antioxidant-rich multivitamin supplement in cystic fibrosis. J Cyst Fibros 2011;10:31–6.

[5] Papas KA, Sontag MK, Pardee C, Sokol RJ, Sagel SD, Accurso FJ, et al. A pilot study on the safety and efficacy of a novel antioxidant rich formulation in patients with cystic fibrosis. J Cyst Fibros 2008;7:60–7.

[6] Park RW, Grant RJ. Gastrointestinal manifestation of cystic fibrosis: a review. Gastroenterology 1981;81(6):1143–61.

[7] Sinaasappel M, Stern M, Littlewood J, Wolfe S, Steinkamp G, Heijermann HGM, et al. Consensus Report, Nutrition in patients with cystic fibrosis: a European Consensus. J Cyst Fibros 2002;1:51–75.

[8] Cantin AM, White TB, Cross CE, Forman HJ, Sokol RJ, Borowitz D. Review article. Antioxidants in cystic fibrosis. Conclusions from the CF Antioxidant Workshop, Bethesda, Maryland, November 11–12, 2003. Free Radical Biol Med 2007;42:15–31.

[9] Sadowska-Woda I, Rachel M, Pazdan J, Bieszczad-Bedrejczuk E, Pawliszak K. Nutritional supplement attenuates selected oxidative stress markers in pediatric patients with cystic fibrosis. Nutr Res 2011;31:509–18.

[10] Carr SB. The role of vitamins in cystic fibrosis. J R Soc Med 2000;93(35):14–9.

[11] Biesalski HK, Stofft E. Biochemical, morphological and functional aspects of systemic and local vitamin A deficiency in the respiratory tract. Ann NY Acad Sci. 1992;669:325–31.

[12] Stern M, Ellemunter H, Palm B, Posselt H-G, Smaczny C. Mukoviszidose (Cystische Fibrose): Ernährung und exokrine Pankreasinsuffizienz.S1-Leitlinie der Gesellschaft für Pädiatrische Gastroenterologie und Ernährung (GPGE); 2011. http://www.awmf.org/uploads/tx_szleitlinien/068-020l_S1_Mukoviszidose_Ernährung_exokrine_Pankreasinsuffizienz_2011-05.pdf. [accessed 01.12.13].

[13] Rees Smith F, Underwood BA, Denning CR, Varma A, Goodman DS. Depressed plasma retinol-binding protein levels in cystic fibrosis. J Clin Lab Med 1972;80:423–33.

[14] Sitrin MD, Liebermann F, Jensen WE, Noronha A, Milburn C, Addington W. Vitamin E deficiency and neurologic disease in adults with cystic fibrosis. Ann Intern Med 1987;107(1):51–4.

[15] James DR, Alfaham M, Goodchild MC. Increased susceptibility to peroxide-induced haemolysis with normal vitamin E concentrations in cystic fibrosis. Clin Chim Acta 1991;204:279–90.

[16] Congden PJ, Bruce G, Rothburn MM, Clarke PCN, Littlewood JM, Kelleher J, et al. Vitamin status in treated patients with cystic fibrosis. Arch Dis Child 1981;56(9):708–14.

[17] Dutta S, Bustin MP, Russel RM, Costa BS. Deficiency in fat-soluble vitamins in treated patients with pancreatic insufficiency. Ann Inter Med 1982;97(4):549–52.

[18] Feranchak AP, Sontag MK, Wagener JS, Hammond KB, Accurso FJ, Sokol RJ. Prospective, long-term study of fat-soluble vitamin status in children with cystic fibrosis identified by newborn screen. J Pediatr 1999;135(5):601–10.

[19] Back EI, Frindt C, Nohr D, Frank J, Ziebach R, Stern M, et al. Antioxidant deficiency in cystic fibrosis: when is the right time to take action. Am J Clin Nutr 2004;80(2):374–84.

[20] Rayner RJ. Fat-soluble vitamins in cystic fibrosis. Proc Nutr Soc 1992;51:245–50.

[21] Dodge JA, Turck D. Cystic fibrosis: nutritional consequences and management. Best Prac Res Clin Gastroenterol 2006;20(3):531–41.

[22] Borowitz D, Robinson KA, Rosenfeld M, Davis SD, Sabadosa KA, Spear SL, et al. Cystic fibrosis foundation evidence-based guidelines for management of infants with cystic fibrosis. J Pediatr 2009;155:S73–93.

[23] Sommer A, Katz J, Tarwotjo L. Increased risk of respiratory disease and diarrhea in children with coexisting mild vitamin A deficiency. Am J Clin Nutr 1984;40:1090–5.

[24] Wood LG, Fitzgerald DA, Lee AK, Garg ML. Improved antioxidant and fatty acid status of patients with cystic fibrosis after antioxidant supplementation is linked to improved lung function. Am J Clin Nutr 2003;77:150–9.

[25] Abello J, Pascaud X, Simones-Nuntes C, Cuber JC, Junien JL, Roze C. Total pancreatic exocrine insufficiency in pigs: a model to study intestinal enzymes and plasma levels of digestive hormones after pancreatic supplementation by a whole pancreas preparation. Pancreas 1989;4(5):556–64.

[26] Gregory PC, Tabeling R, Kamphues J. Growth and digestion in pancreatic duct ligated pigs. Effect of enzyme supplementation. In: Pierzynowski SG, Zabielski R, editors. Biology of the pancreas in growing animals. Elsevier Science B.V.; 1999. pp. 381–94.

[27] Mößeler A, Tabeling R, Gregory PC, Kamphues J. Compensatory digestion of fat, protein and starch (rates and amounts) in the large intestine of minipigs in case of reduced precaecal digestion due to pancreatic duct ligation – a short review. Livest Sci 2007;109:50–2.

[28] Mößeler A, Kramer N, Becker C, Gregory PC, Kamphues J. Prececal digestibility of various sources of starch in minipigs with or without experimentally induced exocrine pancreatic insufficiency. J Anim Sci 2012;90:83–5.

[29] Tabeling R, Gregory PC, Kamphues J. Studies on nutrient digestibilities (pre-caecal and total) in pancreatic duct ligeated pigs and the effects of enzyme substitution. J Anim Physiol Anim Nutr 1999;82:251–63.

[30] Kammlott E, Karthoff J, Stemme K, Gregory P, Kamphues J. Experiments to optimize enzyme substitution therapy in pancreatic duct-ligated pigs. J Anim Physiol Anim Nutr 2005;89(3–6):105–8.

[31] Rahko T, Kalima TV, Salonemi H. Pancreatic duct obstruction in the pig: light microscopy of chronic pancreatitis. Acta Vet Scand 1987;28:285–9.

[32] Boerma D, Straatsburg IH, Offerhaus GJA, Gouma DJ, van Gulik TM. Surgical management of benign biliopancreatic disorders. Chapter 6 Experimental model of chronic pancreatitis in the pig; 2006. 9090-13578-48.

[33] Fedkiv O, Regman S, Weström BR, Pierzynowski SG. Growth is dependent on the exocrine pancreatic function in young weaners biút not in growing-finishing pigs. J Physiol Pharmacol 2009;60(3):55–9.

[34] Rengman S, Fedkiv O, Boltermans J, Svendsen J, Weström B, Pierzynowski S. An elemental diet fed, enteral or parenteral, does not support growth in young pigs with exocrine pancreatic insufficiency. Clin Nutr 2009;28(3):325–30.

[35] Schwarzmaier T. Studies on the effects of substitution of pancreatic enzymes as well as special supplementation of vitamin A and E in growing pancreatic duct ligated pigs as a model for pancreatic insufficiency in children [Doctoral thesis]. Hannover: University of Veterinary Medicine; 2012.
[36] Tabeling R. Studies on the effects of enzyme substitution on nutrient digestibilities (pre-caecal/total) in pancreas ligated pigs [Doctoral thesis]. Hannover: University of Veterinary Medicine; 1998.
[37] GfE, Committee for Requirement Standards of the Society of Nutrition Physiology. Recommendations for the supply of energy and nutrients to pigs. Frankfurt am Main, Germany: DLG-Verlag; 2008978-3-7690-0707-7.
[38] Naumann C, Bassler R. Chemical analysis of feedstuffs. 3rd ed. 5th update, VDLUFA-Verlag; 2004.
[39] Rammel GC, Cunliffe B, Kieboom AJ. Determination of alpha-tocopherol in biological specimens by HPLC. J Liq Chromatogr 1983;6:1123–30.
[40] Mudron P. Plasma and liver α-tocopherol in dairy cows with left abomasal displacement and fatty liver [Doctoral thesis]. Hannover: University of Veterinary Medicine; 1994.
[41] Puls R. Vitamin levels in animal health. Canada: Sherpa International, Clearbrook; 1994.
[42] Rousseau Jr JE, Dicks MW, Teichmann R, Helmboldt CF, Bacon EL, Prouty RM, et al. Relationships between plasma, liver and dietary α-tocopherol in calves, lambs and pigs. J Anim Sci 1957;16:612–22.
[43] Underwood BA, Denning CR. Blood and liver concentrations of vitamins A and E in children with cystic fibrosis. Pediatr Res 1972;6:26–31.
[44] Penniston KL, Tanumihardjo SA. The acute and chronic toxic effects of vitamin A. Am J Clin Nutr 2006;83:137–41.
[45] Chandra RK. Increased bacterial binding to respiratory epithelial cells in vitamin A deficiency. Br Med J 1988;297:834–5.
[46] Myhre AM, Carlsen MH, Bohn SK, Wold HL, Laake P, Blomhoff R. Water-miscible, emulsified, and solid forms of retinol supplements are more toxic than oil-based preparations. Am J Clin Nutr 2003;78:1152–9.
[47] Back EI, Frindt C, Ocenásková E, Nohr D, Stern M, Biesalski HK. Can changes in hydrophobicity increase the bioavailability of alpha-α-tocopherol? Eur J Nutr 2005;45:1–6.
[48] Brei C, Simon A, Krawinkel MB, Naehlich L. Individualized vitamin A supplementation for patients with cystic fibrosis. Clin Nutr 2013;32:805–10.
[49] Maqbool A, Graham-Maar RC, Schall JI, Zemel BS, Stallings VA. Vitamin A intake and elevated serum retinol levels in children and young adults with cystic fibrosis. J Cyst Fibros 2008;7:137–41.
[50] Hathcock JN, Azzi A, Blumberg J, Bray T, Dickinson A, Frei B, et al. Vitamin E and C are safe across a broad range of intakes. Am J Clin Nutr 2005;81:736–45.
[51] Brown RK, Wyatt H, Price JF, Kelly FJ. Pulmonary dysfunction in cystic fibrosis is associated with oxidative stress. Eur Respir J 1996;9:334–9.
[52] Huang SH, Schall JI, Zemel BS, Stallings VA. Vitamin E status in children with cystic fibrosis and pancreatic insufficiency. J Pediatr 2006;148:556–9.

C H A P T E R

39

Unsaturated Fatty Acids in Cystic Fibrosis: Metabolism and Therapy

Adam Seegmiller[1], *Michael O'Connor*[2]

[1]Department of Pathology, Microbiology, and Immunology, Vanderbilt University School of Medicine, Nashville, TN, USA; [2]Department of Pediatrics, Division of Allergy, Immunology, and Pulmonology Medicine, Vanderbilt University School of Medicine, Nashville, TN, USA

39.1 INTRODUCTION

More than 50 years ago, researchers first observed abnormal fatty acid (FA) concentrations in the blood and tissues of patients with cystic fibrosis (CF) when compared with healthy controls [1]. Subsequent studies showed that the most consistent FA abnormalities are in unsaturated FAs, a class of FAs characterized by the presence of one or more double bonds in the hydrocarbon chain [2–4].

These unsaturated FAs fall into three categories based on the position of the most distal double bond. Examples are shown in Figure 39.1(A). In omega-3 or n-3 FAs, such as alpha-linolenate (LNA), and in omega-6 or n-6 FAs, such as linoleate (LA), the most distal double bond is three and six carbons, respectively, from the terminal carbon of the hydrocarbon chain. These FAs contain multiple double bonds and thus are called polyunsaturated FAs. They are considered essential FAs because they cannot be synthesized de novo by mammalian cells and must be obtained from dietary sources. However, mammalian cells can metabolize polyunsaturated FAs to longer and more desaturated forms through the actions of FA elongases and desaturases (see Figure 39.1(B)) [5]. These FAs are important not only for their role in cellular membranes but also because they serve as precursors for eicosanoids (prostaglandins, leukotrienes, and lipoxins) [6,7] and docosanoids (resolvins and protectins) [8], which are paracrine regulators of many physiologic processes.

In omega-7 or n-7 FAs, such as palmitoleate (POA; Figure 39.1(A)), and in omega-9 or n-9 FAs, such as oleate (OA), the most distal double bond is seven and nine carbons, respectively, from the terminal carbon of the hydrocarbon chain. These FAs mostly, but not exclusively, contain a single double bond in the hydrocarbon chain and thus are called monounsaturated FAs. Unlike polyunsaturated FAs, these can be synthesized de novo by elongases and desaturases in mammalian cells , using saturated FAs as substrates (Figure 39.1(B)). Monounsaturated FAs function primarily in membrane structure and energy storage, and there is recent evidence suggesting that they may also have signaling functions [9].

Given the important biochemical and physiologic functions of unsaturated FAs and the consistency of the FA abnormalities observed in CF, some have hypothesized that altered FA metabolism may play a role in disease pathophysiology. This has led to a number of studies investigating the potential of polyunsaturated FA supplementation as a therapy for CF.

In this chapter we examine the patterns of altered FA levels in the blood and tissues of patients CF and CF disease models. We discuss potential mechanisms for these alterations and links to disease pathophysiology. We then consider potential therapeutic approaches that target these metabolic abnormalities and describe the findings of FA-based clinical trials.

39.2 PATTERNS OF POLYUNSATURATED FA ALTERATIONS IN CF

Since the original description, there have been at least 18 additional reports of altered unsaturated FA levels in the serum [10–14], plasma [15–25], erythrocytes [26], or whole blood [27] of patients with CF compared with healthy controls.

Diet and Exercise in Cystic Fibrosis
http://dx.doi.org/10.1016/B978-0-12-800051-9.00039-0

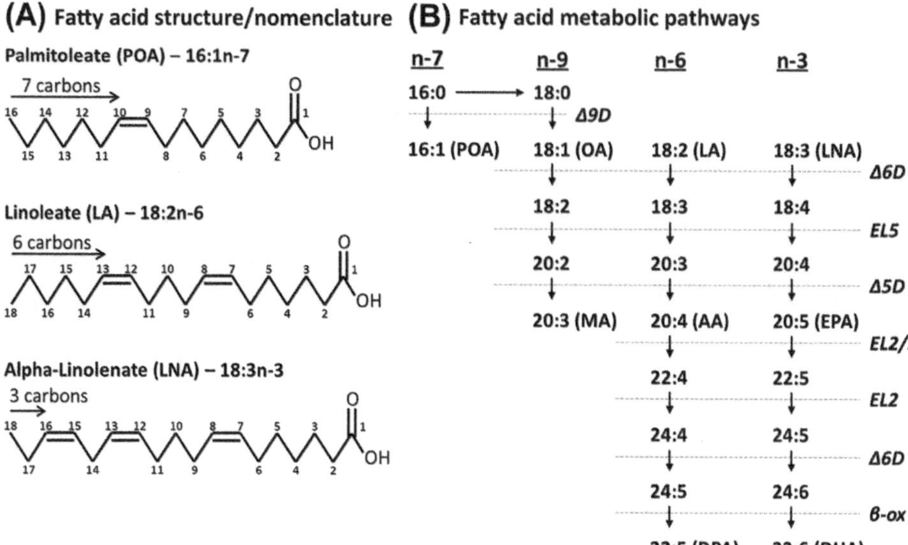

FIGURE 39.1 (A) Examples illustrating the structure and nomenclature of monounsaturated (palmitoleate [POA]) and polyunsaturated fatty acids (linoleate [LA] and alpha-linolenate [LNA]). (B) Metabolic pathways of monounsaturated and polyunsaturated fatty acids. Metabolic steps catalyzed by common enzymes are indicated by horizontal dashed lines. AA, arachidonate; β-ox, beta-oxidation; Δ5D, Δ5-desaturase; Δ6D, Δ6-desaturase; Δ9D, Δ9-desaturase; DHA, docosahexaenoate; DPA, docosapentaenoate EL2, elongase 2; EL5, elongase 5; EPA, eicosapentaenoate; LA, linoleate; LNA, alpha-linolenate; MA, Mead acid; OA, oleate; POA, palmitoleate.

TABLE 39.1 Unsaturated Fatty Acid Alterations in the Blood of Patients with Cystic Fibrosis

Fatty Acid	Change	Frequency	References
18:2n-6 (LA)	Decreased	19/19	[1,10–27]
16:1n-7 (POA)	Increased	11/12	[1,10,11,13, 16–20,23,26,27]
22:6n-3 (DHA)	Decreased	9/11	[15,18–27]
20:3n-9 (MA)	Increased	7/9	[13–15,18–20,22,25,26]
20:4n-6 (AA)	Decreased	5/15	[10,11,13–19,21–23,25–27]
18:3n-3 (LNA)	Increased	2/7	[10,15,16,18,19,25,26]

AA, arachidonate; DHA, docosahexaenoate; LA, linoleate; LNA, alpha-linolenate; MA, Mead acid; POA, palmitoleate.

Common abnormalities in unsaturated FA levels are listed in Table 39.1. The most consistently observed abnormality was in LA, the levels of which were decreased in the blood of patients with CF in all 19 studies. The majority of studies also showed increased POA and Mead acid (MA; 20:3n-9) and decreased docosahexaenoate (DHA; 22:6n-3). Both LNA and arachidonate (AA; 20:4n-6) were unchanged in most studies, although LNA was increased and AA decreased in the blood of patients with CF in a minority of studies.

Because these changes are so consistent, FA analysis has potential as an adjunctive diagnostic test for CF. Batal et al. [24] found that the product of plasma LA × DHA was markedly decreased compared to controls. In open and blinded tests this parameter could distinguish patients with CF from controls with 92–100% sensitivity.

The consistency of the changes in patients with CF raises the important question of how abnormalities in FA levels might be associated with mutations in the *CFTR* gene, the underlying genetic cause of CF. Because the CF transmembrane conductance regulator (CFTR) is primarily expressed in epithelial cells, similar FA changes might be expected in these tissues. To this end, Freedman et al. [23] examined FA levels in biopsies of nasal and rectal epithelia (Table 39.2). Similar to the blood studies, they demonstrated decreased levels of LA and DHA. Unlike in the blood, however, AA levels were significantly increased in the nasal biopsy specimens. A similar increase in AA was noted in another study of bronchoalveolar lavage fluid from patients with CF [28]. Using tissues from autopsy specimens, Farrell et al. [19] showed that the decrease in LA was consistent in fat, skeletal and cardiac muscle, liver, and lung. However, POA was increased in lung only, and there was no difference in AA levels in these tissues between patients with CF and controls.

These findings have been confirmed and extended in animal and cell culture models of CF. Animal studies have been conducted in both homozygous *CFTR* knockout [29,30] and *CFTR*-ΔF508 [31] mice. In general, these studies confirm a consistent decrease in LA levels in tissues expressing CFTR, including the lung, small intestine, and pancreas. They also demonstrated a significant increase in either AA or di-homo-γ-linolenic acid (20:3n-6), the immediate precursor of AA. Freedman et al. [29] also examined heart, brain, and kidney, tissues with little or no CFTR expression, and observed no difference in LA or AA levels between *CFTR*-mutant and wild-type mice. Findings regarding DHA were less consistent: decreased DHA was observed only in mouse tissues in one of the three studies [29].

Cell culture models of CF have been valuable in examining the mechanisms of these FA alterations because they represent a homogenous population of epithelial cells and are not contaminated by mesenchymal cells that do not express CFTR. Accordingly, more

TABLE 39.2 Unsaturated Fatty Acid Alteration in Tissues of Patients with Cystic Fibrosis

Tissue	18:2n-6 (LA)	20:4n-6 (AA)	22:6n-3 (DHA)	16:1n-7 (POA)	References
Adipose	Decreased	No change	n.d.	Increased	[1,17]
Skeletal muscle	Decreased	No change	n.d.	No change	[17]
Cardiac muscle	Decreased	No change	n.d.	No change	[17]
Liver	Decreased	No change	n.d.	No change	[17]
Lung	Decreased	No change	n.d.	Increased	[17]
Nasal biopsy	Decreased	Increased	Decreased	n.d.	[22]
Rectal biopsy	n.d.	No change	Decreased	n.d.	[22]
BAL fluid	n.d.	Increased	n.d.	n.d.	[28]

BAL, bronchoalveolar lavage; n.d., not determined.

extensive differences have been observed that suggest patterns of metabolic disturbance. Studies focused on two bronchial epithelial cell models: 16HBE cells and IB3/C38 cells. 16HBE-antisense cells are immortalized cells that are stably transfected with a transgene expressing the first 131 base pairs of the *CFTR* gene in the antisense orientation, which eliminates both CFTR expression and function [32,33]. Control cells, which expressed the same nucleotides in the sense orientation (sense cells), retained normal CFTR expression and function. IB3 cells are derived from patients with CF with a compound heterozygous genotype (ΔF508/W1282X). This defect is corrected in C38 cells by exogenous expression of wild-type CFTR.

Similar to human and mouse tissues, these models clearly demonstrated decreased LA and increased AA levels in CF versus wild-type cells. However, they also showed parallel changes in the corresponding intermediates of the n-3 pathway (see Figure 39.1(B)), including decreased LNA and increased eicosapentaenoate (20:5n-3; EPA) [32,34]. Similarly, decreased DHA in the n-3 pathway was paralleled by decreased docosapentaenoate (DPA; 20:5n-6) in the n-6 pathway, suggesting that the changes may be due to alterations in the common metabolic enzymes.

These metabolic changes are highly dependent on substrate concentrations. For example, the differences in LA and AA between CF and wild-type cells are appreciated only when the cells are grown in serum containing high concentrations of LA [32,35]. A similar relationship is observed for mouse tissues [36]. Furthermore, there seems to be some degree of competition between the metabolic pathways, such that high levels of LNA (the n-3 pathway) compete with n-6 metabolism and abrogate the differences in metabolism observed in CF and wild-type cells [35]. These observations could have implications for dietary therapy in CF (discussed below).

39.3 RELATIONSHIP BETWEEN FA ALTERATIONS AND CLINICAL FEATURES OF CF

The consistency of FA abnormalities in CF suggests that altered metabolism may play a role in the pathophysiology of this disease. There are three lines of evidence suggesting this connection. First, the degree of FA abnormality correlates with the severity of disease. Second, these FAs participate in pathways with known pathologic significance in CF. Third, reversal of FA abnormalities ameliorated CF-related pathology in a mouse model of the disease.

The severity of CF is determined, to some degree, by the specific *CFTR* mutations carried by the patient. More than 1500 mutations have been identified in patients with CF [37] and have been divided into five classes based on known or predicted effects on CFTR function. Classes I, II, and III are categorized as severe based on their propensity for causing pancreatic insufficiency and malnutrition, liver disease, and meconium ileus, as well as earlier age of onset [38]. Classes IV and V are categorized as mild based on little or no pancreatic disease, later onset, and milder pulmonary disease [38]. Strandvik et al. [12] compared patients with severe genotypes who were had pancreatic insufficiency with patients with mild genotypes and a lower rate of pancreatic insufficiency. They found that whereas patients with severe disease exhibited significantly lower concentrations of LA, those with mild disease had concentrations indistinguishable from healthy controls. This association also was observed in earlier studies in which FA alterations, in particular of LA and POA, were most pronounced in those patients with more severe pancreatic disease [16,18,19,26].

Despite this association with pancreatic insufficiency, FA alterations in CF do not seem to be the sole result of malabsorption. This is evidenced by the fact that these FA abnormalities persist in patients in whom malabsorption has been adequately treated by pancreatic enzyme replacement and nutritional supplementation [22,25]. Furthermore, the presence of essentially identical abnormalities in epithelial cell culture models [32,34] suggests an intrinsic defect in FA metabolism in *CFTR*-mutated cells that is independent of absorption status.

Several studies also have examined the relationship between FA concentrations and pulmonary function [13,20,39]. These have demonstrated weak but statistically significant correlations between levels of particular polyunsaturated FAs, especially LA and DHA, and pulmonary function tests, including forced expiratory flow and forced expiratory volume in 1 s (FEV_1).

Polyunsaturated FAs play significant roles in physiology, both as constituents of cellular membranes and as precursors of bioactive lipid species. Thus it should be no surprise that alterations of polyunsaturated FA metabolism have an effect in disease. This was suggested by two early animal models of essential FA deficiency (EFAD). EFAD is characterized by high MA-to-AA ratios in the blood, similar to the changes observed in CF [40]. Rabbits with EFAD were shown to have defective lung immunity, including impaired macrophage and lymphocyte function, in response to *Pseudomonas* infection [41]. Chickens with EFAD exhibited pulmonary bronchiolitis [42].

One hypothesis regarding the effect of FA alterations on pathophysiology is that changes in FA metabolism alter the composition, and thus the biophysical properties, of epithelial cell membranes, which may regulate membrane protein function. This theory is supported by recent studies showing that polyunsaturated FAs such as AA and LA can modulate the activity of the CFTR and other anion channels [43–46].

FA alterations may also affect cell and organ physiology through their effect on the production of polyunsaturated FA-derived bioactive lipids. These include metabolites of AA, such as prostaglandins, leukotrienes, and lipoxins, and metabolites of EPA and DHA, such as resolvins and protectins.

There is significant evidence that these lipid products play a role in the pathophysiology of CF. Production of prostaglandins, especially PGE_2 and $PGF_{2\alpha}$ and their metabolites, is increased in patients with and models of CF [34,47–52], and the levels of prostaglandins seem to correlate with disease severity [47,51,53]. Leukotrienes are similarly increased [34,54,55], especially in the setting of acute pulmonary exacerbation [56]. However, levels of lipoxins, which decrease inflammation, are reduced [57,58].

The role of increased prostaglandins and leukotrienes in CF also is supported by several studies that demonstrate increased expression of the enzymes that synthesize them. The rate-limiting enzymes for the synthesis of prostaglandins and leukotrienes are cyclooxygenases (COX) 1 and 2 and 5-lipoxygenase. The expression of both of these enzymes is upregulated in the sinonasal tissue of patients with CF [59,60] and in CF cells [34,52] compared with controls.

These changes may influence a number of different physiologic processes that are impaired in CF. Inflammation plays an important role in the pathology of CF. In CF, airways are chronically inflamed, and there is evidence to suggest that this inflammation is out of proportion to or, in some cases, even independent of infectious stimulus [61]. This contributes to the destructive cycle of obstruction, infection, and inflammation that ultimately leads to pulmonary failure in CF. Inflammation may also play a role in intestinal pathology in CF [62,63].

Leukotrienes, in particular LTB_4, are important stimulators of inflammation, whereas lipoxins, such as LXA_4, mediate the resolution of inflammatory episodes [64]. PGE_2, a major subclass of prostaglandins, activates the inflammatory mediator class-switching between LTB_4 and LXA_4 [64,65]. It also stimulates inflammatory cytokine production [66] and inhibits alveolar macrophage function [67–69]. In addition, PGE_2 stimulates mucin secretion, which could contribute to pulmonary obstruction [70], and alters intestinal motility, which could contribute to intestinal obstruction and malabsorption [71]. The role of prostaglandins in CF also is supported by the effects of high-dose ibuprofen, a nonspecific COX inhibitor, which has been shown to reduce the rate of pulmonary decline in patients with CF [72].

Decreased concentrations of DHA could also have an effect on the pathophysiology of CF. DHA and EPA are precursors of docosanoids, a category of oxygenated polyunsaturated FA metabolites that includes resolvins and protectins [8]. These compounds play an important anti-inflammatory role, especially in the resolution of inflammation [73]. Decreased substrate levels of DHA could lead to lower docosanoid production, which might exacerbate the hyperinflammatory condition of airways in CF. In fact, Yang et al. [74] demonstrated that decreased concentrations of resolvin E1 in sputum are associated with poorer pulmonary function in patients with CF.

As these studies indicate, there is significant evidence to suggest that alterations in FA levels contribute to the pathogenesis of CF, although the precise mechanisms are not established. This understanding opens the potential for FA-based dietary and therapeutic intervention in CF.

39.4 MECHANISMS OF ALTERED FA METABOLISM IN CF

An important question arising from the consistent observations of FA alterations in CF involves the mechanistic connection between these changes and *CFTR* mutations. On the surface, it is difficult to intuitively discern how the impaired function of a cell surface ion channel could affect seemingly unrelated metabolic pathways. While this question has yet to be definitively and completely answered, significant progress has been made.

The most obvious potential mechanism of the FA abnormalities in CF is simply intestinal malabsorption. The FA changes in CF are remarkably similar to those of patients with EFAD, in particular, an increased ratio of MA to AA [40,75]. Furthermore, early studies showed that FA abnormalities were primarily seen in patients with CF with pancreatic insufficiency—and thus fat malabsorption [16,18,19,26]. However, more recent studies have demonstrated that in patients with normal nutritional status, FA changes persist because of pancreatic enzyme replacement and intensive nutritional therapy [22,25]. Furthermore, these alterations are seen in isolated cell culture models of bronchial epithelial cells in CF [32,34], in which malabsorption is not a concern. In fact, there is some evidence to suggest that absorption of FAs is actually increased in these cells [76]. In fact, CF-related FA metabolic abnormalities can be induced in wild-type cells in culture using a small-molecule inhibitor of CFTR [34]. These findings suggest an intrinsic connection between CFTR function and FA metabolism that cannot be explained by the malabsorption hypothesis alone.

39.4.1 Mechanisms of Decreased Linoleate

Early studies suggested a defect in n-6 FA metabolism. Working with blood mononuclear cells from patients with CF, Carlstedt-Duke et al. [77] demonstrated that the release of AA could not be suppressed by glucocorticoids in CF cells as it was in control cells. They hypothesized that there is consequent increased turnover of AA in these cells, resulting in the n-6 FA changes observed.

In fact, studies of the human bronchial cell culture model of CF described above demonstrated that there is increased flux of LA to AA in CF cells compared with controls [34,76]. This flux was measured by incubating cells with radiolabeled LA and measuring its conversion to AA. This value was significantly increased in CF cells relative to wild-type controls [34,76]. This correlated with increased messenger RNA and protein expression of the polyunsaturated FA metabolic enzymes Δ5- and Δ6-desaturase in CF cells [34]. The activity of these enzymes is regulated at the transcriptional level [78] and is essential in controlling the rate of this pathway [5]. Thus the low levels of LA and the higher levels of AA in the tissues of patients with CF and mouse models, and in cultured cell systems, is likely due to the increased expression and activity of these enzymes.

This result is supported by studies of DHA supplementation in mice and in cultured cells. When CF mice received supplementation with high doses of oral DHA, not only did their plasma and tissue DHA concentrations increase but the abnormally high AA concentrations in lung and pancreas were reduced to wild-type levels [29]. In cultured cells, DHA supplementation similarly raised LA and reduced AA concentrations, such that they were equivalent to those in control cells [79]. This correlated with a reduction of LA to AA flux and Δ5- and Δ6-desaturase messenger RNA in CF cells to wild-type levels [79].

These studies clearly show that decreased LA can be attributed to upregulation of the LA-to-AA metabolic pathway. However, the connection between this pathway and mutations in CFTR is still unclear, although several hypotheses have been proposed [3]. Understanding this connection is the subject of ongoing studies.

39.4.2 Mechanisms of Increased Palmitoleate and Mead Acid

Unlike LA, POA and MA are not essential FAs. They can be synthesized de novo using palmitate (16:0) as the ultimate substrate (see Figure 39.1(B)). As indicated earlier, both of these FAs are consistently increased in plasma of patients with CF (Table 39.1). These findings were recapitulated in cell culture models of CF [32,80]. Thomsen et al. [80] showed that radiolabeled palmitate is metabolized to POA at an increased rate in CF compared to wild-type cells. Elongation to stearate (18:0) and further metabolism to OA and MA is similarly increased. These changes correlate with increased expression of the relevant metabolic enzymes, namely Δ9-desaturase and elongase-6 (see Table 39.1). Further metabolism of OA to MA is accomplished by Δ5- and Δ6-desaturases, which were previously shown to be increased. Thus the n-7 and n-9 pathway changes in CF seem to be mediated by increased metabolism due to overexpression of metabolic enzymes, similar to what is observed for the n-6 pathway.

39.4.3 Mechanisms of Decreased Docosahexaenoate

Low DHA concentrations are among the most common changes observed in CF. However, the mechanisms of this change have not been as well studied as those of the n-6 pathway. Using cell culture models of CF, Njoroge et al. [34] showed that the metabolism of EPA to DHA in the n-3 metabolic pathway was decreased. However, they also demonstrated no change in the metabolism of 22:5n-3, the immediate product of EPA, to DHA, suggesting that the deficiency is in the first step of that pathway (see Figure 39.1(B)). Thus, one possible mechanism is that the EPA substrate is shunted to another pathway, making it unavailable for metabolism to DHA. A potential candidate for this pathway is metabolism by COX or lipoxygenase to EPA-derived prostaglandins and leukotrienes, although this possibility has not been fully investigated.

A lesser-known DHA metabolic pathway is retroconversion of DHA back to EPA via modified β-oxidation in peroxisomes [81–83]. This pathway is difficult to assay, but its activity can be estimated by comparing the changes in EPA and DHA concentrations in cells supplemented with DHA [84]. This technique, applied to CF cells in culture, showed that the estimated rate of retroconversion is approximately 20 times higher in CF than wild-type cells [79]. This could account for both the lower concentrations of DHA and the disproportionately high concentrations of EPA in the cultured cell model [34]. However, additional studies are required to demonstrate that this abnormality is directly responsible for the lower concentrations of DHA observed in CF.

A final hypothesis to explain DHA abnormalities in CF has to do with alterations in the levels of phospholipid species in CF [3]. Phosphatidylcholine (PC) can be synthesized by one of two mechanisms: (1) de novo, using dietary choline as a substrate, or (2) by multiple methylation steps of phosphatidylethanolamine (PE), using methyltetrahydrofolate as a methyl donor. PE naturally has higher DHA content, and thus PC derived from PE is richer in DHA. However, there seem to be defects in methyl group metabolism that result in a preference for the de novo synthesis pathway of PC, which may result in lower DHA concentrations [85]. This may not only explain differences in DHA levels but also the increased choline uptake that has been observed in CF [86].

39.5 DIETARY AND THERAPEUTIC IMPLICATIONS OF FA METABOLISM IN CF

39.5.1 Diet and FAs

Because of the propensity for malabsorption and malnutrition, it is recommended that even patients receiving pancreatic enzyme replacement therapy consume a high-calorie, high-fat diet [87]. However, there are no known recommendations regarding the specific fat content of that diet.

The FA content of the typical Western diet is very different from historic norms in that the ratio of n-6 to n-3 FAs significantly increased over the past century [88,89]. This may have a particular effect on patients with CF. The studies described earlier demonstrate that the metabolism of LA to AA is increased in CF cells. Furthermore, the abnormal LA and AA concentrations are most pronounced in cells incubated in serum with a high LA-to-LNA ratio [35].

These factors combined suggest that a diet high in LA may result in even greater AA concentrations, increasing the production of potentially harmful prostaglandins and leukotrienes. In fact, Zaman et al. [36] showed that LA supplementation in CF cells and mice increased the production of AA and eicosanoid (PGE_2 and $PGF_{2\alpha}$), causing increases in inflammatory cytokines and airway inflammation. These studies suggest that perhaps attention should be paid not only to the quantity of dietary fat but also to the FA content of that fat and that a lower n-6-to-n-3 FA ratio might be beneficial to patients with CF. However, these hypotheses still need to be tested in human clinical trials.

39.5.2 FA Therapy

Both cell culture [79] and animal studies [29,31,90] have demonstrated that the FA abnormalities in CF can be reversed by supplementation with DHA. Among these, Freedman et al. [29] also showed that high-dose DHA can also reverse CF-related pathology in mice, including reducing the diameter of the pancreatic duct, reversing ileal villus hypertrophy, and decreasing stimulated pulmonary inflammation. A second study using older mice with a different genetic background showed a significant reduction in hepatic inflammation, although effects on the pancreas and ileum could not be confirmed [90]. These studies suggested that supplementation with n-3 FAs, especially DHA, might have a beneficial effect in patients with CF.

To this end, in the past decade, 16 clinical trials involving n-3 essential FA supplementation in CF have been published [91–106] (see Table 39.3). All of the studies used some combination of n-3 FA supplements, with most focusing on DHA or EPA. One study provided the n-3 supplement as an intravenous preparation [97]. The remainder used an enteral n-3 FA preparation. Nearly all of these studies have been small, most with fewer than 20 participants and some with fewer than 10 participants. The single exception involved 30 participants [99]. Most studies were short-term and carried out over a few months. The four longest studies provided supplementation to patients for 1 year [98,103–105].

All of the studies showed that n-3 FA supplementation can partially or completely reverse typical FA abnormalities in serum, erythrocyte membranes, or tissue. However, the clinical benefit has been mixed. Most of the studies used pulmonary function testing as a clinical marker, of which only three showed some statistical evidence of improvement with supplementation. De Vizia et al. [99] gave a combination DHA and EPA supplement to 30 participants over an 8-month period and demonstrated a very small but statistically significant improvement in FEV_1. Olveira et al. [104] gave a DHA and EPA mixture to 17 participants for 1 year and observed an increase in FEV_1 that was significant in absolute terms but not in terms of percent predicted for age and size. Their study also demonstrated significant decreases in

TABLE 39.3 Studies of n-3 Fatty Acid Supplementation in Patients with Cystic Fibrosis

References	Participants (*n*)	Age (years)	Study Type	FA Supplement	Duration	Primary Outcome	Results
Christophe et al. [91]	9	7–20	Double-blind, placebo-controlled crossover	900 mg/day (11.7% DHA; 13.2% EPA)	1 month	Serum lipid n-3 fatty acids	DHA and EPA significantly increased with treatment
Lawrence and Sorrell [92]	19	12–20	Prospective, randomized, double-blind, placebo crossover study	2.7 g/day EPA	12 weeks	LTB_4 production and clinical and other inflammatory markers	EPA treatment caused a statistically significant decrease in sputum production (volume), Shwachman score, and spirometry and an improvement in chemotaxis of neutrophils in response to LTB_4
Henderson et al. [93]	12	12.2 ± 5.4	Double-blind, placebo-controlled crossover	3200 mg/day EPA; 2200 mg/day DHA	6 weeks	1. Safety 2. Change in plasma and RBC EPA and DHA	1. No adverse outcomes 2. Increased DHA and EPA in plasma and RBCs
Kurlandsky et al. [94]	14	6–16	Double-blind crossover study	100 mg/kg/day (44% EPA; 24% DHA)	2 weeks	1. Serum TNF-α and LTB_4 2. Pulmonary function 3. Sweat chloride (Cl^-)	1. Decreased LTB_4; TNF-α was not detectable 2. No significant change in pulmonary function 3. No significant change in sweat chloride (Cl^-)
Lawrence and Sorrell [95]	9	14–22	Prospective study without a placebo but with a control baseline	2.7 g/day EPA	6 weeks	Number and affinity of LTB_4 receptors	Number and affinity of receptors were corrected
Clandinin et al. [96]	23	4–42	Prospective, controlled study without placebo	35 mg/kg of n-3 FA	4 weeks	Lipid fractions of plasma lipoproteins	Increased EPA and DHA
Katz et al. [97]	9	12–29	Prospective study with placebo	150 mg/kg IV of omegavenous (27.6% DHA, 18.3% EPA)	4 weeks	Safety of IV n-3 FA administration in people with CF	1. No adverse side effects 2. Increased serum EPA and DHA 3. No change in lung function
Thies [98]	5	6–16	Uncontrolled pilot study	1.8–2.7 g/day EPA; 1.2–1.8 g/day DHA	1 year	Hospital days and admissions	No significant change in hospital days or admissions
De Vizia et al. [99]	30	0.8–24	Prospective study with a control group baseline and no placebo	400 mg/day EPA; 200 mg/day DHA	8 months	1. Erythrocyte fatty acid membrane change 2. Selected inflammatory markers 3. Lung function and weight	1. Increased DHA and EPA 2. Decreased AA 3. Decreased IgA and α-1 antitrypsin 4. Small but significant improvement in FEV_1
Jumpsen et al. [100]	5	18–43	Prospective study without placebo or control group	70 mg/kg DHA	6 weeks	1. Blood and intestine DHA levels 2. Lung function	1. Increased DHA in blood and intestinal biopsies 2. No change in total AA but decreased AA in plasma PC and PE 3. No significant change in lung function
Lloyd-Still et al. [101]	20	8–20	Prospective study with placebo but no control group	50 mg/kg DHA	6 months	1. Plasma, erythrocyte, and rectal DHA levels 2. Lung function	1. Increased DHA levels in plasma, erythrocyte, and rectal samples 2. Decreased blood AA and AA-to-DHA ratio 3. No detectable change in lung function

Continued

TABLE 39.3 Studies of n-3 Fatty Acid Supplementation in Patients with Cystic Fibrosis—cont'd

References	Participants (*n*)	Age (years)	Study Type	FA Supplement	Duration	Primary Outcome	Results
Panchaud et al. [102]	17	18 ± 9	Prospective, double-blind, placebo-controlled crossover study	390–1170 mg/day n-3 FA	6 months	Biological parameters of inflammation (CRP, albumin, leukocyte count, and orosomucoid acid)	1. Increased EPA in neutrophil membranes without increased DHA 2. Trend toward increased di-homo-γ-linolenate 3. No change in lung function or inflammatory markers
Van Biervliet et al. [103]	9	12.1 (IQR 9.3)	Prospective, double-blind, placebo controlled	3 g/day algal triacylglycerol (40% DHA)	1 year	Plasma fatty acids, pulmonary function, and number of pulmonary infections	1. Increased DHA and EPA 2. Decreased AA and AA-to-DHA ratio 3. No significant change in clinical findings, including lung function
Olveira et al. [104]	17	26.4 ± 10.6	Prospective study with no placebo and no control group	324 mg/day EPA; 216 mg/day DHA	1 year	Clinical markers (inflammatory markers, spirometry, and number of exacerbations) as well as serum fatty acids	1. Increased DHA but no change in EPA 2. Decreased AA and AA-to-DHA ratio 3. Decreased TNF-α 4. Mild but statistically significant increase in FEV_1 5. Decrease in number of exacerbations and number of antibiotic days
Alicandro et al. [105]	41	6–12	Prospective, randomized study with placebo	100 mg/kg DHA × 1 month; 1 g/day × 11 months	1 year	1. Primary: AA-to-DHA ratio 2. Secondary: clinical and biochemical markers	1. Decreased AA-to-DHA ratio 2. No significant changes in clinical or biomarkers, including lung function, lean mass, or inflammatory markers
Leggieri et al. [106]	10	6–18	Prospective study without placebo or control group	1 g/10 kg/day DHA × 1 month; 250 mg/10 kg/day × 5 months	6 months	Respiratory function, nutritional status, and inflammatory markers	1. No significant change in CRP, Igs, or transaminases 2. Decreased IL-8, TNF-α, and fecal calprotectin 3. Increased absolute FEV_1 but no change in percent predicted

AA, arachidonate; CRP, C-reactive protein; DHA, docosahexaenoate; EPA, eicosapentaenoate; FA, fatty acid; FEV_1, forced expiratory volume in 1 s; Ig, immunoglobulin; IL, interleukin; IV, intravenous; IQR, interquartile range; PC, phosphatidylcholine; PE, phosphatidylethanolamine; RBC, red blood cell.

the total number of both respiratory exacerbations and antibiotic days compared to the year before supplementation. Leggieri et al. [106] also showed a significant increase in absolute FEV_1 after 6 months of DHA supplementation. However, there was no change in the percent predicted for age and size.

These studies highlight the difficulty in using pulmonary function testing as a clinical end point. These values change as patients age and vary over the course of study because of pulmonary exacerbations, infectious disease burden, and treatment. Furthermore, the decline in pulmonary function testing due to disease progression often occurs over a period of many years.

Because of this difficulty, several studies focused their end points on biomarkers of inflammation. For example, Kurlandsky et al. [94] showed that n-3 FA supplementation significantly reduced serum LTB_4 levels. The same year, Lawrence and Sorrell [95] showed that EPA normalized the number of high-affinity LTB_4 receptors on neutrophils, indicating downmodulation of the inflammatory response. Olveira et al. [104] more recently demonstrated a significant decrease in tumor necrosis factor-α over 1 year of n-3 FA supplementation, and Leggieri et al. [106] showed a significant decrease in fecal calprotectin, a general marker of inflammation. These studies demonstrate that changes in inflammatory markers may be a better measure of efficacy for studies involving FA supplementation in CF.

Although promising, these studies are not yet definitive. A Cochrane review in 2007 concluded that although n-3 FA supplementation may provide clinical benefit for some patients and that there were few adverse effects, there is not sufficient evidence to recommend routine supplementation for all patients with CF [107]. This conclusion may be due to some common weaknesses in all of these studies. They were all relatively small, and most treated patients with CF for only short periods of time. Most studies were not placebo controlled, but rather compared the status of patients before and after treatment. In addition, the type and amount of n-3 and other FAs was not standardized, making the studies difficult to compare. Finally, the challenge of identifying relevant clinical end points that are measurable over the short term remains a significant obstacle for these types of studies.

39.6 CONCLUSIONS

Abnormalities in FA concentrations, both in plasma and tissues, are a consistent feature of patients with CF. There is significant evidence suggesting that these alterations are not just an effect of the disease but that they may play a role in disease pathogenesis. The mechanism of this effect is not known. However, changes in cellular membrane structure and function and in the production of oxygenated FA metabolites, including eicosanoids and docosanoids, could contribute to the pathophysiology of CF. While the precise connection between *CFTR* mutations and these FA abnormalities is not completely understood, alterations in the expression and activity of FA metabolic enzymes almost certainly play a role. A number of studies of patients and model systems suggest that reversal of these abnormalities through changes in diet or direct supplementation could provide clinical benefits. However, more study is needed before routine supplementation can be recommended.

References

[1] Kuo PT, Huang NN, Bassett DR. The fatty acid composition of the serum chylomicrons and adipose tissue of children with cystic fibrosis of the pancreas. J Pediatr 1962;60:394–403.

[2] Laposata M, Reich EL, Majerus PW. Arachidonoyl-CoA synthetase. Separation from nonspecific acyl-CoA synthetase and distribution in various cells and tissues. J Biol Chem 1985;260:11016–20.

[3] Strandvik B. Fatty acid metabolism in cystic fibrosis. Prostaglandins Leukot Essent Fatty Acids 2010;83:121–9.

[4] Worgall TS. Lipid metabolism in cystic fibrosis. Curr Opin Clin Nutr Metab Care 2009;12:105–9.

[5] Nakamura MT, Nara TY. Essential fatty acid synthesis and its regulation in mammals. Prostaglandins Leukot Essent Fatty Acids 2003;68:145–50.

[6] Funk CD. Prostaglandins and leukotrienes: advances in eicosanoid biology. Science 2001;294:1871–5.

[7] Serhan CN, Takano T, Gronert K, Chiang N, Clish CB. Lipoxin and aspirin-triggered 15-epi-lipoxin cellular interactions anti-inflammatory lipid mediators. Clin Chem Lab Med 1999;37:299–309.

[8] Serhan CN, Gotlinger K, Hong S, Arita M. Resolvins, docosatrienes, and neuroprotectins, novel omega-3-derived mediators, and their aspirin-triggered endogenous epimers: an overview of their protective roles in catabasis. Prostaglandins Other Lipid Mediat 2004;73:155–72.

[9] Cao H, Gerhold K, Mayers JR, Wiest MM, Watkins SM, Hotamisligil GS. Identification of a lipokine, a lipid hormone linking adipose tissue to systemic metabolism. Cell 2008;134:933–44.

[10] Rosenlund ML, Kim HK, Kritchevsky D. Essential fatty acids in cystic fibrosis. Nature 1974;251:719.

[11] Christophe AB, Warwick WJ, Holman RT. Serum fatty acid profiles in cystic fibrosis patients and their parents. Lipids 1994;29:569–75.

[12] Strandvik B, Gronowitz E, Enlund F, Martinsson T, Wahlstrom J. Essential fatty acid deficiency in relation to genotype in patients with cystic fibrosis. J Pediatr 2001;139:650–5.

[13] Olveira G, Dorado A, Olveira C, Padilla A, Rojo-Martinez G, Garcia-Escobar E, et al. Serum phospholipid fatty acid profile and dietary intake in an adult Mediterranean population with cystic fibrosis. Br J Nutr 2006;96:343–9.

[14] Maqbool A, Schall JI, Garcia-Espana JF, Zemel BS, Strandvik B, Stallings VA. Serum linoleic acid status as a clinical indicator of essential fatty acid status in children with cystic fibrosis. J Pediatr Gastroenterol Nutr 2008;47:635–44.

[15] Caren R, Corbo L. Plasma fatty acids in pancreatic cystic fibrosis and liver disease. J Clin Endocrinol Metab 1966;26:470–7.

[16] Hubbard VS, Dunn GD, di Sant'Agnese PA. Abnormal fatty-acid composition of plasma-lipids in cystic fibrosis. A primary or a secondary defect? Lancet 1977;2:1302–4.

[17] Lloyd-Still JD, Johnson SB, Holman RT. Essential fatty acid status in cystic fibrosis and the effects of safflower oil supplementation. Am J Clin Nutr 1981;34:1–7.

[18] Rogiers V, Vercruysse A, Dab I, Baran D. Abnormal fatty acid pattern of the plasma cholesterol ester fraction in cystic fibrosis patients with and without pancreatic insufficiency. Eur J Pediatr 1983;141:39–42.

[19] Farrell PM, Mischler EH, Engle MJ, Brown DJ, Lau SM. Fatty acid abnormalities in cystic fibrosis. Pediatr Res 1985;19:104–9.

[20] Gibson RA, Teubner JK, Haines K, Cooper DM, Davidson GP. Relationships between pulmonary function and plasma fatty acid levels in cystic fibrosis patients. J Pediatr Gastroenterol Nutr 1986;5:408–15.

[21] Thompson GN. Relationships between essential fatty acid levels, pulmonary function and fat absorption in pre-adolescent cystic fibrosis children with good clinical scores. Eur J Pediatr 1989;148:327–9.

[22] Roulet M, Frascarolo P, Rappaz I, Pilet M. Essential fatty acid deficiency in well nourished young cystic fibrosis patients. Eur J Pediatr 1997;156:952–6.

[23] Freedman SD, Blanco PG, Zaman MM, Shea JC, Ollero M, Hopper IK, et al. Association of cystic fibrosis with abnormalities in fatty acid metabolism. N Engl J Med 2004;350:560–9.

[24] Batal I, Ericsoussi MB, Cluette-Brown JE, O'Sullivan BP, Freedman SD, Savaille JE, et al. Potential utility of plasma fatty acid analysis in the diagnosis of cystic fibrosis. Clin Chem 2007;53:78–84.

[25] Aldamiz-Echevarria L, Prieto JA, Andrade F, Elorz J, Sojo A, Lage S, et al. Persistence of essential fatty acid deficiency in cystic fibrosis despite nutritional therapy. Pediatr Res 2009;66:585–9.

[26] Hubbard VS, Dunn GD. Fatty acid composition of erythrocyte phospholipids from patients with cystic fibrosis. Clin Chim Acta 1980;102:115–8.

[27] Campbell IM, Crozier DN, Caton RB. Abnormal fatty acid composition and impaired oxygen supply in cystic fibrosis patients. Pediatrics 1976;57:480–6.

[28] Gilljam H, Strandvik B, Ellin A, Wiman LG. Increased mole fraction of arachidonic acid in bronchial phospholipids in patients with cystic fibrosis. Scand J Clin Lab Invest 1986;46:511–8.

[29] Freedman SD, Katz MH, Parker EM, Laposata M, Urman MY, Alvarez JG. A membrane lipid imbalance plays a role in the phenotypic expression of cystic fibrosis in cftr(−/−) mice. Proc Natl Acad Sci USA 1999;96:13995–4000.

[30] Ollero M, Laposata M, Zaman MM, Blanco PG, Andersson C, Zeind J, et al. Evidence of increased flux to n-6 docosapentaenoic acid in phospholipids of pancreas from cftr−/− knockout mice. Metabolism 2006;55:1192–200.

[31] Mimoun M, Coste TC, Lebacq J, Lebecque P, Wallemacq P, Leal T, et al. Increased tissue arachidonic acid and reduced linoleic acid in a mouse model of cystic fibrosis are reversed by supplemental glycerophospholipids enriched in docosahexaenoic acid. J Nutr 2009;139:2358–64.

[32] Andersson C, Al-Turkmani MR, Savaille JE, Alturkmani R, Katrangi W, Cluette-Brown JE, et al. Cell culture models demonstrate that CFTR dysfunction leads to defective fatty acid composition and metabolism. J Lipid Res 2008;49:1692–700.

[33] Rajan S, Cacalano G, Bryan R, Ratner AJ, Sontich CU, van Heerckeren A, et al. *Pseudomonas aeruginosa* induction of apoptosis in respiratory epithelial cells: analysis of the effects of cystic fibrosis transmembrane conductance regulator dysfunction and bacterial virulence factors. Am J Respir Cell Mol Biol 2000;23:304–12.

[34] Njoroge SW, Seegmiller AC, Katrangi W, Laposata M. Increased Delta5- and Delta6-desaturase, cyclooxygenase-2, and lipoxygenase-5 expression and activity are associated with fatty acid and eicosanoid changes in cystic fibrosis. Biochim Biophys Acta 2011;1811:431–40.

[35] Katrangi W, Lawrenz J, Seegmiller AC, Laposata M. Interactions of linoleic and alpha-linolenic acids in the development of fatty acid alterations in cystic fibrosis. Lipids 2013;48:333–42.

[36] Zaman MM, Martin CR, Andersson C, Bhutta AQ, Cluette-Brown JE, Laposata M, et al. Linoleic acid supplementation results in increased arachidonic acid and eicosanoid production in CF airway cells and in cftr−/− transgenic mice. Am J Physiol Lung Cell Mol Physiol 2010;299:L599–606.

[37] O'Sullivan BP, Freedman SD. Cystic fibrosis. Lancet 2009;373:1891–904.

[38] Rowntree RK, Harris A. The phenotypic consequences of CFTR mutations. Ann Hum Genet 2003;67:471–85.

[39] Ollero M, Astarita G, Guerrera IC, Sermet-Gaudelus I, Trudel S, Piomelli D, et al. Plasma lipidomics reveals potential prognostic signatures within a cohort of cystic fibrosis patients. J Lipid Res 2011;52:1011–22.

[40] Holman RT. The ratio of trienoic:tetraenoic acids in tissue lipids as a measure of essential fatty acid requirement. J Nutr 1960;70:405–10.

[41] Harper TB, Chase HP, Henson J, Henson PM. Essential fatty acid deficiency in the rabbit as a model of nutritional impairment in cystic fibrosis. In vitro and in vivo effects on lung defense mechanisms. Am Rev Respir Dis 1982;126:540–7.

[42] Craig-Schmidt MC, Faircloth SA, Teer PA, Weete JD, Wu CY. The essential fatty acid deficient chicken as a model for cystic fibrosis. Am J Clin Nutr 1986;44:816–24.

[43] Linsdell P. Inhibition of cystic fibrosis transmembrane conductance regulator chloride channel currents by arachidonic acid. Can J Physiol Pharmacol 2000;78:490–9.

[44] Li Y, Wang W, Parker W, Clancy JP. Adenosine regulation of cystic fibrosis transmembrane conductance regulator through prostenoids in airway epithelia. Am J Respir Cell Mol Biol 2006;34:600–8.

[45] Zhou JJ, Linsdell P. Molecular mechanism of arachidonic acid inhibition of the CFTR chloride channel. Eur J Pharmacol 2007;563:88–91.

[46] Dutta AK, Okada Y, Sabirov RZ. Regulation of an ATP-conductive large-conductance anion channel and swelling-induced ATP release by arachidonic acid. J Physiol 2002;542:803–16.

[47] Lemen RJ, Gates AJ, Mathe AA, Waring WW, Hyman AL, Kadowitz PD. Relationships among digital clubbing, disease severity, and serum prostaglandins F2alpha and E concentrations in cystic fibrosis patients. Am Rev Respir Dis 1978;117:639–46.

[48] Rigas B, Korenberg JR, Merrill WW, Levine L. Prostaglandins E2 and E2 alpha are elevated in saliva of cystic fibrosis patients. Am J Gastroenterol 1989;84:1408–12.

[49] Strandvik B, Svensson E, Seyberth HW. Prostanoid biosynthesis in patients with cystic fibrosis. Prostaglandins Leukot Essent Fatty Acids 1996;55:419–25.

[50] De Lisle RC, Meldi L, Flynn M, Jansson K. Altered eicosanoid metabolism in the cystic fibrosis mouse small intestine. J Pediatr Gastroenterol Nutr 2008;47:406–16.

[51] Jabr S, Gartner S, Milne GL, Roca-Ferrer J, Casas J, Moreno A, et al. Quantification of major urinary metabolites of PGE2 and PGD2 in cystic fibrosis: correlation with disease severity. Prostaglandins Leukot Essent Fatty Acids 2013;89:121–6.

[52] Chen J, Jiang XH, Chen H, Guo JH, Tsang LL, Yu MK, et al. CFTR negatively regulates cyclooxygenase-2-PGE(2) positive feedback loop in inflammation. J Cell Physiol 2012;227:2759–66.

[53] Lucidi V, Ciabattoni G, Bella S, Barnes PJ, Montuschi P. Exhaled 8-isoprostane and prostaglandin E(2) in patients with stable and unstable cystic fibrosis. Free Radic Biol Med 2008;45:913–9.

[54] Sampson AP, Spencer DA, Green CP, Piper PJ, Price JF. Leukotrienes in the sputum and urine of cystic fibrosis children. Br J Clin Pharmacol 1990;30:861–9.

[55] Konstan MW, Walenga RW, Hilliard KA, Hilliard JB. Leukotriene B4 markedly elevated in the epithelial lining fluid of patients with cystic fibrosis. Am Rev Respir Dis 1993;148:896–901.
[56] Reid DW, Misso N, Aggarwal S, Thompson PJ, Walters EH. Oxidative stress and lipid-derived inflammatory mediators during acute exacerbations of cystic fibrosis. Respirology 2007;12:63–9.
[57] Karp CL, Flick LM, Park KW, Softic S, Greer TM, Keledjian R, et al. Defective lipoxin-mediated anti-inflammatory activity in the cystic fibrosis airway. Nat Immunol 2004;5:388–92.
[58] Karp CL, Flick LM, Yang R, Uddin J, Petasis NA. Cystic fibrosis and lipoxins. Prostaglandins Leukot Essent Fatty Acids 2005;73:263–70.
[59] Roca-Ferrer J, Pujols L, Gartner S, Moreno A, Pumarola F, Mullol J, et al. Upregulation of COX-1 and COX-2 in nasal polyps in cystic fibrosis. Thorax 2006;61:592–6.
[60] Owens JM, Shroyer KR, Kingdom TT. Expression of cyclooxygenase and lipoxygenase enzymes in sinonasal mucosa of patients with cystic fibrosis. Arch Otolaryngol Head Neck Surg 2008;134:825–31.
[61] Elizur A, Cannon CL, Ferkol TW. Airway inflammation in cystic fibrosis. Chest 2008;133:489–95.
[62] Norkina O, Kaur S, Ziemer D, De Lisle RC. Inflammation of the cystic fibrosis mouse small intestine. Am J Physiol Gastrointest Liver Physiol 2004;286:G1032–41.
[63] Bruzzese E, Raia V, Gaudiello G, Polito G, Buccigrossi V, Formicola V, et al. Intestinal inflammation is a frequent feature of cystic fibrosis and is reduced by probiotic administration. Aliment Pharmacol Ther 2004;20:813–9.
[64] Serhan CN, Savill J. Resolution of inflammation: the beginning programs the end. Nat Immunol 2005;6:1191–7.
[65] Ricciotti E, FitzGerald GA. Prostaglandins and inflammation. Arterioscler Thromb Vasc Biol 2011;31:986–1000.
[66] Caristi S, Piraino G, Cucinotta M, Valenti A, Loddo S, Teti D. Prostaglandin E2 induces interleukin-8 gene transcription by activating C/EBP homologous protein in human T lymphocytes. J Biol Chem 2005;280:14433–42.
[67] Aronoff DM, Canetti C, Peters-Golden M. Prostaglandin E2 inhibits alveolar macrophage phagocytosis through an E-prostanoid 2 receptor-mediated increase in intracellular cyclic AMP. J Immunol 2004;173:559–65.
[68] Kalinski P. Regulation of immune responses by prostaglandin E2. J Immunol 2012;188:21–8.
[69] Serezani CH, Chung J, Ballinger MN, Moore BB, Aronoff DM, Peters-Golden M. Prostaglandin E2 suppresses bacterial killing in alveolar macrophages by inhibiting NADPH oxidase. Am J Respir Cell Mol Biol 2007;37:562–70.
[70] Dharmani P, Srivastava V, Kissoon-Singh V, Chadee K. Role of intestinal mucins in innate host defense mechanisms against pathogens. J Innate Immun 2009;1:123–35.
[71] de Lisle RC, Sewell R, Meldi L. Enteric circular muscle dysfunction in the cystic fibrosis mouse small intestine. Neurogastroenterol Motil 2010;22:341–e87.
[72] Konstan MW, Schluchter MD, Xue W, Davis PB. Clinical use of Ibuprofen is associated with slower FEV1 decline in children with cystic fibrosis. Am J Respir Crit Care Med 2007;176:1084–9.
[73] Kohli P, Levy BD. Resolvins and protectins: mediating solutions to inflammation. Br J Pharmacol 2009;158:960–71.
[74] Yang J, Eiserich JP, Cross CE, Morrissey BM, Hammock BD. Metabolomic profiling of regulatory lipid mediators in sputum from adult cystic fibrosis patients. Free Radic Biol Med 2012;53:160–71.
[75] Rivers JP, Frankel TL. Essential fatty acid deficiency. Br Med Bull 1981;37:59–64.
[76] Al-Turkmani MR, Andersson C, Alturkmani R, Katrangi W, Cluette-Brown JE, Freedman SD, et al. A mechanism accounting for the low cellular level of linoleic acid in cystic fibrosis and its reversal by DHA. J Lipid Res 2008;49:1946–54.
[77] Carlstedt-Duke J, Bronnegard M, Strandvik B. Pathological regulation of arachidonic acid release in cystic fibrosis: the putative basic defect. Proc Natl Acad Sci USA 1986;83:9202–6.
[78] Nakamura MT, Nara TY. Gene regulation of mammalian desaturases. Biochem Soc Trans 2002;30:1076–9.
[79] Njoroge SW, Laposata M, Katrangi W, Seegmiller AC. DHA and EPA reverse cystic fibrosis-related FA abnormalities by suppressing FA desaturase expression and activity. J Lipid Res 2012;53:257–65.
[80] Thomsen KF, Laposata M, Njoroge SW, Umunakwe OC, Katrangi W, Seegmiller AC. Increased elongase 6 and Delta9-desaturase activity are associated with n-7 and n-9 fatty acid changes in cystic fibrosis. Lipids 2011;46:669–77.
[81] Hiltunen JK, Karki T, Hassinen IE, Osmundsen H. beta-Oxidation of polyunsaturated fatty acids by rat liver peroxisomes. A role for 2,4-dienoyl-coenzyme A reductase in peroxisomal beta-oxidation. J Biol Chem 1986;261:16484–93.
[82] Gronn M, Christensen E, Hagve TA, Christophersen BO. Peroxisomal retroconversion of docosahexaenoic acid (22:6(n-3)) to eicosapentaenoic acid (20:5(n-3)) studied in isolated rat liver cells. Biochim Biophys Acta 1991;1081:85–91.
[83] Brossard N, Croset M, Pachiaudi C, Riou JP, Tayot JL, Lagarde M. Retroconversion and metabolism of [13C]22:6n-3 in humans and rats after intake of a single dose of [13C]22:6n-3-triacylglycerols. Am J Clin Nutr 1996;64:577–86.
[84] Stark KD, Holub BJ. Differential eicosapentaenoic acid elevations and altered cardiovascular disease risk factor responses after supplementation with docosahexaenoic acid in postmenopausal women receiving and not receiving hormone replacement therapy. Am J Clin Nutr 2004;79:765–73.
[85] Innis SM, Davidson AG. Cystic fibrosis and nutrition: linking phospholipids and essential fatty acids with thiol metabolism. Annu Rev Nutr 2008;28:55–72.
[86] Ulane MM, Butler JD, Peri A, Miele L, Ulane RE, Hubbard VS. Cystic fibrosis and phosphatidylcholine biosynthesis. Clin Chim Acta 1994;230:109–16.
[87] Borowitz D, Baker RD, Stallings V. Consensus report on nutrition for pediatric patients with cystic fibrosis. J Pediatr Gastroenterol Nutr 2002;35:246–59.
[88] Simopoulos AP. Overview of evolutionary aspects of omega 3 fatty acids in the diet. World Rev Nutr Diet 1998;83:1–11.
[89] Blasbalg TL, Hibbeln JR, Ramsden CE, Majchrzak SF, Rawlings RR. Changes in consumption of omega-3 and omega-6 fatty acids in the United States during the 20th century. Am J Clin Nutr 2011;93:950–62.
[90] Beharry S, Ackerley C, Corey M, Kent G, Heng YM, Christensen H, et al. Long-term docosahexaenoic acid therapy in a congenic murine model of cystic fibrosis. Am J Physiol Gastrointest Liver Physiol 2007;292:G839–48.
[91] Christophe A, Robberecht E, De Baets F, Franckx H. Increase of long chain omega-3 fatty acids in the major serum lipid classes of patients with cystic fibrosis. Ann Nutr Metab 1992;36:304–12.
[92] Lawrence R, Sorrell T. Eicosapentaenoic acid in cystic fibrosis: evidence of a pathogenetic role for leukotriene B4. Lancet 1993;342:465–9.
[93] Henderson Jr WR, Astley SJ, McCready MM, Kushmerick P, Casey S, Becker JW, et al. Oral absorption of omega-3 fatty acids in patients with cystic fibrosis who have pancreatic insufficiency and in healthy control subjects. J Pediatr 1994;124:400–8.
[94] Kurlandsky LE, Bennink MR, Webb PM, Ulrich PJ, Baer LJ. The absorption and effect of dietary supplementation with omega-3 fatty acids on serum leukotriene B4 in patients with cystic fibrosis. Pediatr Pulmonol 1994;18:211–7.

[95] Lawrence RH, Sorrell TC. Eicosapentaenoic acid modulates neutrophil leukotriene B4 receptor expression in cystic fibrosis. Clin Exp Immunol 1994;98:12–6.
[96] Clandinin MT, Zuberbuhler P, Brown NE, Kielo ES, Goh YK. Fatty acid pool size in plasma lipoprotein fractions of cystic fibrosis patients. Am J Clin Nutr 1995;62:1268–75.
[97] Katz DP, Manner T, Furst P, Askanazi J. The use of an intravenous fish oil emulsion enriched with omega-3 fatty acids in patients with cystic fibrosis. Nutrition 1996;12:334–9.
[98] Thies NH. The effect of 12 months' treatment with eicosapentaenoic acid in five children with cystic fibrosis. J Paediatr Child Health 1997;33:349–51.
[99] De Vizia B, Raia V, Spano C, Pavlidis C, Coruzzo A, Alessio M. Effect of an 8-month treatment with omega-3 fatty acids (eicosapentaenoic and docosahexaenoic) in patients with cystic fibrosis. JPEN J Parenter Enteral Nutr 2003;27:52–7.
[100] Jumpsen JA, Brown NE, Thomson AB, Paul Man SF, Goh YK, Ma D, et al. Fatty acids in blood and intestine following docosahexaenoic acid supplementation in adults with cystic fibrosis. J Cyst Fibros 2006;5:77–84.
[101] Lloyd-Still JD, Powers CA, Hoffman DR, Boyd-Trull K, Lester LA, Benisek DC, et al. Bioavailability and safety of a high dose of docosahexaenoic acid triacylglycerol of algal origin in cystic fibrosis patients: a randomized, controlled study. Nutrition 2006;22:36–46.
[102] Panchaud A, Sauty A, Kernen Y, Decosterd LA, Buclin T, Boulat O, et al. Biological effects of a dietary omega-3 polyunsaturated fatty acids supplementation in cystic fibrosis patients: a randomized, crossover placebo-controlled trial. Clin Nutr 2006;25:418–27.
[103] Van Biervliet S, Devos M, Delhaye T, Van Biervliet JP, Robberecht E, Christophe A. Oral DHA supplementation in DeltaF508 homozygous cystic fibrosis patients. Prostaglandins Leukot Essent Fatty Acids 2008;78:109–15.
[104] Olveira G, Olveira C, Acosta E, Espildora F, Garrido-Sanchez L, Garcia-Escobar E, et al. Fatty acid supplements improve respiratory, inflammatory and nutritional parameters in adults with cystic fibrosis. Arch Bronconeumol 2010;46:70–7.
[105] Alicandro G, Faelli N, Gagliardini R, Santini B, Magazzu G, Biffi A, et al. A randomized placebo-controlled study on high-dose oral algal docosahexaenoic acid supplementation in children with cystic fibrosis. Prostaglandins Leukot Essent Fatty Acids 2013;88:163–9.
[106] Leggieri E, De Biase RV, Savi D, Zullo S, Halili I, Quattrucci S. Clinical effects of diet supplementation with DHA in pediatric patients suffering from cystic fibrosis. Minerva Pediatr 2013;65:389–98.
[107] McKarney C, Everard M, N'Diaye T. Omega-3 fatty acids (from fish oils) for cystic fibrosis. Cochrane Database Syst Rev 2007:CD002201.

CHAPTER

40

Essential Fatty Acid Deficiency in Cystic Fibrosis: Malabsorption or Metabolic Abnormality?

S. Van Biervliet[1], *B. Strandvik*[2]

[1]Department of Pediatric Gastroenterology and Nutrition, Ghent University Hospital, Ghent, Belgium;
[2]Department of Biosciences and Nutrition, Karolinska Institutet, Stockholm, Sweden

40.1 INTRODUCTION

Cystic fibrosis (CF) is a chronic disease with high mortality because of progressive pulmonary deterioration related to chronic bacterial colonization with exacerbations. Pancreatic insufficiency is present in 85% of patients with CF and nutritional problems have gained a high priority in the treatment of the disease, as a result of the observed relation between the nutritional status and the pulmonary function [1–4]. Fat, as the most important energy source, has become an important issue in CF nutrition since the description of the survival differences between CF patients from Boston and Toronto [1]. Toronto patients were given a high fat diet in the tradition of Crozier, who found an association between CF symptoms and essential fatty acids (EFA) [5], whereas the rest of the CF world advocated a restricted fat intake because of incomplete restoration of fat malabsorption by the earlier pancreatic enzyme supplements. This resulted in a survival difference during the late 1980s. As a consequence, a general nutritional recommendation of a high-fat, high-energy diet aiming at a caloric intake of 120–150% of the recommended daily allowances (RDA) in association with pancreatic enzyme replacement therapy is currently advocated by all CF centers [6]. However, some reports describe a normal weight gain in pancreatic insufficient patients with a standard energy diet according to RDA, if fat intake was high (mean 37% energy) and EFA supplementation was included [7].

EFA deficiency, or more precisely low linoleic acid (LA, 18:2w6) in serum, was described in CF patients already 50 years ago [8] and confirmed in several studies later on. The low docosahexaenoic acid (DHA, 22:6w3) has only recently been given attention, despite a report in 1972 on low tissue DHA levels in CF [9].

Although the CF gene codes for a chloride transporting channel, the cystic fibrosis transmembrane conductance regulator (CFTR), CF is an inflammatory disease, not only affecting the lungs but also the gastrointestinal tract [10,11]. Similar to results from respiratory cells carrying the d508 mutation [12], studies of transgenic CF mice showed an increased prostaglandin metabolism in the intestine as a result of increased phospholipase A_2 (PLA_2) and cyclooxygenase 2 (Cox 2) activity, indicating increased metabolism of arachidonic acid (AA, 20:4w6). This increased AA metabolism was confirmed in CF patients by increased urinary excretion of Prostaglandin E2 (PGE2) and eicosanoid metabolites [13]. The link between the CFTR and the disturbed EFA metabolism has recently obtained increased interest resulting in several new hypotheses [14]. Furthermore, with the systematic incorporation of the nutritional advice in general CF care and new therapeutic options directed against specific mutations, the proportion of malnourished patients is reducing and the number of overweight and obese patients with CF is increasing [15].

40.2 ESSENTIAL FATTY ACIDS

40.2.1 Essential Fatty Acid Metabolism

The polyunsaturated fatty acids (PUFA) alfa-linolenic acid (ALA, C18:3ω-3) and LA are EFA and have to be derived from the diet in contrast to other fatty acids,

Diet and Exercise in Cystic Fibrosis
http://dx.doi.org/10.1016/B978-0-12-800051-9.00040-7

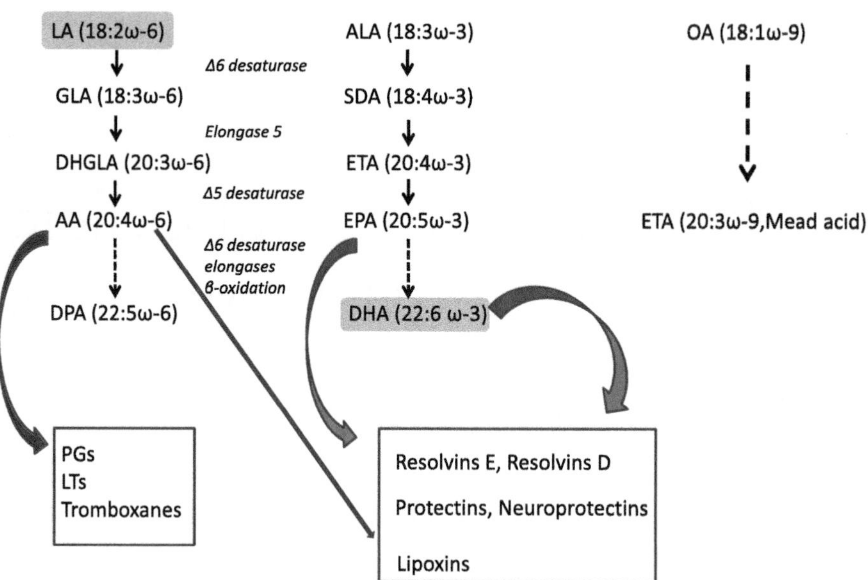

FIGURE 40.1 The major transformation of the essential fatty acids (EFA), linoleic acid (LA, 18:2ω-6) and α-linolenic acid (ALA, 18:3ω-3) to long chain polyunsaturated fatty acids (LCPUFA) and their oxylipin products. The shadowed fatty acids, LA and docosahexaenoic acid (DHA, 22:6ω-3), are those with low concentrations in CF tissues. Proinflammatory oxylipins are indicated by red and antiinflammatory by blue arrows. The competing enzymes in the transformations are indicated by italics. In EFA deficiency oleic acid (OA, 18:1ω-9) increases as does its elongation-desaturation product, the eicosatrienoic acid (Mead acid, 20:3ω-9). Abbreviations: LA: linoleic acid, GLA: gamma-linolenic acid, DHGLA: dihomo-gamma-linolenic acid, AA: arachidonic acid, ALA: alfa-linolenic acid, SDA: stearidonic acid, ETA: eicosatetraenoic acid, EPA: eicosapentaenoic acid, DHA: docosahexaenoic acid, OA: oleic acid, ETA: eicosatrienoic acid, PG: prostaglandin, LT: leukotriene, DPA: docosapentaenoic acid.

which can be synthesized. The EFA are transformed by desaturation and elongation into the long-chain polyunsaturated fatty acid (LCPUFA), metabolically important members of the ω-3 and ω-6 fatty acid series. The dietary balance between LA and ALA is important as the ω-3 and ω-6 FA compete for the same elongases and desaturases (Figure 40.1). The elongation and desaturation to LCPUFA is inefficient, especially in early life when tissue incorporation is the highest, LCPUFA become conditionally essential [16,17].

In case of insufficient LA supply or LCPUFA deficiency the body will try to compensate by producing more oleic acid (OA, 18:1ω-7/ω-9) and eicosatrienoic acid (Mead acid, 20:3w9), the latter is hardly detected during normal conditions except during pregnancy and in neonates. The ratio between Mead acid and AA is often used as an index of EFA deficiency [18].

40.2.2 Function of Polyunsaturated Fatty Acids

Fatty acids are important parts of phospholipids, building the membrane double layer, where they have a structural role. The degree of desaturation of the fatty acids influences significantly membrane viscosity, permeability and membrane protein function [19].

LCPUFA have further important transmitter and signaling functions. AA, DHA and to some extent dihomo-γ-linolenic acid (DHGLA, 20:3ω-6) and eicosapentaenoic acid (EPA, 20:5ω-3) are released from membranes by PLA_2 and transformed into lipid mediators, which have important autocrine and paracrine functions in metabolic and inflammatory regulation [20]. The different fatty acids will give rise to different series of oxylipins (Figure 40.1).

40.3 FATTY ACIDS IN CYSTIC FIBROSIS

40.3.1 Observed Fatty Acids Changes and Association to Clinical Condition

LA deficiency in CF patients has been described since the sixties [8,9,21–25]. After a case report of reversal of pancreatic insufficiency in a newborn with CF receiving Intralipid® in 1974 [26], many authors reported about short-term LA supplementation with varying results [27–29]. A long-term study (3 years) resulted in improvements in liver and renal functions [30,31]. The interest for the lipid abnormality then attenuated in light of the gene discovery 1989 and new interest to the fat abnormality was first obtained after the report by Freedman et al. about a normalization of ileum and pancreas morphology in a CF mice model by supplementation with very high doses of DHA [32]. However, these promising results could not be confirmed in long-term supplementation [33], and most clinical studies with DHA supplementation did not obtain any clinical improvement despite improvement in the DHA profile [34–36].

Most studies describe an increased OA, EPA and DHGLA and a decreased DHA and LA concentration [37,38]. The ratio of AA/DHA is increased suggesting a proinflammatory state [37,38]. This lipid imbalance is not only present in serum of CF patients but also in nasal and rectal biopsies [39]. Clinical manifestations of EFA deficiency, however, are rare except for growth retardation with low body weight and low fat mass. Skin manifestations might sometimes be the presenting symptom in patients with severe deficiencies [40] (Figure 40.2).

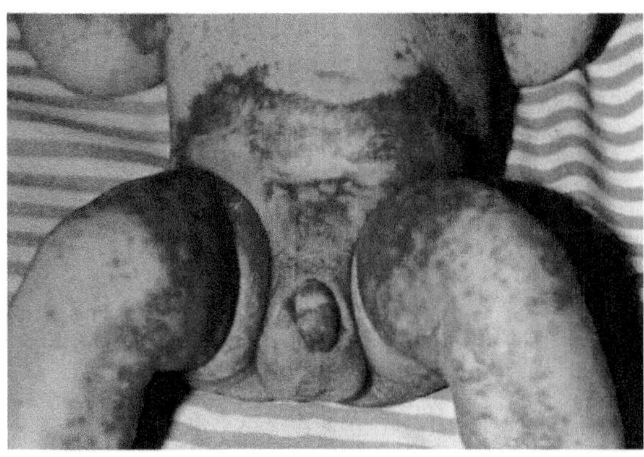

FIGURE 40.2 Dermatosis, as presenting symptoms in a patient with verified classic cystic fibrosis, resistant to ordinary dermatitis treatment, but which resolved after supplementation with Intralipid®.

The serum LA concentration decreases with age [37]. The fatty acid disturbances are associated with genotype [37,38], but not with nutritional status, pulmonary function and presence of CF related diabetes mellitus or pancreatic function [37,38]. However, in selected CF subpopulations (e.g., preadolescent CF children) a relation between LA and pulmonary function and growth as well as an inverse correlation between Mead acid (marker of EFA deficiency) and growth has been described [41]. An association between fatty acid status and bone mineral density [42,43], which often is low even in well-nourished CF patients [44], has also been described. Further on, a normal bone mineral density increase was observed over a 2 year period in CF patients using a general recommendation of daily supplementation with LA rich vegetable oil [45]. The presence of CF related liver disease is associated with lower DHA status [46], and the inflammation in the liver decreased in the CF animal model with long-term DHA therapy [33]. The commonly found steatosis in the CF liver was associated with lower LA concentrations [47].

40.3.2 Etiology of Fatty Acid Disturbances

Initially the EFA deficiencies were exclusively considered to be attributed to the malnutrition, pancreatic insufficiency and low energy intake in severely ill patients [21–25]. The lack of a correlation between nutritional status, pancreatic function and EFA deficiency as well as the persistence of the EFA deficiency despite increased fat intake pleads against this theory [24,25,27,37,42,48]. Further on, the fatty acid abnormalities were commonly present in the first weeks of life [49]. There is furthermore a significant association between EFA deficiency and the genotype severity, suggesting a mutation related problem [37,38]. Also without specific supplementation, some studies were able to demonstrate an association between fatty acid intake and serum fatty acids [50], suggesting that in the context of a high fat intake, EFA's were not used for energy [51].

As described above, eicosanoids are important derivatives of LCPUFA. In CF, in contrast to other EFA deficient patients, the increased systemic eicosanoid production is maintained despite of the LA deficiency [13]. Carlstedt-Duke et al. demonstrated a defective PLA_2 regulation in CF leading to an increased AA turnover [52], which was confirmed by other research groups [53,54]. The increased inflammatory state has been demonstrated in multiple studies not only demonstrating pulmonary inflammation even in absence of bacterial infection [55,56] but also intestinal inflammation [11]. The prostaglandin overproduction was associated to genotype rather than disease severity [57]. The increased expression and activity of fatty acid desaturases further supported this rapid transformation to LCPUFA [58–61]. A rapid turn-over of phospholipids might also be related to changes in thiol and phospholipid metabolism suggested by others [14,62].

Recent studies support the indications for defective PLA_2 inhibition [52]. Bensalem et al. described a low annexin 1 in both patients with CF and CF knockout mice [63]. Annexin 1 is a potent PLA_2 and Cox 2 inhibitor [64]. Although the study results in CF concerning ceramides, important for phagocytosis [65], are not unequivocal, the associations to the plasma fatty acid pattern as well as to the CFTR mutations are interesting [66,67], especially since CF patients are known to display defective pseudomonas internalization [68]. The importance of these pathways in CF have further been demonstrated by Radzioch et al. showing a normalization of LCPUFA and ceramide levels in CF mice by giving fenrenitide, a IL-1 β inhibiting retinoid, which interferes with the ceramide transformations [66,69].

Finally, as PUFAs are very susceptible for peroxidation, this could also play a role in the PUFA deficiency. CF is characterized by an increased oxidative stress, and antioxidative supplementation has been shown to improve lung function [70,71]. Furthermore, a relationship between DHA and vitamin E has been demonstrated [37], which might account for decreased DHA concentrations in CF, since DHA is very susceptible for peroxidation [72].

40.3.3 Influencing Fatty Acid Profiles with Nutrition and Supplements

Ever since PUFA deficiencies in CF were described many reports have tried to reverse the CF symptoms by supplementation with LA, but very little clinical improvement have been obtained. In short term studies improved weight gain and a positive influence on pulmonary function have been reported [41,73,74]. CF care in Sweden, which generally includes supplementation of LA, as well as exercise as additional treatment, has been known to have good clinical results [75, also see the ECFS patient report 2009].

The spectacular effect in CFTR−/− mice after DHA supplementation, initiated many studies with supplementation with DHA in order to shift the fatty acid status toward a less proinflammatory profile, hoping to influence clinical outcome [34–36,75–79]. Studies performed included small study groups, different duration (maximum 1 year), different doses and different oil types. All studies were able to influence the fatty acid status but not the clinical status. Furthermore, some studies describe an influence on inflammatory markers, eicosanoid profile, or nitric oxide production [80]. Only two short studies described an improved clinical outcome [34,76]. Therefore the most recent Cochrane review advises no systematic supplementation of patients with ω3 fatty acids [36]. The one study supplementing both ω6 and ω3 fatty acids for 1 year found improvement in most clinical parameters, including infectious exacerbations and less need for antibiotic treatment [81], in agreement with the Swedish results [82]. Further controlled research remains necessary.

40.4 ISSUES ON OBESITY AND HOW TO INTERVENE

The high energy, high fat intake is the standard nutritional care for CF patients. However, since the discovery of the CF gene, milder mutations have been described that are more likely to be associated with pancreatic sufficiency. The current nutritional guidelines focus on the treatment and prevention of malnutrition in CF as it is correlated with decreased pulmonary function. No recommendations are currently made for the pancreatic sufficient patients and the patients at risk for developing obesity [15]. Controversy persists regarding recommendations of general supplementation with fatty acids.

40.4.1 Relation Obesity and Clinical Parameters

A longitudinal population based cohort study in Toronto demonstrated a 1.4 point increase of the mean body mass index (BMI) in adult CF patients (from 20.7 to 22.3) over the last 3 decades [15]. Besides this trend, they also observed a significant increase in the proportion of overweight (BMI>25) or obese (BMI>30) CF patients (from 7% to 18.4%). The patients with overweight or obesity were more likely to be older, pancreatic sufficient, have better pulmonary function, milder genotype and are less likely to be colonized. Similar results were presented from Greece, where despite increase of overweight or obesity, 22% of patients still showed malnutrition [83].

The positive effect of the BMI on the pulmonary function was more important for pancreatic insufficient patients compared to pancreatic sufficient patients [15]. Finally, the pulmonary function did not increase further in the overweight and obese group of CF patients.

The obese CF patients seem to represent a different subgroup of CF patients. They suffer, like the obese healthy population, from hypercholesterolemia and hypertriglyceridemia [15,84–86], a common finding if the diet is high in carbohydrates [87]. In the Swedish population, being overweight is still no problem despite the regular recommendation of additive vegetable oil (Lindblad and Hjelte, personal communications). One reason might be that physical activity has been a regular part in the treatment of CF for 25 years [88].

40.4.2 Treatment Strategies

The nutritional requirements of pancreatic sufficient CF patients need to be assessed in order to prevent excessive weight gain. The classical high fat, high energy dietary recommendations will need adaptation when the BMI exceeds 25 as this BMI does not further improve the pulmonary function. In that context the quality of the fat recommendations are necessary to consider since also overweight patients can show low EFA concentrations. Obesity in CF might carry the same risk for overweight associated problems as seen in the healthy population. Although these problems are not yet described in CF, it took also several years of follow up in CF-related diabetes mellitus before diabetes associated problems became apparent [89]. Furthermore, one of the effects of CFTR protein repair therapy is an important weight gain [90]. This could also mean we will need to adapt dietary therapy accordingly, which might indicate a stronger awareness of the composition of food, restricting carbohydrates, and increasing essential fatty acids.

Furthermore, pancreatic sufficient patients also display, to a lesser account, the same fatty acid abnormalities as the pancreatic insufficient patients [24,25,91], therefore the fat intake should probably be shifted toward a high LA recommendation in contrast to the healthy population. A high fat, EFA rich diet in CF can reduce the intake of protein and carbohydrate to normal recommendation [92] and thereby reduce the risk of overweight.

Probably, it will be necessary to associate intensified physical exercise to the treatment of the obese CF patient. This will improve the weight but also the body composition of these patients. The lean body mass is, indeed, even more associated with pulmonary function than the BMI explaining why the normal lean body mass but low fat mass, characteristic of the Swedish CF population, might be related to good pulmonary function [93].

40.5 CONCLUSIONS

Future research needs to clarify the physiological relation between the fatty acid abnormality and the gene defect. The possibility of influencing the clinical condition by compensating for different mutations might give new possibilities to investigate such associations. Dietary advice in CF will have to focus on improved dietary fat quality rather than just high energy diets and an adapted dietary advice for those CF patients at risk for developing overweight and obesity.

References

[1] Corey M, McLaughlin FJ, Williams M, Levison H. A comparison of survival, growth and pulmonary function in patients with cystic fibrosis in Boston and Toronto. J Clin Epidemiol 1988;41(6):583–91.

[2] Steinkamp G, Weidemann B. Relationship between nutritional status and lung function in cystic fibrosis: cross sectional and longitudinal analyses from the Germany CF quality assurance (CFQA) project. Thorax 2002;57:596–601.

[3] Kraemer R, Ruedeberg A, Hardorn B, Rossi E. Relative underweight in cystic fibrosis and its prognostic value. Acta Paediatr Scand 1978;67:33–7.

[4] Roulet M. Protein-energy malnutrition in cystic fibrosis patients. Acta Paediatr 1994;83(S395):43–8.

[5] Campbell IM, Crozier DN, Caton RB. Abnormal fatty acid composition and impaired oxygen supply in cystic fibrosis patients. Pediatrics 1976;57:480–6.

[6] Sinaasappel M, Stern M, Littlewood J, Wolfe S, Steinkamp G, Heijerman HG, et al. Nutrition in patients with cystic fibrosis: a European consensus. J Cyst Fibros 2002;1:51–75.

[7] Kindstedt-Arwidson K, Strandvik B. Food intake in patients with cystic fibrosis on an ordinary diet. Scand J Gastroenterol 1988;23(Suppl. 143):160–2.

[8] Kuo PT, Huang NN, Bassett DR. The fatty acid composition of the serum chylomicrons and adipose tissue of children with cystic fibrosis of the pancreas. J Pediatr 1962;60:394–403.

[9] Underwood BA, Denning CR, Navab M. Polyunsaturated fatty acids and tocopherol levels in patients with cystic fibrosis. Ann N Y Acad Sci 1972;203:237–47.

[10] Konstan MW, Berger M. Current understanding of the inflammatory process in cystic fibrosis. Pediatr Pulmonol 1997;24:137–42.

[11] Raia V, Maiuri L, de Ritis G, de Vizia B, Vacca L, Conte R, et al. Evidence of chronic inflammation in morphologically normal small intestine of cystic fibrosis patients. Pediatr Res 2000;47:344–50.

[12] Medjane S1, Raymond B, Wu Y, Touqui L. Impact of CFTR DeltaF508 mutation on prostaglandin E2 production and type IIA phospholipase A2 expression by pulmonary epithelial cells. Am J Physiol Lung Cell Mol Physiol 2005;289:L816–24.

[13] Strandvik B, Svensson E, Seyberth HW. Prostanoid biosynthesis in patients with cystic fibrosis. Prostaglandins Leukot Essent Fatty Acids 1996;55:419–25.

[14] Strandvik B. Fatty acid metabolism in cystic fibrosis. Prostaglandins Leukot Essent Fatty Acids 2010;83:121–9.

[15] Stephenson AL, Mannik LA, Walsh S, Brotherwood M, Robert R, Darling PB, et al. Longitudinal trends in nutritional status and relation between lung function and BMI in cystic fibrosis: a population-based cohort study. Am J Clin Nutr 2013;97:872–7.

[16] Sprecher H. The roles of anabolic and catabolic reactions in the synthesis and recycling of polyunsaturated fatty acids. Prostaglandins Leukot Essent Fatty Acids 2002;67:79–83.

[17] Pawlowsky RJ, Hibbeln JL, Lin Y, Salem Jr N. physiological compartmental analysis of alfa-linolenic acid metabolism in adult humans. J Lipid Res 2001;42:1257–65.

[18] Christiansen K, Marcel Y, Gan MV, Mohrhauer H, Holman RT. Chain elongation of alpha- and gamma-linolenic acids and the effect of other fatty acids on their conversion in vitro. J Biol Chem 1968;243:2969–74.

[19] Tvrzicka E, Kremmyda LS, Stankova B, Zak A. Fatty acids as biocompounds: their role in human metabolism, health and disease – a review. Part 1: classification, dietary sources and biological functions. Biomed Pap Med Fac Univ Palacky Olomouc Czech Repub 2011;155(2):117–30.

[20] Murakami M. Lipid mediators in life and science. Exp Anim 2011;60(1):7–20.

[21] Robinson PG. Essential fatty acids in cystic fibrosis. Lancet 1975;8:919.

[22] Dodge JA, Salter DG, Yassa JG. Essential fatty acid deficiency due to artificial diet in cystic fibrosis. Br Med J 1975;11:192–3.

[23] Farrell PM, Mischler EH, Engle MJ, Brown J, Lau SM. Fatty acid abnormalities in cystic fibrosis. Pediatr Res 1985;19:104–9.

[24] Rogiers V, Dab I, Crokaert R, Vis HL. Long chain non-esterified fatty acid pattern in plasma of cystic fibrosis patients and their parents. Pediatr Res 1980;14:1088–91.

[25] Roulet M, Frascardo P, Pilet M. Essential fatty acid deficiency in well nourished young cystic fibrosis patients. Eur J Pediatr 1997;156:952–6.

[26] Elliott RB, Robinson PG. Unusual clinical course in a child with cystic fibrosis treated with fat emulsion. Arch Dis Child 1975;50:76–8.

[27] Lloyd-Still JD, Johnson SB, Holmart RT. Essential fatty acid status in cystic fibrosis and the effect of safflower supplementation. Am J Clin Nutr 1981;34:1–7.

[28] Mischler EH, Parrell SW, Farrell PM, Raynor WJ, Lemen RJ. Correction of linoleic acid deficiency in cystic fibrosis. Pediatr Res 1986;20:36–41.

[29] Christophe A, Verdonk G, Robberecht E, Mahathanakhun R. Effect of supplementing medium chain triglycerides with linoleic acid-rich monoglycerides on severely disturbed serum lipid fatty acid patterns in patients with cystic fibrosis. Ann Nutr Metab 1985;29:239–45.

[30] Strandvik B, Berg U, Kallner A, Kusoffsky E. Effect on renal function of essential fatty acid supplementation in cystic fibrosis. J Pediatr 1989;115:242–50.

[31] Strandvik B, Hultcrantz R. Liver function and morphology during long term fatty acid supplementation in cystic fibrosis. Liver 1994;14:32–6.

[32] Freedman SD, Katz MH, Parker EM, Laposata M, Urman MY, Alvares JG. A membrane lipid imbalance plays a role in the phenotypic expression of cystic fibrosis in CFTR−/− mice. Proc Natl Acad Sci 1999;96:13995–40000.

[33] Beharry S, Ackerley C, Corey M, Kent G, Heng YM, Christensen H, et al. Long-term docosahexaenoic acid therapy in a congenic murine model of cystic fibrosis. Am J Physiol Gastrointest Liver Physiol 2007;292:G839–48.

[34] De Visia B, Raia V, Spano C, Pavlidis C, Coruzzo A, Alessio M. Effect of an 8-month treatment with ω-3 fatty acids (eicosapentaenoic and docosahexaenoic) in patients with cystic fibrosis. J Parenter Enteral Nutr 2003;27:52–7.

[35] Van Biervliet S, Devos M, Delhaye T, Van Biervliet JP, Robberecht E, Christophe A. Oral DHA supplementation in ΔF508 homozygous cystic fibrosis patients. Prostaglandins Leukot Essent Fatty Acids 2008;78(2):109–15.

[36] Oliver C, Jahnke N. Omega-3 fatty acids for cystic fibrosis. Cochrane Database Syst Rev 2011;10:CD002201.

[37] Van Biervliet S, Vanbillemont G, Van Biervliet JP, Declecq D, Robberecht E, Christophe A. Relation between fatty acid composition an clinical status or genotype in cystic fibrosis patients. Ann Nutr Metab 2007;51:241–9.

[38] Strandvik B, Gronowitz E, Enlund F, Martinsson T, Wahlströrm J. Essential fatty acid deficiency in relation to genotype in patients with cystic fibrosis. J Pediatr 2001;139:650–5.

[39] Freedman SD, Blanco PG, Zaman MM, Shea JC, Ollero M, Hopper IK, et al. Association of cystic fibrosis with abnormalities in fatty acid metabolism. N Engl J Med 2004;350:560–9.

[40] Darmstadt GL, McGuire J, Ziboh VA. Malnutrition associated rash of cystic fibrosis. Pediatr Dermatol 2000;17:337–47.

[41] Maqbool A, Shall JI, Garcia-Espana JF, Zemel BS, Strandvik B, Stallings VA. Serum linoleic acid status as a clinical indicator of essential fatty acid status in children with cystic fibrosis. J Pediatr Gastroenterol Nutr 2008;47(5):635–44.

[42] Gronowitz E, Lorentzon M, Ohlsson C, Mellström D, Strandvik B. Docosahexaenoic acid is associated with endosteal circumference in long bons in young males with cystic fibrosis. Br J Nutr 2008;99(1):160–7.

[43] Gronowitz E, Mellström D, Strandvik B. Serum phospholipid fatty acid pattern is associated with bone mineral density in children, but not adults, with cystic fibrosis. Br J Nutr 2006;95:1159–65.

[44] Gronowitz E, Garemo M, Lindblad A, Mellström D, Strandvik B. Decreased bone mineral density in normal-growing patients with cystic fibrosis. Acta Paediatr 2003;92:688–93.

[45] Gronowitz E, Mellström D, Strandvik B. Normal annual increase of bone mineral density during two years in patients with cystic fibrosis. Pediatrics 2004;114:435–42.

[46] Van Biervliet S, Van Biervliet JP, Robberecht E, Christophe A. Fatty acid composition of serum phospholipids in cystic fibrosis (CF) patients with or without CF related liver disease. Clin Chem Lab Med 2010;48(12):1751–5.

[47] Lindblad A, Glaumann H, Strandvik B. Natural history of liver disease in cystic fibrosis. Hepatology 1999;30:1151–8.

[48] Christophe A, Robberecht E. Current knowledge on fatty acids in cystic fibrosis. Prostaglandins Leukot Essent Fatty Acids 1996;55:129–38.

[49] Lai HC, Kosorok MR, Laxova A, Davis LA, FitzSimmon SC, Farrell PM. Nutritional status of patients with cystic fibrosis with meconium ileus: a comparison with patients without meconium ileus and diagnosed early through neonatal screening. Pediatrics 2000;105:53–61.

[50] Maqbool A, Schall JI, Gallagher PR, Zemel BS, Strandvik B, Stallings VA. Relation between dietary fat intake type and serum fatty acid status in children with cystic fibrosis. J Pediatr Gastroenterol Nutr 2012;55:605–11.

[51] Cunnane SC, Guesnet P. Linoleic acid recommendations–a house of cards. Prostaglandins Leukot Essent Fatty Acids 2011;85:399–402.

[52] Carlstedt-Duke J, Brönnegard M, Strandvik B. Pathological regulation of arachidonic acid release in cystic fibrosis: the putative defect. Proc Natl Acad Sci USA 1986;83:9202–6.

[53] Bhura-Bandali FN, Suh M, Man SFP, Clandinin MT. The F508 mutation in the cystic fibrosis transmembrane conductance regulator alters control of essential fatty acid utilization in epithelial cells. J Nutr 2000;130:2870–5.

[54] Ulane MM, Butler JB, Peri A, Miele L, Ulane RE, Hubbard VS. Cystic fibrosis and phosphatidylcholine biosynthesis. Clin Chim Acta 1994;230:109–16.

[55] Khan TZ, Wagener JS, Bost T, Martinez J, Accurso FJ, Riches DW. Early pulmonary inflammation in infants with cystic fibrosis. Am J Respir Crit Care Med 1995;151:1075–82.

[56] Armstrong DS, Grimwood K, Carzino R, Carlin JB, Olinsky A, Phelan PD. Lower respiratory infection and inflammation in infants with newly diagnosed cystic fibrosis. BMJ 1995;310:1571–2.

[57] Jabr S, Gartner S, Milne GL, Roca-Ferrer J, Casas J, Moreno A, et al. Quantification of major urinary metabolites of PGE2 and PGD2 in cystic fibrosis: correlation with disease severity. Prostaglandins Leukot Essent Fatty Acids 2013;89(2–3):121–6.

[58] Al-Turkmani MRC, Andersson R, Alturkmani R, Katrangi W, Cluette-Brown JE, Freedman SD, et al. A mechanism accounting for the low cellular level of linoleic acid in cystic fibrosis and its reversal by DHA. J Lipid Res 2008;49:1946–54.

[59] Njoroge SW, Seegmiller AC, Katrangi W, Laposata M. Increased delta5- and delta6-desaturase, cycloxygenase-2, and lipoxygenase-5 expression and activity are associated with fatty acid and eicosanoid changes in cystic fibrosis. Biochim Biophys Acta 2011;1811:431–40.

[60] Thomsen KF, Laposata M, Njoroge SW, Umunakwe OC, Katrangi W, Seegmiller AC. Increased elongase-6 and delta9-desaturase activity are associated with n-7 and n-9 fatty acid changes in cystic fibrosis. Lipids 2011;46:669–77.

[61] Ollero M, Laposata M, Zaman MM, Blanco PG, Andersson C, Zeind J, et al. Evidence of increased flux to n-6 docosapentaenoic acid in phospholipids of pancreas from cftr−/− knockout mice. Metabolism 2006;55:1192–200.

[62] Innis SM, Davidson AG. Cystic fibrosis and nutrition: linking phospholipids and essential fatty acids with thiol metabolism. Annu Rev Nutr 2008;28:55–72.

[63] Bensalem N, Ventura AP, Vallée B, Lipecka J, Tondelier D, Davezac N, et al. Down-regulation of the anti-inflammatory protein annexin A1 in cystic fibrosis knock-out mice and patients. Mol Cell Proteomics 2005;4:1591–601.

[64] Perretti M, Gavins NE. Annexin 1: an endogenous anti-inflammatory protein. News Physiol Sci 2003;18:60–4.

[65] Grassmé H, Becker KA, Zhang Y, Gulbins E. Ceramide in bacterial infections and cystic fibrosis. Biol Chem 2008;389:1371–9.

[66] Guilbault C, Wojewodka G, Saeed Z, Hajduch M, Matouk E, De Sanctis JB, et al. Cystic fibrosis fatty acid imbalance is linked to ceramide deficiency and corrected by fenretinide. Am J Respir Cell Mol Biol 2009;41:100–6.

[67] Wojewodka G, De Sanctis JB, Radzioch D. Ceramide in cystic fibrosis: a potential new target for therapeutic intervention. J Lipids 2011;2011:674968.

[68] Schroeder TH, Reiniger N, Meluleni G, Grout M, Coleman FT, Pier GB. Transgenic cystic fibrosis mice exhibit reduced early clearance of Pseudomonas aeruginosa from the respiratory tract. J Immunol 2001;166:7410–8.

[69] Guilbault C, De Sanctis JB, Wojewodka G, Saeed Z, Lachance C, Skinner TA, et al. Fenretinide corrects newly found ceramide deficiency in cystic fibrosis. Am J Respir Cell Mol Biol 2008;38:47–56.

[70] Lezo A, Biasi F, Massarenti P, Calabrese R, Poli G, Santini B, et al. Oxidative stress in stable cystic fibrosis patients: do we need higher antioxidant plasma levels? J Cyst Fibros 2013;12:35–41.

[71] Wood LG, Fitzgerald DA, Lee AK, Garg ML. Improved antioxidant and fatty acid status of patients with cystic fibrosis after antioxidant supplementation is linked to improved lung function. Am J Clin Nutr 2003;77:150–9.

[72] Hulbert AJ, Kelly MA, Abbott SK. Polyunsaturated fats, membrane lipids and animal longevity. J Comp Physiol B 2013;184:149–66.

[73] Lloyd-Still JD, Bibus DM, Powers CA, Johnson SB, Holman RT. Essential fatty acid deficiency and predisposition to lung disease in cystic fibrosis. Acta Paediatr 1996;85:1426–32.

[74] Christophe A, Robberecht E, Franckx H, De Baets F, van de Pas M. Effect of administration of gamma-linolenic acid on the fatty acid composition of serum phospholipids and cholesteryl esters in patients with cystic fibrosis. Ann Nutr Metab 1994;38:40–7.

[75] Keicher U, Koletzko B, Reinhardt D. Omega-3 fatty acids suppress the enhanced production of 5-lipoxygenase products from polymorph neutrophil granolocytes in cystic fibrosis. Eur J Clin Invest 1995;25:915–9.

[76] Lawrence R, Sorell T. Eicosapentaenoic acid in cystic fibrosis: evidence of a pathologenetic role for leukotriene B4. Lancet 1993;342:465–9.

[77] Kurlansky LE, Bennink MR, Webb PM, Ulrich PJ, Baer LJ. The absorption and effect of dietary supplementation with omega-3 fatty acids on serum leukotriene B4 in patients with cystic fibrosis. Pediatr Pulmonol 1994;18:211–7.

[78] Panchaud A, Sauty A, Kernen Y, Decosterd LA, Buclin T, Boulat O, et al. Biolological effects of a dietary omega-3 polyunsaturated fatty acids supplementation in cystic fibrosis patients: a randomized, crossover, placebo-controlled trial. Clin Nutr 2006;3:418–27.

[79] Alicandro G, Faelli N, Gagliardini R, Santini B, Magazzu G, Biffi A, et al. A randomized placebo-controlled study on high dose oral algal docosahexaenoic acid supplementation in children with cystic fibrosis. Prostaglandins Leukot Essent Fatty Acids 2013;88:163–9.

[80] Keen C, Olin AC, Eriksson S, Ekman A, Lindblad A, Basu S, et al. Supplementation with fatty acids influences the airway nitric oxide and inflammatory markers in patients with cystic fibrosis. J Pediatr Gastroenterol Nutr 2010;50:537–44.

[81] Olveira G, Olveira C, Acosta E, Espíldora F, Garrido-Sánchez L, García-Escobar E, et al. Fatty acid supplements improve respiratory, inflammatory and nutritional parameters in adults with cystic fibrosis. Arch Bronconeumol 2010;46:70–7.

[82] Strandvik B. Care of patients with cystic fibrosis. Treatment, screening and clinical outcome. Ann Nestlé 2006;64:131–40.

[83] Panagopoulou P, Fotoulaki M, Nikolaou A, Nousia-Arvanitakis S. Prevalence of malnutrition and obesity among cystic fibrosis patients. Pediatr Int 2014;56:89–94.

[84] Georgiopoulou VV, Denker A, BishopKL, Brown JM, Hirsh B, Wolfenden L, et al. Metabolic abnormalities in adults with cystic fibrosis. Respirology 2010;15:823–9.

[85] Rhodes B, Nash EF, Tullis E, Pencharz PB, Brotherwood M, Dupuis A, et al. Prevalence of dyslipidemia in adults with cystic fibrosis. J Cyst Fibros 2010;9:24–8.

[86] Ishimo MC, Belson L, Ziai S, Levy E, Berthiaume Y, Coderre L, et al. Hypertriglyceridemia is associated with insulin levels in adult cystic fibrosis patients. J Cyst Fibros 2013;12:271–6.

[87] Rajaie S, Azadbakht L, Khazaei M, Sherbafchi M, Esmaillzadeh A. Moderate replacement of carbohydrates by dietary fats affects features of metabolic syndrome: a randomized crossover clinical trial. Nutrition 2014;30:61–8.

[88] Sahlberg ME, Svantesson U, Thomas EM, Strandvik B. Muscular strength and function in patients with cystic fibrosis. Chest 2005;127:1587–92.

[89] Konrad K, Scheuing N, Badenhoop K, Borkenstein MH, Gohlke B, Schöfl C, et al. Cystic fibrosis-related diabetes compared with type 1 and type 2 diabetes in adults. Diabetes Metab Res Rev 2013;29:568–75.

[90] Harrison MJ, Murphy DM, Plant BJ. Ivacaftor in a G551D homozygote with cystic fibrosis. N Engl J Med 2013;369:1280–2.

[91] Lloyd-Still JD, Johnson SB, Holman RT. Essential fatty acid status and fluidity of plasma phospholipids in cystic fibrosis infants. Am J Clin Nutr 1991;54:1029–35.

[92] Kindstedt-Arfwidson K, Strandvik B. Food intake in patients with cystic fibrosis on an ordinary diet. Scand J Gastroenterol 1988;23:160–2.

[93] Sahlberg M, Eriksson BO, Sixt R, Strandvik B. Cardiopulmonary data in response to 6 months of training in physically active adult patients with classic cystic fibrosis. Respiration 2008;76:413–20.

41

Persistent Fat Malabsorption in Cystic Fibrosis

Frank A.J.A. Bodewes, Marjan Wouthuyzen-Bakker, Henkjan J. Verkade

University Medical Center Groningen, University of Groningen, Groningen, The Netherlands

41.1 INTRODUCTION

One of the most striking features of the gastrointestinal phenotypes in cystic fibrosis (CF) is fat malabsorption [1]. The fatty stools (steatorrhoea) are a clinical sign of fat malabsorption. Fat malabsorption in patients causes several severe problems. Due to the high energy content of fat in general, dietary fat malabsorption can contribute to malnutrition and poor growth. A secondary effect of intestinal fat malabsorption is the reduced absorption of the fat-soluble vitamins A, D, E, and K. Vitamin malabsorption can lead to hypovitaminosis and vitamin K dependent coagulopathy [2–4]. Therefore, CF patients with intestinal fat malabsorption are usually dependent on oral vitamin ADEK supplementation.

In this chapter we review the impaired intestinal fat absorption in CF. In CF the intestinal fat malabsorption can be divided into two major causal categories. The first is the intestinal fat malabsorption as a consequence of the CF-related exocrine pancreatic insufficiency (EPI) and the subsequent intestinal deficiency of pancreatic enzymes, in particular pancreatic lipase, and/or of the optimal pH milieu to operate. This causal category has been well addressed in some recent excellent reviews [5–7]. In the present chapter we will focus on the second causal category of intestinal fat malabsorption, which is independent of the presence or activity of pancreatic enzymes: it causes (some degree of) fat malabsorption to remain in CF patients despite pancreatic enzyme replacement therapy, or PERT. This therapy-resistant form of intestinal fat malabsorption has several CF-related causes that we will discuss individually, based on data from studies in experimental animal models and humans. To be able to discuss the subject properly, we will begin with a short structured review of the mechanisms and phases involved in the physiology of intestinal fat absorption.

41.2 PHYSIOLOGY OF INTESTINAL DIETARY FAT ABSORPTION

Dietary fat intake mainly consists of long- and medium-chain triglycerides. The mechanism of digestion and absorption of these lipids can be dissected into several sequential steps. In contrast to long-chain triglycerides, medium-chain triglycerides are known to circumvene several steps in fat digestion and absorption that are essential for long-chain triglycerides. For example, unlike long-chain triglycerides, medium-triglycerides are less dependent upon solubilization, escape the re-esterification into triglycerides, and are directly absorbed into the portal system without being assembled into chylomicrons [8]. Since the majority of dietary fat and energy intake consists of long-chain triglycerides (92–96%), we exclusively focused on their mechanism of intestinal digestion and absorption. The mechanism of digestion and absorption of long-chain triglycerides can be dissected into several sequential processes.

41.2.1 Emulsification

Emulsification of triglycerides is a process in which (water-insoluble) fat droplets are suspended in an aqueous environment. Emulsification can be achieved by mechanical as well as biochemical means. Mechanical emulsification is attained by chewing and forcing dietary fat through a small opening (e.g., the pylorus) with high pressure, thus dispersing large fat droplets into smaller droplets. Biochemical emulsification is attained by the action of bile and gastric lipases and prevents the emulsion from recoalescing. Emulsification results in a fine, relatively stable oil-in-water emulsion with an increased surface area. Emulsification aids in the efficiency of fat absorption because the water soluble digestive/lipolytic

Diet and Exercise in Cystic Fibrosis
http://dx.doi.org/10.1016/B978-0-12-800051-9.00041-9

enzymes are active at the site of the water–oil interface. Increasing the surface area will thus increase the rate of lipolysis.

41.2.2 Lipolysis

During lipolysis, the triglycerides are broken down (hydrolyzed) into fatty acids and monoglycerides. Around 10–30% of the triglycerides are hydrolyzed in the stomach into diglycerides and free fatty acids, catalyzed by lingual and gastric lipase. Because the resulting molecules are surface-active compounds, they induce a further emulsification into smaller particles (a process also known as biochemical/lipolytic emulsification [8]). Under physiological conditions, pancreatic lipase completes the lipolysis in the proximal part of the small intestine, by hydrolyzing the remaining triglycerides and diglycerides into monoglycerides and free fatty acid molecules. Bile salts are amphipathic molecules; i.e., they possess a hydrophilic and a hydrophobic site. Upon exposure to dietary fat emulsion at the level of the proximal small intestine, the hydrophobic site orients itself toward the (hydrophobic) fat droplets, whereas their hydrophilic site exposes itself to the aqueous (water) phase. Fat droplets whose surface is thus covered with bile salts are not accessible to pancreatic lipase. Pancreatic colipase allows the anchoring of the pancreas lipase to the oil–water interface and is essential for lipolytic activity. Under physiological conditions, pancreatic lipase is present in vast excess. Thus, in CF, fat maldigestion occurs only during severe pancreatic malfunction [9]. During exocrine pancreatic insufficiency, when pancreatic lipase is severely reduced, lipolysis is more dependent on the activity of the preduodenal gastric and lingual lipases. In humans, gastric lipase is the predominant preduodenal lipase species [10]. In CF conditions, preduodenal lipase remains fully active in the intestine, where it can account for more than 90% of lipase activity, even when the intestinal pH drops below the optimal level for bile salt dependent lipolysis [11].

41.2.3 Solubilization

The process in which fat molecules are dissolved in water in the form of (mixed) micelles is called solubilization. Solubilization enhances the aqueous solubility of fatty acids by several orders of magnitude (100–1000 fold) [8]. Solubilization is necessary for monoglycerides and free fatty acids to efficiently overcome the diffusion barrier of the so-called unstirred water layer of the enterocytes [8]. The unstirred water layer separates the enterocytes from the luminal contents of the intestine. Solubilization is achieved by the formation of mixed micelles, mainly consisting of phospholipids and bile salts derived from biliary secretion. The diameter of the fat droplets ranges from 100 to 1000 nm, whereas that of mixed micelles ranges from 3 to 5 nm. The hydrophobic part of the fat molecules, such as the acyl-chains, will be oriented inwards, whereas the hydrophilic parts (such as the carboxylic head groups) orient toward the aqueous outside of the micelle. Saturated fatty acids are more dependent on solubilization than unsaturated fatty acids, due to the higher hydrophobicity of the former. The difference in bile salt dependency can be derived from absorption studies in rats with chronic bile diversion: in such intestinal bile-deficient conditions, absorption of saturated fatty acids is below 30% of the ingested amount, while the absorption of unsaturated fatty acids is relatively maintained (~80%) [12]. Just like the digestion of fat, the process of micelle formation is also pH sensitive. Low intestinal pH levels can severely inhibit micelle formation or induce premature release of lipolytic products out of micelles [8].

41.2.4 Translocation

Once the mixed micelles diffuse through the unstirred water layer and arrive at the proximity of the (apical) enterocyte membrane, the free fatty acids and monoglycerides dissociate from the micelles. It has been postulated that the acidic microclimate near the apical membrane of intestinal mucosal cells induces micelle disintegration and favors the translocation of fatty acid molecules across the enterocyte membrane [8]. It has remained unclear whether intestinal membrane transporters are essential for the translocation of fatty acids and monoglycerides. It has been suggested that free fatty acid uptake is concentration-dependent, in which a high intra-luminal concentration drives passive diffusion. Two putative intestinal transporters have been proposed to be involved fatty acid uptake: the fatty acid binding protein (FABP) and the fatty acid translocase/cluster determinant 36 (FAT/Cd36) [13]. However, both transporters are not likely to be involved in fatty acid uptake into the enterocytes. FABP is expressed only in a small area of the crypt-villus of the intestine, and knockout mice for either FABP or CD36 do not exhibit impairments in fatty acid uptake [14–16].

41.2.5 Intracellular Processing

After being absorbed into the enterocytes, the lipolytic products migrate to the endoplasmic reticulum, possibly mediated via the fatty acid binding proteins [13,14]. At the cytoplasmic surface of the endoplasmic reticulum, the fatty acids and monoglycerides are re-esterified into triglycerides. Under physiological conditions, re-esterification mainly occurs via the monoacylglycerol pathway, i.e., the sequential acylation of monoacylglycerol by acyl-CoA [8].

41.2.6 Chylomicron Production

Newly synthesized triglycerides are transferred into the smooth endoplasmic reticulum and are assembled into lipoprotein particles called chylomicrons. Intestinal phospholipids are required for chylomicron production in order to prevent accumulation in the enterocyte [17,18]. Maturation of chylomicrons (i.e., assembly of fat particles with a phospholipid-cholesterol-apolipoprotein surface) takes place in the Golgi apparatus. Chylomicron formation is followed by exocytosis via the secretory pathway at the basolateral surface of the enterocyte. The Chylomicrons are released into the circulation via the mesenteric lymph system, via the thoracic duct into the venous system, after which their contents are systemically delivered.

41.3 IMPAIRED INTESTINAL DIETARY FAT ABSORPTION IN CYSTIC FIBROSIS

41.3.1 Exocrine Pancreatic Insufficiency

The leading cause of intestinal fat malabsorption in CF is EPI [19]. CF causes fibrotic degeneration of the acinar tissue of the pancreas secondary to destruction of the ductular structures due to loss of Cystic fibrosis transmembrane conductance regulator (CFTR) function [20]. The fibrotic pancreas is no longer able to excrete pancreatic enzymes, including lipases and proteases essential for intestinal fat and protein absorption. The pancreatic destruction, partly based on autodigestion, starts in utero and, in most patients with a severe genotype, this develops into complete PI already during infancy [21].

EPI is treated with PERT [22]. These products contain pancreatic enzymes exclusively of animal origin [5]. However, bioengineered products based on human lipases are currently developed and coming to the market [23]. PERT is individually dosed based on the dietary fat intake and on its effects on intestinal fat absorption.

41.3.2 PERT Therapy Persistent Intestinal Fat Malabsorption in CF

Most pancreatic-insufficient CF patients display a degree of intestinal fat malabsorption despite optimal PERT. Rather than the physiological absorption (i.e., 95% of the dietary fat ingested), most CF patients show a decreased intestinal fat absorption of ±85–90% of dietary fat intake [24,25].

Optimizing of PERT for individual patients toward the physiological range of fat absorption percentage has proven to be rather difficult. In daily clinical practice it is reasonable to assume that at least part of the failure to approach a physiological degree of fat absorption could still be attributed to inefficient or inaccurate dosing of PERT. PERT formulations, based on the use of various acid resistant coatings and adjuvants, differ in their intestinal pharmokinetic properties. Individual variations between patients in intestinal milieu such as pH and transit time make it difficult to predict how effective a specific PERT formulation will behave and function in individual patients. Additional factors such as non-synchronous entrance of PERT and dietary stomach contents into the duodenum probably also play a role in the CF-related persistent fat malabsorption [26].

In CF, apart from EPI- and PERT-related issues, multiple intestinal pathophysiological mechanisms are, to a greater or lesser extent, involved in the impaired intestinal fat malabsorption localized in various gastrointestinal organs (Figure 41.1). The factors involved in persistent intestinal fat malabsorption in CF (despite PERT) can be divided into several main groups based on their causal mechanisms:

1. Intestinal pH and bicarbonate secretion
2. Bile salt metabolism and enterohepatic circulation
3. Intestinal mucosal abnormalities
4. Small intestinal bacterial overgrowth

41.4 INTESTINAL PH AND BICARBONATE SECRETION

Neutralization of gastric acid is essential for functional intestinal fat absorption. Intestinal pH is guiding the efficiency of triglyceride lipolysis and bile salt-induced mixed micelle formation [8,27]. The activity of all involved enzymes is affected by changes in pH. The optimum pH for maximal pancreatic lipase enzyme activity is 6–7. For optimal pancreatic lipase activity to occur, it is physiologically important that the low gastric pH of the stomach contents (pH ~2) is buffered adequately when passing on into the duodenum and the proximal intestine.

Physiologically, gastric acid is buffered by secretion of bicarbonate by the pancreas and by the enterocytes of the duodenum where the acid load from the stomach is the highest. CFTR plays a pivotal role in pancreatic and duodenal bicarbonate secretion. CFTR is highly expressed in the pancreas and duodenum. In CF, the pancreatic and duodenal bicarbonate secretion is insufficient to neutralize the gastric acid load [28,29]. Hence, the duodenal pH is (on average) 1–2 units lower in CF patients compared with healthy controls. Accordingly, CF patients have significantly longer postprandial periods in which the duodenal pH drops below 4 [30]. More distally in the small intestine, the pH values in jejunal and ileal contents from CF patients vary from lower to similar pH values compared with healthy controls [30]. Using a wireless motility and pH-detecting intestinal capsule, Gelfond et al. recently showed in CF patients in vivo a deficient buffering capacity required to neutralize the gastric acid in the proximal small bowel [31].

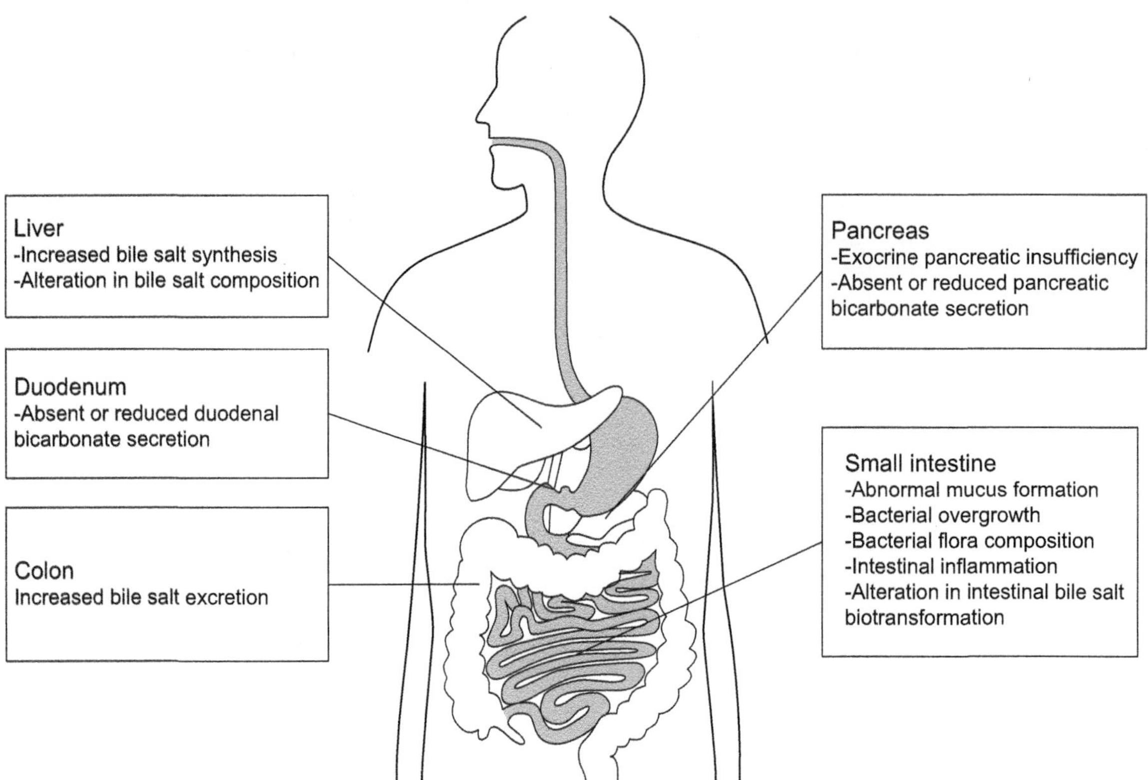

FIGURE 41.1 Gasto-intestinal organs possibly involved in CF patients with persistent fat malabsorption despite pancreatic enzyme replacement therapy.

A relation has been described between low postprandial duodenal pH levels and the degree of fat malabsorption in CF patients treated with pancreatic enzymes [32]. These data suggest that interventions to increase intestinal pH values might target persistent fat malabsorption in cystic fibrosis. Two studies evaluated fat absorption in pediatric and adult CF patients after the addition of bicarbonate to enteric-coated pancreatic enzymes. In the first study, the average fat absorption increased from 75% to 82% of the ingested amount [33]. In this study, a beneficial effect on fat absorption was observed in 75% of the patients. In the second study, the average fat absorption did not increase, but 50% of patients did show an improvement in fat absorption (>5%) after bicarbonate supplementation [34]. Due to the overall limited effects on fat absorption, pancreatic enzymes buffered with bicarbonate have not become standard treatment in CF care. It is unclear to what extent the activity of the bicarbonate supplementation is already decreased in the stomach.

In a systematic review, Jones evaluated the effect of acid suppressant therapy on fat absorption or fecal fat excretion and nutritional status [35]. While multiple studies reported improved fat absorption or reduced fecal fat excretion after acid suppressive therapy, other studies reported no improvements. Improved fat absorption was not accompanied with an improved nutritional status. The performed studies showed high variability in the dosage of pancreatic enzymes used, and in the choice of acid suppressive medication, and differed in the inclusion criteria for the degree of fat malabsorption. Due to these large inter-study differences, it remains difficult to draw an overall definite conclusion about the effectiveness of acid suppressive drugs. Upon evaluation of individual data in these intervention studies, a subgroup of patients clearly showed improved fat absorption. The results indicate that some CF patients do benefit from acid suppressive therapy. Why some patients respond, but others show no improvement in fat absorption, illustrates that an altered pH is not the only factor responsible for CF-related intestinal fat malabsorption. In addition to this, acid suppressive drugs can induce small intestinal bacterial overgrowth (SIBO) and can alter bile salt metabolism, with a potentially negative effect on fat absorption [36]. Therefore, acid suppressive therapy should be imposed with caution, especially in CF patients who are relatively more prone to SIBO. A trial of proton pump inhibitors could be considered to evaluate its effectiveness in the individual CF patient, but in that case a predefined evaluation of its effect is warranted.

Bijvelds et al. investigated the effect of the acid suppressive drug omeprazole on lipolysis and uptake of lipolytic products in CFTR^{tm1CAM} knockout mice by using radioisotope labeled triglycerides and fatty acids [37]. The researchers showed that lipolytic activity and fat absorption improved after omeprazole treatment [37]. It is likely that increasing the intestinal pH has a generally positive effect on intestinal fat absorption, since lipolytic and post-lipolytic activity also partially improved in the omeprazole-treated wild type mice [37].

41.5 INTRALUMINAL BILE SALTS

Bile salts are polarized steroids that play a vital role in intestinal fat absorption [38]. In the intestine, bile salts function as essential surfactants used to solubilize dietary fats in the hydrophilic milieu of gut [39]. Bile salts are synthesized in the hepatocytes from cholesterol. Bile salts are excreted into the bile and transported, to the intestine, via the intra- and extrahepatic bile ducts. In the bile and the gut, bile salts form water-solvable aggregates, so-called micelles, together with the fatty acids originating from the dietary fats. The formation of micelles is essential to transport the dietary fats toward the enterocytes across the aqueous intestinal lumen and the unstirred water layer at the intestinal epithelium. Absence of bile salts in the gut results in severe intestinal fat malabsorption. In recent years, it has become clear that bile salts are not only involved in dietary food digestion, but also play vital roles in a variety of systemic metabolic regulatory processes, which are, however, outside the scope of this chapter [40].

Bile salts are efficiently recycled via the portal system back to the liver in the so-called enterohepatic circulation [41]. Bile salts are to a large extent (>95% per cycle) absorbed in the terminal ileum, the final section of the small intestine. The total amount of bile salts in the body is balanced and is kept in a tight, steady state [42]. Under steady-state conditions, the fecal loss of bile salts is entirely compensated by de novo bile salt synthesis of primary bile salts in the liver. The primary bile salts in humans, i.e., synthesized de novo from cholesterol in the liver, are cholate and chenodeoxycholate. The primary bile salts are excreted via de bile into the intestine. In the intestinal lumen, the bile salts can be metabolized by the gut flora. Bacteria are capable of deconjugating bile salts and subsequently transforming them into a variety of different, so-called secondary, bile salts. Bile salts differ in their water solubility and their hydrophobic–hydrophilic balance. Hydrophobic bile salts have a high capability for solubilizing fats and lipids [43]. As a result, hydrophobic bile salts also have the ability to solubilize the lipid structures of cell membranes.

CF patients have an increased fecal loss of bile salts compared with non-CF individuals [44,45]. It is speculated that the loss of bile salts is due to impaired bile salt uptake, secondary to alterations in the intestinal mucosa in CF patients, like thickening of the mucus barrier or SIBO [46]. Because the biosynthesis of taurine is limited in humans, the fecal loss of bile salts induces an increased glycine/taurine ratio of conjugated bile salts in CF patients [47]. As a consequence, CF patients have an altered bile salt composition in gallbladder bile. Due to a quantitative increase of the primary bile salt cholate, its percent contribution is higher in the bile of CF patients, at the expense of the percent contribution of chenodeoxycholate and deoxycholate [48]. Theoretically, the increased glycine/taurine ratio may impair fat absorption in an acidic intestinal lumen. Due to the higher pKa of glycine, glycine-conjugated bile salts are less able to remain in micellar solution [49]. In addition, part of the glycoconjugates are passively absorbed in the proximal part of the intestine and are, in comparison to tauroconjugates, less resistant to bacterial degradation [50,51].

It has been proposed that excessive fecal loss of bile salts diminishes the bile salt pool in CF patients and consequently impairs fat absorption by reducing the solubilization capacity of bile [44]. However, a subsequent study indicated that the amount of fecal bile salt excretion was not related to the degree of fat malabsorption in CF patients [45]. Strandvik et al. showed that adult CF patients have normal to large bile salt pool sizes and similar amounts of duodenal bile salts as healthy controls [48]. Bile salt synthesis was normal or even increased, indicating that CF patients adequately compensate for the fecal bile salt loss. In CF mouse models, that fecal loss of bile salts does not influence the absorption of fat [37]. Homozygous ΔF508 mice and $CFTR^{tm1CAM}$ knockout mice both exhibit, to the same extent, an increased fecal loss of bile salts, but only the $CFTR^{tm1CAM}$ knockout mice had fat malabsorption [37]. In conclusion, bile salt malabsorption in itself, at the levels observed in CF patients and CF mice, has not been proven to contribute to persistent fat malabsorption.

Several studies showed that taurine supplementation reduces fecal fat excretion and improves the nutritional status of CF patients, particularly in patients with severe steatorrhoea [52–54]. However, the beneficial effect of taurine supplementation on fat absorption is not unequivocally demonstrated [55–58]. Furthermore, the degree of fat absorption did not relate to changes in the serum glycine/taurine ratio in CF children [58]. Altogether, the use of taurine supplementation remains controversial and is not implemented in nutritional CF care.

41.6 INTESTINAL MUCOSAL ABNORMALITIES

Several intestinal mucosal abnormalities are described in CF patients. These abnormalities include accumulation of viscous and sticky mucus, SIBO, and (chronic) inflammation of the small intestine [59–66].

In addition to the direct effect of bicarbonate on efficiency of lipolysis and micelle formation, it has been suggested that bicarbonate is involved in the intestinal mucus formation. Quinton et al. proposed in 2008 that HCO_3^- is crucial to normal mucin expansion in different organs, including the intestine [67]. Quinton et al. stated that because HCO_3^- secretion is defective in CF, mucins in organs affected by CF tend to remain aggregated, poorly solubilized, and less transportable [68]. When mucin expansion is disturbed by inadequate release of bicarbonate in CF conditions, viscous and sticky mucus might impair translocation of lipids as a causal mechanism for intestinal fat malabsorption.

41.7 SMALL INTESTINAL BACTERIAL OVERGROWTH

In CF patients we find an altered bacterial composition in the gastrointestinal tract. Several factors, like difference in intestinal pH, transit times, bile salt composition, and fluid secretion are suggested to be involved in this process. Based on these findings, CF patients are more prone to the development of SIBO. The clinical entity of bacterial overgrowth in general (i.e., independent of the concurrence with CF) has been related to intestinal malabsorption and potentially steatorrhoea.

In *Cftr*$^{-/-tm1Unc}$ mice, we found Cftr-specific alterations in the bacterial production of secondary bile salts and, accordingly, in the fecal and biliary bile salt composition. These results provide a clear example of the way in which alterations in the interaction between intestinal flora and bile salt formation can significantly affect the enterohepatic circulation of bile salt. The changes in bile salt metabolism in CF could be related to phenotypical characteristics that are described in CF patients and CF mice models. These abnormalities include accumulation of viscous and sticky mucus, small intestinal bacterial overgrowth, increased intestinal permeability, and inflammation of the small intestine [59–66,69,70].

All the above-mentioned factors may theoretically contribute to fat malabsorption in CF patients, as they might impair adequate translocation of fatty acids in(to) the enterocyte. In addition, SIBO might impair micelle formation by the bacterial deconjugation of bile salts [71]. In CF patients with proven SIBO, it has been shown that antibiotic treatment results in improved intestinal fat digestion as represented by improved ^{13}C mixed triglyceride breath test [72].

The group of De Lisle et al. evaluated the effect of antibiotic treatment on intestinal inflammation, mucus accumulation, and SIBO in CFTR^{tm1Unc} knockout mice [73]. CFTR^{tm1Unc} knockout mice display a phenotype of SIBO with a 400 fold increase in small intestinal bacterial content compared to wild type littermates [69]. Broad spectrum antibiotic treatment with ciprofloxacin and metronidazole did not only reduce the bacterial load in the small intestine, but also decreased intestinal mucus accumulation and inflammation. More important, 3 weeks of treatment substantially improved the body weight of these mice. The same effect on the bacterial load, intestinal mucus accumulation, inflammation, and body weight was observed after laxative treatment [70]. It remained to be elucidated whether the growth benefit was due to improved fat absorption, reduced competition for nutrients by intestinal bacteria, or reduced intestinal inflammation [74]. Wouthuyzen et al. subsequently demonstrated that oral antibiotics reduced the fecal excretion of cholate by ~50% in both CF mouse models, indicating improved intestinal bile salt absorption [74]. Interestingly, antibiotic treatment did not improve total fat absorption in CF mice. Finally, antibiotics accelerated the absorption of isotope-labeled fats in ΔF 508/508 mice, but also in wild type littermate controls. O'Brien et al. showed that a 7-day treatment with metronidazole reduced fecal fat excretion in four CF patients [45], compatible with the hypothesis that the increased weight is due to improved fat absorption. Although the effect on fat absorption was not assessed in either of these studies, the results suggest that metronidazole might be a potential therapeutic option for increasing fat absorption or improving nutritional status in CF patients.

One study evaluated the effect of probiotics on intestinal inflammation in CF patients [62]. It has been suggested that probiotics improve intestinal barrier function and modify the immune response [75]. Bruzzese et al. reported that a 4 week treatment with *Lactobacillus rhamnosus* GG reduced fecal calprotectin levels (marker for intestinal inflammation) in 8 out of 10 treated CF patients [62]. The long-term effect of probiotics on intestinal inflammation was not evaluated, nor the possible effects on fat absorption or nutritional status. Only one study in mice evaluated fat absorption after probiotics; *Lactobacillus* supplementation increased intestinal absorption of dietary fats in germ-free mice colonized with human baby flora [76]. Until now, the clinical value of probiotics and its relation with fat absorption in CF patients was not clear.

41.8 SMALL INTESTINAL TRANSIT TIME

Different studies have revealed the prolonged small intestinal transit time in adult and pediatric CF patients with exocrine pancreatic insufficiency compared to healthy controls [77–79]. The prolonged small intestinal transit time was not corrected by PERT. None of these studies related the small intestinal transit time to the intestinal fat absorption. In older studies, the oro-cecal transit time (OCTT) has been used to determine intestinal motility [30]. The lactulose/hydrogen breath test was most commonly applied to determine the OCTT. Eighty-five percent of CF patients had a prolonged OCTT (at least +50%) in the fasted state [80,81]. Another method to evaluate intestinal motility is based on measuring the intestinal muscle activity during the inter-digestive state, also known as the migrating motor complex. The migrating motor complex indirectly reflects the ability to transit food through the intestine [82]. In experimental animals it has been shown that CF mice have a slower small intestinal transit time in comparison to wild type littermates on the same diet [70,75,76,80,81,83–85]. An underlying mechanism may be found in intestinal smooth muscle activity of CF mice models that shows an erratic pattern and is unresponsive to cholinergic stimulation [83]. The group of De Lisle et al. performed several (intervention) studies on gastrointestinal transit in CFTR^{tm1Unc} knockout mice [70,83,85]. Oral laxative treatment improved circular smooth muscle function in the small intestine and normalized intestinal transit time [83]. Moreover, as earlier described, laxative treatment reduced intestinal mucus accumulation, eradicated overgrowth of bacteria in the small intestine, and improved body weight [70].

Intestinal fat absorption partly depends on the time fat is in contact with the absorptive epithelium of the intestine. A classic view states that intestinal fat malabsorption prolongs the intestinal transit time (as a feedback mechanism) in order to enhance its ability to absorb fat, a phenomenon also known as the 'ileal brake' [86]. It is reasonable to assume that this compensatory mechanism also occurs in persistent fat malabsorption in CF, as intestinal transit time is prolonged in CF. Additionally, a prolonged intestinal transit time is a risk factor for the occurrence of SIBO [87]. As a consequence, a prolonged intestinal transit time might actually induce or worsen the persistence of intestinal fat malabsorption rather than alleviate it. The relation between intestinal transit time and persistent intestinal fat malabsorption is yet to be elucidated.

41.9 CONCLUSION

Persistent intestinal fat malabsorption is a frequent occurrence in exocrine pancreatic-insufficient CF patients despite adequate pancreatic enzyme replacement therapy. In this chapter a number of CF-related intestinal factors like intestinal pH and bicarbonate secretion, bile salt metabolism, and enterohepatic circulation, intestinal mucosal abnormalities and small intestinal bacterial overgrowth are proposed to be involved (Figure 41.1). It is likely that not a single one of these factors determines the persistence of fat malabsorption. All factors can interact with one another and can individually increase or decrease the effect of the others. It is most likely that the persistent fat malabsorption in CF is a complex, multifactorial condition. It is, however, to be expected that in the current phase of evolving CFTR modulation therapies, new insights will become apparent. Trials with CFTR modulators have already shown significant increases in body mass index as a result of the new treatments. It is likely that part of this increase in body weight is related to improved intestinal fat absorption after CFTR correction. It will therefore be of great interest to see what intestinal factors are associated with these improvements. As a consequence, in the future, these intestinal factors could potentially be used as new clinical outcome parameters in future clinical therapeutic trails.

References

[1] Andersen DH. Cystic fibrosis of the pancreas and its relation to celiac diseasea clinical and pathologic study. Am J Dis Child 1938;56(2):344–99.

[2] Hahn TJ, Squires AE, Halstead LR, Strominger DB. Reduced serum 25-hydroxyvitamin D concentration and disordered mineral metabolism in patients with cystic fibrosis. J Pediatr 1979;94(1):38–42.

[3] Walters TR, Koch CHF. Hemorrhagic diathesis and cystic fibrosis in infancy. Arch Pediatr Adolesc Med 1972;124(5):641.

[4] Farrell PM, Bieri JG, Fratantoni JF, Wood RE, di Sant'Agnese PA. The occurrence and effects of human vitamin E deficiency: a study in patients with cystic fibrosis. J Clin Invest 1977;60(1):233.

[5] Fieker A, Philpott J, Armand M. Enzyme replacement therapy for pancreatic insufficiency: present and future. Clin Exp Gastroenterol 2011;4:55.

[6] Taylor J, Gardner T, Waljee A, Dimagno M, Schoenfeld P. Systematic review: efficacy and safety of pancreatic enzyme supplements for exocrine pancreatic insufficiency. Aliment Pharmacol Ther 2010;31(1):57–72.

[7] Stallings VA, Stark LJ, Robinson KA, Feranchak AP, Quinton H. Evidence-based practice recommendations for nutrition-related management of children and adults with cystic fibrosis and pancreatic insufficiency: results of a systematic review. J Am Diet Assoc 2008;108(5):832–9.

[8] Verkade HJ, Tso P. Biophysics of intestinal luminal lipids. Intestinal lipid metabolism. Springer; 2001. p. 1–18.

[9] Zoppi G, Andreotti G, Pajno-Ferrara F, Njai D, Gaburro D. Exocrine pancreas function in premature and full term neonates. Pediatr Res 1972;6(12):880–6.

[10] Moreau H, Laugier R, Gargouri Y, Ferrato F, Verger R. Human preduodenal lipase is entirely of gastric fundic origin. Gastroenterology November 1988;95(5):1221–6.

[11] Abrams CK, Hamosh M, Hubbard VS, Dutta SK, Hamosh P. Lingual lipase in cystic fibrosis. Quantitation of enzyme activity in the upper small intestine of patients with exocrine pancreatic insufficiency. J Clin Invest 1984;73(2):374–82.

[12] Minich DM, Voshol PJ, Havinga R, Stellaard F, Kuipers F, Vonk RJ, et al. Biliary phospholipid secretion is not required for intestinal absorption and plasma status of linoleic acid in mice. Biochim Biophys Acta (BBA)-Mol Cell Biol Lipids 1999;1441(1):14–22.

[13] Peretti N, Marcil V, Drouin E, Levy E. Mechanisms of lipid malabsorption in cystic fibrosis: the impact of essential fatty acids deficiency. Nutr Metab (Lond) 2005;2(1):11–9.

[14] Vassileva G, Huwyler L, Poirier K, Agellon LB, Toth MJ. The intestinal fatty acid binding protein is not essential for dietary fat absorption in mice. FASEB J 2000;14(13):2040–6.

[15] Levy E, Ménard D, Delvin E, Montoudis A, Beaulieu J, Mailhot G, et al. Localization, function and regulation of the two intestinal fatty acid-binding protein types. Histochem Cell Biol 2009;132(3): 351–67.

[16] Goudriaan JR, Dahlmans VE, Febbraio M, Teusink B, Romijn JA, Havekes LM, et al. Intestinal lipid absorption is not affected in CD36 deficient mice. Cellular lipid binding proteins. Springer; 2002. p. 199–202.
[17] Lightfoot FG, Lauretta Grau MSS, Satchitanandum M. Lipid accumulation in jejunal and colonic mucosa following chronic cholestyramine (Questran) feeding. Dig Dis Sci 1985;30(5):468–76.
[18] Werner A, Havinga R, Perton F, Kuipers F, Verkade HJ. Lymphatic chylomicron size is inversely related to biliary phospholipid secretion in mice. Am J Physiol Gastrointest Liver Physiol 2006;290(6):G1177–85.
[19] Shwachman H, Dooley RR, Guilmette F, Patterson PR, Weil C, Leubner H. Cystic fibrosis of the pancreas with varying degrees of pancreatic insufficiency. AMA J Dis Child 1956;92(4):347–68.
[20] Marino CR, Matovcik LM, Gorelick FS, Cohn JA. Localization of the cystic fibrosis transmembrane conductance regulator in pancreas. J Clin Invest 1991;88(2):712.
[21] Couper R, Corey M, Moore D, Fisher L, Forstner G, Durie P. Decline of exocrine pancreatic function in cystic fibrosis patients with pancreatic sufficiency. Pediatr Res 1992;32(2):179–82.
[22] Harper A, Raper HS. Pancreozymin, a stimulant of the secretion of pancreatic enzymes in extracts of the small intestine. J Physiol (Lond) 1943;102(1):115–25.
[23] Borowitz D, Stevens C, Brettman LR, Campion M, Chatfield B, Cipolli M. International phase III trial of liprotamase efficacy and safety in pancreatic-insufficient cystic fibrosis patients. J Cyst Fibros 2011.
[24] Littlewood JM, Wolfe SP, Conway SP. Diagnosis and treatment of intestinal malabsorption in cystic fibrosis. Pediatr Pulmonol 2006;41(1):35–49.
[25] FitzSimmons SC, Burkhart GA, Borowitz D, Grand RJ, Hammerstrom T, Durie PR, et al. High-dose pancreatic-enzyme supplements and fibrosing colonopathy in children with cystic fibrosis. N Engl J Med 1997;336(18):1283–9.
[26] Taylor C, Hillel P, Ghosal S, Frier M, Senior S, Tindale W, et al. Gastric emptying and intestinal transit of pancreatic enzyme supplements in cystic fibrosis. Arch Dis Child 1999;80(2):149–52.
[27] Carey MC, Small DM, Bliss CM. Lipid digestion and absorption. Annu Rev Physiol 1983;45(1):651–77.
[28] Tang L, Fatehi M, Linsdell P. Mechanism of direct bicarbonate transport by the CFTR anion channel. J Cyst Fibros 2009;8(2):115–21.
[29] Kaur S, Norkina O, Ziemer D, Samuelson LC, De Lisle RC. Acidic duodenal pH alters gene expression in the cystic fibrosis mouse pancreas. Am J Physiol Gastrointest Liver Physiol 2004;287(2):G480–90.
[30] Gregory P. Gastrointestinal pH, motility/transit and permeability in cystic fibrosis. J Pediatr Gastroenterol Nutr 1996;23(5):513–23.
[31] Gelfond D, Ma C, Semler J, Borowitz D. Intestinal pH and gastrointestinal transit profiles in cystic fibrosis patients measured by wireless motility capsule. Dig Dis Sci 2012:1–7.
[32] Robinson PJ, Smith AL, Sly PD. Duodenal pH in cystic fibrosis and its relationship to fat malabsorption. Dig Dis Sci 1990;35(10):1299–304.
[33] Brady MS, Garson JL, Krug SK, Kaul A, Rickard KA, Caffrey HH, et al. An enteric-coated high-buffered pancrelipase reduces steatorrhea in patients with cystic fibrosis: a prospective, randomized study. J Am Diet Assoc 2006;106(8):1181–6.
[34] Kalnins D, Ellis L, Corey M, Pencharz PB, Stewart C, Tullis E, et al. Enteric-coated pancreatic enzyme with bicarbonate is equal to standard enteric-coated enzyme in treating malabsorption in cystic fibrosis. J Pediatr Gastroenterol Nutr 2006;42(3):256–61.
[35] Jones DE, Palmer JM, Kirby JA, De Cruz DJ, McCaughan GW, Sedgwick JD, et al. Experimental autoimmune cholangitis: a mouse model of immune-mediated cholangiopathy. Liver 2000;20(5):351–6.
[36] Shindo K, Machida M, Fukumura M, Koide K, Yamazaki R. Omeprazole induces altered bile acid metabolism. Gut 1998;42(2):266–71.
[37] Bijvelds MJC, Bronsveld I, Havinga R, Sinaasappel M, de Jonge HR, Verkade HJ. Fat absorption in cystic fibrosis mice is impeded by defective lipolysis and post-lipolytic events. Am J Physiol Gastrointest Liver Physiol April 2005;288(4):G646–53.
[38] Hofmann AF. Bile acid secretion, bile flow and biliary lipid secretion in humans. Hepatology September 1990;12(3):17S–22S.
[39] Hofmann A. Fat digestion: the interaction of lipid digestion products with micellar bile acid solutions. Lipid absorption: biochemical and clinical aspects. Springer; 1976. p. 3–21.
[40] Trauner M, Claudel T, Fickert P, Moustafa T, Wagner M. Bile acids as regulators of hepatic lipid and glucose metabolism. Dig Dis 2010;28(1):220–4.
[41] Hofmann AF. Enterohepatic circulation of bile acids; 1969.
[42] Small DM, Dowling RH, Redinger RN. The enterohepatic circulation of bile salts. Arch Intern Med 1972;130(4):552.
[43] Hofmann A, Small D. Detergent properties of bile salts: correlation with physiological function. Annu Rev Med 1967;18(1):333–76.
[44] Weber AM. Relationship between bile acid malabsorption and pancreatic insufficiency in cystic fibrosis. Gut 1976;17(4):295.
[45] O'Brien S, Mulcahy H, Fenlon H, O'Broin A, Casey M, Burke A, et al. Intestinal bile acid malabsorption in cystic fibrosis. Gut 1993;34(8):1137.
[46] Fondacaro JD, Heubi JE, Kellogg FW. Intestinal bile acid malabsorption in cystic fibrosis: a primary mucosal cell defect. Pediatr Res 1982;16(6):494–8.
[47] Roy CC, Weber AM, Morin CL, Combes JC, Nusslé D, Mégevand A, et al. Abnormal biliary lipid composition in cystic fibrosis. N Engl J Med 1977;297(24):1301–5.
[48] Strandvik B, Einarsson K, Lindblad A, Angelin B. Bile acid kinetics and biliary lipid composition in cystic fibrosis. J Hepatol July 1996;25(1):43–8.
[49] Regan PT, Malagelada J-, Dimagno EP, Go VLW. Reduced intraluminal bile acid concentrations and fat maldigestion in pancreatic insufficiency: correction by treatment. Gastroenterology August 01, 1979;77(2):285–9.
[50] Krag E, Phillips SF. Active and passive bile acid absorption in man. Perfusion studies of the ileum and jejunum. J Clin Invest 1974;53(6):1686.
[51] Hepner GW, Sturman JA, Hofmann AF, Thomas PJ. Metabolism of steroid and amino acid moieties of conjugated bile acids in man III. cholyltaurine (taurocholic acid). J Clin Invest 1973;52(2):433–40.
[52] Belli DC, Levy E, Darling P, Leroy C, Lepage G, Giguère R, et al. Taurine improves the absorption of a fat meal in patients with cystic fibrosis. Pediatrics 1987;80(4):517–23.
[53] Darling PB, Lepage G, Leroy C, Masson P, Roy CC. Effect of taurine supplements on fat absorption in cystic fibrosis. Pediatr Res 1985;19(6):578–82.
[54] Smith LJ, Lacaille F, Lepage G, Ronco N, Lamarre A, Roy CC. Taurine decreases fecal fatty acid and sterol excretion in cystic fibrosis: a randomized double-blind trial. Arch Pediatr Adolesc Med 1991;145(12):1401–4.
[55] Colombo C, Arlati S, Curcio L, Maiavacca R, Garatti M, Ronchi M, et al. Effect of taurine supplementation on fat and bile acid absorption in patients with cystic fibrosis. Scand J Gastroenterol 1988;23(S143):151–6.
[56] Merli M, Bertasi S, Servi R, Diamanti S, Martino F, De Santis A, et al. Effect of a medium dose of ursodeoxycholic acid with or without taurine supplementation on the nutritional status of patients with cystic fibrosis: a randomized, placebo-controlled, crossover trial. J Pediatr Gastroenterol Nutr 1994;19(2):198–203.
[57] De Curtis M, Santamaria F, Ercolini P, Vittoria L, De Ritis G, Garofalo V, et al. Effect of taurine supplementation on fat and energy absorption in cystic fibrosis. Arch Dis Child 1992;67(9):1082–5.

[58] Thompson GN, Robb TA, Davidson GP. Taurine supplementation, fat absorption, and growth in cystic fibrosis. J Pediatr 1987;111(4):501–6.
[59] Van Elburg RM, Uil JJ, Van Aalderen WMC, Mulder CJJ, Heymans HSA. Intestinal permeability in exocrine pancreatic insufficiency due to cystic fibrosis or chronic pancreatitis. Pediatr Res 1996;39 (6):985–91.
[60] Lewindon P, Robb T, Moore DJ, Davidson GP, Martin AJ. Bowel dysfunction in cystic fibrosis: importance of breath testing. J Paediatr Child Health 1998;34(1):79–82.
[61] Lisowska A, Wójtowicz J, Walkowiak J. Small intestine bacterial overgrowth is frequent in cystic fibrosis: combined hydrogen and methane measurements are required for its detection. Acta Biochim Pol 2009;56(4):631.
[62] Bruzzese E, Raia V, Gaudiello G, Polito G, Buccigrossi V, Formicola V, et al. Intestinal inflammation is a frequent feature of cystic fibrosis and is reduced by probiotic administration. Aliment Pharmacol Ther 2004;20(7):813–9.
[63] Raia V, Maiuri L, de Ritis G, de Vizia B, Vacca L, Conte R, et al. Evidence of chronic inflammation in morphologically normal small intestine of cystic fibrosis patients. Pediatr Res 2000;47(3):344–50.
[64] Sbarbati A, Bertini M, Catassi C, Gagliardini R, Osculati F. Ultrastructural lesions in the small bowel of patients with cystic fibrosis. Pediatr Res 1998;43(2):234–9.
[65] Smyth RL, Croft NM, O'Hea U, Marshall TG, Ferguson A. Intestinal inflammation in cystic fibrosis. Arch Dis Child 2000;82(5):394–9.
[66] Quigley EM, Quera R. Small intestinal bacterial overgrowth: roles of antibiotics, prebiotics, and probiotics. Gastroenterology 2006;130(2):S78–90.
[67] Quinton PM. Birth of mucus. Am J Physiol Lung Cell Mol Physiol 2010;298(1):L13–4.
[68] Quinton PM. Cystic fibrosis: impaired bicarbonate secretion and mucoviscidosis. Lancet 2008;372(9636):415–7.
[69] Norkina O, Burnett TG, De Lisle RC. Bacterial overgrowth in the cystic fibrosis transmembrane conductance regulator null mouse small intestine. Infect Immun 2004;72(10):6040.
[70] De Lisle RC, Roach E, Jansson K. Effects of laxative and *N*-acetylcysteine on mucus accumulation, bacterial load, transit, and inflammation in the cystic fibrosis mouse small intestine. Am J Physiol Gastrointest Liver Physiol 2007;293(3):G577–84.
[71] Martin FJ, Dumas M, Wang Y, Legido-Quigley C, Yap IK, Tang H, et al. A top-down systems biology view of microbiome-mammalian metabolic interactions in a mouse model. Mol Syst Biol 2007;3(1):1–16.
[72] Lisowska A, Pogorzelski A, Oracz G, Siuda K, Skorupa W, Rachel M, et al. Oral antibiotic therapy improves fat absorption in cystic fibrosis patients with small intestine bacterial overgrowth. J Cyst Fibros 2011;10(6):418–21.
[73] De Lisle RC, Roach EA, Norkina O. Eradication of small intestinal bacterial overgrowth in the cystic fibrosis mouse reduces mucus accumulation. J Pediatr Gastroenterol Nutr 2006;42(1):46–52.
[74] Wouthuyzen-Bakker M, Bijvelds MJC, de Jonge HR, De Lisle RC, Burgerhof JGM, Verkade HJ. Effect of antibiotic treatment on fat absorption in mice with cystic fibrosis. Pediatr Res 2011;71(1):4–12.
[75] Perdigón G, Fuller R, Raya R. Lactic acid bacteria and their effect on the immune system. Curr Issues Intestinal Microbiol 2001;2(1):27–42.
[76] Martin FJ, Wang Y, Sprenger N, Yap IK, Lundstedt T, Lek P, et al. Probiotic modulation of symbiotic gut microbial–host metabolic interactions in a humanized microbiome mouse model. Mol Syst Biol 2008;4(1):1–14.
[77] Hedsund C, Gregersen T, Joensson IM, Olesen HV, Krogh K. Gastrointestinal transit times and motility in patients with cystic fibrosis. Scand J Gastroenterol 2012:1–7.
[78] Rovner AJ, Schall JI, Mondick JT, Zhuang H, Mascarenhas MR. Delayed small bowel transit in children with cystic fibrosis and pancreatic insufficiency. J Pediatr Gastroenterol Nutr July 2013;57(1):81–4.
[79] Borowitz D, Gelfond D, Maguiness K, Heubi JE, Ramsey B. Maximal daily dose of pancreatic enzyme replacement therapy in infants with cystic fibrosis: a reconsideration. J Cyst Fibros 2013;12(6):784–5.
[80] Dalzell A, Freestone N, Billington D, Heaf D. Small intestinal permeability and orocaecal transit time in cystic fibrosis. Arch Dis Child 1990;65(6):585–8.
[81] Bali A, Stableforth DE, Asquith P. Prolonged small-intestinal transit time in cystic fibrosis. Br Med J (Clin Res Ed) 1983;287(6398):1011–3.
[82] Deloose E, Janssen P, Depoortere I, Tack J. The migrating motor complex: control mechanisms and its role in health and disease. Nat Rev Gastroenterol Hepatol 2012;9(5):271–85.
[83] De Lisle R, Sewell R, Meldi L. Enteric circular muscle dysfunction in the cystic fibrosis mouse small intestine. Neurogastroenterol Motil 2010;22(3):341–e87.
[84] Lin HC. Ileal brake: neuropeptidergic control of intestinal transit. Curr Gastroenterol Rep 2006;8(5):367–73.
[85] De Lisle RC. Altered transit and bacterial overgrowth in the cystic fibrosis mouse small intestine. Am J Physiol Gastrointest Liver Physiol 2007;293(1):G104–11.
[86] Freedman SD, Blanco PG, Zaman MM, Shea JC, Ollero M, Hopper IK, et al. Association of cystic fibrosis with abnormalities in fatty acid metabolism. N Engl J Med 2004;350(6):560–9.
[87] Borowitz D, Durie PR, Clarke LL, Werlin SL, Taylor CJ, Semler J, et al. Gastrointestinal outcomes and confounders in cystic fibrosis. J Pediatr Gastroenterol Nutr 2005;41(3):273–85.

NOTE ADDED IN PROOF

Part of this chapter were published in "Persistent fat malabsorption in cystic fibrosis; lessons from patients and mice." Journal of Cystic Fibrosis 10.3 (2011): 150-158., Vol number, Wouthuyzen-Bakker, M., F. A. J. A. Bodewes, and H. J. Verkade. Copyright Elsevier, Permission for re-use was granted.

CHAPTER

42

Omega-3 Fatty Acids and Cystic Fibrosis

Gaurav Paul[1], *Ronald Ross Watson*[2]

[1]College of Public Health, University of Arizona, Tucson, AZ, USA; [2]Health Sciences Center, School of Medicine, Mel and Enid Zuckerman College of Public Health, University of Arizona, Tucson, AZ, USA

42.1 INTRODUCTION

42.1.1 Omega-3 Fatty Acids in Biology and Health

Omega-3 fatty acids, which are polyunsaturated fatty acids with a double bond after the third carbon atom in the carbon chain, are fats that are commonly found in marine and plant oils. It is a nutrient that the body needs but can't produce, and can be consumed through different foods. Omega-3 fatty acids can be found in many types of seafoods, including some types of fish such as salmon, tuna, and halibut, as well as algae and krill. Furthermore, omega-3 fatty acids can be found in some plants and various nut oils. By consuming omega-3 fatty acids as a supplement or through the diet, one can reduce the risk of getting cancer, cardiovascular disease, inflammation, and many other disorders.

A ratio of omega-6/omega-3 polyunsaturated fatty acids of about 1:1 that is favorable can have a protective effect against cancer [1]. Although this is true, the effect of omega-3 in childrens' diets is not known because clinical studies were only performed on adults. It is very likely that omega-3 fatty acids can play a positive role in preventing cancer in children, but more tests would be needed to verify this.

Furthermore, omega-3 fatty acids are known to play a role in cardiovascular disease. There is a large amount of evidence that shows omega-3 fatty acid supplementation can help to prevent cardiovascular disease or stroke. For example, eating a diet that is high in seafood reduces the risk of stroke [2]. In other studies, a high consumption of omega-3 fatty acids helped to increase the amount of LDL cholesterol in the body, lower blood pressure, and reduce the risk of heart attacks [2].

42.1.2 Cystic Fibrosis

Cystic fibrosis (CF), which is also known as mucoviscidosis, is a disease that is autosomal recessive and affects mostly the lungs, and also the pancreas, liver, and intestine. Its main effect is an abnormal transport of chloride and sodium across an epithelium, which leads to thick, viscous secretions [3]. The actual name of the disorder itself refers to the scarring and formation of cysts that are usually seen within the pancreas. However, the symptom that causes the most concern is difficulty breathing because of a higher susceptibility of lung infections. Other symptoms include sinus infections, stunted growth, infertility, and many others.

The cause of CF is genetic; specifically, a mutation in a gene that encodes for the protein cystic fibrosis transmembrane conductance regulator (CFTR). This protein is responsible for many different functions in the human body. It regulates sweat, digestive fluids, and mucus. It also controls the movement of chloride and sodium ions across the epithelial membranes throughout the body. People generally have two copies of CFTR when they are healthy, but CF develops when both copies of CFTR are missing, because the disorder is autosomal recessive [4].

The most common signs and symptoms of cystic fibrosis are skin that develops a salty taste, stunted growth, weight loss even with a normal diet, lots of mucus that develops, and reoccurring chest infections [5–8]. Loss of weight that is associated with CF comes about because of pancreatic insufficiency. Thus, people with CF need extra nutrition just to maintain their normal body weight. A high-calorie, high-fat diet is recommended for those with CF. A pancreatic enzyme replacement therapy can also be administered in severe cases. Furthermore, since CF is a genetic disorder, signs and symptoms appear in early infancy/childhood as well. Because of the buildup

Diet and Exercise in Cystic Fibrosis
http://dx.doi.org/10.1016/B978-0-12-800051-9.00042-0

of mucus, children with CF have to exercise a lot beginning from an early age [9]. CF is a disease that has many symptoms associated with it, but cures and remedies have only recently been proposed to treat it.

42.2 OMEGA-3 FATTY ACID: A ROLE IN CYSTIC FIBROSIS?

Omega-3 fatty acids may play a role in CF. Since omega-3 fatty acids have many anti-inflammatory effects, this can help those with CF. In CF, infection and inflammation are the two main symptoms of the disease that can impact lung function severely over time. In humans, the polyunsaturated fatty acids (PUFA) linoleic acid and alpha-linolenic acid can only be obtained from the diet. The best way to incorporate PUFAs into your diet is by consuming omega-3 fatty acids, and the easiest way to do this is by consuming fish oils. Fish oils are also the best source for omega-3 fatty acid derivatives, known as eicosapentaenoic acid (EPA) and docosahexaenoic acid (DHA). The mechanism behind omega-3 fatty acids is that they can strengthen cellular membranes, which will largely increase the anti-inflammatory response of the human body. This is partially explained by the decrease in pro-inflammatory metabolites from the omega-6 fatty acid family [10].

Something characteristic of CF in all patients is abnormalities with their PUFAs, such as a low amount of linoleic acid and a high amount of arachidonic acid, a member of the polyunsaturated omega-6 fatty acid family. These changes in the amount of PUFAs are thought to play a role when it comes to the pathophysiology of the disease. Recent studies have shown this to be true. A study that increased the amount of parallel n-3 and n-6 PUFAs (linoleic acid and alpha-linoleic acid) actually showed that the amount of products in that pathway was increased, although the formation of products in the parallel pathway was decreased. Therefore, by increasing the amount of n-3 and n-6 PUFAs in the diet, the amount of linoleic acid and alpha linoleic acid will remain high. Since CF is linked to low levels of linoleic acid, it makes sense that increasing the amount of omega-3 fatty acids can possibly help treat the disease.

Alterations of PUFA levels are found in CF patients, regardless of the level of nutrition. Even with pancreatic enzyme replacement therapy, which is used to maintain nutrition, there is still an irregular amount of PUFAs [11]. This has been seen in mouse studies as well as cell culture models [12–14]. Most often, the alterations of PUFAs that are seen in CF are high levels of linoleic acid (LNA) and docosahexaenoic acid (DHA), as well as increases in palmitoleate and Mead acid. Sometimes, increased arachidonic acid levels can be seen. The theory behind the arachidonic acid is that increased levels can also increase the amount of pro-inflammatory eicosanoids produced [15]. Generally, the severity of the alterations in the levels of PUFAs corresponds with the severity of CF [16,17].

Laboratory work has shown that most of the fluctuations in PUFA levels that are associated with CF are because of changes in the fatty acid metabolism. n-3 and n-6 PUFAs are created by the same biochemical pathways that both use desaturase and elongase enzymes [18]. Compared to normal cells, (wild-type), cells affected by CF tend to have increased expression of many fatty acid enzymes such as Δ5 desaturase, Δ6 desaturase, Δ9 desaturase, and elongase 6 which are all used in the production of arachidonic acid, palmitoleate, and Mead acid [19,20]. In further tests also done in a lab, it was shown that DHA and eicosapentaenoate can actually suppress the activity of Δ5 and Δ6 desaturases and normalize the levels of linoleic acid and arachidonic acid [21]. However, in cell culture studies done on linoleic acid supplementation, there was an increased amount of arachidonic acid being produced, which could lead to increased inflammation in the airways of humans with the disease [22]. Thus, a large concern of patients with CF is that the high-calorie, high-fat diets they have to consume to maintain nutrition (typical western diets) are high in linoleic acid but low in alpha-linoleate [23].

A cell culture model was performed in which human bronchial cells expressing CFTR (wild-type) and human bronchial cells not expressing CFTR (CF) were used. Specifically, in this study those bronchial cells were actually transfected with a segment of the CFTR gene in either the sense or antisense orientation. The cells with the sense gene (wild-type cells) had normal expression and activity of CFTR, whereas the cells with the antisense gene (CF) did not have either [14,24]. Both wild-type and CF cells were spread onto plates and grown for 7 days. On the 7th day, the cells were supplemented with varying amounts of linoleic acid or alpha linoleic acid for 4 h, either 0, 5, 10, or 20 μM. Afterward, the cells that were ready for analysis went through a process that involved centrifugation, incubation, and methylation. Results showed that there was not a big difference in the amount of linoleic acid between the wild-type and CF cells. However, these results were compared to two other cell cultures that had a higher concentration of linoleic acid supplemented, and it became clear that the difference between wild-type and CF cells in terms of linoleic acid levels only becomes apparent when the concentrations of linoleic acid are moderate to high. The main piece of data to take from this study is that as linoleic acid and alpha linoleic acid levels increased, EPA and DHA levels actually dropped, showing that they are inversely related [25].

To better measure the effect of alpha linoleic acid on linoleic acid to arachidonic acid metabolism, cells radiolabeled with linoleic acid were incubated in the presence of an increasing amount of arachidonic acid for 4h, and then the levels of both were measured. As expected, the CF cells converted linoleic acid to arachidonic acid much more than wild-type cells, evidenced by the low amounts of labeled linoleic acid and high amounts of arachidonic acid in the CF cells. Adding alpha-linoleic acid to this mixture decreased the metabolism of linoleic acid to arachidonic acid, which was shown by the large increases in linoleic and small increases in arachidonic.

Essentially, all of these results show that there are many different interactions between the substrates of the n-3 and n-6 pathways, and that a substrate of one pathway can suppress the substrate of the other pathway. However, how this is just a cell culture model. What can animal trials tell us?

42.3 OMEGA FATTY ACIDS AND CF: ANIMAL MODELS

Increased conversion of linoleic acid to arachidonic acid led to more proinflammatory metabolites in CF patients, and the process gets worse with increased amounts of linoleic acid. The hypothesis was tested by using an in vivo model with *cftr*$^{-/-}$ transgenic mice [22]. These mice were housed in an animal facility and were an average of 23days of age. Both CF mice, as well as their wild-type counterparts, were fed normal Purina mouse food and drinking water while being supplemented with either 100mg/day of linoleic acid or a control fatty acid supplement of oleic acid. Examination of these mice after a given period of time demonstrates that the linoleic and arachidonic acid levels in the lungs of the mice showed a large increase in the linoleic acid levels after supplementation in both the CF and wild-type mice. However, the arachidonic acid levels only increased in the CF mice and not in the wild-type mice.

It is known that previous studies have shown that mice with CF that are exposed to *Pseudomonas* LPS, which CFTR recognizes, have increased levels of neutrophils in their Bronchoalveolar Lavage (BAL) fluid. Furthermore, these same mice with CF that were then fed linoleic acid (100mg/day) for 10days had a fivefold increase in the amount of neutrophils in their BAL fluid compared to the CF mice that were given the control oleic acid. Do these results carry over to cases where humans have CF as well?

42.4 OMEGA FATTY ACIDS AND CF: CLINICAL TRIALS

Randomized, controlled clinical trials were performed to see if supplementation of omega-3 fatty acids in the form of eicosapentaenoic acid (EPA) or docosahexaenoic acid (DHA) would have any effect on patients with CF. The control group used for this study was a placebo that had a low omega-3 or omega-6 fatty acid amount such as olive oil. In total, this study included four different trials [26–29]. The variations of the trial length were from 6weeks [26] all the way up to 6months [27]. Furthermore, the number of participants differed from 12 [26] to 43 [28]. Out of the four studies, two (the Henderson, 1994 and Keen, 2010) were similar in design. The Henderson study included four groups of people, two with CF and two without. However, information from the groups without CF was not analyzed. Out of the two groups with CF, one group received the omega supplementation and the other received placebo. The Keen study included three groups, one given a mix of EPA and DHA, one given mostly omega-6 fatty acids (linoleic acid and arachidonic acid), and the last was a control given a high amount of saturated fatty acid. Furthermore, the group receiving the omega-6 fatty acids in the Keen study did not have their data considered as it was not part of the whole trial. Lawrence only took into account results after 12weeks, and Panchaud took in results from the first 6weeks.

Participants of the trials included children and adults, with only one study including adults above the age of 41years (Keen, 2010). Three of the studies had patients that were said to have pancreatic insufficiency (Henderson, Keen, and Panchaud), and two of the studies had patients that had chronic infections of *Pseudomonas aeruginosa* (Keen and Lawrence). Furthermore, the Keen group was said to have severe mutations. Henderson and Lawrence compared omega-3 fatty acids to the olive oil control in a 6-week period, whereas Panchaud compared omega-3 fatty acids to a placebo in a 6-month period, and Keen compared essential fatty acids to a placebo in a 3-month period. Dosage varied between each study as well. The Henderson study used four 1g capsules of fish oil two times daily, which contained 3.2g EPA and 2.2g DHA. The Lawrence study administered fish oil capsules daily, which contained 2.7g EPA. Panhaud used a mixture of PUFAs that contained 0.2g EPA and 0.1g DHA. Finally, Keen used a blend of fatty acids that contained both EPA and DHA, with participants of the study receiving 50mg of the mixture per kg of their body weight per day.

There were a variety of outcomes from these studies. All four included adverse results, and only one (Keen) did not report any deaths. Primary outcomes included the number of respiratory exacerbations, which was not actually measured in any of the studies; adverse events and dropouts; and lung function. Adverse events and dropouts included steatorrhoea, diarrhea, asthma, and stomach pains. The change in lung function was measured in each study as well. Secondary outcomes included the quality of life, not measured in any study;

the number of antibiotics given, which was only measured in the Keen study; and the number of deaths.

The results from these four studies show that it is not enough to change the way CF is treated in clinical practice. These studies are too small, and only three of the studies showed any benefits with omega-3 supplementation (Lawrence, Panchaud, and Keen). Therefore, there is not enough evidence here to make a statement that people with CF can use fatty acid supplementation in order to help themselves. In order to further investigate the effect of fatty acid supplementation on CF, a larger long-term study needs to be performed [10].

42.5 CONCLUSION

These studies provide great information to further the knowledge of CF. It is known that CF patients exhibit low levels of linoleic acid [30,31]. However, explanations for these low levels include fat malabsorption, decreased dietary fat intake, and altered fatty acid metabolism [16,30]. Furthermore, this study shows that supplementation with linoleic acid is connected to increased levels of arachidonic acid in CF cells and mice, but not in wild-type cells or in animals. Elevated linoleic acid levels are also connected to increased production of the neutrophil interleukin chemokine IL-8. Another way inflammatory metabolites are produced more with elevated linoleic acid levels is from the increased arachidonic acid production that comes with it.

Increased linoleic acid intake does not always lead to an inflammatory response. This only occurs in conditions when there the regulation of inflammation is altered for a certain reason [32]. The most important thing to note is that linoleic acid will only have an impact on the amount of proinflammatory metabolites released when NF-kB is activated, such as in CF when NF-kB activity is greatly increased because of the loss of negative regulation by CFTR [33]. The risk-benefit of dietary linoleic acid and other omega fatty acids needs to be examined for each patient's situation.

To further investigate whether omega-3 fatty acids are a viable treatment or prevention method for CF, much more research needs to be done. Specifically, stage II and stage III clinical trials on humans need to be on a larger scale to truly observe the vast effect of omega-3 fatty acids on patients with CF. Currently, research only supports omega-3 fatty acids as a way to reduce inflammation, but not to directly prevent or treat CF.

References

[1] Cole GM, Lim GP, Yang F, Teter B, Begum A, Ma Q, et al. Prevention of Alzheimer's disease: omega-3 fatty acid and phenolic antioxidant interventions. National Library of Medicine; 2013.

[2] Dangour A. The OPAL study: older people and n-3 long chain polyunsaturated fatty acids. Current Controlled Trials; 2005.

[3] Yankaskas JR, Marshall BC, Sufian B, Simon RH, Rodman D. Cystic fibrosis adult care consensus conference report. Chest 2004;125(90010):1–39.

[4] Andersen DH. Cystic fibrosis of the pancreas and its relation to celiac disease: a clinical and pathological study. Am J Dis Child 1938;56:344–99.

[5] Quinton PM. Cystic fibrosis: lessons from the sweat gland. Physiology (Bethesda) June 2007;22(3):212–25.

[6] Hardin DS. GH improves growth and clinical status in children with cystic fibrosis – a review of published studies. Eur J Endocrinol August 2004;151(Suppl. 1):S81–5.

[7] De Lisle RC. Pass the bicarb: the importance of HCO_3^- for mucin release. J Clin Invest September 2009;119(9):2535–7.

[8] O'Malley CA. Infection control in cystic fibrosis: cohorting, cross-contamination, and the respiratory therapist. Respir Care May 2009;54(5):641–57.

[9] Ratjen FA. Cystic fibrosis: pathogenesis and future treatment strategies. Respir Care May 2009;54(5):595–605.

[10] Oliver C, Watson H. Omega-3 fatty acids for cystic fibrosis. Cochrane Database Syst Rev November 27, 2013;11:CD002201.

[11] Aldamiz-Echevarria L, Prieto JA, Andrade F, Elorz J, Sojo A, Lage S, et al. Persistence of essential fatty acid deficiency in cystic fibrosis despite nutritional therapy. Pediatr Res 2009;66:585–9.

[12] Freedman SD, Katz MH, Parker EM, Laposata M, Urman MY, Alvarez JG. A membrane lipid imbalance plays a role in the phenotypic expression of cystic fibrosis in cftr(−/−) mice. Proc Natl Acad Sci USA 1999;96:13995–4000.

[13] Mimoun M, Coste TC, Lebacq J, Lebecque P, Wallemacq P, Leal T, et al. Increased tissue arachidonic acid and reduced linoleic acid in a mouse model of cystic fibrosis are reversed by supplemental glycerophospholipids enriched in docosahexaenoic acid. J Nutr 2009;139:2358–64.

[14] Andersson C, Al-Turkmani MR, Savaille JE, Alturkmani R, Katrangi W, Cluette-Brown JE, et al. Cell culture models demonstrate that CFTR dysfunction leads to defective fatty acid composition and metabolism. J Lipid Res 2008;49:1692–700.

[15] Strandvik B. Fatty acid metabolism in cystic fibrosis. Prostaglandins Leukot Essent Fatty Acids 2010;83:121–9.

[16] Strandvik B, Gronowitz E, Enlund F, Martinsson T, Wahlstrom J. Essential fatty acid deficiency in relation to genotype in patients with cystic fibrosis. J Pediatr 2001;139:650–5.

[17] Van Biervliet S, Vanbillemont G, Van Biervliet JP, Declercq D, Robberecht E, Christophe A. Relation between fatty acid composition and clinical status or genotype in cystic fibrosis patients. Ann Nutr Metab 2007;51:541–9.

[18] Nakamura MT, Nara TY. Essential fatty acid synthesis and its regulation in mammals. Prostaglandins Leukot Essent Fatty Acids 2003;68:145–50.

[19] Njoroge SW, Seegmiller AC, Katrangi W, Laposata M. Increased Delta5- and Delta6-desaturase, cyclooxygenase-2, and lipoxygenase-5 expression and activity are associated with fatty acid and eicosanoid changes in cystic fibrosis. Biochim Biophys Acta 2011;1811:431–40.

[20] Thomsen KF, Laposata M, Njoroge SW, Umunakwe OC, Katrangi W, Seegmiller AC. Increased elongase 6 and Delta9-desaturase activity are associated with n-7 and n-9 fatty acid changes in cystic fibrosis. Lipids 2011;46:669–77.

[21] Njoroge SW, Laposata M, Katrangi W, Seegmiller AC. DHA and EPA reverse cystic fibrosis-related FA abnormalities by suppressing FA desaturase expression and activity. J Lipid Res 2012;53:257–65.

[22] Zaman MM, Martin CR, Andersson C, Bhutta AQ, Cluette-Brown JE, Laposata M, et al. Linoleic acid supplementation results in increased arachidonic acid and eicosanoid production in CF airway cells and in cftr−/− transgenic mice. Am J Physiol Lung Cell Mol Physiol 2010;299:L599–606.

[23] Kris-Etherton PM, Taylor DS, Yu-Poth S, Huth P, Moriarty K, Fishell V, et al. Polyunsaturated fatty acids in the food chain in the United States. Am J Clin Nutr 2000;71:179S–88S.

[24] Rajan S, Cacalano G, Bryan R, Ratner AJ, Sontich CU, van Heerckeren A, et al. *Pseudomonas aeruginosa* induction of apoptosis in respiratory epithelial cells: analysis of the effects of cystic fibrosis transmembrane conductance regulator dysfunction and bacterial virulence factors. Am J Respir Cell Mol Biol 2000;23:304–12.

[25] Katrangi W, Lawrenz J, Seegmiller AC, Laposata M. Interactions of linoleic and alpha-linolenic acids in the development of fatty acid alterations in cystic fibrosis. Lipids 2013;48:333–42.

[26] Henderson WR. Omega-3 supplementation in cystic fibrosis. In: Proceedings of the 6th North American Cystic Fibrosis Conference; 1992. Abstract edition:S21.2.

[27] Keen C, Olin A, Erikkson S, Ekman A, Lindblad A, Basu S, et al. Supplementation with fatty acids influences the airway nitric oxide and inflammatory markers in patients with cystic fibrosis. J Pediatr Gastroenterol Nutr 2010;50(5):537–44.

[28] Lawrence R, Sorrell T. Eicosapentaenoic acid in cystic fibrosis: evidence of a pathogenetic role for leukotriene B4. Lancet 1993;342(8869):465–9.

[29] Panchaud A, Sauty A, Kernan Y, Decosterd LA, Buclin T, Boulat O, et al. Biological effects of a dietary omega-3 polyunsaturated fatty acids supplementation in cystic fibrosis patients: a randomized, crossover placebo-controlled trial. Clin Nutr 2006;25(3):418–27.

[30] Farrell PM, Mischler EH, Engle MJ, Brown DJ, Lau SM. Fatty acid abnormalities in cystic fibrosis. Pediatr Res 1985;19:104–9.

[31] Freedman SD, Blanco PG, Zaman MM, Shea JC, Ollero M, Hopper IK, et al. Association of cystic fibrosis with abnormalities in fatty acid metabolism. N Engl J Med 2004;350:560–9.

[32] Fritsche KL. Too much linoleic acid promotes inflammation – doesn't it? Prostaglandins Leukot Essent Fatty Acids 2008;79:173–5.

[33] Vij N, Mazur S, Zeitlin PL. CFTR is a negative regulator of NFkappaB mediated innate immune response. PLoS One 2009;4:e4664.

Index

Note: Page numbers with "*f*" denote figures; "*t*" tables.

D

Printed and bound by CPI Group (UK) Ltd, Croydon, CR0 4YY
08/05/2025
01865029-0002